Ergebnisse der Mathematik und ihrer Grenzgebiete

3. Folge · Band 17

A Series of Modern Surveys in Mathematics

G. A. Margulis

Discrete Subgroups of Semisimple Lie Groups

Springer-Verlag
Berlin Heidelberg New York
London Paris Tokyo
Hong Kong Barcelona

Gregori Aleksandrovitch Margulis
Institute for Problems of Transmission of Information
Academie of Sciences, ul. Ermolovoi 19
Moscow, 103 151
USSR

Mathematics Subject Classification (1989):
22E40 20G30 22D10 28D15 60B15 20G25

ISBN 3-540-12179-X Springer-Verlag Berlin Heidelberg New York
ISBN 0-387-12179-X Springer-Verlag New York Berlin Heidelberg

Printed in the United States of America

2141/3140-543210 Printed on acid-free paper

Preface

A detailed treatment of the geometric aspects of discrete groups was carried out by Raghunathan in his book "Discrete subgroups of Lie Groups" which appeared in 1972. In particular he covered the theory of lattices in nilpotent and solvable Lie groups, results of Mal'cev and Mostow, and proved the Borel density theorem and local rigidity theorem of Selberg-Weil. He also included some results on unipotent elements of discrete subgroups as well as on the structure of fundamental domains. The chapters concerning discrete subgroups of semi-simple Lie groups are essentially concerned with results which were obtained in the 1960's.

The present book is devoted to lattices, i.e. discrete subgroups of finite covolume, in semi-simple Lie groups. By "Lie groups" we not only mean real Lie groups, but also the sets of k-rational points of algebraic groups over local fields k and their direct products. Our results can be applied to the theory of algebraic groups over global fields. For example, we prove what is in some sense the best possible classification of "abstract" homomorphisms of semi-simple algebraic group over global fields.

Our work deals mainly with problems of structure, classification, and the description of discrete subgroups of Lie groups. Many of our statements are formulated in terms of algebraic group theory and of course make extensive use of methods from this subject. In addition we take advantage of techniques which at first sight might appear to be unrelated to the theory of algebraic groups, e.g. measure theory, ergodic theory, the theory of infinite-dimensional unitary representations, and the theory of amenable groups.

A number of important topics have been omitted. The most significant of these is the theory of Kleinian groups and Thurston's theory of 3-dimensional manifolds; these two theories can be united under the common title "Theory of discrete subgroups of $\mathbf{SL}_2(\mathbb{C})$". Discrete subgroups of rank 1 semi-simple Lie groups are considered only in Appendix C where we give examples of non-arithmetic lattices in such groups.

A number of topics covered here are closely related to those in Zimmer's book "Ergodic Theory and Semisimple Groups" which appeared in 1985. However, in our opinion the two books are in many ways complementary.

We would like to thank J. Tits who encouraged us to write this book and whose advice and suggestions were extremely valuable. We also greatly appreciate the constructive remarks of A.T. Huckleberry and G. Prasad. Finally, we would like to express our gratitude to Ya. G. Sinai who, as superviser and mentor, profoundly influenced our approach to mathematics.

Moscow, Autumn 1990

Table of Contents

Introduction

1. Statement of Main Results

Let A be a nonempty finite set. Suppose that for each $\alpha \in A$ a local (i.e. non-discrete locally compact) field k_α and a connected semisimple k_α-group $\mathbf{G}_\alpha$ have been chosen. Denote by $\mathbf{G}_\alpha(k_\alpha)$ the group of k_α-rational points of $\mathbf{G}_\alpha$ and let $G = \prod_{\alpha \in A} \mathbf{G}_\alpha(k_\alpha)$.

In this book the main objects of study are lattices (i.e. discrete subgroups with finite covolume) in the group G. We mainly consider the so called irreducible lattices (a precise definition of this concept will be given in 5.9 of Chapter III; here we only note that if for every $\alpha \in A$ the group $\mathbf{G}_\alpha$ has no nontrivial k_α-anisotropic factors, then a lattice $\Gamma \subset G$ is irreducible if and only if no subgroup of finite index in Γ can be represented in the form of a direct product of two infinite subgroups).

We now provide an important example of an irreducible lattice. Let K be a global field, let $\mathscr{R}$ be the set of all (inequivalent) valuations of the field K, and let $\mathscr{R}_\infty \subset \mathscr{R}$ be the set of archimedean valuations of K. Denote by K_v the completion of the field K with respect to the valuation $v \in \mathscr{R}$. If $S \subset \mathscr{R}$, then the ring $\{x \in K \mid |x|_v \leq 1 \text{ for all } v \in \mathscr{R} - \mathscr{R}_\infty - X\}$, where $|x|_v$ is the value of the valuation v at x, of S-integral elements of the field K, will be denoted by $K(S)$. Now let $\mathbf{H}$ be a connected non-commutative absolutely almost simple K-subgroup of the group $\mathbf{SL}_n$ of unimodular $n \times n$ matrices. We set $\mathscr{T} = \mathscr{T}(\mathbf{H}) = \{v \in \mathscr{R} \mid$ the group $\mathbf{H}$ is anisotropic over K_v, or equivalently, the group $\mathbf{H}(K_v)$ is compact$\}$. Let $S \subset \mathscr{R}$ be a finite set such that $S \supset \mathscr{R}_\infty - \mathscr{T}$ and let $\mathbf{H}(K(S))$ denote the subgroup of $\mathbf{H}$ consisting of matrices with entries in $K(S)$.

We set $H_S = \prod_{v \in S} \mathbf{H}(K_v)$ and identify the group $\mathbf{H}(K(S))$ with its image under the diagonal embedding in H_S. By the Borel-Harish-Chandra-Behr-Harder reduction theorem $\mathbf{H}(K(S))$ is a lattice in H_S. It can easily be verified that the lattice $\mathbf{H}(K(S))$ is irreducible. Let φ: $H_S \to G$ be a continuous isomorphism. Then $\varphi(\mathbf{H}(K(S)))$ is an irreducible lattice in G. The irreducible lattices obtained in this way and all the lattices commensurable with them will be called *arithmetic*. This definition is the final one in the case where for each $\alpha \in A$ the group $\mathbf{G}_\alpha$ has no nontrivial k_α-anisotropic factors and the groups $\mathbf{G}_\alpha$ are either all simply connected or all adjoint. In the general case one should take as φ not only isomorphisms but also continuous homomorphisms from a certain more general class (for details we refer to 1.3 and 1.4 of Chapter IX). If $G = \mathbf{G}(\mathbb{R})$, where $\mathbf{G}$ is a connected semisimple $\mathbb{R}$-group, then an irreducible lattice $\Gamma \subset G$ is arithmetic if and only if there exist a connected non-commutative almost $\mathbb{Q}$-simple $\mathbb{Q}$-group $\mathbf{F}$ and an $\mathbb{R}$-epimorphism $\tau: \mathbf{F} \to \mathbf{G}$ such that the Lie group $(\operatorname{Ker} \tau)(\mathbb{R})$ is compact and the subgroups $\tau(\mathbf{F}(\mathbb{Z}))$ and Γ are commensurable.

We can now state the arithmeticity theorem (although not in full generality). First, recall that $\operatorname{char} k$ denotes the characteristic of the field k and $\operatorname{rank}_k \mathbf{F}$ denotes the k-rank (i.e. the dimension of a maximal k-split torus) of the k-group $\mathbf{F}$. Let us set $\operatorname{rank} G = \sum_{\alpha \in A} \operatorname{rank}_{k_\alpha} \mathbf{G}_\alpha$.

(1) Theorem. (See Theorems A and B and Remark 2 in 1.9 of Chapter IX.) *Let Γ be an irreducible lattice in G. Suppose that either* $\operatorname{rank} G \geq 2$ *or Γ is finitely generated and is of infinite index in the commensurability subgroup* $\operatorname{Comm}_G(\Gamma) \stackrel{\text{def}}{=} \{g \in G \mid$ *the subgroups $g\Gamma g^{-1}$ and Γ are commensurable*$\}$. *Assume in addition that for no $\alpha \in A$ the group $\mathbf{G}_\alpha$ has a nontrivial k_α-anisotropic factor. Then the lattice Γ is arithmetic.*

It should be noted that an irreducible lattice $\Gamma \subset G$ is finitely generated if at least one of the following conditions is satisfied:

(a) the lattice Γ is cocompact (i.e. the quotient space $\Gamma \setminus G$ is compact);
(b) there exists $\alpha \in A$ such that $\operatorname{char} k_\alpha = 0$;
(c) $\operatorname{rank} G \geq 2$.

The proof of Theorem 1 makes fundamental use of the superrigidity theorems, more precisely Theorem 2 below. Before stating this theorem we set $G^+ = \prod_{\alpha \in A} \mathbf{G}_\alpha(k_\alpha)^+$, where $\mathbf{G}_\alpha(k_\alpha)^+$ denotes the subgroup of $\mathbf{G}_\alpha(k_\alpha)$ generated by the sets of k_α-rational points of unipotent radicals of parabolic k_α-subgroups of the group $\mathbf{G}_\alpha$. Note that

1) if for each $\alpha \in A$ the group $\mathbf{G}_\alpha$ is simply connected and has no nontrivial k_α-anisotropic factors, then $G^+ = G$;
2) if the groups $\mathbf{G}_\alpha$ have no nontrivial k_α-anisotropic factors and the fields k_α are archimedean, then G^+ coincides with the identity component of the Lie group G.

(2) Theorem. (See Theorems 5.4 and 5.6 of Chapter VII.) *Let Γ be a lattice in G, let Λ be a subgroup of* $\operatorname{Comm}_G(\Gamma)$ *containing Γ, k a local field, $\mathbf{H}$ a connected adjoint k-simple k-group, and let $\tau: \Lambda \to \mathbf{H}(k)$ be a homomorphism. Suppose that for no $\alpha \in A$ the group $\mathbf{G}_\alpha$ has a nontrivial k_α-anisotropic factor and the following conditions are satisfied:*

(i) either $\operatorname{rank} G \geq 2$ *and the lattice Γ is irreducible or the closure of the subgroup Λ contains G^+;*

(ii) the subgroup $\tau(\Gamma)$ is Zariski dense in $\mathbf{H}$ but is not relatively compact in $\mathbf{H}(k)$ *(in the topology induced from that of the field k).*

Then τ extends, uniquely, to a continuous homomorphism $\tilde{\tau}: G \to$ $\mathbf{H}(k)$.

In Theorem 2 we actually consider homomorphisms τ of the lattice Γ to algebraic groups over local fields k such that the Zariski closure of the subgroup $\tau(\Gamma)$ is a k-simple k-group. Nevertheless, we may consider homomorphisms for which this condition is not satisfied and prove for them assertions similar to that of Theorem 2. Making use of these assertions we shall prove the following.

(3) Theorem. (See Theorem 5.8 and Corollary 5.9 of Chapter IX.) *Let $\Gamma \subset G$ be an irreducible lattice, l a field, $\mathbf{F}$ an algebraic l-group, and let $\delta \colon \Gamma \to \mathbf{F}(l)$ be a homomorphism. Suppose that* $\operatorname{rank} G \geq 2$ *and for no $\alpha \in A$ the group $\mathbf{G}_\alpha$ has a nontrivial k_α-anisotropic factor. Then*

(i) in case $\operatorname{char} l = 0$, *the Zariski closure of the group $\delta(\Gamma)$ is a semisimple l-group;*

(ii) in case $\operatorname{char} l \neq \operatorname{char} k_\alpha$ *for all $\alpha \in A$, the group $\delta(\Gamma)$ is finite;*

(iii) the 1*-dimensional cohomology group $H^1(\Gamma, \rho)$ is trivial for every representation ρ of the group Γ on a finite-dimensional vector space over a field of characteristic* 0.

In connection with (iii) above, note that if ρ is a continuous non-trivial absolutely irreducible representation of the group G on a finite-dimensional vector space over a local field k of arbitrary characteristic, then, under certain assumptions on G and Γ, the restriction map of the cohomology groups $H^1_{\mathrm{cont}}(G, \rho) \to H^1(\Gamma, \rho)$ is an isomorphism (see Corollary 5.21 of Chapter VII).

Another group of results is related to the finiteness of factor groups of lattices $\Gamma \subset G$. These results can be stated as follows:

(4) Theorem. (See Proposition 5.3 and Theorem 5.4 of Chapter IX.) *Let $\Gamma \subset G$ be an irreducible lattice, let* $\operatorname{rank} G \geq 2$, *and assume that for no $\alpha \in A$ the group $\mathbf{G}_\alpha$ has a nontrivial k_α-anisotropic factor. Then*

(i) for every normal subgroup N of Γ, either N is contained in the center $\mathscr{Z}(G)$ of G or the group Γ/N is finite;

(ii) the factor group of the group Γ modulo its commutator subgroup is finite.

In case k_α is archimedean for every $\alpha \in A$, Theorems 1–4 can be restated in terms of Lie group theory. Before doing this we shall give some definitions.

Let H be a connected semisimple Lie group. A connected commutative subgroup $T \subset H$ is said to be an $\mathbb{R}$-*split torus* if for each $x \in T$, $\operatorname{Ad} x$ is semisimple and all its eigenvalues are real. Here Ad denotes as usual the adjoint representation. The dimension of a maximal $\mathbb{R}$-split torus of H is called the *rank* of H and denoted by $\operatorname{rank} H$. We remark that if H is the connected component of the identity in the group of isometries of a symmetric space X of non-positive curvature, then $\operatorname{rank} H$ coincides with the rank of the space X, i.e. with the dimension of a maximal totally geodesic flat submanifold $M \subset X$.

We call a lattice $\Gamma \subset H$ *reducible* if there exist two connected infinite normal subgroups H' and H'' in H such that $H = H' \cdot H''$, $H' \cap H'' \subset \mathscr{Z}(H)$ and the subgroup $(\Gamma \cap H') \cdot (\Gamma \cap H'')$ is of finite index in Γ; otherwise Γ is called *irreducible*.

Now we can present the Lie-theoretic restatements.

(1') Theorem. (See Theorem 6.5 of Chapter IX.) *Let H be a connected semisimple Lie group with trivial center and no non-trivial compact factor groups, and let Γ be an irreducible lattice in H. Assume that either* $\operatorname{rank} H \geq 2$ *or Γ is of infinite index in* $\operatorname{Comm}_H(\Gamma)$. *Then the lattice Γ is arithmetic, i.e., there exist a connected non-commutative $\mathbb{Q}$-simple $\mathbb{Q}$-group $\mathbf{F}$ and a continuous epimorphism $\sigma\colon \mathbf{F}(\mathbb{R})^0 \to H$ such that the kernel* $\operatorname{Ker}\sigma$ *is compact and the subgroups $\sigma(\mathbf{F}(\mathbb{Z}) \cap \mathbf{F}(\mathbb{R})^0)$ and Γ are commensurable, where $\mathbf{F}(\mathbb{R})^0$ denotes the connected component of the identity in the Lie group $\mathbf{F}(\mathbb{R})$.*

(2') Theorem. (See Theorem 6.16 of Chapter IX.) *Let H be a connected semisimple Lie group without non-trivial compact factor groups. Let $\Gamma \subset H$ be a lattice, $\Lambda \supset \Gamma$ a subgroup of* $\operatorname{Comm}_H(\Gamma)$, *$k$ a local field, $\mathbf{F}$ a connected semisimple k-group, and $\delta\colon \Lambda \to \mathbf{F}(k)$ a homomorphism such that the subgroup $\delta(\Gamma)$ is Zariski dense in $\mathbf{F}$. Assume that either* $\operatorname{rank} H \geq 2$ *and the lattice Γ is irreducible or the subgroup Λ is dense in H. Then,*

(a) for k isomorphic neither to $\mathbb{R}$ nor to $\mathbb{C}$, i.e. for non-archimedean k, the subgroup $\delta(\Gamma)$ is relatively compact in $\mathbf{F}(k)$;
(b) for archimedean k, if the subgroup $\delta(\Gamma)$ is not relatively compact in $\mathbf{F}(k)$ and the group $\mathbf{F}$ is adjoint and k-simple, δ extends, uniquely, to a continuous homomorphism $\tilde{\delta}\colon H \to \mathbf{F}(k)$;
(c) for $k = \mathbb{R}$, if the group $\mathbf{F}$ is adjoint and has no nontrivial $\mathbb{R}$-anisotropic factors, δ extends, uniquely, to a continuous homomorphism $\tilde{\delta}\colon H \to \mathbf{F}(\mathbb{R})$.

(3') Theorem. (See Theorem 6.15 of Chapter IX.) *Let H be a connected semisimple Lie group with finite center, and no non-trivial compact factor groups, let Γ be an irreducible lattice in H, l a field, $\mathbf{F}$ an algebraic l-group, and let $\delta\colon \Gamma \to \mathbf{F}(l)$ be a homomorphism. Assume* $\operatorname{rank} H \geq 2$. *Then*

(i) if $\operatorname{char} l = 0$, *then the Zariski closure of the subgroup $\delta(\Gamma)$ is a semisimple l-group;*
(ii) if $\operatorname{char} l \neq 0$, *then the group $\delta(\Gamma)$ is finite;*
(iii) the one-dimensional cohomology group $H^1(\Gamma, \rho)$ is trivial for every representation ρ of the group Γ on a finite-dimensional vector space over a field of characteristic 0.

(4') Theorem. (See Theorem 6.14 of Chapter IX.) *With H and Γ as in Theorem $3'$, assume* $\operatorname{rank} H \geq 2$. *Then*

(i) for every normal subgroup N of the group Γ, either N is contained in the center of the group H or the factor group Γ/N is finite;
(ii) the factor group of the group Γ modulo its commutator subgroup is finite.

In Sections 4–7 of Chapter IX we present a number of consequences of the results mentioned above. Among them are Theorems 7.20 and 7.23 of Chapter IX which may be viewed as a strengthening of Mostow's theorem on the strong rigidity of locally symmetric spaces (for spaces of rank ≥ 2).

We now turn to some applications of the superrigidity theorems and the theorems on finiteness of factor groups of discrete subgroups to the study of algebraic groups over global fields. With K, $\mathscr{R}$, $\mathscr{R}_\infty$, K_v, $K(S)$, $\mathbf{H}$, $\mathscr{T} = \mathscr{T}(\mathbf{H})$ and $\mathbf{H}(K(S))$ as in the above definition of an arithmetic lattice, we fix a subset $S \subset \mathscr{R}$ such that $S \supset \mathscr{R}_\infty - \mathscr{T}$. We call a subgroup of $\mathbf{H}$ *S-arithmetic* if it is commensurable with $\mathbf{H}(K(S))$. We set $\operatorname{rank}_S \mathbf{H} = \sum_{v \in S} \operatorname{rank}_{K_v} \mathbf{H}$.

Given an l-group $\mathbf{F}$ and a field homomorphism $\sigma\colon l \to l'$, we denote by ${}^\sigma\mathbf{F}$ the l'-group obtained from $\mathbf{F}$ via σ, and by σ^0 the homomorphism of the group $\mathbf{F}(l)$ into ${}^\sigma\mathbf{F}(l')$ induced from σ. (If $\mathbf{F}$ is a matrix l-group, then ${}^\sigma\mathbf{F}$ is obtained by applying the homomorphism σ to the equations defining $\mathbf{F}$; furthermore, if $x = (x_{ij}) \in \mathbf{F}(l)$, then $\sigma^0(x) = (\sigma(x_{ij}))$.)

(5) Theorem. (See Theorem C of Chapter VIII.) *Let Λ be an S-arithmetic subgroup of the group $\mathbf{H}$. Let l be a field, $\mathbf{F}$ a connected non-commutative absolutely almost simple l-group, and let $\delta\colon \Lambda \to \mathbf{F}(l)$ be a homomorphism such that the subgroup $\delta(\Lambda)$ is Zariski dense in $\mathbf{F}$. Assume that* $\operatorname{rank}_S \mathbf{H} \geq 2$ *and that either $\mathbf{H}$ is simply connected or $\mathbf{F}$ is adjoint. Then there exist a homomorphism $\sigma\colon K \to l$, an l-epimorphism $\eta\colon {}^\sigma\mathbf{H} \to \mathbf{F}$, and a homomorphism $\nu\colon \Lambda \to \mathscr{Z}(\mathbf{F})$ such that $\delta(\lambda) = \nu(\lambda) \cdot \eta(\sigma^0(\lambda))$ for all $\lambda \in \Lambda$.*

(6) Theorem. (See Theorem B of Chapter VIII.) *Let Λ be an S-arithmetic subgroup of the group $\mathbf{H}$. Let l be a field, $\mathbf{F}$ an algebraic l-group, and $\delta\colon \Lambda \to \mathbf{F}(l)$ a homomorphism. Assume that* $\operatorname{char} K = 0$ *and* $\operatorname{rank}_S \mathbf{H} \geq 2$. *Then*

(i) for $\operatorname{char} l \neq 0$ *and $\mathbf{H}$ simply connected, the group $\delta(\Lambda)$ is finite;*

(ii) for $\operatorname{char} l = 0$, *the Zariski closure of the group $\delta(\Lambda)$ is a semisimple l-group;*

(iii) for $\operatorname{char} l = 0$ *(i.e. $l \supset \mathbb{Q}$) and $\mathbf{H}$ simply connected, there exist a uniquely determined l-morphism $\varphi\colon R_{K/\mathbb{Q}}\mathbf{H} \to \mathbf{F}$ and a homomorphism $\nu\colon \Lambda \to \mathbf{F}$ such that the subgroup $\nu(\Lambda)$ is finite and commutes with $\varphi(R_{K/\mathbb{Q}}\mathbf{H})$ and $\lambda(\gamma) = \nu(\lambda) \cdot \varphi(R^0_{K/\mathbb{Q}}(\lambda))$ for each $\lambda \in \Lambda$, where $R_{K/\mathbb{Q}}$ denotes the restriction of scalars functor and $R^0_{K/\mathbb{Q}}\colon \mathbf{H}(K) \to (R_{K/\mathbb{Q}}\mathbf{H})(\mathbb{Q})$ denotes the natural isomorphism (for precise definitions see 1.7 of Chapter I).*

(7) Theorem. (See Theorem A of Chapter VIII.) *Let Λ be an S-arithmetic subgroup of the group $\mathbf{H}$ and let N be a normal subgroup in Λ. Assume that* $\operatorname{rank}_S \mathbf{H} \geq 2$ *and that either $\mathbf{H}$ is simply connected or S is finite. Then either N is contained in the center of $\mathbf{H}$ or Λ/N is finite.*

It should be noted that Theorems 5–7 are deduced from Theorems 2–4 (more precisely, from the weak versions of these theorems obtained in Chapters IV and VII) via the Borel-Harish-Chandra-Behr-Harder reduction theorem.

2. Synopsis of the Chapters

Chapter I is preliminary. This chapter contains the results from different areas of mathematics used in the succeeding chapters. Here we state without proofs almost all the results from other books and papers which are used in Chapters II–IX.

The results presented in Chapter II can be divided into three parts:

1) density theorems for subgroups with property (*S*) and, in particular, a version of the Borel-Wang theorem on Zariski density of certain subgroups in algebraic groups;
2) density theorems for projections of discrete subgroups onto direct factors;
3) ergodicity theorems for actions on quotient spaces.

Furthermore we prove a series of assertions concerning the theory of unitary representations and iterations of linear transformations. We make use of these assertions in Chapter II and in Chapters III and V.

Chapter III is devoted to property (*T*). For a locally compact group this means that the trivial one-dimensional representation is isolated in the space of irreducible unitary representations. It turns out that property (*T*) for discrete groups is closely connected with their structure. In particular, if a discrete group Γ has property (*T*), then Γ is finitely generated and the factor group of the group Γ modulo its commutator subgroup is finite. Applications of property (*T*) in the theory of discrete subgroups are based on the fact that lattices in a sufficiently large class of semisimple groups have this property.

In Chapter IV we prove a weak version of above mentioned Theorem 4 on finiteness of factor groups of discrete subgroups. The proofs are largely based on the results of Chapter III and on the study of invariant algebras of measurable sets.

In Chapters V and VI the theorems on existence of equivariant measurable maps are proved. These are used in Chapter VII in the proof of the superrigidity theorems. In Chapter V we present an approach based on the multiplicative ergodic theorem and in Chapter VI an approach based on boundary theory. Chapter VI, or more precisely its first three sections, may be considered as an introduction to the Furstenberg boundary theory.

In Chapter VII the superrigidity theorems for discrete subgroups are established. In particular we prove Theorem 2 stated above. A series of consequences of these theorems are obtained as well. At the end of Chapter VII, namely in Section 8, certain results on rigidity of ergodic actions of semisimple groups are stated.

In Chapter VIII a series of assertions on normal subgroups and "abstract" homomorphisms of algebraic groups over global fields are established with the help of the results of Chapters VI and VIII. In particular, we prove Theorems 5–7 stated above.

In Chapter IX the arithmeticity theorems are stated in full generality and proved. A series of consequences of these theorems are obtained and, in particular, it is shown how with the help of these theorems certain results of Chapters IV and VII can be strengthened. Sections 6 and 7 are largely devoted to the restatement of some previous results in terms of Lie group theory, symmetric space theory, and the theory of complex manifolds.

In Appendices A and B we prove two theorems which are used in Chapter V. In Appendix C we provide some examples of non-arithmetic lattices in real rank 1 semisimple Lie groups.

Finally, we would like to indicate how one might use this book. The reader who is only interested in the arithmeticity theorems may pass, after Chapter I, immediately to Chapter VI, then to Chapter VII till 5.17, and finally to Chapter IX. Doing this the reader would occasionally have to consult other chapters, in particular Sections 2, 4 and 7 of Chapter II. The reader who is mainly interested in the statements of the results should consult Sections 2, 4, 6 and 7 of Chapter II, Section 5 of Chapter III, Sections 5–8 of Chapter VII, Sections 2, 3 of Chapter VIII, and Sections 1, 3–7 of Chapter IX. It should be noted that in Sections 1, 2 of Chapter II, 1–3 of Chapter III, 1, 3 of Chapter IV, 1–3 of Chapter V, 1–4 of Chapter VII and in Chapter VI the machinery of algebraic group theory has almost never been used.

3. Remarks on the Structure of the Book, References and Notation

In this book chapters are divided into sections and sections into paragraphs. In Chapter I, §§1–5, we use a triple spaced numeration, in other chapters only a double spaced one. Cross references to results in another chapter are preceded by the chapter number denoted by a Roman numeral, e.g. Lemma II.3.5, while cross references to results within the same chapter are given without indication of chapter number, e.g. Lemma 3.5

The notation and the terminology used throughout the book are introduced in Section 0 of Chapter I. For the most part, they agree with generally accepted usage. The notation used within a chapter appears at the beginning of that chapter before Section 1. The notation used within a section is usually introduced either at the beginning of that section or before the paragraph with the last number 1 or in the paragraph with the last number 0, e.g. the notation used in Section 3 appears either before paragraph 3.1 or in 3.0.

Chapter I. Preliminaries

0. Notation, Terminology and Some Basic Facts

(0.1) The letters $\mathbb{C}$, $\mathbb{R}$, $\mathbb{R}^+$, $\mathbb{Q}$, $\mathbb{Z}$, $\mathbb{N}$, $\mathbb{N}^+$ and $\mathbb{Q}_p$ will as usual denote the set of complex numbers, reals, positive reals, rationals, integers, non-negative integers, positive integers, and p-adic numbers. Sometimes we write $\mathbb{Q}_\infty$ instead of $\mathbb{R}$ (i.e. $\mathbb{Q}_p$ coincides with $\mathbb{R}$ for $p = \infty$). In particular a p-adic Lie group is a real one for $p = \infty$.

(0.2) We denote, respectively, by $\mathscr{Z}_H(F)$ and $\mathscr{N}_H(F)$ the centralizer and the normalizer of a subset F in a group H. By $\mathscr{Z}(H)$ and $\mathscr{D}(H)$ we denote respectively the center and the commutator subgroup of a group H. The identity element of a group will as usual be denoted by e. If A and B are two subsets of a group, we set $A \cdot B = \{ab \mid a \in A,\ b \in B\}$ and $A^{-1} = \{a^{-1} \mid a \in A\}$. Let $\operatorname{Int} h$ denote the inner automorphism of a group H corresponding to $h \in H$, i.e. $(\operatorname{Int} h)(h') = hh'h^{-1}$, for all $h' \in H$.

Two subgroups of a group are called *commensurable* if their intersection has finite index in both of them. Let G be a group and let Γ be a subgroup of G. We call the set $\operatorname{Comm}_G(\Gamma) = \{g \in G \mid g\Gamma g^{-1}$ and Γ are commensurable$\}$ the *commensurability subgroup* of Γ in G. We remark that $\operatorname{Comm}_G(\Gamma)$ is actually a subgroup of G and that $\Gamma \subset \mathscr{N}_G(\Gamma) \subset \operatorname{Comm}_G(\Gamma)$. If two subgroups Γ_1 and Γ_2 of G are commensurable, then $\operatorname{Comm}_G(\Gamma_1) = \operatorname{Comm}_G(\Gamma_2)$.

We denote, respectively, by hf and fh the left and the right translation of a function (map) f on a group H defined by $h \in H$, i.e., $(hf)(x) = f(h^{-1}x)$ and $(fh)(x) = f(xh)$, $x \in H$.

By $[x, y]$ we denote as usual the commutator of two elements x and y in a Lie algebra. If X and Y are two subsets of a Lie algebra, then we set $[X, Y] = \{[x, y] \mid x \in X,\ y \in Y\}$. The center and the commutator subalgebra of a Lie algebra $\mathfrak{G}$ will be denoted by $\mathscr{Z}(\mathfrak{G})$ and $\mathscr{D}(\mathfrak{G})$ respectively.

(0.3) We denote by $\operatorname{card} A$ or by $|A|$ the number of elements in a finite set A.

(0.4) We denote by $\operatorname{char} k$ the characteristic of a field k.

(0.5) We denote as usual by $f \circ g$ the composition of two maps f, g and by $\varphi|_A$ the restriction of a map φ to a set A.

If f is a real-valued function on a set X, then we denote by f^+ its positive part, i.e. $f^+(x) = \max\{0, f(x)\}$, $x \in X$.

(0.6) If $G = G_1 \times \ldots \times G_n$ is a product of groups $G_1, \ldots, G_n$, then we shall always identify these groups G_i, $1 \leq i \leq n$, with their images under the natural embeddings into G (i.e. with $\{e\} \times \ldots \times \{e\} \times G_i \times \{e\} \times \ldots \times \{e\}$).

(0.7) For standard definitions and results of algebraic geometry and algebraic group theory we refer the reader to [Bo 6] and [Hum 1].

(0.8) We consider the dimension *dim*, unless explicitly stated otherwise, to be defined in the sense of algebraic geometry.

(0.9) Algebraic varieties. By the *Zariski topology* on an algebraic variety we mean as usual the topology with closed subsets being algebraic subvarieties. We say that a subset of an algebraic variety is *Zariski closed* (resp. *open*, *dense*, and so forth) if it is closed (resp. open, dense, and so forth) in its Zariski topology. By the *Zariski closure* of a set we mean its closure in the Zariski topology.

An algebraic variety is said to be *irreducible*, if it is not empty and can not be represented as a union of two proper algebraic subvarieties. Every algebraic variety $\mathbf{M}$ is a union of finitely many maximal algebraic irreducible subvarieties which are called *irreducible components* of $\mathbf{M}$.

Every algebraic variety is Noetherian in the Zariski topology, i.e. its Zariski closed subsets satisfy the descending chain condition. In other words, every descending chain of algebraic subvarieties stabilizes.

A subset of a topological space is said to be *locally closed* if it is representable in the form of the intersection of a closed and an open set. A subset of an algebraic variety is said to be *constructible* if it can be represented in the form of the union of finitely many Zariski locally closed subsets.

Let $\mathbf{X}$ and $\mathbf{Y}$ be two algebraic varieties. A map $\alpha\colon \mathbf{X} \to \mathbf{Y}$ is said to be *regular* or an *algebraic variety morphism* if it is defined by locally regular functions or, in other words, if it satisfies the following two conditions:

(I) α is continuous in the Zariski topology;
(II) for any Zariski open subset U of $\mathbf{Y}$ and for each regular function f on U the function $f \circ \alpha$ is regular on $\alpha^{-1}(U)$.

Let $\alpha\colon \mathbf{X} \to \mathbf{Y}$ be an algebraic variety morphism. Then (see [Bo 6], Chapter AG, Corollary 10.2 or [Hum 1], 4.4) the images under α of all constructible sets are constructible. In particular, if $\alpha(\mathbf{X})$ is Zariski dense in $\mathbf{Y}$, then $\alpha(\mathbf{X})$ contains a Zariski dense and open subset of $\mathbf{Y}$.

If $V \subset \mathbf{X}$ and $U \subset \mathbf{Y}$ are Zariski open, $\alpha\colon \mathbf{X} \to \mathbf{Y}$ is an algebraic variety morphism, and $\alpha(V) \subset U$, then the correspondence $f \mapsto f \circ \alpha|_V$ is a homomorphism of the algebra of regular functions on U into the algebra of regular functions on V. This homomorphism is called the *comorphism* of the morphism α. Strictly speaking a comorphism is a homomorphism of the sheaves of regular functions.

In case $\mathbf{X}$ and $\mathbf{Y}$ are algebraic subvarieties of affine spaces, a mapping $\alpha: \mathbf{X} \to \mathbf{Y}$ is a morphism if and only if the coordinates of the point $\alpha(x)$ are regular functions of $x \in \mathbf{X}$. In other words, in the above-mentioned case, the map $\alpha: \mathbf{X} \to \mathbf{Y}$ is a morphism if and only if it is defined by polynomials. The comorphism of an affine variety morphism $\alpha: \mathbf{X} \to \mathbf{Y}$ may be viewed as a morphism of the algebra of regular functions on $\mathbf{Y}$ to the algebra of regular functions on $\mathbf{X}$.

The morphism $\alpha: \mathbf{X} \to \mathbf{Y}$ is called an *isomorphism* if it is bijective and $\alpha^{-1}: \mathbf{X} \to \mathbf{Y}$ is also a morphism. The isomorphisms are called *biregular maps* as well. Note that not every bijective algebraic variety morphism is an isomorphism. But (see [Bo 6], Chapter AG, 18.2), if $\mathbf{X}$ and $\mathbf{Y}$ are normal irreducible algebraic varieties over a field of characteristic 0, then every bijective morphism $\alpha: \mathbf{X} \to \mathbf{Y}$ is an isomorphism (for the definition of a normal variety we refer to 0.12 below).

Let $\mathbf{X}$ be an algebraic variety. On every constructible subset $A \subset \mathbf{X}$ there is a natural structure of an algebraic variety. A constructible set $A \subset \mathbf{X}$ is called *affine*, if it is isomorphic to an affine algebraic variety.

(0.10) k-structures, k-varieties and k-morphisms. (See [Bo 6], Chapter AG, §§11–14.) Let K be an algebraically closed field and let k be a subfield of K. To define a *k-structure* on a vector space W (not necessarily finite-dimensional) over a field K is the same thing as to distinguish in W a k-submodule $W_k \subset W$ such that the natural homomorphism $K \bigotimes_k W_k \to W$ is an isomorphism. A linear subspace $U \subset W$ is said to be *defined over* k if $U \cap W_k$ is a k-structure on U, i.e. if U is spanned by $U \cap W_k$ over K.

Let $f: V \to W$ be a K-linear map of vector spaces equipped with k-structures. The map f is said to be *defined over* k if $f(V_k) \subset W_k$.

By a *k-structure* on a K-algebra A we mean a k-structure A_k on the underlying space which is simultaneously a k-subalgebra. In other words, a k-structure on a K-algebra A is a k-subalgebra $A_k \subset A$ such that $K \bigotimes_k A_k = A$.

Let $n \in \mathbb{N}^+$ and let $\mathbf{M}$ be an algebraic subvariety of the affine space K^n. The subvariety $\mathbf{M}$ is called *k-closed* or *closed over* k if $\mathbf{M}$ is the set of common zeros of a finite system of polynomials with coefficients in k. A subvariety $\mathbf{M}$ is k-closed if and only if it is invariant under the natural action on K^n of the Galois group of the field K over k. Let us denote by $J(\mathbf{M}) \subset K[x_1, \ldots, x_n]$ the ideal of polynomials vanishing on $\mathbf{M}$, and put $J_k(\mathbf{M}) = J(\mathbf{M}) \cap k[x_1, \ldots, x_n]$. A subvariety $\mathbf{M}$ is said to be *k-defined*, or *defined over* k, if $J_k(\mathbf{M})$ is a k-structure on the space $J(\mathbf{M})$, i.e., if $J(\mathbf{M}) = K \bigotimes_k J_k(\mathbf{M})$. A subvariety $\mathbf{M}$ is closed over k if and only if it is defined over some purely inseparable field extension of k. If k is a perfect field (for instance, if $\operatorname{char} k = 0$), then a subvariety $\mathbf{M}$ is k-closed if and only if it is defined over k. Denote by $K[\mathbf{M}] = K[x_1, \ldots, x_n]/J(\mathbf{M})$ the algebra of regular functions on $\mathbf{M}$. If $\mathbf{M}$ is defined over k, then $K[\mathbf{M}] = K \bigotimes_k k[\mathbf{M}]$, where $k[\mathbf{M}] = k[x_1, \ldots, x_n]/J_k(\mathbf{M})$. In other words, if $\mathbf{M}$ is defined over k, then the k-algebra $k[\mathbf{M}]$ is a k-structure on the K-algebra $K[\mathbf{M}]$. A regular function $f \in K[\mathbf{M}]$ is said to be *defined over* k or *k-regular* if $f \in k[\mathbf{M}]$.

To define a *k-structure* on an affine algebraic variety $\mathbf{X}$ is the same thing as to define an isomorphism $\alpha\colon \mathbf{X} \to \mathbf{M}$ of the variety $\mathbf{X}$ onto an algebraic subvariety $\mathbf{M}$ of an affine space which is defined over k. An affine variety equipped with a k-structure is called an *affine k-variety*. The image of the k-algebra $k[\mathbf{M}]$ under the comorphism of the morphism α is called the algebra of *k-regular functions* on the k-variety $\mathbf{X}$ and is denoted by $k[\mathbf{X}]$. Note that $k[\mathbf{X}]$ is a k-structure of the K-algebra $K[\mathbf{X}]$ of regular functions on $\mathbf{X}$. Two k-structures on an affine variety are called *equivalent* if the corresponding algebras of k-regular functions coincide.

We say that a morphism $\alpha\colon \mathbf{X} \to \mathbf{X}'$ between two affine k-varieties is *defined over k* or that α is a *k-morphism* if the image of $k[\mathbf{X}']$ under the comorphism of α is contained in $k[\mathbf{X}]$. If $\mathbf{X}$ and $\mathbf{X}'$ are subvarieties of affine spaces defined over k, then a morphism $\alpha\colon \mathbf{X} \to \mathbf{X}'$ is defined over k if and only if α is defined by polynomials with coefficients in k.

Let $\mathbf{X}$ be an algebraic variety over K. For any Zariski open subset $U \subset \mathbf{X}$, denote by $K[U]$ the K-algebra of regular functions on U. A k-structure on $\mathbf{X}$ is a topology on $\mathbf{X}$ (called the *k-topology*) which is coarser than the Zariski topology and which has the following properties:

(i) for every k-open (i.e. open in the k-topology) subset $U \subset \mathbf{X}$ a k-structure on $K[U]$ is given;
(ii) for every pair of k-open sets $U \supset V$ the restriction homomorphism $K[U] \to K[V]$ is defined over k;
(iii) there exists a finite cover of $\mathbf{X}$ by k-open affine subsets;
(iv) for every k-open affine subset $U \subset \mathbf{X}$ a k-structure is given on U so that the algebra of k-regular functions coincides with the k-structure on $K[U]$.

A variety $\mathbf{X}$ endowed with a k-structure is called a *k-variety*.

A subvariety $\mathbf{M}$ of a k-variety $\mathbf{X}$ is k-closed (i.e. closed in the k-topology) if and only if, for every k-open affine subset $U \subset \mathbf{X}$, $\mathbf{M} \cap U$ is the set of common zeros of a finite system of k-regular functions on U.

A subvariety $\mathbf{M}$ of a k-variety $\mathbf{X}$ is said to be *defined over* k or *k-defined*, if for each k-open affine subset $U \subset \mathbf{X}$ the intersection $\mathbf{M} \cap U$ is a subvariety of U defined over k (i.e. the ideal $J(\mathbf{M} \cap U)$ of regular functions, which vanish on $\mathbf{M} \cap U$ is defined over k).

Like for the case of subvarieties of K^n, a subvariety $\mathbf{M}$ of $\mathbf{X}$ is k-closed if and only if it is defined over a purely inseparable field extension of k. Subvarieties defined over k are called *k-subvarieties*.

Let $\alpha\colon \mathbf{X} \to \mathbf{Y}$ be a morphism of k-varieties. We say that α is *defined over k* or is a *k-morphism* if

(I) α is continuous with respect to the k-topologies,
(II) for every pair of k-open sets $V \subset \mathbf{X}$ and $U \subset \mathbf{Y}$ such that $\alpha(V) \subset U$, the comorphism $\alpha_V^U\colon K[U] \to K[V]$, $f \mapsto f \circ \alpha|_V$ is defined over k.

If a morphism α is defined over k and is an isomorphism, then α^{-1} is also defined over k.

If $\mathbf{M}$ is a k-subvariety of K^n, we denote by $\mathbf{M}(k) = \mathbf{M} \cap k^n$ the set of its *k-rational points*. If $\mathbf{X}$ is an affine k-variety and $\alpha\colon \mathbf{X} \to \mathbf{M}$ is an isomorphism defined over k of the variety $\mathbf{X}$ onto a k-subvariety $\mathbf{M}$ of an affine space, then we set $\mathbf{X}(k) = \alpha^{-1}(\mathbf{M}(k))$. For an arbitrary k-variety $\mathbf{X}$ we set $\mathbf{X}(k) = \{x \in \mathbf{X} \mid$ there exists a k-open affine neighbourhood $\mathbf{M}$ of x such that $x \in \mathbf{M}(k)\} = \{x \in \mathbf{X} \mid$ for each k-open affine neighbourhood $\mathbf{M}$ of x, $x \in \mathbf{M}(k)\}$.

In case $\mathbf{X}$ is affine, $\mathbf{X}(k)$ may be identified with the set of homomorphisms of the K-algebra $K[\mathbf{X}]$ to K which are defined over k.

In the sequel we shall as usual identify the k-variety $\mathbf{X}$ with the set $\mathbf{X}(K)$.

(0.11) Let K be an algebraically closed field, k a subfield of K, $\mathbf{M}$ a k-variety and $B \subset \mathbf{M}(k)$. We denote by $\overline{B}$ the Zariski closure of B in $\mathbf{M}$. Then the following propositions hold:

(I) *The subvariety $\overline{B}$ is defined over k* (see [Bo 6], Chapter AG, Theorem 14.4).

(II) *If $f\colon \mathbf{M} \to \mathbf{M}'$ is a k-variety morphism, $\overline{B} = \mathbf{M}$, and $f(B) \subset \mathbf{M}'(k)$, then f is defined over k* (see [Bo-T4], 1.4).

(0.12) Tangent spaces. Smooth and normal varieties. Separable morphisms. (See [Bo 6], Chapter AG, §§15–18 and [Hum 1], §5.) Let $\mathbf{M}$ be an algebraic variety defined over an algebraically closed field K and let $x \in \mathbf{M} = \mathbf{M}(K)$. Let us denote by $\mathcal{O}_x$ the local ring of the point x and by m_x the unique maximal ideal in $\mathcal{O}_x$ (we recall that if $\mathbf{M}$ is affine and irreducible, then $\mathcal{O}_x$ is the ring of rational functions on $\mathbf{M}$ representable in the form g/h, where $g, h \in k[\mathbf{M}]$ and $h(x) \neq 0$). The quotient ring $\mathcal{O}_x/m_x$ is naturally isomorphic to K. We denote the image of the element $f \in \mathcal{O}_x$ under the natural epimorphism $\mathcal{O}_x \to \mathcal{O}_x/m_x = K$ by $f(x)$. *A derivation at the point* x is defined as a K-linear map $\delta\colon \mathcal{O}_x \to \mathcal{O}_x/m_x$, satisfying the condition $\delta(fg) = f(x) \cdot \delta(g) + g(x) \cdot \delta(f)$, $f, g \in \mathcal{O}_x$. The derivations at the point x form a finite-dimensional vector space over K which is called the *tangent space* of $\mathbf{M}$ at x and denoted by $\mathcal{T}_x(\mathbf{M})$. The vector space $\mathcal{T}_x(\mathbf{M})$ is naturally isomorphic to the dual space of m_x/m_x^2.

If $\alpha\colon \mathbf{M} \to \mathbf{M}'$ is a morphism of algebraic varieties, then one can define a natural map $(d\alpha)_x\colon \mathcal{T}_x(\mathbf{M}) \to \mathcal{T}_{\alpha(x)}(\mathbf{M}')$ which is called the *differential* of α at the point x. This mapping is defined as follows: $(d\alpha)_x(X)(f) = X(\alpha_0(f))$, where $X \in \mathcal{T}_x(\mathbf{M})$, $f \in \mathcal{O}_{\alpha(x)}$, and α_0 is the comorphism of α.

If $\mathbf{M}$ is defined over k and $x \in \mathbf{M}(k)$, then there exists a natural k-structure $\mathcal{O}_{x,k}$ on $\mathcal{O}_x$. This k-structure induces the k-structure $\mathcal{T}_{x,k}(\mathbf{M})$ on the space $\mathcal{T}_x(\mathbf{M})$, consisting of the derivations which map $\mathcal{O}_{x,k}$ into k. If $\alpha\colon \mathbf{M} \to \mathbf{M}'$ is a k-morphism and $x \in \mathbf{M}(k)$, then $(d\alpha)_x(\mathcal{T}_{x,k}(\mathbf{M})) \subset \mathcal{T}_{\alpha(x),k}(\mathbf{M}')$.

A point $x \in \mathbf{M}$ is called *simple* if its local ring $\mathcal{O}_x$ is regular. For an irreducible variety $\mathbf{M}$ this means that $\dim \mathcal{T}_x(\mathbf{M}) = \dim \mathbf{M}$ (we recall that if $\mathbf{M}$ is irreducible, then $\dim \mathcal{T}_x(\mathbf{M}) \geq \dim \mathbf{M}$ for all $x \in \mathbf{M}$). The set of simple points of $\mathbf{M}$ is Zariski dense and open in $\mathbf{M}$. If this set coincides with $\mathbf{M}$, we call $\mathbf{M}$ *smooth*.

We say that a point $x \in \mathbf{M}$ is *normal* in $\mathbf{M}$ if $\mathcal{O}_x$ is an integral domain, integrally closed in its quotient field. A variety $\mathbf{M}$ is called *normal* if all of its points are normal in $\mathbf{M}$. Every smooth algebraic variety is normal.

A k-variety morphism $\alpha\colon \mathbf{M} \to \mathbf{M}'$ is called *separable* if for every irreducible component $\mathbf{V}$ of $\mathbf{M}$ the field of rational functions on $\mathbf{V}$ is a separable extension of the image of the field of rational functions on $\alpha(\mathbf{V})$ under the comorphism $\alpha|_{\mathbf{V}}$. If char $k = 0$, then every k-variety morphism is separable.

A morphism α is called *dominant* if the image $\alpha(\mathbf{M})$ is Zariski dense in $\mathbf{M}'$ and the image under α of every irreducible component of $\mathbf{M}$ is a Zariski dense subset of an irreducible component of $\mathbf{M}'$.

Let $\alpha: \mathbf{M} \to \mathbf{M}'$ be a dominant morphism. Then the following conditions are equivalent:

(1) the morphism α is separable;
(2) the map $(d\alpha)_x$ is surjective for each point x of a Zariski dense and open subset $V \subset \mathbf{M}'$;
(3) every irreducible component of $\mathbf{M}$ contains a simple point x of $\mathbf{M}$ such that $\alpha(x)$ is a simple point of $\mathbf{M}$ and the map $(d\alpha)_x$ is surjective.

(0.13) Algebraic groups and their morphisms. Let $\mathbf{G}$ be an algebraic variety with group structure and suppose that both maps $\mu: \mathbf{G} \times \mathbf{G} \to \mathbf{G}$ and $\tau: \mathbf{G} \to \mathbf{G}$, where $\mu(x, y) = xy$ and $\tau(x) = x^{-1}$ are morphisms of algebraic varieties. Then $\mathbf{G}$ is called an *algebraic group.* If the algebraic group $\mathbf{G}$ is a k-variety and the morphisms μ and τ are defined over k, then $\mathbf{G}$ is said to be *defined over k.* Algebraic groups defined over a field k are also called *algebraic groups over k* or *algebraic k-groups,* or simply *k-groups.* Every algebraic group is a smooth, and hence normal, algebraic variety (see [Bo 6], Chapter I, Proposition 1.2).

A subgroup $\mathbf{H}$ of an algebraic group $\mathbf{G}$ is called *algebraic* if $\mathbf{H}$ is an algebraic subvariety of $\mathbf{G}$. Algebraic subgroups defined over k (as algebraic subvarieties) are called *k-subgroups.* An algebraic subgroup of an algebraic group is called *k-closed* or *closed over k* (resp. *k-defined* or *defined over k*) if it is k-closed (resp. k-defined) as an algebraic subvariety.

The commutator subgroup of a k-group is a k-subgroup (see [Bo 6], Chapter I, Proposition 2.3). If H and F are two subgroups of an algebraic group $\mathbf{G}$, and H normalizes F, then $\overline{(H,F)} = (\overline{H},\overline{F})$, where $\overline{M}$ denotes the Zariski closure of $M \subset \mathbf{G}$ and (M_1, M_2) denotes the mutual commutator subgroup of the subgroups M_1 and M_2 of $\mathbf{G}$ (see [Bo 6], Chapter I, 2.3). We obtain as a consequence that the Zariski closure of a solvable (resp. nilpotent) group is a solvable (resp. nilpotent) group as well. The center of a k-group $\mathbf{G}$ is an algebraic subgroup of $\mathbf{G}$ which is k-closed (but not necessarily defined over k).

An algebraic group is called *connected* if it is connected in the Zariski topology. Let $\mathbf{G}$ be an algebraic k-group. Let us denote by $\mathbf{G}^0$ the connected component of the identity in the group $\mathbf{G}$, that is the maximal connected algebraic subgroup of $\mathbf{G}$. Then the factor group $\mathbf{G}/\mathbf{G}^0$ is finite, $\mathbf{G}^0$ is defined over k and every algebraic subgroup of finite index in $\mathbf{G}$ contains $\mathbf{G}^0$ (see [Bo 6], Chapter I, Proposition 1.2). Therefore, if H is a Zariski dense subgroup of $\mathbf{G}$ and F is a subgroup of finite index in H, then the Zariski closure $\overline{F}$ of F contains $\mathbf{G}^0$. In particular, if $\mathbf{G}$ is connected, then F is Zariski dense in $\mathbf{G}$. This implies in turn that if H_1 and H_2 are two commensurable subgroups of a connected algebraic group $\mathbf{G}$, then H_1 is Zariski dense in $\mathbf{G}$ if and only if H_2 is Zariski dense in $\mathbf{G}$.

If $\mathbf{G}$ is an algebraic k-group and H is a subgroup of the group $\mathbf{G}(k)$, then the Zariski closure of H is a k-subgroup of $\mathbf{G}$ (see [Bo 6], Chapter I,

Proposition 1.3). An algebraic group *morphism* (resp. *epimorphism, automorphism, isomorphism* etc.) is defined as a group homomorphism which at the same time is an algebraic variety morphism (resp. epimorphism, automorphism, isomorphism etc.). The assumption "α: $\mathbf{G} \to \mathbf{G}'$ is a *k-group k-morphism* (resp. *k-epimorphism, k-automorphism, k-isomorphism,* etc.)" means that $\mathbf{G}$ and $\mathbf{G}'$ are k-groups and α is a morphism (resp. epimorphism, automorphism, isomorphism, etc.) defined over k. A homomorphism α: $\mathbf{G} \to \mathbf{G}'$ is said to be *rational* (resp. *k-rational* or a *k-homomorphism*) if α is an algebraic group morphism (resp. k-morphism). The k-groups $\mathbf{G}$ and $\mathbf{G}'$ are said to be *isomorphic over k* or *k-isomorphic* if there exists a k-isomorphism α: $\mathbf{G} \to \mathbf{G}'$. If α: $\mathbf{G} \to \mathbf{G}'$ is a k-group morphism, then $\alpha(\mathbf{G})$ is a k-subgroup in $\mathbf{G}'$, $\alpha(\mathbf{G}^0) = \alpha(\mathbf{G})^0$, and $\dim \mathbf{G} = \dim \operatorname{Ker} \alpha + \dim \alpha(\mathbf{G})$ (see [Bo 6], Chapter I, Corollary 1.4).

An algebraic group is called *affine* if it is affine viewed as an algebraic variety. If $\mathbf{G}$ is an affine k-group, then $\mathbf{G}$ is k-isomorphic to a k-subgroup of the group $\mathbf{GL}_n$ (for a suitable n), where $\mathbf{GL}_n$ denotes as usual the group of invertible $n \times n$ matrices (see [Bo 6], Chapter I, Proposition 1.10). Algebraic subgroups of $\mathbf{GL}_n$ are called *linear algebraic groups.*

The algebra $k[\mathbf{GL}_n]$ of k-regular functions on $\mathbf{GL}_n$ has the form $k[\mathbf{GL}_n] = k[T_{11}, T_{12}, \ldots, T_{nn}, D^{-1}]$, where T_{ij} are the entries of a matrix $g \in \mathbf{GL}_n$ and $D = \det g$. Therefore, every k-regular function on a k-subgroup $\mathbf{H} \subset \mathbf{GL}_n$ is defined by a polynomial in the entries of the matrices of h and h^{-1}, where $h \in \mathbf{H}$. If $\mathbf{H} \subset \mathbf{SL}_n$, then every regular function on $\mathbf{H}$ is defined by a polynomial in the entries of a matrix $h \in \mathbf{H}$. Here $\mathbf{SL}_n$ denotes as usual the group of unimodular $n \times n$ matrices.

In what follows the term "algebraic group" will always mean an affine algebraic group realized as an algebraic subgroup of $\mathbf{GL}_n$.

(0.14) We denote by End_n the vector space of all linear transformations of an n-dimensional vector space, by $\mathbf{P}_n$ the projective space of dimension n, by $\mathbf{Gr}_{n,l}$ the Grassman variety of l-dimensional linear subspaces in an n-dimensional vector space, and by $\mathbf{Gr}_n$ the union $\bigcup_{0 \le l \le n} \mathbf{Gr}_{n,l}$. On $\mathbf{P}_n$ and on $\mathbf{Gr}_{n,l}$ (and hence on $\mathbf{Gr}_n$) one can define the structure of an algebraic variety (see [Bo 6], Chapter III, 10.3). By $\mathrm{End}(W)$ we denote the vector space of all linear transformations of a finite-dimensional vector space W, by $\mathbf{GL}(W)$ (resp. $\mathbf{SL}(W)$) the group of invertible (resp. of determinant 1) linear transformations of the space W, by $\mathbf{P}(W)$ the projective space corresponding to W (i.e. the set of lines in W), by $\mathbf{Gr}_l(W)$ the Grassman variety of l-dimensional linear subspaces in W and by $\mathbf{Gr}(W)$ the union $\bigcup_{0 \le l \le \dim W} \mathbf{Gr}_l(W)$.

Let K be an algebraically closed field, $k \subset K$ a subfield, and let W be a finite-dimensional vector space over K with k-structure W_k. Then we can define the natural k-structures on $\mathbf{GL}(W)$, $\mathbf{SL}(W)$, $\mathbf{Gr}_l(W)$, $\mathbf{P}(W)$, etc. We shall identify the sets of k-points of k-groups and k-varieties $\mathbf{GL}(W)$, $\mathbf{SL}(W)$, $\mathbf{Gr}_l(W)$, $\mathbf{P}(W)$, etc., with $\mathbf{GL}(W_k)$, $\mathbf{SL}(W_k)$, $\mathbf{Gr}_l(W_k)$, $\mathbf{P}(W_k)$, etc. We shall similarly identify $\mathbf{GL}_n(k)$ with $\mathbf{GL}(k^n)$, $\mathbf{SL}_n(k)$ with $\mathbf{SL}(k^n)$, $\mathbf{P}_{n-1}(k)$ with $\mathbf{P}(k^n)$, $\mathbf{Gr}_{n,l}(k^n)$ with $\mathbf{Gr}_l(k^n)$, $\mathbf{Gr}_n(k)$ with $\mathbf{Gr}(k^n)$ and also $\mathrm{End}_n(k)$ with $\mathrm{End}(k^n)$.

A representation ρ: $\mathbf{G} \to \mathbf{GL}(W)$ of a k-group $\mathbf{G}$ on the space W is called *rational* (resp. *k-rational* or *defined over k*) if ρ is an algebraic group morphism (resp. k-morphism).

A representation ρ: $G \to \mathbf{GL}(W_k)$ of an abstract group G (resp. k-rational representation ρ: $\mathbf{G} \to \mathbf{GL}(W)$ of a k-group $\mathbf{G}$) is called *k-irreducible* or *irreducible over k*, if W does not contain $\rho(G)$-invariant (resp. $\rho(\mathbf{G})$-invariant) non-trivial subspaces defined over k. A representation ρ is called *absolutely irreducible* if it is irreducible over K.

(0.15) Lie algebra of an algebraic group. (See [Bo 6], Chapter I and [Hum 1], Chapter III.) Let K be an algebraically closed field, k a subfield of K, and let $\mathbf{G}$ be an algebraic k-group. The *Lie algebra* of the group $\mathbf{G}$, i.e. the Lie algebra of left-invariant derivations of the algebra $k[\mathbf{G}]$, is denoted by $\mathrm{Lie}(\mathbf{G})$. The set $\{D \in \mathrm{Lie}(\mathbf{G}) \mid D(k[\mathbf{G}]) \subset k[\mathbf{G}]\}$ will be denoted by $\mathrm{Lie}(\mathbf{G})_k$. We remark that $\mathrm{Lie}(\mathbf{G})_k$ is a k-structure on $\mathrm{Lie}(\mathbf{G})$. If $D \in \mathrm{Lie}(\mathbf{G})$, then by setting $D_e(f) = (Df)(e)$ we obtain a derivation D_e at the identity element of the group $\mathbf{G}$. The correspondence $D \mapsto D_e$ is a linear isomorphism of the Lie algebra $\mathrm{Lie}(\mathbf{G})$ onto the tangent space $\mathscr{T}_e(\mathbf{G})$ defined over k.

The Lie algebra $\mathrm{Lie}(\mathbf{GL}_m)$ of the group $\mathbf{GL}_m$ is canonically identified with the Lie algebra End_m of all $m \times m$ matrices. If $\mathbf{G}$ is a k-subgroup in $\mathbf{GL}_m$, then, denoting by E the identity matrix, we can set

$$\mathrm{Lie}(\mathbf{G}) = \{A \in \mathrm{End}_m \mid \frac{df(E+tA)}{dt}(0) = 0 \text{ for all } f \in J(\mathbf{G})\},$$

where $J(\mathbf{G}) \subset K[\mathrm{End}_m]$ is the ideal of polynomials vanishing on $\mathbf{G}$.

Each k-group morphism f: $\mathbf{G} \to \mathbf{G}'$ induces the Lie algebra morphism $\mathrm{Lie}(\mathbf{G}) \to \mathrm{Lie}(\mathbf{G}')$, which is called the *differential* of the morphism f and is denoted by df (note that $df(D)_e = (df)_e(D)$, where $D \in \mathrm{Lie}(\mathbf{G})$, and $(df)_e$ is defined in 0.12 above). If f is defined over k, then so is df, that is $df(\mathrm{Lie}(\mathbf{G})_k) \subset \mathrm{Lie}(\mathbf{G}')_k$.

Let f: $\mathbf{G} \to \mathbf{G}'$ be a k-epimorphism. Then the following conditions are equivalent:

(1) viewed as a morphism of algebraic varieties, f is separable;
(2) the morphism df is surjective.

The differential of the morphism $\mathrm{Int}\, g$, $g \in \mathbf{G}$, is denoted by $\mathrm{Ad}\, g$. If $\mathbf{G}$ is an algebraic subgroup in $\mathbf{GL}_m$, then $(\mathrm{Ad}\, g)x = gxg^{-1}$ for all $x \in \mathrm{Lie}(\mathbf{G}) \subset \mathrm{End}_m$. The mapping Ad: $\mathbf{G} \to \mathbf{GL}(\mathrm{Lie}(\mathbf{G}))$ is a k-group morphism and hence $\mathrm{Ad}\,\mathbf{G}$ is a k-subgroup in $\mathbf{GL}(\mathrm{Lie}(\mathbf{G}))$. We call Ad the *adjoint representation* of the group $\mathbf{G}$. The differential of the morphism Ad is the adjoint representation ad: $\mathrm{Lie}(\mathbf{G}) \to \mathrm{End}(\mathrm{Lie}(\mathbf{G}))$ of the Lie algebra $\mathrm{Lie}(\mathbf{G})$. Since for each $g \in \mathbf{G}(k)$ the automorphism $\mathrm{Int}\, g$ is defined over k, it follows that $\mathrm{Lie}(\mathbf{G})_k$ is invariant relative to $\mathrm{Ad}\,\mathbf{G}(k)$. If α: $\mathbf{G} \to \mathbf{G}'$ is an algebraic group morphism, then $d\alpha \circ \mathrm{Ad}\, g = \mathrm{Ad}(\alpha(g)) \circ d\alpha$ for all $g \in \mathbf{G}$.

For every $A \subset \mathfrak{G} = \mathrm{Lie}(\mathbf{G})$ we set

$$\mathscr{Z}_{\mathbf{G}}(A) = \{g \in \mathbf{G} \mid \mathrm{Ad}\, g(a) = a \text{ for each } a \in A\} \text{ and}$$
$$\mathscr{Z}_{\mathfrak{G}}(A) = \{f \in \mathfrak{G} \mid \mathrm{ad}\, f(a) = a \text{ for each } a \in A\},$$

and call $\mathscr{Z}_{\mathbf{G}}(A)$ (resp. $\mathscr{Z}_{\mathfrak{G}}(A)$) the *centralizer* of the subset A in $\mathbf{G}$ (resp. in $\mathfrak{G}$).

If $\operatorname{char} k = 0$, then the following assertions hold (see [Bo 6], Chapter II, §7 and [Hum 1], §13):

(i) *The map* $\mathbf{H} \mapsto \operatorname{Lie}(\mathbf{H})$ *is one-to-one and preserves the inclusion between connected algebraic subgroups* $\mathbf{H}$ *of the group* $\mathbf{G}$ *and their Lie algebras, viewed as subalgebras of the Lie algebra* $\operatorname{Lie}(\mathbf{G})$.

(ii) *Let* $\mathbf{G}$ *be a connected algebraic group and let* $\mathbf{H}$ *be a connected algebraic subgroup of* $\mathbf{G}$. *Then* $\operatorname{Lie}(\mathbf{H})$ *is an ideal in* $\operatorname{Lie}(\mathbf{G})$ *if and only if* $\mathbf{H}$ *is a normal subgroup of* $\mathbf{G}$.

(iii) *If* $\mathbf{G}$ *is connected,* $\mathfrak{G} = \operatorname{Lie}(\mathbf{G})$ *and* $A \subset \mathfrak{G}$, *then*
   (a) $\operatorname{Lie}(\mathscr{Z}_{\mathbf{G}}(A)) = \mathscr{Z}_{\mathfrak{G}}(A)$;
   (b) $\operatorname{Ker}\operatorname{Ad} = \mathscr{Z}(\mathbf{G})$;
   (c) $\operatorname{Lie}(\mathscr{Z}(\mathbf{G})) = \mathscr{Z}(\mathfrak{G})$.

(iv) *If* $\mathbf{G}$ *is connected, then* $\operatorname{Lie}(\mathscr{D}(\mathbf{G})) = \mathscr{D}(\operatorname{Lie}(\mathbf{G}))$;

(v) *If a Lie subalgebra* $\mathfrak{H} \subset \operatorname{Lie}(\mathbf{G})$ *coincides with its own commutator subalgebra, then* $\mathfrak{H}$ *is a Lie algebra of an algebraic subgroup* $\mathbf{H}$ *of* $\mathbf{G}$.

If φ_1 and φ_2 are algebraic group morphisms of an algebraic group $\mathbf{H}$ to $\mathbf{G}$ and $d\varphi_1 = d\varphi_2$, then the Lie algebras of the subgroups

$$\{(\varphi_1(h),\ \varphi_2(h)) \mid h \in \mathbf{H}\} \text{ and } \{(\varphi_1(h),\ \varphi_1(h)) \mid h \in \mathbf{H}\}$$

of the group $\mathbf{G} \times \mathbf{G}$ coincide. Therefore (i) implies that

(vi) *If* $\operatorname{char} k = 0$, *then each morphism of a connected k-group is uniquely determined by its differential.*

(0.16) If $\alpha\colon\ \mathbf{G} \times \mathbf{V} \to \mathbf{V}$, $(g, v) \mapsto gv = \alpha(g, v)$, is a morphism with $ev = v$ and $g(hv) = (gh)v$ for all $v \in \mathbf{V}$ and $g, h \in \mathbf{G}$, we say that the algebraic group $\mathbf{G}$ *acts rationally* on the algebraic variety $\mathbf{V}$. If $\mathbf{G}$, $\mathbf{V}$ and α are defined over k, then we say that $\mathbf{G}$ *acts k-rationally* on $\mathbf{V}$. The following proposition (see [Bo 6], Chapter I, Proposition 1.8 or [Hum 1], 8.3) gives some basic information on orbits of rational actions.

Proposition. *Let* $\mathbf{G}$ *be an algebraic group acting rationally on an algebraic variety* $\mathbf{V}$. *Then every orbit (i.e. a set of the form* $\{gv \mid g \in \mathbf{G}\}$, $v \in \mathbf{V}$*) is a smooth variety which is open in its Zariski closure in* $\mathbf{V}$. *The boundary of each orbit is the union of orbits of strictly lower dimension. In particular the orbits of minimal dimension are Zariski closed.*

(0.17) Quotient spaces and factor groups. (See [Bo 6], Chapter II, §6 or [Hum 1], Chapter IV.) Let $\pi\colon\ \mathbf{V} \to \mathbf{W}$ be a morphism of k-varieties defined over k. We say that π is a *quotient morphism* (over k) if

(a) π is surjective and Zariski open;

(b) for every Zariski open subset U of the variety $\mathbf{V}$ the comorphism of π induces an isomorphism of the algebra of regular functions on $\pi(U)$ onto the set of all regular functions on U which are constant on the fibers of the restriction $\pi|_U$.

Quotient morphisms satisfy the following universality property.

Let π: $\mathbf{V} \to \mathbf{W}$ *be a quotient k-morphism. If* α: $\mathbf{V} \to \mathbf{M}$ *is an arbitrary morphism which is constant on the fibers of* π, *then there exists a unique morphism* β: $\mathbf{W} \to \mathbf{M}$ *such that* $\alpha = \beta \circ \pi$. *If* α *is a k-morphism, then so is* β.

If $\mathbf{G}$ is an (affine) algebraic k-group and $\mathbf{H}$ a k-subgroup of $\mathbf{G}$, then one can introduce on the homogeneous space $\mathbf{G}/\mathbf{H}$ a k-variety structure so that the natural map $\mathbf{G} \to \mathbf{G}/\mathbf{H}$ is a separable quotient k-morphism. More precisely, there exist a k-variety $\mathbf{M}$ and a separable quotient k-morphism π: $\mathbf{G} \to \mathbf{M}$ with fibers being the cosets modulo $\mathbf{H}$. The variety $\mathbf{G}/\mathbf{H}$ is smooth and quasi-projective. If the subgroup $\mathbf{H}$ is normal, then $\mathbf{G}/\mathbf{H}$ is a k-group and the canonical homomorphism $\mathbf{G} \to \mathbf{G}/\mathbf{H}$ is a k-group morphism. The proof of the above-mentioned assertions are based on the following theorem of Chevalley (see [Bo 6], Chapter II, Theorem 5.1 or [Hum 1], Theorem 11.2):

Theorem. *Let* $\mathbf{H}$ *be a k-subgroup of a k-group* $\mathbf{G}$. *Then there exist a faithful, finite-dimensional representation* α: $\mathbf{G} \to \mathbf{GL}(W)$ *defined over k and a* 1*-dimensional k-subspace* $L \subset W$ *such that* $\mathbf{H} = \{g \in \mathbf{G} \mid \alpha(g)L = L\}$ *and* $\mathrm{Lie}(\mathbf{H}) = \{y \in \mathrm{Lie}(\mathbf{G}) \mid d\alpha(y)L \subset L\}$.

(0.18) A surjective group homomorphism with finite kernel is called an *isogeny*. Since every finite normal subgroup of a connected algebraic group $\mathbf{G}$ is central (i.e. contained in $\mathscr{Z}(\mathbf{G})$), the kernel of every isogeny of connected algebraic groups is central. A k-group epimorphism (resp. isomorphism, isogeny etc.) which is a k-morphism will be called a *k-epimorphism* (resp. *k-isomorphism*, *k-isogeny* etc.).

(0.19) An algebraic group $\mathbf{G}$ is said to be a *direct product* of the algebraic normal subgroups $\mathbf{G}_1, \ldots, \mathbf{G}_n$, provided that the multiplication map $\mathbf{G}_1 \times \ldots \times \mathbf{G}_n \to \mathbf{G}$ is an algebraic group isomorphism. If this map is an isogeny, $\mathbf{G}$ is called an *almost direct product* of the subgroups $\mathbf{G}_i$. An algebraic group $\mathbf{G}$ is called a *semi-direct product* of a normal subgroup $\mathbf{H} \subset \mathbf{G}$ and an algebraic subgroup $\mathbf{F} \subset \mathbf{G}$ if the multiplication map induces an isomorphism of the algebraic variety $\mathbf{H} \times \mathbf{F}$ onto $\mathbf{G}$. In that case we shall write $\mathbf{G} = \mathbf{H} \rtimes \mathbf{F}$ or $\mathbf{G} = \mathbf{F} \ltimes \mathbf{H}$. The symbols $\rtimes$ and $\ltimes$ will be used for semi-direct products of not only algebraic groups but also of other groups, e.g. topological, Lie groups etc., for which a semi-direct product is defined analogously.

(0.20) Semisimple, unipotent, and nilpotent elements. Jordan decomposition. (See [Bo 6], Chapter I, §4 and [Hum 1], §15.) Let K be an algebraically closed field and $x \in \mathrm{End}(W)$, where W is a finite-dimensional vector space over K. An endomorphism x is said to be *semisimple* if W is spanned by the eigen-vectors of x, i.e. if x is diagonalizable. An endomorphism x is called *nilpotent* if $x^n = 0$ for some $n \in \mathbb{N}^+$, and *unipotent* if $x-1$ is nilpotent. In other words, an endomorphism x is nilpotent (resp. unipotent) if all the eigenvalues of x equal 0 (resp. 1). The following propositions are well known (see [Bo 6], Chapter I, 4.2 or [Hum 1], 15.1).

(a) *Let* $x \in \mathrm{End}(W)$. *Then there exist uniquely determined elements* $x_s, x_n \in \mathrm{End}(W)$, *such that:*

$$x = x_s + x_n,\ x_s \text{ is semisimple,}$$
$$x_n \text{ is nilpotent and}$$
$$x_s x_n = x_n x_s.$$

(b) *Let* $x \in \mathbf{GL}(W)$. *Then there exist uniquely determined elements* $x_s, x_u \in \mathbf{GL}(W)$ *such that:*

$$x = x_s x_u,\ x_s \text{ is semisimple,}$$
$$x_u \text{ is unipotent and}$$
$$x_s x_u = x_u x_s.$$

These decompositions $x = x_s + x_n$ and $x = x_s x_u$ are called, the additive and the multiplicative *Jordan decompositions.* We shall call x_s and x_u, the *semisimple part* and the *unipotent part* of the endomorphism $x \in \mathbf{GL}(W)$.

Let k be a subfield of the field K, and let $\mathbf{G} \subset \mathbf{GL}_n$ be an algebraic k-group. The set of all unipotent (resp. semisimple) elements of the group $\mathbf{G}$ will be denoted by $\mathbf{G}^{(u)}$ (resp. $\mathbf{G}^{(s)}$). Similarly, the set of all nilpotent (resp. semisimple) elements of the Lie algebra $\mathrm{Lie}(\mathbf{G})$ viewed as a subspace of End_n will be denoted by $\mathrm{Lie}(\mathbf{G})^{(n)}$ (resp. $\mathrm{Lie}(\mathbf{G})^{(s)}$). The set $\mathbf{G}^{(u)}$ is a k-closed algebraic subvariety of $\mathbf{G}$. If $\mathbf{G} = \mathbf{G}^{(u)}$, then the group $\mathbf{G}$ is called *unipotent*. For connected groups $\mathbf{G}$ the following conditions are equivalent:

(a) $\mathbf{G}$ is unipotent;
(b) $\mathrm{Lie}(\mathbf{G})^{(n)} = \mathrm{Lie}(\mathbf{G})$.

Every unipotent k-subgroup of the group $\mathbf{GL}_n$ is conjugate by some $g \in \mathbf{GL}_n(k)$ to a subgroup of the group of upper triangular matrices with ones in the diagonal. Hence, if the group $\mathbf{G}$ is unipotent, then $\mathbf{G}$ is nilpotent (see [Bo 6], Chapter V, Corollary 15.5).

We say that a unipotent k-group $\mathbf{G}$ is *split over* k or k-*split*, if it has a composition series

$$\mathbf{G} = \mathbf{G}_0 \supset \mathbf{G}_1 \supset \ldots \supset \mathbf{G}_m = \{e\}$$

of connected k-subgroups such that the factor groups $\mathbf{G}_i/\mathbf{G}_{i+1}$ are isomorphic over k to the additive group $\mathbf{G}_a$ of dimension 1. If the field k is perfect, in particular if $\mathrm{char}\, k = 0$, then every unipotent k-group is k-split (see [Bo 6], Chapter V, Corollary 15.5).

The following propositions hold:

(i) *Let* $g \in \mathbf{G}$ *and* $x \in \mathrm{Lie}(\mathbf{G})$. *Then the semisimple and unipotent (resp. nilpotent) parts of* g *(resp.* x*) belong to* $\mathbf{G}$ *(resp.* $\mathrm{Lie}(\mathbf{G})$*). If* $g \in \mathbf{G}(k)$ *(resp.* $x \in \mathrm{Lie}(\mathbf{G})_k$*) and* $\mathrm{char}\, k = 0$, *then* $g_s, g_u \in \mathbf{G}(k)$ *(resp.* $x_s, x_n \in \mathrm{Lie}(\mathbf{G})_k$*). If* $g \in \mathbf{G}(k)$ *(resp.* $x \in \mathrm{Lie}(\mathbf{G})_k$*) and* $p = \mathrm{char}\, k > 0$, *then* $g_s, g_u \in \mathbf{G}(k^{p^{-\infty}})$ *(resp.* $x_s, x_n \in \mathrm{Lie}(\mathbf{G})_{k^{p^{-\infty}}}$*), where* $k^{p^{-\infty}} = \{y \in K \mid \text{there exists } n \in \mathbb{N}^+ \text{ such that } y^{p^n} \in k\}$ *denotes the maximal purely inseparable field extension of* k.

(ii) *If* $\varphi\colon \mathbf{G} \to \mathbf{G}_1$ *is an algebraic group morphism,* $g \in \mathbf{G}$ *and* $x \in \mathrm{Lie}(\mathbf{G})$, *then*

$$\varphi(g)_s = \varphi(g_s),\ \varphi(g)_u = \varphi(g_u),$$
$$(d\varphi)(x_s) = (d\varphi)(x)_s \text{ and}$$
$$(d\varphi)(x_n) = (d\varphi)(x)_n.$$

In particular, under an algebraic group morphism the image of any semisimple (resp. unipotent) element is semisimple (resp. unipotent).

(iii) *Let* $\mathbf{G}$ *be a commutative k-group. Then* $\mathbf{G}^{(u)}$ *and* $\mathbf{G}^{(s)}$ *are algebraic subgroups of* $\mathbf{G}$ *and the map* $\mathbf{G}^{(u)} \times \mathbf{G}^{(s)} \to \mathbf{G}$ *is an algebraic group morphism (i.e. the algebraic group* $\mathbf{G}$ *is the direct product of* $\mathbf{G}^{(u)}$ *and* $\mathbf{G}^{(s)}$*).*

(iv) *Let* $\mathrm{char}\, k = 0$. *Then* (see [Bo 6], Chapter II, 7.3) $x \in \mathrm{Lie}(\mathbf{G})^{(n)}$ *implies that*

$$\exp x \overset{\text{def}}{=} \sum_{i \geq 0} (i!)^{-1} x^i$$

belongs to $\mathbf{G}^{(u)}$. *Conversely, if* $g \in \mathbf{G}^{(u)}$, *then the logarithm*

$$\ln g \overset{\text{def}}{=} \sum_{i > 0} (-i)^{-1} (1-g)^i$$

belongs to $\mathrm{Lie}(\mathbf{G})^{(n)}$. *The set* $\mathrm{Lie}(\mathbf{G})^{(n)}$ *(resp.* $\mathbf{G}^{(u)}$*) is a k-subvariety in* $\mathrm{Lie}(\mathbf{G})$ *(resp. in* $\mathbf{G}$*), the maps* $\exp : \mathrm{Lie}(\mathbf{G})^{(n)} \to \mathbf{G}^{(u)}$ *and* $\ln : \mathbf{G}^{(u)} \to \mathrm{Lie}(\mathbf{G})^{(n)}$ *are inverses of each other, biregular and defined over k.*

If $\mathrm{char}\, k = 0$, then (iv) implies that the order of each element $g \in \mathbf{G}^{(u)} - \{e\}$ is infinite, the variety $\mathbf{G}^{(u)}$ is connected, and the set $\mathbf{G}(k)^{(u)}$ is Zariski dense in $\mathbf{G}^{(u)}$. In particular, every unipotent algebraic group over a field of characteristic 0 is connected. From (ii) we deduce that

(v) *if* $\varphi\colon \mathbf{G}_1 \to \mathbf{G}$ *is an algebraic group morphism and* $\mathrm{Ker}\,\varphi$ *is unipotent, then* $\varphi^{-1}(\mathbf{G}^{(u)}) = \mathbf{G}_1^{(u)}$.

In the sequel, for $M \subset \mathbf{G}$, the set $M \cap \mathbf{G}^{(u)}$ will be denoted by $M^{(u)}$.

(0.21) Let $\mathbf{G}$ be an algebraic group. An algebraic group morphism of $\mathbf{G}$ to $\mathbf{GL}_1$ is called a *character* or a *rational character* of $\mathbf{G}$. The set of all characters, endowed with the composition law defined by the product of the values of the characters, is a finitely generated commutative group. It is free if $\mathbf{G}$ is connected. This character group will be denoted by $\mathrm{X}(\mathbf{G})$.

(0.22) Tori and commutative algebraic groups. (See [Bo 6], Chapter III, §8 and [Hum 1], §16.) A commutative algebraic group $\mathbf{T}$ is said to be a *torus* if it is connected and the following equivalent conditions are satisfied:

(i) $\mathbf{T}$ consists of semisimple elements;

(ii) $\mathbf{T}$ is diagonalizable, i.e., $\mathbf{T}$ can be conjugated to a subgroup of the group of diagonal matrices;

(iii) $\mathbf{T}$ is isomorphic to the product of $\dim \mathbf{T}$ many copies of the group $\mathbf{GL}_1$.

If $\mathbf{T}$ is a torus, then $\mathrm{X}(\mathbf{T})$ is a free Abelian group of rank $\dim \mathbf{T}$. A torus $\mathbf{T}$ is called *k-split* or *split over k*, if it is defined over k and is k-isomorphic to the direct product of $\dim \mathbf{T}$ many copies of $\mathbf{GL}_1$. If $\mathbf{T} \subset \mathbf{GL}_n$ is a torus defined over k, then the following conditions are equivalent:

(a) $\mathbf{T}$ is split over k;

(b) $\mathbf{T}$ is diagonalizable over k, i.e., there exists a matrix $g \in \mathbf{GL}_n(k)$ such that $g\mathbf{T}g^{-1}$ consists of diagonal matrices;

(c) each character of $\mathbf{T}$ is defined over k.

Every torus defined over k is split over some finite *separable* field extension of k.

Every connected algebraic subgroup $\mathbf{S}$ of a torus $\mathbf{T}$ is a torus and is a direct factor (i.e. there exist a subtorus $\mathbf{S}' \subset \mathbf{T}$ such that $\mathbf{T} = \mathbf{S} \times \mathbf{S}'$). Every connected algebraic subgroup of a k-split torus is a k-split torus defined over k. All morphisms of k-split tori are defined over k, and in particular every morphism of $\mathbf{GL}_1$ into a k-split torus is defined over k. The image of a torus (resp. k-split torus) under an algebraic group morphism (resp. k-group morphism) is a torus (resp. k-split torus).

Let p be the characteristic exponent of a field k (i.e. $p = \operatorname{char} k$ if $\operatorname{char} k > 0$ and $p = 1$ if $\operatorname{char} k = 0$) and let $\mathbf{T}$ be a torus defined over k. For $m \in \mathbb{N}^+$ we define the morphism $\alpha_m \colon \mathbf{T} \to \mathbf{T}$ by setting $\alpha_m(x) = x^m$, $x \in \mathbf{T}$. Then

(a) the morphism α_m is surjective for all $m \in \mathbb{N}^+$;

(b) if $m = p^i$, where $i \in \mathbb{N}^+$, then the morphism α_m is bijective.

There is only a finite number of elements in $\mathbf{T}$ of a fixed order m. If $m = p^i$, $i \in \mathbb{N}^+$, then the identity e is the unique element in $\mathbf{T}$ of order m.

Let $\mathbf{G}$ be a commutative algebraic group which consists of semisimple elements. Then $\mathbf{G}$ is a direct product $\mathbf{G} = \mathbf{F} \times \mathbf{G}^0$ of a finite group $\mathbf{F}$ and the torus $\mathbf{G}^0$.

If $\mathbf{H}$ is a connected algebraic group and h is a semisimple element of $\mathbf{H}$, then h belongs to a torus $\mathbf{T} \subset \mathbf{H}$ (see [Bo 6], Chapter IV, Corollary 11.12 or [Hum 1], 22.3). As a torus containing h one can choose some maximal torus of $\mathbf{H}$. Therefore, by Proposition 0.20(i) every connected non-unipotent algebraic group contains a torus of some positive dimension. All the maximal tori of the group $\mathbf{H}$ are mutually conjugate (see [Bo 6], Chapter IV, Corollary 11.3 or [Hum 1], 21.3).

Every connected k-group $\mathbf{H}$ contains a maximal torus defined over k (see [Bo 6], Chapter V, Theorem 18.2). Therefore, it follows from the above observations that every connected non-unipotent k-group $\mathbf{H}$ contains a k-torus (i.e. a torus defined over k) of positive dimension.

(0.23) We recall that the *radical* (resp. *unipotent radical*) of a k-group $\mathbf{G}$ is the maximal connected algebraic solvable (resp. unipotent) normal subgroup in $\mathbf{G}$. These subgroups are k-closed, because they are invariant relative to the action of the Galois group of the field K over k, where, as above, K denotes an algebraically closed field containing k. In particular, if k is a perfect field, e.g., if $\operatorname{char} k = 0$, then the radical and the unipotent radical of every k-group are defined over k.

(0.24) Reductive, semisimple and almost simple groups. Let $\mathbf{G}$ be an algebraic k-group. We say that $\mathbf{G}$ is *semisimple* (resp. *reductive*), if its radical (resp. unipotent radical) is $\{e\}$. The factor group of every algebraic group modulo its radical (resp. unipotent radical) is semisimple (resp. reductive). Every semisimple group is reductive. A group $\mathbf{G}$ is called (absolutely) *simple* (resp. (absolutely) *almost simple*), if $\{e\}$ is the only proper algebraic normal subgroup of $\mathbf{G}$ (resp. all such subgroups are finite), and $\mathbf{G}$ is called *k-simple* or *simple over k* (resp. *almost k-simple* or *almost simple over k*) if this condition holds for k-closed normal subgroups. If $\mathbf{G}$ is a connected group, then every finite normal subgroup is contained in $\mathscr{Z}(\mathbf{G})$. Therefore, if $\mathbf{G}$ is connected and almost k-simple, then every normal k-closed subgroup of $\mathbf{G}$ is either equal to $\mathbf{G}$ or is contained in $\mathscr{Z}(\mathbf{G})$. Since $\mathscr{D}(\mathbf{G})$ is a k-subgroup of $\mathbf{G}$ and $\mathscr{D}(\mathbf{G})$ is connected for $\mathbf{G}$ connected, $\mathscr{D}(\mathbf{G}) = \mathbf{G}$ if $\mathbf{G}$ is non-commutative, and is either k-simple or connected and almost k-simple.

By definition a group $\mathbf{G}$ is semisimple (resp. reductive) if the identity component of $\mathbf{G}$ is semisimple (resp. reductive). If $\mathbf{G}$ is connected, non-commutative and almost k-simple, then $\mathbf{G}$ is semisimple. If $\mathbf{G}$ is an almost direct product of semisimple (resp. reductive) subgroups, then $\mathbf{G}$ is semisimple (resp. reductive). If $\mathbf{G} \neq \{e\}$ is connected and semisimple, then $\mathbf{G}$ decomposes uniquely (up to permutation of the factors) into an almost direct product of connected non-commutative almost simple algebraic subgroups $\mathbf{G}_1, \ldots, \mathbf{G}_i$ and also into an almost direct product of connected non-commutative almost k-simple k-subgroups $\mathbf{G}'_1, \ldots, \mathbf{G}'_j$ (see [Bo-T 1], 2.15). These groups $\mathbf{G}_1, \ldots, \mathbf{G}_i$ are called *almost simple factors* of the group $\mathbf{G}$, and the groups $\mathbf{G}'_1, \ldots, \mathbf{G}'_j$ are called *almost k-simple factors* of $\mathbf{G}$. The decomposition of $\mathbf{G}$ into almost k-simple factors is invariant relative to k-isogenies. More precisely, if $f\colon \mathbf{G} \to \mathbf{H}$ is a k-isogeny of connected semisimple k-groups, then the images of almost k-simple factors of $\mathbf{G}$ are almost k-simple factors of $\mathbf{H}$. This proposition can easily be deduced from the fact that every connected algebraic normal subgroup of a semisimple k-group is defined over the separable closure of the field k (see 0.25 below).

A connected k-group $\mathbf{G}$ is reductive if and only if it is an almost direct product of a k-torus and the connected semisimple k-group $\mathscr{D}(\mathbf{G})$ (see [Bo-T 1], 2.2 and 2.15). It follows from the properties of Jordan decomposition given above in 0.20 that if $\mathbf{G}$ is a connected reductive k-group, then $\mathbf{G}^{(u)} \subset \mathscr{D}(\mathbf{G})$.

It follows from the above description of semisimple and reductive groups that the image of a semisimple (resp. reductive) group under an algebraic group morphism is semisimple (resp. reductive).

In a connected reductive group $\mathbf{G}$ the set of semisimple elements contains a Zariski open and dense subset of $\mathbf{G}$ (see [Bo 6], Chapter IV, Theorem 12.3 and Corollary 2(c) in 13.17).

A connected algebraic group $\mathbf{G}$ is reductive if and only if $\mathbf{G}$ admits a fully reducible rational representation with finite kernel (see [Bo-T 1], Proposition 2.2). If $\operatorname{char} k = 0$, then every rational representation of a reductive k-group is fully reducible (see [C 3], Chapter IV, Theorem 4(a)). It should be noted that the analogous proposition is false if $\operatorname{char} k \neq 0$.

If $\operatorname{char} k = 0$, then (see [Hum 1], Theorem 13.5) a k-group is semisimple if and only if its Lie algebra is semisimple. For fields of positive characteristic this fact is in general false , e.g. $\mathbf{SL}_2$ over a field of characteristic 2.

Let $\mathbf{G}$ be a connected reductive k-group. Then (see [Bo-T 1], 2.14 and 2.15)

(a) if k is infinite, then the subgroup $\mathbf{G}(k)$ is Zariski dense in $\mathbf{G}$;
(b) the center of the group $\mathbf{G}$ is defined over k.

For infinite k, (a) implies that $\mathscr{Z}(\mathbf{G}(k)) = \mathscr{Z}(\mathbf{G})(k)$ (for finite k, the same equality is also true but does not follow from (a)). We remark that, if k is a perfect field, then (a) and (b) are also valid without the assumption of reductivity of the group $\mathbf{G}$.

As to Proposition 0.15(iii)(b), observe that if $\mathbf{G}$ is connected and reductive, then the equality $\operatorname{Ker} \operatorname{Ad} = \mathscr{Z}(\mathbf{G})$ is also valid in case $\operatorname{char} k \neq 0$ (see [Hum 1], §27, Exercise 5).

(0.25) Let $\mathbf{G}$ be an algebraic k-group. If $\mathbf{G}$ is reductive or the field k is perfect, then (see [Bo-T 1], 4.21 and 8.2) the maximal k-split tori of $\mathbf{G}$ are conjugate by elements of $\mathbf{G}(k)$ and, hence, all have the same dimension. In these two cases (i.e. if $\mathbf{G}$ is reductive or k is perfect) we denote by $\operatorname{rank}_k \mathbf{G}$ the *k-rank* of the group $\mathbf{G}$, i.e., the common dimension of maximal k-split tori in $\mathbf{G}$. If $\operatorname{rank}_k \mathbf{G} > 0$, then $\mathbf{G}$ is said to be *k-isotropic* or *isotropic over k*. Otherwise $\mathbf{G}$ is said to be *k-anisotropic* or *anisotropic over k*. An almost k-simple factor of $\mathbf{G}$ which is k-isotropic (resp. k-anisotropic) will be called a *k-isotropic* (resp. *k-anisotropic*) *factor*. A reductive k-group $\mathbf{G}$ is called *k-split* or *split over k*, if $\mathbf{G}$ contains a k-split torus which is a maximal torus of the group $\mathbf{G}$, in other words, if $\operatorname{rank}_k \mathbf{G} = \operatorname{rank}_K \mathbf{G}$, where K is the algebraic closure of k. The following propositions hold.

(i) *Every connected reductive k-group is split over some finite separable field extension of k* (see [Bo 6], Corollary 18.8).
(ii) *Every connected algebraic normal subgroup of a connected k-split reductive k-group is defined over k* (this is a consequence of Theorem 18.7 in [Bo 6]).
(iii) *Every connected algebraic normal subgroup of a connected reductive k-group is defined over a finite separable field extension of k* (this is a consequence of (i) and (ii)).

(0.26) Root systems. We mention here certain properties of root systems. For proofs and more detailed information the reader is referred to Chapter VI of [Bou 6].

Given a finite-dimensional vector space V over $\mathbb{R}$, we define a *reflection* relative to a non-zero vector $\alpha \in V$ as a linear transformation s sending α to $-\alpha$ and leaving invariant every element of a subspace of codimension 1; this subspace is called the *hyperplane of the reflection s*.

A *root system* in the space V is a subset $\Psi \subset V$ satisfying the following three conditions:

(R_1) The set Ψ is finite, V is spanned by Ψ, and Ψ does not contain 0 (elements of Ψ are called *roots*).
(R_2) If $\alpha \in \Psi$, then there exists a reflection s_α relative to α, leaving invariant the set Ψ.
(R_3) If $\alpha, \beta \in \Psi$, then the vector $s_\alpha(\beta) - \beta$ is integrally proportional to α.

Let Ψ be a root system in V. The group $W(\Psi) \subset \mathbf{GL}(V)$, generated by reflections s_α, $\alpha \in \Psi$, is called the *Weyl group* of the root system Ψ. Since the set Ψ is finite and V is spanned by Ψ, the group $W(\Psi)$ is finite. There exists a positive definite inner product on V invariant under $W(\Psi)$. A root system is called *irreducible*, if it can not be represented in the form of a union of two mutually orthogonal proper subsystems. If Ψ is irreducible, then a $W(\Psi)$-invariant scalar product is unique up to a scalar multiple.

If two roots are proportional, then the proportionality coefficient can be equal only to ± 1, $\pm 1/2$ and ± 2. A root $\alpha \in \Psi$ for which $(1/2) \cdot \alpha \notin \Psi$ (resp. $2 \cdot \alpha \notin \Psi$) is called *indivisible* (resp. *non-multipliable*). A root system Ψ is called *reduced* if every root $\alpha \in \Psi$ is indivisible. The set of non-multipliable roots of a root system Ψ is a reduced root system having the same Weyl group as Ψ.

We call *Weyl chambers* (of a root system Ψ) the connected components of the complement of the union of hyperplanes of reflections s_α, $\alpha \in \Psi$. The group $W(\Psi)$ operates simply transitively on the set of Weyl chambers of Ψ. Let $C \subset V$ be a Weyl chamber of Ψ. Then there exists a uniquely determined set $B(C) = \{\alpha_1, \ldots, \alpha_l\} \subset \Psi$ such that $B(C)$ is a basis of the space V, each α_i, $1 \leq i \leq l$, is indivisible, and $C = \bigcap_{1 \leq i \leq l} Y_i$, where Y_i is a half-space containing α_i whose boundary is the hyperplane of the reflection s_{α_i}. The set $B(C)$ is called the *base* of the root system Ψ corresponding to the Weyl chamber C. An ordering $\geq$ on V compatible with the linear structure of V such that the elements ≥ 0 are linear combinations of the roots α_i with non-negative coefficients will be called an *ordering compatible with the Weyl chamber* C. A root ≥ 0 (≤ 0) is called *positive* (*negative*). Every root of the system Ψ is either positive or negative. Furthermore, every root $\alpha \in \Psi$ can be represented uniquely in the form $\alpha = \sum_i c_i \alpha_i$, where c_i are integers (possibly zeroes) of same sign. A positive root is said to be *simple* (with respect to the ordering defined above) if it is not representable as a sum of two positive roots. It is easily seen that the set of simple roots coincides with $B(C)$.

We now fix a Weyl group invariant positive definite inner product on V. For each root $\alpha \in \Psi$ we define the linear form α^* on V by setting

$$\alpha^*(v) = 2\langle \alpha, v \rangle \langle \alpha, \alpha \rangle^{-1}, \quad v \in V,$$

where $\langle x, y \rangle$ denotes the inner product of the vectors x, $y \in V$. Every pair α, β of distinct elements of $B(C)$ satisfies (possibly after a transposition of α and β) one of the following systems of relations:

(a) $\quad \alpha^*(\beta) = \beta^*(\alpha) = 0;$

(b) $\quad \alpha^*(\beta) = \beta^*(\alpha) = -1;$

(c) $\quad \alpha^*(\beta) = -1, \ \beta^*(\alpha) = -2;$

(d) $\quad \alpha^*(\beta) = -1, \ \beta^*(\alpha) = -3.$

We can define now the *Dynkin diagram of a root system* Ψ. To this end we represent the elements of $B(C)$ by points and join those corresponding to α and β in the way indicated below subject to one of relations (a)–(d) holding for this pair α, β.

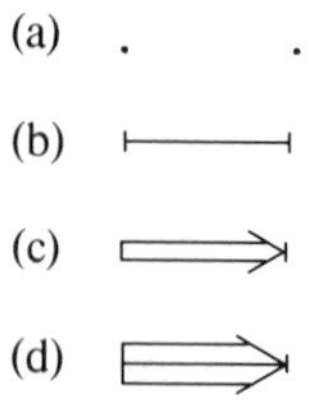

Fig. 1

Since $W(\Psi)$ acts transitively on the set of Weyl chambers, the Dynkin diagram does not depend upon the choice of the chamber C. The Dynkin diagram of a root system Ψ is connected if and only if Ψ is irreducible. Every reduced root system is determined up to isomorphism by its Dynkin diagram. The Dynkin diagram of any irreducible root system is one of the following:

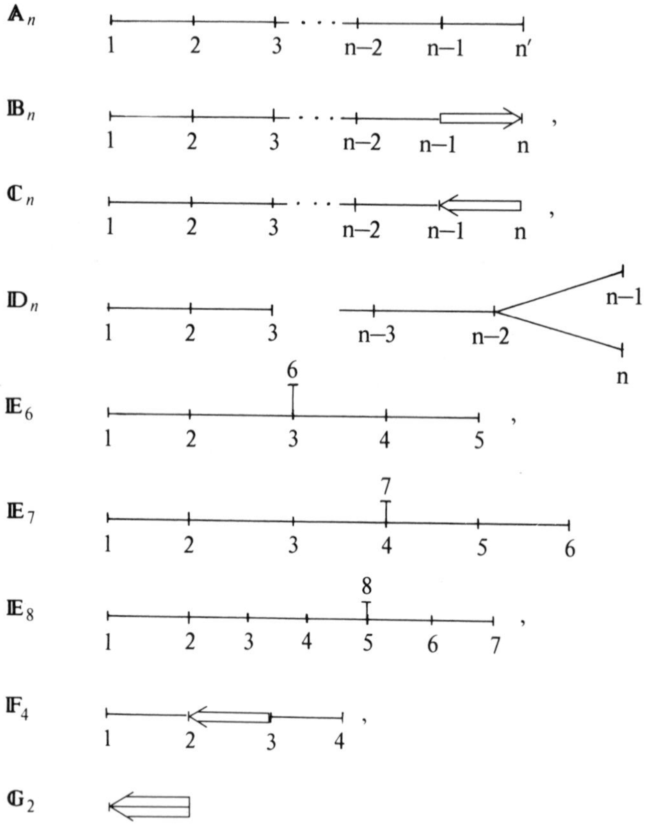

Fig. 2

The *type* of an irreducible root system Ψ is by definition the type of its Dynkin diagram.

(0.27) Roots of an algebraic group relative to a torus. Let **G** be an algebraic group defined over k, $\mathbf{S} \subset \mathbf{G}$ a torus, and ρ a rational representation of **G** in a finite-dimensional vector space V. A character $\chi \in \mathrm{X}(\mathbf{S})$ is said to be a *character* or a *weight of the torus* **S** *in the representation* ρ if there exists a non-zero vector $v \in V$ such that $\rho(s)v = \chi(x)v$ for all $s \in \mathbf{S}$. Non-trivial characters of the torus **S** in the adjoint representation of the group **G** are said to be *roots of the group* **G** *relative to* **S**. The set of these roots will be denoted by $\Phi(\mathbf{S},\mathbf{G})$.

The factor group $\mathscr{N}_{\mathbf{G}}(\mathbf{S})/\mathscr{Z}_{\mathbf{G}}(\mathbf{S})$ will be called the *Weyl group of* **G** *relative to* **S** and will be denoted by $W(\mathbf{S},\mathbf{G})$. The group $W(\mathbf{S},\mathbf{G})$ operates naturally on **S**, namely, $\pi(g)s = gsg^{-1}$, where $g \in \mathscr{N}_{\mathbf{G}}(\mathbf{S})$ and $\pi\colon \mathscr{N}_{\mathbf{G}}(\mathbf{S}) \to W(\mathbf{S},\mathbf{G})$ is the natural epimorphism. This action induces the action of the group $W(\mathbf{S},\mathbf{G})$ on $\mathrm{X}(\mathbf{S})$, and hence, also on $\mathrm{X}(\mathbf{S}) \otimes \mathbb{R}$ by the formula

$$(w\chi)(s) = \chi(w^{-1}s),\ w \in W(\mathbf{S},\mathbf{G}),\ \chi \in \mathrm{X}(\mathbf{S}),\ s \in \mathbf{S}.$$

Notice $\Phi(\mathbf{S},\mathbf{G})$ is $W(\mathbf{S},\mathbf{G})$-invariant.

Let **G** be connected and semisimple and let **S** be a maximal k-split torus in **G**. Then (see [Bo-T 1], §5) the following assertions hold: (a) $\Phi(\mathbf{S},\mathbf{G})$ is a root system in $\mathrm{X}(\mathbf{S}) \otimes \mathbb{R}$; it is irreducible if **G** is almost k-simple; (b) $W(\mathbf{S},\mathbf{G})$ as the automorphism group of the space $\mathrm{X}(\mathbf{S}) \otimes \mathbb{R}$ coincides with the Weyl group of the root system $\Phi(\mathbf{S},\mathbf{G})$; (c) the image of the group $\mathscr{N}_{\mathbf{G}}(\mathbf{S})(k)$ under the natural epimorphism $\mathscr{N}_{\mathbf{G}}(\mathbf{S}) \to W(\mathbf{S},\mathbf{G})$ is all of $W(\mathbf{S},\mathbf{G})$. And this implies, in particular, that if $\mathrm{rank}_k\,\mathbf{G} = 1$, then there exists $x \in \mathscr{N}_{\mathbf{G}}(\mathbf{S})(k)$, such that $xsx^{-1} = s^{-1}$ for each $s \in \mathbf{S}$. As the *ordering* on $\mathrm{X}(\mathbf{S})$ we shall consider an ordering compatible with a Weyl chamber of the system $\Phi(\mathbf{S},\mathbf{G})$.

Let **T** be a maximal torus in a connected semisimple algebraic group **G**. Then (see [Bo 6], Theorem 14.8 or [Hum 1], 27.1) the root system $\Phi(\mathbf{T},\mathbf{G})$ is reduced. The *Dynkin diagram of the group* **G** is by definition the Dynkin diagram of the root system $\Phi(\mathbf{T},\mathbf{G})$. The root system $\Phi(\mathbf{T},\mathbf{G})$ is irreducible if and only if **G** is absolutely almost simple. The *type* of an absolutely almost simple group **G** is the type of its Dynkin diagram. Two connected semisimple k-groups **G** and $\mathbf{G}'$ are strictly isogenous (for the definition see 1.4.3 below) if and only if they have the same Dynkin diagram (see [Ti 2], Theorem 1.2.2). In particular, two absolutely almost simple k-groups **G** and $\mathbf{G}'$ are strictly isogenous if and only if they are of same type. We shall say that **G** is of type $\mathbb{A}$ (resp. $\mathbb{B}$, $\mathbb{C}$, $\mathbb{D}$), if the group **G** is of type $\mathbb{A}_n$ (resp. $\mathbb{B}_n$, $\mathbb{C}_n$, $\mathbb{D}_n$) for some n.

(0.28) Levi subgroups. (See [Bo-T 1], 0.8.) Let **G** be an algebraic k-group. A *Levi subgroup* of **G** is any algebraic *reductive* subgroup $\mathbf{H} \subset \mathbf{G}$ such that **G**, viewed as an algebraic group, is the semi-direct product of **H** and the unipotent radical $R_u(\mathbf{G})$. The decomposition $\mathbf{G} = \mathbf{H} \ltimes R_u(\mathbf{G})$ is called a *Levi decomposition* of **G**. If $\mathrm{char}\,k = 0$, then **G** always has a Levi k-subgroup **H** and every reductive k-subgroup in **G** is conjugate to a subgroup of the group **H** by an element of $R_u(\mathbf{G})(k)$. However, in case $\mathrm{char}\,k \neq 0$, this assertion is false.

(0.29) Borel and parabolic subgroups. Let **G** be a connected algebraic group. A *Borel subgroup* of the group **G** is a connected algebraic solvable subgroup of

G which is contained in no other such subgroup. An algebraic subgroup **P** of **G** is called *parabolic* if the algebraic variety **G**/**P** is complete, equivalently, is projective. The following propositions (see [Bo 6], §51 and [Hum 1], §§21, 23) hold.

(a) *Any two Borel subgroups are conjugate and the union of all Borel subgroups of the group* **G** *is equal to* **G**;

(b) *An algebraic subgroup of the group* **G** *is parabolic if and only if it contains a Borel subgroup of* **G**;

(c) *Every parabolic subgroup of the group* **G** *is connected and coincides with its own normalizer in* **G**.

Two parabolic subgroups of a reductive algebraic group are said to be *opposite* if their intersection is a Levi subgroup in both of them. If **G** is a connected reductive k-group, then (see [Bo-T 1], 4.8 and 4.18) the intersection of any two parabolic k-subgroups **P** and **P**′ of **G** contains the centralizer of a maximal k-split torus **S** of **G**, and coincides with this centralizer (i.e. $\mathbf{P} \cap \mathbf{P}' = \mathscr{Z}_{\mathbf{G}}(\mathbf{S})$) if and only if **P** and **P**′ are opposite.

A connected reductive k-group contains a proper parabolic k-subgroup if and only if it contains a non-central k-split torus (see [Bo-T 1], Corollary 4.17). In particular, a connected semisimple k-group contains a proper parabolic k-subgroup if and only if it is isotropic over k.

(0.30) A topological space is called *σ-compact*, if it is a countable union of compact subsets. A topological space is called *semi-separated* or a *T_0-space* if for each pair of distinct points at least one of them has a neighbourhood not containing the other.

We say that a sequence $\{x_n\}_{n \in \mathbb{N}}$ in a locally compact topological space X *converges to infinity in* X, if for every compact subset $K \subset X$ we have $x_n \in X - K$ for all but finitely many $n \in \mathbb{N}$.

If $f\colon X \to \mathbb{C}$ is a function on a topological space, then the closure in X of the set $\{x \in X \mid f(x) \neq 0\}$ is called the *support* of f and is denoted by $\operatorname{supp} f$.

A continuous map $f\colon X \to Y$ of topological spaces is said to be *proper* if the preimage $f^{-1}(K)$ of every compact set $K \subset Y$ is compact in X.

A subset of a topological space is called *Borel* if it belongs to the σ-algebra generated by the family of its closed subsets.

A mapping $f\colon X \to Y$ of locally compact spaces is called *Borel* if the inverse image $f^{-1}(U)$ of every open set $U \subset Y$ is Borel in X.

A topological group is called *compactly generated* if it is generated by a compact subset.

(0.31) Local fields. (See [Weil 4], Chapter I.) By a *local field* we will always mean a commutative non-discrete locally compact field. A local field is isomorphic either to $\mathbb{C}$ or $\mathbb{R}$ or a finite extension of $\mathbb{Q}_p$ or a field of formal power series in one variable over a finite field. (Here and in what follows by formal power series we mean a series of the form $\sum_{n \le i \le \infty} a_i t^i$ with n an integer, possibly negative.) A local field is of finite degree over any of its local subfield. A finite field extension of a local field is a local field as well. A local field contains no infinite discrete subfields; in other words, the closure of every infinite subfield of a local field is also a local field. Let k be a local field and let $x \in k$. We denote by $\mathrm{mod}_k(x)$ the modulus of the automorphism $y \mapsto yx$ of the additive group k^+ of the field k (i.e. $\mathrm{mod}_k(x)$ is defined by $\mu(xB) = \mathrm{mod}_k(x)\mu(B)$, where $B \subset k^+$ and μ is Haar measure on k^+). We call $\mathrm{mod}_k(x)$ the *modulus* of x. An *absolute value* or a *valuation* on k is a continuous function $|\ |\colon k \to \mathbb{R}^+ \cup \{0\}$, such that

(A) $|x| = 0$ is equivalent to $x = 0$;
(B) $|xy| = |x|\,|y|$ for all $x, y \in k$;
(C) $|x + y| \leq |x| + |y|$ for all $x, y \in k$.

If k is isomorphic to $\mathbb{R}$ or $\mathbb{C}$, then we can take as a valuation $|\ |$ the standard absolute value. If k is isomorphic neither to $\mathbb{R}$ nor to $\mathbb{C}$, then one can take as a valuation $|\ |$, mod_k. If $|\ |$ is an absolute value on k, then there exists a number $c > 0$, such that $\mathrm{mod}_k(x) = |x|^c$ for all $x \in k$. Thus on every local field there exists a valuation which is unique up to a positive power. Any valuation on k can be extended uniquely to any finite field extension of k and, hence, to its algebraic closure (see [Wae 2], §144).

If $|\ |$ is an absolute value on k, then one can define a metric ρ on k by setting $\rho(x, y) = |x - y|$. The topology of the field k coincides with the topology defined by this metric.

A local field k is called *archimedean* if it is isomorphic to $\mathbb{R}$ or $\mathbb{C}$; otherwise k is *non-archimedean*. A field k is non-archimedean if and only if it is totally disconnected. Let k be non-archimedean, and let $|\ |$ be an absolute value on k. Then $|k - \{0\}|$ is an infinite cyclic group and $|\ |$ satisfies the *ultrametric inequality*:

$$|x + y| \leq \max\{|x|, |y|\}, \quad x, y \in k.$$

The set $\mathrm{Ker}|\ | = \{x \in k \mid |x| = 1\}$ is called the *group of units* of the local field k. This group is compact. An element x of k is called a *uniformizer*, if $|x|$ is a generator of the cyclic group $|k - \{0\}|$ and $|x| < 1$. The set of uniformizers is not empty. If $x \in k$ is a uniformizer, then the multiplicative group of the field k is the direct product of the group of units and the infinite cyclic group $\{x^n \mid n \in \mathbb{Z}\}$. The set $\{x \in k \mid |x| \leq 1\}$ is closed relative to the operations of addition and multiplication and is called the *ring of integers* of the field k. This ring is the unique maximal compact subring of the field k. The set $\{x \in k \mid |x| < 1\}$ is the unique maximal ideal of the ring of integers of the field k.

Let $\varphi\colon k \to k'$ be a continuous morphism of local fields. Then φ is a proper map, and hence, $\varphi(k)$ is a closed subfield of k' (for non-archimedean k this follows from the fact that for each uniformizer x of k the absolute value of $\varphi(x)$ is less than 1). In particular, if $\varphi(k)$ is dense in k, then $\varphi(k) = k'$ and the isomorphism φ^{-1} is continuous.

The local fields k and k' are said to be of *common type*, provided that k and k' contain isomorphic local subfields; otherwise they are of *different type*. It is easily seen that k and k' are of common type if and only if there exists a continuous homomorphism of the field k into some finite field extension of k'.

From the classification of local fields it follows that k and k' are of common type if and only if one of the following conditions is satisfied:

(a) $\operatorname{char} k = \operatorname{char} k' \neq 0$;

(b) k and k' are isomorphic to finite field extensions of $\mathbb{Q}_p$, for some prime p;

(c) k and k' are isomorphic to $\mathbb{R}$ or $\mathbb{C}$.

In particular, if p and p' are two elements of the set of prime numbers, supplemented by ∞, and the field k (resp. k') is isomorphic to a finite field extension of $\mathbb{Q}_p$ (resp. $\mathbb{Q}_{p'}$) then k and k' are of common type if and only if $p = p'$.

Every time we work with a finite-dimensional vector space V over a local field k we assume that V is equipped with the structure of a Hausdorff topological vector space over k. It is well known (see [Bou 2], Chapter 1, §2, Theorem 2) that every n-dimensional Hausdorff topological vector space over k is isomorphic to k^n.

If $\mathbf{M}$ is a k-variety, then $\mathbf{M}(k)$ is equipped with a natural topology induced from the topology of the local field k, in case $\mathbf{M}$ is a k-subvariety of the n-dimensional affine space, this topology is the restriction to $\mathbf{M}(k)$ of the natural topology of the space k^n. From now on, if not stated otherwise, the closure of a subset $X \subset \mathbf{M}(k)$ will be considered in the topology indicated above. Since k is locally compact, σ-compact and metrizable, $\mathbf{M}(k)$ is also locally compact, σ-compact and metrizable. Note that the projective space $\mathbf{P}_n(k) = \mathbf{P}(k^{n+1})$ is compact. If k is isomorphic to $\mathbb{R}$ or $\mathbb{C}$, then (see [Wh]) the number of connected components of the space $\mathbf{M}(k)$ is finite. If k is isomorphic neither to $\mathbb{R}$ nor to $\mathbb{C}$, then the space $\mathbf{M}(k)$ is totally disconnected.

If H is a compact subgroup of the group $\mathbf{GL}_n(\mathbb{R})$, then $H = \mathbf{H}(\mathbb{R})$, where $\mathbf{H}$ is an $\mathbb{R}$-subgroup of the group $\mathbf{GL}_n$ (see [Vi 2], Theorem 6). This proposition is known as the *theorem on the algebraicity of a compact real linear group.* It can be restated as follows. Given an algebraic $\mathbb{R}$-group $\mathbf{H}$, if $\mathbf{H}(\mathbb{R})$ contains a relatively compact subgroup, Zariski dense in $\mathbf{H}$, then the group $\mathbf{H}(\mathbb{R})$ is compact. We remark that analogous proposition for an arbitrary local field k in place of $\mathbb{R}$ is false.

(0.32) Global fields. (See [Wae 2], Chapter 18 and [Weil 4], Chapter III.) By a *global field* we mean a finite field extension either of the field of rational numbers or of a field of rational functions in one variable over a finite field. Let K be a global field. A *valuation* or an *absolute value* on K is, by definition, a function $v\colon K \to \mathbb{R}^+ \cup \{0\}$ satisfying conditions (A), (B), (C) of 0.31 with $|x| = v(x)$. Everywhere in the sequel all the valuations v are considered to be non-trivial, i.e., we assume that $v(x) \neq 1$ for some $x \in K$, $x \neq 0$. Any valuation v of the field K defines a metric d_v on K, namely, $d_v(x, y) = v(x - y)$, $x \in K$. We may consider the completion of the field K relative to this metric. This completion is a local field which is denoted by K_v. Two valuations v and v' of the field K are called *equivalent* if the corresponding metrics d_v and $d_{v'}$ define the same topology on K. The value of v at $x \in K$ will often be denoted by $|x|_v$.

Let λ be an embedding of the field K into a local field k. The pair (λ, k) is called a *completion* of the field K, if $\lambda(K)$ is dense in k. Two completions (λ, k) and (λ', k') of the field K are said to be *equivalent*, if there exists a continuous isomorphism ρ of the field k onto k', such that $\lambda' = \rho \cdot \lambda$. If (λ, k) is a completion of the field K, then one can define a valuation v on k by setting $v(x) = |\lambda(x)|$, where $| \, |$ denotes an absolute value on k. Furthermore, the embedding $\lambda \colon K \to k$ can be extended to a continuous isomorphism of the field K_v onto k. Therefore the set of equivalence classes of valuations of the field K is naturally identified with the set of equivalence classes of their completions. In what follows we shall not distinguish a valuation and its equivalence class.

The set of inequivalent valuations of the field K and hence, the set of equivalent classes of completions of the field K are countable.

If K' is a finite field extension of K, then every valuation of the field K can be extended to a valuation of the field K' and the number of such extensions does not exceed the degree of the field extension K' over K.

A valuation is said to be *non-archimedean*, if it satisfies the ultrametric inequality and otherwise *archimedean*. It should be noted that in many books and articles a valuation means only a non-archimedean valuation. A valuation v is archimedean if and only if the field K_v is archimedean. A field K has only a finite number of inequivalent archimedean valuations. The set of archimedean valuations is non-empty if and only if $\operatorname{char} K = 0$ (i.e. if K is a finite field extension of $\mathbb{Q}$). If a valuation v is non-archimedean, then the ring of integral elements of the field K_v is denoted by $\mathcal{O}_v$.

Let S be a set of valuations of the field K. An element $x \in K$ is called *S-integral*, if $|x|_v \le 1$ for each non-archimedean valuation $v \notin S$. The set of S-integral elements is a subring of the field K, which will be denoted by $K(S)$. If $\operatorname{card} K(S) = \infty$ (i.e. if $\operatorname{char} K = 0$ or $\operatorname{card} S \ge 1$), then K is a quotient field of the ring $K(S)$. The quotient ring of the ring $K(S)$ modulo any of its non-zero ideal is finite. We remark that if $\operatorname{char} K = 0$ and S is contained in the set of archimedean valuations, the $K(S)$ coincides with the ring of algebraic integers belonging to the field K.

If $x \in K$, then the set of valuations v of the field K with $|x|_v > 1$ is finite. In other words, K coincides with the union of subrings $K(S)$, where S runs through all finite sets of valuations of the field K.

We shall clarify the concepts introduced above by considering the example of the field $\mathbb{Q}$ of rationals. We denote by $\mathcal{P}$ the set of prime numbers, supplemented by ∞. If $p = \infty$, then we set $v_p(x) = |x|$, where $|x|$ is the standard absolute value. If p is a prime number, then we set $v_p(x) = p^{-l}$, where $x = p^l m/n$, m and n are prime to p. The functions v_p are the valuations of the field $\mathbb{Q}$. A valuation v_p is archimedean if and only if $p = \infty$. If $p \neq p'$, then the valuations v_p and $v_{p'}$ are inequivalent. Every valuation of the field $\mathbb{Q}$ is equivalent to one of the valuations v_p, $p \in \mathcal{P}$. Thus, the set of valuations of the field $\mathbb{Q}$ is naturally identified with $\mathcal{P}$. Under this identification an element $x \in \mathbb{Q}$ is S-integral if and only if x is representable in the form $x = m/n$, where $m \in \mathbb{Z}$ and n is the product of powers of prime numbers belonging to $S \subset \mathcal{P}$. In particular, if $S = \{\infty\}$, then $\mathbb{Q}(S) = \mathbb{Z}$.

(0.33) Adèles. (See [Hum 2], §§3 and 14 and [Weil 2], Chapter I.) Let $\{X_\lambda\}_{\lambda\in\Lambda}$ be a family of topological spaces with open sets $Y_\lambda \subset X_\lambda$ given for all but finitely many $\lambda \in \Lambda$. We put $X = \{(x_\lambda) \in \prod_{\lambda\in\Lambda} X_\lambda \mid x_\lambda \in Y_\lambda$ for all but finitely many $\lambda \in \Lambda\}$. We introduce on X the topology for which the products $\prod_{\lambda\in\Lambda} Z_\lambda$ form a base of open sets, where Z_λ is open in X_λ and $Z_\lambda = Y_\lambda$ for all but finitely many λ. The space X with this topology is said to be the *restricted topological product* of the spaces X_λ relative to the subsets Y_λ, $\lambda \in \Lambda$.

Let K be a global field, let $\mathscr{R}$ be the set of all (inequivalent) valuations of K, and let K_v and $\mathcal{O}_v$ denote the same as in 0.32. The ring $\mathcal{O}_v$ is defined for all non-archimedean valuations v and is a compact open subring of the locally compact field K_v. Therefore, the restricted topological product of the spaces K_v relative to the subsets $\mathcal{O}_v$, where v runs through $\mathscr{R}$, is a locally compact topological ring, which is denoted by $\mathbb{A}_K$ (ring operations being defined component-wise). The ring $\mathbb{A}_K$ is called the *adèle ring* of the field K. If $x \in K$, then $x \in \mathcal{O}_v$ for all but finitely many $v \in \mathscr{R}$. Therefore the field K can be embedded into $\mathbb{A}_K$ as the set of *principal adèles* $(\alpha, \alpha, \ldots)$, $\alpha \in K$. Furthermore, K is a discrete subset of $\mathbb{A}_K$.

Let $\mathbf{M}$ be an affine K-variety and let $\alpha\colon \mathbf{M} \to \mathbf{M}'$ be a K-isomorphism of $\mathbf{M}$ onto a (Zariski closed) K-subvariety $\mathbf{M}'$ of an affine space. Let us denote by $\mathbf{M}'(\mathcal{O}_v)$ the subset of the variety $\mathbf{M}'$ consisting of the points with coordinates in $\mathcal{O}_v$. Let us consider the restricted topological product $\mathbf{M}(\mathbb{A}_K) \subset \prod_{v\in\mathscr{R}} \mathbf{M}(K_v)$ of the spaces $\mathbf{M}(K_v)$ relative to the subsets $\alpha^{-1}(\mathbf{M}'(\mathcal{O}_v))$. It is easily seen that $\mathbf{M}(\mathbb{A}_K)$, as a subset of $\prod_{v\in\mathscr{R}} \mathbf{M}(K_v)$, and the topology on $\mathbf{M}(\mathbb{A}_K)$ do not depend on the choice of the K-isomorphism α. We call $\mathbf{M}(\mathbb{A}_K)$ the *adèle space corresponding to the variety* $\mathbf{M}$. One could define the adèle space for an arbitrary K-variety making use of its covers by affine K-open subsets. Since the sets $\mathbf{M}'(\mathcal{O}_v)$, and hence the sets $\alpha^{-1}(\mathbf{M}'(\mathcal{O}_v))$, are compact and the spaces $\mathbf{M}(K_v)$ are locally compact, the space $\mathbf{M}(\mathbb{A}_K)$ is locally compact. If $x \in \mathbf{M}(K)$, then $x \in \alpha^{-1}(\mathbf{M}'(\mathcal{O}_v))$ for all but finitely many $v \in \mathscr{R}$. Therefore $\mathbf{M}(K)$ is naturally identified with the set of *principal adèles* of the space $\mathbf{M}(\mathbb{A}_K)$ via the diagonal embedding $x \mapsto (x, x, \ldots)$, $x \in \mathbf{M}(K)$.

Let now $\mathbf{G}$ be a linear algebraic K-group. For $\mathbf{G}$ viewed as a K-variety, one can define the space $\mathbf{G}(\mathbb{A}_K)$. Let us endow $\mathbf{G}(\mathbb{A}_K)$ with the component-wise group operation. Then $\mathbf{G}(\mathbb{A}_K)$ turns into a locally compact topological group which is called the *adèle group of the group* $\mathbf{G}$ or the *adèle group associated with* $\mathbf{G}$. If $\mathbf{G}$ is a K-subgroup of $\mathbf{GL}_n$, then $\mathbf{G}(\mathbb{A}_K)$ may be viewed as the group consisting of invertible matrices whose entries belong to $\mathbb{A}_K$ and satisfy the polynomial conditions defining $\mathbf{G}$ in $\mathbf{GL}_n$. The group $\mathbf{G}(K)$, identified with the set of principal adèles of the space $\mathbf{G}(\mathbb{A}_K)$, is a discrete subgroup in $\mathbf{G}(\mathbb{A}_K)$.

(0.34) Measures. (See [Ha 1].) By a *measure* on a set X we mean a non-negative countably additive function μ defined on a σ-algebra Ω of subsets of X and taking finite or infinite values. The subsets of X belonging to Ω are called *μ-measurable* or simply *measurable*. A measure μ on X is said to be *finite* if $\mu(X) < \infty$, *σ-finite* if X is representable in the form of a countable union of measurable subsets of finite measure, and a *normalized* or a *probability* measure if $\mu(X) = 1$.

We denote by δ_x, $x \in X$, the normalized measure concentrated at the point x, i.e.

$$\delta_x(Y) = \begin{cases} 1, & \text{if } x \in Y \\ 0, & \text{if } x \notin Y \,. \end{cases}$$

If μ_1 is a measure on X_1 and μ_2 is a measure on X_2, then by $\mu_1 \times \mu_2$ we denote the product of measures μ_1 and μ_2 (i.e. $(\mu_1 \times \mu_2)(Y_1 \times Y_2) = \mu_1(Y_1)\mu_2(Y_2)$ for measurable $Y_1 \subset X_1$ and $Y_2 \subset X_2$).

For a given measure space the expressions "almost everywhere" and "for almost all" mean, respectively, "everywhere but a null set" and "for all but belonging to a null set" (a *null set* is a set of measure 0).

A map of a space X with measure μ to a topological space (resp. measure space without any topology) Y is called *μ-measurable* or simply *measurable* if it is defined almost everywhere on X and the preimage of every open (resp. measurable) subset of Y is μ-measurable. We say that a measurable map φ: $X \to X$ is *measure preserving* if $\mu(\varphi^{-1}(Y)) = \mu(Y)$ for every measurable $Y \subset X$. A map φ: $X \to X$ of a measure space X with measure μ into itself is said to be a measure space *automorphism* provided that φ is almost everywhere bijective, μ-measurable, and measure preserving. A measurable map of (or a function on) a measure space with measure μ is said to be *μ-constant* if it is constant almost everywhere.

Two measures μ and ν on X are said to be *equivalent* if

$$\{Y \subset X \mid \mu(Y) = 0\} = \{Y \subset X \mid \nu(Y) = 0\}.$$

The *image* $f(\mu)$ of a measure μ on X under a map f: $X \to Y$ is defined by $f(\mu)(B) = \mu(f^{-1}(B))$, for measurable $B \subset Y$.

Let X be a locally compact Hausdorff space. A measure μ on X is said to be *Borel* if it is defined on the σ-algebra of Borel subsets of X and such that $\mu(C) < \infty$ for every compact $C \subset X$. A Borel measure μ on X is called *regular* if for every Borel subset $Y \subset X$ and for each $\varepsilon > 0$ there exist an open set $U \subset X$ and a closed set $F \subset X$ such that $U \supset Y \supset F$ and $\mu(U - F) < \varepsilon$. The *support* of a Borel measure μ is the complement of the union of all open null sets; we denote it by $\operatorname{supp}\mu$. The space of continuous complex-valued functions on X with compact supports will be denoted by $\mathcal{K}(X)$; we denote by $\mathcal{M}(X)$ the set of regular Borel measures on X, and by $\mathcal{P}(X)$ the subset of $\mathcal{M}(X)$ consisting of probability measures on X. For each $\mu \in \mathcal{M}(X)$ we define the linear functional on $\mathcal{K}(X)$ by the formula

$$(*) \qquad \mu(f) = \int_X f(x)d\mu(x).$$

By the F. Riesz theorem on the general form of linear functionals on $\mathcal{K}(X)$, we can identify $\mathcal{M}(X)$ via (*) with the set of positive (i.e. $l(f) \geq 0$, if $f \geq 0$) linear functionals l on $\mathcal{K}(X)$. We endow $\mathcal{M}(X)$ with the *weak topology*, i.e., the topology for which the finite intersections of the sets of type

$$U_{f,V} = \{\mu \in \mathcal{M}(X) \mid \mu(f) \in V\},$$

where $f \in \mathscr{K}(X)$, and V is an open subset of $\mathbb{C}$, form a base of open sets. A subset $M \subset \mathscr{M}(X)$ is said to be *bounded* if there exist relatively compact open sets $X_i \subset X$ and $b_i > 0$, $i \in \mathbb{N}^+$, such that

$$X = \bigcup_{i \in \mathbb{N}^+} X_i \text{ and } \mu(X_i) < b_i$$

for each measure $\mu \in M$. By the Tychonov theorem on compactness of a product of compact spaces, every weakly closed bounded subset of $\mathscr{M}(X)$ is weakly compact. In particular, if X is compact, then $\mathscr{P}(X)$ is weakly compact.

In what follows, unless explicitly stated otherwise, all measures on locally compact spaces are considered to be regular and Borel.

(0.35) ***G*-spaces.** Let G be a topological group. A *left* (resp. *right*) *action* of G on a set X is a map

$$\varphi\colon\ G \times X \to X$$

(resp. $\varphi\colon\ X \times G \to X$) such that, with the notation $\varphi(g, x) = gx$ (resp. $\varphi(x, g) = xg$),

1) $ex = x$ (resp. $xe = x$) for all $x \in X$;
2) $(g_1 g_2)x = g_1(g_2 x)$ (resp. $x(g_1 g_2) = (xg_1)g_2$) for all $x \in X$, and $g_1, g_2 \in G$.

If X is a topological space, then φ is assumed to be continuous. If X is a measure space, then φ is assumed to be measurable in the following strong sense: for every measurable subset $A \subset X$ the preimage $\varphi^{-1}(A)$ belongs to the σ-algebra in $G \times X$ generated by the sets of the form $Y \times B$, where $Y \subset G$ is Borel and $B \subset X$ is measurable.

If a left (resp. right) action of a group G on a set X is given, we say that X is a *left* (resp. *right*) *G-space.*

One can turn a right (resp. left) G-space into a left (resp. right) G-space by setting $gx = xg^{-1}$ (resp. $xg = g^{-1}x$). Usually, it is more convenient to consider left G-spaces, for in this case the map $g \mapsto T_g$, assigning to each $g \in G$ the transformation $T_g\colon\ x \mapsto \varphi(g, x)$ is a homomorphism, while for right G-spaces the map $T_g\colon\ x \mapsto \varphi(x, g)$ is an anti-homomorphism. Therefore a left action and a left G-space will, for simplicity, be often called an *action* and a *G-space.* However, sometimes it is more convenient to consider right G-spaces. The following is an example of such a situation. Let X be a right G-space and let Y be a set. Let us denote by $F(X, Y)$ the set of all maps of X into Y and for each $g \in G$ we define the transformation $\rho(g)$ of the set $F(X, Y)$ by setting $(\rho(g)f)(x) = f(xg)$, $x \in X$, $g \in G$, $f \in F(X, Y)$. Then the map $g \mapsto \rho(g)$ is a homomorphism, while for left G-spaces the similar map $g \mapsto \rho'(g)$ is an anti-homomorphism (where $(\rho'(g)f)(x) = f(gx)$).

If X is a locally compact Hausdorff G-space, then $\mathscr{K}(X)$, $\mathscr{M}(X)$ and $\mathscr{P}(X)$ are also G-spaces:

$$(gf)(x) = f(g^{-1}x) \text{ and } (g\mu)(A) = \mu(g^{-1}A),$$

where $x \in X$, $f \in \mathscr{K}(X)$, $\mu \in \mathscr{M}(X)$ or $\mu \in \mathscr{P}(X)$, $A \subset X$, $g \in G$. It is straight forward that

$$(g\mu)(gf) = \mu(f), \text{ for all } g \in G,\ \mu \in \mathcal{M}(X) \text{ and } f \in \mathcal{K}(X).$$

Let X and Y be two G-spaces. A map $\varphi\colon X \to Y$ is said to be *G-equivariant* or simply *equivariant* if $\varphi(gx) = g\varphi(x)$ for all $x \in X$ and $g \in G$. Furthermore, if a measure μ is given on X and φ is μ-measurable, then φ will be called equivariant if for each $g \in G$ the equality $\varphi(gx) = g\varphi(x)$ holds for almost all $x \in X$.

A measure μ on a left (resp. right) G-space is said to be *G-invariant* or simply *invariant* if $\mu(gA) = \mu(A)$ (resp. $\mu(Ag) = \mu(A)$) for each $g \in G$ and any measurable $A \subset X$, and *G-quasi-invariant* or simply *quasi-invariant* if $\mu(gA) = 0$ (resp. $\mu(Ag) = 0$) is equivalent to $\mu(A) = 0$.

(0.36) Haar measure. (See [Bou 3], Chapter VII and [Hal 1], Chapter XI.) Let G be a locally compact σ-compact group. The *left* (resp. *right*) *Haar measure* on G is a left-invariant (resp. right-invariant) regular Borel measure on G. The left (resp. right) Haar measure on G is unique up to a scalar multiple. The left Haar measure on a group G will be denoted by μ_G. The Haar measure μ_G is right-quasi-invariant, i.e., if $\mu_G(A) = 0$, then $\mu_G(Ag) = 0$ for each $g \in G$. Furthermore, there exists a continuous homomorphism $\Delta_G\colon G \to \mathbb{R}^+$ such that $\mu_G(Ag) = \Delta_G(g)^{-1}\mu_G(A)$ for each measurable $A \subset G$ and every $g \in G$. We call Δ_G the *modular function* of G and remark that $\Delta_G \cdot \mu_G$ is a right Haar measure on G. If $\Delta_G(g) = 1$, i.e., if the measure μ_G is also right-invariant, then the group G is called *unimodular*. If G is compact or $\mathcal{D}(G) = G$, then G is unimodular. If $\mu_G(A) = 0$, then $\mu_G(A^{-1}) = 0$. For every discrete subgroup Γ of the group G the measure on $\Gamma \setminus G$ induced from μ_G will also be denoted by μ_G (if $A \subset G$ and $A \cap \gamma A = \varnothing$ for each $\gamma \in \Gamma - \{e\}$, then $\mu_G(\pi(A)) = \mu_G(A)$, where $\pi\colon G \to \Gamma \setminus G$ is the natural projection).

Let H be a closed subgroup of G. A measure on G/H is called *G-invariant* or *invariant* (resp. *G-quasi-invariant* or *quasi-invariant*), if it is invariant (resp. quasi-invariant) relative to the left translations by all $g \in G$. Replacing the left by the right translations, we similarly define invariant and quasi-invariant measures on $H \setminus G$. We shall denote by $\mu_{G/H}$ (resp. $\mu_{H\setminus G}$) a non-zero quasi-invariant measure on G/H (resp. $H \setminus G$); such a measure always exists and it is unique up to equivalence. The measure $\mu_{G/H}$ (resp. $\mu_{H\setminus G}$) has the following property: $\mu_{G/H}(Y) = 0$ (resp. $\mu_{H\setminus G}(Y) = 0$) if and only if $\mu_G(\pi^{-1}(Y)) = 0$, where $\pi\colon G \to G/H$ (resp. $\pi\colon G \to H \setminus G$) is the natural projection. There exists a non-zero G-invariant measure on G/H (equivalently on $H \setminus G$) if and only if $\Delta_G(h) = \Delta_H(h)$ for each $h \in H$ (see [Bou 3], Chapter VII, §2, Theorem 3, Corollary 2).

If F and H are closed subgroups of G and $H \supset F$, then there exists a finite invariant measure on $F \setminus G$ if and only if there exist finite invariant measures on both $H \setminus G$ and $F \setminus H$ (see [Rag 5], Lemma 1.6).

(0.37) If x and y are two vectors in a Hilbert space, then we denote by $\langle x, y\rangle$ their inner product, and by $\|x\|$ the norm of x. The space of complex-valued functions on a measure space (X, μ) whose p-th power is integrable will be denoted by $L_p(X, \mu)$.

(0.38) Let G be a locally compact group. By a *representation* of G on a vector space V we mean as usual a homomorphism U of the group G into the group of invertible linear transformations of V. We call V the *space of the representation* U. A representation U of the group G on a Hilbert space V is said to be *unitary*, if the operators $U(g)$ are unitary for all $g \in G$. Unless the contrary is stated explicitly unitary representations are supposed to be continuous (i.e. $U(g)v$ depends continuously on $(g, v) \in G \times V$). If U is a unitary representation of a group G and $f \in L_1(G, \mu_G)$, then we set $U(f) = \int_G f(g)U(g)d\mu_G(g)$.

Let U and U' be two unitary representations of a group G on Hilbert spaces V and V' respectively. The representations U and U' are called *equivalent* if there exists a unitary operator A: $V \to V'$ such that $U'(g) = AU(g)A^{-1}$ for each $g \in G$. If a closed subspace $V_1 \subset V$ is invariant under the operators $U(g)$, $g \in G$, then the restriction of U to V_1 is called *subrepresentation* of the representation U. A representation U is called *irreducible*, if it has no non-trivial subrepresentations. We say that a representation U' *is contained in* U, if it is equivalent to a subrepresentation of U.

For each $g \in G$ we define the operator $\lambda(g)$ in $L_2(G, \mu_G)$ by setting for $f \in L_2(G, \mu_G)$ and $x \in G$,

$$(\lambda(g)f)(x) = f(g^{-1}x).$$

It is straightforward that λ is a continuous representation of the group G on the space $L_2(G, \mu_G)$; it is called the *left regular representation*. Similarly we define the *right regular representation* ρ of the group G on the space $L_2(G, \mu'_G)$ by setting

$$(\rho(g)f)(x) = f(xg),$$

where μ'_G is a right Haar measure on G, $f \in L_2(G, \mu'_G)$, x, $g \in G$. The right regular representation will often be called *regular representation*.

(0.39) Let φ be a topological automorphism of a Hausdorff topological space X. A point $x \in X$ is said to be *wandering* (relative to φ) if there exists a neighbourhood U of x such that $\varphi^n(U) \cap U = \emptyset$ for each $n \in \mathbb{N}^+$; otherwise x is *non-wandering*. It can easily be shown that a point x is wandering if and only if it has a neighbourhood W such that $\varphi^n(W) \cap W = \emptyset$ for all n greater then a positive integer $N(W)$. The homeomorphism φ is called *recurrent* if every point of X is non-wandering relative to φ.

Let X be a locally compact Hausdorff topological space, φ: $X \to X$ a topological automorphism, A a closed and B an open subset of X. We say that φ *attracts* B towards A if for every compact set $K \subset B$ and for every open set $U \subset X$ containing A there exists a positive integer $N = N(K, U)$ such that $\varphi^n(K) \subset U$ for all $n > N$. If φ attracts B towards A, then every point $x \in B - A$ is wandering relative to φ.

Let G be a locally compact group and let φ be a continuous automorphism of G. The automorphism φ is said to be *contracting* if it attracts G towards $\{e\}$, i.e., if for every compact set $K \subset G$ and for every neighbourhood U of the identity, there exists an integer $m = m(K, U) \in \mathbb{N}^+$ such that $\varphi^n(K) \subset U$ for all

$n > m$. We call an automorphism φ *non-expanding* if the following conditions are satisfied:

(a) for every neighbourhood W of the identity, there exists a neighbourhood V of the identity such that

$$\bigcup_{n\in\mathbb{N}^+} \varphi^n(V) \subset W;$$

(b) for every compact set $K \subset G$ the set

$$\bigcup_{n\in\mathbb{N}^+} \varphi^n(K)$$

is relatively compact in G.

It is easily seen that every contracting automorphism is non-expanding.

(0.40) Let G be a locally compact group. A discrete subgroup Γ of G is called a *lattice*, if $\mu_G(\Gamma \setminus G) < \infty$. A lattice Γ is said to be *cocompact* or *uniform* if the quotient space $\Gamma \setminus G$ is compact; otherwise Γ is *non-cocompact* or *non-uniform*.

If H_1 and H_2 are closed subgroups of G, $H_1 \supset H_2$ and there exists a finite invariant measure on G/H_2, then (see 0.36) there exists a finite invariant measure on H_1/H_2. Therefore, if Γ is a lattice in G and a closed subgroup G' of G contains Γ, then Γ is a lattice in G'.

Let Γ be a discrete subgroup of G. A subset $X \subset G$ is said to be a *left* (resp. *right*) *fundamental domain for* Γ, if $G = \Gamma \cdot X$ and $(\gamma_1 X) \cap (\gamma_2 X) = \varnothing$ (resp. $G = X \cdot \Gamma$ and $(X\gamma_1) \cap (X\gamma_2) = \varnothing$) for all $\gamma_1, \gamma_2 \in \Gamma$, $\gamma_1 \neq \gamma_2$. If X is a left (resp. right) fundamental domain, then X^{-1} is a right (resp. left) fundamental domain. If the group G is σ-compact, then (see [Bou 3], Chapter VII, §2, Exercise 12) there exists a left (resp. right) Borel fundamental domain X such that

1) $\mu_G(\partial X) = 0$, where ∂X is the boundary of X;
2) for every compact set $K \subset G$ the set $\{\gamma \in \Gamma \mid (\gamma X) \cap K \neq \varnothing$ (resp. $(X\gamma) \cap K \neq \varnothing)\}$ is finite.

Furthermore, if $\Gamma \setminus G$ is compact, then the domain X can be chosen to be relatively compact.

If the group G is compactly generated and $\Gamma \subset G$ is a cocompact lattice, then Γ is finitely generated. Indeed, since G is compactly generated and $\Gamma \setminus G$ is compact, there exists a compact set $Y \subset G$, such that $\Gamma \cdot Y = G$ and G is generated by Y. Since Γ is discrete, Y is compact and $\Gamma \cdot Y = G$, there exists a finite set $L \subset \Gamma$ such that $Y \cdot Y \subset L \cdot Y$ and $Y \cap \Gamma \subset L$. Let us denote by Γ_0 the subgroup of Γ generated by L. By induction on $m \in \mathbb{N}^+$ we have that $Y_m \subset \Gamma_0 \cdot Y$, where $Y_1 = Y$, $Y_m = Y_{m-1} \cdot Y$. Since G is generated by Y, we have that $G = \bigcup_{m\in\mathbb{N}^+} Y_m$. Therefore $G \subset \Gamma_0 \cdot Y$ and hence

$$\Gamma \subset \Gamma_0 \cdot (Y \cap \Gamma) \subset \Gamma_0 \cdot L = \Gamma_0.$$

This implies that Γ coincides with the finitely generated group Γ_0.

(0.41) Let N be a group, Aut N its automorphism group, and let H be a subgroup of Aut N. Then one can define the *semi-direct product* $N \rtimes H$, by introducing on the cartesian product $N \times H$ a multiplication as follows

$$(x_1, y_1) \cdot (x_2, y_2) = (x_1 \cdot (y_1 x_2), y_1 \cdot y_2).$$

If N and H are topological groups (resp. algebraic groups, k-groups) and H acts continuously (resp. rationally, k-rationally) on N, then there exists a topological group (resp. algebraic group, k-group) structure on $N \rtimes H$.

Let G be a group and let $\rho\colon G \to H$ be a homomorphism. We say that a homomorphism $\sigma\colon G \to N \rtimes H$ *covers* ρ if $\rho = \pi \circ \sigma$, where $\pi\colon N \rtimes H \to H$ is the natural epimorphism defined by $\pi(x, y) = y$, $x \in N$, $y \in H$. Two homomorphisms σ and σ' of G into $N \rtimes H$ covering ρ are called *equivalent*, if there exists $x \in N$ such that $\sigma'(g) = x\sigma(g)x^{-1}$ for all $g \in G$.

For any map $f\colon G \to N$ we define the map $\rho_f\colon G \to N \rtimes H$ by setting $\rho_f(g) = (f(g), \rho(g)) \in N \rtimes H$. It is easy to check that the condition "ρ_f is a homomorphism" is equivalent to the condition

(a) $f(g_1 g_2) = f(g_1) \cdot (\rho(g_1) f(g_2))$ for all $g_1, g_2 \in G$.

Furthermore, two homomorphisms ρ_f and $\rho_{f'}$ are equivalent if and only if

(b) there exists $x \in N$ such that for each $g \in G$, $f'(g) = x \cdot f(g) \cdot (\rho(g)x^{-1})$.

A map $f\colon G \to N$ is said to be a *cocycle* (relative to ρ), if condition (a) is satisfied. Two cocycles f and f' are said to be *equivalent* or *cohomologous* if (b) holds. The set of equivalence classes of cocycles will be denoted by $H^1(G, \rho)$. If N, H and G are topological (resp. algebraic) groups, H operates continuously (resp. rationally) on N, and the homomorphism ρ is continuous (resp. rational), then, considering only continuous (resp. regular) maps $f\colon G \to N$, we may define the set $H^1_{\mathrm{cont}}(G, \rho)$ (resp. $H^1_{\mathrm{rat}}(G, \rho)$) that may be viewed as a subset of $H^1(G, \rho)$. Thus, there exists a bijection between $H^1(G, \rho)$ (resp. $H^1_{\mathrm{cont}}(G, \rho)$, $H^1_{\mathrm{rat}}(G, \rho)$) and the set of equivalence classes of all (resp. continuous, rational) homomorphisms of the group G into $N \rtimes H$ covering ρ.

If F is a subgroup of the group G, then restricting the maps of G into N to F we obtain the natural map $H^1(G, \rho) \to H^1(F, \rho)$ which is called the *restriction map*. Similarly, if F is a closed (resp. algebraic) subgroup of a topological (resp. algebraic) group G, then there exists the natural restriction map

$$H^1_{\mathrm{cont}}(G, \rho) \to H^1_{\mathrm{cont}}(F, \rho)$$

(resp. $H^1_{\mathrm{rat}}(G, \rho) \to H^1_{\mathrm{rat}}(F, \rho)$).

Suppose now that the group N is commutative. With additive notation for the group operation on N we can rewrite conditions (a) and (b) as follows:

(a′) $f(g_1 g_2) = f(g_1) + \rho(g_1) f(g_2)$ for all $g_1, g_2 \in G$;
(b′) there exists $x \in N$ such that $h(g) = f(g) + x - \rho(g)x$ for all $g \in G$.

A cocycle $f\colon G \to N$ is called a *coboundary* if it is cohomologous to zero, i.e., if there exists $x \in N$ such that $f(g) = \rho(g)x - x$ for all $g \in G$. The operation of point-wise addition defines an Abelian group structure on the set Ω of maps of the group G to N. It is easily seen that the set of cocycles and the set of coboundaries are subgroups of the group Ω and two cocycles are cohomologous if and only if their difference is a coboundary. Therefore if N is commutative, then $H^1(G, \rho)$ is the factor group of the group of cocycles modulo the subgroup of coboundaries. Similarly, one can define the Abelian group structure on $H^1_{\text{cont}}(G, \rho)$ and on $H^1_{\text{rat}}(G, \rho)$. The group $H^1(G, \rho)$ is said to be the *first cohomology group* of the group G relative to ρ.

1. Algebraic Groups Over Arbitrary Fields

Throughout this section k denotes a field and $\mathbf{G}$ denotes a connected reductive k-group.

1.1 Closed and Quasi-Closed Sets of Roots and Subgroups Associated with Them

(See [Bo-T 1], 3.8.) Let $\mathbf{S}$ be a torus in $\mathbf{G}$ and let $\mathbf{T}$ be a maximal torus in $\mathbf{G}$ containing $\mathbf{S}$. For each $b \in \Phi(\mathbf{T}, \mathbf{G})$ we denote by $\mathbf{U}_b$ the one-parameter root subgroup associated with b. This is the unique subgroup (see [Bo 6], Chapter V, 18.6 or [Hum 1], 26.3) characterized by the existence of an isomorphism $\vartheta_b \colon \mathbf{G}_a \to \mathbf{U}_b$ such that

$$t\vartheta_b(x)t^{-1} = \vartheta_b(b(t)x), \quad t \in \mathbf{T}, \ x \in \mathbf{G}_a, \tag{1}$$

where $\mathbf{G}_a$ is the additive group of dimension 1. Let us observe incidentally that

(i) $$\operatorname{Lie}(\mathbf{U}_b) = \mathfrak{U}_b \overset{\text{def}}{=} \{v \in \operatorname{Lie}(\mathbf{G}) \mid \operatorname{Ad} t(v) = b(t)v, \ t \in \mathbf{T}\},$$

(ii) $$\dim \mathfrak{U}_b = 1,$$

(iii) $$\operatorname{Lie} \mathbf{G} = \bigoplus_{b \in \Phi(\mathbf{T}, \mathbf{G})} \mathfrak{U}_b \oplus \operatorname{Lie}(\mathbf{T}).$$

If the torus $\mathbf{T}$ is defined and split over k, then the subgroups $\mathbf{U}_b$ are defined over k and the isomorphisms ϑ_b can be chosen to be defined over k (see [Bo 6], Chapter V, 18.7).

A subset $\Psi \subset \Phi(\mathbf{T}, \mathbf{G})$ is called *quasi-closed* if for each $b \in \Phi(\mathbf{T}, \mathbf{G})$ such that $\mathbf{U}_b$ is contained in the subgroup generated by all $\mathbf{U}_a$, $a \in \Psi$, we have $b \in \Psi$.

Let $\Psi \subset \Phi(\mathbf{S}, \mathbf{G}) \subset X(\mathbf{S})$ and let $\sigma = \sigma(\Psi)$ (resp. $v = v(\Phi)$) be the set of those elements of $\Phi(\mathbf{T}, \mathbf{G})$ whose restrictions to $\mathbf{S}$ belong to $\Psi \cup \{0\}$ (resp. Ψ). A set $\Psi \subset \Phi(\mathbf{S}, \mathbf{G})$ is called *closed* if $a, b \in \Psi$ and $a + b \in \Phi(\mathbf{S}, \mathbf{G})$ imply $a + b \in \Psi$, and is called *quasi-closed* if σ is quasi-closed. If Ψ is closed, then it is quasi-closed (this follows from the corresponding proposition for σ which, in turn, is a consequence of Proposition 3.4 in [Bo-T 1]). If Ψ is quasi-closed, then by $\mathbf{G}_\Psi^{(\mathbf{S})}$ or $\mathbf{G}_\Psi$ we denote the subgroup generated by $\mathbf{T}$ and the subgroups $\mathbf{U}_a$, $a \in \sigma$, and by $\mathbf{G}_\Psi^{*(\mathbf{S})}$ or $\mathbf{G}_\Psi^*$ the subgroup generated by all the subgroups

$\mathbf{U}_a$, $a \in \nu$. The subgroups $\mathbf{G}_\Psi$ and $\mathbf{G}^*_\Psi$ are algebraic and do not depend on the choice of a maximal torus $\mathbf{T}$ containing $\mathbf{S}$. Furthermore, $\mathscr{Z}_\mathbf{G}(\mathbf{S}) \subset \mathscr{N}_\mathbf{G}(\mathbf{G}^*_\Psi)$ and $\mathbf{G}_\Psi = \mathscr{Z}_\mathbf{G}(\mathbf{S}) \cdot \mathbf{G}^*_\Psi$. If the subgroup $\mathbf{G}^*_\Psi$ is unipotent, then it will often be denoted by $\mathbf{U}_\Psi$ and the set Ψ in this case will be called *unipotent*. If Ψ is quasi-closed and consists of positive roots with respect to some suitable ordering on $\mathrm{X}(\mathbf{S})$, then Ψ is unipotent and ν is quasi-closed and unipotent.

An algebraic group $\mathbf{H}$ is said to be *directly generated* by its algebraic subgroups $\mathbf{H}_1, \ldots, \mathbf{H}_n$ (in the order given) if the multiplication map $\mathbf{H}_1 \times \ldots \times \mathbf{H}_n \to \mathbf{H}$ is an isomorphism of the variety $\mathbf{H}_1 \times \ldots \times \mathbf{H}_n$ onto $\mathbf{H}$.

(1.1.1) Proposition. (See [Bo 6], Chapter IV, 14.3 and [Hum 1], 28.1.) *Let $\mathbf{B}$ be a Borel subgroup of $\mathbf{G}$ containing $\mathbf{T}$ and let $\mathbf{W}$ be a subgroup of $R_u(\mathbf{B})$ normalized by the torus $\mathbf{T}$ and let $C = \{a \in \Phi(\mathbf{T}, \mathbf{G}) \mid \mathbf{U}_a \subset \mathbf{W}\}$. Then $\mathbf{W}$ is directly generated (in any order) by the subgroups $\mathbf{U}_a$, $a \in C$, and hence* $\mathrm{Lie}(\mathbf{W}) = \bigoplus_{a \in C} \mathrm{Lie}(\mathbf{U}_a)$. *In particular, if $\Psi \subset \Phi(\mathbf{S}, \mathbf{G})$ is quasi-closed and unipotent, then $\mathbf{U}_\Psi$ is directly generated by the subgroups $\mathbf{U}_a$, $a \in \nu(\Psi)$, and* $\mathrm{Lie}(\mathbf{U}_\Psi) = \bigoplus_{a \in \nu(\Psi)} \mathrm{Lie}(\mathbf{U}_a)$.

(1.1.2) Proposition. (See [Bo-T 1], Proposition 3.11.) *Let $\Psi \subset \Phi(\mathbf{S}, \mathbf{G})$ be a quasi-closed unipotent set, let $\Psi_1, \ldots, \ldots \Psi_n$ be a partition of Ψ into quasi-closed subsets, and let $\mathbf{H}$ be a connected unipotent algebraic subgroup of $\mathbf{U}_\Psi$ (resp. $\mathbf{G}_\Psi$) normalized by the torus $\mathbf{S}$. Then $\mathbf{H}$ is directly generated by the subgroups $\mathbf{H} \cap \mathbf{U}_{\Psi_1}, \ldots, \mathbf{H} \cap \mathbf{U}_{\Psi_n}$ (resp. $\mathbf{H} \cap \mathscr{Z}_\mathbf{G}(\mathbf{S})$, $\mathbf{H} \cap \mathbf{U}_{\Psi_1}, \ldots, \mathbf{H} \cap \mathbf{U}_{\Psi_n}$).*

Let us note that Proposition 1.1.1 is a special case of Proposition 1.1.2 (in fact it is used in the proof of Proposition 1.1.2).

(1.1.3) Proposition. (See [Bo-T 1], Corollary 3.18.) *Let $\mathbf{S}(\subset \mathbf{G})$ be a torus defined and split over k and let Ψ be a quasi-closed unipotent subset of $\Psi(\mathbf{S}, \mathbf{G})$. Then the unipotent subgroup $\mathbf{U}_\Psi$ is defined and split over k.*

1.2 Subgroups Associated with Sets of Simple Roots

(See [Bo-T 1], 4.2, 4.3, 4.8 and 5.12.) Let $\mathbf{G}$ be a semisimple group and let $\mathbf{S}$ be a maximal k-split torus in $\mathbf{G}$. Then (see 0.27) the set $\Phi = \Phi(\mathbf{S}, \mathbf{G})$ is a root system in $\mathrm{X}(\mathbf{S}) \otimes \mathbb{R}$. Let us fix an ordering on $\mathrm{X}(\mathbf{S}) \otimes \mathbb{R}$ and denote, respectively, by

$$\Phi^+ = \Phi^+(\mathbf{S}, \mathbf{G}), \ \Phi^- = \Phi^-(\mathbf{S}, \mathbf{G}),$$

and

$$\Delta = \Delta(\mathbf{S}, \mathbf{G})$$

the set of positive, negative and simple roots with respect to the ordering. For every $\vartheta \subset \Delta$ we denote by $[\vartheta]$ the set of the roots expressible as linear combinations of elements from ϑ with integral coefficients, and we define the following closed subsets of Φ:

$$\pi_\vartheta = [\vartheta] \cup \Phi^+,$$
$$\pi_\vartheta^- = [\vartheta] \cup \Psi^-,$$
$$\beta_\vartheta = \Phi^+ - [\vartheta],$$
$$\beta_\vartheta^- = \Phi^- - [\vartheta].$$

For every $\Psi \subset \mathrm{X}(\mathbf{S})$ denote by $\mathbf{S}_\Psi$ the identity component in the intersection of the kernels of the characters $\chi \in \Psi$. For the sake of simplicity, we shall denote by $\mathbf{P}_\vartheta$, $\mathbf{P}_\vartheta^-$, $\mathbf{V}_\vartheta$, $\mathbf{V}_\vartheta^-$, the subgroups $\mathbf{G}_{\pi_\vartheta}$, $\mathbf{G}_{\pi_\vartheta^-}$, $\mathbf{U}_{\beta_\vartheta}$, $\mathbf{U}_{\beta_\vartheta^-}$. The subgroups $\mathbf{P}_\vartheta$, $\mathbf{P}_\vartheta^-$, $\mathbf{V}_\vartheta$, $\mathbf{V}_\vartheta^-$ and $\mathscr{Z}_\mathbf{G}(\mathbf{S}_\vartheta)$ are connected and defined over k. The following Levi decompositions hold:

(1) $$\mathbf{P}_\vartheta = \mathscr{Z}_\mathbf{G}(\mathbf{S}_\vartheta) \ltimes \mathbf{V}_\vartheta, \ \mathbf{P}_\vartheta^- = \mathscr{Z}_\mathbf{G}(\mathbf{S}_\vartheta) \ltimes \mathbf{V}_\vartheta^-.$$

The subgroups $\mathbf{P}_\vartheta(\vartheta \subset \Delta)$ are called *standard parabolic k-subgroups* of $\mathbf{G}$ associated with $\mathbf{S}$ and Φ^+. These are precisely the k-subgroups of $\mathbf{G}$ containing $\mathbf{P}_\varnothing$, and every parabolic k-subgroup of $\mathbf{G}$ is conjugate by an element of $\mathbf{G}(k)$ to a unique standard parabolic k-subgroup of $\mathbf{G}$ (see [Bo-T 1], Proposition 5.14). In particular, any minimal parabolic k-subgroup of $\mathbf{G}$ is conjugate by an element of $\mathbf{G}(k)$ to $\mathbf{P}_\varnothing$. For every $\vartheta \subset \Delta$ we have

$$\mathbf{V}_\vartheta^- \cap \mathbf{P}_\vartheta = \{e\} \text{ and } \mathbf{P}_\vartheta \cap \mathbf{P}_\vartheta^- = \mathscr{Z}_\mathbf{G}(\mathbf{S}_\vartheta),$$

and so $\mathbf{P}_\vartheta$ and $\mathbf{P}_\vartheta^-$ are opposite parabolic k-subgroups. The multiplication map induces an isomorphism of the k-variety $\mathbf{V}_\vartheta^- \times \mathbf{P}_\vartheta$ and also of the k-variety $\mathbf{V}_\vartheta \times \mathbf{P}_\vartheta^-$ onto a Zariski dense and open subset of $\mathbf{G}$.

(1.2.1) Proposition. *For every $\vartheta \subset \Delta$, the group $\mathbf{G}(k)$ is generated by the subgroups $\mathbf{P}_\vartheta(k)$ and $\mathbf{V}_\vartheta^-(k)$.*

Proof. As was observed above, the multiplication map induces an isomorphism of the k-variety $\mathbf{V}_\vartheta^- \times \mathbf{P}_\vartheta$ onto a Zariski dense and open subset W of $\mathbf{G}$. Then $W \cap \mathbf{G}(k) = \mathbf{V}_\vartheta^-(k) \cdot \mathbf{P}_\vartheta(k)$. On the other hand, since W is Zariski dense and open in $\mathbf{G}$ and $\mathbf{G}(k)$ is Zariski dense in $\mathbf{G}$, we have that $W \cap gW^{-1} \cap \mathbf{G}(k) \neq \varnothing$ for each $g \in \mathbf{G}(k)$. Hence every $g \in \mathbf{G}(k)$ is contained in $(W \cap \mathbf{G}(k)) \cdot (W \cap \mathbf{G}(k))$. Therefore

$$\mathbf{G}(k) \subset \mathbf{V}_\vartheta^-(k) \cdot \mathbf{P}_\vartheta(k) \cdot \mathbf{V}_\vartheta^-(k) \cdot \mathbf{P}_\vartheta(k).$$

This completes the proof. □

(1.2.2) Proposition. *If* $\dim \mathbf{S} = \mathrm{rank}_k \mathbf{G} \geq 2$, *then there exist a positive integer n and 1-dimensional subtori $\mathbf{S}_1, \ldots, \mathbf{S}_n$ of the torus $\mathbf{S}$ such that*

$$\mathbf{G}(k) = \mathscr{Z}_{\mathbf{G}(k)}(\mathbf{S}_1(k)) \cdot \ldots \cdot \mathscr{Z}_{\mathbf{G}(k)}(\mathbf{S}_n(k)).$$

Proof. For each indivisible root $a \in \Phi = \Phi(\mathbf{S}, \mathbf{G})$ we denote by $\mathbf{W}_a$ the subgroup $\mathbf{G}^\bullet_{\Psi(a)} = \mathbf{U}_{\Psi(a)}$, where $\Psi(a) = \{na \mid n \in \mathbb{N}^+\} \cap \Phi$. By Proposition 1.1.3 the subgroups $\mathbf{W}_a$ are defined over k. Since $\dim \mathbf{S} \geq 2$, for each indivisible $a \in \Phi$ there exists a 1-dimensional subtorus $\mathbf{S}_a$ of $\mathbf{S}$ such that $\mathbf{S}_a \subset \operatorname{Ker} a$, and hence we have $\mathbf{W}_a \subset \mathscr{Z}_{\mathbf{G}}(\mathbf{S}_a)$. It follows from Proposition 1.1.2 that the subgroup $\mathbf{P}_\varnothing$ is directly generated by the k-subgroups $\mathscr{Z}_{\mathbf{G}}(\mathbf{S})$, $\mathbf{W}_{a_1}, \ldots, \mathbf{W}_{a_r}$, where $a_1, \ldots, a_r \in \Phi^+$ are positive indivisible roots, and hence

$$\mathbf{P}_\varnothing(k) = \mathscr{Z}_{\mathbf{G}(k)}(\mathbf{S}(k)) \cdot \mathbf{W}_{a_1}(k) \cdot \ldots \cdot \mathbf{W}_{a_r}(k).$$

But $\mathbf{W}_a \subset \mathscr{Z}_{\mathbf{G}}(\mathbf{S}_a)$ and $\mathscr{Z}_{\mathbf{G}}(\mathbf{S}) \subset \mathscr{Z}_{\mathbf{G}}(\mathbf{S}_a)$, so

$$\mathbf{P}_\varnothing(k) \subset \mathscr{Z}_{\mathbf{G}(k)}(\mathbf{S}_{a_1}(k)) \cdot \ldots \cdot \mathscr{Z}_{\mathbf{G}(k)}(\mathbf{S}_{a_r}(k)) \tag{2}$$

and similarly

$$\mathbf{P}^-_\varnothing(k) \subset \mathscr{Z}_{\mathbf{G}(k)}(\mathbf{S}_{a_1}(k)) \cdot \ldots \cdot \mathscr{Z}_{\mathbf{G}(k)}(\mathbf{S}_{a_r}(k)). \tag{3}$$

Since $\mathbf{G}(k)$ is generated by the subgroups $\mathbf{P}_\varnothing(k)$ and $\mathbf{P}^-_\varnothing(k)$ (see Proposition 1.2.1), inclusions (2) and (3) imply the required equality. □

(1.2.3) Proposition. (See [Bo-T 1], Corollary 5.18.)
(i) *The groups* $\mathbf{P}_\vartheta(k)$ *are precisely the subgroups of* $\mathbf{G}(k)$ *containing* $\mathbf{P}_\varnothing(k)$;
(ii) *Let* $\vartheta, \vartheta' \subset \Delta$ *and* $g \in \mathbf{G}(k)$. *Then* $g\mathbf{P}_{\vartheta'}(k)g^{-1} \subset \mathbf{P}_\vartheta(k)$ *if and only if* $\vartheta' \subset \vartheta$ *and* $g \in \mathbf{P}_\vartheta(k)$. *In particular,* $\mathbf{P}_\vartheta(k) \subset \mathbf{P}_{\vartheta'}(k)$ *(resp.* $\mathbf{P}_\vartheta(k) = \mathbf{P}_{\vartheta'}(k)$*) if and only if* $\vartheta \subset \vartheta'$ *(resp.* $\vartheta = \vartheta'$*).*

(1.2.4) Corollary. *Let* $\vartheta \subset \Delta$. *Then* $\mathbf{P}_\vartheta(k)$ *is generated by the subgroups* $\mathbf{P}_{\{b\}}(k)$, $b \in \vartheta$.

Proof. If H is the subgroup of $\mathbf{G}(k)$ generated by the subgroups $\mathbf{P}_{\{b\}}(k)$, $b \in \vartheta$, then by Proposition 1.2.3(i) there exists $\vartheta' \subset \Delta$ such that $H = \mathbf{P}_{\vartheta'}(k)$. Since $H \subset \mathbf{P}_\vartheta(k)$ and $\mathbf{P}_{\{b\}}(k) \subset H$ for each $b \in \vartheta$, by Proposition 1.2.3(ii) we have $\vartheta' = \vartheta$. □

1.3 Equivariant Mappings of Unipotent Groups onto Lie Algebras

Let $\operatorname{char} k = 0$ and let $\mathbf{H}$ be an algebraic k-group, with $\exp : \operatorname{Lie}(\mathbf{H})^{(n)} \to \mathbf{H}^{(u)}$ and $\ln : \mathbf{H}^{(u)} \to \operatorname{Lie}(\mathbf{H})^{(n)}$ as in 0.20. Before stating the following assertion, we recall that if $\alpha \colon \mathbf{H} \to \mathbf{F}$ is an algebraic group morphism, then $\alpha(\mathbf{H}^{(u)}) \subset \mathbf{F}^{(u)}$ (see 0.20).

(1.3.1) Proposition. *If* $\operatorname{char} k = 0$ *and* $\alpha \colon \mathbf{H} \to \mathbf{F}$ *is a* k*-group morphism, then for each* $u \in \mathbf{H}^{(u)}$ *we have* $\alpha(u) = \exp((d\alpha)(\ln u))$, *where* $d\alpha \colon \operatorname{Lie}(\mathbf{H}) \to \operatorname{Lie}(\mathbf{F})$ *is the differential of* α.

Proof. Let $u \in \mathbf{H}^{(u)}$. Let us set $u^t = \exp(t \cdot \ln u)$ and $(\alpha(u))^t = \exp(t \cdot \ln \alpha(u))$. Then the mapping $u^t \mapsto (\alpha(u))^t$ is a morphism of the one-parameter group $\{u^t\}$ onto the one-parameter group $\{(\alpha(u))^t\}$ coinciding with α on the Zariski dense subgroup $\{u^n \mid n \in \mathbb{Z}\}$. Therefore, $\alpha(u^t) = (\alpha(u))^t$ for each $t \in k$, which implies that $d\alpha(\ln u) = \ln \alpha(u)$. This proves the proposition. □

It follows from the preceding proposition that if $\operatorname{char} k = 0$, then for every unipotent k-group $\mathbf{U}$ the logarithmic mapping $\ln : \mathbf{U} \to \operatorname{Lie}(\mathbf{U})$ is equivariant in the following sense: if α is a biregular automorphism of $\mathbf{U}$, then $\ln \circ \alpha = d\alpha \circ \ln$. But, if $\operatorname{char} k \neq 0$, then in general there is no such equivariant mapping. Nevertheless, the following assertion is valid.

(1.3.2) Proposition. (See [Bo-Sp], Corollary 9.12.) *Let $\mathbf{H}$ be a connected solvable k-group, let $\mathbf{U} = R_u(\mathbf{H})$, and let $\mathbf{S}$ be a torus in $\mathbf{H}$ defined over k. Assume that either $\mathscr{Z}_{\mathbf{H}}(\mathbf{S}) \cap \mathbf{U} = \{e\}$ or k is a perfect field. Then the unipotent group $\mathbf{U}$ is defined and split over k and there exists an $\mathbf{S}$-equivariant k-isomorphism $f\colon \mathbf{U} \to \operatorname{Lie}(\mathbf{U})$ of algebraic varieties; here $\mathbf{S}$-equivariance means that $f(sus^{-1}) = \operatorname{Ad} s(f(u))$ for all $s \in \mathbf{S}$ and $u \in \mathbf{U}$.*

(1.3.3) Proposition. *Let $\mathbf{S}$ be a maximal k-split torus in $\mathbf{G}$.*

(i) *If $\mathbf{P}$ is a parabolic k-subgroup of $\mathbf{G}$ containing $\mathbf{S}$, then the unipotent radical $R_u(\mathbf{P})$ is defined and split over k and there exists an $\mathbf{S}$-equivariant k-isomorphism of algebraic varieties $R_u(\mathbf{P}) \to \operatorname{Lie}(R_u(\mathbf{P}))$;*

(ii) *If $\mathbf{G}$ is semisimple and $\vartheta \subset \Delta = \Delta(\mathbf{S}, \mathbf{G})$, then the unipotent groups $\mathbf{V}_\vartheta$ and $\mathbf{V}_\vartheta^-$ (see 1.2) are defined and split over k and there exist $\mathbf{S}$-equivariant k-isomorphisms $\mathbf{V}_\vartheta \to \operatorname{Lie}(\mathbf{V}_\vartheta)$ and $\mathbf{V}_\vartheta^- \to \operatorname{Lie}(\mathbf{V}_\vartheta^-)$.*

To show (i) it suffices to observe that $\mathscr{Z}_{\mathbf{G}}(\mathbf{S}) \cap R_u(\mathbf{P}) = \{e\}$ (see [Bo-T 1], Theorem 4.15) and to apply Proposition 1.3.2 for $\mathbf{H} = \mathbf{S} \cdot R_u(\mathbf{P})$.

(ii) is a particular case of (i). Since $\mathscr{Z}_{\mathbf{G}}(\mathbf{S}) \cap \mathbf{V}_\vartheta = \mathscr{Z}_{\mathbf{G}}(\mathbf{S}) \cap \mathbf{V}_\vartheta^- = \{e\}$ (which follows, for instance, from decomposition (1) in 1.2), (ii) can also be deduced from Proposition 1.3.2.

1.4 Central Isogenies. Simply Connected Adjoint Groups

(See [Bo-T 3], §2.)

(1.4.1) Definition. *A morphism $\varphi\colon \mathbf{H} \to \mathbf{H}'$ of algebraic groups is called* quasi-central *if* $\operatorname{Ker} \varphi \subset \mathscr{Z}(\mathbf{H})$. *In other words, φ is quasi-central if there exists a mapping $\chi\colon \varphi(\mathbf{H}) \times \varphi(\mathbf{H}) \to \mathbf{H}$ such that $\chi(\varphi(x), \varphi(y)) = xyx^{-1}y^{-1}$ for all $x, y \in \mathbf{H}$.*

(1.4.2) Definition. *A morphism $\varphi\colon \mathbf{H} \to \mathbf{H}'$ of algebraic groups is said to be* central *if it is quasi-central and the mapping χ in Definition 1.4.1 (which is obviously unique) is a morphism of algebraic varieties.*

(1) *Remark.* If $\varphi\colon \mathbf{H} \to \mathbf{H}'$ is a central k-group morphism, then the mapping χ is defined over k (see [Bo-T 3], Remark 2.6). Therefore if $\varphi(x) \in \mathbf{H}'(k)$ and $\varphi(y) \in \mathbf{H}'(k)$, then $xyx^{-1}y^{-1} = \chi(\varphi(x), \varphi(y)) \in \mathbf{H}(k)$. Thus, $\mathscr{D}(\varphi^{-1}(\mathbf{H}'(k))) \subset \mathbf{H}(k)$ and hence, if $\varphi(\mathbf{H}) = \mathbf{H}'$, then $\mathscr{D}(\mathbf{H}'(k)) \subset \varphi(\mathbf{H}(k))$.

(2) *Remark.* Let $\beta\colon \mathbf{G} \to \mathbf{G}'$ be a central k-isogeny of reductive k-groups and let d be the degree of the isogeny (i.e. d is the degree of the field $K(\mathbf{G})$ of rational functions on $\mathbf{G}$ over the field $\beta_0(K(\mathbf{G}'))$, where β_0 is the comorphism of β). Then (see [Bo-T 4], Proposition 3.16) there exists a k-morphism of algebraic varieties μ: $\mathbf{G}' \to \mathbf{G}$ such that $\mu(\beta(g)) = g^d$ for each $g \in \mathbf{G}$. If $\beta(g) \in \mathbf{G}'(k)$, then $g^d = \mu(\beta(g)) \in \mathbf{G}(k)$. But in view of Remark 1, $\mathscr{D}(\beta^{-1}(\mathbf{G}'(k))) \subset \mathbf{G}(k)$. Therefore $\mathbf{G}(k)$ is a normal subgroup of $\beta^{-1}(\mathbf{G}'(k))$ and $\beta^{-1}(\mathbf{G}'(k))/\mathbf{G}(k)$ is a commutative torsion group whose exponent divides d. It follows that $\beta(\mathbf{G}(k))$ is a normal subgroup of $\mathbf{G}'(k)$ and $\mathbf{G}'(k)/\beta(\mathbf{G}(k))$ is a commutative torsion group whose exponent divides d (see [Bo-T 4], Corollary 3.17).

(3) *Remark.* If a morphism is quasi-central and separable, then it is central (the converse is in general false). In particular, in case $\operatorname{char} k = 0$, every isogeny of connected k-groups is central.

(1.4.3) Definition. *Two algebraic groups* $\mathbf{H}$ *and* $\mathbf{H}'$ *are called* isogenous *(resp.* strictly isogenous*), if there exist an algebraic group* $\mathbf{F}$ *and two isogenies (resp. central isogenies)* $\mathbf{F} \to \mathbf{H}$ *and* $\mathbf{F} \to \mathbf{H}'$.

(1.4.4) Definition. *Two algebraic k-groups* $\mathbf{H}$ *and* $\mathbf{H}'$ *are called* k-isogenous *(resp.* strictly k-isogenous*), if there exist a k-group* $\mathbf{F}$ *and two k-isogenies (resp. central k-isogenies)* $\mathbf{F} \to \mathbf{H}$ *and* $\mathbf{F} \to \mathbf{H}'$.

(1.4.5) Proposition. (See [Bo-T 3], Theorem 2.20.) *Let* $\mathbf{G}'$ *be a reductive k-group and let* f: $\mathbf{G} \to \mathbf{G}'$ *be a central k-isogeny. Then*

(i) *The maximal k-split tori in* $\mathbf{G}'$ *(resp.* $\mathbf{G}$*) are the images (resp. the connected components of the identity in the inverse images) under* f *of maximal k-split tori in* $\mathbf{G}$ *(resp.* $\mathbf{G}'$*);*

(ii) *The parabolic k-subgroups of* $\mathbf{G}'$ *(resp.* $\mathbf{G}$*) are the images (resp. the inverse images) under* f *of parabolic k-subgroups in* $\mathbf{G}$ *(resp.* $\mathbf{G}'$*).*

(1.4.6) Corollary.

(a) *Strictly k-isogenous reductive k-groups have common k-rank* (this follows from Proposition 1.4.5(i)).

(b) *Let n be a non-negative integer and* $\mathbf{G}$, $\mathbf{G}'$ *be two strictly k-isogenous connected semisimple k-groups. Then* $\mathbf{G}$ *has an almost k-simple factor of k-rank n if and only if* $\mathbf{G}'$ *also has such a factor. In particular,* $\mathbf{G}$ *has no k-anisotropic factors if and only if* $\mathbf{G}'$ *also has no such factors.*

(c) *Let* f: $\mathbf{G} \to \mathbf{G}_0$ *be a central k-isogeny of connected semisimple k-groups. Let us denote by* $\mathbf{G}^{is} \subset \mathbf{G}$ *(resp.* $\mathbf{G}_0^{is} \subset \mathbf{G}_0$*) the subgroup which is the almost direct product of the k-isotropic factors of* $\mathbf{G}$ *(resp.* $\mathbf{G}_0$*). Then* $f(\mathbf{G}^{is}) = \mathbf{G}_0^{is}$.

Assertions (b) and (c) are consequences of (a) and the fact that the decomposition into k-simple factors is invariant under k-isogenies.

(1.4.7) Proposition. (See [Bo-T 3], 2.13, 2.15, 2.17.) *Let* $\mathbf{G}$ *be a reductive algebraic group,* f: $\mathbf{G} \to \mathbf{G}'$ *an isogeny,* $\mathbf{T}$ *a maximal torus in* $\mathbf{G}$, *and let* $f_{\mathbf{T}}^{*}$: $\mathrm{X}(f(\mathbf{T})) \to \mathrm{X}(\mathbf{T})$

be the homomorphism sending $\chi \in \mathrm{X}(f(\mathbf{T}))$ *to* $\chi \circ f \in \mathrm{X}(\mathbf{T})$. *Then the following conditions are equivalent:*

(i) *the isogeny* f *is central;*

(ii) $f_{\mathbf{T}}^{*}(\Phi(f(\mathbf{T}), \mathbf{G}')) = \Phi(\mathbf{T}, \mathbf{G})$*:*

(iii) *the kernel of the differential* $df\colon \mathrm{Lie}(\mathbf{G}) \to \mathrm{Lie}(\mathbf{G}')$ *is contained in the center of the Lie algebra* $\mathrm{Lie}(\mathbf{G})$ *(this center coincides with* $\mathscr{Z}_{\mathrm{Lie}(\mathbf{G})}(\mathbf{G})$ *and is contained in* $\mathrm{Lie}(\mathbf{T})$*).*

(iv) $df(\mathrm{Lie}(\mathbf{G})^{(n)}) = \mathrm{Lie}(\mathbf{G}')^{(n)}$, *where as in 0.20* $\mathrm{Lie}(\mathbf{G})^{(n)}$ *and* $\mathrm{Lie}(\mathbf{G}')^{(n)}$ *denote the set of nilpotent elements in* $\mathrm{Lie}(\mathbf{G})$ *and* $\mathrm{Lie}(\mathbf{G}')$ *respectively;*

(v) $\mathrm{Lie}(\mathbf{G})^{(n)} \cap \mathrm{Ker}\, df = \{0\}$*;*

(vi) *for each* $b \in \Phi(\mathbf{T}, \mathbf{G})$ *the map* $f\colon \mathbf{U}_b \to f(\mathbf{U}_b)$ *is an algebraic group isomorphism, where* $\mathbf{U}_b$ *denotes the root subgroup associated with* b.

Remark. It follows from the equivalence of (i) and (iv) above that the composition of central isogenies of connected reductive algebraic groups is a central isogeny (see [Bo-T 3], Corollary 2.18).

(1.4.8) Corollary. *Let* $\mathbf{G}$ *be a reductive algebraic group, let* $f\colon \mathbf{G} \to \mathbf{G}'$ *be a central isogeny, and let* $g \in \mathbf{G}$. *Then the characteristic polynomials of the transformations* $\mathrm{Ad}\, g$ *and* $\mathrm{Ad}\, f(g)$ *coincide and, in particular,* $\mathrm{Tr}\,\mathrm{Ad}\, g = \mathrm{Tr}\,\mathrm{Ad}\, f(g)$.

Proof. Since the characteristic polynomial of a linear transformation and of its semisimple part coincide and in view of (ii) in 0.20, $(\mathrm{Ad}\, g)_s = \mathrm{Ad}\, g_s$ and $(\mathrm{Ad}\, f(g))_s = \mathrm{Ad}\, f(g_s)$, we may assume that g is semisimple. All that remains now is to observe that every semisimple element of $\mathbf{G}$ belongs to a maximal torus of $\mathbf{G}$ (see 0.22) and to apply Proposition 1.4.7. □

(1.4.9) Definition. *A connected semisimple algebraic group* $\mathbf{H}$ *is called* simply connected *(resp.* adjoint*) if every central isogeny* $\varphi\colon \mathbf{H}' \to \mathbf{H}$ *(resp.* $\varphi\colon \mathbf{H} \to \mathbf{H}'$*), for* $\mathbf{H}'$ *connected, is an algebraic group isomorphism.*

If a connected semisimple k-group $\mathbf{H}$ is adjoint, then $\mathscr{Z}(\mathbf{H}) = \{e\}$. In case $\mathrm{char}\, k = 0$, the converse is also true. But if $\mathrm{char}\, k \neq 0$, a connected semisimple k-group with trivial center is not necessarily adjoint.

As examples of simply connected groups we mention $\mathbf{SL}_n$ and the group $\mathbf{Sp}_{2n}$ of symplectic matrices of order $2n$ as well.

(1.4.10) Proposition. (See [Ti 2], 3.1.2.) *A connected simply connected (resp. adjoint) semisimple k-group decomposes uniquely into a direct product of simply connected (resp. adjoint) almost k-simple k-groups.*

(1.4.11) Proposition. (See [Bo-T 3], Propositions 2.24 and 2.26 and [Ti 2], 2.6.1.) *Let* $\mathbf{G}$ *and* $\mathbf{G}'$ *be connected semisimple k-groups. Then*

(i) *there exists an exact sequence* $\tilde{\mathbf{G}} \xrightarrow{\tilde{p}} \mathbf{G} \xrightarrow{\bar{p}} \bar{\mathbf{G}}$, *where* $\tilde{\mathbf{G}}$ *is a simply connected k-group,* $\bar{\mathbf{G}}$ *is an adjoint k-group,* $\tilde{p}$ *and* $\bar{p}$ *are central k-isogenies. The groups* $\tilde{\mathbf{G}}$ *and* $\bar{\mathbf{G}}$ *and the isogenies* $\tilde{p}$ *and* $\bar{p}$ *are determined uniquely up to k-isomorphism;*

(ii) *if* $f\colon \mathbf{G} \to \mathbf{G}'$ *is a central k-isogeny and* $h\colon \mathbf{G}_1 \to \mathbf{G}'$ *is a k-homomorphism of a semisimple simply connected k-group* $\mathbf{G}_1$, *then there exists a unique k-homomorphism* $h'\colon \mathbf{G}_1 \to \mathbf{G}$ *such that* $h = f \circ h'$;

(iii) *if* $f\colon \mathbf{G} \to \mathbf{G}'$ *is a central k-isogeny and* $h\colon \mathbf{G} \to \mathbf{G}_1$ *is a k-epimorphism to a semisimple adjoint k-group* $\mathbf{G}_1$, *then there exists a unique k-homomorphism* $h'\colon \mathbf{G}' \to \mathbf{G}_1$ *such that* $h = h' \circ f$;

(iv) *the isogeny* $\mathrm{Ad}\colon \mathbf{G} \to \mathrm{Ad}\,\mathbf{G}$ *is central and* $\mathrm{Ad}\,\mathbf{G}$ *is an adjoint group (i.e. one can take* $\mathrm{Ad}\,\mathbf{G}$ *and* Ad *as* $\bar{\mathbf{G}}$ *and* $\bar{\mathrm{p}}$ *in (i) above).*

(1.4.12) Definition. *The groups* $\tilde{\mathbf{G}}$ *and* $\bar{\mathbf{G}}$ *in (i) of Proposition 1.4.11 are called, respectively, the* simply connected covering *and the* adjoint group *of the connected semisimple k-group* $\mathbf{G}$.

(1.4.13) Proposition. *Let* $\mathbf{G}'$ *be an algebraic k-group and let* $\beta\colon \mathrm{Lie}(\mathbf{G}) \to \mathrm{Lie}(\mathbf{G}')$ *be a Lie algebra morphism defined over k. Suppose* $\mathrm{char}\, k = 0$ *and the group* $\mathbf{G}$ *is semisimple and simply connected. Then there exists a unique morphism* $\pi\colon \mathbf{G} \to \mathbf{G}'$, *whose differential is* β. *Furthermore,* π *is defined over k.*

Proof. Let us set $\mathfrak{G} = \mathrm{Lie}(\mathbf{G})$ and denote by $\mathfrak{H} = \{(x, \beta(x)) \mid x \in \mathfrak{G}\} \subset \mathfrak{G} \times \mathrm{Lie}(\mathbf{G}')$ the graph of β. Since $\mathbf{G}$ is semisimple (see 0.24), the Lie algebra $\mathfrak{G}$, and hence the Lie algebra $\mathfrak{H}$ which is isomorphic to $\mathfrak{G}$, are semisimple. Therefore $\mathfrak{H}$ is the Lie algebra of a connected semisimple algebraic subgroup $\mathbf{H} \subset \mathbf{G} \times \mathbf{G}'$ (see assertion 0.15(v) and 0.24). Let $\alpha\colon \mathbf{G} \times \mathbf{G}' \to \mathbf{G}$ and $d\alpha\colon \mathfrak{G} \times \mathrm{Lie}(\mathbf{G}') \to \mathfrak{G}$ be natural projections. Since $\dim \mathbf{H} = \dim \mathfrak{H} = \dim \mathfrak{G} = \dim \mathbf{G}$ and $d\alpha(\mathfrak{H}) = \mathfrak{G}$, the restriction of the morphism α to $\mathbf{H}$ is an isogeny. But $\mathrm{char}\, k = 0$ and $\mathbf{G}$ is simply connected. Therefore $\alpha|_{\mathbf{H}}$ is an isomorphism and hence $\mathbf{H}$ is the graph of a morphism $\pi\colon \mathbf{G} \to \mathbf{G}'$. Since $\mathbf{H}$ is the graph of π, $\mathfrak{H}$ is the graph of β, and $\mathfrak{H} = \mathrm{Lie}(\mathbf{H})$, we have that $\beta = d\pi$. The uniqueness of the morphism π is a consequence of (vi) in 0.15. Since β is defined over k and π is unique, it follows that ${}^{\sigma}\pi = \pi$ for every k-automorphism σ of any field extension of k (see the definition of ${}^{\sigma}\pi$ below at the beginning of 1.7). Since k is perfect, it then follows that π is defined over k. □

(1.4.14) Let $f\colon \mathbf{G} \to \mathbf{G}'$ be an epimorphism of connected semisimple k-groups. If f is an isogeny, then by Proposition 1.4.7 it is central if and only if $df(\mathrm{Lie}(\mathbf{G})^{(n)}) = \mathrm{Lie}(\mathbf{G}')^{(n)}$. The epimorphism f is said to be *special* if $df(\mathrm{Lie}(\mathbf{G})^{(n)}) \cap \mathrm{Lie}(\mathbf{G}'_1)^{(n)} \neq \{0\}$ for every almost simple factor $\mathbf{G}'_1$ of $\mathbf{G}'$. Obviously every central isogeny is special. The converse is in general false. However, if $\mathrm{char}\, k \geq 5$ and f is a special isogeny, then f is central. More precisely (see [Bo-T 4], 3.3), if $f\colon \mathbf{G} \to \mathbf{G}'$ is a special non-central isogeny and $\mathbf{G}, \mathbf{G}'$ are almost simple, then one of the following conditions is satisfied:

$\mathrm{char}\, k = 2$, $\mathbf{G}$ is either the group of type $\mathbb{B}_n$, $\mathbb{C}_n$ or $\mathbb{F}_4$ and $\mathbf{G}'$ is the group of dual type $\mathbb{C}_n$, $\mathbb{B}_n$ or $\mathbb{F}_4$;

$\mathrm{char}\, k = 3$, $\mathbf{G}$ and $\mathbf{G}'$ are the groups of type $\mathbb{G}_2$.

It follows from Proposition 1.4.7 that if $f_0\colon \mathbf{G}_0 \to \mathbf{G}$ (resp. $f_1\colon \mathbf{G}' \to \mathbf{G}_1$) is a rational epimorphism and $f\colon \mathbf{G} \to \mathbf{G}'$ is a central isogeny, then the epimorphism $f \circ f_0$ (resp. $f_1 \circ f$) is special if and only if f_0 (resp. f_1) is likewise special.

1.5 Unipotent Subgroups and the Group $\mathbf{H}(k)^+$

(1.5.1) Proposition. (See [Bo-T 2], Corollary 3.7.) *Let us assume that the field k is perfect, and let $\mathbf{H}$ be a connected k-group. Then any unipotent subgroup of $\mathbf{H}(k)$ (resp. unipotent k-subgroup of $\mathbf{H}$) is contained in the unipotent radical of a parabolic k-subgroup of $\mathbf{H}$. In particular, maximal unipotent subgroups of $\mathbf{H}(k)$ (resp. maximal unipotent k-subgroups of $\mathbf{H}$) are of the form $R_u(\mathbf{P})(k)$ (resp. $R_u(\mathbf{P})$), where $\mathbf{P}$ denotes a minimal parabolic k-subgroup of $\mathbf{H}$. Thus all such unipotent subgroups are conjugate to each other by elements of $\mathbf{H}(k)$.*

(1.5.2) If $\mathbf{H}$ is a connected k-group, then by $\mathbf{H}(k)^+$ we denote the normal subgroup of $\mathbf{H}(k)$ generated by the subgroups $R_u(\mathbf{P})(k)$, where $\mathbf{P}$ runs through the set of all parabolic k-subgroups of $\mathbf{H}$. It follows from Proposition 1.5.1 that if k is a perfect field, then $\mathbf{H}(k)^+$ coincides with the subgroup generated by the set $\mathbf{H}(k)^{(u)}$ of unipotent elements of $\mathbf{H}(k)$. If $\mathbf{H}$ is semisimple and k is algebraically closed, then (see [Hum 1], Theorem 27.5(d)) $\mathbf{H}(k)^{(u)}$ generates $\mathbf{H}(k)$ and hence $\mathbf{H}(k)^+ = \mathbf{H}(k)$.

(1.5.3) Proposition. *Let $\mathbf{G}$ be semisimple. Then the subgroup $\mathbf{G}(k)^+$ coincides with $\{e\}$ if and only if $\mathbf{G}$ is anisotropic over k.*

To prove this proposition it suffices to observe that

(a) a connected semisimple k-group contains a proper parabolic k-subgroup if and only if it is isotropic over k (see 0.29) and
(b) if $\mathbf{S}$ is a maximal k-split torus in a semisimple group $\mathbf{G}$ and $\vartheta \subsetneq \Delta = \Delta(\mathbf{S}, \mathbf{G})$, then by Proposition 1.3.3(ii) the subgroup $R_u(\mathbf{P}_\vartheta)(k) = \mathbf{V}_\vartheta(k)$ is different from $\{e\}$.

(1.5.4) Proposition. (See [Bo-T 4], 6.2, 6.9, 6.11.) *Let $\mathbf{G}_i$, $1 \leq i \leq r$, be the connected non-commutative almost k-simple k-isotropic normal k-subgroups of $\mathbf{G}$. Then*

(i) *$\mathbf{G}(k)^+$ coincides with the subgroup of $\mathbf{G}(k)$ generated by the subgroups $\mathbf{U}(k)$, where $\mathbf{U}$ runs through the set of unipotent k-split subgroups of $\mathbf{G}$;*
(ii) *if $\mathbf{P}$ and $\mathbf{P}^-$ are opposite parabolic k-subgroups of $\mathbf{G}$ not containing any of the $\mathbf{G}_i$, $1 \leq i \leq r$, then $\mathbf{G}(k)^+$ is generated by $R_u(\mathbf{P})(k)$ and $R_u(\mathbf{P}^-)(k)$. In particular, if $\mathbf{P}$ and $\mathbf{P}^-$ are opposite minimal parabolic k-subgroups of $\mathbf{G}$, then $\mathbf{G}(k)^+$ is generated by $R_u(\mathbf{P})(k)$ and $R_u(\mathbf{P}^-)(k)$;*
(iii) (follows from (ii)) *Assume that $\mathbf{G}$ is almost k-simple and non-commutative. Then, if $\mathbf{P}$ and $\mathbf{P}^-$ are opposite parabolic k-subgroups of $\mathbf{G}$ and $\mathbf{P} \neq \mathbf{G}$, then $\mathbf{G}(k)^+$ is generated by $R_u(\mathbf{P})(k)$ and $R_u(\mathbf{P}^-)(k)$. In particular, if $\mathbf{S}$ is a maximal k-split torus of $\mathbf{G}$, $\vartheta \subsetneq \Delta = \Delta(\mathbf{S}, \mathbf{G})$ and $\mathbf{V}_\vartheta$, $\mathbf{V}_\vartheta^-$ denote the same as in 1.2, then $\mathbf{G}(k)^+$ is generated by $\mathbf{V}_\vartheta(k)$ and $\mathbf{V}_\vartheta^-(k)$;*
(iv) *the group $\mathbf{G}(k)^+$ is the almost direct product of the subgroups $\mathbf{G}_i(k)^+$, $1 \leq i \leq r$;*

(v) *if k is infinite, then the Zariski closure of* $\mathbf{G}(k)^+$ *in G coincides with the product of the* $\mathbf{G}_i$, $1 \leq i \leq r$. *In particular, if* $\mathbf{G}$ *is almost k-simple, k-isotropic and non-commutative, then* $\mathbf{G}(k)^+$ *is Zariski dense in* $\mathbf{G}$;
(vi) *if* $\mathbf{S}$ *is a k-split torus in* $\mathbf{G}$, *then* $\mathbf{G}(k) = \mathbf{G}(k)^+ \cdot \mathscr{Z}_{\mathbf{G}}(\mathbf{S})(k)$.

(1.5.5) Proposition. (See [Bo-T 4], Corollary 6.3.) *Let* $f\colon \mathbf{F} \to \mathbf{G}'$ *be a central k-isogeny of reductive k-groups. Then* $f(\mathbf{G}(k)^+) = \mathbf{G}'(k)^+$.

(1.5.6) Theorem. (See [Ti 1], the main theorem and Corollary 6.4.) *Assume that the field k contains at least four elements. Then*

(i) *if the group* $\mathbf{G}$ *is almost k-simple, any subgroup of* $\mathbf{G}(k)$ *normalized by* $\mathbf{G}(k)^+$ *is either contained in* $\mathscr{Z}(\mathbf{G})$ *or contains* $\mathbf{G}(k)^+$;
(ii) *the group* $\mathbf{G}(k)^+$ *coincides with its commutator subgroup.*

(1.5.7) Corollary. (see [Bo-T 4], Corollary 6.7.) *Assume k is infinite. Then* $\mathbf{G}(k)^+$ *does not contain any proper subgroup of finite index and hence every subgroup of finite index in* $\mathbf{G}(k)$ *contains* $\mathbf{G}(k)^+$.

1.6 Split Semisimple Subgroups

We assume in this paragraph that the group $\mathbf{G}$ is semisimple, and we denote by $\mathbf{S}$ a maximal k-split torus in $\mathbf{G}$. Let us denote by Φ' the system of non-multipliable roots in $\Phi = \Phi(\mathbf{S},\mathbf{G})$. Making use of the notation in 1.1 we set for each $a \in \Phi$, $\mathbf{U}_{(a)} = \mathbf{U}_{\{a\}}$, where $\{a\} = \{na \mid n \in \mathbb{N}^+\} \cap \Phi$. Notice that if a root a is non-multipliable, then $\mathrm{Lie}(\mathbf{U}_{(a)})$ coincides with the root subspace $\{v \in \mathrm{Lie}(\mathbf{G}) \mid \mathrm{Ad}\, s(v) = a(s)v \text{ for each } s \in \mathbf{S}\}$.

(1.6.1) Theorem. (See [Bo-T 1], Theorem 7.2.) *The group* $\mathbf{G}$ *contains a connected semisimple k-split k-subgroup* $\mathbf{H}$ *such that* $\mathbf{H} \supset \mathbf{S}$, $\Phi' = \Phi(\mathbf{S},\mathbf{H})$, *and for each* $a \in \Phi' = \Phi(\mathbf{S},\mathbf{H})$ *the subgroup* $\mathbf{U}_{(a)} \cap \mathbf{H}$ *coincides with the (1-dimensional k-split) root subgroup* $\mathbf{U}_a$ *of* $\mathbf{H}$. *If* $\mathbf{G}$ *is almost k-simple, then* $\mathbf{H}$ *is almost k-simple as well.*

It should be noted that the subgroup $\mathbf{H}$ above is not uniquely determined, even up to conjugacy (see [Bo-T 1], 7.3).

Let $\mathbf{Sp}_{2n} = \{g \in \mathbf{GL}_{2n} \mid {}^t g J g = J\}$, where $J = \begin{pmatrix} 0 & E \\ -E & 0 \end{pmatrix}$ and, for a matrix g, ${}^t g$ denotes its transpose.

(1.6.2) Proposition. *Let* $\mathbf{G}$ *be almost k-simple and* $\mathrm{rank}_k\, \mathbf{G} \geq 2$. *Then* $\mathbf{G}$ *contains an almost k-simple k-subgroup* $\mathbf{F}$ *whose simply connected covering is isomorphic over k either to* $\mathbf{SL}_3$ *or* $\mathbf{Sp}_4$.

To prove this proposition it suffices to apply Theorem 1.6.1 and to observe that for a k-split group $\mathbf{G}$ the subgroup $\mathbf{F}$ can be constructed as follows: since $\mathbf{G}$ is almost k-simple and $\mathrm{rank}_k\, \mathbf{G} \geq 2$, we can find two non-proportional roots

$a, b \in \Phi = \Phi(\mathbf{S}, \mathbf{G})$ which are not orthogonal under a $W(\Phi)$-invariant scalar product, where $W(\Phi)$ is the Weyl group of the system Φ. Furthermore, if Φ is of type $\mathbb{G}_2$, then we may assume that a and b have equal length (i.e. $a \in W(\Phi)b$). It then follows from the classification of semisimple split groups (see [Hum 1], Chapters XI, XII and [Ti 2]) that the simply connected covering of the subgroup $\mathbf{F}$ generated by the root subgroups $\mathbf{U}_a$, $\mathbf{U}_{-a}$, $\mathbf{U}_b$ and $\mathbf{U}_{-b}$ is either $\mathbf{SL}_3$ or $\mathbf{Sp}_4$.

(1.6.3) Proposition. (See [Ti 1], 3.1, Proposition 13.) *Let*

$$U = \left\{ \begin{pmatrix} 1 & x \\ 0 & 1 \end{pmatrix} \mid x \in k \right\}, \; D = \left\{ \begin{pmatrix} c & 0 \\ 0 & c^{-1} \end{pmatrix} \mid c \in k, \; c \neq 0 \right\},$$

and let $a \in \Phi(\mathbf{S}, \mathbf{G})$ be a non-multipliable root. Then there exists a k-morphism $\sigma\colon \mathbf{SL}_2 \to \mathbf{G}$ with finite kernel such that $\sigma(U) \subset \mathbf{U}_{(a)}(k)$ and $\sigma(D) \subset \mathbf{S}(k)$.

1.7 The Restriction of Scalars

(See [Bo-T 1], 6.17–6.21 and [Weil 2], 1.3.) Let K be a finite separable field extension of k of degree d. If $\varphi\colon K \to K'$ is a field homomorphism and $p\colon \mathbf{M} \to \mathbf{M}'$ is a K-morphism of K-varieties, then we denote, respectively, by ${}^{\varphi}\mathbf{M}$, ${}^{\varphi}\mathbf{M}'$ and ${}^{\varphi}p\colon {}^{\varphi}\mathbf{M} \to {}^{\varphi}\mathbf{M}'$ the K'-varieties and the K'-morphism obtained from $\mathbf{M}$, $\mathbf{M}'$ and p by φ. If $\mathbf{M}$ and $\mathbf{M}'$ are subvarieties of affine spaces defined over K, then ${}^{\varphi}\mathbf{M}$, ${}^{\varphi}\mathbf{M}'$ and ${}^{\varphi}p$ can be obtained by applying φ to the coefficients of the polynomials defining $\mathbf{M}$, $\mathbf{M}'$, and p respectively. Let $\{\sigma_1 = \mathrm{Id}, \sigma_2, \ldots, \sigma_d\}$ be the set of all distinct k-embeddings of the field K into the algebraic closure of k. Then for each K-variety $\mathbf{W}$ there exist a k-variety $\mathbf{V}$ and a K-morphism $p\colon \mathbf{V} \to \mathbf{W}$ such that the map

$$({}^{\sigma_1}p, \ldots, {}^{\sigma_d}p)\colon \mathbf{V} \to {}^{\sigma_1}\mathbf{W} \times \ldots \times {}^{\sigma_d}\mathbf{W}$$

is an isomorphism of algebraic varieties; furthermore, the pair $(\mathbf{V}, p)$ is defined uniquely up to a k-isomorphism. The variety $\mathbf{V}$ is denoted by $\mathbf{R}_{K/k}\mathbf{W}$. If $\mathbf{W}$ is a K-group, then $R_{K/k}\mathbf{W}$ is a k-group and p is a K-group morphism. The correspondence $\mathbf{W} \mapsto R_{K/k}\mathbf{W}$ is a functor from the category of K-varieties (resp. affine K-varieties, K-groups) into the category of k-varieties (resp. affine k-varieties, k-groups) which is called the *restriction of scalars functor* (from K to k). The map $p\colon R_{K/k}\mathbf{W} \to \mathbf{W}$ has the following *universality property*: for a k-variety $\mathbf{X}$ and a K-morphism $f\colon \mathbf{X} \to \mathbf{W}$ there exists a unique k-morphism $\varphi\colon \mathbf{X} \to R_{K/k}\mathbf{W}$ such that $f = p \circ \varphi$. Notice that if f is a K-group morphism, then φ is a k-group morphism. The restriction of the projection p to $\mathbf{V}(k)$ induces a bijection of the set $\mathbf{V}(k)$ onto $\mathbf{W}(K)$. The inverse map to $p|_{\mathbf{V}(k)}$ will be denoted by $R^0_{K/k}$. If k is a local field and K is endowed with the topology of a Hausdorff topological vector space over k and if $\mathbf{V}(k)$ and $\mathbf{W}(K)$ are endowed with topologies induced, respectively, from that of the fields k and K, then $p|_{\mathbf{V}(k)}$ and $R^0_{K/k}$ are homeomorphisms. If $\mathbf{W}$ is a K-group, then $R^0_{K/k}\colon \mathbf{W}(K) \to \mathbf{V}(k)$ is a group isomorphism.

If $\mathbf{H}$ is a K-group, then the functor $R_{K/k}$ defines a bijection of the set of parabolic K-subgroups of $\mathbf{H}$ onto the set of parabolic k-subgroups of the k-group $R_{K/k}\mathbf{H}$ and sends $R_u(\mathbf{H})$ to $R_u(R_{K/k}\mathbf{H})$. It follows that $(R_{K/k}\mathbf{H})(k)^+ = R^0_{K/k}(\mathbf{H}(K)^+)$

for any connected K-group $\mathbf{H}$. If $\mathbf{T}$ is a K-torus, then $R_{K/k}\mathbf{T}$ is a k-torus; if $\mathbf{T}$ is a maximal K-split torus in $\mathbf{H}$, then the maximal k-split subtorus of the k-torus $R_{K/k}\mathbf{T}$ is a maximal k-split torus of the group $R_{K/k}\mathbf{H}$. If $\mathbf{H}$ is reductive, then $\operatorname{rank}_K \mathbf{H} = \operatorname{rank}_k R_{K/k}\mathbf{H}$. If the K-group $\mathbf{H}$ is almost simple over K (resp. K-simple, semisimple, reductive, unipotent), then the group $R_{K/k}\mathbf{H}$ is almost simple over k (resp. k-simple, semisimple, reductive, unipotent).

If $\mathbf{G}$ is a simply connected (adjoint) almost k-simple k-group, then there exist a finite separable field extension k' of k and a connected simply connected (adjoint) absolutely almost simple k'-group $\mathbf{G}'$ such that $\mathbf{G} = R_{k'/k}\mathbf{G}'$ (see [Ti 2], 3.1.2).

1.8 "Abstract" Homomorphisms of Isotropic Algebraic Groups

Let k' be a field. If $\varphi\colon k \to k'$ is a homomorphism, we define $\varphi_{\mathbf{G}}$ as in 1.7 above and we denote by φ^0 the canonical homomorphism of the group $\mathbf{G}(k)$ into ${}^{\varphi}\mathbf{G}(k')$ (which is obtained by applying φ to the entries of the matrices in $\mathbf{G}(k)$).

(1.8.1) Theorem. (See [Bo-T4], Theorem (A).) *Let H be a subgroup of $\mathbf{G}(k)$ containing $\mathbf{G}(k)^+$, let $\mathbf{G}'$ be a connected non-commutative absolutely almost simple k-group, and let $\delta\colon H \to \mathbf{G}'(k')$ be a homomorphism. Assume that the fields k and k' are both infinite and that*

(a) *$\mathbf{G}$ is k-isotropic, almost k-simple and non-commutative;*
(b) *either $\mathbf{G}$ is simply connected or $\mathbf{G}'$ is adjoint;*
(c) *$\mathbf{G}' \neq \{e\}$ and the subgroup $\delta(\mathbf{G}(k)^+)$ is Zariski-dense in $\mathbf{G}'$.*

Then there exist a uniquely determined homomorphism $\varphi\colon k \to k'$, a special (in the sense of 1.4.14) k-epimorphism $\beta\colon {}^{\varphi}\mathbf{G} \to \mathbf{G}'$, and a homomorphism $\tau\colon H \to \mathscr{Z}(\mathbf{G}'(k')) = \mathscr{Z}(\mathbf{G}')(k')$ such that $\delta(g) = \tau(g)\cdot\beta(\varphi^0(g))$ for each $g \in H$. If $\mathbf{G}$ is absolutely simple, then β is an isogeny.

(1.8.2) *Remarks on Theorem 1.8.1.*

(I) It is superfluous to assume that the fields k and k' are infinite, for this is a consequence of (c) and the connectedness of the group $\mathbf{G}'$.

(II) In [Bo-T 4] $\mathbf{G}$ is assumed to be absolutely almost simple. But this assumption is used only for proving that the k-epimorphism β is an isogeny. It should be noted that by using the restriction of scalars functor, one can easily reduce the general case to the special case where $\mathbf{G}$ is absolutely almost simple.

(III) If k and k' are local fields and the homomorphism δ is continuous, then φ is continuous. The assertion follows from the construction of the homomorphism φ given in [Bo-T 4], §8.3, but it can also be deduced from the equality $\delta(g) = \tau(g)\cdot\beta(\varphi^0(g))$, $g \in H$, if we note that:

(a) since $\mathbf{G}$ is isotropic over k, by Proposition 1.5.3 and Proposition 1.5.4(i) $\mathbf{G}$ contains a 1-dimensional unipotent k-group $\mathbf{U}$ such that $\mathbf{U}(k) \subset \mathbf{G}(k)^+$ and $\mathbf{U}$ is isomorphic over k to the additive group $\mathbf{G}_a$;

(b) Since $\mathscr{Z}(\mathbf{G}')$ is finite and by Corollary 1.5.7 $\mathbf{G}(k)^+$ does not contain proper subgroups of finite index, $\tau(\mathbf{G}(k)^+) = \{e\}$.

Finally, we observe that if k' is not isomorphic to $\mathbb{C}$, then any homomorphism $\vartheta\colon k \to k'$ is continuous (see [Bo-T 4], 2.3).

(IV) Assume k and k' are local fields, $\mathbf{G}$ and $\mathbf{G}'$ are adjoint absolutely simple groups, $\operatorname{Ker}\delta = \{e\}$ and $\delta(H) \supset \mathbf{G}'(k')^+$. Then $\varphi\colon k \to k'$ is a field isomorphism and $\beta\colon {}^{\varphi}\mathbf{G} \to \mathbf{G}'$ is an algebraic group isomorphism. This is a special case of Corollary 8.13 in [Bo-T 4].

(1.8.3) Proposition. *Let $\mathbf{G}'$ be an algebraic k'-group, H a subgroup of $\mathbf{G}(k)$ containing $\mathbf{G}(k)^+$, $\alpha\colon H \to \mathbf{G}'(k')$ a homomorphism and let $\mathbf{P}$ be a parabolic k-subgroup of the group $\mathbf{G}$. Assume that the field k is infinite. Then*

(i) *the group $\alpha(R_u(\mathbf{P})(k))$ is unipotent;*
(ii) *if $\alpha(\mathbf{G}(k)^+) \neq \{e\}$, then $\operatorname{char} k = \operatorname{char} k'$;*
(iii) *if k is a perfect field, then $\alpha(H^{(u)}) \subset \mathbf{G}'(k')^{(u)}$.*

Since $\mathbf{G}^{(u)} \subset \mathscr{D}(\mathbf{G})$ and the group $\mathscr{D}(\mathbf{G})$ is semisimple (see 0.24), we may assume that $\mathbf{G}$ is semisimple. But in this case one can find assertions (i) and (ii) in [Bo-T 4] (Proposition 7.2). Assertion (iii) follows from (i) and Proposition 1.5.1.

2. Algebraic Groups Over Local Fields

In this section we denote by k a local field with absolute value $|\ |$, by K an algebraic closure of the field k and by $\mathbf{G}$ a connected semisimple k-group. Every k-variety $\mathbf{M}$ will be equipped with the topology induced from that of the field k (see 0.31).

2.1 The Behaviour of the Set of k-rational Points Under k-morphisms

(2.1.1) Proposition. *Let $\alpha\colon \mathbf{V} \to \mathbf{W}$ be a bijective k-morphism of irreducible normal k-varieties. Then*

(i) *the subset $\alpha(\mathbf{V}(k))$ is closed in $\mathbf{W}(k)$ and the map $\alpha\colon \mathbf{V}(k) \to \alpha(\mathbf{V}(k))$ is a homeomorphism;*
(ii) *there exists a finite purely inseparable field extension k' of k such that $\alpha^{-1}(\mathbf{W}(k)) \subset \mathbf{V}(k')$.*

Proof. The morphism α is induced from the comorphism $K(\mathbf{W}) \to K(\mathbf{V})$, which will be considered as an embedding, where $K(\mathbf{M})$ denotes as usual the field of rational functions on an irreducible K-variety $\mathbf{M}$. Since α is a bijective morphism of irreducible normal varieties, $K(\mathbf{V})$ (see [Bo 6], Chapter AG, 18.2) is a purely inseparable field extension of $K(\mathbf{W})$ and for every Zariski open affine subset U of the variety $\mathbf{W}$, the set $U' = \alpha^{-1}(U)$ coincides with the space $\operatorname{spec}_K(K[U]')$ of maximal ideals of the integral closure $K[U]'$ of the ring $K[U]$ in $K(\mathbf{V})$ and the morphism α is induced from the embedding $K[U] \hookrightarrow K[U]' = K[U']$. Thus we may assume that $\mathbf{V}$ and $\mathbf{W}$ are affine and $k[\mathbf{W}] \supset k[\mathbf{V}]^{p^n} \overset{\text{def}}{=} \{f^{p^n} \mid f \in k[\mathbf{V}]\}$ for some positive integer n, where $p = \operatorname{char} k$ (in particular $k[\mathbf{W}] = k[\mathbf{V}]$, if $\operatorname{char} k = 0$, which proves the proposition in this case). Let $f_1, \dots, f_i$ be the generators of the k-algebra $k[\mathbf{V}]$. We set $h_j = f_j^{p^n} \in k[\mathbf{W}]$, $1 \leq j \leq i$. Then $f_j(x)^{p^n} = h_j(\alpha(x))$ for each $x \in \mathbf{V}$ and hence a point $y \in \mathbf{W}(k)$ belongs to $\alpha(\mathbf{V}(k))$ if and only if $h_j(y) \in k^{p^n}$ for each j, $l \leq j \leq i$. On the other hand, it follows from the description of local fields given in 0.31 that the set k^{p^n} is closed in k and the map $x \mapsto x^{p^n}$, $x \in k$, is a homomorphism of the field k onto k^{p^n}. Thus

(a) the set $\alpha(\mathbf{V}(k))$ is closed in $\mathbf{W}(k)$ and $f_j(\alpha^{-1}(y))$ depends continuously on $y \in \alpha(\mathbf{V}(k))$, hence (i);

(b) $f_j(\alpha^{-1}(y)) \in k^{p^{-n}}$ for each $y \in \mathbf{W}(k)$, and hence $\alpha^{-1}(\mathbf{W}(k)) \subset \mathbf{V}(k^{p^{-n}})$, this proves (ii). □

(2.1.2) *Remarks.*

(i) If $\operatorname{char} k = 0$, then, as was observed in 0.9, every bijective k-morphism of irreducible normal k-varieties is an isomorphism. Hence, in case $\operatorname{char} k = 0$, we have $\alpha(\mathbf{V}(k)) = \mathbf{W}(k)$.

(ii) In Proposition 2.1.1 the assumption of irreducibility of the varieties $\mathbf{V}$ and $\mathbf{W}$ can be replaced by the assumption that all irreducible components of the variety $\mathbf{V}$ have the same dimension.

(2.1.3) Corollary. *Let $f\colon \mathbf{H} \to \mathbf{H}'$ be a k-group morphism with trivial kernel. Then*

(i) the subgroup $f(\mathbf{H}(k))$ is closed in $\mathbf{H}'(k)$ and the homomorphism $f\colon \mathbf{H}(k) \to f(\mathbf{H}(k))$ is a topological group isomorphism.

(ii) there exists a finite purely inseparable field extension k' of k such that $f^{-1}(\mathbf{H}'(k)) \subset \mathbf{H}(k')$.

This corollary is a consequence of Proposition 2.1.1 and Remark 2.1.2(ii), because every algebraic group, viewed as an algebraic variety, is normal (see 0.13), irreducible components of the group $\mathbf{H}$ are cosets of the subgroup $\mathbf{H}^0$, and the image of a k-group under a k-morphism is a k-group. Furthermore, we need not use Remark (ii) of 2.1.2. Indeed, since $\mathbf{H}^0$ is of finite index in $\mathbf{H}$, we may assume that $\mathbf{H}$ is connected. But one can then apply Proposition 2.1.1.

(2.1.4) Proposition. (See [Bo-T 4], 3.18.) *Suppose that a k-group $\mathbf{H}$ acts k-rationaly on a k-variety $\mathbf{M}$ and x is an element of $\mathbf{M}(k)$ such that the map $h \mapsto hx$, $h \in \mathbf{H}$, of the group $\mathbf{H}$ onto the orbit $\mathbf{H}x$ is separable. Then*

(i) The subset $\mathbf{H}(k)x$ is closed and open in $(\mathbf{H}x)(k)$ and hence is locally closed in $\mathbf{M}(k)$;

(ii) the natural map $\mathbf{H}(k)/\mathbf{H}(k)_x \to \mathbf{H}(k)x$ is a homeomorphism, where $\mathbf{H}(k)_x = \{h \in \mathbf{H}(k) \mid hx = x\}$.

(2.1.5) Corollary. *If $\mathbf{P}$ is a parabolic k-subgroup of a k-group $\mathbf{H}$, then the quotient space $\mathbf{H}(k)/\mathbf{P}(k)$ is compact.*

Proof. Let $\pi\colon \mathbf{H} \to \mathbf{H}/\mathbf{P}$ be the natural k-morphism. The morphism is separable (see 0.17). Thus, in view of Proposition 2.1.4, the space $\mathbf{H}(k)/\mathbf{P}(k)$ is homeomorphic to the closed subset $\mathbf{H}(k)\pi(e)$ of the space $(\mathbf{H}/\mathbf{P})(k)$. On the other hand, since $\mathbf{H}/\mathbf{P}$ is a projective variety, $(\mathbf{H}/\mathbf{P})(k)$ is compact. □

2.2 Cartan and Iwasawa Decompositions

If k is (isomorphic to) $\mathbb{R}$ or $\mathbb{C}$, we set

$$\hat{k} = \{x \in \mathbb{R} \mid x \geq 1\} \text{ and } k_0 = \{x \in \mathbb{R} \mid x > 0\}.$$

Otherwise we set

$$\hat{k} = \{\beta^n \mid n \in \mathbb{N}\} \text{ and } k_0 = \{\beta^n \mid n \in \mathbb{Z}\},$$

where β is a fixed uniformizer of the non-archimedean local field k.

Let $\mathbf{S}$ be a maximal k-split torus in $\mathbf{G}$. Define on $\mathrm{X}(\mathbf{S}) \otimes \mathbb{R}$ the ordering compatible with a Weyl chamber of the system $\Phi(\mathbf{S},\mathbf{G})$ (see 0.26, 0.27) and denote by $\mathrm{X}^+ \subset \mathrm{X}(\mathbf{S})$ the set of positive characters with respect to the order, and by $\mathbf{V}$ the unipotent k-subgroup $\mathbf{V}_\varnothing$ defined in 1.2. We set $S^+ = \{s \in \mathbf{S}(k) \mid \chi(s) \in \hat{k}$ for each $\chi \in \mathrm{X}^+\}$ and $S' = \{s \in \mathbf{S}(k) \mid \chi(s) \in k_0$ for each $\chi \in \mathrm{X}(\mathbf{S})\}$.

(2.2.1) Theorem. *Assume* $\mathbf{G}$ *is simply connected. Then there exists a compact subgroup* M *of* $\mathbf{G}(k)$ *such that*

(1) $$\mathscr{N}_{\mathbf{G}}(\mathbf{S})(k) \subset M \cdot \mathbf{S}(k),$$

and the following decompositions hold (called, respectively, a Cartan and an Iwasawa decomposition)

(2) $$\mathbf{G}(k) = M \cdot S^+ \cdot M \text{ and } \mathbf{G}(k) = M \cdot S' \cdot \mathbf{V}(k),$$

furthermore,

(i) if $Ms_1M = Ms_2M$, $s_1, s_2 \in S^+$, *then* $s_1 = s_2$, *and*
(ii) if $Ms_1\mathbf{V}(k) = Ms_2\mathbf{V}(k)$, $s_1, s_2 \in S'$, *then* $s_1 = s_2$.

If k is isomorphic to $\mathbb{R}$, then this theorem is a classical fact of semisimple Lie group theory (see Theorem 1.1 and 5.1 of Chapter VI in [He]). In view of the connectedness of the group $\mathbf{G}(k)$ (see Remark 2 in 2.3), the remarks in 14.7 of [Bo-T 1], and the results of §6 of Chapter V in [He], the above formulations are equivalent to that given in [He]. As to inclusion (1) for $k = \mathbb{R}$, see 14.7 of [Bo-T 1]. The case $k = \mathbb{C}$ reduces to the case $k = \mathbb{R}$ after the replacement of the group $\mathbf{G}$ by $R_{\mathbb{C}/\mathbb{R}}\mathbf{G}$, where $R_{\mathbb{C}/\mathbb{R}}$ is the restriction of scalars from $\mathbb{C}$ to $\mathbb{R}$ (see 1.7). For the case where k is non-archimedean and $\mathbf{G}$ is almost k-simple, see, for instance, 2.5 and Theorem 2.6.11 in [Macd] (inclusion (1) in this case follows from the construction given in [Macd] of the subgroup M and from (IV) of 2.5 in the same book). All that remains now is to observe that any semisimple simply connected k-group is a direct product of almost k-simple simply connected k-groups (see Proposition 1.4.10) and a maximal k-split torus of a direct product of k-groups is a direct product of maximal k-split tori of the factors (see [Bo-T 1], Proposition 4.27).

(2.2.2) Corollary. *Assume* $\operatorname{rank}_k \mathbf{G} = 1$ *and* $\mathbf{G}$ *is simply connected. Then there exists a compact subgroup* M *of* $\mathbf{G}(k)$ *such that* $g^{-1} \in MgM$ *for each* $g \in \mathbf{G}(k)$.

Proof. Let M be a compact subgroup of $\mathbf{G}(k)$ as in Theorem 2.2.1 and let $Y = \{g \in \mathbf{G}(k) \mid g^{-1} \in MgM\}$. Since $\operatorname{rank}_k \mathbf{G} = 1$ and $\mathbf{G}$ is semisimple, there exists $x \in \mathscr{N}_{\mathbf{G}}(\mathbf{S})(k)$ such that $xsx^{-1} = s^{-1}$ for each $s \in \mathbf{S}$ (see 0.27). It follows, in view of (1) and the commutativity of the group $\mathbf{S}$, that $\mathbf{S}(k) \subset Y$. But $\mathbf{G}(k) = M \cdot \mathbf{S}(k) \cdot M$ (in view of (2)), and, as one easily sees, $M \cdot Y \cdot M = Y$. Hence $Y = \mathbf{G}(k)$. □

2.3 Certain Properties of the Group $\mathbf{G}(k)$ and Its Subgroup $\mathbf{G}(k)^+$

(2.3.1) Theorem.

(a) *If* $\mathbf{G}$ *is simply connected, k-isotropic and almost k-simple, then* $\mathbf{G}(k)^+ = \mathbf{G}(k)$ (see [Pl 2] and [Pr-R 4]).

(a') *If* $\mathbf{G}$ *is simply connected and does not have k-anisotropic factors, then* $\mathbf{G}(k)^+ = \mathbf{G}(k)$ (this is a consequence of (a), Proposition 1.4.10, and Proposition 1.5.4(iv)).

(b) $\mathbf{G}(k)^+$ *is a closed normal subgroup of* $\mathbf{G}(k)$ *and the factor group* $\mathbf{G}(k)/\mathbf{G}(k)^+$ *is compact* (see [Bo-T 4], 6.14).

(c) *Assume* $\mathbf{G}$ *has no non-trivial k-anisotropic factors and denote by q the inseparability degree of the central isogeny* $f\colon \tilde{\mathbf{G}} \to \mathbf{G}$, *where* $\tilde{\mathbf{G}}$ *is the simply connected covering of the group* $\mathbf{G}$ *(i.e. q is the inseparability degree of the field* $K(\tilde{\mathbf{G}})$ *over the field* $f_0(K(\mathbf{G}))$*, where f_0 is the comorphism of the morphism f; if* $\operatorname{char} k = 0$, *then* $q = 1$*). Then the factor group* $\mathbf{G}(k)/\mathbf{G}(k)^+$ *is commutative and contains an open torsion subgroup (having finite index because the group* $\mathbf{G}(k)/\mathbf{G}(k)^+$ *is compact) whose exponent divides q. In particular,*

  1) *there exists* $r \in \mathbb{N}^+$ *such that* $g^r \in \mathbf{G}(k)^+$ *for all* $g \in \mathbf{G}(k)$;
  2) *if* $\operatorname{char} k = 0$, *then* $\mathbf{G}(k)^+$ *is an open normal subgroup of finite index in* $\mathbf{G}(k)$. *Furthermore, if* $k = \mathbb{R}$, *then* $\mathbf{G}(k)^+$ *coincides with the identity component of the Lie group* $\mathbf{G}(k)$ (see [Bo-T 4], 6.14, 6.15).

(1) *Remark.* In [Pl 2] the proof of (a) was given only for the fields of characteristic 0 but, as was remarked there, this proof can be carried over to the case of positive characteristic. The proof of (a) given in [Pr-R 4] is based on a reduction to the groups of k-rank 1.

(2) *Remark.* If $k = \mathbb{R}$ and $\mathbf{G}$ is isotropic and almost k-simple, then (see (c)) $\mathbf{G}(k)^+$ coincides with the identity component of the Lie group $\mathbf{G}(k)$. Thus, if $k = \mathbb{R}$, then assertion (a) follows from the connectedness of the group of k-rational points of any connected simply connected semisimple k-group. (This fact is well-known and follows from a theorem, already known to E. Cartan, on the connectedness of the set of fixed points of an involutory automorphism of a connected simply connected semisimple Lie group.)

(3) *Remark.* Since the field $\mathbb{C}$ is algebraically closed, we have $\mathbf{G}(k)^+ = \mathbf{G}(k)$, if $k = \mathbb{C}$ (see 1.5.2).

(2.3.2) Corollary.

(a) *If* $\mathbf{G}$ *is simply connected, k-isotropic, and almost k-simple, then every normal subgroup of* $\mathbf{G}(k)$ *is either contained in* $\mathscr{Z}(\mathbf{G})$ *or coincides with* $\mathbf{G}(k)$ (this is a consequence of Theorem 2.3.1(a) and Theorem 1.5.6(i)).

(b) *If* $\mathbf{G}$ *is simply connected and has no k-anisotropic almost k-simple factors, then* $\mathscr{D}(\mathbf{G}(k)) = \mathbf{G}(k)$ *and every subgroup of finite index in* $\mathbf{G}(k)$ *coincides with* $\mathbf{G}(k)$ (this is a consequence of Theorem 2.3.1(a), Theorem 1.5.6(ii), and Corollary 1.5.7).

(2.3.3) Corollary. *The locally compact group* $\mathbf{G}(k)$ *is unimodular.*

Proof. Since $\mathscr{D}(\mathbf{G}(k)^+) = \mathbf{G}(k)^+$ (see Theorem 1.5.6(ii)), we have that the group $\mathbf{G}(k)^+$ is unimodular. But the factor group $\mathbf{G}(k)/\mathbf{G}(k)^+$ is compact (see Theorem 2.3.1(b)) and hence is unimodular. Therefore $\mathbf{G}(k)$ is unimodular. □

(2.3.4) Proposition. (See [Bo-T 4], Proposition 3.19 and Corollary 3.20.) *Let* $\beta\colon \mathbf{H} \to \mathbf{H}'$ *be a central k-isogeny of connected reductive k-groups. Then*

(i) *the subgroup* $\beta(\mathbf{H}(k))$ *is closed and normal in* $\mathbf{H}'(k)$ *and the factor group* $\mathbf{H}'(k)/\beta(\mathbf{H}(k))$ *is compact, commutative and torsion. If the isogeny* β *is separable, in particular, if* $\operatorname{char} k = 0$, *then the subgroup* $\beta(\mathbf{H}(k))$ *is open and of finite index in* $\mathbf{H}'(k)$;

(ii) *the isomorphism* $\mathbf{H}(k)/\operatorname{Ker}\beta(k) \to \beta(\mathbf{H}(k))$ *is topological and hence the restriction of* β *to* $\mathbf{H}(k)$ *is proper.*

(2.3.5) Corollary. *The groups* $\mathbf{G}(k)$ *and* $\mathbf{G}(k)^+$ *are compactly generated.*

Proof. Let $\tilde{\mathbf{G}}$ be the simply connected covering of the group $\mathbf{G}$, let $\tilde{p}\colon \tilde{\mathbf{G}} \to \mathbf{G}$ be the central k-isogeny, and let $\mathbf{S}$ be a maximal k-split torus in $\tilde{\mathbf{G}}$. Since $\mathbf{S}$ is split over k, the group $\mathbf{S}(k)$ can be represented as a direct product of a number of subgroups isomorphic to the multiplicative group k^* of the field k. But from the properties of local fields given in 0.31 it follows that the group k^* is compactly generated. Therefore $\mathbf{S}(k)$ is compactly generated. It then follows from the Cartan decomposition (see Theorem 2.3.1) that the group $\tilde{\mathbf{G}}(k)$, and hence the group $\tilde{p}(\tilde{\mathbf{G}}(k))$, is compactly generated. But by Proposition 2.3.4 the factor group $\mathbf{G}(k)/\tilde{p}(\tilde{\mathbf{G}}(k))$ is compact. Therefore $\mathbf{G}(k)$ is compactly generated. Let $\tilde{\mathbf{G}}_i$, $1 \le i \le r$, be the connected almost k-simple k-isotropic normal k-subgroups of $\tilde{\mathbf{G}}$. Since $\tilde{\mathbf{G}}(k)^+$ is the product of the groups $\tilde{\mathbf{G}}_i(k)^+$, $1 \le i \le r$, (see Proposition 1.5.4(iv)), $\tilde{\mathbf{G}}_i(k)^+ = \tilde{\mathbf{G}}_i(k)$ for each i, $1 \le i \le r$ (see Theorem 2.3.1(a)), and the groups $\tilde{\mathbf{G}}_i(k)$ have already been proved to be compactly generated, we see that $\tilde{\mathbf{G}}(k)^+$ is compactly generated. But, by Proposition 1.5.5, $\mathbf{G}(k)^+ = \tilde{p}(\tilde{\mathbf{G}}(k)^+)$, and hence $\mathbf{G}(k)^+$ is compactly generated. □

Remark. In case char $k = 0$ the compact generation of the group $\mathbf{H}(k)$ has been proved by Borel and Tits not only for semisimple but also for reductive groups $\mathbf{H}$, and what is more, without using the Cartan decomposition (see [Bo-T 1], Proposition 13.4). The proof given in [Bo-T 1] can be carried over to the case char $k \neq 0$. But we shall not do this, for the compact generation of the group $\mathbf{H}(k)$ will be used below only in the special case where $\mathbf{H}$ is semisimple.

(2.3.6) Proposition. *Let $\mathbf{H}$ be a reductive k-group and let $\mathbf{S}$ be maximal k-split torus in $\mathbf{H}$. Then the topological group $\mathscr{Z}_{\mathbf{H}}(\mathbf{S})(k)/\mathbf{S}(k)$ is compact. In particular, the group $\mathbf{H}(k)$ is compact if and only if* $\operatorname{rank}_k \mathbf{H} = 0$.

This assertion was obtained in [Bo-T 4] in the course of the proof of Proposition 3.19. We remark that a simple proof of the assertion "the group $\mathbf{H}(k)$ is compact if and only if $\operatorname{rank}_k \mathbf{H} = 0$" is contained in [Pr 7].

(2.3.7) Corollary. *The group $\mathbf{G}(k)^+$ is compact if and only if* $\operatorname{rank}_k \mathbf{G} = 0$.

This follows immediately from Proposition 2.3.6 and Theorem 2.3.1(b).

(2.3.8) Proposition. *Let $\mathbf{G}' \subset \mathbf{G}$ be the subgroup of $\mathbf{G}$ which is the almost direct product of (almost k-simple) k-isotropic factors of $\mathbf{G}$. Then the factor group $\mathbf{G}(k)/\mathbf{G}'(k)$ is compact* (this is a straightforward consequence of Theorem 2.3.1(b) and Proposition 1.5.4(iv)).

2.4 Split Tori Over Local Fields

(2.4.1) Proposition. *Let $\mathbf{S}$ be a torus of positive dimension defined and split over k and let A be a finite set of non-trivial characters of $\mathbf{S}$. Then there exists $s \in \mathbf{S}(k)$ such that $|\chi(s)| \neq 1$ for all $\chi \in A$.*

Proof. Let $X_*(\mathbf{S}) = \operatorname{Mor}(\mathbf{GL}_1, \mathbf{S})$ be the free $\mathbb{Z}$-module of multiplicative one-parameter subgroups of the torus $\mathbf{S}$ and let $X(\mathbf{S}) \times X_*(\mathbf{S}) \to \mathbb{Z} = X(\mathbf{GL}_1)$ be the duality over $\mathbb{Z}$ defined by the formula $\langle \chi, \lambda \rangle = m$, if $(\chi \cdot \lambda)(y) = y^m$ (see [Bo 6], Chapter III, 8.6). Since A is finite and consists of non-trivial characters, there exists $\lambda \in X_*(S)$ such that $\langle \chi, \lambda \rangle \neq 0$, whenever $\chi \in A$. Let us identify $\mathbf{GL}_1(k)$ with the multiplicative group k^* of the field k and pick a uniformizer x of the field k. The element $s = \lambda(x)$ has the desired property. □

(2.4.2) Proposition. *Let $\mathbf{S}$ be a torus defined and split over k, let $A \subset B \subset X(\mathbf{S})$, and let $\mathbf{S}_A$ be the identity component in the intersection of the kernels of the characters $a \in A$. Assume that the characters in B are linearly independent. Then there exists $s \in \mathbf{S}_A$ such that $|\chi(s)| < 1$ for all $\chi \in B - A$.*

Proof. Let $X_*(\mathbf{S})$, $\langle \chi, \lambda \rangle$, and $x \in \mathbf{GL}_1(k) = k^*$ be as in 2.4.1. We deduce from the linear independence of the characters in B that there exists $\lambda \in X_*(\mathbf{S})$ such that $\langle \chi, \lambda \rangle = 0$ for all $\chi \in A$ and $\langle \chi, \lambda \rangle < 0$ for all $\chi \in B - A$. Then $s = \lambda(x)$ has the desired property. □

2.5 Analytic Manifolds Over Local Fields

A Hausdorff space X is said to be a *k-analytic manifold* if for each $x \in X$ there exist a positive integer $n(x)$ and a homeomorphism φ_x of a neighbourhood U_x of x onto an open subset of $k^{n(x)}$ such that for all $x, x' \in X$ the map $\varphi_x \circ \varphi_{x'}^{-1}$ is analytic on $\varphi_{x'}(U_x \cap U_{x'})$ (for more details see §5 in [Bou 4]). By the *dimension* of the manifold X we mean the least upper bound of the numbers $n(x)$, $x \in X$. If $n(x) = n$ for all $x \in X$, we say that X is a *pure* k-analytic manifold of dimension n. A subset $Y \subset X$ is called an *analytic submanifold* of the analytic manifold X if for each $y \in Y$ there exists a neighbourhood $\tilde{U}_y \subset U_y$ of y and a bianalytic homeomorphism f_y of $\varphi_y(\tilde{U}_y)$ onto an open subset in $k^{n(y)}$ such that $f_y(\varphi_y(Y \cap \tilde{U}_y))$ coincides with the intersection of the set $f_y(\varphi_y(Y))$ and a linear subspace of the space $k^{n(y)}$.

If X is a σ-compact k-analytic manifold, then (see [Bou 4], 10.1.4) there exists on X a unique class M of equivalent measures such that for every measure $\nu \in M$ and for each $x \in X$ the image under φ_x of the restriction of ν to U_x is equivalent to the restriction of the measure μ_{k^n} to $\varphi_x(U_x)$. The set M is said to be the *canonical class* of measures on X.

(2.5.1) Lemma. *Let* $\mathbf{V}$ *be a smooth k-variety and let* $\mathbf{W}$ *be a k-subvariety of* $\mathbf{V}$. *Then*

(i) *The space* $\mathbf{V}(k)$ *equipped with topology induced from that of the field k is a k-analytic manifold. If all the irreducible components of the variety* $\mathbf{V}$ *have the same dimension, then* $\mathbf{V}(k)$ *is pure k-analytic manifold of dimension* $\dim \mathbf{V}$.

(ii) *If we set* $Y = \{w \in \mathbf{W}(k) \mid w$ *is a simple point of the variety* $\mathbf{W}\}$, *then* Y *is an analytic submanifold of the k-analytic manifold* $\mathbf{V}(k)$ *of dimension at most* $\dim \mathbf{W}$.

(iii) *The set* $\mathbf{W}(k)$ *is representable in the form of a finite union of disjoint locally closed subsets each of which is an analytic submanifold of* $\mathbf{V}(k)$ *of dimension* $\dim \mathbf{W}$.

(iv) *If all irreducible components of the variety* $\mathbf{V}$ *have the same dimension and* $\dim \mathbf{W} < \dim \mathbf{V}$, *then for every* ν *from the canonical class of measures on* $\mathbf{V}(k)$ *we have* $\nu(\mathbf{W}(k)) = 0$ (this is a consequence of (i), (ii) and (iii)).

For assertions (i) and (ii), see 5.8.10 and 5.8.11 in [Bou 4]. Since in every algebraic variety the set of simple points is Zariski open and dense (see [Bo 6], Chapter AG, Corollary 17.2) and the Zariski closure of every $A \subset \mathbf{W}(k)$ is defined over k, there exists a k-subvariety $\mathbf{W}' \subset \mathbf{W}$ such that $\dim \mathbf{W}' < \dim \mathbf{W}$ and $\mathbf{W}(k) - \mathbf{W}'(k) = Y$. Therefore (iii) is deduced from (ii) by induction on $\dim \mathbf{W}$.

(2.5.2) Proposition. *Let* $\mathbf{H}$ *be a k-group. Then*

(i) $\mathbf{H}(k)$ *is a pure k-analytic manifold of dimension* $\dim \mathbf{H}$.

(ii) *The Haar measures* $\mu_{\mathbf{H}(k)}$ *belongs to the canonical class of measures on* $\mathbf{H}(k)$ (see [Bou 4], 10.1.7).

Assertion (i) follows from Proposition 2.5.1(i) and from the fact that every algebraic group is a smooth variety (see 0.13).

(2.5.3) Proposition. *Let* $\mathbf{H}$ *be a connected* k*-group. Then*

(i) *for a proper algebraic* k*-subvariety* $\mathbf{W}$ *in* $\mathbf{H}$, $\mu_{\mathbf{H}(k)}(\mathbf{W}(k)) = 0$ *and hence* $\mathbf{W}(k)$ *is nowhere dense in* $\mathbf{H}(k)$*;*

(ii) $\mathbf{H}(k)$ *is Zariski dense in* $\mathbf{H}$ (follows from (i)).

Proof. (i) By Proposition 2.5.2(i) and Lemma 2.5.1(iii), $\mathbf{H}(k)$ is a pure k-analytic manifold of dimension $\dim \mathbf{H}$ and $\mathbf{W}(k)$ is representable in the form of a finite union of analytic submanifolds of $\mathbf{H}(k)$ of dimension at most $\dim \mathbf{W}$. Since the k-group $\mathbf{H}$ is connected and $\mathbf{W}$ is a proper subvariety in $\mathbf{H}$, $\dim \mathbf{W} < \dim \mathbf{H}$. To prove (i) it remains now to apply Proposition 2.5.2(ii) and Proposition 2.5.1(iv). □

(2.5.4) Proposition. (See [Bou 4], 5.8.10, 5.8.11 and 5.9.1.) *Let* $\varphi\colon \mathbf{V} \to \mathbf{V}'$ *be a* k*-morphism of* k*-varieties and* $x \in \mathbf{V}(k)$*. Assume* x *is a simple point of* $\mathbf{V}$ *and the differential* $(d\varphi)_x$ *of the morphism* φ *at* x *is surjective. Then*

(i) $\varphi(\mathbf{V}(k))$ *contains a neighbourhood* $U' \subset \mathbf{V}'(k)$ *of the point* $\varphi(x)$*;*

(ii) *if* $(d\varphi)_x$ *is an isomorphism, then there exists a neighbourhood* U *of* x *such that the set* $\varphi(U)$ *is open in* $\mathbf{V}'(k)$ *and the restriction* $\varphi|_u\colon U \to \varphi(U)$ *is an analytic isomorphism (i.e.* $\varphi|_U$ *is bijective and bianalytic).*

(2.5.5) Proposition. *Let* l *be a proper closed subfield of the field* k *and let* $\mathbf{H}$ *be an* l*-group of positive dimension. Then* $\mu_{\mathbf{H}(k)}(\mathbf{H}(l)) = 0$ *and hence the subgroup* $\mathbf{H}(l)$ *is of infinite index in* $\mathbf{H}(k)$.

Proof. Let us set $n = \dim \mathbf{H}$. Since $\mathbf{H}$ is a smooth affine l-variety, for each $h \in \mathbf{H}(l)$ there exists a defined over l regular map φ of $\mathbf{H}$ to the n-dimensional affine space, such that $(d\varphi)_h$ is an isomorphism (see 0.13). By Proposition 2.5.4(ii) there exists then a neighbourhood U_h of h in $\mathbf{H}(k)$ such that the set $\varphi(U_h)$ is open in k^n and the map $\varphi|_{U_h}\colon U_h \to \varphi(U_h)$ is an analytic isomorphism. Since l is a proper closed subfield of k and $n > 0$, we have that $\mu_{k^n}(l^n) = 0$. On the other hand, since $\varphi|_{U_h}$ is an analytic isomorphism and according to Proposition 2.5.2(ii), $\mu_{\mathbf{H}(k)}$ belongs to the canonical class of measures on $\mathbf{H}(k)$, we conclude that for every $Y \subset U_h$ the conditions $\mu_{\mathbf{H}(k)}(Y) = 0$ and $\mu_{k^n}(\varphi(Y)) = 0$ are equivalent. Therefore,

$$\mu_{\mathbf{H}(k)}(U_h \cap \varphi^{-1}(l^n)) = 0.$$

But, since φ is defined over l, we have that $U_h \cap \mathbf{H}(l) \subset U_h \cap \varphi^{-1}(l^n)$. Hence, $\mu_{\mathbf{H}(k)}(U_h \cap \mathbf{H}(l)) = 0$. Now covering $\mathbf{H}(l)$ by countably many open sets of the form $U_h \cap \mathbf{H}(l)$, $h \in \mathbf{H}(l)$, we obtain that $\mu_{\mathbf{H}(k)}(\mathbf{H}(l)) = 0$. □

2.6 Continuous "Abstract" Homomorphisms of Algebraic Groups Over Local Fields

Denote by $\mathscr{B}$ the set of prime numbers supplemented by ∞. If char $k = 0$, then (see 0.31) there exists $p(k) \in \mathscr{B}$ such that k is (isomorphic to) a finite extension of $\mathbb{Q}_{p(k)}$ (we recall that $\mathbb{Q}_\infty = \mathbb{R}$). If char $k \neq 0$, we put $p(k) = 0$.

(2.6.1) Proposition. *Let $p \in \mathscr{B}$, $\mathbf{H}$ a reductive $\mathbb{Q}_p$-group, $\mathbf{H}'$ an algebraic k-group, and let $\delta\colon \mathbf{H}(\mathbb{Q}_p) \to \mathbf{H}'(k)$ be a continuous homomorphism.*

(i) *If $\delta(\mathbf{H}(\mathbb{Q}_p)^{(u)}) \neq \{e\}$, then $p(k) = p$ and the restriction of δ to $\mathbf{H}(\mathbb{Q}_p)^{(u)}$ can be extended to a regular map $\delta'\colon \mathbf{H}^{(u)} \to \mathbf{H}'$ defined over k.*

(ii) *If the group $\mathbf{H}$ is connected, semisimple, simply connected, without $\mathbb{Q}_p$-anisotropic factors, and $\delta(\mathbf{H}(\mathbb{Q}_p)) \neq \{e\}$, then $p(k) = p$ and δ can be extended to a k-morphism $\tilde{\delta}\colon \mathbf{H} \to \mathbf{H}'$.*

(iii) *If the group $\mathbf{H}$ is connected, semisimple, without $\mathbb{Q}_p$-anisotropic factors, the subgroup $\delta(\mathbf{H}(\mathbb{Q}_p))$ is Zariski dense in $\mathbf{H}'$, the group $\mathbf{H}'$ is connected, different from $\{e\}$, and $\mathscr{Z}(\mathbf{H}') = \{e\}$, then $p(k) = p$ and δ extends to a k-morphism $\tilde{\delta}\colon \mathbf{H} \to \mathbf{H}'$.*

Proof. (i) Since $\delta(\mathbf{H}(\mathbb{Q}_p)^{(u)}) \neq \{e\}$ and $\mathbf{H}(\mathbb{Q}_p)^{(u)} \subset \mathbf{H}(\mathbb{Q}_p)^+$ (see Proposition 1.5.1), we have, in view of Proposition in 1.8.3(ii), that char $k = 0$. Replacing then $\mathbf{H}'$ by $R_{k/\mathbb{Q}_{p(k)}}\mathbf{H}'$, where $R_{k/\mathbb{Q}_{p(k)}}$ is the restriction of scalars (see 1.7), we may assume that $k = \mathbb{Q}_{p'}$, where $p' = p(k)$. By Corollary 2.5.2 $\mathbf{H}(\mathbb{Q}_p))$ (resp. $\mathbf{H}'(\mathbb{Q}_{p'})$) is a $\mathbb{Q}_p$ (resp. $\mathbb{Q}_{p'}$)-analytic manifold and hence is a p (resp. p')-adic Lie group. The Lie algebras $\mathrm{Lie}(\mathbf{H}(\mathbb{Q}_p))$ and $\mathrm{Lie}(\mathbf{H}'(\mathbb{Q}_{p'}))$ of the Lie groups $\mathbf{H}(\mathbb{Q}_p)$ and $\mathbf{H}'(\mathbb{Q}_{p'})$ are naturally identified with $\mathrm{Lie}(\mathbf{H})_{\mathbb{Q}_p}$ and $\mathrm{Lie}(\mathbf{H}')_{\mathbb{Q}_{p'}}$ (this follows from the results of Section 11 of §3 of Chapter III in [Bou 5]). Let $\exp : U \to \mathbf{H}(\mathbb{Q}_p)$ and $\exp : U' \to \mathbf{H}'(\mathbb{Q}_{p'})$ be exponential maps, where U and U' are neighbourhoods of zero in $\mathrm{Lie}(\mathbf{H}(\mathbb{Q}_p))$ and $\mathrm{Lie}(\mathbf{H}'(\mathbb{Q}_{p'}))$ (see [Bou 5], Chapter III, §3). Observe that the exponential maps agree with the exponential maps introduced earlier on $\mathrm{Lie}(\mathbf{H})^{(n)}$ and $\mathrm{Lie}(\mathbf{H}')^{(n)}$. For every $t \in \mathbb{Q}_p$ and $u \in \mathbf{H}(\mathbb{Q}_p)^{(n)}$ (resp. $t \in \mathbb{Q}_{p'}$ and $u \in \mathbf{H}'(\mathbb{Q}_{p'})^{(u)}$) we set $u^t = \exp(t \ln u) \in \mathbf{H}(\mathbb{Q}_p)^{(u)}$ (resp. $\mathbf{H}'(\mathbb{Q}_{p'})^{(u)}$). If $t = m/n$, $m, n \in \mathbb{Z} - \{0\}$, then the unipotent element u^t is uniquely determined via the equality $(u^t)^n = u^m$.On the other hand, according to Proposition 1.8.3(iii), $\delta(\mathbf{H}(\mathbb{Q}_p)^{(u)}) \subset \mathbf{H}'(\mathbb{Q}_{p'})^{(u)}$. Therefore we have

$$(1) \qquad \delta(u^r) = (\delta(u))^r,\ r \in \mathbb{Q},\ u \in \mathbf{H}(\mathbb{Q}_p)^{(u)}.$$

Let $p \neq p'$. Then, by Proposition 1 of §8 of Chapter III in [Bou 5], there exists a neighbourhood V of the identity in the p-adic Lie group $\mathbf{H}(\mathbb{Q}_p)$ such that $\delta(V) = \{e\}$. Since for each $u \in \mathbf{H}(\mathbb{Q}_p)^{(u)}$, $u^t \to e$ as t tends to zero in $\mathbb{Q}_p$, for each $u \in \mathbf{H}(\mathbb{Q}_p)^{(u)}$ there exists $r \in \mathbb{Q}$ such that $u^r \in V$. Therefore, in view of (1) above, we obtain a contradiction, for $\delta(\mathbf{H}(\mathbb{Q}_p)^{(u)}) \neq \{e\}$. Thus $p = p'$ and by Theorem of §8 of Chapter III in [Bou 5], there exist a neighbourhood $\tilde{U} \subset U$ of zero in $\mathrm{Lie}(\mathbf{H}(\mathbb{Q}_p))$ and a Lie algebra morphism $L(\delta)\colon \mathrm{Lie}(\mathbf{H}) \to \mathrm{Lie}(\mathbf{H}')$ defined over $\mathbb{Q}_p$ such that the set $V \overset{\text{def}}{=} \exp(\tilde{U})$ is open in $\mathbf{H}(\mathbb{Q}_p)$ and $\delta(\exp x) = \exp(L(\delta)(x))$ for all $x \in \tilde{U}$. Thus $u = \exp(L(\delta)(\ln u))$ for all $u \in \mathbf{H}^{(u)} \cap V$. But for each $u \in \mathbf{H}(\mathbb{Q}_p)^{(u)}$, as we have seen above, there exists $r \in \mathbb{Q}$ such that $u^r \in V$. Therefore (1) implies that $\delta(u) = \exp(L(\delta)(\ln u))$ for all $u \in \mathbf{H}(\mathbb{Q}_p)^{(u)}$ and, since the maps exp, $L(\delta)$, and ln are regular and defined over $\mathbb{Q}_p$, this proves assertion (i).

(ii) According to Theorem 2.3.1(a), $\mathbf{H}(\mathbb{Q}_p) = \mathbf{H}(\mathbb{Q}_p)^+$. But $\delta(\mathbf{H}(\mathbb{Q}_p)) \neq \{e\}$. Therefore $\delta(\mathbf{H}(\mathbb{Q}_p)^{(u)}) \neq \{e\}$, and hence (see (i)) $p(k) = p$. Replacing $\mathbf{H}'$ by $R_{k/\mathbb{Q}_p}\mathbf{H}'$ we may assume that $k = \mathbb{Q}_p$. Let $L(\delta)$: $\mathrm{Lie}(\mathbf{H}) \to \mathrm{Lie}(H')$ be the Lie algebra morphism defined over $\mathbb{Q}_p$ as in the proof of (i). Then by Proposition 1.4.13 there exists a k-morphism $\tilde{\delta}$: $\mathbf{H} \to \mathbf{H}'$ whose differential is $L(\delta)$. Since $\delta(u) = \exp(L(\delta)(\ln u))$ for all $u \in \mathbf{H}(\mathbb{Q}_p)^{(u)}$ (see the proof of (i) above and Proposition 1.3.1) we have that $\delta(u) = \tilde{\delta}(u)$ for each $u \in \mathbf{H}(\mathbb{Q}_p)^{(u)}$. But in view of Theorem 2.3.1(a), $\mathbf{H}(\mathbb{Q}_p)^{(u)}$ generates $\mathbf{H}(\mathbb{Q}_p)$. Therefore $\delta(h) = \tilde{\delta}(h)$ for all $h \in \mathbf{H}(\mathbb{Q}_p)$, i.e. $\tilde{\delta}$ is the desired extension of δ.

(iii) Let $\tilde{\mathbf{H}}$ be the simply connected covering of the group $\mathbf{H}$ and let $\pi: \tilde{\mathbf{H}} \to \mathbf{H}$ be the (central) $\mathbb{Q}_p$-isogeny. By Proposition 2.3.4(i) the subgroup $\pi(\tilde{\mathbf{H}}(\mathbb{Q}_p))$ is open, normal, and of finite index in $\mathbf{H}(\mathbb{Q}_p)$. But $\mathbf{H}$ is connected, $\mathbf{H} \neq \{e\}$, and the subgroup $\delta(\mathbf{H}(\mathbb{Q}_p))$ is Zariski dense in $\mathbf{H}'$. Therefore the subgroup $\delta(\pi(\tilde{\mathbf{H}}(\mathbb{Q}_p)))$ is non-trivial and Zariski dense in $\mathbf{H}'$. Now, applying (ii) to the homomorphism $\delta \circ \pi$, we obtain that $p(k) = p$ and $\delta \circ \pi$ extends to a k-morphism δ': $\tilde{\mathbf{H}} \to \mathbf{H}'$. Since the subgroup $\delta'(\tilde{\mathbf{H}}(\mathbb{Q}_p))$ containing $\delta(\pi(\tilde{\mathbf{H}}(\mathbb{Q}_p)))$ is Zariski dense in $\mathbf{H}'$ and $\mathscr{Z}(\mathbf{H}') = \{e\}$, we have that $\delta'(\mathrm{Ker}\,\pi) \subset \delta'(\mathscr{Z}(\tilde{\mathbf{H}})) = \{e\}$. It follows that there exists a k-morphism $\tilde{\delta}$ of the group $\mathbf{H} = \tilde{\mathbf{H}}/\mathrm{Ker}\,\pi$ to $\mathbf{H}'$ such that $\delta' = \tilde{\delta} \circ \pi$. Since $\delta \circ \pi = \delta'|_{\tilde{\mathbf{H}}(\mathbb{Q}_p)}$, we have that $\tilde{\delta}(h) = \delta(h)$ for all $h \in \pi(\tilde{\mathbf{H}}(\mathbb{Q}_p))$. Let us set $v(x) = \tilde{\delta}(x)^{-1}\delta(x)$ and let $x \in \mathbf{H}(\mathbb{Q}_p)$, $h \in \pi(\tilde{\mathbf{H}}(\mathbb{Q}_p))$. Since the subgroup $\pi(\tilde{\mathbf{H}}(\mathbb{Q}_p))$ is normal in $\mathbf{H}(\mathbb{Q}_p)$ and $\tilde{\delta}$ agrees with δ on $\pi(\tilde{\mathbf{H}}(\mathbb{Q}_p))$, we have that

$$\begin{aligned}\tilde{\delta}(x)\delta(h)\tilde{\delta}(x)^{-1} &= \tilde{\delta}(xhx^{-1}) = \\ &= \delta(xhx^{-1}) = \delta(x)\delta(h)\delta(x)^{-1},\end{aligned}$$

and hence, $v(x)$ and $\delta(h)$ commute. But the subgroup $\delta(\pi(\mathbf{H}(\mathbb{Q}_p)))$ is Zariski dense in $\mathbf{H}'$ and $\mathscr{Z}(\mathbf{H}') = \{e\}$. Thus $\delta = \tilde{\delta}|_{\mathbf{H}(\mathbb{Q}_p)}$ and the proof is complete. □

(2.6.2) Corollary. *Let l be a local field of characteristic* 0, *let* $\mathbf{H}$ *be a connected semisimple l-group,* $\mathbf{H}'$ *a k-group, and let δ: $\mathbf{H}(l) \to \mathbf{H}'(k)$ be a continuous homomorphism. Denote by* $\mathbf{F}$ *the Zariski closure of the subgroup $\delta(\mathbf{H}(l))$ in* $\mathbf{H}'$. *Assume the group* $\mathbf{H}$ *has no l-anisotropic factors. Then the group* $\mathbf{F}$ *is semisimple.*

Proof. In view of the results in 1.7 the group $R_{l/\mathbb{Q}_{p(l)}}\mathbf{H}$ is connected, semisimple and does not have $\mathbb{Q}_{p(l)}$-anisotropic factors and

$$R^0_{l/\mathbb{Q}_{p(l)}}: \mathbf{H}(l) \to (R_{l/\mathbb{Q}_{p(l)}}\mathbf{H})(\mathbb{Q}_{p(l)})$$

is a topological group isomorphism. Therefore we may assume, replacing $\mathbf{H}$ by $R_{l/\mathbb{Q}_{p(l)}}\mathbf{H}$ and δ by $\delta \cdot (R^0_{l/\mathbb{Q}_{p(l)}})^{-1}$, that $l = \mathbb{Q}_p$, where $p \in \mathscr{P}$.

Let $\tilde{\mathbf{H}}$ be the simply connected covering of the group $\mathbf{H}$, let $\pi: \tilde{\mathbf{H}} \to \mathbf{H}$ be the $\mathbb{Q}_p$-isogeny, and let $\tilde{\mathbf{F}}$ be the Zariski closure of the subgroup $\delta(\pi(\tilde{\mathbf{H}}(\mathbb{Q}_p)))$ in $\mathbf{H}$. By Proposition 2.6.1(ii) the morphism $\delta \circ \pi$ extends to a k-morphism $\tilde{\delta}$: $\tilde{\mathbf{H}} \to \mathbf{H}'$. Since the subgroup $\tilde{\mathbf{H}}(\mathbb{Q}_p)$ is Zariski dense in $\tilde{\mathbf{H}}$ (see 0.24 or Proposition 2.5.3), we have that $\tilde{\mathbf{F}} = \tilde{\delta}(\tilde{\mathbf{H}})$. But the image of a semisimple group under an algebraic group morphism is semisimple (see 0.24). Therefore the group $\tilde{\mathbf{F}}$ is semisimple. Since the subgroup $\pi(\tilde{\mathbf{H}}(\mathbb{Q}_p))$ is of finite index in $\mathbf{H}(\mathbb{Q}_p)$ (see Proposition 2.3.4(i)), the semisimple group $\tilde{\mathbf{F}}$ is of finite index in $\mathbf{F}$ and hence $\mathbf{F}$ is semisimple. □

(2.6.3) *Remark.* It is not hard to show that the assumption in Corollary 2.6.2 concerning the absence of l-anisotropic factors can be dropped.

(2.6.4) Proposition. *Let A be a finite set. For each $\alpha \in A$ let k_α be a local field and let $\mathbf{G}_\alpha$ be a connected semisimple k_α-group with no k_α-anisotropic factors. Let $G = \prod_{\alpha \in A} \mathbf{G}_\alpha(k_\alpha)$, let $\mathbf{H}$ be a k-group and let $\delta : G \to \mathbf{H}(k)$ be a continuous homomorphism. Assume that for $\alpha \in A$, with $\operatorname{char} k_\alpha \neq 0$ the group $\mathbf{G}_\alpha$ is simply connected.*

(a) If $\operatorname{char} k \neq \operatorname{char} k_\alpha$ for each $\alpha \in A$, then the group $\delta(G)$ is finite.

(b) If $\operatorname{char} k = 0$ and $\mathbf{F}$ is the Zariski closure of the subgroup $\delta(G)$ in $\mathbf{H}$, then the subgroup $\mathbf{F}$ is semisimple.

Proof. (a) According to Proposition 1.8.3(ii), $\delta(\mathbf{G}_\alpha(k_\alpha)^+) = \{e\}$ for all $\alpha \in A$. But by assertions (a′) and (c) of Theorem 2.3.1 the subgroup $\mathbf{G}_\alpha(k_\alpha)^+$ is of finite index in $\mathbf{G}_\alpha(k_\alpha)$, $\alpha \in A$. Therefore all the groups $\delta(\mathbf{G}_\alpha(k_\alpha))$ are finite, and hence the group $\delta(G)$ is finite as well.

(b) Let us denote by $\mathbf{F}_\alpha$ the Zariski closure of the subgroup $\delta(\mathbf{G}_\alpha(k_\alpha))$ in $\mathbf{H}$. In view of (a) and Corollary 2.6.2 the groups $\mathbf{F}_\alpha$, $\alpha \in A$, are semisimple. But $\mathbf{F}$ coincides with the image of the group $\prod_{\alpha \in A} \mathbf{F}_\alpha$ under the multiplication map $\prod_{\alpha \in A} \mathbf{F}_\alpha \to \mathbf{H}$ and the image of a semisimple algebraic group under an algebraic group morphism is also a semisimple algebraic group (see 0.24). Thus $\mathbf{F}$ is semisimple. □

(2.6.5) Proposition. *Let $\mathbf{G}$ and $\mathbf{G}'$ be two connected semisimple $\mathbb{R}$-groups. Denote by $\mathbf{G}(\mathbb{R})^0$ and $\mathbf{G}'(\mathbb{R})^0$ the identity components of the Lie groups $\mathbf{G}(\mathbb{R})$ and $\mathbf{G}'(\mathbb{R})$. Assume that $\mathbf{G}'$ is an adjoint group, Then every continuous epimorphism φ: $\mathbf{G}(\mathbb{R})^0 \to \mathbf{G}'(\mathbb{R})^0$ is rational, i.e., extends to an $\mathbb{R}$-epimorphism $\tilde{\varphi}$: $\mathbf{G} \to \mathbf{G}'$.*

Proof. Let $\tilde{\mathbf{G}}$ be the simply connected covering of $\mathbf{G}$, let p: $\tilde{\mathbf{G}} \to \mathbf{G}$ be the (central) $\mathbb{R}$-isogeny, and let β: $\operatorname{Lie}(\mathbf{G}) \to \operatorname{Lie}(\mathbf{G}')$ be the Lie algebra epimorphism defined over $\mathbb{R}$ and induced from the continuous epimorphism φ. According to Proposition 1.4.13 there exists an $\mathbb{R}$-epimorphism π: $\tilde{\mathbf{G}} \to \mathbf{G}'$ whose differential is β (we identify $\operatorname{Lie}(\mathbf{G})$ with $\operatorname{Lie}(\tilde{\mathbf{G}})$ here). Since $\mathbf{G}'$ is an adjoint group, there exists an $\mathbb{R}$-epimorphism $\tilde{\varphi}$: $\mathbf{G} \to \mathbf{G}'$ such that $\tilde{\varphi} \circ p = \pi$ (see Proposition 1.4.11(iii)). Then the differential of the morphism $\tilde{\varphi}$ equals β and hence $\tilde{\varphi}$ is an extension of the epimorphism φ. □

3. Arithmetic Groups

In this section K denotes a global field, $\mathscr{R}$ denotes the set of all non-equivalent valuations of the field K and $\mathscr{R}_\infty \subset \mathscr{R}$ denotes the set of archimedean valuations.

We shall use the notation introducted in 0.32 and 0.33. In particular, K_v denotes the completion of the field K relative to the valuation $v \in \mathscr{R}$ and for $S \subset \mathscr{R}$ we denote by $K(S)$ the ring of S-integral elements of the field K. We shall only consider the case where K is the quotient field of the ring $K(S)$, that is either $\operatorname{char} K = 0$ or $\operatorname{card} S \geq 1$.

3.1 The Group $\mathbf{H}(K(S))$, Congruence Subgroups and S-arithmetic Subgroups

For $n \in \mathbb{N}^+$ and for any commutative ring L with 1, we denote by $\mathbf{GL}_n(L)$ the group of matrices with entries in L whose determinants are invertible elements of the ring L. It is easily seen that for $S \subset \mathscr{R}$, $\mathbf{GL}_n(K(S)) = \{A \in \mathbf{GL}_n(K) \mid$ the entries of A and A^{-1} belong to $K(S)\}$. If $\mathfrak{a}$ is a non-zero ideal in $K(S)$, we denote by $\mathbf{GL}_n(\mathfrak{a})$ the set of elements of $\mathbf{GL}_n(K(S))$ congruent to the unit matrix E modulo $\mathfrak{a}$, i.e., $\mathbf{GL}_n(\mathfrak{a}) = \{A \in \mathbf{GL}_n(K(S)) \mid$ the entries of the matrix A–E belong to $\mathfrak{a}\}$. The natural epimorphism $K(S) \to K(S)/\mathfrak{a}$ induces a group homomorphism

$$\sigma_{\mathfrak{a}}\colon \mathbf{GL}_n(K(S)) \to \mathbf{GL}_n(K(S)/\mathfrak{a})$$

whose kernel is $\mathbf{GL}_n(\mathfrak{a})$. Therefore $\mathbf{GL}_n(\mathfrak{a})$ is a normal subgroup of the group $\mathbf{GL}_n(K(S))$. Since the quotient ring $K(S)/\mathfrak{a}$, and hence the group $\mathbf{GL}_n(K(S)/\mathfrak{a})$, are finite, the subgroup $\mathbf{GL}_n(\mathfrak{a}) = \operatorname{Ker} \sigma_{\mathfrak{a}}$ is of finite index in $\mathbf{GL}_n(K(S))$.

For every K-subgroup $\mathbf{H}$ of the group $\mathbf{GL}_n$, we set $\mathbf{H}(K(S)) = \mathbf{H} \cap \mathbf{GL}_n(K(S))$ and $\mathbf{H}(\mathfrak{a}) = \mathbf{H} \cap \mathbf{GL}_n(\mathfrak{a})$. Since $\mathbf{GL}_n(\mathfrak{a})$ is a normal subgroup of finite index in $\mathbf{GL}_n(K(S))$, $\mathbf{H}(\mathfrak{a})$ is a normal subgroup of finite index in $\mathbf{H}(K(S))$. The subgroups of the form $\mathbf{H}(\mathfrak{a})$ are called the *principle S-congruence subgroups* of the group $\mathbf{H}(K(S))$. An *S-congruence subgroup* of $\mathbf{H}(K(S))$ is a subgroup containing a principle S-congruence subgroup.

(3.1.1) Lemma. *Let $S \subset \mathscr{R}$, $n, m \in \mathbb{N}^+$, let $\mathbf{H} \subset \mathbf{GL}_n$ and $\mathbf{H}' \subset \mathbf{GL}_m$ be K-subgroups, let E be the identity matrix and let $f\colon \mathbf{H} \to \mathbf{H}'$ be a regular map defined over K with $f(E) = E$. Then*

(i) *for every non-zero ideal $\mathfrak{a}$ of the ring $K(S)$, there exists a non-zero ideal $\mathfrak{a}' \subset K(S)$ such that if a matrix $h \in \mathbf{H}(K(S))$ is congruent to E modulo $\mathfrak{a}'$, then the matrix $f(h)$ is congruent to E modulo $\mathfrak{a}$.*

(ii) *Assume that either $\mathbf{H}' \subset \mathbf{SL}_m$ or f is a K-homomorphism. Then the inverse image under f of every S-congruence subgroup of the group $\mathbf{H}'(K(S))$ contains an S-congruence subgroup of the group $\mathbf{H}(K(S))$. In particular, there exists an S-congruence subgroup D of the group $\mathbf{H}(K(S))$ such that $f(D) \subset \mathbf{H}'(K(S))$.*

(iii) *Every S-congruence subgroup of the group $\mathbf{H}(K(S))$ is of finite index in $\mathbf{H}(K(S))$.*

(iv) *Assume f is a K-isomorphism. Then the subgroups $f(\mathbf{H}(K(S))$ and $\mathbf{H}'(K(S))$ are commensurable.*

(v) *For each $h \in \mathbf{H}(K)$ the subgroups $h\mathbf{H}(K(S))h^{-1}$ and $\mathbf{H}(K(S))$ are commensurable. In other words, $\mathbf{H}(K) \subset \operatorname{Comm}_{\mathbf{H}}(\mathbf{H}(K(S)))$.*

Proof. Every regular function on $\mathbf{H}$ is defined by a polynomial in entries of the matrices h and h^{-1}, $h \in \mathbf{H}$ (see 0.13). Therefore the entries of the matrices $f(h) - E$ are defined via m^2 polynomials $P_1, \ldots, P_{m^2}$ in entries of the matrices $h - E$ and $h^{-1} - E$, $h \in \mathbf{H}$. Since $f(E) = E$, the constant terms of these polynomials equal zero. On the other hand, since K is the quotient field of the ring $K(S)$, there exists $a \in K(S)$, $a \neq 0$, such that the coefficients of the polynomials $P_l, \ldots, P_{m^2}$ belong to $a^{-1}K(S)$. Therefore, if a matrix $h \in \mathbf{H}(K(S))$, and hence h^{-1} are congruent to E modulo $a\mathfrak{a}$, then the matrix $f(h)$ is congruent to E modulo $\mathfrak{a}$ and this completes the proof of (i).

If $\mathbf{H}' \subset \mathbf{SL}_m$, then (ii) follows immediately from (i). Since $\mathbf{H}(\mathfrak{a}) = \{h \in \mathbf{H}(k) \mid$ the entries of the matrices $h-E$ and $h^{-1}-E$ belong to $\mathfrak{a}\}$ and $\mathbf{H}'(\mathfrak{a}) = \{h \in \mathbf{H}'(K) \mid$ the entries of the matrices $h - E$ and $h^{-1} - E$ belong to $\mathfrak{a}\}$, if f is a group homomorphism, (ii) follows from (i) and from the equality $f(h^{-1}) = f(h)^{-1}$ for all $h \in \mathbf{H}$. Assertion (iii) is true, because $\mathbf{H}(\mathfrak{a})$ is of finite index in $\mathbf{H}(K(S))$ for every non-zero ideal $\mathfrak{a} \subset K(S)$. Assertion (iv) is a consequence of (ii) and (iii). Assertion (v) follows from (iv) applied to the map $\operatorname{Int} h$. □

(3.1.2) Given a K-isomorphism $f\colon \mathbf{H} \to \mathbf{H}'$ of K-groups $\mathbf{H} \subset \mathbf{GL}_n$ and $\mathbf{H}' \subset \mathbf{GL}_m$, in general $f(\mathbf{H}(K(S))) \neq \mathbf{H}'(K(S))$. Therefore, for a given K-group $\mathbf{H}$ whenever, we speak about the groups $\mathbf{H}(K(S))$ and $\mathbf{H}(\mathfrak{a})$, we assume $\mathbf{H}$ to be a K-subgroup of the group $\mathbf{GL}_n$. Notice that the group $\mathbf{H}(K(S))$ can be determined intrinsically if we suppose that $\mathbf{H}$ is equipped with a "$K(S)$-structure", in other words, if an affine group $K(S)$-scheme is given whose generic fiber is $\mathbf{H}$.

Let $S \subset \mathscr{R}$ and let $\mathbf{H}$ be a K-group. We call a subgroup of the group $\mathbf{H}$ *S-arithmetic* if it is commensurable with $\mathbf{H}(K(S))$. In view of Lemma 3.1.1(iv) this notion is intrinsic with respect to the K-structure on $\mathbf{H}$, i.e., if $f\colon \mathbf{H} \to \mathbf{H}'$ is a K-isomorphism, then a subgroup X of $\mathbf{H}$ is S-arithmetic if and only if $f(X)$ is an S-arithmetic subgroup of $\mathbf{H}$. If $K = \mathbb{Q}$ and $S = \{\infty\}$ is the only archimedean valuation of the field $\mathbb{Q}$, we simply call an S-arithmetic subgroup *arithmetic*.

(3.1.3) Lemma. *Let $S \subset \mathscr{R}$, $f\colon \mathbf{H} \to \mathbf{H}'$ be a K-homomorphism, and let $X \subset \mathbf{H}$, $X' \subset \mathbf{H}'$ be S-arithmetic subgroups. Then*

(a) *The subgroup $X \cap f^{-1}(X')$ is of finite index in X;*
(b) *for each $h \in \mathbf{H}(k)$ the subgroups hXh^{-1} and X are commensurable. In other words, $\mathbf{H}(K) \subset \operatorname{Comm}_{\mathbf{H}}(X)$.*

Assertion (a) follows from (ii) and (iii) of Lemma 3.1.1 and (b) is a consequence of Lemma 3.1.1(v).

(3.1.4) Lemma. *Let K' be a finite separable field extension of K, let $\mathbf{H}$ be a K'-group and let $S \subset \mathscr{R}$. For each $v \in \mathscr{R}$, we denote by v' the set of valuations of the field K' extending the valuation v. Let $S' = \bigcup_{v \in S} v'$. We shall use the notation $R_{l/k}$ and $R^0_{l/k}$ introduced in 1.7.*

(i) *For each* $v \in \mathscr{R}$ *there exists a natural* K_v*-isomorphism*

$$f_v\colon R_{K'/K}\mathbf{H} \to \prod_{w \in v'} R_{K'_w/K_v}\mathbf{H}.$$

The isomorphisms $f_v^{-1} \circ R^0_{K'_w/K_v}$, $v \in R$, $w \in v'$ *induce a topological group isomorphism of the adèle groups associated with* $\mathbf{H}$ *and* $R_{K'/K}\mathbf{H}$, *whose restriction to* $\mathbf{H}(K')$ *agrees with* $R^0_{K'/K}$.

(ii) $\sum_{w \in S'} \operatorname{rank}_{K'_w} \mathbf{H} = \sum_{v \in S} \operatorname{rank}_{K_v}(R_{K'/K}\mathbf{H})$.

(iii) *The subgroups* $R^0_{K'/K}\mathbf{H}(K'(S'))$ *and* $(R_{K'/K}\mathbf{H})(K(S))$ *are commensurable.*

Assertion (i) is a reformulation of some results in [Weil 2] (see Theorems 1.3.1 and 1.3.3). Since for all $v \in \mathscr{R}$ and $w \in v'$, $\operatorname{rank}_{K_v}(R_{K'_w/K_v}\mathbf{H}) = \operatorname{rank}_{K'_w}\mathbf{H}$ (see 1.7), (ii) follows from (i). Assertion (iii) is a consequence of (i), but it can also be easily derived from Lemma 3.1.1(iv).

3.2 The Finiteness of the Volume of Quotient Spaces

Let $\mathbf{G} \subset \mathbf{GL}_n$ be a connected K-group. Let us denote by $\mathrm{X}_K(\mathbf{G})$ the set of rational characters of the group $\mathbf{G}$ defined over K. As in 0.33 we denote by $\mathbf{G}(\mathbb{A}_K)$ the adèle group of the group $\mathbf{G}$ and identify $\mathbf{G}(K)$ with the subgroup of the principal adèles of the group $\mathbf{G}(\mathbb{A}_K)$. The subgroup $\mathbf{G}(K)$ is discrete in $\mathbf{G}(\mathbb{A}_K)$. It often turns out that $\mathbf{G}(K)$ is a lattice in $\mathbf{G}(\mathbb{A}_K)$.

(3.2.1) Theorem. *Let* $\mathbf{G}$ *be a connected reductive* K*-group. Then*

(a) the subgroup $\mathbf{G}(K)$ *is a lattice in* $\mathbf{G}(\mathbb{A}_K)$ *if and only if* $\mathrm{X}_K(\mathbf{G}) = 1$;

(b) the quotient space $\mathbf{G}(K) \setminus \mathbf{G}(\mathbb{A}_K)$ *is compact if and only if* $\mathbf{G}$ *is anisotropic over* K.

In case $\operatorname{char} K = 0$, this theorem was proved by Borel in [Bo 3] and in case $\operatorname{char} K \neq 0$ by Behr [Be 2] and Harder [Har 2].

(1) *Remark.* For the fields K of characteristic 0 the assumption of reductivity of the group $\mathbf{G}$ can be dropped.

(2) *Remark.* In case $\mathrm{X}_K(\mathbf{G}) \neq 1$ the following analogue of (a) holds. We set $\mathbf{G}^1(\mathbb{A}_K) = \bigcap_{\chi \in \mathrm{X}_K(\mathbf{G})} \operatorname{Ker} \psi_\chi$, where the homomorphism $\psi_\chi\colon \mathbf{G}(\mathbb{A}_K) \to \mathbb{R}^+$ is defined as follows: for $g = (g_v) \in \mathbf{G}(\mathbb{A}_K)$, $\psi_\chi(g) = \prod_{v \in \mathscr{R}} v(\chi(g_v))$. Then $\mathbf{G}(K)$ is a lattice in $\mathbf{G}^1(\mathbb{A}_K)$ (the subgroup $\mathbf{G}(K)$ is contained in $\mathbf{G}^1(\mathbb{A}_K)$, because $\prod_{v \in \mathscr{R}} v(x) = 1$ for all $x \in K - \{0\}$).

The following assertion is a special case of (a).

(3.2.2) Theorem. *If* $\mathbf{G}$ *is a connected semisimple* K*-group, then* $\mathbf{G}(K)$ *is a lattice in* $\mathbf{G}(\mathbb{A}_K)$.

(3.2.3) For every subset $S \subset \mathscr{R}$ we denote by G_S the subgroup in $\mathbf{G}(\mathbb{A}_K)$ consisting of the adèles whose v-component is equal to the identity for all $v \in \mathscr{R} - S$. Observe that $G_{\mathscr{R}} = \mathbf{G}(\mathbb{A}_K)$ and $G_S = \prod_{v \in \mathscr{R}} \mathbf{G}(K_v)$, if S is finite. It is good to keep in mind that

$$(*) \qquad \mathbf{G}(K(S)) = \mathbf{G}(K) \cap (G_{S \cup \mathscr{R}_\infty} \cdot \prod_{v \in \mathscr{R} - \mathscr{R}_\infty - S} \mathbf{G}(\mathscr{O}_v)),$$

where $\mathscr{O}_v$ is the ring of integral elements of the field K_v and $\mathbf{G}(\mathscr{O}_v) = \mathbf{G} \cap \mathbf{GL}_n(\mathscr{O}_v)$. Assume $\mathbf{G}$ is reductive and set $\mathscr{T} = \mathscr{T}(\mathbf{G}) = \{v \in \mathscr{R} \mid$ the group $\mathbf{G}(K_v)$ is compact or, equivalently, the group $\mathbf{G}$ is anisotropic over $K_v\}$. We recall the well-known fact that the set $\mathscr{T}(G)$ is finite (see [Spr], Lemma 4.9). Since $\mathbf{G}(\mathscr{O}_v)$ is open in $\mathbf{G}(K_v)$ for all $v \in \mathscr{R} - \mathscr{R}_\infty$, $\mathbf{G}(\mathscr{O}_v)$ is of finite index in $\mathbf{G}(K_v)$ for each $v \in \mathscr{T} - \mathscr{R}_\infty$. It follows, in view of (*), that $\mathbf{G}(K(S))$ is of finite index in $\mathbf{G}(K(S \cup \mathscr{T}))$ and that the group $\mathbf{G}(K(S))$, identified with its image under the diagonal embedding into G_S, is discrete in G_S whenever $\mathscr{R}_\infty - \mathscr{T}(\mathbf{G}) \subset S \subset \mathscr{R}$. The following two theorems follow from (*) and Theorems 3.2.1 and 3.2.2.

(3.2.4) Theorem. *Let* $\mathbf{G}$ *be a connected reductive* K*-group and let* $\mathscr{R}_\infty - \mathscr{T}(G) \subset S \subset \mathscr{R}$. *Then*

(a) the subgroup $\mathbf{G}(K(S))$ *is a lattice in* G_S *if and only if* $\mathrm{X}_K(\mathbf{G}) = 1$;

(b) the quotient space $\mathbf{G}(K(S)) \backslash G_S$ *is compact if and only if* $\mathbf{G}$ *is anisotropic over* K.

(3.2.5) Theorem. *If* $\mathbf{G}$ *is a connected semisimple* K*-group and* $\mathscr{R}_\infty - \mathscr{T}(\mathbf{G}) \subset S \subset \mathscr{R}$, *then* $\mathbf{G}(K(S))$ *is a lattice in* G_S.

(3.2.6) If $\operatorname{char} K = 0$, then, as in Theorem 3.2.1, it is unnecessary to assume in Theorem 3.2.4 that the group $\mathbf{G}$ is reductive. Applying now Theorem 3.2.4 to the special case where $S \subset \mathscr{R}_\infty$ and dropping the assumption that $\mathbf{G}$ is reductive, we obtain the following

(3.2.7) Theorem. (See [Bo-Hari], Theorem 12.3.) *Let* K *be a finite field extension of* $\mathbb{Q}$, *let* $L = K(\mathscr{R}_\infty) \subset K$ *be the ring of integers in the field* K, *let* $\mathbf{G}$ *be a connected* K*-group, and let* $\mathscr{R}_\infty - \mathscr{T}(\mathbf{G}) \subset S \subset \mathscr{R}$. *Let us embed* $\mathbf{G}(L)$ *in* $\prod_{v \in S} \mathbf{G}(K_v)$ *via the diagonal embedding.*

(a) *The subgroup* $\mathbf{G}(L)$ *is a lattice in* $\prod_{v \in S} \mathbf{G}(K_v)$ *if and only if* $\mathrm{X}_K(\mathbf{G}) = 1$. *In particular, if* $\mathbf{G}$ *is semisimple, then* $\mathbf{G}(L)$ *is a lattice in* $\prod_{v \in S} \mathbf{G}(K_v)$.

(b) *The quotient space* $\mathbf{G}(L) \backslash \prod_{v \in S} \mathbf{G}(K_v)$ *is compact if and only if* $\mathbf{G}$ *is anisotropic over* K.

Remark. The condition "$\mathbf{G}$ is anisotropic over K" is equivalent to "$\mathrm{X}_K(\mathbf{G}) = 1$ and every unipotent element of the group $\mathbf{G}(L)$ (or equivalently of $\mathbf{G}(K)$) belongs to $R_u(\mathbf{G})$".

The following theorem is a special case of Theorem 3.2.7.

(3.2.8) Theorem. *Let* $\mathbf{G}$ *be a connected* $\mathbb{Q}$*-group.*

(a) $\mathbf{G}(\mathbb{Z})$ *is a lattice in* $\mathbf{G}(\mathbb{R})$ *if and only if* $X_{\mathbb{Q}}(\mathbf{G}) = 1$. *In particular, if* $\mathbf{G}$ *is semisimple, then* $\mathbf{G}(\mathbb{Z})$ *is a lattice in* $\mathbf{G}(\mathbb{R})$ (see [Bo-Hari], Theorem 9.4).

(b) *The quotient space* $\mathbf{G}(\mathbb{Z}) \setminus \mathbf{G}(\mathbb{R})$ *is compact if and only if* $\mathbf{G}$ *is anisotropic over* $\mathbb{Q}$ (see [Bo-Hari], Theorem 11.8 of [M-T]).

We shall need the following corollary of Theorem 3.2.5.

(3.2.9) Corollary. *Let* $S \subset \mathscr{R}$ *be a finite set of valuations of the field* K *and let* $f: \mathbf{G} \to \mathbf{G}'$ *be a central* K*-isogeny of connected semisimple* K*-groups. Then the subgroups* $f(\mathbf{G}(K(S)))$ *and* $\mathbf{G}'(K(S))$ *are commensurable and hence the preimage* $f^{-1}(\Lambda)$ *of every* S*-arithmetic subgroup* $\Lambda \subset \mathbf{G}'$ *is an* S*-arithmetic subgroup in* $\mathbf{G}$.

Proof. First observe that, since $K(S) = K(S \cup \mathscr{R}_\infty)$, we may assume that $S \supset \mathscr{R}_\infty$. The isogeny f induces a homomorphism $f_S: G_S \to G'_S$ which sends $(g_v) \in G_S$ to $(f(g_v)) \in G'_S$. In view of the finiteness of S, it follows from Proposition 2.3.4 that $f_S(G_S)$ is a closed subgroup of G'_S and the factor group $G'_S/f_S(G_S)$ is compact, moreover the induced isomorphism $G_S/\operatorname{Ker} f_S \to f_S(G_S)$ is an isomorphism of topological groups. On the other hand, $\operatorname{Ker} f_S$ is finite and by Theorem 3.2.5 $\mathbf{G}(K(S))$ is a lattice in G_S. Therefore, $f(\mathbf{G}(K(S)))$ is a lattice in G'_S and since, by (ii) and (iii) of Lemma 3.1.1, the subgroup $\mathbf{G}'(K(S)) \cap f(\mathbf{G}(K(S)))$ is of finite index in $f(\mathbf{G}(K(S)))$, it is a lattice in G'_S as well. All that remains now is to make use of the discreteness of the subgroup $\mathbf{G}'(K(S)) \subset G'_S$ and to observe that a lattice is of finite index in any discrete subgroup in which it is contained. □

Remark. In the statement of Corollary 3.2.9 one can replace "semisimple K-groups" by "reductive K-groups". Furthermore, if $\operatorname{char} K = 0$, then the semisimplicity assumption is superfluous, i.e., one can replace "connected semisimple K-groups" by "connected K-groups".

Applying Theorem 3.2.5 we shall prove the following

(3.2.10) Proposition. *Let* $\mathbf{G}$ *be a connected semisimple* K*-group,* $\mathscr{R}_\infty \subset S \subset \mathscr{R}$*,* Λ *an* S*-arithmetic subgroup of the group* $\mathbf{G}(K)$ *and let* k *be a subfield of the field* K *such that* K *is a finite separable field extension of* k. *Assume that* $\sum_{v \in S} \operatorname{rank}_{K_v} \mathbf{G}' > 0$ *for every almost* K*-simple factor* $\mathbf{G}'$ *of the group* $\mathbf{G}$. *Then the subgroup* $R^0_{K/k}(\Lambda)$ *is Zariski dense in* $R_{K/k}\mathbf{G}$.

Proof. It suffices to consider the case where $\mathbf{G}$ is almost K-simple. Denote by $\mathbf{H}$ the Zariski closure of the subgroup $R^0_{K/k}(\Lambda)$ in $R_{K/k}\mathbf{G}$ and by $\mathbf{H}^0$ the identity component of the group $\mathbf{H}$. Since for each $g \in \mathbf{G}(K)$ the subgroups $g\Lambda g^{-1}$ and Λ are commensurable (see Lemma 3.1.3(b)), the map $R^0_{K/k}: \mathbf{G}(K) \to (R_{K/k}\mathbf{G})(k)$ is an isomorphism and $\mathbf{H}^0$ coincides with the intersection of all algebraic subgroups of finite index in the group $\mathbf{H}$, we see that $(R_{k/K}\mathbf{G})(k)$ normalizes $\mathbf{H}^0$. On the other hand, since the group $\mathbf{G}$, and hence the group $R_{K/k}\mathbf{G}$, are connected and semisimple, the subgroup $R_{K/k}\mathbf{G})(k)$ is Zariski dense in $R_{K/k}\mathbf{G}$ (see 0.24).

Therefore the subgroup $\mathbf{H}^0$ is normal in $R_{K/k}\mathbf{G}$. Since the group $\prod_{v \in S} \mathbf{G}(K_v)$ is not compact and in a non-compact locally compact group any lattice is infinite, it then follows by Theorem 3.2.5 that the group $\mathbf{G}(K(S))$, and hence the groups Λ and $\mathbf{H}$, are infinite. Furthermore, since $R^0_{K/k}(\Lambda) \subset (R_{K/k}\mathbf{G})(k)$, the subgroup $\mathbf{H}^0$ is defined over k (see 0.11 and 0.13). Thus $\mathbf{H}^0$ is an infinite normal k-subgroup in $R_{K/k}\mathbf{G}$. But $\mathbf{G}$ is almost K-simple, and hence $R_{K/k}\mathbf{G}$ is almost k-simple. Thus $\mathbf{H}^0 = R_{K/k}\mathbf{G}$. □

The following assertion is a special case of Proposition 3.2.10.

(3.2.11) Proposition. *Let* $\mathbf{G}$ *be a connected semisimple* $\mathbb{Q}$*-group. Assume* $\operatorname{rank}_{\mathbb{R}} \mathbf{G}' > 0$ *for each almost* $\mathbb{Q}$*-simple factor* $\mathbf{G}'$ *of* $\mathbf{G}$*. Then every arithmetic subgroup of* $\mathbf{G}$ *is Zariski dense in* $\mathbf{G}$.

3.3 Unipotent Arithmetic Groups and Their Unipotent Homomorphisms

We call a homomorphism $\rho\colon H \to \mathbf{F}$ of a group H into an algebraic group $\mathbf{F}$ *unipotent*, if $\rho(H) \subset \mathbf{F}^{(u)}$. Below in Lemma 3.3.4 we shall state certain properties of unipotent homomorphisms of unipotent arithmetic groups. But we shall quote three other lemmas beforehand.

(3.3.1) Lemma. (See [Rag 5], Theorem 2.11.) *Let N and V be two connected simply connected nilpotent Lie groups and let H be a closed subgroup of the group N such that the quotient space $H \setminus N$ is compact. Then every continuous homomorphism $\rho\colon H \to V$ can be extended uniquely to a continuous homomorphism $\tilde{\rho}\colon N \to V$.*

(3.3.2) Lemma. (See [Mal], Lemma 4.) *Let N be a connected simply connected nilpotent Lie group and let Γ be a cocompact lattice in N. Then there exist cyclic subgroups Γ_i, $1 \leq i \leq n = \dim N$, such that every element $\gamma \in \Gamma$ is representable in the form $\gamma = \gamma_1 \cdot \ldots \cdot \gamma_n$, $\gamma_i \in \Gamma_i$, $1 \leq i \leq n$.*

(3.3.3) Lemma. *Let* $\mathbf{W}$ *be a unipotent* $\mathbb{R}$*-group.*

(i) *The nilpotent Lie group* $\mathbf{W}(\mathbb{R})$ *is connected and simply connected.*

(ii) *If* $\mathbf{W}'$ *is a unipotent* $\mathbb{R}$*-group and* $f\colon \mathbf{W}(\mathbb{R}) \to \mathbf{W}'(\mathbb{R})$ *is a continuous homomorphism, then f extends uniquely to an* $\mathbb{R}$*-morphism* $\tilde{f}\colon \mathbf{W} \to \mathbf{W}'$.

(iii) *If* $\mathbf{W}$ *is defined over* $\mathbb{Q}$*, then every arithmetic (i.e. commensurable with* $\mathbf{W}(\mathbb{Z})$*) subgroup of* $\mathbf{W}(\mathbb{Q})$ *is Zariski dense in* $\mathbf{W}$.

This lemma is a consequence of the existence, for every unipotent $\mathbb{R}$-group $\mathbf{V}$, of the $\mathbb{R}$-isomorphism $\ln\colon \mathbf{V} \to \operatorname{Lie}(\mathbf{V})$ of algebraic varieties which is defined over $\mathbb{Q}$ whenever $\mathbf{V}$ is defined over $\mathbb{Q}$.

(3.3.4) Lemma. *Let* $\mathbf{W}$ *be a unipotent* $\mathbb{Q}$*-group, let Γ be an arithmetic (i.e. commensurable with* $\mathbf{W}(\mathbb{Z})$*) subgroup of the group* $\mathbf{W}(\mathbb{Q})$*, k a local field,* $\mathbf{F}$ *an algebraic k-group, and let $\rho\colon \Gamma \to \mathbf{F}(k)$ be a unipotent homomorphism.*

(a) *If k is $\mathbb{R}$ or $\mathbb{C}$, then ρ extends uniquely to a rational homomorphism $\tilde{\rho}\colon \mathbf{W} \to \mathbf{F}$.*
(b) *If k is isomorphic neither to $\mathbb{R}$ nor to $\mathbb{C}$, i.e., if k is totally disconnected, then the subgroup $\rho(\Gamma)$ is relatively compact in $F(k)$.*

Proof. By Theorem 3.2.7(b) the quotient space $\Gamma \setminus \mathbf{W}(\mathbb{R})$ is compact. So we see that in case $k = \mathbb{R}$, (a) follows from Lemmas 3.3.1 and 3.3.3. The case $k = \mathbb{C}$ reduces to the case $k = \mathbb{R}$ by passing from $\mathbf{F}$ to $R_{\mathbb{C}/\mathbb{R}}\mathbf{F}$. Assertion (b) follows from the compactness of the quotient space $\Gamma \setminus \mathbf{W}(\mathbb{R})$, Lemma 3.3.2, and from the fact that for a totally disconnected field k any cyclic unipotent subgroup of $\mathbf{F}(k)$ is relatively compact. □

4. Measure Theory and Ergodic Theory

In this section we shall use the notation and the notions of 0.34, 0.35 and 0.36.

4.1 Invariance of Sets and Maps Relative to Translations on Locally Compact Groups

(4.1.1) Lemma. *Let H be a locally compact σ-compact group, B a μ_H-measurable subset of H, where μ_H is a Haar measure on H, and let f be a measurable map of H into a semi-separated second countable topological space X. If we put*

$$H_1(B) = \{h \in H \mid \mu_H((hB)\triangle B) = 0\},$$
$$H_2(B) = \{h \in H \mid \mu_H((Bh)\triangle B) = 0\},$$
$$H_1(f) = \{h \in H \mid f(hh') = f(h') \text{ for almost all (relative to } \mu_H)h' \in H\},$$
$$H_2(f) = \{h \in H \mid f(h'h) = f(h') \text{ for almost all } h' \in H\},$$

then

(i) *the subgroups $H_1(B)$ and $H_2(B)$ are closed in H;*
(ii) *there exist two subsets B_1 and B_2 of H such that*

$$\mu_H(B\triangle B_1) = \mu_H(B\triangle B_2) = 0,$$
$$H_1(B)\cdot B_1 = B_1, \text{ and}$$
$$B_2 \cdot H_2(B) = B_2$$

(iii) *if $H_1(B) = H$ or $H_2(B) = H$, then either $\mu_H(B) = 0$ or $\mu_H(H - B) = 0$;*
(iv) *the subgroups $H_1(f)$ and $H_2(f)$ are closed in H;*
(v) *there exist two maps f_1, f_2 of H into X such that f_1 and f_2 agree almost everywhere with f, $f_1(hh') = f_1(h')$ for all $h \in H_1(f)$ and $h' \in H$, and $f_2(h'h) = f_2(h')$ for all $h \in H_2(f)$, $h' \in H$;*
(vi) *if $H_1(f) = H$ or $H_2(f) = H$, then the map f is μ_H-constant.*

Proof. (i) Since H is σ-compact, $H - B$ can be represented in the form of a countable union of sets C_j, $1 \leq j < \infty$, of finite measure. Then $H_1(B) = \{h \in H \mid$

$\mu_H((hB) \cap C_j) = 0$ for all $j \in \mathbb{N}^+\}$ and, since the functions $f_j(h) = \mu_H((hB) \cap C_j)$ are continuous (see [Bou 3], Chapter VII, §1, Exercise 20), the subgroup $H_1(B)$ is closed. The fact that $H_2(B)$ is closed can be proved similarly.

(iv) Let $\{U_i \mid i \in \mathbb{N}^+\}$ be a countable base of open sets in X. Then, since X is semi-separated,

$$H_1(f) = \bigcap_{i \in \mathbb{N}^+} H_1(f^{-1}(U_i)).$$

It then follows from (i) that $H_1(f)$ is closed. Similarly we obtain that $H_2(f)$ is closed.

(v) Let us consider the following two sets

$$D = \{(h, h') \in H_1(f) \times H \mid f(hh') \neq f(h')\} \text{ and}$$
$$C = \{h' \in H \mid \mu_{H_1(f)}\{h \in H_1(f) \mid f(hh') \neq f(h')\} = 0\}.$$

Since X is semi-separated, we have that

$$D = \{(h, h') \in H_1(f) \times H \mid$$
$$(hh', h') \in \bigcup_{i \in \mathbb{N}^+} [(f^{-1}(U_i) \times (H - f^{-1}(U_i))) \cup ((H - f^{-1}(U_i)) \times f^{-1}(U_i))]\}.$$

So D is $\mu_{H_1(f)} \times \mu_H$-measurable, for f is measurable. But for each $h \in H_1(f)$ we have $\mu_H\{h' \in H \mid (h, h') \in D\} = 0$. Therefore, by Fubini's theorem $\mu_H(H - C) = 0$. It follows from the definition of the set C that $f(h') = f(h'')$ for all $h', h'' \in C$, $h'' \in H_1(f) \cdot h'$. Therefore for each $h \in H_1(f) \cdot C$ the element $f_1(h) = f(C \cap (H_1(f) \cdot h)) \in X$ is well-defined. Thus, since $\mu_H(H - C) = 0$, we have that f_1 is the desired map. The existence of the map f_2 is proved analogously.

Assertion (ii) is a consequence of (v) applied to the characteristic function of the set B. Assertion (iii) and (vi) follows from (ii) and (v). □

(4.1.2) Lemma. *Let H be a locally compact σ-compact group, D a closed subgroup of H, π: $H \to H/D$ the natural projection, let R be a subset of H/D and let p be a map of the space H/D into a topological space. Then*

(i) *the set $\pi^{-1}(R)$ is μ_H-measurable if and only if R is $\mu_{H/D}$-measurable;*
(ii) *the map $p \circ \pi$ is μ_H-measurable if and only if p is $\mu_{H/D}$-measurable* (see [Bou 3], Chapter VII, §2, Proposition 6 and Lemma 4).

Assertion (i) is a consequence of (ii) applied to the characteristic function of the set R.

(4.1.3) Lemma. *With H, D and π as in Lemma 4.1.2, let $B \subset H$ be a μ_H-measurable set such that $\mu_H((Bd) \triangle B) = 0$ for all $d \in D$ and let f be a μ_H-measurable map of the group H into a second countable semi-separated topological space X such that for every $d \in D$ the equality $f(hd) = f(h)$ holds for almost all $h \in H$. Then*

(i) *there exists a $\mu_{H/D}$-measurable subset $R \subset H/D$ such that $\mu_H(B \triangle (\pi^{-1}(R))) = 0$;*

(ii) *there exists a* $\mu_{H/D}$*-measurable map* $p\colon H/D \to X$ *such that* $p \circ \pi$ *agrees almost everywhere with* f.

Proof. In view of (ii) and (v) of Lemma 4.1.1 we may assume, modifying B and f on a null set, that $BD = B$ and $f(hd) = f(h)$ for all $h \in H$ and $d \in D$. Then $B = \pi^{-1}(R)$ and $f = p \circ \pi$, where $R = \pi(B)$ and $p(y) = f(\pi^{-1}(y)) \in X$, $y \in H/D$. But by Lemma 4.1.2, R and p are $\mu_{H/D}$-measurable. □

4.2 Algebras of Measurable Sets

(See [Rohli 1] and [Rohli 2].) Let X be a measure space with a σ-finite measure μ. We shall divide the family of all measurable subsets of X into the classes of sets which differ from one another on a null set and denote the set of these classes by $\mathfrak{M}(X, \mu)$. The operations of countable union, countable intersection, and complementation can be carried over from sets to classes and turn $\mathfrak{M}(X, \mu)$ into an *algebra.* Any subset of this algebra which is closed under these operations is said to be a *subalgebra* of the algebra $\mathfrak{M}(X, \mu)$. The subalgebra of the class of null sets and the class of their complements is called *trivial.* The map of the set of measurable subsets of X sending every measurable subset of X to its class will be denoted by β. We set for each $B \in \mathfrak{M}(X, \mu)$, $\mu(B) = \mu(B')$, where $B' \in \beta^{-1}(B)$.

We say that a sequence $B_1, \dots, B_n, \dots$ of measurable subsets of X (resp. elements of the algebra $\mathfrak{M}(X, \mu)$) *converges in measure* μ to a measurable subset $B \subset X$ (resp. to $B \in \mathfrak{M}(X, \mu)$) if

$$\lim_{n \to \infty} \mu(Y \cap (B_n \triangle B)) = 0$$

(resp. $\lim_{n\to\infty} \mu(\beta(Y) \cap (B_n \Delta B)) = 0$) for every subset $Y \subset X$ of finite μ-measure. If X is a locally compact σ-compact space and μ is a regular Borel measure on X, then $\{B_n\}_{n \in \mathbb{N}^+}$ converges to B if and only if the indicated limits are equal to zero for all compact sets $Y \subset X$. It can easily be shown that if two σ-finite measures μ and ν on X are equivalent and the sequence $\{B_n\}$ converges in measure μ to B, then it also converges in measure ν; we shall not make use of this fact, and so we omit the proof. Every subalgebra $\mathscr{B}$ of the algebra $\mathfrak{M}(X, \mu)$ is closed relative to the convergence defined above. (Indeed, let $B_n \in \mathscr{B}$, $n \in \mathbb{N}^+$ and the sequence $\{B_n\}$ converges in measure μ to $B \in \mathfrak{M}(X, \mu)$. Since the measure μ is σ-finite, $X = \bigcup_{i \in \mathbb{N}^+} Y_i$, where $\mu(Y_i) < \infty$. By passing to a subsequence of $\{B_n\}$, we then may assume that $\mu(\beta(Y_i) \cap (B_n \triangle B)) < 2^{-n}$ for all $i \in \mathbb{N}^+$ and for all $n > i$. But we then have $B \cap \beta(Y_i) = \bigcap_{m \in \mathbb{N}^+} (\bigcup_{n > m} B_n \cap \beta(Y_i))$ for all $i \in \mathbb{N}^+$ and, since $X = \bigcup_{i \in \mathbb{N}^+} Y_i$, we have $B = \bigcap_{m \in \mathbb{N}^+} (\bigcup_{n > m} B_n) \in \mathscr{B}$).

Let X' be a measure space with a σ-finite measure μ' and let $f\colon X \to X'$ be a measurable map such that the preimage of any null set is a null set. For each $B \in \mathfrak{M}(X', \mu')$ one can then define $f^*(B) = \beta(f^{-1}(\beta^{-1}(B))) \in \mathfrak{M}(X', \mu')$. The subalgebra $f^*(\mathscr{B})$ of the algebra $\mathfrak{M}(X, \mu)$ is said to be the *inverse image of the subalgebra* $\mathscr{B} \subset \mathfrak{M}(X', \mu')$ *under the map* f.

If X is a locally compact σ-compact group and $H \subset H'$ are two closed subgroups in X, we set $\mathfrak{M}(X) = \mathfrak{M}(X, \mu_X)$ and $\mathfrak{M}(X/H) = \mathfrak{M}(X/H, \mu_{X/H})$

and denote by $\mathfrak{M}(X,H')$ and $\mathfrak{M}(X/H,H')$ the inverse images of the algebra $\mathfrak{M}(X/H')$ under the natural mappings $X \to X/H'$ and $X/H \to X/H'$. Let us denote, respectively, by $xM_1 = \beta(x\beta^{-1}(M_1)) \in \mathfrak{M}(X/H)$ and $xM \in \mathfrak{M}(X)$ the left translations of the classes $M_1 \in \mathfrak{M}(X/H)$ and $M \in \mathfrak{M}(X)$. We define similarly $\mathfrak{M}(H \setminus X) = \mathfrak{M}(H \setminus X), \mu_{H\setminus X})$ and the right translations M_1x and Mx of the classes $M_1 \in \mathfrak{M}(H \setminus X)$ and $M \in \mathfrak{M}(X)$. Clearly $\mathfrak{M}(X,H')$ is the inverse image of the subalgebra $\mathfrak{M}(X/H,H')$ under the natural map $X \to X/H$. It follows from Lemma 4.1.1(ii) that $\mathfrak{M}(X,H) = \{M \in \mathfrak{M}(X) \mid MH = M\}$.

4.3 Certain Properties of Haar Measure

(4.3.1) Proposition. (See [Bou 3], Chapter VII, §2, Proposition 13.) *Let H be a unimodular locally compact, σ-compact group and let X, Y be two closed subgroups of H such that the subgroup $X \cap Y$ is compact and the set $\Omega = X \cdot Y$ is open in H. Then the restriction to Ω of the measure μ_H is, up to a constant multiple, the image of the measure $\mu_X \times \mu_Y$ under the map $(x,y) \mapsto xy^{-1}$ of the product $X \times Y$ onto Ω.*

We shall use in the sequel the following simple

(4.3.2) Lemma. *Let H be a locally compact group and let $F \subset H$ be a closed subgroup such that H/F (resp. $F \setminus H$) has an H-invariant measure μ. Then for every compact set $K \subset H$ there exists a constant $c(K) > 0$ such that $\mu_H(\pi^{-1}(A) \cap K) \le c(K)\mu(A)$ for each measurable set $A \subset H/F$ (resp. $A \subset F \setminus H$), where $\pi\colon H \to H/F$ (resp. $\pi\colon H \to F \setminus H$) is the natural projection.*

Proof. Since $\Delta_H \cdot \mu_H$ is the right Haar measure, and the modular function $\Delta_H\colon H \to \mathbb{R}^+$ is continuous, if suffices to consider the case where $A \subset H/F$. Then (see [Bou 3], Chapter VII, §2, Lemma 6)

$$\mu_H(\pi^{-1}(A) \cap K) = a \int_A f(x)d\mu(x),$$

where $a > 0$ does not depend on A and K, $f(\pi(y)) = q(y)$, and $q(y) = \mu_F(F \cap (y^{-1}K))$ (observe that $q(yz) = q(y)$ for all $y \in H$ and $z \in F$, since the measure μ_F is left-invariant). Since K is compact,

$$b \stackrel{\text{def}}{=} \sup_{x \in A} f(x) < \infty.$$

Thus, $c(K) = ab$ as required. □

4.4 Borel Sections

A Borel map $f\colon X \to X'$ of locally compact spaces is said to be *regular* if the image $f(K)$ of every compact set $K \subset X$ is relatively compact in X'. A map f is called a *regular Borel isomorphism* if f is bijective and both f and f^{-1} are regular Borel.

(4.4.1) Theorem. (Existence of Borel sections, see [F-G], Theorem 1 and [Mack], Lemma 1.1.) *Let G be a locally compact second countable group, H a closed subgroup of G, and let $\pi\colon G \to G/H$ and $\pi'\colon G \to H \setminus G$ be the natural maps. There exists a regular Borel section $\varphi\colon G/H \to G$ (resp. $\varphi'\colon H \setminus G \to G$) such that $\pi \circ \varphi = \mathrm{Id}$ (resp. $\pi' \circ \varphi' = \mathrm{Id}$) and $\varphi(G/H)$ (resp. $\varphi(H \setminus G)$) is a Borel subset in G.*

(4.4.2) Corollary. *With G, H, π and π' as in Theorem 4.4.1, we set $X = G/H$ and $X' = H \setminus G$. There exist regular Borel maps $\vartheta, \vartheta'\colon G \to H$ such that*

(a) *$\vartheta(gh) = \vartheta(g)h$ and $\vartheta'(hg) = h\vartheta'(g)$ for all $g \in G$, $h \in H$ and*

(b) *the maps $f\colon G \to X \times H$ and $f'\colon G \to H \times X'$ sending $g \in G$ to $(\pi(g), \vartheta(g)) \in X \times H$ and $(\vartheta'(g), \pi'(g)) \in H \times X'$ respectively are regular Borel isomorphisms.*

To prove this corollary it suffices to set $\vartheta(g) = \varphi(\pi(g))^{-1}g$ and $\vartheta'(g) = g\varphi'(\pi(g))^{-1}$, where φ and φ' are from Theorem 4.4.1.

(4.4.3) Corollary. *Let G be a locally compact second countable group and let H, F be two closed subgroups of G with $G = HF$.*

(a) Let $C \subset G$ be a μ_G-measurable subset such that $CF = C$. Then the set $C \cap H$ is μ_H-measurable, and furthermore, if $\mu_G(C) = 0$, then $\mu_H(C \cap H) = 0$.

(b) If $\omega\colon G \to X$ is a μ_G-measurable map into a topological space X such that $\omega(gx) = \omega(g)$ for all $g \in G$ and $x \in F$, then $\omega|_H$ is μ_H-measurable.

Proof. By Theorem 4.4.1 there exists a Borel subset $Y \subset F$ such that $F = (H \cap F)Y$ and for each $x \in F$ the representation $x = hy$, $h \in H \cap F$, $y \in Y$, is unique. Since $G = HF$ we then have, that $G = HY$, and for each $g \in G$ the representation $g = hy$, $h \in H$, $y \in Y$, is unique. Let us identify Y with $H \setminus G$ associating to $y \in Y$ the coset $Hy \in H \setminus G$. Since the measures μ_G and μ_H are left-invariant, there exists a measure μ on $Y = H \setminus G$ such that μ_G is the image of the measure $\mu_H \times \mu$ under the map $(h, y) \mapsto hy$, $h \in H$, $y \in Y$ (notice that $\mu(B)$, for $B \subset Y$, can be defined as the proportionality coefficient to the left-invariant measures μ_B and μ_H, where $\mu_B(A) = \mu(AB)$, $A \subset H$). This proves (a) and (b) is a straightforward consequence of (a). □

4.5 Ergodic Actions

Given a left action of a topological group G on a measure space X with σ-finite G-quasi-invariant measure μ we say that the action of G on X is *ergodic* if the following equivalent conditions are satisfied:

(a) if $Y \subset X$ is measurable and $\mu(Y \triangle (gY)) = 0$ for all $g \in G$, then either $\mu(Y) = 0$ or $\mu(X - Y) = 0$;

(a′) if $B \in \mathfrak{M}(X, \mu)$ and $gB = B$ for all $g \in G$, then either $\mu(B) = 0$ or $\mu(X - B) = 0$.

The ergodicity of a right action is defined analogously.

An automorphism L of the space X is said to be *ergodic* if the action of the group $\{L^m \mid m \in \mathbb{Z}\}$ on X is ergodic. It is easily seen that the ergodicity of the automorphism L is equivalent to the following property: if $Y \subset X$ is measurable and $L(Y) = Y$, then either $\mu(Y) = 0$ or $\mu(X - Y) = 0$.

(4.5.1) Proposition. *Let G be a locally compact σ-compact group and let Λ be a dense subgroup of G. Let G be viewed as a measure space with measure μ_G. Then the action of the group Λ by left (resp. right) translations on G is ergodic, i.e., if $B \in \mathfrak{M}(G)$ and $\Lambda B = B$ (resp. $B\Lambda = B$), then either $\mu_G(B) = 0$ or $\mu_G(G - B) = 0$.*

Proof. Let $B \in \mathfrak{M}(G)$ and $\Lambda B = B$ (resp. $B\Lambda = B$). Then, since Λ is dense in G, by Lemma 4.1.1(i) we have $GB = B$ (resp. $BG = B$). All that remains now is to apply Lemma 4.1.1(iii). □

4.6 Poincaré Recurrence Theorem and Ergodic Theorems

(See [Bi], Chapter I, [Hal 2] and [K-S-F], Chapter I.)

(4.6.1) Theorem. (Poincaré recurrence theorem.) *Let $L\colon X \to X$ be an automorphism of a finite measure space X, and let $U \subset X$ be a measurable set of positive measure. Then almost all $x \in U$ return to U infinitely often, i.e., for almost all $x \in U$ the set $\{n \in \mathbb{N}^+ \mid L^n x \in U\}$ is infinite.*

(4.6.2) Corollary. *Let L be an automorphism of a measure space X with finite measure and let f be a non-negative measurable function on X. Then for almost all $x \in X$*

$$\liminf_{m\to+\infty}(1/m)f(L^m x) = 0 \textit{ and } \liminf_{m\to-\infty}(-1/m)f(L^m x) = 0.$$

To prove this corollary it suffices to apply the Poincaré recurrence theorem to the sets

$$U_m = \{x \in X \mid f(x) \leq m\},\ m \in \mathbb{N}^+.$$

(4.6.3) Corollary. *Let G be a locally compact, σ-compact group, $d \in G$, H a closed subgroup of G such that $H \setminus G$ has a finite G-invariant measure μ, and let f be a measurable function on G such that $f(hg) = f(g)$ for all $g \in G$ and $h \in H$. Then*

$$\liminf_{m\to+\infty}(1/m)f(gd^m) = 0$$

and

$$\liminf_{m\to-\infty}(-1/m)f(gd^m) = 0$$

for almost all $g \in G$.

This corollary follows from Corollary 4.6.2 applied to the measure space automorphism $x \mapsto xd$ of $(H \setminus G, \mu)$ and to the function $\hat{f} = f \circ \pi^{-1}$, where $\pi\colon G \to H \setminus G$ is the natural map.

(4.6.4) Theorem. (Statistical ergodic theorem.) *Let X be a measure space with finite measure μ, $L\colon X \to X$ a measure space automorphism, and $f \in L_1(X,\mu)$. For $n \in \mathbb{N}^+$, let $f_n(x) = (1/n)\sum_{i=1}^{n} f(L^i x)$. Then as $n \to \infty$ the sequence $\{f_n\}$ converges in $L_1(X,\mu)$ to an L-invariant integrable function $\tilde{f}$ (the L-invariance of $\tilde{f}$ means that $\tilde{f}(Lx) = \tilde{f}(x)$ for almost all $x \in X$). If L is ergodic, then $\tilde{f}(x) = (1/\mu(X))\int_X f(x)d\mu(x)$ for almost all $x \in X$.*

(4.6.5) Theorem. (Birkhoff individual ergodic theorem.) *Let L be an automorphism of a measure space X with finite measure μ and let $f \in L_1(X,\mu)$. Then there exists an L-invariant function $\tilde{f} \in L_1(X,\mu)$ such that*

$$\int_X \tilde{f}(x)d\mu(x) = \int_X f(x)d\mu(x)$$

and

$$\lim_{n\to\infty}(1/n)\sum_{i=1}^{n} f(L^i x) = \tilde{f}(x)$$

for almost all $x \in X$.

If the automorphism L is ergodic, then $\tilde{f}(x) = (1/\mu(X))\int_X f(x)d\mu(x)$ for almost all $x \in X$.

Remark. A part of this (and the preceding theorem) is true for σ-finite measures μ.

5. Unitary Representations and Amenable Groups

In this section G denotes a locally compact second countable group.

5.1 The Dual Space of the Group G and Decomposition of Unitary Representations

(See [Ki], 7.3 and 8.4.)

(5.1.1) We denote by $\tilde{G}$ (resp. $\hat{G}$) the set of equivalence classes of unitary (resp. irreducible unitary) representations of the group G on separable Hilbert spaces. The set $\hat{G}$ is said to be *dual space* of the group G.

One can define a topology on $\tilde{G}$ as follows. Let ρ be a unitary representation of the group G on a Hilbert space V. Let us consider for a compact set $K \subset G$, a finite collection of vectors $\xi_1,\ldots,\xi_n \in V$ and a number $\varepsilon > 0$ the subset $U(K,\xi_1,\ldots,\xi_n,\varepsilon)$ of $\tilde{G}$ consisting of the classes $[\rho']$ of those representations ρ' in the space of which there exist the vectors $\eta_1,\ldots,\eta_n$ such that

$$|\langle\rho(g)\xi_i,\xi_j\rangle - \langle\rho'(g)\eta_i,\eta_j\rangle| < \varepsilon$$

for all $g \in K$, $1 \leq i, j \leq n$. We may consider the family of the sets $U(K, \xi_1, \dots, \xi_n, \varepsilon)$ as a neighbourhood base at the point $[\rho] \in \tilde{G}$. If $[\rho'] \in \tilde{G}$ belongs to the closure of the singleton $\{[\rho]\}$, $[\rho] \in \tilde{G}$, we say that the representation ρ' is *weakly contained* in ρ. We remark that if the representation ρ' is contained in ρ, then it is weakly contained in ρ. So the space $\tilde{G}$ is not Hausdorff.

If the group G is commutative, then each of its irreducible unitary representations is 1-dimensional. In this case the tensor product operation defines a commutative group structure on $\hat{G}$, and this group is naturally identified with the group of characters of the group G.

If the group G is compact, then the dual space $\hat{G}$ is discrete (see [Ki], §9, Corollary 2 to Theorem 2).

(5.1.2) Let X be a measure space with measure μ and V be a separable Hilbert space. Assume that for each $x \in X$ a closed subspace H_x of V has been chosen so that the map $x \mapsto H_x$ is measurable, i.e., for all $v, w \in V$ the real-valued function $x \mapsto \langle P_x v, w\rangle$ is measurable, where P_x is the orthogonal projector of the space V onto H_x. Let us consider the space of measurable functions $f\colon X \to V$ such that

$$\int_X \|f(x)\|^2 d\mu(x) < \infty$$

and $f(x) \in H_x$ for all $x \in X$. We define on this space an inner product as follows

$$\langle f_1, f_2\rangle = \int_X \langle f_1(x), f_2(x)\rangle d\mu(x),$$

and identify functions which agree almost everywhere. Thus we obtain a Hilbert space H, which is denoted by $\int_X H_x d\mu(x)$ and is called the *continuous sum* of the Hilbert spaces H_x.

Assume that for each $x \in X$ a unitary representation ρ_x of the group G on the space H_x is given. We say that a unitary representation ρ of the group G on the space $H = \int_X H_x d\mu(x)$ is a *continuous sum* (or an *integral*) of the representations ρ_x in measure μ, if for each $g \in G$ and each $f \in H$, $(\rho(g)f)(x) = \rho_x(g)f(x)$ for almost all $x \in X$. In this case we write $\rho = \int_X \rho_x d\mu(x)$. If X is finite (or countable) and the measure μ is discrete, then a continuous sum of representations is just a direct sum of finitely (or countably) many representations.

(5.1.3) Theorem. (See [Ki], 8.4, the Corollary to Theorem 2.) *Every unitary representation ρ of the group G on a separable Hilbert space can be decomposed into a continuous sum $\int_X \rho_x d\mu(x)$ of irreducible unitary representations.*

(5.1.4) We say that a unitary representation ρ' is a *multiple* of a unitary representation ρ, if it is a direct sum of finitely or countably many unitary representations equivalent to ρ. For commutative groups the following decomposition theorem for unitary representations holds.

Theorem. (See [Ki], 8.4, Theorem 3 and Problem 4.) *If G is commutative and $[\rho] \in \tilde{G}$, then the unitary representation ρ can be decomposed into a continuous sum $\rho = \int_{\hat{G}} \rho_\chi d\mu(\chi)$, where ρ_χ is a multiple of the character $\chi \in \hat{G}$ (i.e. $\rho_\chi(g)x = \chi(g)x$).*

5.2 Unitary Induced Representations

(See [Ki], 13.2.) Let H be a closed subgroup of the group G and let U be a unitary representation of the group H on a space V. Let us set for each $h \in H$,

$$U_0(h) = (\Delta_H(h)/\Delta_G(h))^{1/2} U(h),$$

where, as in 0.36, Δ_G and Δ_H denote the modular functions of the groups G and H respectively. Let us consider the space $L(G, H, U_0)$ of μ_G-measurable vector-functions F on G taking values in V such that

$$F(hg) = U_0(h)F(g), \quad h \in H, \quad g \in G. \tag{1}$$

Denote by μ'_G (resp. μ'_H) a right Haar measure on G (resp. H). Pick a non-negative continuous function q on G such that for each $g \in G$, $Hg \cap \operatorname{supp} q$ is compact and $\int_H q(hg)d\mu'_H(h) = 1$. Let $L_2(G, H, U)$ be the subspace of $L(G, H, U_0)$ consisting of F such that $\int_G \|F(g)\|^2 q(g) d\mu'_G(g) < \infty$. On $L_2(G, H, U)$ we introduce an inner product as follows:

$$\langle F_1, F_2 \rangle = \int_G \langle F_1(g), F_2(g) \rangle q(g) d\mu'_G(g), \tag{2}$$

where $F_i \in L_2(G, H, U)$ $(i = 1, 2)$. We remark that the subspace $L_2(G, H, U)$ and the integral on the right side of (2) do not depend upon the choice of the function q. Now define a representation of the group G on the space $L_2(G, H, U)$ as follows:

$$(\rho(g)F)(x) = F(xg), \quad F \in L_2(G, H, U) \text{ and } x, g \in G. \tag{3}$$

The representation ρ is unitary and is called the representation of the group G *induced in the sense of Mackey* from the representation U and is denoted by $\operatorname{Ind}(G, H, U)$ or simply by $\operatorname{Ind} U$.

The definition of the space $L_2(G, H, U)$ and the inner product on it can be simplified, if $H \setminus G$ carries a G-invariant measure μ. To this end choose a regular Borel map $\varphi\colon H \setminus G \to G$ such that $\pi \circ \varphi = \mathrm{Id}$, where $\pi\colon G \to H \setminus G$ is the natural projection. We set $X = \varphi(H \setminus G)$ and define the measure μ' on X as the image of the measure μ under the map φ. Then $L_2(G, H, U)$ consists of μ_G-measurable functions $F\colon G \to V$ with

$$F(hg) = U(h)F(g), \quad h \in H, \quad g \in G \text{ and}$$

$$\int_X \|F(x)\|^2 d\mu'(x) < \infty.$$

The inner product on $L_2(G, H, U)$ is defined by the formula

$$\langle F_1, F_2\rangle = \int_X \langle F_1(x), F_2(x)\rangle d\mu'(x).$$

If the group G is unimodular and its subgroup H is discrete, we may take as X the left Borel fundamental domain for H and as μ' the restriction of the measure μ_G to X.

If $H \setminus G$ has a G-invariant measure μ and U stands for the trivial 1-dimensional representation I_H of the group H, then $\operatorname{Ind} U$ is the so called *quasi-regular representation* τ of the group G in the space $L_2(H \setminus G, \mu)$. This representation can also be defined by

$$(\tau(g)f)(x) = f(xg), \text{ for } f \in L_2(H \setminus G, \mu),\ x \in H \setminus G,\ g \in G.$$

In particular, the regular representation of the group G can be written as follows: $\operatorname{Ind}(G, \{e\}, I_{\{e\}})$.

(5.2.1) The following "inducing through" principle holds. Let H be a closed subgroup of G and let K be a closed subgroup of H. Then for every unitary representation U of the group K, the following equivalence holds:

$$\operatorname{Ind}(G, H, \operatorname{Ind}(H, K, U)) \sim \operatorname{Ind}(G, K, U)$$

or symbolically

$$\operatorname{Ind}_H^G \operatorname{Ind}_K^H = \operatorname{Ind}_K^G,$$

where Ind_H^G stands for the operation which sends U to $\operatorname{Ind}(G, H, U)$.

Denote by R_G the regular representation of the group G. Since $R_G = \operatorname{Ind}(G, \{e\}, I_{\{e\}})$ and $R_H = \operatorname{Ind}(H, \{e\}, I_{\{e\}})$, by the "inducing through" principle we have $R_G = \operatorname{Ind}(G, H, R_H)$.

(5.2.2) If two unitary representations U_1 and U_2 of the subgroup H are equivalent, then the representations $\operatorname{Ind} U_1$ and $\operatorname{Ind} U_2$ of the group G are also equivalent. Therefore the inducing operation can be transferred from representations to their equivalence classes, i.e., a map Ind: $\tilde{H} \to \tilde{G}$ can be defined. This map is continuous relative to the topology defined in 5.1.1 (see [Fe], Theorem 4.1). Therefore, if U_1 is weakly contained in U_2, then $\operatorname{Ind} U_1$ is weakly contained in $\operatorname{Ind} U_2$.

(5.2.3) Theorem. (Mackey theorem, see [Ki], 13.3, Theorem 1.) *Let N be a closed commutative normal subgroup of the group G. We define the natural action of G on the group of characters $\hat{N}$ of the group N by setting*

$$(g\chi)(n) := \chi(g^{-1}ng),\ g \in G,\ \chi \in \hat{N},\ n \in N.$$

Assume that every orbit $G\chi$, $\chi \in \hat{N}$, is locally closed in $\hat{N}$. Then every irreducible unitary representation ρ of the group G is equivalent to a representation of the form $\operatorname{Ind}(G, H, U)$, *where H is the stabilizer of a character $\chi \in \hat{N}$, $U \subset \hat{H}$, and the restriction of the unitary representation U to N is a multiple of the character χ (i.e.*

$U(n)x = \chi(n)x$ for all $n \in N$ and for all vectors x in the space of the representation U).

5.3 Group Algebras of Locally Compact Groups

(See [Bou 3], Chapter VIII, §§4, 5 and [Ki], 10.2.)

(5.3.1) We recall that the *convolution* $f_1 * f_2$ of two functions $f_1, f_2 \in L_1(G, \mu_G)$ is defined by the formula

$$(f_1 * f_2)(g) = \int_G f_1(h) f_2(h^{-1}g) d\mu_G(h).$$

With the operation of convolution of functions as the product, $L_1(G, \mu_G)$ becomes an associative algebra. For all $f_1, f_2 \in L_1(G, \mu_G)$, we have

$$\int_G (f_1 * f_2)(g) d\mu_G(g) = \int_G f_1(g) d\mu_G(g) \int_G f_2(g) d\mu_G(g).$$

(5.3.2) Define for a complex valued function $f \in L_1(G, \mu_G)$, the function $f^* \in L_1(G, \mu_G)$ by setting $f^*(g) = \Delta_g(g) \cdot \overline{f(g^{-1})}$, where Δ_G denotes the modular function of the group G, and the bar denotes complex conjugation. The map $f \mapsto f^*$ is an involution of the algebra $L_1(G, \mu_G)$, i.e., for all $x, y \in L_1(G, \mu_G)$ and λ, $\mu \in \mathbb{C}$ we have

1) $$(\lambda x + \mu y)^* = \lambda, \mu \in \mathbb{C},$$
2) $$(xy)^* = y^* x^*,$$
3) $$(x^*)^* = x.$$

If ρ is a unitary representation of the group G, then

$$\rho(f_1 * f_2) = \rho(f_1)\rho(f_2) \text{ and } \rho(f^*) = \rho(f)^*,$$

where $f, f_1, f_2 \in L_1(G, \mu_G)$, and $\rho(f)^*$ denotes the adjoint of $\rho(f)$.

5.4 Positive Definite Functions

(5.4.1) A continuous complex-valued function φ on G is called *positive definite* provided that for each $n \in \mathbb{N}^+$ and for all $g_1, \ldots, g_n \in G$ the matrix $(\varphi(g_i^{-1} g_j))_{1 \le i,j \le n}$ is positive definite. In other words, the function φ is positive definite, if

$$\sum_{i,j=1}^{n} \alpha_i \bar{\alpha}_j \varphi(g_i^{-1} g_j) \ge 0$$

for all $n \in \mathbb{N}^+$, $g_1, \ldots, g_n \in G$ and $\alpha_1, \ldots, \alpha_n \in \mathbb{C}$.

(5.4.2) Let U be a continuous representation of the group G on a topological vector space V. A vector $\xi \in V$ is called *cyclic* with respect to U, if finite linear combinations of the vectors $U(g)\xi$, $g \in G$, form a dense subset of V.

(5.4.3) Theorem. (See [Dix], Theorem 13.4.5 (II).) *A complex function $\psi\colon G \to \mathbb{C}$ is continuous and positive definite if and only if there exist a continuous unitary representation ρ of the group G on a Hilbert space V and a vector $\xi \in V$ (which we may assume to be cyclic) such that*

$$\psi(g) = \langle \rho(g)\xi, \xi \rangle$$

for all $g \in G$.

5.5 Amenable Groups

(5.5.1) A group G is called *amenable* if one of the following equivalent conditions is satisfied:

(i) the space Ω of continuous bounded functions on G has a left-invariant mean (i.e. a linear functional m such that
  a) $m(f) \geq 0$, whenever $f \geq 0$,
  b) $m(1) = 1$ and
  c) $m(gf) = m(f)$ for all $g \in G$ and $f \in \Omega$);

(ii) for every continuous affine action of the group G on a compact convex subset W of a locally convex topological vector space there exists a G-invariant element $w \in W$ (an action of G on W is called affine, if $g(\lambda w_1 + (1-\lambda)w_2) = \lambda g w_1 + (1-\lambda) g w_2$ for all $w_1, w_2 \in W$, $0 \leq \lambda \leq 1$ and $g \in G$);

(iii) for every non-empty compact G-space X, there exists a G-invariant measure on X belonging to $\mathscr{P}(X)$.

In view of Theorem 3.3.1 in [Gre], conditions (i) and (ii) are equivalent. Since the set $\mathscr{P}(X)$ is convex and compact, (iii) is a consequence of (ii). Finally, since the map assigning to each measure $\nu \in \mathscr{P}(W)$ its centroid $\int_W w\,d\nu(w) \in W$ is G-equivariant, (iii) implies (ii).

(5.5.2) If a group G is amenable, then so is any closed subgroup of G (see [Gre], Theorem 2.3.2). If H is a closed normal subgroup of G and the groups H and G/H are amenable, then G is amenable (see [Gre], Theorem 2.2.3). If the group G is commutative, then G is amenable (see [Gre], Theorem 1.2.1). Finally, if G is compact, then G is amenable with $m(f) = \int_G f(g)\,d\mu_G(g)$. From the results stated above, it follows that if there exists a closed solvable normal subgroup H of G such that G/H is compact, then the group G is amenable.

(5.5.3) Theorem. (See [Gre], Theorem 3.5.2 and remarks following it.) *The following conditions for a group G are equivalent:*

(i) *The group G is amenable;*

(ii) *The trivial* 1*-dimensional representation of* G *is weakly contained in the regular representation of* G;

(iii) *Every irreducible unitary representation of the group* G *is weakly contained in the regular representation of* G.

Remark. The equivalence of conditions (ii) and (iii) was proved for the first time in [Go] and the equivalence of condition (i) to (ii) and (iii) in [Hul].

Chapter II. Density and Ergodicity Theorems

In Section 2, with the help of the results of Section 1, we shall prove the following

(2.5) Theorem. *Let G be a locally compact group and k a local field. Let $\mathscr{W}$ be a finite-dimensional vector space over k and let $T\colon\ G \to \mathbf{GL}(\mathscr{W})$ be a continuous representation. Let H be a subgroup of G and $g \in G$. Assume that the pair (H, g) has property (S) in G (we refer to 2.2 for the definition of property (S)). Denote by Ψ_g the closure of the subgroup generated by the set $\{x \in G \mid$ the closure of $\{g^i x g^{-i} \mid i \in \mathbb{Z}\}$ contains $e\}$. Then any $T(H)$-linear subspace $\mathscr{W}' \subset \mathscr{W}$ is also $T(\Psi_g)$-invariant.*

Theorem 2.5 is equivalent to the assertion that for every continuous homomorphism φ of the group G into the set of k-rational points of a k-group, the Zariski closure of the subgroup $\varphi(H)$ contains $\varphi(\Psi_g)$ (see Corollary 2.6).

In Section 4 the results of Section 2 are applied to the case where

$$G = \prod_{\alpha \in A} \mathbf{G}_\alpha(k_\alpha),$$

and A is a finite set, k_α is a local field, and $\mathbf{G}_\alpha$ is a connected semisimple k_α-group. The main result of Section 6 (Theorem 6.7(a)) asserts that, for a lattice Γ in

$$G = \prod_{\alpha \in A} \mathbf{G}_\alpha(k_\alpha),$$

$B \subsetneq A$, and the natural map

$$\mathrm{pr}_B\colon\ G \to G_B = \prod_{\alpha \in B} \mathbf{G}_\alpha(k_\alpha),$$

the closure in G_B of the subgroup $\mathrm{pr}_B(\Gamma)$ under certain hypotheses contains $\prod_{\alpha \in B} G_\alpha(k_\alpha)^+$. It is not hard to deduce the strong approximation theorem from this result (see 6.8). Theorem 6.7 is a simple consequence of a theorem on non-discrete closed subgroups with finite volume of quotient space proved in Section 5. In the proof we make use of certain results of Section 3, namely Corollary 3.10 which can be interpreted as the assertion that certain actions on homogeneous spaces have the property of "topological mixing". The results of Section 3 are also used in Section 7 to prove some propositions concerning the ergodicity of actions on quotient spaces.

1. Iterations of Linear Transformations

(1.0) Let k be a local field, equipped with a valuation $|\ |$, K an algebraic closure of k, and let k_s be the separable closure of the field k (i.e. the set of elements of K which are separable over k). The unique extension of $|\ |$ to the valuation on K will also be denoted by $|\ |$. Let $n \in \mathbb{N}^+$ and let $\mathscr{W}$ be a vector space over k of dimension n. For each extension l of k we set $\mathscr{W}_l = l \otimes_k \mathscr{W}$, identify $\mathscr{W}$ and $\mathscr{W}_k$, and assume $\mathscr{W}_l$ is naturally embedded in $\mathscr{W}_{l'}$, whenever $l \subset l'$. We say that a linear subspace $\mathscr{W}' \subset \mathscr{W}_k$ is defined over a field $l \subset K$ if $\mathscr{W}'$ is spanned by $\mathscr{W}'_l \overset{\text{def}}{=} \mathscr{W}' \cap \mathscr{W}_l$. Given a system $y = (y_1, \ldots, y_n)$ of linear coordinates in $\mathscr{W}$ viewed as a system of linear coordinates in $\mathscr{W}_K$ we define

$$\|w\| = \|w\|_y = \sum_{1 \le j \le n} |\, y_j(w)\, |, \ w \in \mathscr{W}_K$$

and

$$\|B\| = \|B\|_y = \sup_{w \in \mathscr{W}_K, w \neq 0} \|Bw\|_y / \|w\|_y, \ B \in \operatorname{End}(\mathscr{W}_K).$$

The function $w \mapsto \|w\|$ is a norm on $\mathscr{W}_K$, i.e., for all w, w_1, $w_2 \in \mathscr{W}_K$ and $\lambda \in K$

1) $\|w\| = 0$ if and only if $w = 0$;
2) $\|w_1 + w_2\| \le \|w_1\| + \|w_2\|$ and
3) $\|\lambda w\| = |\lambda| \|w\|$.

The function $B \mapsto \|B\|$ is a norm on $\operatorname{End}(\mathscr{W}_K)$, furthermore $\|BC\| \le \|B\| \, \|C\|$ for all B, $C \in \operatorname{End}(\mathscr{W}_K)$. It is easily seen that

(*) *if y and z are two coordinate systems in $\mathscr{W}$, then the norms $\|w\|_y$ and $\|w\|_z$ are equivalent in the usual sense: there exist two positive numbers c_1, c_2 such that*

$$c_1 \|w\|_z \le \|w\|_y \le c_2 \|w\|_z \text{ for all } w \in \mathscr{W}_K.$$

For a linear subspace $\mathscr{W}' \subset \mathscr{W}$, $\mathbf{P}(\mathscr{W}') \subset \mathbf{P}(\mathscr{W})$ will denote its projective space. We endow $\mathscr{W}$, $\operatorname{End}(\mathscr{W})$, and $\mathbf{P}(\mathscr{W})$ with the topologies induced from the topology of the field k, and denote by π the natural projection $\mathscr{W} - \{0\} \to \mathbf{P}(\mathscr{W})$. A transformation $B \in \mathbf{GL}(\mathscr{W})$ induces the transformation $\pi B \pi^{-1}$ of the space $\mathbf{P}(\mathscr{W})$, which will be denoted by B_π. By $E \in \operatorname{End}(\mathscr{W})$ we denote the identity transformation (defined by the unit matrix).

For a transformation $B \in \mathbf{GL}(\mathscr{W})$, we denote by $\Omega(B) \subset K$ the set of its eigenvalues and by $\mathscr{W}_\lambda(B)$ we denote the weight subspace corresponding to $\lambda \in \Omega(B)$, i.e.,

$$\mathscr{W}_\lambda(B) = \{w \in \mathscr{W}_K \mid (B - \lambda E)^n w = 0\}.$$

Since $k_s = K$, if $\operatorname{char} k = 0$, and $K = \{x^{p^{-m}} \mid x \in k_s,\ m \in \mathbb{N}^+\}$, if $\operatorname{char} k = p > 0$, there exists $m \in \mathbb{N}^+$ such that $\lambda^m \in k_s$ and hence, the subspaces $\mathscr{W}_{\lambda^m}(B^m)$ are defined over k_s for all $\lambda \in \Omega(B)$. But $\mathscr{W}_\lambda(B) = \mathscr{W}_{\lambda^m}(B^m)$. Therefore the subspaces $\mathscr{W}_\lambda(B)$ are defined over k_s. The space $\mathscr{W}$ is the direct sum of the subspaces $\mathscr{W}_\lambda(B)$. Let $p_{\lambda,B}\colon \mathscr{W} \to \mathscr{W}_\lambda(B)$ be the natural projection, i.e, $p_{\lambda,B}(w) \in \mathscr{W}_\lambda(B)$ and

$$w = \sum_{\lambda \in \Omega(B)} p_{\lambda,B}(w)$$

for each $w \in \mathscr{W}$. We put for $d \in \mathbb{R}$,

$$\begin{aligned}
\Omega_d(B) &= \{\lambda \in \Omega(B) \mid \ln|\lambda| = d\},\\
\mathscr{W}_d(B) &= [\bigoplus_{\lambda \in \Omega_d(B)} \mathscr{W}_\lambda(B)]_k,\\
\mathscr{W}_d^+(B) &= \bigoplus_{d' \geq d} \mathscr{W}_{d'}(B) \text{ and}\\
\mathscr{W}_d^-(B) &= \bigoplus_{d' \leq d} \mathscr{W}_{d'}(B).
\end{aligned}$$

Since the subspaces $\mathscr{W}_\lambda(B)$ are defined over k_s, and the set $\Omega_d(B)$ is invariant under all the automorphisms of the field K over k, the subspaces

$$\bigoplus_{\lambda \in \Omega_d(B)} \mathscr{W}_\lambda(B)$$

are defined over k. Therefore, $\mathscr{W}$ is a direct sum of the subspaces $\mathscr{W}_d(B)$. If $\Omega_d(B) \neq \emptyset$, we say that d is a *characteristic exponent* of the transformation B and $\mathscr{W}_d(B)$ is the *characteristic subspace* of B corresponding to d. We further set $\Delta_B = \bigcup_{d \in \mathbb{R}} \mathscr{W}_d(B)$ and call Δ_B the *characteristic cross* of the transformation B.

(1.1) Lemma. *Let $B \in \mathbf{GL}(\mathscr{W})$.*

(a) *There exist positive numbers $c_1 = c_1(B)$ and $c_2 = c_2(B)$ such that for all $\lambda \in \Omega(B)$, $w \in \mathscr{W}_\lambda(B)$, and $i \in \mathbb{Z}$, $i \neq 0$, we have that*

$$c_1|i|^{-n}|\lambda|^i\|w\| \leq \|B^i w\| \leq c_2|i|^n|\lambda|^i\|w\|. \tag{1}$$

(b) *If $\lambda \in \Omega(B)$, $w \in \mathscr{W}$ and $p_{\lambda,B}(w) \neq 0$, then there exists a number $c(w) > 0$, such that $\|B^i w\| \geq c(w)|\lambda|^i$, for each $i \in \mathbb{Z}$.*

(c) *If $|\lambda| = 1$ for all $\lambda \in \Omega(B)$, then for $w \in \mathscr{W} - \{0\}$ the set $\{B^i w \mid i \in \mathbb{Z}\}$ does not contain 0 in its closure.*

(d) *There exists a number $c = c(B) > 1$, such that for all $d \in \mathbb{R}$, $v \in \mathscr{W}_d^+(B)$, $w \in \mathscr{W}_d^-(B)$ and $i \in \mathbb{N}^+$ we have*

$$\begin{aligned}
&\|B^i v\| \geq c^{-1} i^{-n} \exp(di)\|v\| \text{ and}\\
&\|B^i w\| \leq c i^n \exp(di)\|w\|.
\end{aligned}$$

Proof. Expanding $B^i = (\lambda E + (B - \lambda E))^i$ by the binomial theorem and using the fact that $(B - \lambda E)^n w = 0$ for all $w \in \mathscr{W}_\lambda$ we obtain the right-hand inequality of (1). Its left-hand inequality follows from the right-hand one applied to B^{-1} and $B^i w$.

In view of (*) in 1.0, we can assume that, in the base of the coordinate system y, the transformation B can be reduced to Jordan normal form. Therefore, if $p_{\lambda,B}(w) \neq 0$, then there exists a natural number $j = j(w)$, $1 \leq j \leq n$, such that $y_j(B^i w) = \lambda^i y_j(w)$ for each $i \in \mathbb{Z}$. Thus we obtain (b). Now (c) is an immediate consequence of (b). Since

$$\mathscr{W}_d^+(B) = [\bigoplus_{\ln|\lambda|\geq d} \mathscr{W}_\lambda(B)]_k \text{ and } \mathscr{W}_d^-(B) = [\bigoplus_{\ln|\lambda|\leq d} \mathscr{W}_\lambda(B)]_k,$$

(a) implies (d). □

(1.2) Proposition. *Let $B \in \mathbf{GL}(\mathscr{W})$ and assume that $d \in \mathbb{R}$ is not a characteristic exponent of B. Then B attracts $P(\mathscr{W}) - \pi(\mathscr{W}_d^-(B))$ towards $\pi(\mathscr{W}_d^+(B))$.*

Proof. Since $\Omega_d(B) = \varnothing$, $\mathscr{W} = \mathscr{W}_d^+(B) \oplus \mathscr{W}_d^-(B)$ and there exists $\varepsilon > 0$ such that

$$\mathscr{W}_{d+\varepsilon}^+(B) = \mathscr{W}_d^+(B) \text{ and } \mathscr{W}_{d-\varepsilon}^-(B) = \mathscr{W}_d^-(B). \tag{1}$$

Let $p^+ \colon \mathscr{W} \to \mathscr{W}_d^+(B)$ and $p^- \colon \mathscr{W} \to \mathscr{W}_d^-(B)$ be the natural projections (i.e. $w = p^+(w) + p^-(w)$, $w \in \mathscr{W}$). Choose an arbitrary compact set $M \subset \mathbf{P}(\mathscr{W}) - \pi(\mathscr{W}_d^-(B))$. Then

$$\sup_{w \in \pi^{-1}(M)} (\|p^-(w)\| / \|p^+(w)\|) < \infty. \tag{2}$$

In view of (1), (2), and Lemma 1.1(d) we have that

$$\lim_{m \to +\infty} \sup_{w \in \pi^{-1}(M)} (\|p^-(B^m w)\| / \|p^+(B^m w)\|) = 0$$

and this completes the proof. □

(1.3) Proposition. *Let $B \in \mathbf{GL}(\mathscr{W})$ and let $w \in \mathscr{W} - \Delta_B$. Then the point $\pi(w)$ is wandering relative to B_π.*

Proof. Since $w \notin \Delta_B$, one can choose a number $d \in \mathbb{R}$ such that $\pi(w) \notin Y_d^+ \cup Y_d^-$, where $Y_d^+ = \pi(\mathscr{W}_d^+(B))$, and $Y_d^- = \pi(\mathscr{W}_d^-(B))$. Then $\pi(w) \in \mathbf{P}(\mathscr{W}) - Y_d^-$ and $\pi(w) \notin Y_d^+$. But B_π attracts $\mathbf{P}(\mathscr{W}) - Y_d^-$ towards Y_d^+ (see Proposition 1.2). Therefore $\pi(w)$ is a wandering point relative to B_π. □

(1.4) Lemma. *Let $B \in \mathbf{GL}(\mathscr{W})$ and let $Y \in \mathrm{End}(\mathscr{W})$, $Y \neq E$. Assume that $\Omega(B) = \Omega_d(B)$ for some $d \in \mathbb{R}$. Then the set $\{B^i Y B^{-i} \mid i \in \mathbb{Z}\}$ does not contain E in its closure.*

Proof. Define the linear transformation $\tilde{B}$ of the space $\mathrm{End}(\mathscr{W})$ by setting $\tilde{B}(X) = BXB^{-1}$, $X \in \mathrm{End}(\mathscr{W})$. The eigenvalues of the transformation $\tilde{B}$ have the form λ_1/λ_2, where $\lambda_1, \lambda_2 \in \Omega(B)$. Therefore, in view of Lemma 1.1(c), the set $\{\tilde{B}^i(Y-E) \mid$

$i \in \mathbb{Z}\}$ does not contain 0 in its closure and hence E does not belong to the closure of the set $\{B^i Y B^{-i} \mid i \in \mathbb{Z}\}$. □

(1.5) Proposition. *Let $B \in \mathbf{GL}(\mathscr{W})$ and let $Y \in \operatorname{End}(\mathscr{W})$, $Y \neq E$. Assume that there exists a Y-invariant subset L of the characteristic cross Δ_B such that $\mathscr{W}$ is spanned by L. Then the closure of the set $\{B^i Y B^{-i} \mid i \in \mathbb{Z}\}$ does not contain E.*

Proof. Let $L_d = L \cap \mathscr{W}_d(B)$ and let us consider the following two cases:

(a) $Y L_d \subset L_d$ for all $d \in \mathbb{R}$ and
(b) $Y L_d \not\subset L_d$ for some $d \in \mathbb{R}$.

Since L is contained in Δ_B, $\mathscr{W}$ is spanned by L, and since $\mathscr{W}$ is a direct sum of the subspaces $\mathscr{W}_d(B)$, it follows that $\mathscr{W}_d(B)$ is spanned by L_d. Therefore in the case (a), the subspaces $\mathscr{W}_d(B)$ are invariant relative to Y and the proposition follows from Lemma 1.4 applied to the restrictions of the transformations B and Y to the subspaces $\mathscr{W}_d(B)$. Since $L \subset \Delta_B$ and $YL = L$, in the case (b) there exist c, $d \in \mathbb{R}$ and $w \in L_c - \{0\}$ such that $c \neq d$ and $Yw \in L_d$. Since $\mathscr{W}_c(B)$ and $\mathscr{W}_d(B)$ are invariant relative to B, we have $B^i w \in \mathscr{W}_c(B) - \{0\}$ and $B^i Y B^{-i}(B^i w) = B^i Y w \in \mathscr{W}_d(B)$. On the other hand, since the intersection of linear subspaces $\mathscr{W}_c(B)$ and $\mathscr{W}_d(B)$ reduces to 0, there exists a neighbourhood U of the transformation E in $\mathbf{GL}(\mathscr{W})$ such that $\mathscr{W}_d(B) \cap g(\mathscr{W}_c(B) - \{0\}) = \varnothing$ for all $g \in U$. Therefore in the case (b) the closure of the set $\{B^i Y B^{-i} \mid i \in \mathbb{Z}\}$ does not contain E as well. □

2. Density Theorems for Subgroups with Property (S)I

Let G be a locally compact group, H a subgroup of G, and $g \in G$. We set $\Theta_g = \{x \in G \mid e$ belongs to the closure of the set $\{g^i x g^{-i} \mid i \in \mathbb{Z}\}\}$ and denote by Ψ_g the closure of the subgroup in G generated by the set Θ_g.

(2.1) Lemma. *The following conditions are equivalent:*

(a) *For every neighbourhood Ω of the identity in G and for each $x \in G$ there exists a positive integer $n = n(\Omega, x)$ such that*
$$x g^n x^{-1} \in \Omega H \Omega;$$
(b) *The homeomorphism of the space $\bar{H} \setminus G$ sending $x \in \bar{H} \setminus G$ to xg is recurrent (in the sense of I.0.39), where $\bar{H}$ denotes the closure of the subgroup H in G.*

Proof. Let $\pi\colon G \to \bar{H} \setminus G$ be the natural map. The equivalence of (a) and (b) is a consequence of the fact that $x g^n x^{-1} \in \Omega H \Omega$ if and only if $\pi(\Omega x) g^n \cap \pi(\Omega x) \neq \varnothing$, where $x \in G$, $n \in \mathbb{Z}$, and Ω is a neighbourhood of the identity in G such that $\Omega = \Omega^{-1}$. □

(2.2) Definition. *We say that a pair (H, g) has* property (S) *in G, provided that the conditions (a) and (b) of Lemma 2.1 are satisfied. A subgroup H in G is said to have* property (S) *in G if for every $g \in G$, the pair (H, g) has property (S) in G.*

From the Poincaré recurrence theorem (see Theorem I.4.6.1) we deduce

(2.3) Lemma. *If H is closed and there exists a finite G-invariant measure on $H \backslash G$, then H has property (S) in G.*

(2.4) Lemma.

(a) *Let φ_1 be a topological automorphism of a topological space X_1, φ_2 a topological automorphism of a space X_2, and let $f\colon X_1 \to X_2$ be a continuous map such that $f \circ \varphi_1 = \varphi_2 \circ f$. If a point $x \in X_1$ is non-wandering relative to φ_1, then $f(x)$ is non-wandering relative to φ_2.*

(b) *Let $f\colon G \to F$ be a continuous epimorphism of G onto a locally compact group F. If a pair (H, g) has property (S) in G, then so does the pair $(f(H), f(g))$ in F. Hence, if H has property (S) in G, then $f(H)$ has property (S) in F.*

This lemma is a straightforward consequence of the definitions.

(2.5) Theorem. *Let k be a local field, $\mathscr{W}$ a finite-dimensional vector space over k, and let $T\colon G \to \mathbf{GL}(\mathscr{W})$ be a continuous representation. Suppose a pair (H, g) has property (S) in G. If $\mathscr{W}' \subset \mathscr{W}$ is a $T(H)$-invariant linear subspace of $\mathscr{W}$, then $\mathscr{W}'$ is $T(\Psi_g)$-invariant.*

Proof. Let us consider the exterior power $\Lambda^d \mathscr{W}$ of $\mathscr{W}$, where $d = \dim \mathscr{W}'$. Let $\mathscr{D} = \Lambda^d \mathscr{W}' \subset \Lambda^d \mathscr{W}$. Since

$$\{h \in \mathbf{GL}(\mathscr{W}) \mid h\mathscr{W}' = \mathscr{W}'\} = \{h \in \mathbf{GL}(\mathscr{W}) \mid (\Lambda^d h)\mathscr{D} = \mathscr{D}\}$$

(see [Bo 6], Lemma 5.1), we may assume, replacing T by $\Lambda^d T$ and $\mathscr{W}'$ by $\mathscr{D}$, that $\dim \mathscr{W}' = 1$. Furthermore, we can assume that $\mathscr{W}$ is spanned by $T(G)\mathscr{W}'$. Let $B = T(g)$ and with π, B_π, and Δ_B as in 1.0 we put $w = \pi(\mathscr{W}') \in \mathbf{P}(\mathscr{W})$. Since $T(H)\mathscr{W}' = \mathscr{W}'$, one can define a continuous map $f\colon G/H \to T(G)_\pi w$, by setting $f(yH) = T(y)_\pi w$, $y \in G$. Since T is a homomorphism, $f(gx) = B_\pi(f(x))$ for all $x \in G/H$. But the pair (H, g) has property (S) in G. Therefore by Lemma 2.4(a) none of the points of the set $T(G)_\pi w = \pi(T(G)\mathscr{W}')$ is wandering relative to the transformation B_π. It follows by Proposition 1.3 that $T(G)\mathscr{W}' \subset \Delta_B$. On the other hand, $\mathscr{W}$ is spanned by $T(G)\mathscr{W}'$ and the set $T(G)\mathscr{W}'$ is $T(G)$-invariant. Therefore, by Proposition 1.5, for any transformation $Y \in T(G)$, $Y \neq E$, the closure of the set $\{B^i Y B^{-i} \mid i \in \mathbb{Z}\}$ does not contain E and hence, for each $y \in G$, either $T(y) = E$ or the closure of the set $\{B^i T(y) B^{-i} = T(g^i y g^{-i}) \mid i \in \mathbb{Z}\}$ does not contain E. It then follows from the continuity of the representation T that $T(\Theta_g) = \{E\}$. Thus $T(\Psi_g) = \{E\}$ and this completes the proof. □

(2.6) Corollary. *Let k be a local field, $\mathbf{F}$ an algebraic k-group, and let $\varphi\colon G \to \mathbf{F}(k)$ be a continuous homomorphism. Assume a pair (H, g) has property (S) in G. Then the Zariski closure $\overline{\varphi(H)}$ of the subgroup $\varphi(H)$ in $\mathbf{F}(k)$ contains $\varphi(\Psi_g)$.*

Proof. By Chevalley's theorem (see I.0.17) there is a faithful finite-dimensional representation $\alpha\colon \mathbf{F} \to \mathbf{GL}(\mathscr{W})$ defined over k, and a 1-dimensional subspace $\mathscr{W}' \subset \mathscr{W}$ over k such that

$$\overline{\varphi(H)} = \{x \in \mathbf{F} \mid \alpha(x)\mathscr{W}' = \mathscr{W}'\}.$$

Now, to prove the inclusion $\varphi(\Psi_g) \subset \overline{\varphi(H)}$ it suffices to apply Theorem 2.5 to the homomorphism $T = \alpha \circ \varphi\colon G \to \mathbf{GL}(\mathscr{W})$. □

(2.7) Corollary. *Let k be a local field, $\mathbf{F}$ an algebraic k-group, $\varphi_1\colon G \to \mathbf{F}(k)$ and $\varphi_2\colon G \to \mathbf{F}(k)$ two continuous homomorphisms. Assume a pair (H, g) has property (S) in G and $\varphi_1|_H = \varphi_2|_H$. Then $\varphi_1|_{\Psi_g} = \varphi_2|_{\Psi_g}$.*

Proof. Let us define the continuous homomorphism

$$\varphi\colon G \to \mathbf{F}(k) \times \mathbf{F}(k) = (\mathbf{F} \times \mathbf{F})(k)$$

by setting $\varphi(s) = (\varphi_1(s), \varphi_2(s))$. Since $\varphi_1(h) = \varphi_2(h)$ for all $h \in H$, $\varphi(H)$ is contained in the diagonal $D = \{(x, x) \mid x \in \mathbf{F}\}$. But the diagonal D is Zariski closed in $\mathbf{F} \times \mathbf{F}$ and the pair (H, g) has property (S) in G. Therefore, in view of Corollary 2.6, we have that $\varphi(\Psi_g) \subset D$. □

(2.8) Corollary. *Let k be a local field, l a finite field extension of k, $\mathbf{F}$ an algebraic k-group, and let $\varphi\colon G \to \mathbf{F}(l)$ be a continuous homomorphism. If a pair (H, g) has property (S) in G and $\varphi(H) \subset \mathbf{F}(k)$, then $\varphi(\Psi_g) \subset \mathbf{F}(k)$.*

Proof. We may assume that $\mathbf{F} = \mathbf{GL}_m$, $m \in \mathbb{N}^+$. Let us set $\mathscr{V} = l^m$, $\mathscr{W} = k^m \subset \mathscr{V}$, and consider $\mathscr{V}$ as a vector space over k. Since $\mathbf{GL}_m(k) = \{g \in \mathbf{GL}_m(l) \mid g\mathscr{W} = \mathscr{W}\}$, the inclusion $\varphi(\Psi_g) \subset \mathbf{F}(k)$ follows from Theorem 2.5. □

3. The Generalized Mautner Lemma and the Lebesgue Spectrum

(3.0) Let A be a finite set. For $\alpha \in A$, let k_α be a local field with valuation $|\ |_\alpha$ and let $\mathbf{G}_\alpha$ be a nontrivial connected semi-simple group defined over k_α. The locally compact group $\mathbf{G}_\alpha(k_\alpha)$ (equipped with the topology induced from that of the field k_α) will be denoted by G_α. For $B \subset A$, let $G_B = \prod_{\alpha \in B} G_\alpha$; we denote G_A by G. Here, by convention, $G_\varnothing = \{e\}$.

The groups G_α are σ-compact, metrizable, and compactly generated (see I.0.31 and Corollaries I.2.3.3 and I.2.3.5). Therefore G is σ-compact, metrizable, and compactly generated. The natural projections $G \to G_B$ and $G \to G_\alpha$, where $B \subset A$ and $\alpha \in A$, will be denoted by pr_B and pr_α.

Let us choose for each $\alpha \in A$ a maximal k_α-split torus $\mathbf{S}_\alpha$ in $\mathbf{G}_\alpha$ and set $\Phi_\alpha = \Phi(\mathbf{S}_\alpha, \mathbf{G}_\alpha)$. Fix an ordering on the root system Φ_α and denote, respectively, by Φ_α^+, Φ_α^-, and Δ_α the set of positive, negative, and simple roots with respect to this ordering. With notation as in I.1.2, define for every $\vartheta \subset \Delta_\alpha$ the subgroups $\mathbf{S}_\vartheta$, $\mathbf{P}_\vartheta$, $\mathbf{P}_\vartheta^-$, $\mathbf{V}_\vartheta = R_u(\mathbf{P}_\vartheta)$, and $\mathbf{V}_\vartheta^- = R_u(\mathbf{P}_\vartheta^-)$ of the group $\mathbf{G}_\alpha$. The subgroups $\mathbf{P}_\varnothing$, $\mathbf{P}_\varnothing^-$, $\mathbf{V}_\varnothing$, and $\mathbf{V}_\varnothing^-$ of $\mathbf{G}_\alpha$ will be denoted, respectively, by $\mathbf{P}_\alpha$, $\mathbf{P}_\alpha^-$, $\mathbf{V}_\alpha$, and $\mathbf{V}_\alpha^-$. We set

$$S = \prod_{\alpha \in A} \mathbf{S}_\alpha(k_\alpha),\ P = \prod_{\alpha \in A} \mathbf{P}_\alpha(k_\alpha),\ P^- = \prod_{\alpha \in A} \mathbf{P}_\alpha^-(k_\alpha),$$
$$V = \prod_{\alpha \in A} \mathbf{V}_\alpha(k_\alpha), \text{ and } V^- = \prod_{\alpha \in A} \mathbf{V}_\alpha^-(k_\alpha).$$

We set $\Delta = \cup_{\alpha \in A} \Delta_\alpha$ and define for every $\vartheta \subset \Delta$ the subgroups

$$S_\vartheta = \prod_{\alpha \in A} \mathbf{S}_{\vartheta \cap \Delta_\alpha}(k_\alpha),\ P_\vartheta = \prod_{\alpha \in A} \mathbf{P}_{\vartheta \cap \Delta_\alpha}(k_\alpha),\ P_\vartheta^- = \prod_{\alpha \in A} \mathbf{P}_{\vartheta \cap \Delta_\alpha}^-(k_\alpha),$$
$$V_\vartheta = \prod_{\alpha \in A} \mathbf{V}_{\vartheta \cap \Delta_\alpha}(k_\alpha), \text{ and } V_\vartheta^- = \prod_{\alpha \in A} \mathbf{V}_{\vartheta \cap \Delta_\alpha}^-(k_\alpha).$$

We see that

1) $S_\varnothing = S$, $P_\varnothing = P$, $V_\varnothing = V$, and $V_\varnothing^- = V^-$;
2) $V_{\vartheta_1} \subset V_{\vartheta_2} \subset P_{\vartheta_2} \subset P_{\vartheta_1}$ and $V_{\vartheta_1}^- \subset V_{\vartheta_2}^- \subset P_{\vartheta_2}^- \subset P_{\vartheta_1}^-$ whenever $\vartheta_2 \subset \vartheta_1 \subset \Delta$.

Since $\mathscr{Z}_{\mathbf{G}_\alpha}(\mathbf{S}_\vartheta) = \mathbf{P}_\vartheta \cap \mathbf{P}_\vartheta^-$ is a Levi subgroup of both $\mathbf{P}_\vartheta$ and $\mathbf{P}_\vartheta^-$ for all $\alpha \in A$ and $\vartheta \subset \Delta_\alpha$ (see I.1.2), we have for every $\vartheta \subset \Delta$

(1) $$\mathscr{Z}_G(S_\vartheta) = P_\vartheta \cap P_\vartheta^-,\ P_\vartheta = \mathscr{Z}_G(S_\vartheta) \ltimes V_\vartheta, \text{ and } P_\vartheta^- = \mathscr{Z}_G(S_\vartheta) \ltimes V_\vartheta^-.$$

For every $\vartheta \subset \Delta$ we put $R_\vartheta = \{s \in S \mid |b_1(\mathrm{pr}_\alpha(s))|_\alpha \leq 1$ and $|b_2(\mathrm{pr}_\alpha(s))|_\alpha < 1$ for all $\alpha \in A$, $b_1 \in \Delta_\alpha$, $b_2 \in \Delta_\alpha - \vartheta\}$ and $D_\vartheta = S_\vartheta \cap R_\vartheta = \{s \in S_\vartheta \mid |b(\mathrm{pr}_\alpha(s))|_\alpha < 1$ for all $\alpha \in A$ and $b \in \Delta_\alpha - \vartheta\}$. It follows from Proposition I.2.4.2 that $D_\vartheta \neq \varnothing$. For each $s \in S$ we set $A(s) = \{\alpha \in A \mid |b(\mathrm{pr}_\alpha(s))|_\alpha \neq 1$ for at least one $b \in \Phi_\alpha\}$. Denote the set $\{s \in S \mid A(s) = A\}$ by $\tilde{S}$.

We recall that $\mathbf{H}(k)^+$ denotes the subgroup of the group $\mathbf{H}(k)$ generated by the sets of k-rational points of unipotent radicals of all parabolic k-subgroups of a k-group $\mathbf{H}$. We put $G_\alpha^+ = \mathbf{G}_\alpha(k_\alpha)^+$, $G_B^+ = \prod_{\alpha \in B} G_\alpha^+$, where $B \subset A$, and $G^+ = G_A^+$.

We shall clarify the notation above by considering the special case where $G = \mathbf{SL}_n(\mathbb{R})$, $n \geq 2$. In this case we may assume that S is the group of diagonal matrices and P is the group of upper triangular ones (of determinant 1). Then P^- is the group of lower triangular matrices and V (resp. V^-) is the group of upper (resp. lower) triangular matrices with all diagonal entries equal to one. The set Δ is naturally identified with the set $\{1, 2, \ldots, n-1\}$. Let $\vartheta \subset \Delta$, where $\Delta - \vartheta = \{i_1, \ldots, i_t\}$ and $1 \leq i_1 < i_2 < \ldots < i_t \leq n-1$. We then have

$$S_\vartheta = \left\{ \begin{pmatrix} \lambda_1 E_{i_1} & 0 & \ldots & 0 \\ 0 & \lambda_2 E_{i_2 - i_1} & \ldots & 0 \\ \ldots & \ldots & \ldots & \ldots \\ 0 & 0 & \ldots & \lambda_t E_{i_t - i_{t-1}} \end{pmatrix} \middle| \ \lambda_1^{i_1} \lambda_2^{i_2 - i_1} \ldots \lambda_t^{i_t - i_{t-1}} = 1 \right\},$$

where E_j denotes the $j \times j$ identity matrix. The group P_ϑ is the group of block triangular matrices of determinant 1 of the form

$$\begin{pmatrix} A_{11} & A_{12} & \ldots & A_{1t} \\ 0 & A_{22} & \ldots & A_{2t} \\ \ldots & \ldots & \ldots & \ldots \\ 0 & 0 & \ldots & A_{tt} \end{pmatrix},$$

where A_{ss} is a square matrix of order $i_s - i_{s-1}$, and we assume $i_0 = 0$. The group V_ϑ consists of those block triangular matrices having the identity matrices in the diagonal, i.e.

$$V_\vartheta = \left\{ \begin{pmatrix} E_{i_1} & A_{12} & \dots & A_{1t} \\ 0 & E_{i_2-i_1} & \dots & A_{2t} \\ \dots & \dots & \dots & \dots \\ 0 & 0 & \dots & E_{i_t-i_{t-1}} \end{pmatrix} \right\}.$$

The group P_ϑ^- (resp. V_ϑ^-) results from P_ϑ (resp. V_ϑ) by transposition. Further we have

$$R_\vartheta = \left\{ \begin{pmatrix} \chi_1 & 0 & & 0 \\ 0 & \chi_2 & \dots & 0 \\ \dots & \dots & \dots & \dots \\ 0 & 0 & & \chi_n \end{pmatrix} \middle| \begin{array}{l} |\chi_1| \leq \dots \leq |\chi_{i_1}| < \\ < |\chi_{i_1+1}| \leq \dots \leq |\chi_{i_2}| < \\ < |\chi_{i_2+1}| \leq \dots \leq |\chi_{i_3}| < \\ < |\chi_{i_3+1}| \leq \dots \leq |\chi_n|, \ \chi_1\chi_2\cdots\chi_n = 1 \end{array} \right\}$$

and

$$D_\vartheta = \left\{ \begin{pmatrix} \lambda_1 E_{i_1} & 0 & \dots & 0 \\ 0 & \lambda_2 E_{i_2-i_1} & \dots & 0 \\ \dots & \dots & \dots & \dots \\ 0 & 0 & \dots & \lambda_t E_{i_t-i_{t-1}} \end{pmatrix} \middle| \begin{array}{l} |\lambda_1| < |\lambda_2| < \dots < |\lambda_t|; \\ \lambda_1^{i_1}\lambda_2^{i_2-i_1}\dots\lambda_t^{i_t-i_{t-1}} = 1 \end{array} \right\}.$$

Finally,

$$\tilde{S} = \left\{ \begin{pmatrix} \chi_1 & 0 & \dots & 0 \\ 0 & \chi_2 & \dots & 0 \\ \dots & \dots & \dots & \dots \\ 0 & 0 & & \chi_n \end{pmatrix} \middle| \begin{array}{l} \chi_1\chi_2\cdots\chi_n = 1 \text{ and there is} \\ \text{an integer } j,\ 1 \leq j \leq n, \\ \text{such that } |\chi_j| \neq 1 \end{array} \right\}.$$

Notice also that if $G = \mathbf{SL}_n(\mathbb{R})$, then $G^+ = G$.

(3.1) Lemma.

(a) *For every* $\vartheta \subset \Delta$ *and* $s \in R_\vartheta$ *the automorphisms* $\operatorname{Int} s|_{V_\vartheta}$ *and* $\operatorname{Int} s^{-1}|_{V_\vartheta^-}$ *are contracting.*

(b) *For each* $s \in R_\Delta$ *the automorphism* $\operatorname{Int} s|_P$ *is non-expanding.*

Proof. By Proposition I.1.1.1 we have, with notation as in I.1.1 and I.1.2, the following equality

$$\operatorname{Lie}(\mathbf{V}_{\vartheta\cap\Delta_\alpha}) = \bigoplus_{a\in v(\beta_{\vartheta\cap\Delta_\alpha})} \operatorname{Lie}(\mathbf{U}_a), \ \alpha \in A.$$

Therefore, if $s \in R_\vartheta$, then $|\lambda|_\alpha < 1$ for all $\alpha \in A$ and for all eigenvalues λ of the transformation $\operatorname{Ad}\operatorname{pr}_\alpha(s)|_{\operatorname{Lie}(\mathbf{V}_{\vartheta\cap\Delta_\alpha})}$. It then follows from Lemma 1.1(d) and the existence of $\mathbf{S}_\alpha$-equivariant k_α-isomorphisms of algebraic varieties

$$\mathbf{V}_{\vartheta\cap\Delta_\alpha} \to \operatorname{Lie}(\mathbf{V}_{\vartheta\cap\Delta_\alpha})$$

and

$$\mathbf{V}_{\vartheta\cap\Delta_\alpha}^- \to \operatorname{Lie}(\mathbf{V}_{\vartheta\cap\Delta_\alpha}^-)$$

(see Proposition I.3.3(ii)) that (a) is true. Since

$$P = V \rtimes \mathscr{Z}_G(S)$$

(see equalities (1) in 3.0) and since s commutes with $\mathscr{Z}_G(S)$, we see that (a) implies (b). $\square$

(3.2) Lemma. (Generalized Mautner lemma.) *Let H be a topological group and let $x, y \in H$ be the elements such that the sequence $\{x^n y x^{-n}\}$ converges to e as $n \to +\infty$. If ρ is a continuous unitary representation of the group H on a Hilbert space W, $w \in W$, and $\rho(x)w = w$, then $\rho(y)w = w$.*

Proof. Since $\rho(x)w = w$ and ρ is unitary, we have for each $n \in \mathbb{Z}$ that

$$\|\rho(y)w - w\| = \|\rho(y)\rho(x^{-n})w - \rho(x^{-n}w\| = \|\rho(x^n y x^{-n})w - w\|.$$

But $\{x^n y z^{-n}\} \to e$ as $n \to +\infty$, and ρ is continuous. Therefore $\|\rho(y)w - w\| = 0$, and hence $\rho(y)w = w$. $\square$

(3.3) Proposition.

(a) *Let $s \in S$, H a subgroup of the group G containing $\{s\} \cup G^+$, and let ρ be a unitary representation of the group H on a Hilbert space W. Suppose for each $\alpha \in A$, the group $\mathbf{G}_\alpha$ is almost k_α-simple. If $w \in W$ and $\rho(s)w = w$, then $\rho(G^+_{A(s)})w = w$.*

(b) *Let $s \in S$ and let ρ be a unitary representation of G on a Hilbert space W. Suppose for each $\alpha \in A$, the group $\mathbf{G}_\alpha$ is simply connected, k_α-isotropic, and almost k_α-simple. If $w \in W$ and $\rho(s)w = w$, then $\rho(G_{A(s)})w = w$.*

Proof. (a) Let $\alpha \in A(s)$. Modifying, if necessary, the ordering on Φ_α we may assume that $|b(\mathrm{pr}_\alpha(s))|_\alpha \leq 1$ for each $b \in \Phi^+_\alpha$. Since $\alpha \in A(s)$, we have that

$$\vartheta \overset{\text{def}}{=} \{b \in \Delta_\alpha \mid |b(\mathrm{pr}_\alpha(s))|_\alpha < 1\} \neq \varnothing.$$

Therefore by Proposition I.1.5.4(iii) the subgroups $V_{\Delta-\vartheta}$ and $V^-_{\Delta-\vartheta}$ both generate G^+_α. On the other hand, since $\rho(s)w = \rho(s^{-1})w = w$, it follows from Lemmas 3.1(a) and 3.2 that

$$\rho(V_{\Delta-\vartheta})w = \rho(V^-_{\Delta-\vartheta})w = w.$$

Therefore $\rho(G^+_\alpha)w = w$ for all $\alpha \in A(s)$ and hence $\rho(G^+_{A(s)})w = w$.

(b) Since the groups $\mathbf{G}_\alpha$ are simply connected, k_α-isotropic, and almost k_α-simple, by virtue of Theorem I.2.3.1(a) one has $G^+_\alpha = G_\alpha$ for each $\alpha \in A$. Thus $G^+_{A(s)} = G_{A(s)}$. It then follows from (a) that $\rho(G_{A(s)})w = w$. $\square$

(3.4) Lemma. *Let k be a local field and let ρ be a unitary representation of the group $\mathbf{SL}_2(k)$ on a Hilbert space W and let*

$$U = \left\{ \begin{pmatrix} 1 & x \\ 0 & 1 \end{pmatrix} \mid x \in k \right\}.$$

If w is an element of W such that $\rho(U)w = w$, then $\rho(\mathbf{SL}_2(k))w = w$.

Proof. We may assume that $\|w\| = 1$. Let us consider the continuous function $\varphi(g) = \langle \rho(g)w, w\rangle$, $g \in \mathbf{SL}_2(k)$. Since ρ is unitary and $\rho(U)w = w$, the function φ is constant on a double coset modulo U. But if $\begin{pmatrix} a & b \\ c & d \end{pmatrix} \in \mathbf{SL}_2(k)$ and $c \neq 0$, we have that

$$\begin{pmatrix} 1 & c^{-1}(1-a) \\ 0 & 1 \end{pmatrix}\begin{pmatrix} a & b \\ c & d \end{pmatrix}\begin{pmatrix} 1 & c^{-1}(1-d) \\ 0 & 1 \end{pmatrix} = \begin{pmatrix} 1 & 0 \\ c & 1 \end{pmatrix}.$$

Therefore, one has for $c \neq 0$ that

$$\varphi\left(\begin{pmatrix} a & 0 \\ c & a^{-1} \end{pmatrix}\right) = \varphi\left(\begin{pmatrix} 1 & 0 \\ c & 1 \end{pmatrix}\right).$$

Passing to the limit, as $c \to 0$, in this equality we obtain that

$$\varphi\left(\begin{pmatrix} a & 0 \\ 0 & a^{-1} \end{pmatrix}\right) = 1.$$

It then follows from the property of ρ being unitary that $\rho(g)w = w$ for all $g = \begin{pmatrix} a & 0 \\ 0 & a^{-1} \end{pmatrix}$. Indeed, for such g we have

$$\|\rho(g)w - w\|^2 = \langle \rho(g)w - w, \rho(g)w - w\rangle = 2 - 2\operatorname{Re}\varphi(g) = 0.$$

Applying now Proposition 3.3(b) to $s = \begin{pmatrix} \pi & 0 \\ 0 & \pi^{-1} \end{pmatrix}$, where π is a uniformizer of the field k, we obtain $\rho(\mathbf{SL}_2(k))w = w$. □

(3.5) Definition. *We say that a unitary operator B on a Hilbert space W has the* Lebesgue spectrum *if there exists a subspace L of W such that the subspaces $B^m L$, $m \in \mathbb{Z}$, are mutually orthogonal and their sum equals W.*

Remark. It is not hard to show that a unitary operator B on a separable complex Hilbert space W has the Lebesgue spectrum if and only if the measure μ in the spectral decomposition

$$W = \int_{|z|=1} W_z d\mu(z)$$

of the space W relative to the operator B is equivalent to the Lebesgue measure on the circle $\{s \in \mathbb{C} \mid |z| = 1\}$ and the multiplicity function $n(z) = \dim W_z$ is constant.

(3.6) Lemma. *If a unitary operator B on a Hilbert space W has the Lebesgue spectrum, then for every pair $w_1, w_2 \in W$*

$$\lim_{n\to\infty} \langle B^n w_1, w_2\rangle = 0. \tag{1}$$

Proof. Denote by X the set of those pairs $(w_1, w_2) \in W \times W$ satisfying (1) and put $L_n = \bigoplus_{-n\le i\le n} B^i L$, where L is the subspace in Definition 3.5. Since B is unitary, the set X is closed. On the other hand, the union

$$\bigcup_{n \in \mathbb{N}^+} L_n$$

is dense in W and it is easily seen that $X \supset L_n \times L_n$ for all $n \in \mathbb{N}^+$. Therefore $X = W \times W$ and the proof is complete. □

Remark. The converse is in general false, i.e., one can find a unitary operator B on a Hilbert space W such that (1) is satisfied for all $w_1, w_2 \in W$ but B does not have the Lebesgue spectrum.

(3.7) Lemma. *Let the group F be a semi-direct product of a discrete cyclic group C generated by an element q and a commutative locally compact second countable normal subgroup N, and let ρ be a unitary representation of the group F on a separable Hilbert space W. Suppose 1) there is no non-zero $\rho(N)$-invariant vector in W and 2)* $\operatorname{Int} q|_N$ *is a contracting automorphism. Then the operator $\rho(q)$ has the Lebesgue spectrum.*

Proof. The restriction of ρ to N can be written in the form of a continuous sum

(1) $$\rho|_N = \int_{\hat{N}} T_\chi d\mu(\chi), \; W = \int_{\hat{N}} W_\chi d\mu(\chi),$$

where T_χ is the one-dimensional representation determined by the character $\chi \in \hat{N}$ of the group N on a space W_χ. For every Borel set $X \supset \hat{N}$ we put

(2) $$W_X = \int_X W_\chi d\mu(\chi).$$

We define the action of the group F on $\hat{N}$ in the following manner:

$$(f\chi)(n) = \chi(f^{-1}nf), \; f \in F, \; \chi \in \hat{N}, \; n \in N.$$

It is straightforward to verify that

(3) $$\rho(f)W_X = W_{fX} \text{ for all } f \in F \text{ and } X \subset \hat{N}.$$

(Indeed, if $w \in W_\chi$ is viewed as a usual and not "generalized" eigenvector of the operators $\rho(n)$, $n \in N$, then we have that

$$\rho(n)\rho(f)w = \rho(f)\rho(f^{-1}nf)w = \chi(f^{-1}nf)\rho(f)w = (f\chi)(n)\rho(f)w$$

and hence $\rho(f)w \in W_{f\chi}$.) Since W does not have a non-zero $\rho(N)$-invariant vector, one has

(4) $$W_{\{\chi_0\}} = 0,$$

where $\chi_0 \in \hat{N}$ is the trivial character of the group N. Since $\operatorname{Int} q|_N$ is a contracting automorphism, the automorphism $\chi \mapsto q^{-1}\chi$ of the group $\hat{N}$ is contracting as well. Therefore there exists a Borel subset $Y \subset \hat{N} - \{\chi_0\}$ such that the subsets $q^i Y$, $i \in \mathbb{Z}$, are mutually disjoint and form a cover of $\hat{N} - \{\chi_0\}$. It then follows from

(1), (2), (3), and (4) that the subspaces $\rho(q^i)W_Y$, $i \in \mathbb{Z}$, are mutually orthogonal and their sum equals W. □

(3.8) *Remark.* One could prove Lemma 3.7 by decomposing ρ into a continuous sum of irreducible representations and making use of the following argument: if ρ is irreducible and the conditions of Lemma 3.7 are satisfied, then by Mackey's theorem (see I.5.2.3) the representation ρ is induced from a non-trivial character of the group N. In the demonstration above, we have in fact taken advantage of some arguments from the proof of Mackey's theorem.

(3.9) Proposition. *Let ρ be a unitary representation of the group G on a Hilbert space W and $s \in \tilde{S}$. Suppose the groups $\mathbf{G}_\alpha$ are almost k_α-simple and there is no non-zero $\rho(G^+)$-invariant vector in W. Then the operator $\rho(s)$ has the Lebesgue spectrum.*

Proof. Let us consider for each $\alpha \in A$ the groups

$$Y_\alpha = \left\{\begin{pmatrix} 1 & x \\ 0 & 1 \end{pmatrix} \,\middle|\, x \in k_\alpha\right\} \text{ and } D_\alpha = \left\{\begin{pmatrix} c & 0 \\ 0 & c^{-1} \end{pmatrix} \,\middle|\, c \in k_\alpha,\ c \neq 0\right\}.$$

Let $c_\alpha \in k_\alpha$, $c_\alpha \neq 0$, $|c_\alpha|_\alpha \neq 1$, and let $d_\alpha = \begin{pmatrix} c_\alpha & 0 \\ 0 & c_\alpha^{-1} \end{pmatrix} \in D_\alpha$. We set

$$\Psi_\alpha = \{b \in \Phi_\alpha \mid |b(\mathrm{pr}_\alpha(s))|_\alpha < 1\}.$$

Modifying, if necessary, the ordering on Φ_α, we may assume that $\Psi_\alpha \subset \Phi_\alpha^+$. Put $\Delta(s) = \bigcup_{\alpha \in A}(\Delta_\alpha \cap \Psi_\alpha)$. Since $s \in \tilde{S}$, one has that $\Delta_\alpha \cap \Psi_\alpha \neq \varnothing$ and hence $\mathbf{V}_{\Delta_\alpha \cap \Psi_\alpha} \supset \mathbf{U}_{\{b\}}$ for a non-multipliable root $b \in \Phi_\alpha^+$, where $\mathbf{U}_{\{b\}}$ is a unipotent k_α-subgroup corresponding to the quasi-closed subset $\{b\}$ (see I.1.1). It then follows from Theorem I.1.6.1 and Proposition I.1.6.3 that for each $\alpha \in A$ there exists a k_α-morphism $\sigma_\alpha \colon \mathbf{SL}_2 \to \mathbf{G}_\alpha$ with finite kernel such that

$$(1) \qquad \sigma_\alpha(Y_\alpha) \subset \mathbf{V}_{\Delta_\alpha \cap \Psi_\alpha}(k_\alpha),\ \sigma_\alpha(D_\alpha) \subset \mathbf{S}_\alpha(k_\alpha),$$

and $\sigma_\alpha(Y_\alpha)$ is normalized by S. Let s_0 be an element of S such that $\mathrm{pr}_\alpha(s_0) = \sigma_\alpha(d_\alpha)$ for all $\alpha \in A$. Since $|c_\alpha|_\alpha \neq 1$, the groups $\mathbf{G}_\alpha$ are semisimple, and the kernels of k_α-morphisms σ_α are finite, we have that $s_0 \in \tilde{S}$. But there are no $\rho(G^+)$-invariant vectors in W, so by virtue of Proposition 3.3(a) there are no $\rho(s_0)$-invariant vectors in W. It then follows from (1) and Lemma 3.4 applied to the representations $\rho \cdot \sigma_\alpha$ that there are no $\rho(F)$-invariant vectors in W, where $F = \prod_{\alpha \in A} \sigma_\alpha(Y_\alpha)$. Now it suffices to observe that $\mathrm{Int}\, s|_{V_{\Delta(s)}}$ is a contracting automorphism (see Lemma 3.1(a)) and to apply Lemma 3.7 to the group $C \cdot F$, where $C = \{s^n | n \in \mathbb{Z}\}$. □

(3.10) Corollary. *Let H be a closed subgroup of the group G such that $H \setminus G$ carries a finite non-zero G-invariant measure μ, and let $s \in \tilde{S} \cap G^+$. Then for every open non-empty subset $Y \subset G$ and for each $g_0 \in G^+$ there exists a positive integer $N(Y, g_0)$ such that $H \cap (Y \cdot g_0 \cdot s^{-m} \cdot Y^{-1}) \neq \varnothing$ for all $m > N(Y, g_0)$, $m \in \mathbb{N}^+$.*

Proof. We define a representation ρ of the group G on the space $W = L_2(H \setminus G, \mu)$ as follows:

$$(\rho(g)f)(q) = f(qg), \ q \in H \setminus G, \ g \in G, \ f \in W.$$

Since the measure μ is G-invariant, the representation ρ is unitary. We set $T = \{w \in W \mid \rho(G^+)w = w\}$ and denote by $T^\perp$ the orthogonal complement of T in W. Since the subgroup G^+ is normal in G, the subspace T, and hence $T^\perp$ are $\rho(G)$-invariant. It follows from the definition of the subspace T that there are no $\rho(G^+)$-invariant vectors in $T^\perp$. This implies by Proposition 3.9 that the restriction of $\rho(s)$ to $T^\perp$ has the Lebesgue spectrum, and hence (see Lemma 3.6)

$$\lim_{n\to\infty} \langle \rho(s^m)f_1, f_2\rangle = 0 \text{ for all } f_1, f_2 \in T^\perp.$$

But since $s \in G^+$, we have that $\rho(s)t = t$ for all $t \in T$. Therefore

(1) $$\lim_{m\to\infty} \langle \rho(s^m)f, f+t\rangle = \langle J(f), J(f)\rangle \text{ for all } f \in W \text{ and } t \in T^\perp,$$

where J is the orthogonal projector of the space W onto T. Let $\pi\colon G \to H \setminus G$ be the natural map. We set $\hat{Y} = \pi(Y)$ and denote by χ the characteristic function of the set $\hat{Y}$. Since $g_0 \in G^+$, we have that $\rho(g_0^{-1})\chi - \chi \in T^\perp$. Therefore (1) implies that

(2) $$\lim_{n\to\infty} \mu((\hat{Y}s^m) \cap (\hat{Y}g_0)) = \lim_{m\to\infty} \langle \rho(s^{-m})\chi, \rho(g_0^{-1})\chi\rangle =$$
$$= \lim_{n\to\infty} \langle \rho(s^{-m})\chi, \chi + (\rho(g_0^{-1})\chi - \chi)\rangle = \langle J(\chi), J(\chi)\rangle.$$

Since Y, and hence $\hat{Y}$, are open and the non-zero measure μ is G-invariant, we have $\mu(\hat{Y}) > 0$. On the other hand, the space of constants is contained in T, because μ is finite. Therefore $\langle J(\chi), J(\chi)\rangle > 0$. This implies, in view of (2), that $(\hat{Y}s^m) \cap (\hat{Y}g_0) \neq \varnothing$ for all sufficiently large m. But

$$(\hat{Y}s^m) \cap (\hat{Y}g_0) = \pi(H \cdot Y \cdot s^m) \cap (Y \cdot g_0)).$$

Thus we have $(H \cdot Y \cdot s^m) \cap (Y \cdot g_0) \neq \varnothing$ and hence $H \cap (Y \cdot g_0 \cdot s^{-m} \cdot Y^{-1}) \neq \varnothing$ for all sufficiently large m. □

4. Density Theorems for Subgroups with Property (S)II

In this section we preserve the notation introduced in 3.0. As in Section 2 we set for each $g \in G$, $\Theta_g = \{g \in G \mid$ the set $\{g^i x g^{-i} \mid i \in \mathbb{Z}\}$ contains e in its closure$\}$ and denote by Ψ_g the closure of the subgroup generated by the set Θ_g.

(4.1) Lemma. *There exists $s \in S$ such that $\Psi_s = G^+$.*

Proof. Pick an element $s \in D_\varnothing = R_\varnothing$. Then by Lemma 3.1(a) the automorphisms $\operatorname{Int} s|_V$ and $\operatorname{Int} s^{-1}|_{V^-}$ are contracting. Therefore $V \cup V^- \subset \Theta_s$. On the other hand, by virtue of Proposition I.1.5.4(ii) the subgroups V and V^- generate G^+, hence

$\Psi_s \supset G^+$. It remains to observe that, since by Theorem I.2.3.1(b) the factor group G/G^+ is compact and in a compact group any class of conjugated elements is closed, the subgroup G^+ contains Θ_s, and hence Ψ_s. □

(4.2) Theorem. *Let k be a local field, $\mathscr{W}$ a finite-dimensional vector space over k. Let $T: G \to \mathbf{GL}(\mathscr{W})$ be a continuous representation of G and let H be a subgroup of G with property (S) in G. Let $\mathscr{W}'$ be a linear subspace of $\mathscr{W}$ which is invariant under $T(H)$. Then*

(a) *the subspace $\mathscr{W}'$ is $T(G^+)$-invariant and hence $T(G')$-invariant, where G' is the closure in G of the subgroup $H \cdot G^+$;*
(b) *if all the groups $\mathbf{G}_\alpha$, $\alpha \in A$, are simply connected and have no k_α-anisotropic factors, then $T(G)\mathscr{W}' = \mathscr{W}'$.*

Assertion (a) is a consequence of Theorem 2.5 and Lemma 4.1 and (b) is a consequence of (a) and Theorem I.2.3.1 (a′).

From Corollary 2.6 and Lemma 4.1 we obtain the following

(4.3) Proposition. *Let k be a local field, $\mathbf{F}$ an algebraic k-group, φ: $G \to \mathbf{F}(k)$ a continuous homomorphism, and let H be a subgroup of G with property (S). Then*

(a) *the Zariski closure of the subgroup $\varphi(H)$ contains $\varphi(G^+)$;*
(b) *if all the groups $\mathbf{G}_\alpha$, $\alpha \in A$, are simply connected and have no k_α-anisotropic factors , then $\varphi(G)$ is contained in the Zariski closure of the subgroup $\varphi(H)$* (this follows from (a) and Theorem I.2.3.1 (a′)).

(4.4) Corollary. (Borel-Wang density theorem.) *Let k be a local field, $\mathbf{G}$ a connected semisimple k-group, $\mathbf{G}' \subset \mathbf{G}$ the product of the k-isotropic factors of the group $\mathbf{G}$, and let H be a subgroup of $\mathbf{G}(k)$ with property (S). Then $\mathbf{G}'$ is contained in the Zariski closure of H in $\mathbf{G}$. In particular, if $\mathbf{G}$ has no k-anisotropic factors (i.e. if $\mathbf{G} = \mathbf{G}'$), then H is Zariski dense in $\mathbf{G}$.*

Proof. Since, by virtue of Proposition I.1.5.4(v), the Zariski closure of the subgroup $G(k)^+$ coincides with $\mathbf{G}'$, the desired assertion follows from Proposition 4.3 applied to the identity homomorphism. □

(4.5) The Borel-Wang density theorem is, in fact, a more general assertion than that of Corollary 4.4. We shall give the appropriate statement, but for that we need the following definition.

We say that a closed subgroup H of a locally compact group F has *property (NP)* if for each $g \in F$ and for every dense and open subset $U \subset H \backslash F$ which is invariant under right multiplication by g, there exists a compact set $K \subset U$ such that the set $\{n \in \mathbb{N}^+ \mid K \cap Kg^n \neq \varnothing\}$ is infinite.

Theorem. (See [WS 8], Theorem A.) *Let k be a local field, $\mathbf{G}$ a connected k-group, and let H be a closed subgroup of the group $\mathbf{G}(k)$ with property (NP). Then the Zariski closure $\bar{H}$ of the subgroup H contains all k-split tori of the group $\mathbf{G}$.*

Furthermore, if $k = \mathbb{R}$ *or* $\mathbb{C}$, *then* $\bar{H}$ *contains all unipotent k-subgroups of the group* $\mathbf{G}$.

Corollary. (See [WS 8], Corollary 1.4.) *Let k be a local field and let* $\mathbf{G}$ *be a connected semisimple k-group having no k-anisotropic factors, then every closed subgroup of* $\mathbf{G}(k)$ *having property (NP) is Zariski dense in* $\mathbf{G}$.

(4.6) Proposition. *Let there be given a local field k, an algebraic k-group* $\mathbf{F}$, *and two continuous homomorphisms* $\varphi_1, \varphi_2 \colon G \to \mathbf{F}(k)$ *agreeing on a subgroup H of G with property (S). Then*

(a) $\varphi_1(g) = \varphi_2(g)$ *for all* $g \in G^+$;

(b) *if, in addition, all the groups* $\mathbf{G}_\alpha$, $\alpha \in A$, *are simply connected and have no* k_α*-anisotropic factors, then* $\varphi_1 = \varphi_2$.

Proof. Assertion (a) is a consequence of Corollary 2.7 and Lemma 4.1 and (b) follows from (a) and Theorem I.2.3.1 (a′). □

(4.7) Proposition. *Let k be a local field, l a finite field extension of k,* $\mathbf{F}$ *an algebraic k-group, and let* $\varphi \colon G \to \mathbf{F}(l)$ *be a continuous homomorphism. If H is a subgroup of G with property (S) and* $\varphi(H) \subset \mathbf{F}(k)$, *then*

(a) $\varphi(G^+) \subset \mathbf{F}(k)$;

(b) *in case all the groups* $\mathbf{G}_\alpha$, $\alpha \in A$, *are simply connected and have no* k_α*-anisotropic factors,* $\varphi(G) \subset \mathbf{F}(k)$.

Proof. Assertion (a) is a consequence of Corollary 2.8 and Lemma 4.1 and (b) follows from (a) and Theorem I.2.3.1 (a′). □

(4.8) Proposition. *Let H be a subgroup with property (S) in the group G and let* $\varphi \colon G \to G$ *be a continuous homomorphism. Suppose that for each* $\alpha \in A$, *the group* $\mathbf{G}_\alpha$ *has no* k_α*-anisotropic factors and* $\mathscr{Z}(\mathbf{G}_\alpha) = \{e\}$. *If* $\varphi(h) = h$ *for each* $h \in H$, *then* $\varphi(g) = g$ *for all* $g \in G$.

Proof. From Proposition 4.6(a) applied to the homomorphisms $\varphi_1 = \mathrm{pr}_\alpha$ and $\varphi_2 = \mathrm{pr}_\alpha \circ \varphi$, $\alpha \in A$, it follows that $\varphi(x) = x$ for all $x \in G^+$. Let $g \in G$. We set $v(g) = \varphi(g)g^{-1}$. Since $\varphi|_{G^+} = \mathrm{Id}$ and the subgroup G^+ is normal in G, we have for each $x \in G^+$, $gxg^{-1} = \varphi(gxg^{-1}) = \varphi(g)\varphi(x)\varphi(g)^{-1} = \varphi(g)x\varphi(g)^{-1}$ and hence

$$v(g) \in \mathscr{Z}_G(G^+).$$

Since $\mathscr{Z}(\mathbf{G}_\alpha) = \{e\}$ and by virtue of Proposition I.1.5.4(v) the subgroup $\mathbf{G}_\alpha(k_\alpha)^+$ is Zariski dense in $\mathbf{G}_\alpha$, $\alpha \in A$, we have that $\mathscr{Z}_G(G^+) = \{e\}$. Thus, $v(g) = e$, and hence $\varphi(g) = g$. □

5. Non-Discrete Closed Subgroups of Finite Covolume

A closed subgroup E of a locally compact group F is said to be a *subgroup of finite covolume* if $E \setminus F$ has a finite non-zero F-invariant measure. It is the purpose of the present section to prove the following.

(5.1) Theorem. *Let k be a local field, $\mathbf{G}$ a connected semisimple k-group, isotropic and almost simple over k, and let H be a closed (in the topology induced from the topology of the field k) non-discrete subgroup of finite covolume in the group $\mathbf{G}(k)$. Then*

(a) $H \supset \mathbf{G}(k)^+$;

(b) *if $\mathbf{G}$ is simply connected, then $H = \mathbf{G}(k)$.*

Proof. Let $\mathbf{S}$ be a maximal k-split torus in $\mathbf{G}$. The group $\mathbf{G}$ may be viewed as a k-subgroup of $\mathbf{GL}_n$ for some $n \in \mathbb{N}^+$. We set for each character $a \in \chi(\mathbf{S})$, $W_a = \{w \in \operatorname{End}_n \mid sws^{-1} = a(s)w$ for all $s \in \mathbf{S}\}$. Let $\Omega = \{a \in \chi(S) \mid W_\alpha \neq 0\}$. Since the torus $\mathbf{S}$ is split over k, End_n is the direct sum of the subspaces W_a, $a \in \Omega$. Let $w_1, \ldots, w_{n^2}$ be a basis of $\operatorname{End}_n(k)$, such that every w_i, $1 \leq i \leq n$, belongs to some W_{a_i}, $a_i \in \Omega$. The coefficient of w_i in the expansion $w = \sum_{1 \leq i \leq n^2} c_i w_i$ of w with respect to this basis will be denoted by $c_i(w)$. Let $|\ |$ be a valuation on k. We call a sequence $\{g_t\}_{t \in \mathbb{N}^+}$ of elements of $\mathbf{G}(k)$ which differ from e a *tolerance sequence*, provided that $\lim_{t \to \infty} g_t = e$ and the following condition is satisfied:
(*) for every $\varepsilon > 0$, there is a positive integer $N(\varepsilon)$ and a neighbourhood $Y(\varepsilon)$ of the identity in $\mathbf{G}(k)$ such that

$$|1 - c_i(yg_ty^{-1})/c_i(g_t)| < \varepsilon,$$

for all $t \geq N(\varepsilon)$, $y \in Y(\varepsilon)$, and $1 \leq i \leq n^2$ (here, by convention, $|1 - 0/0| = 0$ and $|1 - a/0| = \infty > \varepsilon$ for all $a \in k$, $a \neq 0$).

We shall state two lemmas on tolerance sequences.

(A) Lemma. *If there is a tolerance sequence in H, then $H \supset \mathbf{G}(k)^+$.*

(B) Lemma. *There exists $g \in \mathbf{G}(k)$ such that there is a tolerance sequence in gHg^{-1}.*

If there is a tolerance sequence in gHg^{-1}, then after the replacement of the torus $\mathbf{S}$ by $g^{-1}\mathbf{S}g$ the subgroup H will contain a tolerance sequence. Therefore assertion (a) follows from Lemmas A and B which will be proved, respectively, in 5.3 and 5.4. Finally, we note that (b) is a consequence of (a) and Proposition I.2.3.1(a). □

(5.2) Lemma. *Let $\{g_t\}_{t \in \mathbb{N}^+}$ be a tolerance sequence in $\mathbf{G}(k)$ and let χ_0 be the trivial character of the torus $\mathbf{S}$. Then there exists a positive integer i such that $1 \leq i \leq n^2$, $a_i \neq \chi_0$, and the set $\{t \in \mathbb{N}^+ \mid c_i(g_t) \neq 0\}$ is infinite.*

Proof. Suppose the contrary, that is $c_j(g_t) = 0$ for all sufficiently large t and every $j \in J$, where $J = \{j \in \mathbb{N}^+ \mid a_j \neq \chi_0\}$. Then, since $\{g_t\}$ is a tolerance sequence, there exists a neighbourhood Y of the identity in $\mathbf{G}(k)$ such that $c_j(yg_ty^{-1}) = 0$ for all $y \in Y$ and $j \in J$ and all sufficiently large t. But Y is Zariski dense in $\mathbf{G}$ (see Proposition I.2.5.3(i)) and $c_j(gg_tg^{-1})$ viewed as a function of $g \in \mathbf{G}$ is regular. Therefore $c_j(gg_tg^{-1}) = 0$ for all $g \in G$, $j \in J$, and sufficiently large t. On the other hand, it is clear that $\mathscr{Z}_{\mathbf{G}}(\mathbf{S}) = W_{\chi_0} \cap \mathbf{G}$. Thus for $t \in \mathbb{N}^+$ sufficiently large we have that

$$\{gg_tg^{-1} \mid g \in \mathbf{G}\} \subset \mathscr{Z}_{\mathbf{G}}(\mathbf{S})$$

and hence the minimal k-closed normal subgroup of the group $\mathbf{G}$ containing g_t is contained in $\mathscr{Z}_{\mathbf{G}}(\mathbf{S})$. But $\mathbf{G}$ is almost k-simple and $\mathbf{G} \neq \mathscr{Z}_{\mathbf{G}}(\mathbf{S})$ (because $\mathbf{G}$ is isotropic over k). Thus $g_t \in \mathscr{Z}(\mathbf{G})$ for all sufficiently large $t \in \mathbb{N}^+$, a contradiction, for $\mathscr{Z}(\mathbf{G})$ is finite (because $\mathbf{G}$ is semisimple), $g_t \neq e$, and $\lim_{t\to\infty} g_t = e$. □

(5.3) *Proof of Lemma A.* Let $\{h_t\}_{t\in\mathbb{N}^+}$ be a tolerance sequence in H and let χ_0 be the trivial character of the torus $\mathbf{S}$. By virtue of Lemma 5.2 there exists i_0, $i \leq i_0 \leq n^2$, such that $a_{i_0} \neq \chi_0$ and, by passing to a subsequence of the sequence $\{h_t\}$, that

(1) $$c_{i_0}(h_t) \neq 0 \text{ for all } t \in \mathbb{N}^+.$$

We set $\tilde{S} = \{s \in \mathbf{S}(k) \mid$ at least one of the eigenvalues of the transformation $\operatorname{Ad} s$ is different from 1 in absolute value$\}$. Since the group $\mathbf{G}$ is semisimple and isotropic over k and $\mathbf{S}$ is a maximal k-split torus in $\mathbf{G}$, by Proposition I.2.4.1 there exists $s_0 \in \tilde{S}$ such that $|a_{i_0}(s_0)| \neq 1$. Replacing, if necessary, s_0 by s_0^{-1} we may assume that

(2) $$|a_{i_0}(s_0)| < 1 \text{ and } s_0 \in \tilde{S}.$$

By virtue of Theorem I.2.3.1(c) there exists $r \in \mathbb{N}^+$ such that $g^r \in \mathbf{G}(k)^+$ for every $g \in \mathbf{G}(k)$. So replacing s_0 by s_0^r we may assume that

(3) $$s_0 \in \mathbf{G}(k)^+.$$

Since $\mathscr{Z}(\mathbf{G})$ is finite, there exist relatively compact neighbourhoods U' and U'' of the identity in $\mathbf{G}(k)$ such that

(4) $$U' \subset U'',\ \overline{U''} \cap \mathscr{Z}(\mathbf{G}) = \{e\},\ \text{and } s_0^{-1}U's_0 \subset U'',$$

where $\overline{U''}$ is the closure of U''. From (1) and (2) we see that for each $t \in \mathbb{N}^+$

(5) $$\{s_0^{-m} \cdot h_t \cdot s_0^m \mid m \in \mathbb{N}^+\} \text{ is not relatively compact in } \mathbf{G}(k) \subset \operatorname{End}_n(k).$$

Since $\lim_{t\to\infty} h_t = e$, dropping a finite number of elements in the sequence $\{h_t\}$, we may assume that $h_t \in U'$ for all $t \in \mathbb{N}^+$. It then follows from (4), (5), and from the relative compactness of U'' that for each $t \in \mathbb{N}^+$ there exists a number $m(t) \in \mathbb{N}^+$ such that

(6) $$\tilde{h}_t \overset{\text{def}}{=} s_0^{-m(t)} \cdot h_t \cdot s_0^{m(t)} \in U'' - U'.$$

Since $\lim_{t\to\infty} h_t = e$, one has

(7) $$\lim_{t\to\infty} m(t) = \infty.$$

Since U'' is relatively compact in $\mathbf{G}(k)$, by passing, if necessary, to a subsequence of $\{h_t\}$, we may assume that

(8) $$\lim_{t\to\infty} \tilde{h}_t = \tilde{h} \in \overline{U''} - U'.$$

Fix an element $g_0 \in \mathbf{G}(k)$. Since H is of finite covolume and (in view of (2) and (3)) $s_0 \in \tilde{S} \cap \mathbf{G}(k)^+$, we infer from Corollary 3.10 that for every neighbourhood Y of the identity in $\mathbf{G}(k)$ there is a positive integer $N(Y)$ such that $H \cap (Y \cdot g_0 \cdot s_0^{-m} \cdot Y^{-1}) \neq \varnothing$ for all $M > N(Y)$. It then follows from (7) that there exist two sequences $\{y_t\}$ and $\{z_t\}$ in $\mathbf{G}(k)$ such that

(9) $$\lim_{t\to\infty} y_t = e, \ \lim_{t\to\infty} z_t = e$$

and

(10) $$u_t \overset{\text{def}}{=} y_t g_0 s_0^{-m(t)} z_t \in H.$$

Since $\{h_t\}$ is a tolerance sequence, $\lim_{t\to\infty} z_t = e$, and $c_i(s_0^m w s_0^{-m}) = a_i(s_0^m) c_i(w)$ for all $m \in \mathbb{Z}$, $1 \leq i \leq n^2$, and $w \in \mathrm{End}_n$, we have that

$$\lim_{t\to\infty} |1 - c_i(s_0^{-m(t)} z_t h_t z_t^{-1} s_0^{m(t)})/c_i(s_0^{-m(t)} h_t s_0^{m(t)})| = 0$$

for all i, $1 \leq i \leq n^2$. Therefore we obtain in view of (6) and (8)

(11) $$\lim_{t\to\infty} s_0^{-m(t)} z_t h_t z_t^{-1} s_0^{m(t)} = \tilde{h}.$$

Since

$$u_t = y_t g_0 s_0^{-m(t)} z_t \text{ and } \lim_{t\to\infty} y_t = e,$$

it follows from (11) that

$$\lim_{t\to\infty} u_t h_t u_t^{-1} = g_0 \tilde{h} g_0^{-1}.$$

But the subgroup H is closed and $u_t, h_t \in H$, hence

(12) $$g_0 \tilde{h} g_0^{-1} \in H \text{ for all } g_0 \in \mathbf{G}(k)^+.$$

By virtue of Theorem I.1.5.6(i) any subgroup of the group $\mathbf{G}(k)$ normalized by the subgroup $\mathbf{G}(k)^+$ is either contained in $\mathscr{Z}(\mathbf{G})$ or contains $\mathbf{G}(k)^+$. On the other hand, since $\tilde{h} \notin \mathscr{Z}(\mathbf{G})$ (see (4) and (8)), it follows from (12) that H contains a subgroup (namely, the subgroup generated by the set $\{g\tilde{h}g^{-1} \mid g \in \mathbf{G}(k)^+\}$) which is normalized by the subgroup $\mathbf{G}(k)^+$ and not contained in $\mathscr{Z}(\mathbf{G})$. Thus $H \supset \mathbf{G}(k)^+$. □

(5.4) *Proof of Lemma B.* We set

(1) $$d_{i,w}(g) = c_i(g^{-1} w g), \ w \in \mathrm{End}_n(k), \text{ and } g \in \mathbf{G}(k), \ 1 \leq i \leq n^2.$$

Since $\mathbf{G}$ is connected and semisimple, one has $\mathbf{G} \subset \mathbf{SL}_n$. Therefore the functions $d_{i,w}(g)$ are the restrictions to $\mathbf{G}(k)$ of the polynomials on $\mathrm{End}_n(k)$ of degree at most n^2, and hence the linear space D of all linear combinations of the functions $d_{i,w}$, $1 \le i \le n^2$, $w \in \mathrm{End}_n(k)$ with coefficients in k is finite-dimensional. Fix a compact set $M \subset \mathbf{G}(k)$ having e in its interior and set

$$\|f\| = \sup_{g \in M} |f(g)|, \quad f \in D.$$

Since M is Zariski dense in $\mathbf{G}$ (see Proposition I.2.5.3(i)), one has $\|f\| = 0$ if and only if $f = 0 (f \in D)$. Let us define the set $D_1 = \{f \in D \mid \|f\| = 1\}$. Since D is finite-dimensional, D_1 is compact. But M is compact and D consists of continuous functions. Hence the set D_1 is equicontinuous on M, i.e., for each $\varepsilon > 0$ there exists a neighbourhood of the identity $X(\varepsilon) \subset \mathbf{G}(k)$ such that $|f(gx) - f(g)| < \varepsilon$ for all $f \in D_1$, $g \in M$, and $x \in X(\varepsilon)$.

For every compact set $Z \subset \mathbf{G}(k)$ and for each $q = (q_1, \dots, q_{n^2}) \in D_1^{n^2}$ we set

(2) $$\varphi_Z(q) = \sup_{g \in Z} \inf_{1 \le i \le n^2} |q_i(g)|.$$

Since M is Zariski dense in $\mathbf{G}$, the function φ_M is not vanishing on $D_1^{n^2}$. On the other hand, φ_M is continuous and D_1 is compact. Therefore

(3) $$a \stackrel{\text{def}}{=} \inf_{q \in D_1^{n^2}} \varphi_M(q) > 0.$$

Since D_1 is equicontinuous on the compact set M, there is a finite set $R \subset M$ such that $|\varphi_M(q) - \varphi_R(q)| < a/2$ for all $q \in D_1^{n^2}$. It then follows from (3) that

(4) $$\inf_{q \in D_1^{n^2}} \varphi_R(q) \ge a/2.$$

It follows from (2) and (4) that for each $q = (q_1, \dots, q_{n^2}) \in D_1^{n^2}$ there exists $r(q) \in R$ such that

(5) $$\inf_{1 \le i \le n^2} |q_i(r(q))| \ge a/2.$$

Since H is not discrete in $\mathbf{G}(k)$, there exists a sequence $\{h_t\}_{t \in \mathbb{N}^+}$ of non-identity elements in H converging to the identity. Choose $b_{i,h_t} \in k$, so that $|b_{i,h_t}| = \|d_{i,h_t}\|$. Put for $t \in \mathbb{N}^+$ and $1 \le i \le n^2$

(6) $$\tilde{d}_{i,t} = d_{i,h_t}/b_{i,h_t} \in D_1 \text{ and } \tilde{d}_t = (\tilde{d}_{1,t}, \dots, \tilde{d}_{n^2,t}) \in D_1^{n^2}.$$

Since the set R is finite, by passing, if necessary, to a subsequence of the sequence $\{h_t\}$, we may assume that there is $r \in R$ such that $r(\tilde{d}_t) = r$ for all $t \in \mathbb{N}^+$. Then, according to (5), for each $t \in \mathbb{N}^+$

(7) $$\inf_{1 \le i \le n^2} |\tilde{d}_{i,t}(r)| \ge a/2.$$

In view of (1) and (6), for each $g \in \mathbf{G}(k)$ we have

$$(8) \qquad d'_{i,t}(g) \overset{\text{def}}{=} \tilde{d}_{i,t}(g)/\tilde{d}_{i,t}(r) = d_{i,h_t}(g)/d_{i,h_t}(r) = \\ = c_i(g^{-1}h_t g)/c_i(r^{-1}h_t r).$$

Since $a > 0$, $\tilde{d}_{i,t} \in D_1$, and D_1 is equicontinuous on M, it follows from (7) that the set $\{d'_{i,t} \mid t \in \mathbb{N}^+,\ 1 \leq i \leq n^2\}$ is equicontinuous on the compact set M having e in its interior. In view of (8) this implies that $\{r^{-1}h_t r \in r^{-1}Hr\}$ is a tolerance sequence. □

(5.5) *Remark 1.* In case $\operatorname{char} k = 0$ Theorem 5.1 is an easy consequence of the Borel-Wang density theorem. Indeed, if $\operatorname{char} k = 0$, then k is a finite field extension of $\mathbb{Q}_p$, where p is prime or ∞. Replacing $\mathbf{G}$ by $R_{k/\mathbb{Q}_p}\mathbf{G}$, where $R_{k/\mathbb{Q}_p}$ is the restriction of scalars from k to $\mathbb{Q}_p$, we may assume that $k = \mathbb{Q}_p$. Since the subgroup H is closed, H is a (p-adic) Lie subgroup of the group $\mathbf{G}(k)$. Let $\mathfrak{H} \subset \operatorname{Lie}(\mathbf{G})_k$ be the Lie algebra of the subgroup H. Since by the Borel-Wang density theorem H is Zariski dense in $\mathbf{G}$, $\mathfrak{H}$ is $(\operatorname{Ad}\mathbf{G})$-invariant. On the other hand, $\mathbf{G}$ is almost simple over k and $\dim \mathfrak{H} > 0$ (because H is not discrete). Thus, $\mathfrak{H} = \operatorname{Lie}(\mathbf{G})_k$ and hence H is open in $\mathbf{G}(k)$. It follows from the finiteness of the covolume of H that H is a subgroup of finite index in $\mathbf{G}(k)$. Thus, by Corollary I.1.5.7, we have $H \supset \mathbf{G}(k)^+$.

Remark 2. In the proof of Theorem 5.1 above (i.e. in 5.1–5.4) we did not make use of the Borel-Wang density theorem. While, in the special case where the subgroups are of finite covolume, this theorem (in the form presented in 4.4) follows easily from Theorem 5.1.

6. Density of Projections and the Strong Approximation Theorem

With A, k_α, $\mathbf{G}_\alpha$, G_α, G_B, $G = G_A$, G_α^+, G_B^+, $G^+ = G_A^+$, $\operatorname{pr}_B\colon G \to G_B$ and, $\operatorname{pr}_\alpha\colon G \to G_\alpha$ throughout this section as in 3.0, we set $A_0 = \{a \in A \mid G_\alpha \text{ is not compact}\}$ and observe that

1) by Proposition I.2.3.6 $A_0 = \{\alpha \in A \mid \mathbf{G}_\alpha \text{ is isotropic over } k_\alpha\}$ and
2) the group G_{A-A_0} is compact.

In this section the group $\mathbf{G}_\alpha$ is assumed to be almost k_α-simple for each $\alpha \in A$. By $\bar{Y} \subset G$ will be denoted the closure of the subset $Y \subset G$. We shall make use of the notion of a subgroup of finite covolume defined at the beginning of Section 5.

(6.1) Lemma. *Let D be a locally compact group, let F be a closed normal subgroup of D, and let $\pi\colon D \to D/F$ be the natural epimorphism. Then for every closed subgroup Λ of finite covolume of the group D the closure $\overline{\pi(\Lambda)}$ of the subgroup $\pi(\Lambda)$ of D/F is a subgroup of finite covolume in D/F.*

Proof. The epimorphism π induces the map $\pi_0\colon \Lambda/D \to \overline{\pi(\Lambda)} \setminus (D/F)$. Let μ be a finite non-zero D-invariant measure on $\Lambda \setminus D$ and let μ_0 be the image of μ

under π_0 (i.e. $\mu_0(U) = \mu(\pi_0^{-1}(U))$, $U \subset \overline{\pi(\Lambda)} \setminus (D/F)$). Then μ_0 is a finite non-zero (D/F)-invariant measure on $\overline{\pi(\Lambda)} \setminus (D/F)$. □

(6.2) Theorem. *Let H be a closed subgroup of finite covolume of the group G. Then*

(a) *there exists a subset $B \subset A_0$ such that $H \supset G_B^+$ and $\mathrm{pr}_{A_0-B}(H)$ is a lattice in G_{A-B};*

(b) *in the case that $\mathbf{G}_\alpha$ is simply connected for each $\alpha \in A_0$ there exists a subset $B \subset A_0$ such that $H \supset G_B$ and $\mathrm{pr}_{A_0-B}(H)$ is a lattice in G_{A_0-B}.*

Proof. (a) We set

$$B = \{\alpha \in A_0 \mid H \supset \mathbf{G}_\alpha(k_\alpha)^+\}.$$

First let us consider the special case where $B = \varnothing$. Let $M = \{\alpha \in A \mid k_\alpha$ is isomorphic to $\mathbb{R}$ or $\mathbb{C}\}$ and let $R_{\mathbb{C}/\mathbb{R}}$ be the restriction of scalars (from $\mathbb{C}$ to $\mathbb{R}$). Replacing $\mathbf{G}_\alpha$ by $R_{\mathbb{C}/\mathbb{R}}\mathbf{G}_\alpha$ for all those $\alpha \in M$ for which $k_\alpha = \mathbb{C}$, we may assume that $k_\alpha = \mathbb{R}$ for all $\alpha \in M$. Since H is of finite covolume in G, by Lemma 2.3 H has property (S) in G, and hence by Lemma 2.4 $\mathrm{pr}_M(H)$ has property (S) in G_M. It follows from Corollary 4.4 that $G_{A_0 \cap M}$ is contained in the Zariski closure of the subgroup $\mathrm{pr}_M(H)$. Since $\mathrm{pr}_M(H)$ normalizes $H \cap G_M$, the Lie algebra L of the group $H \cap G_M$ is invariant under $\mathrm{Ad}\,\mathrm{pr}_M(H)$. Therefore L is $(\mathrm{Ad}\, G_{A_0 \cap M})$-invariant and, since $B = \varnothing$ and for each $\alpha \in A_0 \cap M$ the subgroup $\mathbf{G}_\alpha$ is almost k_α-simple and $G_\alpha^+ = \mathbf{G}_\alpha(\mathbb{R})^0$ (see Theorem I.2.3.1(c)), one has that $(H \cap G_M)^0 \subset G_{M-A_0}$ (from now on in the proof F^0 denotes the identity component of a Lie group F). It then follows from the compactness of the Lie group G_{M-A_0} that

(1) the subgroup $\mathrm{pr}_{A_0}(H \cap G_M) \subset G_{A_0 \cap M}$ is discrete.

Since every local field isomorphic neither to $\mathbb{R}$ nor to $\mathbb{C}$ is totally disconnected, the group G_{A-M} is totally disconnected. Therefore there exists a decreasing sequence $\{U_i\}_{i \in \mathbb{N}^+}$ of open compact subgroups in G_{A-M} such that $\bigcap_{i \in \mathbb{N}^+} U_i = \{e\}$. We shall show that

(2) the subgroup $\mathrm{pr}_{A_0}(H \cap G_{A-M}) \subset G_{A_0-M}$ is discrete.

Suppose the contrary. Since the subgroups $G_{A_0-M} \cdot U_i$ have finite index in G_{A-M} (because G_{A-A_0} is compact and U_i is open, $i \in \mathbb{N}^+$), the subgroups $\mathrm{pr}_{A_0}(H \cap (G_{A_0-M} \cdot U_i))$ are non-discrete, $i \in \mathbb{N}^+$. Therefore one can find $\alpha \in A_0 - M$ such that the subgroups

$$H_i \overset{\text{def}}{=} \mathrm{pr}_\alpha(H \cap (G_\alpha \cdot U_i))$$

are nondiscrete, $i \in \mathbb{N}^+$. We set $H' = \overline{\mathrm{pr}_{A-M}(H)}$ and $H_i' = \mathrm{pr}_\alpha(H' \cap (G_\alpha \cdot U_i))$, $i \in \mathbb{N}^+$. Then 1) since H is a closed subgroup of G of finite covolume and the subgroups U_i are compact and open, by Lemma 6.1 the subgroups H_i', $i \in \mathbb{N}^+$, are closed and of finite covolume in G_α; 2) since $H' \supset H \cap G_{A-M}$, we have that $H_i' \supset H_i$ and hence the subgroups H_i', $i \in \mathbb{N}^+$, are not discrete. It then follows from Theorem 5.1 that $H_i' \supset G_\alpha^+$ for all $i \in \mathbb{N}^+$. But H' is closed, all the subgroups U_i, $i \in \mathbb{N}^+$, are compact, $U_{i+1} \subset U_i$ and $\bigcap_{i \in \mathbb{N}^+} U_i = \{e\}$. Therefore we have that

$H' \supset G_\alpha^+$. On the other hand, 1) since H is closed, H' normalizes $H \cap G_{A-M}$; 2) since $\mathrm{pr}_\alpha(H \cap G_{A-M}) \not\subset \mathscr{Z}(\mathbf{G}_\alpha)$ (because $\mathscr{Z}(\mathbf{G}_\alpha)$ is finite and H_i is not discrete, $i \in \mathbb{N}^+$), making use of Theorem I.1.5.6 and denoting by $[F_1, F_2]$ the mutual commutator subgroup of the subgroups F_1 and F_2, we obtain that

$$[G_\alpha^+, H \cap G_{A-M}] = [G_\alpha^+, \mathrm{pr}_\alpha(H \cap G_{A-M})] \not\subset \mathscr{Z}(\mathbf{G}_\alpha).$$

Thus, the subgroup $[G_\alpha^+, H \cap G_{A-M}]$ is contained in H, non-central in G_α and normalized by the subgroup G_α^+; hence by virtue of Theorem I.1.5.6(i) we have $H \supset G_\alpha^+$. This contradicts $B = \emptyset$ and proves (2).

Let $V_i = \mathrm{pr}_M(H \cap (G_M \cdot U_i))$, $i \in \mathbb{N}^+$. Since H is closed and U_i is compact, one has that V_i is a closed subgroup of the Lie group G_M and hence is a Lie subgroup in G_M. But any decreasing sequence of connected Lie subgroups stabilizes. Therefore one can find a number $j \in \mathbb{N}^+$ such that $V_j^0 = V_m^0$ for all $m \geq j$. Since H is closed, U_i is compact, $U_{i+1} \subset U_i$, and $\bigcap_{i \in \mathbb{N}^+} U_i = \{e\}$, we have that $(H \cap G_M)^0 = V_j^0$. But 1) U_j is open in G_{A-M}; 2) since V_j is a Lie group, V_j^0 is open in V_j. Therefore the subgroup $(H \cap G_M) \cdot (H \cap G_{A-M})$ is open in H. From this, (1), (2), and from the compactness of the group G_{A-A_0} we deduce that

(3) $$\mathrm{pr}_{A_0}(H) \text{ is discrete in } G_{A_0}, \text{ if } B = \emptyset.$$

Let us now consider the general case dropping the assumption that $B = \emptyset$. Since H is closed and of finite covolume in G, G_B/G_B^+ is compact (see Theorem I.2.3.1(b)), and $H \supset G_B^+$, we have that the subgroup $\hat{H} \stackrel{\text{def}}{=} \mathrm{pr}_{A-B}(H)$ is closed and of finite covolume in G_{A-B}. Suppose the subgroup $\mathrm{pr}_{A_0-B}(H) = \mathrm{pr}_{A_0-B}(\hat{H})$ is not discrete. It then follows from (3) that $\hat{H} \supset G_{\alpha_0}^+$ for some $\alpha_0 \in A_0 - B$. But 1) $H \supset G_B^+$; and 2) for each $\alpha \in A_0$ the group G_α/G_α^+ is commutative and $\mathscr{D}(G_\alpha)^+ = G_\alpha^+$ (see Theorems I.2.3.1(c) and I.1.5.6(ii)). Thus, $\mathscr{D}(H) \cap G_B = G_B^+$, $\mathscr{D}(H) \subset G_{A-B} \cdot G_B^+$, and $\mathrm{pr}_{A-B}(\mathscr{D}(H)) = \mathscr{D}(\hat{H}) \supset G_{\alpha_0}^+$. It follows that $\mathscr{D}(H) \supset G_{\alpha_0}^+$, i.e., $\alpha_0 \in B$, a contradiction, for $\alpha_0 \in A - B$. Thus the subgroup $\mathrm{pr}_{A_0-B}(H)$ is discrete and by Lemma 6.1 this completes the proof of assertion (a). Assertion (b) is a direct consequence of (a) and Theorem I.2.3.1(a). □

(6.3) Lemma. *Let $G_0 = G_{A_0}$ and let Λ be a lattice in G_0. Then*

(I) $\mathscr{Z}_{G_0}(\Lambda) = \mathscr{Z}(G_0)$;

(II) *$\mathscr{N}_{G_0}(\Lambda)$ is discrete in G_0 and Λ is of finite index in $\mathscr{N}_{G_0}(\Lambda)$;*

(III) *if $A_0 \neq \emptyset$, then the commutator subgroup $\mathscr{D}(\Lambda)$ is infinite.*

Proof. (I) Since Λ is a lattice in G_0, by Lemmas 2.3 and 2.4 the subgroup $\mathrm{pr}_\alpha(\Lambda)$ has property (S) in G for each $\alpha \in A_0$. Therefore, by Corollary 4.4, $\mathrm{pr}_\alpha(\Lambda)$ is Zariski dense in $\mathbf{G}_\alpha$ for each $\alpha \in A_0$. It follows that $\mathrm{pr}_\alpha(\mathscr{Z}_{G_0}(\Lambda)) \subset \mathscr{Z}(\mathbf{G}_\alpha)$, $\alpha \in A_0$, and hence $\mathscr{Z}_{G_0}(\Lambda) = \mathscr{Z}(G_0)$.

(II) Clearly $\mathscr{Z}_{\mathbf{G}_\alpha}(D)$ is an algebraic subgroup in $\mathbf{G}_\alpha$ and

$$\mathscr{Z}_{\mathbf{G}_\alpha}(D) = \bigcap_{d \in D} \mathscr{Z}_{\mathbf{G}_\alpha}(d)$$

for all $\alpha \in A$ and $D \subset \mathbf{G}_\alpha$. On the other hand, the groups $\mathbf{G}_\alpha$ are Noetherian in the Zariski topology and

$$\mathscr{Z}_{G_0}(F) = \prod_{\alpha \in A_0} \mathscr{Z}_{G_\alpha}(\mathrm{pr}_\alpha(F))$$

for all $F \subset G_0$. Therefore, there is a finite subset $M \cap \Lambda$ such that $\mathscr{Z}_{G_0}(M) = \mathscr{Z}_{G_0}(\Lambda)$. From this and the countability of the group Λ we deduce that the factor group $\mathscr{N}_{G_0}(\Lambda)/\mathscr{Z}_{G_0}(\Lambda)$ is countable. Since $\mathscr{Z}_{G_0}(\Lambda) = \mathscr{Z}(G_0)$ and the groups $\mathbf{G}_\alpha$ are semisimple, $\mathscr{Z}_{G_0}(\Lambda)$ is finite and hence $\mathscr{N}_{G_0}(\Lambda)$ is countable. On the other hand, since Λ is closed in G_0, $\mathscr{N}_{G_0}(\Lambda)$ is closed in G_0. By the Baire category theorem every countable closed subgroup of G_0 is discrete in G_0. Therefore $\mathscr{N}_{G_0}(\Lambda)$ is discrete and, since $\mathscr{N}_{G_0}(\Lambda) \supset \Lambda$ and Λ is a lattice in G_0, the group $\mathscr{N}_{G_0}(\Lambda)/\Lambda$ is finite.

(III) Let $\alpha \in A_0$. As in the proof of (I) it can easily be verified that the subgroup $\mathrm{pr}_\alpha(\Lambda)$ is Zariski dense in $\mathbf{G}_\alpha$. Since $\mathbf{G}_\alpha$ is connected and semisimple, $\mathscr{D}(\mathbf{G}_\alpha) = \mathbf{G}_\alpha$. Thus the group $\mathscr{D}(\mathrm{pr}_\alpha(\Lambda)) = \mathrm{pr}_\alpha(\mathscr{D}(\Lambda))$ is Zariski dense in the infinite group $\mathbf{G}_\alpha$ and hence the group $\mathscr{D}(\Lambda)$ is infinite. □

(6.4) Lemma. *Let $G_0 = G_{A_0}$, Λ a lattice in G_0, and $B \subset A_0$. If* $\mathrm{pr}_B(\Lambda)$ *is discrete in G_B, then $(\Lambda \cap G_B) \cdot (\Lambda \cap G_{A_0-B})$ is of finite index in Λ.*

Proof. Since Λ is a lattice in G_0 and $\mathrm{pr}_B(\Lambda)$ is discrete in G_B, we have that $\mathrm{pr}_B(\Lambda)$ and $\Lambda \cap G_{A_0-B}$ are lattices in G_B and G_{A_0-B}, respectively. Therefore $\Lambda' \overset{\mathrm{def}}{=} \mathrm{pr}_B(\Lambda) \cdot (\Lambda \cap G_{A_0-B})$ is a lattice in G_0 and hence (see Lemma 6.3) Λ' is of finite index in $\mathscr{N}_{G_0}(\Lambda')$. On the other hand, since Λ normalizes both $\mathrm{pr}_B(\Lambda)$ and $\Lambda \cap G_{A_0-B}$, one has $\Lambda \subset \mathscr{N}_{G_0}(\Lambda')$. Therefore $\Lambda' \cap \Lambda$ is of finite index in Λ. Since Λ and Λ' are lattices in G_0, they are commensurable and hence $\Lambda \cap G_B$ and $\mathrm{pr}_B(\Lambda) = \Lambda' \cap G_B$ are commensurable as well. Thus, $(\Lambda \cap G_B) \cdot (\Lambda \cap G_{A_0-B})$ is of finite index in Λ. □

(6.5) Definition. *We call a lattice Λ in G_C, $C \subset A$,* irreducible *if for every $B \subset C$, $B \neq \varnothing$, $B \neq C$, the subgroup $(\Lambda \cap G_B) \cdot (\Lambda \cap G_{C-B})$ is of infinite index in Λ.*

(6.6) *Remark.* Since the group G_{A-A_0} is compact, for every lattice Λ in G the subgroup $\mathrm{pr}_{A_0}(\Lambda)$ is a lattice in G_{A_0}.

(6.7) Theorem. *Let Γ be a lattice in G and let B a non-empty subset of A_0. Suppose* $\mathrm{pr}_{A_0}(\Gamma)$ *is an irreducible lattice in G_{A_0}. Then*

(a) $\overline{\mathrm{pr}_{A-B}(\Gamma)} \supset G^+_{A-B}$;

(a') $\overline{\Gamma \cdot G^+_B} \supset G^+$;

(b) *in case all the $\mathbf{G}_\alpha$, $\alpha \in A_0$, are simply connected,* $\overline{\mathrm{pr}_{A-B}(\Gamma)} \supset G_{A_0-B}$ *and hence* $\overline{\Gamma \cdot G_B} \supset G_{A_0}$;

(c) $\Gamma \cap G_{A-B} \subset G_{A-A_0} \cdot \mathscr{Z}(G)$, *and hence the subgroup $\Gamma \cap G_{A-B}$ is finite;*

(c') $\Gamma \cap (\mathscr{Z}(G_B) \cdot G_{A-B}) \subset G_{A-A_0} \cdot \mathscr{Z}(G)$;

(d) *if $A_0 = A$, then $\Gamma \cap G_{A-B} \subset \mathscr{Z}(G)$;*

(d') *if* $A_0 = A$, *then* $\Gamma \cap (\mathscr{Z}(G_B) \cdot G_{A-B}) \subset \mathscr{Z}(G)$.

Proof. (a) Since Γ is a lattice in G, by virtue of Lemma 6.1, $\overline{\mathrm{pr}_{A-B}(\Gamma)}$ is a subgroup of finite covolume in G_{A-B}. Therefore by Theorem 6.2(a) there exists a subset $C \subset A_0 - B$ such that $\overline{\mathrm{pr}_{A-B}(\Gamma)} \supset G_C^+$ and $\mathrm{pr}_{A_0-B-C}(\overline{\mathrm{pr}_{A-B}(\Gamma)})$ is a lattice in G_{A_0-B-C}. Since the lattice $\mathrm{pr}_{A_0}(\Gamma)$ is irreducible and $A_0 - B - C \neq A_0$ (for $B \neq \varnothing$), it follows from the discreteness of the subgroup $\mathrm{pr}_{A_0-B-C}(\mathrm{pr}_{A-B}(\Gamma)) = \mathrm{pr}_{A_0-B-C}(\mathrm{pr}_{A_0}(\Gamma))$ and Lemma 6.4 that $A_0 - B - C = \varnothing$. Thus, $C = A_0 - B$ and $\overline{\mathrm{pr}_{A-B}(\Gamma)} \supset G^+_{A_0-B} = G^+_{A-B}$.

(a′) It follows from (a) that $\overline{\Gamma \cdot G_B} \supset G^+$. On the other hand, 1) since by Theorem I.2.3.1(c) the factor group G_α/G_α^+ is commutative for each $\alpha \in A_0$, one has $\mathscr{D}(G_B) \subset G_B^+$; 2) since by Theorem I.1.5.6 (ii) the group G_α^+ coincides with its commutator subgroup for each $\alpha \in A$, one has $\mathscr{D}(G^+) = G^+$. Therefore

$$\overline{\Gamma \cdot G_B^+} \supset \overline{\Gamma \cdot \mathscr{D}(G_B)} \supset \mathscr{D}(\overline{\Gamma \cdot G_B}) \supset \mathscr{D}(G^+) = G^+.$$

(b) follows from (a) and Theorem I.2.3.1(a).

(c) We set $\Gamma_0 = \Gamma \cap G_{A-B}$. It is easily seen that $\mathrm{pr}_{A-B}(\Gamma)$ normalizes Γ_0. Since the group Γ_0 is discrete, and hence closed, its normalizer is closed as well. Therefore, in view of assertion (a), $G^+_{A_0-B}$ normalizes Γ_0. Hence, G^+ normalizes $\mathrm{pr}_\alpha(\Gamma_0)$ for all $\alpha \in A_0 - B$. But Γ_0 is countable and according to Theorem I.1.5.6 (i) if $\alpha \in A_0$, then any subgroup of G_α normalized by G_α^+ is either contained in $\mathscr{Z}(\mathbf{G}_\alpha)$ or has the cardinality of the continuum. Thus $\mathrm{pr}_\alpha(\Gamma_0) \subset \mathscr{Z}(\mathbf{G}_\alpha)$ for all $\alpha \in A_0 - B$ and hence $\Gamma \cap G_{A-B} = \Gamma_0 \subset G_{A-A_0} \cdot \mathscr{Z}(G)$.

(c′) follows from (a) applied to the lattice $\Gamma \cdot \mathscr{Z}(G)$.

Assertions (d) and (d′) follow from (c) and (c′), respectively. □

(6.8) In this paragraph we shall state and prove the strong approximation theorem for semisimple groups over global fields. Let K be a global field. We denote by $\mathscr{R}$ the set of all (inequivalent) valuations of K and by $\mathscr{R}_\infty \subset \mathscr{R}$ the (finite) set of archimedean valuations. Let K_v and $\mathscr{O}_v$ be as in I.0.32. Let $\mathbf{G}$ be a connected simply connected almost K-simple K-subgroup of $\mathbf{SL}_n$. As in I.0.33 we denote by $\mathbf{G}(\mathbb{A}_K)$ the adèle group of the group $\mathbf{G}$ and identify $\mathbf{G}(K)$ via the diagonal embedding with the subgroup of the principal adèles of the group $\mathbf{G}$. We set $\mathscr{T} = \{v \in \mathscr{R} \mid$ the group $\mathbf{G}$ is anisotropic over K_v, or, equivalently, $\mathbf{G}(K_v)$ is compact$\}$. We observe as in I.3.2.3 that $\mathscr{T}$ is finite.

We denote by G_B, where $B \subset \mathscr{R}$, the subgroup in $\mathbf{G}(\mathbb{A}_K)$ consisting of the adèles whose v-components, for all $v \notin B$, are equal to the identity. We denote by π_B the natural projection $\mathbf{G}(\mathbb{A}_K) \to G_B$.

Theorem. (Strong approximation theorem.) *Let* $\mathscr{E} \subset \mathscr{R}$. *Suppose the group* $G_{\mathscr{E}}$ *is not compact or, equivalently,* $\mathscr{E} \not\subset \mathscr{T}$. *Then the subgroup* $\mathbf{G}(K)G_{\mathscr{E}}$ *is dense in* $\mathbf{G}(\mathbb{A}_K)$ *or, in other words, the subgroup* $\pi_{\mathscr{R}-\mathscr{E}}(\mathbf{G}(K))$ *is dense in* $G_{\mathscr{R}-\mathscr{E}}$.

Proof. Since $\mathbf{G}$ is representable in the form $R_{K'/K}\mathbf{G}'$, where $\mathbf{G}'$ is absolutely almost simple and simply connected (see I.1.7), replacing $\mathbf{G}$ by $\mathbf{G}'$ and K by K' we may assume that $\mathbf{G}$ is absolutely almost simple (and simply connected). If $B \subset \mathscr{R}$ and $B \supset \mathscr{R}_\infty$, we set

$$U_B = \prod_{v \in B} \mathbf{G}(K_v) \cdot \prod_{v \notin B} \mathbf{G}(\mathcal{O}_v)$$

and observe that U_B is an open subgroup of $\mathbf{G}(\mathbb{A}_K)$. By virtue of Theorem I.3.2.2 $\mathbf{G}(K)$ is a lattice in $\mathbf{G}(\mathbb{A}_K)$. Therefore $\mathbf{G}(K) \cap U_B$ is a lattice in the open subgroup U_B and hence $\Gamma_B \stackrel{\text{def}}{=} \pi_B(\mathbf{G}(K) \cap U_B)$ is a lattice in G_B. Clearly the lattice Γ_B is irreducible (for any finite B). Applying Theorem 6.7(b) to the lattice Γ_B we obtain that G_{B-C-T} is contained in the closure of the subgroup $\pi_{B-C}(\Gamma_B)$ for every finite set B such that $\mathscr{R}_\infty \subset B \subset \mathscr{R}$ and for every non-empty set $C \subset B - T$. But $\mathscr{E} \not\subset \mathscr{R}$. Hence the closure of the subgroup $\pi_{B-E}(\Gamma_B)$ containes G_{B-E-T} for every finite set B such that $\mathscr{R}_\infty \subset B \subset \mathscr{R}$. This is equivalent (by the definition of the topology in $\mathbf{G}(\mathbb{A}_K)$) to the following

(*) the closure of the subgroup $\mathbf{G}(K)G_{\mathscr{E}}$ contains $G_{\mathscr{R}-\mathscr{T}}$.

According to the theorem on weak approximation (see [Kn 2], [Har 1], and [Pl-J]) the subgroup $\mathbf{G}(K)G_{\mathscr{R}-B}$ is dense in $\mathbf{G}(\mathbb{A}_K)$ for every finite $B \subset \mathscr{R}$. But $\mathscr{T}$ is finite. So

(**) $\mathbf{G}(K)G_{\mathscr{R}-\mathscr{T}}$ is dense in $\mathbf{G}(\mathbb{A}_K)$.

The assertion of the theorem now follows from (*) and (**). □

(1) *Remark.* In the case in which $\mathscr{T}$ consists entirely of archimedean valuations (**) follows easily from the following observations:

1) in any connected $\mathbb{Q}$-group $\mathbf{F}$ the set $\mathbf{F}(\mathbb{Q})$ is Zariski dense;
2) $G_{\mathscr{T}}$ is compact;
3) any closed subgroup of a compact real algebraic group is algebraic.

On the other hand (see [B-T 3]), any anisotropic almost simple group over a non-archimedean local field is an "inner form" of type $\mathbb{A}$. Therefore to obtain (**) it suffices to prove the weak approximation property in case $\mathbf{G}$ is a K-anisotropic group of type $\mathbb{A}$. In this case (see [Ti 2]) $\mathbf{G}(K)$ is either (a) $\mathbf{SL}_1(D) = \{x \in D \mid v(x) = 1\}$, where D is a finite-dimensional central simple associative algebra over K and v is the reduced norm on D, or (b) $\{x \in D \mid xx^\sigma = 1$ and $v(x) = 1\}$, where D is a finite-dimensional central simple associative algebra over a quadratic field extension K' of K with reduced norm v and involution σ of the second kind such that $K = \{x \in K' \mid x^\sigma = x\}$. But for such groups the weak approximation property follows easily from the strong approximation property for the additive group (see [Hum 2], 6.3) and the fact (see [Pl-J]) that the commutator subgroup of the group of invertible elements of a finite-dimensional central simple associative algebra over a local field coincides with the group of elements of reduced norm 1; in case (b) it is also necessary to make use of the following consequence of Hilbert's theorem 90: if $x \in D$ and $xx^\sigma = 1$, then $x = y/y_\sigma$, where $y \in K(x)$. Here we also use the fact that the K-rational points are dense in $\{z \in D \mid z$ commutes with $z^\sigma\}$. This easily follows from the observation that the K-rational points are dense in $\{z \in D \mid \sigma(z) = -z\}$, i.e. in the set of the imaginary points with respect to σ.

(2) *Remark.* In view of I.3.2.3 (*) the strong approximation theorem may be restated as follows. With $\mathscr{E} \subset S \subset \mathscr{R}$ and $\mathbf{G}(K(S))$ as in Section 3 of Chapter I suppose that $\mathscr{E} \not\subset \mathscr{T}$. Then under the diagonal embedding into $G_{S-\mathscr{E}}$ the subgroup $\mathbf{G}(K(S))$ is dense in $G_{S-\mathscr{E}}$.

7. Ergodicity of Actions on Quotient Spaces

With A, k_α, $\mathbf{G}_\alpha$, G_α, G_B, $G = G_A$, G_α^+, G_B^+, $G^+ = G_A^+$, $\mathrm{pr}_B\colon G \to G_B$, $\mathrm{pr}_\alpha\colon G \to G_\alpha$, $\mathbf{S}_\alpha$, S, and $A(s)$ throughout this section as in 3.0 we set as in Section 6 $A_0 = \{\alpha \in A \mid \text{the group } G_\alpha \text{ is not compact}\}$. Besides, we shall use the notation introduced in I.4.2. In particular, for every locally compact group H and every discrete subgroup $\Lambda \subset H$ we denote, respectively, by $\mathfrak{M}(H) = \mathfrak{M}(H, \mu_H)$ and $\mathfrak{M}(\Lambda \setminus H) = \mathfrak{M}(\Lambda \setminus H, \mu_H)$ the algebras of classes of measurable sets in H and $\Lambda \setminus H$.

(7.1) Let H be a locally compact σ-compact group, Λ a discrete subgroup in H, and $h \in H$. We say that h *acts ergodically* on $\Lambda \setminus H$ if the transformation $x \mapsto xh$, $x \in \Lambda \setminus H$, of the space $\Lambda \setminus H$ with measure μ_H is ergodic, i.e., if the following condition is satisfied: for each $Y \in \mathfrak{M}(\Lambda \setminus H)$ with $Yh = Y$ either $\mu_H(Y) = 0$ or $\mu_H((\Lambda \setminus H) - Y) = 0$. The ergodicity of the action of the element h on $\Lambda \setminus H$ is equivalent (in view of Lemma I.4.1.3(i)) to the following condition: if $B \in \mathfrak{M}(H)$ and $\Lambda \cdot B \cdot h = B$, then either $\mu_H(B) = 0$ or $\mu_H(H - B) = 0$.

(7.2) Theorem. *Let Γ be a lattice in G and $s \in S \cap G^+$. Suppose the group $\mathbf{G}_\alpha$ is almost k_α-simple for each $\alpha \in A$. Denote by G' the closure in G of the subgroup $\Gamma \cdot G^+$.*

(a) *If the subgroup $\Gamma \cdot G_{A(s)}^+$ is dense in G', then s acts ergodically on $\Gamma \setminus G'$;*
(b) *If $A(s) \neq \varnothing$ and $\mathrm{pr}_{A_0}(\Gamma)$ is an irreducible lattice in G_{A_0}, then s acts ergodically on $\Gamma \setminus G'$.*

Proof. (a) Let π be the natural map $G' \to \Gamma \setminus G'$ and let ρ be a quasi-regular representation of the group G on the space $L_2(\Gamma \setminus G', \mu_{G'})$ (i.e. $(\rho(g)f)(x) = f(xg)$, $g \in G'$, $x \in \Gamma \setminus G'$, $f \in L_2(\Gamma \setminus G', \mu_{G'})$. Let $Y \in \mathfrak{M}(\Gamma \setminus G')$ be a class such that $Ys = Y$. We set $B = \pi^{-1}(Y) \in \mathfrak{M}(G')$ and denote by χ_Y the characteristic function of the class Y. Since Γ is a lattice in G and the subgroup G' is closed and contains Γ, the subgroup Γ is a lattice in G' (see I.0.40). Therefore $\chi_Y \in L_2(\Gamma \setminus G', \mu_{G'})$. Since $Ys = Y$, we have that $\rho(s)\chi_Y = \chi_Y$, and hence (see Proposition 3.3(a)) $\rho(G_{A(s)}^+)\chi_Y = \chi_Y$. Thus, $B \cdot G_{A(s)}^+ = B$. It then follows from Lemma I.4.1.1(ii) that there exists a measurable set B' in the class B such that $B' \cdot G_{A(s)}^+ = B'$. Since $G_{A(s)}^+$ is normal in G, we have $G_{A(s)}^+ \cdot B' = B'$, and hence $G_{A(s)}^+ \cdot B = B$. But $\Gamma \cdot B = B$, the subgroup $\{g \in G \mid gB = B\}$ is closed in G (see Lemma I.4.1.1(i)), and the subgroup $\Gamma \cdot G^+_{A(s)}$ is dense in G'. Therefore $G' \cdot B = B$. It then follows from Lemma I.4.1.1(iii) that either $\mu_{G'}(B) = 0$ or $\mu_{G'}(G' - B) = 0$, and hence either $\mu_{G'}(Y) = 0$ or $\mu_{G'}((\Gamma \setminus G') - Y) = 0$.

(b) is a consequence of (a) and Theorem 6.7 (a'). □

(7.3) Corollary. *Let Γ be a lattice in G and let $s \in S$. Suppose the group $\mathbf{G}_\alpha$ is simply connected and almost k_α-simple for each $\alpha \in A$.*

(a) *If $\Gamma \cdot G_{A(s)}$ is dense in G, then s acts ergodically on $\Gamma \setminus G$;*

(b) *If $A(s) \neq \varnothing$, the subgroup $\mathrm{pr}_{A-A_0}(\Gamma)$ is dense in G_{A-A_0}, and $\mathrm{pr}_{A_0}(\Gamma)$ is an irreducible lattice in G_A, then s acts ergodically on $\Gamma \setminus G$;*

(c) *If $A(s) \neq \varnothing$, the groups $\mathbf{G}_\alpha$ are isotropic over k_α, $\alpha \in A$, and the lattice Γ is irreducible, then s acts ergodically on $\Gamma \setminus G$.*

Assertions (a) and (b) follows from the appropriate parts of Theorem 7.2 and from Theorem I.2.3.1(a), and (c) is a special case of (b).

Chapter III. Property (T)

A locally compact group is said to have property (T) if the trivial one-dimensional representation is isolated in the space of its irreducible unitary representations. This notion is due to Kazhdan (see [Kaz 1]). It was also established in that paper that

(1) if a discrete group Γ has property (T), then Γ is finitely generated and the factor group $\Gamma/\mathscr{D}(\Gamma)$ is finite;
(2) if Γ is a lattice in a simple Lie group of rank greater than 2, then Γ has property (T).

These results of Kazhdan are presented in Sections 2 and 5 in somewhat greater generality than in the original paper [Kaz 1]. In Sections 1 and 4 we provide some results from representation theory which are used in sections 2 and 5. In Section 3 a theorem of Watatani asserting that the groups with property (T) are not amalgams has been proved. There we present not only the original proof of Watatani but also a sketch of the proof due to Serre. In Section 6 a series of results on the structure of closed subgroups of a locally compact group H are obtained in the case where H, possibly without property (T), contains a "sufficiently large" normal subgroup with this property.

Throughout this chapter H will denote a locally compact second countable group, μ_H a fixed left invariant Haar measure on H, and $\tilde{H}$ (resp. $\hat{H}$) the set of equivalence classes of (continuous) unitary (resp. irreducible unitary) representations of the group H on separable Hilbert spaces. We shall not distinguish a unitary representation from its equivalence class. The space of a representation $\rho \in \tilde{H}$ will be denoted by $L(\rho)$. For $\rho \in \tilde{H}$, we set

$$L(\rho)^H = \{x \in L(\rho) \mid \rho(H)x = x\}.$$

Let us denote by $I_H \in \hat{H}$ the trivial one-dimensional representation of the group H. We say that a representation $\rho \in \tilde{H}$ *contains* I_H and denote this by $\rho \geq I_H$ if $L(\rho)^H \neq \varnothing$. We write $\mathfrak{A}(H)$ for the set of non-negative continuous functions f on H with compact support and $\int_H f\, d\mu_H = 1$.

1. Representations Which Are Isolated from the Trivial One-Dimensional Representation

(1.0) For a compact set $K \subset H$ and $\varepsilon > 0$, we set $W(\varepsilon, K) = \{\rho \in \tilde{H} \mid$ there is $y \in L(\rho)$ such that $\|\rho(h)y - y\| < \varepsilon\|y\|$ for all $h \in K\}$. Clearly, the sets $W(\varepsilon, K)$ form a local neighbourhood base at I_H relative to the standard topology defined in I.5.1.1. Since $\tilde{H}$ consists of unitary representations, we have that

(1) $$\|\rho(f)\| \leq \int_H |f| d\mu_H \text{ for all } f \in L_1(H, \mu_H) \text{ and } \rho \in \tilde{H},$$

and hence

(2) $$\|\rho(f)\| \leq 1 \text{ for all } f \in \mathfrak{A}(H) \text{ and } \rho \in \tilde{H}.$$

For $f \in \mathfrak{A}(H)$ and $\varepsilon > 0$, we set $W(\varepsilon, f) = \{\rho \in \tilde{H} \mid \|\rho(f)\| > 1 - \varepsilon\}$.

(1.1) Lemma.
(a) *$W(\varepsilon, f) \supset W(\varepsilon, \operatorname{supp} f)$ for all $\varepsilon > 0$ and $f \in \mathfrak{A}(H)$.*
(b) *For any $\varepsilon > 0$ and any compact set $K \subset H$ there exist $\delta > 0$ and a function $f \in \mathfrak{A}(H)$ such that $W(\varepsilon, K) \supset W(\delta, f)$.*
(c) *The sets $W(\varepsilon, f)$, $\varepsilon > 0$, $f \in \mathfrak{A}(H)$ form a local neighbourhood base at I_H in $\tilde{H}$.*

Proof. If $\rho \in \tilde{H}$, $u \in L(\rho)$, and $f \in \mathfrak{A}(H)$, then

(1) $$\|\rho(f)y - y\| = \|\int_H f(h)\rho(h)y\, d\mu_H(h) - \int_H f(h)y\, d\mu_H(h)\| =$$
$$= \|\int_H f(h)(\rho(h)y - y)\, d\mu_H(h)\| \leq \int_H f(h)\|\rho(h)y - y\|\, d\mu_H(h) \leq$$
$$\leq \sup_{h \in \operatorname{supp} f} \|\rho(h)y - y\|$$

This implies (a). We will show (b). Pick an element $d \in \mathfrak{A}(H)$. There exist $f \in \mathfrak{A}(H)$ and $\lambda > 0$ such that $f \geq \lambda((hd) + d)$ for all $h \in K$, where the function hd is defined by the equality $(hd)(h') = d(h^{-1}h')$, $h' \in H$. Now let $\rho \notin W(\varepsilon, K)$ and $y \in L(\rho)$. Then

$$\|\rho(h_0)\rho(d)y - \rho(d)y\| \geq \varepsilon\|\rho(d)y\|$$

for some $h_0 \in K$. But

$$\|u + v\|^2 = 2\|u\|^2 + 2\|v\|^2 - \|u - v\|^2$$

for all elements u, v of the Hilbert space $L(\rho)$, the operator $\rho(h_0)$ is unitary and $\rho(h_0)\rho(d) = \rho(h_0 d)$. Therefore, in view of inequality (2) in 1.0, one has

$$\|\rho((h_0 d) + d)y\| \leq \sqrt{4 - \varepsilon^2}\|\rho(d)y\| \leq \sqrt{4 - \varepsilon^2}\|y\|.$$

Since the function

$$p \stackrel{\text{def}}{=} f \dot{-} \lambda((h_0 d) + d)$$

is non-negative, making use of inequality (1) in 1.0 and the left invariance of μ_H, we obtain that

$$\begin{aligned}\|\rho(f)y\| &\le \lambda\|\rho((h_0 d) + d)y\| + \|\rho(p)y\| \le \\ &\le \lambda\sqrt{4-\varepsilon^2}\|y\| + (1-2\lambda)\|y\| = \\ &= [1 - \lambda(2 - \sqrt{4-\varepsilon^2})]\|y\|\end{aligned}$$

for all $y \in L(\rho)$. Hence $\|\rho(f)\| \le 1 - \delta$ for each $\rho \notin W(\varepsilon, K)$, where $\delta = \lambda(2 - \sqrt{4-\varepsilon^2}) > 0$. Thus $W(\varepsilon, K) \supset W(\delta, f)$ and this implies (b). Assertion (c) is a straightforward consequence of (a) and (b). □

(1.2) We say that a representation $\rho \in \tilde{H}$ is *close to the trivial one-dimensional representation* I_H if I_H belongs to the closure of the singleton $\{\rho\}$, i.e., if $\rho \in W(\varepsilon, K)$ for each $\varepsilon > 0$ and for every compact $K \subset H$. Otherwise (i.e. if $\rho \notin W(\varepsilon, K)$ for a compact subset $K \subset H$ and some $\varepsilon > 0$)we say that ρ is *isolated from the trivial one-dimensional representation* I_H.

Let $\rho \in \tilde{H}$. We say that a sequence $\{y_i\}_{i \in \mathbf{N}^+}$ in $L(\rho)$ is *asymptotically* $\rho(H)$-*invariant* if $y_i \neq 0$ for all sufficiently large i and

$$\overline{\lim_{i\to\infty}} \sup_{h \in K} \|\rho(h)y_i - y_i\| / \|y_i\| = 0$$

for every compact $K \subset H$. Since H is σ-compact, ρ is close to I_H if and only if there exists an asymptotically $\rho(H)$-invariant sequence in $L(\rho)$.

Lemma 1.1(c) implies

(1.3) Proposition. *For each* $\rho \in \tilde{H}$ *the following conditions are equivalent:*

(I) ρ *is isolated from the trivial one-dimensional representation;*
(II) *there is a function* $f \in \mathfrak{A}(H)$ *such that* $\|\rho(f)\| < 1$.

(1.4) Proposition. *Let* $\rho \in \tilde{H}$ *and let* $\rho = \int_X \rho_x d\mu(x)$ *be the decomposition of* ρ *into a continuous sum of irreducible representations* $\rho_x \in \hat{H}$. *Then the following conditions are equivalent:*

(I) ρ *is isolated from the trivial one-dimensional representation;*
(II) *there is a neighbourhood* W *of* I_H *such that* $\rho_x \notin W$ *for almost all (with respect to the measure* μ*)* $x \in X$.

Proof. Since $\rho(f) = \int_X \rho_x(f) d\mu(x)$, one has

$$\|\rho(f)\| = \operatorname{vrai}_\mu \sup \|\rho_x(f)\|$$

for each $f \in \mathfrak{A}(H)$, where $\operatorname{vrai}_\mu \sup$ denotes the essential least upper bound relative to the measure μ. This implies, in view of Lemma 1.1(c), the equivalence of (I) and (II). □

(1.5) Let X be a complex normed linear space and let $\{j_i\}_{i \in \mathbf{N}^+}$ be a sequence of μ_H-measurable maps of H into X. We say that the sequence $\{j_i\}_{i \in \mathbf{N}^+}$ is *equicontinuous on compact sets* if for every compact $K \subset H$ and every $\varepsilon > 0$ there exists a neighbourhood R of the identity in H such that

$$\|j_i(rh) - j_i(h)\| < \varepsilon$$

for all $h \in K$, $r \in R$, and $i \in \mathbb{N}^+$. We say that the sequence $\{j_i\}$ is *uniformly bounded on compact sets* if

$$\sup_{i \in \mathbb{N}^+, h \in K} \|j_i(h)\| < \infty,$$

for every compact $K \subset H$. We say that the sequence $\{j_i\}$ is *mean uniformly bounded* if

$$\sum_{i \in \mathbb{N}^+} \int_K \|j_i(h)\| d\mu_H(h) < \infty$$

for every compact $K \subset H$. It follows from Hölder's inequality applied to the characteristic functions χ_K of compact sets $K \subset H$ and the functions $\|j_i(h)\|$ that the sequence $\{j_i\}$ is mean uniformly bounded whenever

$$\sup_{i \in \mathbb{N}^+} \int_K \|j_i(h)\|^p d\mu_H(h) < \infty$$

for some integer $p > 1$ and every compact $K \subset H$.

(1.6) Lemma. *If a sequence $\{i_j\}_{i \in \mathbb{N}^+}$ of complex-valued measurable functions $j_i \colon H \to \mathbb{C}$ is equicontinuous and uniformly bounded on compact sets, then it has a subsequence converging uniformly on every compact set to a continuous function j.*

The assertion of this lemma is a generalization of the classical Ascoli theorem and is a special case of Theorem 2 in paragraph 5 of §2 of Chapter X in [Bou 1]. Notice that the proof of the lemma is a direct generalization of that of Ascoli's theorem.

(1.7) Lemma. *Let f be a continuous function on H with compact support and let $\{j_i\}_{i \in \mathbb{N}^+}$ be a sequence of measurable maps on H with values in a complex normed linear space X. Suppose the sequence $\{j_i\}$ is mean uniformly bounded and set*

$$p_i(h) = \int_H f(x) j_i(hx)\, d\mu_H(x), \quad h \in H.$$

Then the sequence $\{p_i\}_{i \in \mathbb{N}^+}$ is equicontinuous and uniformly bounded on compact sets.

Since this assertion is, in fact, a special case of standard theorems on smoothing operators, we shall confine ourselves to a brief argument. It can easily be verified that for all $x, y \in H$ and $i \in \mathbb{N}^+$ one has

$$\|p_i(xy) - p_i(y)\| \leq \int_H |f(y^{-1}x^{-1}h - f(y^{-1}h)| \, \|j_i(h)\| d\mu_H(h) \tag{1}$$

and

$$\|p_i(x)\| \le \int_H |f(x^{-1}h)| \|j_i(h)\| d\mu_H(h). \tag{2}$$

Since f is continuous with compact support and $\{j_i\}$ is mean uniformly bounded, we obtain from (1) that the sequence $\{p_i\}$ is equicontinuous on compact sets and from (2) that $\{p_i\}$ is uniformly bounded on compact sets.

(1.8) Let F be a closed subgroup of the group H such that there exists a (nonzero) H-invariant measure μ on $F \setminus H$ and let $\pi\colon H \to F \setminus H$ be the natural map. For every unitary representation Y of the group F on a Hilbert space Y we denote by $\operatorname{Ind}(H, F, U)$ or simply by $\operatorname{Ind} U$ the unitary representation of H induced in the sense of Mackey from the representation U. We recall that the space $L_2(H, F, U)$ of the representation $\operatorname{Ind} U$ consists of measurable vector valued functions $j\colon H \to Y$ with

$$j(zh) = U(z)j(h), \; z \in F, \; h \in H, \tag{1}$$

and with

$$\|j\|^2 \stackrel{\text{def}}{=} \int_{F\setminus H} \|j(\pi^{-1}(x)\|^2 d\mu(x) < \infty.$$

The representation $\rho = \operatorname{Ind} U$ is defined by the formula

$$(\rho(h)j)(x) = j(xh), \tag{2}$$

where $j \in L_2(H, F, U)$ and $h, x \in H$. We now denote by τ the quasi-regular representation of the group H on the space $L_2(F \setminus H, \mu)$, i.e., $(\tau(h)f)(x) = f(xh)$, $x \in F \setminus H$, $h \in H$, and $f \in L_2(F \setminus H, \mu)$. There is the natural isometry $\tilde{\pi}\colon L_2(F \setminus H, \mu) \to L_2(H, F, I_F)$, $\tilde{\pi}(p) = p \circ \pi$, $p \in L_2(F \setminus F, \mu)$ which identifies τ with $\operatorname{Ind} I_F$. If the measure μ is finite, we set

$$L_2^0 = \{f \in L_2(F \setminus H, \mu) \mid \int_{F\setminus H} f(x)d\mu(x) = 0\}.$$

If μ is infinite, we set $L_2^0 = L_2(F \setminus H, \mu)$. The linear subspace L_2^0 is closed in $L_2(F \setminus H, \mu)$. Since μ is H-invariant, it follows that τ is unitary and the subspace L_2^0 is $\tau(H)$-invariant. The subgroup F is called *weakly cocompact* if the restriction of the representation τ to L_2^0 is isolated from I_H.

(1.9) Lemma. *With F, μ, and τ as in 1.8 the following conditions are equivalent:*

(I) *the subgroup F is not weakly cocompact;*

(II) *there is an asymptotically $\tau(H)$-invariant (in the sense of 1.2) sequence $\{q_i\}_{i\in\mathbb{N}^+}$ in $L_2(F \setminus H, \mu)$ such that*

$$\inf_{i\in\mathbb{N}^+} \|q_i\| > 0 \text{ and}$$

$$\lim_{i\to\infty} \int_K |q_i(x)|^2 \, d\mu(x) = 0$$

for any compact $K \subset F \setminus H$.

Proof. (I) $\Rightarrow$ (II). Choose an asymptotically $\tau(H)$-invariant sequence $\{p_i\}_{i\in\mathbb{N}^+}$ in $L_2^0 - \{0\}$ (see 1.2). Let $\pi\colon H \to F \setminus H$ be the natural map and let $f \in \mathfrak{A}(H)$. By virtue of Lemma I.4.3.2 for every compact $K \subset H$, there exists a constant $c(K) > 0$ such that $\mu_H(\pi^{-1}(A) \cap K) \leq c(K)\mu(A)$ for every measurable set $A \subset F \setminus H$. It follows that the sequence $\{(p_i/\|p_i\|) \circ \pi\}_{i\in\mathbb{N}^+}$ is mean uniformly bounded. On the other hand,

1) since $\{p_i\}$ is asymptotically $\tau(H)$-invariant and $f \in \mathfrak{A}(H)$, if follows from inequality (1) in 1.1 that

$$\lim_{i\to\infty} \|\tau(f)p_i - p_i\|/\|p_i\| = 0;$$

2) it follows from the definition of τ that for every $p \in L_2(F \setminus H, \mu)$ and $h \in H$,

$$(\tau(f)p) \circ \pi(h) = \int_H f(x)(p \circ \pi)(hx) d\mu_H(x).$$

Therefore, replacing p_i by $\tau(f)p_i/\|\tau(f)p_i\|$, we may assume (see Lemma 1.7) that $\|p_i\| = 1$ and the sequence $\{p_i \circ \pi\}$ is equicontinuous and uniformly bounded on compact sets. Hence, by passing to a subsequence (see Lemma 1.6) we can assume that $\{p_i\}$ converges uniformly on any compact set to a continuous function p. Since $\|p_i\| = 1$ and the sequence $\{p_i\}$ is asymptotically $\tau(H)$-invariant, we have that $p \in L_2(F \setminus H, \mu)$ and $\tau(H)p = p$. Therefore, p is constant, and hence p is orthogonal to L_2^0. But $p_i \in L_2^0$ and $\|p_i\| = 1$. Thus, $\|p_i - p\| \geq 1$ and, since $\{p_i\}$ converges to p uniformly on any compact subset, the sequence $\{q_i = p_i - p\}$ has the desired properties.

(II) $\Rightarrow$ (I). Let $\tilde{q}_i$ be the projection of the vector q_i to L_2^0. Since $\inf_{i\in\mathbb{N}^+} \|q_i\| > 0$,

$$\lim_{i\to\infty} \int_K |q_i(x)|^2 d\mu(x) = 0$$

for any compact set $K \subset F \setminus H$, and the function $p(x) \equiv 1$, $x \in F \setminus H$, is approximated in $L_2(F \setminus H, \mu)$ by the characteristic functions of compact sets, it follows that

$$\lim_{i\to\infty} \|\tilde{q}_i - q_i\|/\|q_i\| = 0.$$

This implies, since $\{q_i\}$ is asymptotically $\tau(H)$-invariant, that the sequence $\{\tilde{q}_i\}$ is asymptotically $\tau(H)$-invariant and hence (I). □

(1.10) Corollary. *With F and μ as in 1.8, assume $F \setminus H$ is compact. Then F is weakly cocompact.*

To prove this assertion it suffices to apply Lemma 1.9 and to recall that

$$\|f\|^2 = \int_{F\setminus H} |f(x)|^2 d\mu(x), \quad f \in L_2(F \setminus H, \mu).$$

(1.11) Proposition. *With F and μ as in 1.8, let U be a continuous unitary representation of the subgroup F and let $\rho = \mathrm{Ind}(H, F, U)$. Then*

(a) *if* $\mu(F \setminus H) < \infty$ *and* U *is close to* I_F, *then* ρ *is close to* I_H;
(b) *if the subgroup* F *is weakly cocompact and* U *is isolated from* I_F, *then* ρ *is isolated from* I_H.

Proof. (a) Since $\mu(F\backslash H) < \infty$, the representation $\operatorname{Ind} I_F$ contains I_H. It then follows from the closeness of U to I_F and the continuity of the inducing operation (see I.5.2.2) that ρ is close to I_H.

(b) Suppose the contrary. Then (see 1.2) there is an asymptotically $\rho(H)$-invariant sequence $\{j_i\}_{i\in\mathbf{N}^+}$ in $L_2(H,F,U) - \{0\}$. Take $f \in \mathfrak{A}(H)$. Using the same argument as at the beginning of the proof of Lemma 1.9 and replacing j_i by $\rho(f)j_i/\|\rho(f)j_i\|$ we may assume that $\|j_i\| = 1$ and the sequence $\{j_i\}$ is equicontinuous and uniformly bounded on compact sets. Then so are the sequences $\{c_i\}$ and $\{d_i\}$, where $c_i(h) = \|j_i(h)\|$ and

$$d_i(h_1, h_2) = \|j_i(h_1) - j_i(h_2)\|, \quad h, h_1, h_2 \in H.$$

Therefore, according to Lemma 1.6, by passing to a subsequence we can assume that the sequences $\{c_i\}$ and $\{d_i\}$ converge on every compact set to continuous functions $c(h)$ and $d(h_1, h_2)$ respectively. Then, since $\|j_i\| = 1$ and $\{j_i\}$ is asymptotically $\rho(H)$-invariant, $d(h_1, h_2) \equiv 0$. But, since U is isolated from I_F, it follows from equality (1) in 1.8 that there are a compact set $K \subset F$ and $\varepsilon > 0$ such that $\sup_{x\in K} d_i(zh, h) \geq \varepsilon c_i(h)$, and hence $\sup_{z\in K} d(zh, h) \geq \varepsilon c(h)$ for all $h \in H$. Thus, $c(h) \equiv 0$. Let π, $\tilde{\pi}$, and τ be as in 1.8. Since $j_i \in L_2(H, F, U)$, we have that $c_i \in L_2(H, F, I_F)$ and $\|c_i\| = \|j_i\| = 1$. Let us set

$$\tilde{c}_i = c_i \circ \pi^{-1} = \tilde{\pi}^{-1}(c_i) \in L_2(F \setminus H, \mu).$$

Since $\{j_i\}$ is almost $\rho(H)$-invariant, $\{\tilde{c}_i\}$ is almost $\tau(H)$-invariant. On the other hand,

1) $\|\tilde{c}_i\| = \|c_i\| = 1$;
2) since $c = 0$ and $\{c_i\}$ converges uniformly on compact sets to c, it follows that

$$\lim_{i\to\infty} \int_K |\tilde{c}_i(x)|^2 d\mu(x) = 0$$

for any compact $K \subset F \setminus H$. Thus (see Lemma 1.9), the subgroup F is not weakly cocompact. This contradiction completes the proof of (b). □

(1.12) *Remark 1.* With F, μ, U, and ρ as in Proposition 1.11, it is not known to the author if the following assertion is true.

(*) *If* $\mu(F \setminus H) < \infty$ *and* U *is isolated from* I_F, *then* ρ *is isolated from* I_H.

In view of Proposition 1.11(b) assertion (*) could be deduced from the following:

(**) *if* $\mu(F \setminus H) < \infty$, *then the subgroup* F *is weakly cocompact.*

Assertion (**) is very probable in the case where H is a Lie group. In any case (**) can be demonstrated if H is a connected semisimple Lie group and F is a discrete subgroup of H. (The proof is not simple. It is based on the results concerning property (T) which will be presented in the next section and on the study of fundamental domains for discrete subgroups).

(2) *Remark.* One can prove assertion (a) of Proposition 1.11 directly without using the theorem on the continuity of the inducing operation in its full generality. Indeed, let $\{y_i\}_{i\in\mathbb{N}^+}$ be an asymptotically $U(F)$-invariant sequence in the space of the representation $U \in \tilde{F}$. Let us choose a regular Borel map $\varphi\colon F\setminus H \to H$ such that $\pi\circ\varphi = \mathrm{Id}$ (see I.4.4.1), where $\pi\colon H \to F\setminus H$ is the natural map. Then every $h \in H$ can be written in the form $h = z(h)x(h)$, where $x(h) \in \varphi(F\setminus H)$, $z(h) \in F$. We set $j_i(h) = U(z(h))y_i$, $h \in H$. Then $j_i \in L(\rho)$, where $\rho = \mathrm{Ind}(H, F, U)$. It follows easily from the asymptotic $U(F)$-invariance of $\{y_i\}$ and the finiteness of $\mu(F\setminus H)$, that the sequence $\{j_i\}$ is asymptotically $\rho(H)$-invariant. Thus ρ is close to I_H.

2. Property (T) and Some of Its Consequences. Relationship Between Property (T) for Groups and for Their Subgroups

In this section we shall use the concepts defined in 1.2.

(2.1) Lemma. *The following conditions are equivalent*

(i) *I_H is isolated in $\hat{H}$ (relative to the topology defined in I.5.1.1).*

(ii) *If $\rho \in \tilde{H}$ is close to I_H, then $\rho \geq I_H$.*

(iii) *If a sequence $\{\rho_n\}_{n\in\mathbb{N}^+}$ in $\tilde{H}$ converges to I_H, then there exists a number $n_0 \in \mathbb{N}^+$ such that $\rho_n \geq I_H$ for all $n \geq n_0$.*

Proof. (i) $\Rightarrow$ (ii). Suppose $\rho \in \tilde{H}$ is close to I_H and let

$$\rho = \int_X \rho_x d\mu(x)$$

be the decomposition of ρ into the continuous sum of irreducible representations $\rho_x \in \hat{H}$. Since I_H is isolated in $\hat{H}$, there is a neighbourhood W of I_H such that $W\cap\hat{H} = \{I_H\}$. We set $X_0 = \{x \in X \mid \rho_x \in W\}$ and $L_0 = \int_{X_0}\rho_x d\mu(x)$. Since ρ is close to I_H, by Proposition 1.4 $\mu(X_0) > 0$, and hence $L_0 \neq \{0\}$. On the other hand, since $W\cap\hat{H} = \{I_H\}$, the subspace L_0 consists of $\rho(H)$-invariant vectors. Thus, ρ contains I_H.

(ii) $\Rightarrow$ (iii). Suppose the contrary. Then there exists a sequence $\{\rho_n\}_{n\in\mathbb{N}^+}$ of elements of $\tilde{H}$ converging to I_H such that no ρ_n contains I_H. Let

$$\rho = \sum_{n\in\mathbb{N}^+} \rho_n$$

be the direct sum of these representations. Since $\{\rho_n\}$ converges to I_H, ρ is close to I_H. On the other hand, since no ρ_n contains I_H, neither does ρ. This contradicts (ii).

(iii) $\Rightarrow$ (i). Since H is σ-compact, the space $\hat{H}$ is second countable. Therefore, if I_H is not isolated in $\hat{H}$, there is a sequence $\{\rho_n\}_{n \in \mathbb{N}^+}$ of distinct elements of $\hat{H}$ (which differ from I_H) converging to I_H. Since every ρ_n is irreducible, we have that $\rho_n \cong I_H$, if $\rho_n \geq I_H$, contradicting (iii). □

(2.2) Definition. *We say that a locally compact second countable group* H *has* property (T) *if the three equivalent conditions of Lemma 2.1 are satisfied.*

Every compact group H has property (T). Indeed, if H is compact, $\rho \in \tilde{H}$, and $\{y_i\}_{i \in \mathbb{N}^+}$ is an asymptotically $\rho(H)$-invariant sequence in $L(\rho) - \{0\}$, then for all sufficiently large i, the $\rho(H)$-invariant vectors $\int_H \rho(h) y_i d\mu_H(h)$ are different from zero. If H is not compact, then the regular representation R_H of the group H on the space $L_2(H, \mu_H)$ does not contain I_H. But by Theorem I.5.5.3 R_H is close to I_H if and only if H is amenable. Thus, if a group H is amenable and has property (T), then it is compact. In particular, if H is discrete, amenable, and has property (T), then it is finite.

Since the character group of a locally compact commutative group F is discrete if and only if F is compact, the following is true

(2.3) Lemma. *A commutative group has property* (T) *if and only if it is compact.*

(2.4) Lemma. *If* F *is a closed normal subgroup of* H *and* H *has property* (T), *then so does* $F \setminus H$.

This lemma follows from the fact that $\rho \in \widetilde{F \setminus H}$ is close to $I_{F \setminus H}$ (resp. contains $I_{F \setminus H}$) if and only if $\rho \circ \pi$ is close to I_H (resp. contains I_H), where $\pi \colon H \to F \setminus H$ is the natural map.

(2.5) Theorem. *Let* $\overline{\mathscr{D}(H)}$ *be the closure of the commutator subgroup of the group* H. *If* H *has property* (T), *then the factor group* $H/\overline{\mathscr{D}(H)}$ *is compact. In particular, if* H *is discrete and has property* (T), *then* $H/\mathscr{D}(H)$ *is finite.*

Proof. If H has property (T), then by Lemma 2.4 the commutative group $H/\overline{\mathscr{D}(H)}$ has property (T), and hence by Lemma 2.3, it is compact. □

(2.6) Corollary. *A group* H *with property* (T) *is unimodular.*

Proof. Suppose H has property (T) and let $\Delta_H \colon H \to \mathbb{R}^+$ be the modular function of the locally compact group H. By virtue of Theorem 2.3 $H/\overline{\mathscr{D}(H)}$ is compact. On the other hand, since Δ_H is a continuous homomorphism into a commutative group, Δ_H is trivial on $\overline{\mathscr{D}(H)}$. Therefore, $\Delta_H(H)$ is a compact subgroup of the group $\mathbb{R}^+$, and hence, $\Delta_H(h) = 1$ for all $h \in H$. □

(2.7) Theorem. *If* H *has property* (T), *then* H *is compactly generated. In particular, if* H *is discrete and has property* (T), *then* H *is finitely generated.*

Proof. Assume that H has property (T). Since H is σ-compact, there is a sequence $\{M_n\}_{n \in \mathbb{N}^+}$ of relatively compact open subsets such that $H = \bigcup_{n \in \mathbb{N}^+} M_n$ and

$M_n \subset M_{n+1}$ for all $n \in \mathbb{N}^+$. We denote by H_n the subgroup of H generated by the set M_n. Since M_n is open, the subgroup H_n is open in H, and hence the space $H_n \setminus H$ is discrete. Assuming the measure of every singleton to be equal to 1, we obtain an H-invariant measure μ_n on the discrete space $H_n \setminus H$. Let us denote by ρ_n the quasi-regular unitary representation of the group H on the space $L_2(H_n \setminus H, \mu_n)$, i.e.,

$$(\rho_n(h)f)(x) = f(xh), \ x \in H_n \setminus H, \ h \in H, \ f \in L_2(H_n \setminus H, \mu_n).$$

Let $\pi_n\colon H \to H_n \setminus H$ be the natural projection and let f_n be the characteristic function of the singleton $\{\pi_n(e)\}$. Clearly, $\rho_n(H_n)f_n = f_n$. On the other hand, every compact subset K of H is contained in H_n for all n greater han some $n(K)$. Therefore the sequence $\{\rho_n\}$ converges to I_H. But H has property (T). Hence there exists an integer $n_0 \in \mathbb{N}^+$ such that $\rho_n \geq I_H$ for all $n \geq n_0$. On the other hand, it follows from the definition of ρ_n that $\rho_n \geq I_H$ if and only if the discrete space $H_n \setminus H$ is finite. Therefore $H_n \setminus H$ is finite for all $n \geq n_0$, and hence the increasing sequence $\{H_n\}$ of subgroups stabilizes. But

$$H = \bigcup_{n \in \mathbb{N}^+} H_n.$$

So, there exists a positive integer n such that H coincides with the compactly generated subgroup H_n. □

(2.8) Proposition. *For $\rho \in \tilde{H}$, let P_ρ denote the orthogonal projector of the space $L(\rho)$ onto the subspace $L(\rho)^H$. Assume H has property (T). Then there exists a compact set $K \subset H$ such that*

(A) *for any $\alpha > 0$ there exists $\delta > 0$ such that if $\rho \in \tilde{H}$, $x \in L(\rho)$, and $|\rho(h)x - x| \leq \delta \|x\|$ for all $h \in K$, then $\|P_\rho x - x\| \leq \alpha \|x\|$;*
(B) *for any $\alpha > 0$ there is $\delta' > 0$ such that if $\rho \in \tilde{H}$, $x \in L(\rho)$, and $\|\rho(h)x - x\| \leq \delta' \|x\|$ for all $h \in K$, then $\|\rho(h)x - x\| \leq \alpha \|x\|$ for all $h \in H$;*
(C) *for any $\alpha > 0$ there is $\delta'' > 0$ such that if $\rho \in \tilde{H}$, $x \in L(\rho)$, and $\operatorname{Re}\langle \rho(h)x, x\rangle \geq (1 - \delta'')\|x\|^2$ for all $h \in K$, then $\operatorname{Re}\langle \rho(h)x, x\rangle \geq (1 - \alpha)\|x\|^2$ for all $h \in H$.*

Proof. Since H has property (T), there exist $\varepsilon > 0$ and a compact set $K \subset H$ such that

(1) $$\rho \notin W(\varepsilon, K), \text{ if } \rho \text{ does not contain } I_H,$$

where $W(\varepsilon, K)$ was defined in 1.0. Let $\rho \in \tilde{H}$. We denote by $L' \subset L(\rho)$ the orthogonal complement of the subspace $L(\rho)^H$ and by P' the orthogonal projector onto L'. Since ρ is unitary, $\rho(H)L' = L'$. By ρ' we denote the restriction $\rho|_{L'}$. Since L' is orthogonal to $L(\rho)^H$, ρ' does not contain I_H. From this and (1) we deduce that for each $x \in L(\rho)$

(2) $$\sup_{h \in K} \|\rho(h)x - x\| \geq \varepsilon \|P'x\| = \varepsilon \|P_\rho x - x\|.$$

It follows from (2) that K has property (A) (with $\delta = \alpha\varepsilon$). Since ρ is unitary, $\|\rho(h)x - x\| \le 2\|P_\rho x - x\|$ for all $x \in L(\rho)$ and $h \in H$. Therefore (B) is a consequence of (A). Finally, since

$$2\,\mathrm{Re}\langle\rho(h)x, x\rangle = 2\|x\|^2 - \|\rho(h)x - x\|^2$$

for all $x \in L(\rho)$ and $h \in H$ (because ρ is unitary), (B) implies (C). □

(2.9) Proposition. *Let F be a closed normal subgroup of the group H. If F and $F \setminus H$ have property (T), then so does H.*

Proof. Let $\rho \in \tilde{H}$ be close to I_H. We set

$$W = \{x \in L(\rho) \mid \rho(F)x = x\}.$$

Since the subgroup F is normal in H, W is $\rho(H)$-invariant. Therefore there exists a representation $\rho_1 \in \widetilde{F \setminus H}$ such that $\rho_1 \circ \pi = \rho|_W$, where $\pi\colon H \to F \setminus H$ is the natural map. Since F has property (T), it follows by Proposition 2.8 (A) that the image under the orthogonal projection onto W of any asymptotically $\rho(H)$-invariant sequence is also asymptotically $\rho(H)$-invariant. But ρ is close to I_H. Therefore $W \neq \{0\}$, $\rho|_W$ is close to I_H, and hence ρ_1 is close to $I_{F\setminus H}$. On the other hand, since $F \setminus H$ has property (T), $\rho_1 \geq I_{F\setminus H}$, and hence $\rho \geq I_H$. □

The following assertion is a straightforward consequence of the proposition just proved.

(2.10) Corollary. *Let $H = H_1 \times H_2$ be a direct product of locally compact groups. If H_1 and H_2 have property (T), then so does H.*

(2.11) Lemma. *Let F be a closed subgroup of the group H and $U \in \tilde{F}$. Assume* $\mathrm{Ind}(H, F, U) \geq I_H$. *Then $U \geq I_F$.*

This lemma is an immediate consequence of the definition of an induced representation and Lemma I.4.1.1(vi).

(2.12) Theorem. *Let F be a closed subgroup of the group H such that $F \setminus H$ has a finite H-invariant Borel measure μ. Then the following conditions are equivalent*

(i) *H has property (T).*
(ii) *F has property (T).*

Proof. (i) ⇒ (ii). Let $U \in \tilde{F}$ be close to I_F. Then, since $\mu(F \setminus H) < \infty$, by Proposition 1.11(a) the representation $\rho = \mathrm{Ind}(H, F, U) \in \tilde{H}$ is close to I_H, and hence by Lemma 2.11 $U \geq I_F$.

(ii) ⇒ (i). Let $\rho \in \tilde{H}$ be close to I_H. Since F has property (T), by Proposition 2.8 (A) the orthogonal projection onto the subspace $\{x \in L(\rho) \mid \rho(F)x = x\}$ of any asymptotically $\rho(H)$-invariant sequence is asymptotically $\rho(H)$-invariant. But ρ is close to I_H. Therefore there exist $y_i \in L(\rho)$, $i \in \mathbb{N}^+$, such that $\rho(F)y_i = y_i$ and the sequence $\{y_i\}_{i\in\mathbb{N}^+}$ is asymptotically $\rho(H)$-invariant. Since $\rho(F)y_i = y_i$, one can define for all $x = hF \in H/F$ and $i \in \mathbb{N}^+$ the element $xy_i \in L(\rho)$, by setting $xy_i = \rho(hF)y_i = \rho(h)y_i$. The anti-automorphism $h \mapsto h^{-1}$, $h \in H$, induces the map $F \setminus H \to H/F$. Under this map the measure μ is transformed into a finite H-invariant measure $\tilde{\mu}$ on H/F. Since ρ is continuous and unitary, for each $i \in \mathbb{N}^+$, the vector valued function $x \mapsto xy_i$, $x \in H/F$, is continuous and bounded. Therefore, for each $i \in \mathbb{N}^+$, one can define $z_i \in L(\rho)$ by setting

$$z_i = (1/\tilde{\mu}(H/F)) \int_{H/F} xy_i d\tilde{\mu}(x).$$

We then have that

$$(h_1 x)y_i = \rho(h_1 hF)y_i = \rho(h_1)\rho(hF)y_i = \rho(h_1)xy_i,$$

for all $h_1 \in H$ and $x = hF \in H/F$. From this and the invariance of $\tilde{\mu}$ we deduce that for all $h \in H$ and $i \in \mathbb{N}^+$,

$$\begin{aligned}\tilde{\mu}(H/F)\rho(h)z_i &= \rho(h) \int_{H/F} xy_i d\tilde{\mu}(x) = \int_{H/F} \rho(h)xy_i d\tilde{\mu}(x) = \\ &= \int_{H/F} (hx)y_i d\tilde{\mu}(x) = \int_{H/F} (hx)y_i d\tilde{\mu}(hx) = \tilde{\mu}(H/F)z_i.\end{aligned}$$

Hence, $\rho(H)z_i = z_i$ for all $i \in \mathbb{N}^+$. Replacing y_i by $y_i/\|y_i\|$ we may assume that $\|y_i\| = 1$. Since the sequence $\{y_i\}$ is asymptotically $\rho(H)$-invariant,

$$\lim_{i\to\infty} \langle xy_i, y_i \rangle = 1$$

for all $x \in H/F$. On the other hand, since ρ is unitary and $\|y_i\| = 1$, $|\langle xy_i, y_i\rangle| \leq 1$ for all $x \in H/F$ and $i \in \mathbb{N}^+$. Therefore by the classical Lebesgue dominated convergence Theorem we have

$$\begin{aligned}\lim_{i\to\infty} \langle z_i, y_i \rangle &= (1/\tilde{\mu}(H/F)) \lim_{i\to\infty} \int_{H/F} \langle xy_i, y_i \rangle d\tilde{\mu}(x) = \\ &= (1/\tilde{\mu}(H/F)) \int_{H/F} \lim_{i\to\infty} \langle xy_i, y_i \rangle d\tilde{\mu}(x) = 1.\end{aligned}$$

Hence, $z_i \neq 0$ for all sufficiently large i. But $\rho(H)z_i = z_i$. Thus $\rho \geq I_H$. □

Corollaries 2.13 and 2.15 are special cases of Theorem 2.12.

(2.13) Corollary. *If F is a closed normal subgroup of H with $F \setminus H$ compact, then the following are equivalent*

(i) *H has property (T).*
(ii) *F has property (T).*

(2.14) *Remark.* Since every compact group has property (T) (see 2.2), implication (ii) ⇒ (i) in Corollary 2.13 can also be obtained from Proposition 2.9.

(2.15) Corollary. *Let F be a countable discrete group and D a subgroup of finite index in F. Then the following are equivalent*

(i) *F has property (T).*
(ii) *D has property (T).*

(2.16) As was observed above in 2.14 to prove implication (ii) ⇒ (i) in Corollary 2.13 it is not necessary to make use of the more difficult implication (ii) ⇒ (i) in Theorem 2.12. That is why a direct deduction of Corollary 2.15 from Corollary 2.13 is worthwhile.

Deduction of Corollary (2.15) from Corollary (2.13). Since D is of finite index in F, the subgroup

$$D' = \bigcap_{z \in F} zDz^{-1}$$

is of finite index in F. On the other hand, the subgroup D' is normal in F. By Corollary 2.13 the condition "D' has property (T)" is equivalent both to (i) and to (ii) of Corollary 2.15.

3. Property (T) and Decompositions of Groups into Amalgams

In this section we shall prove Watatani's theorem asserting that groups with property (T) are not amalgams. The proof is based on the study of negative definite functions.

(3.1) Definition. *A continuous complex-valued function f on a group H is said to be* negative definite, *provided that $f(h^{-1}) = \overline{f(h)}$ for all $h \in H$ and the inequality*

$$\sum_{i,j=1}^{n} f(h_i h_j^{-1}) z_i \bar{z}_j \le 0$$

holds for $n \in \mathbb{N}^+$, for all $h_1, \ldots, h_n \in H$, and for all $z_1, \ldots, z_n \in \mathbb{C}$ with

$$\sum_{i=1}^{n} z_i = 0.$$

It is obvious that a linear combination with positive coefficients of negative definite functions is negative definite. To avoid ambiguity it should be noted that for a negative definite function f the function $-f$ is not necessarily positive definite.

(3.2) Theorem. *A function f on the group H is negative definite if and only if the function* $\exp(-tf)$ *is positive definite for all positive $t \in \mathbb{R}$.*

In fact, this theorem was established by Schoenberg while studying the problem of embedding of metric spaces into a Hilbert space (see [Scho]). In case $H = \mathbb{R}$ the proof is given in [G-V] (Chapter 3, §4, Theorem 4). That proof can verbatim be carried over to the general case. In fact, the following assertion is proved. Let $A = (a_{ij})$ be an Hermitian $n \times n$ matrix. If the inequality

$$\sum_{i,j=1}^{n} a_{ij} z_i \bar{z}_j \leq 0$$

holds for all $z_1, \ldots, z_n \in \mathbb{N}$ with $\sum_{i=1}^{n} z_i = 0$, then the matrix $(\exp(a_{ij}))$ is positive definite.

(3.3) Theorem. *If there is an unbounded continuous negative definite function on the group H, then H does not have property (T).*

Proof. Let f be an unbounded continuous negative definite function on H. By virtue of Theorem 3.2, for each $t \in \mathbb{R}$, $t > 0$, the function $\exp(-tf)$ is positive definite. Therefore, by Theorem I.5.4.3, for each $t > 0$ there exist $\rho_t \in \tilde{H}$ and $x_t \in L(\rho_t)$ such that

(1) $$\langle \rho_t(h) x_t, x_t \rangle = \exp(-tf(h)) \text{ for all } h \in H.$$

Since f is continuous, and hence locally bounded, it follows from (1) that for every compact $K \subset H$

(2) $$\lim_{t \to 0} \sup_{h \in K} (1 - \operatorname{Re}\langle \rho_t(h) x_t, x_t \rangle / \|x_t\|^2) = 0.$$

On the other hand, since f is unbounded, it follows from (1) that

(3) $$\overline{\lim_{t \to 0}} \sup_{h \in H} (1 - \operatorname{Re}\langle \rho_t(h) x_t, x_t \rangle / \|x_t\|^2) > 0.$$

Making use of Proposition 2.8 (C) we obtain from (2) and (3) that H does not have property (T). □

(3.4) *Remark.* The converse to Theorem 3.3 is also true. Namely, one can prove (see [A-W], Theorem 3) that if H does not have property (T), then there exists an unbounded continuous negative definite function on H.

(3.5) Here we assume familiarity with standard notions of graph theory. Let X be a non-oriented graph and let V be the set of vertices of X. Two vertices are called *adjacent* if they are joint by an edge. An *automorphism* of the graph X is a bijection $f\colon V \to V$ such that the vertices $f(x)$ and $f(y)$ are adjacent if and only if x and y are likewise adjacent, for all $x, y \in V$. The automorphism group of the graph X will be denoted by $\operatorname{Aut} X$. Regard $\operatorname{Aut} X$ as a totally disconnected topological group by taking the family of the sets of the form $U_B = \{h \in \operatorname{Aut} X \mid hb = b \text{ for all } B\}$, where $B \subset V$ is finite, as a neighbourhood base of the identity. By an action of a group G on X we mean an action of

G on V such that for each $g \in G$ the map $\varphi(g)\colon V \to V$, $\varphi(g)x = gx$, $x \in V$, is a graph automorphism. If G is a topological group and the homomorphism $\varphi\colon G \to \operatorname{Aut} X$ is continuous, then we call this action *continuous*. We define the standard distance function on V as follows: $d(x,y) =$ the number of edges in the shortest path joining the vertices x and y. Clearly, $d(x,x) = 0$ and the equality $d(x,y) = 1$ holds if and only if x and y are adjacent. This distance function is $(\operatorname{Aut} X)$-invariant, i.e., $d(hx,hy) = d(x,y)$ for all $x, y \in V$ and $h \in \operatorname{Aut} X$.

By a *tree* we mean a connected non-oriented graph without cycles. Any two vertices x and y of a tree X can be joined by a unique reduced path (i.e. the path without recurrence). The length of this path equals $d(x,y)$.

(3.6) Proposition. *Let X be a tree, P_0 a vertex of X, and suppose that H acts continuously on X. For $h \in H$, let $l(h) = d(P_0, hP_0)$. Then l is a negative definite function on H.*

Proof. Let us denote by Y the set of edges of the tree X, and let $\mathscr{L}$ be a complex Hilbert space with an orthonormal basis $\{e(y) \mid y \in Y\}$. Define the map $f\colon H \to \mathscr{L}$ by setting

$$f(h) = e(u_1) + \ldots + e(u_n), \quad h \in H,$$

where $u_1, \ldots, u_n$ are the edges of the reduced path from P_0 to $h^{-1}P_0$. Since the vectors $e(u_1), \ldots, e(u_n)$ are mutually orthogonal and $\|e(u_i)\| = 1$, we have $\|f(h)\|^2 = n = l(h)$. We shall show that

(1) $$\|f(g) - f(h)\|^2 = l(gh^{-1}), \text{ for all } g, h \in H.$$

Let $u_1, \ldots, u_n$ be the edges of the reduced path from P_0 to $g^{-1}P_0$ and $v_1, \ldots, v_n$ the edges of the reduced path from P_0 to $h^{-1}P_0$. Let $a = \max\{k \mid u_k = v_k\}$. Thus we obtain the following diagram

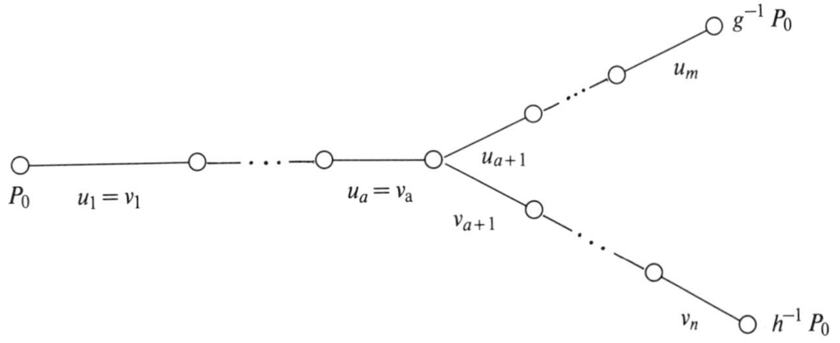

We see in this diagram that $d(g^{-1}P_0, h^{-1}P_0) = (m-a) + (n-a) = m+n-2a$. But the distance function d is $(\operatorname{Aut} X)$-invariant. Therefore

(2) $$l(gh^{-1}) = d(P_0, gh^{-1}P_0) = d(g^{-1}P_0, h^{-1}P_0) = m + n - 2a.$$

Since

$$u_1 = v_1, \dots, u_a = v_a,$$

we have

$$f(g) - f(h) = e(u_{a+1}) + \dots + e(u_m) - (e(v_{a+1}) + \dots + e(v_n)).$$

But the vectors

$$e(u_{a+1}), \dots, e(u_m), e(v_{a+1}), \dots, e(v_n)$$

are mutually orthogonal with norm 1. Therefore

$$\begin{aligned} \|f(g) - f(h)\|^2 &= \|e(u_{a+1})\|^2 + \dots + \|e(u_m)\|^2 + \|e(v_{a+1})\|^2 + \dots + \|e(v_n)\|^2 = \\ &= (m-a) + (n-a) = m + n - 2a, \end{aligned}$$

which in view of (2) proves (1). Since $\langle e(y_1), e(y_2)\rangle \in \{0,1\}$ for all $y_1, y_2 \in Y$, it follows that

(3) $$\langle f(g), f(h)\rangle \in \mathbb{N} \subset \mathbb{R} \text{ for all } g, h \in H.$$

Let now $h_1, \dots, h_n \in H$ and let $z_1, \dots, z_n \in \mathbb{C}$ with $\sum_{i=1}^n z_i = 0$. Then in view of (1) and (3) we have that

$$\begin{aligned} \sum_{i,j=1}^n l(h_i h_j^{-1}) z_i \bar{z}_j &= \sum_{i,j=1}^n \|f(h_i) - f(h_j)\|^2 z_i \bar{z}_j = \\ &= \sum_{i,j=1}^n ((\|f(h_i)\|^2 + \|f(h_j)\|^2 - 2\langle f(h_i), f(h_j)\rangle) z_i \bar{z}_j) = \\ &= \sum_{i=1}^n (\|f(h_i)\|^2 z_i \cdot \sum_{j=1}^n \bar{z}_j) + \sum_{j=1}^n (\|f(h_j)\|^2 \bar{z}_j \cdot \sum_{i=1}^n z_i) - \\ &\qquad - 2 \sum_{i,j=1}^n (\langle f(h_i), f(h_j)\rangle z_i \bar{z}_j) = \\ &= -2 \sum_{i,j=1}^n (\langle f(h_i), f(h_j)\rangle z_i \bar{z}_j) = \\ &= -2 \|\sum_{i=1}^n z_i f(h_i)\|^2 \le 0 \end{aligned}$$

(the next to the last equality follows from the assumption that $\sum_{i=1}^n z_i = 0$.) Thus, l is negative definite and the proof is complete. □

(3.7) Proposition. *As in Proposition 3.6 let X be a tree on which a topological group H operates continuously, let P_0 be a vertex of X, and let $l(h) = d(P_0, hP_0)$. Assume H has property (T). Then $\sup_{h \in H} l(h) < \infty$.*

This proposition is a straightforward consequence of Theorem 3.3 and Proposition 3.6. We note that J.-P. Serre has suggested a proof of Proposition

3.7 which does not use negative definite functions. A sketch of his proof will be given in 3.11-12.

(3.8) Let G be a group, G_1 and G_2 two subgroups of G, and $A = G_1 \cap G_2$. We associate with the triple (G, G_1, G_2) the graph $X = X(G, G_1, G_2)$, whose set of vertices is $G/G_1 \cup G/G_2$, with various pairs $\{gG_1, gG_2\}$, $g \in G$, as edges. The action of G on G/G_1 and on G/G_2 by left translations induces the action of G on the graph X. If G is a topological group and the subgroups G_1, G_2 are open in G, then this action is continuous. Obviously the group G acts transitively on the set of edges of X and the set of vertices falls into two orbits G/G_1 and G/G_2. Choosing a representative in every coset $x \in G/G_1$ (resp. $x \in G/G_2$) we obtain a subset Y_1 (resp. Y_2)of G. If the union $G_1 \cup G_2$ generates G, then each $g \in G$ can be represented in the form $g = y_n \cdot \ldots \cdot y_1 a$, where $a \in A$, $y_i \in Y_1$ for i odd, $y_i \in Y_2$ for i even, and $y_i \notin A$ for $i > 1$. We call this representation of g *canonical.* The group G is said to be the *free product* of the subgroups G_1 and G_2 with amalgamated subgroup A (we denote this by $G = G_1 *_A G_2$ or simply by $G = G_1 * G_2$) if

(a) G is generated by $G_1 \cup G_2$ and
(b) a canonical representation of each $g \in G$ is unique.

It can easily be verified that (b) is equivalent to the following: $e \in G$ can not be represented in the form $e = g_n \ldots g_1$, where $g_i \in G_{j(i)} - A$, $j(i) = 1, 2$ $(1 \leq i \leq n)$ and $j(i+1) \neq j(i)$ for all i with $1 \leq i \leq n-1$. It is not hard to show (see [Ser 3], Theorem 7) that $G = G_1 *_A G_2$ if and only if $X(G, G_1, G_2)$ is a tree.

We say that the group H is an *amalgam* if $H = H_1 *_A H_2$, where H_1 and H_2 are open subgroups of H, $A = H_1 \cap H_2$, and $H_1 \neq A \neq H_2$.

(3.9) Theorem. *If a group H has property (T), then H is not an amalgam.*

Proof. Suppose the contrary, i.e., $H = H_1 *_A H_2$, where H_1 and H_2 are open subgroups of H, $H_1 \cap H_2 = A$, and $H_1 \neq A \neq H_2$. Then (see 3.8) the graph $X = X(H, H_1, H_2)$ is a tree. Let $P_0 = eH_1 \in H/H_1$. As in 3.6 we set, for $h \in H$, $l(h) = d(P_0, hP_0)$. Since $H = H_1 *_A H_2$ and $H_1 \neq A \neq H_2$, the function l is not bounded (indeed, $l((h_1 h_2)^n) = 2n$ for all $n \in \mathbb{N}^+$ and $h_1 \in H - H_1$, $h_2 \in H - H_2$). On the other hand, since H has property (T), by Proposition 3.7 the function l is bounded. A contradiction which completes the proof. □

(3.10) A group G is said to be *acting without inversions* on a tree X if for no pair of adjacent vertices x and y of X one can find an element $g \in G$ with $gx = y$ and $gy = x$. Following Serre we say that the group G has *property (FA)* if for every tree X and for every action of G on X without inversions, there exists a G-invariant vertex of X. A countable group G has property (FA) if and only if the following three conditions are satisfied:

(I) G is not an amalgam;
(II) the factor group $G/\mathcal{D}(G)$ is finite;
(III) the group G is finitely generated

(see [Ser 3], Theorem 15); although in [Ser 3] instead of (ii) another condition (II′) asserting that no factor group of G is isomorphic to $\mathbb{Z}$ was considered, these two conditions are equivalent in the presence of (III), because every finitely generated Abelian group is a direct sum of finitely many cyclic groups). Therefore it follows from Theorems 2.5, 2.7, and 3.9 that a countable group with property (T) has also property (FA). We remark that in [Wat] an example of a countable group with property (FA), but without property (T) has been constructed. As an example of such a group in [Wat] a group with two generators x and y and defining relations $x^a = y^b = (xy)^c = e$ has been given, where

$$a, b, c \geq 2 \text{ and } (1/a) + (1/b) + (1/c) = 1.$$

(3.11) We now provide the promised sketch of Serre's proof of Proposition 3.7. Let us denote by Y (resp. V) the set of edges (resp. vertices) of the tree X. Let $\mathscr{L}$ be a complex Hilbert space with an orthonormal basis $\{e(y) \mid y \in Y\}$. We embed V into $\mathscr{L}$ by sending $P \in V$ to $e(u_1) + \ldots + e(u_m)$, where $u_1, \ldots, u_m$ are the edges of the reduced path from P_0 to P and the image of P_0 is 0. It can easily be checked that under this embedding every tree automorphism of X extends to an isometry of the space $\mathscr{L}$. Therefore we may assume that H operates on $\mathscr{L}$ by isometries. It is not hard to show that $\|P_1 - P_2\|^2 = d(P_1, P_2)$ for all $P_1, P_2 \in V$. In particular,

$$l(h) = d(P_0, hP_0) = \|hP_0 - P_0\|^2$$

for all $h \in H$. Thus, it suffices to prove the following

(3.12) Proposition. *Let $\mathscr{L}$ be a Hilbert space and assume that the group H operates continuously by isometries on $\mathscr{L}$. If H has property (T), then*

(1) $$\sup_{h \in H} \|hx - x\| < \infty, \text{ for all } x \in \mathscr{L}.$$

Proof. We denote by $\mathscr{B}$ the vector space of finite formal series of the form $\Sigma \lambda_y y$, $y \in \mathscr{L}$, $\lambda_y \in \mathbb{C}$. The elements of $\mathscr{L}$ form a basis of the space $\mathscr{B}$. For each $t > 0$ we define the Hermitian form $\langle\ ,\ \rangle_t$ on $\mathscr{B}$ by setting

(2) $$\langle y, z \rangle_t = e^{-t\|y-z\|^2}, \quad y, z \in \mathscr{L}.$$

For each $t > 0$ the function $e^{-t\|u\|^2}$, $u \in \mathscr{L}$, is positive definite (this is an easy consequence of the classical Bochner theorem on Fourier transform of positive definite functions applied to the restrictions of the function $e^{-t\|u\|^2}$ to finite-dimensional linear subspaces of $\mathscr{L}$). Therefore, the form $\langle\ ,\ \rangle_t$ is positive definite or, in other words, it is a scalar product. We denote by $\mathscr{B}_t$ the Hilbert space obtained by completing the space $\mathscr{B}$ relative to the scalar product $\langle\ ,\ \rangle_t$. Since $\|hy - hz\| = \|y - z\|$ for all $h \in H$ and $y, z \in \mathscr{L}$, it follows from (2) that the action of the group H on $\mathscr{L}$ induces, for each $t > 0$, a unitary representation ρ_t of H on $\mathscr{B}_t$.

Let $x \in \mathscr{L}$ and let K be a compact subset as in Proposition 2.8. Since K is compact and H operates continuously on $\mathscr{L}$,

$$\sup_{h\in K} \|hx - x\| < \infty.$$

But $\rho_t(h)x = hx$. It then follows from (2) that

$$\liminf_{t\to 0}\,\inf_{h\in K} \langle \rho_t(h)x, x\rangle_t = \langle x, x\rangle_t = 1. \tag{3}$$

(3) and property (C) (Proposition 2.8) of the compact subset K imply that

$$\lim_{t\to 0}\,\inf_{h\in H} \langle \rho_t(h)x, x\rangle_t = 1.$$

But in view of (2) and the equality $\rho_t(h)x = hx$, this implies (1). □

4. Property (R)

In this section R_H denotes the right regular representation of the group H on the space $L_2(H, \mu'_H)$, where μ'_H is right Haar measure on H.

(4.1) Definition. *We say that H has* property (R) *if R_H is close to I_H. The following conditions are equivalent (see Theorem I.5.5.3).*

(i) *H has property (R).*
(ii) *The closure of the representation R_H in $\tilde{H}$ contains $\hat{H}$.*
(iii) *H is amenable.*

It then follows from the appropriate properties of amenable groups (see I.5.5.2) that

(4.2) Lemma.
(a) *If H has property (R) and F is a closed subgroup of the group H, then F has property (R);*
(b) *Let F be a closed normal subgroup of H. If F and $F \setminus H$ have property (R), then so does H;*
(c) *If H is solvable, then H has property (R).*

(4.3) Proposition. *If H is a free discrete group with at least two generators, then H does not have property (R).*

Proof. In view of Lemma 4.2(a), it suffices to consider the case where H is the group with two generators which we denote by a, b. Since

$$\|R_H(h)f - f\|^2 = 2\|f\|^2 - 2\,\mathrm{Re}\langle R_H(h)f, f\rangle$$

for all $f \in L_2(H, \mu'_H)$, $h \in H$, it suffices to show that there is $\varepsilon > 0$ such that for every $f \in L_2(H, \mu'_H)$ the following three conditions can not be satisfied simultaneously

(a) $$\|f\| = 1;$$
(b) $$|\langle R_H(a^{-1})f, f\rangle - 1| \leq \varepsilon^2/2;$$
(c) $$|\langle R_H(b^{-1})f, f\rangle - 1| \leq \varepsilon^2/2.$$

Assume there exist $f \in L_2(H, \mu'_H)$ and $\varepsilon > 0$ satisfying conditions (a), (b), and (c) and consider the measure μ on H defined by the equality $\mu(X) = \int_X |f(h)|^2 d\mu'_H(h)$. It then follows that

$$\begin{aligned}|\mu(aX) - \mu(X)| &= |\int_X |f(ah)|^2 - |f(h)|^2 d\mu'_H(h)| \leq \\ &\leq 2(\int_H |f(ah) - f(h)|^2 d\mu'_H(h))^{1/2} = 2\|R_H(a^{-1})f - f\| = \\ &= 2(2\|f\|^2 - 2\operatorname{Re}\langle R_H(a^{-1})f, f\rangle)^{1/2} \leq 2\varepsilon,\end{aligned}$$

and hence

(1) $$\mu(aX) \geq \mu(X) - 2\varepsilon.$$

Let P be the set of those elements of the free group H whose reduced word expression in terms of the generators $\{a, b\}$ begins with a nonzero power of a, and let N be the complement of P in H. It then follows from (1) that

$$1 \geq \mu(N) + \mu(aN) + \mu(a^2N) \geq 3\mu(N) - 6\varepsilon,$$

and hence

$$\mu(N) \leq (1/3) + 2\varepsilon.$$

Analogously, $\mu(P) \leq (1/3) + 2\varepsilon$. As a result we obtain that

$$1 = \mu(N) + \mu(P) \leq (2/3) + 4\varepsilon.$$

Thus, for $\varepsilon = 1/13$ we have the desired contradiction. □

(4.4) Corollary. *The group* $\mathbf{SL}_2(k)$, *where* k *is a local field, does not have property* (R).

Proof. In view of Lemma 4.2(a) and Proposition 4.3 it suffices to show that $\mathbf{SL}_2(k)$ contains a free discrete subgroup with two generators. But this fact is a special case of the theorem, proved in Appendix B, on free discrete subgroups of linear groups (the reader can prove, as an exercise, that for a suitable $\lambda \in k$ the matrices

$$A = \begin{pmatrix} \lambda & 0 \\ 0 & \lambda^{-1} \end{pmatrix} \text{ and } B = CAC^{-1}, \text{ where } C = \begin{pmatrix} 2 & 1 \\ 1 & 1 \end{pmatrix},$$

generate a free discrete subgroup in $\mathbf{SL}_2(k)$). □

(4.5) Proposition. *Let* $H = F \ltimes N$ *be a semi-direct product of locally compact groups, the group* N *being commutative. We consider the natural action of the group* H *on the character group* $\hat{N}$ *of* N *defined by*

$$(h\chi)(n) = \chi(h^{-1}nh), \ h \in H, \ \chi \in \hat{N}, \ n \in N.$$

Assume the following conditions are satisfied:

(i) *every orbit* F_χ, $\chi \in \hat{N}$, *is locally closed in* $\hat{N}$;

(ii) *for each* $\chi \in \hat{N}$, $\chi \neq 0$, *the stabilizer* $\{f \in F \mid f\chi = \chi\}$ *has property* (R);

(iii) F *does not have property* (R).

If $\rho \in \tilde{H}$ *is close to* I_H, *then* $\rho|_N \geq I_N$.

Proof. Suppose the contrary. Let $\rho = \int_X \rho_x d\mu(x)$ be the decomposition of ρ into a continuous sum of irreducible representations $\rho_x \in \hat{H}$. Since $\rho|_N$ does not contain I_N, we may assume that $\rho_x|_N$ does not contain I_N for all $x \in X$. It then follows from (i) by Mackey's theorem (see I.5.2.3) that $\rho_x = \mathrm{Ind}(H, H_x, \pi_x)$, where H_x is the stabilizer $\{h \in H \mid h\chi_x = \chi_x\}$ of a non-trivial character $\chi_x \in \hat{N}$ and $\pi_x \in \hat{H}_x$. Lemma 4.2 ((b) and (c)) and (ii) imply that H_x has property (R). Therefore (see 4.1) π_x is contained in the closure of the regular representation R_{H_x} of the group H_x. But the restriction operation is continuous and $R_H = \mathrm{Ind}(H, H_x, R_{H_x})$, hence ρ_x is contained in the closure of the representation R_H. On the other hand, since ρ is close to I_H, by Proposition 1.4 I_H is an adherent point of the set $\{\rho_x \mid x \in X\}$ in $\tilde{H}$. Thus, R_H is close to I_H which contradicts (iii), according to Lemma 4.2 (a). □

(4.6) Corollary. *Let k be a local field. Let us consider the following two subgroups of* $\mathbf{SL}_3(k)$:

$$H = \left\{ X = \begin{pmatrix} a & b & c \\ c & d & y \\ 0 & 0 & 1 \end{pmatrix} \middle| X \in \mathbf{SL}_3(k) \right\} \text{ and}$$

$$N = \left\{ \begin{pmatrix} 1 & 0 & x \\ 0 & 1 & y \\ 0 & 0 & 1 \end{pmatrix} \middle| x, y \in k \right\}.$$

If $\rho \in \tilde{H}$ *is close to* I_H, *then* $\rho|_N \geq I_N$.

Proof. Let us consider the subgroup

$$F = \left\{ X = \begin{pmatrix} a & b & 0 \\ c & d & 0 \\ 0 & 0 & 1 \end{pmatrix} \middle| X \in \mathbf{SL}_3(k) \right\} \subset H.$$

The group F has a natural identification with $\mathbf{SL}_2(k)$ and the group N with the vector space $k \times k$. It is well known (see [Weil 4], Chapter II, §5, Theorem 3) that if V is a finite-dimensional vector space over k and λ is a non-trivial character of the additive group of k, then the formula $\langle v, \hat{v} \rangle = \lambda([v, v'])$ for all $v \in V$, defines a bijective map $v' \mapsto \hat{v}$ of the conjugate space V' onto the character group $\hat{V}$ of V, where $\langle\ ,\ \rangle$ is the natural pairing between V and $\hat{V}$, and $[\ ,\ ]$ is the natural one between V and V'. Therefore $\hat{N}$ can be identified with $N' \cong k \times k$. Under these identifications the natural actions of the group F on N and $\hat{N}$ pass, respectively, to the standard and the conjugate contragradient of the standard representation of F on $k \times k$. This implies that

1) $F\chi = \{0\}$, whenever $\chi = 0$, and $F\chi = \hat{N} - \{0\}$, whenever $\chi \in \hat{N} - \{0\}$;

2) for each $\chi \in \hat{N}$, $\chi \neq 0$, the stabilizer $\{f \in F \mid f\chi = \chi\}$ is solvable, and hence, by Lemma 4.2(c) has property (R).

On the other hand, by Corollary 4.4 the group $F = \mathbf{SL}_2(k)$ does not have property (R). Thus, the semi-direct product $H = F \ltimes N$ satisfies all the conditions of Proposition 4.5, completing the proof. □

(4.7) Corollary. *Let k be a local field. Let us consider the following subgroups of the group $\mathbf{SL}_4(k)$:*

$$H = \left\{ \begin{pmatrix} A & B \\ 0 & {}^tA^{-1} \end{pmatrix} \middle| A{}^tB = B{}^tA \right\} \text{ and}$$

$$N = \left\{ \begin{pmatrix} E & B \\ 0 & E \end{pmatrix} \middle| B = {}^tB \right\},$$

where A and B are 2×2-matrices with entries in k, $\det A = 1$, and ${}^tX$ denotes the transpose of a matrix X. If $\rho \in \tilde{H}$ is close to I_H, then $\rho|_N \geq I_N$.

Proof. Let us consider the subgroup

$$F = \left\{ \begin{pmatrix} A & 0 \\ 0 & {}^tA^{-1} \end{pmatrix} \middle| A \in \mathbf{SL}_2(k) \right\} \subset H.$$

Using the same argument as in the proof above, we identify F with $\mathbf{SL}_2(k)$ and $\hat{N}$ with a vector space over k. Since

$$\begin{pmatrix} A & 0 \\ 0 & {}^tA^{-1} \end{pmatrix} \begin{pmatrix} E & B \\ 0 & E \end{pmatrix} \begin{pmatrix} A & 0 \\ 0 & {}^tA^{-1} \end{pmatrix}^{-1} = \begin{pmatrix} E & AB{}^tA \\ 0 & E \end{pmatrix},$$

under this identification the natural action of the group F on $\hat{N}$ induces a representation equivalent to the standard representation of the group $\mathbf{SL}_2(k)$ on the space W of symmetric bilinear forms of order 2 over k. Now

1) for each non-zero form $w \in W$ the stabilizer $\{g \in \mathbf{SL}_2(k) \mid gw = w\}$ is solvable, and hence, has property (R);
2) by virtue of Corollary 4.4 the group $F = \mathbf{SL}_2(k)$ does not have property (R).

Therefore, in view of Proposition 4.5, it remains to prove the following assertion:

$(*)$ for each form $w \in W$ the orbit $\mathbf{SL}_2(k)w$ is locally closed in W.

We set $\varphi_w(f) = fw$, $f \in F$. If $\operatorname{char} k \neq 2$, then it is straightforward that the differential of the morphism $\varphi_w \colon \mathbf{SL}_2 \to \mathbf{SL}_2 w$ is surjective at any point $h \in H$, hence the morphism φ_w is separable. This and Proposition I.2.1.4(i) imply $(*)$ in case $\operatorname{char} k \neq 2$. The case in which $\operatorname{char} k = 2$, $w = ax_1x_2 + b(x_1y_2 + x_2y_1) + cy_1y_2$, and $b \neq 0$ can be examined analogously. Thus, we may assume that $\operatorname{char} k = 2$ and $w = ax_1x_2 + cy_1y_2$. Then the set $k^2 \overset{\text{def}}{=} \{x^2 \mid x \in k\}$ is a closed subfield in k and

$$\mathbf{SL}_2(k)w = \left\{ (ap + cq)x_1x_2 + (ar + cs)y_1y_2 \middle| \begin{pmatrix} p & r \\ q & s \end{pmatrix} \in \mathbf{SL}_2(k^2) \right\}$$

which implies $(*)$. □

5. Semisimple Groups with Property (T)

(5.1) Lemma. *Let k be a local field.*

(a) *The group $\mathbf{SL}_3(k)$ has property (T).*

(b) *The group $\mathbf{Sp}_4(k)$ has property (T), where $\mathbf{Sp}_4$ is the symplectic group ($\mathbf{Sp}_4 = \{g \in \mathbf{GL}_4 \mid {}^t g J g = J\}$, ${}^t g$ is the transpose of g, and $J = \begin{pmatrix} 0 & E \\ -E & 0 \end{pmatrix}$).*

Proof. (a) Let $G = \mathbf{SL}_3(k)$. Let H and N be as in Corollary 4.6. We set

$$Y = \left\{ X = \begin{pmatrix} a & 0 & b \\ 0 & 1 & 0 \\ c & 0 & d \end{pmatrix} \middle| X \in \mathbf{SL}_3(k) \right\} \text{ and}$$

$$s = \begin{pmatrix} \lambda & 0 & 0 \\ 0 & 1 & 0 \\ 0 & 0 & \lambda^{-1} \end{pmatrix} \in Y,$$

where $\lambda = 2$, if k is $\mathbb{R}$ or $\mathbb{C}$, and λ is a uniformizer of the field k, if k is non-archimedean. Let now $\rho \in \tilde{G}$ be close to I_G. Then, by Corollary 4.6, there exists $w \in L(\rho)$, $w \neq 0$, such that $\rho(N)w = w$. But (see Lemma II.3.4) if $U = \left\{ \begin{pmatrix} 1 & x \\ 0 & 1 \end{pmatrix} \middle| x \in k \right\}$, σ is a (continuous) unitary representation of the group $\mathbf{SL}_2(k)$, and a vector z is $\sigma(U)$-invariant, then $\sigma(\mathbf{SL}_2(k))z = z$. Therefore $\rho(Y)w = w$ and so $\rho(s)w = w$. It follows from the definition of s that at least one of the eigenvalues of the transformation $\operatorname{Ad} s$ is different from 1 in absolute value. On the other hand, the group $\mathbf{SL}_3$ is simply connected and almost simple. Therefore, according to Proposition II.3.3, if σ is a unitary representation of the group G and the vector z is invariant relative to $\sigma(s)$, then $\rho(G)z = z$. This implies, since $\rho(s)w = w$, that $\rho(G)w = w$, and hence $\rho \geq I_G$.

Assertion (b) is proved in exactly the same way as (a). It suffices to take as H and N the subgroups defined in the statement of Corollary 4.7 and to set

$$Y = \left\{ X = \begin{pmatrix} a & 0 & b & 0 \\ 0 & 1 & 0 & 0 \\ c & 0 & d & 0 \\ 0 & 0 & 0 & 1 \end{pmatrix} \middle| X \in \mathbf{SL}_4(k) \right\} \subset \mathbf{SP}_4(k) \text{ and}$$

$$s = \begin{pmatrix} \lambda & 0 & 0 & 0 \\ 0 & 1 & 0 & 0 \\ 0 & 0 & \lambda^{-1} & 0 \\ 0 & 0 & 0 & 1 \end{pmatrix} \in Y,$$

with λ as in the proof of (a). □

(5.2) Lemma. *Let k be a local field, $\mathbf{G}$ a connected semisimple k-group and $\tilde{\mathbf{G}}$ the simply connected covering of $\mathbf{G}$. Then the following conditions are equivalent*

(i) *$\mathbf{G}(k)$ has property (T);*

(ii) *$\tilde{\mathbf{G}}(k)$ has property (T).*

Proof. Let $\pi\colon \tilde{\mathbf{G}} \to \mathbf{G}$ be the central k-isogeny. By virtue of Proposition I.2.3.4 the subgroup $\pi(\tilde{\mathbf{G}}(k))$ is closed in $\mathbf{G}(k)$ and the isomorphism $\tilde{\mathbf{G}}(k)/(\operatorname{Ker}\pi)(k) \to \pi(\tilde{\mathbf{G}}(k))$ is one of topological groups. But the group $(\operatorname{Ker}\pi)(k)$ is finite and hence has property (T). Therefore by Lemma 2.4 and Proposition 2.9 $\tilde{\mathbf{G}}(k)$ has property (T) if and only if so does $\pi(\tilde{\mathbf{G}}(k))$. On the other hand, by virtue of Proposition II.2.3.4(i) the subgroup $\pi(\tilde{\mathbf{G}}(k))$ is normal in $\mathbf{G}(k)$ and the factor group $\mathbf{G}(k)/\pi(\tilde{\mathbf{G}}(k))$ is compact. Therefore the equivalence of (i) and (ii) is a consequence of Corollary 2.13. □

(5.3) Theorem. *Let k be a local field, $\mathbf{G}$ a connected non-commutative almost k-simple k-group, and $\operatorname{rank}_k \mathbf{G} \neq 1$. Then $\mathbf{G}(k)$ has property (T).*

Proof. It follows from Corollary I.1.4.6(a) that the group $\mathbf{G}$ and its simply connected covering have the same k-rank. Therefore, in view of Lemma 5.2, we may assume that $\mathbf{G}$ is simply connected. If $\operatorname{rank}_k \mathbf{G} = 0$, then by Proposition I.2.3.6 the group $\mathbf{G}(k)$ is compact and hence (see 2.2) has property (T). Thus, we may assume that $\operatorname{rank}_k \mathbf{G} \geq 2$. It then follows from Proposition I.1.6.2 that $\mathbf{G}$ contains a semisimple k-subgroup $\mathbf{F}$ of k-rank 2 the simply connected covering of which is isomorphic over k either to $\mathbf{SL}_3$ or $\mathbf{Sp}_4$. Lemmas 5.1 and 5.2 then imply that $\mathbf{F}(k)$ has property (T). Therefore, if $\rho \in \widetilde{\mathbf{G}(k)}$ is close to $I_{\mathbf{G}(k)}$, then there is $w \in L(\rho)$, $w \neq 0$, such that $\rho(\mathbf{F}(k))w = w$. Let $\mathbf{S}$ be the maximal k-split torus in $\mathbf{F}$. Since $\operatorname{rank}_k \mathbf{F} > 0$ and $\mathbf{G}$ is semisimple, there exists an element $s \in \mathbf{S}(k)$ such that at least one of the eigenvalues of the transformation $\operatorname{Ad} s$ is different from 1 in absolute value. It then follows, in view of the almost k-simplicity and simple connectedness of the group $\mathbf{G}$ and the equality $\rho(s)w = w$ (see Proposition II.3.3), that $\rho(\mathbf{G}(k))w = w$. Thus, $\rho \geq I_{\mathbf{G}(k)}$. □

(5.4) Corollary. *Let A be a finite non-empty set and for each $\alpha \in A$ let k_α be a local field and $\mathbf{G}_\alpha$ a connected semisimple group defined over k_α and containing no connected normal k_α-subgroup of k_α-rank 1. Then the group*

$$G = \prod_{\alpha \in A} \mathbf{G}_\alpha(k_\alpha)$$

has property (T).

Proof. Let $\tilde{\mathbf{G}}_\alpha$ be the simply connected covering of $\mathbf{G}_\alpha$. Since k-rank is invariant under central k-isogenies (see Corollary I.1.4.6), the k_α-rank of any k_α-simple factor of the group $\tilde{\mathbf{G}}_\alpha$ is different from 1. On the other hand, the simply connected k_α-group $\tilde{\mathbf{G}}_\alpha$ is the direct product of its almost k_α-simple factors. Therefore, Corollary 2.10 and Theorem 5.3 imply that $\tilde{\mathbf{G}}_\alpha(k_\alpha)$ has property (T), and hence, by Lemma 5.2 $\mathbf{G}_\alpha(k_\alpha)$ has property (T). This implies by Corollary 2.10 the desired assertion. □

(5.5) We shall now analyze what will happen if we drop in the statement of Theorem 5.3 the assumption that $\operatorname{rank}_k \mathbf{G} \neq 1$. Let $\operatorname{rank}_k \mathbf{G} = 1$. We shall consider separately the cases (a) k is non-archimedean and (b) k is $\mathbb{R}$ or $\mathbb{C}$.

(a) Since $\operatorname{rank}_k \mathbf{G} = 1$, the Bruhat-Tits building of the group $\mathbf{G}$ over k is a tree. This implies by Proposition 3.7 that $\mathbf{G}(k)$ does not have property (T).

(b) If k is $\mathbb{R}$ or $\mathbb{C}$, then (see [Ko], Remark 10) $\mathbf{G}(k)$ does not have property (T) only in the following two cases:
(i) $k = \mathbb{C}$ and $\mathbf{G}$ is locally isomorphic (i.e. isogenous) to $\mathbf{SL}_2$;
(ii) $k = \mathbb{R}$ and $\mathbf{G}$ is locally isomorphic (i.e. $\mathbb{R}$-isogenous) either to $\mathbf{SO}_{n,1}$ or $\mathbf{SU}_{n,1}$, where $\mathbf{SO}_{n,1}$ (resp. $\mathbf{SU}_{n,1}$) denotes as usual the group of linear transformations of determinant 1 preserving the quadratic form $x_1^2 + \ldots + x_n^2 - x_{n+1}^2$ (resp. Hermitian form $z_1\bar{z}_1 + \ldots + z_n\bar{z}_n - z_{n+1}\bar{z}_{n+1}$)

From Theorem 5.3 and (a) and (b) above we deduce

(5.6) Theorem. *Let k be a local field and let $\mathbf{G}$ be a connected semisimple almost k-simple k-group. Then $\mathbf{G}(K)$ fails to have property (T) only in the following three cases:*

(a) *k is non-archimedean and $\operatorname{rank}_k \mathbf{G} = 1$;*
(b) *$k = \mathbb{C}$ and $\mathbf{G}$ is locally isomorphic to $\mathbf{SL}_2$;*
(c) *$k = \mathbb{R}$ and $\mathbf{G}$ is locally isomorphic either to $\mathbf{SO}_{n,1}$ or $\mathbf{SU}_{n,1}$.*

(5.7) Theorem. *Let A be a finite non-empty set and for $\alpha \in A$ let k_α be a local field and $\mathbf{G}_\alpha$ a connected semisimple group defined over k_α. We set $G = \prod_{\alpha \in A} \mathbf{G}_\alpha(k_\alpha)$. Let Γ be a lattice in G. Assume*

(∗) *for each $\alpha \in A$ the k_α-rank of any almost k_α-simple factor of $\mathbf{G}_\alpha$ is different from 1. Then*

(a) *Γ has property (T);*
(b) *the factor group of Γ modulo its commutator subgroup is finite;*
(c) *Γ is finitely generated;*
(d) *Γ is not an amalgam.*

Proof. (a) is a consequence of Corollary 5.4 and Theorem 2.12. Assertions (b), (c), and (d) are obtained from (a) and, respectively, Theorems 2.5, 2.7, and 3.9. □

(5.8) Theorem. *Let K be a global field, S a finite set of (inequivalent) valuations of K containing all the archimedean ones, $\mathbf{G}$ a connected semisimple K-group, and Γ an S-arithmetic (i.e. commensurable with $\mathbf{G}(K(S))$) subgroup of $\mathbf{G}$. Assume*

(∗) *for each $v \in S$, the K_v-rank of any almost K_v-simple factor of the group $\mathbf{G}$ is different from 1, where as before K_v denotes the completion of the field K relative to v.*

Then

(a) *Γ has property (T);*
(b) *the factor group of Γ modulo its commutator subgroup is finite;*
(c) *Γ is not an amalgam.*

Proof. Let $\Gamma_0 = \Gamma \cap \mathbf{G}(K)$. By virtue of Theorem I.3.2.5 the image of the group Γ_0 under the diagonal embedding of the group $\mathbf{G}(K)$ into $\prod_{v \in S} \mathbf{G}(K_v)$ is a lattice. On the other hand, condition (∗) is satisfied and Γ_0 is of finite index in Γ.

Therefore, by virtue of Theorem 5.7 and Corollary 2.15, Γ has property (T). All that remains now is to make use of Theorems 2.5, 2.7, and 3.9. □

(5.9) Conditions $(*)$ in the statements of Theorem 5.7 and 5.8 are not necessary for the validity of assertions (a), (b), (c), and (d) of these theorems. According to Theorem 5.6 one can replace $(*)$ in Theorem 5.7 by the following condition:

$(**)$ if k_α is totally disconnected, then the group G_α does not have almost k_α-simple factors of k_α-rank 1 and if $k_\alpha = \mathbb{R}$ (resp. $k_\alpha = \mathbb{C}$), then $\mathbf{G}_\alpha$ contains no almost k_α-simple factors locally isomorphic to $\mathbf{SO}_{n,1}$ or $\mathbf{SU}_{n,1}$ (resp. $\mathbf{SL}_2$).

As to $(*)$ in Theorem 5.8, it can be replaced by the following condition

$(**)'$ if $v \in S$ is non-archimedean, then $\mathbf{G}$ contains no almost K_v-simple factors of K_v-rank 1 and if, for $v \in S$, $K_v = \mathbb{R}$ (resp. $K_v = \mathbb{C}$), then the group $\mathbf{G}$ contains no almost K_v-simple factors locally isomorphic to $\mathbf{SO}_{n,1}$ or $\mathbf{SU}_{n,1}$ (resp. $\mathbf{SL}_2$).

Let A, k_α, $\mathbf{G}_\alpha$, G, and Γ be as in Theorem 5.7 and denote by pr_α the natural projection $G \to \mathbf{G}_\alpha(k_\alpha)$. Related with Theorem 5.7(d) we have the following

(A) Theorem. (See [Mar 15], Theorem 2.) *Suppose for each $\alpha \in A$ the group $\mathbf{G}_\alpha$ is simply connected, k_α-isotropic, and almost k_α-simple. Assume in addition that* $\mathrm{pr}_\alpha(\Gamma)$ *is dense in* $\mathbf{G}_\alpha(k_\alpha)$, *whenever* $\mathrm{rank}_{k_\alpha} \mathbf{G}_\alpha = 1$, $\alpha \in A$.

(a) *Let Γ_1 and Γ_2 be two subgroups of Γ with* $\mathrm{card}(\Gamma_1/(\Gamma_1 \cap \Gamma_2)) \geq 2$, $\mathrm{card}(\Gamma_2/(\Gamma_1 \cap \Gamma_2)) \geq 3$, *and let, with notation of 3.8, $\Gamma = \Gamma_1 * \Gamma_2$. Then there exist $\alpha_0 \in A$ and two open subgroups G_1 and G_2 of $\mathbf{G}_{\alpha_0}(k_{\alpha_0})$ such that* $\mathrm{rank}_{k_{\alpha_0}} \mathbf{G}_{\alpha_0} = 1$, $\Gamma_i = \mathrm{pr}_{\alpha_0}^{-1}(G_i) \cap \Gamma\ (i = 1, 2)$, *and* $G = \mathrm{pr}_{\alpha_0}^{-1}(G_1) * \mathrm{pr}_{\alpha_0}^{-1}(G_2)$.

(b) *Assume* $\mathrm{rank}_{k_\alpha} \mathbf{G}_\alpha \geq 2$, *whenever k_α is non-archimedean, $\alpha \in A$. Then there exist no subgroups Γ_1 and Γ_2 of Γ such that* $\mathrm{card}(\Gamma_1/(\Gamma_1 \cap \Gamma_2)) \geq 2$, $\mathrm{card}(\Gamma_2/(\Gamma_1 \cap \Gamma_2)) \geq 3$, *and* $\Gamma = \Gamma_1 * \Gamma_2$.

The proof of Theorem A given in [Mar 15] is essentially different from that of Theorem 5.7(d) and is based on the study of Γ-equivariant measurable maps of the group $\mathbf{G}$ to the space of finitely additive measures on the set of ends of the tree $X(\Gamma, \Gamma_1, \Gamma_2)$.

A normal subgroup G' of the group G is called *standard* if G' can be represented in the form

$$G' = \prod_{\alpha \in A} \mathbf{G}'_\alpha(k_\alpha),$$

where $\mathbf{G}'_\alpha$ is a connected normal k_α-subgroup of $\mathbf{G}_\alpha$. A standard normal subgroup $G' = \prod_{\alpha \in A} \mathbf{G}'_\alpha(k_\alpha)$ is infinite if and only if $\mathbf{G}'_\alpha \neq \{e\}$ for some $\alpha \in A$. We call two standard normal subgroups $G' = \prod_{\alpha \in A} \mathbf{G}'_\alpha(k_\alpha)$ and $G'' = \prod_{\alpha \in A} \mathbf{G}''_\alpha(k_\alpha)$ *complementary* if for each $\alpha \in A$ the group $\mathbf{G}_\alpha$ is an almost direct product of the subgroups $\mathbf{G}'_\alpha$ and $\mathbf{G}''_\alpha$. We now give the definition of an irreducible lattice which coincides with Definition II.6.5 in case all the $\mathbf{G}_\alpha$ are almost k_α-simple.

Definition. *A lattice Γ is said to be* reducible *if there are two infinite standard complementary normal subgroups G' and G'' of G such that the subgroup $(\Gamma \cap G') \cdot (\Gamma \cap G'')$ is of finite index in Γ; otherwise the lattice Γ is called* irreducible.

In 6.9 the following theorem will be proved

(B) Theorem. *Suppose the lattice Γ is irreducible and the set $\hat{A} \overset{\text{def}}{=} \{\alpha \in A \mid$ at least one of almost k_α-simple factors of $\mathbf{G}_\alpha$ has k_α-rank greater than $1\}$ is not empty. Assume in addition that for no $\alpha \in A$, the group $\mathbf{G}_\alpha$ contains a nontrivial k_α-anisotropic factor. Then*

(a) *the factor group Γ/N of Γ modulo any normal subgroup $N \not\subset \mathbf{Z}(G)$ has property (T), and hence (see 2.2) is either non-amenable or finite;*
(b) *the factor group $\Gamma/\mathscr{D}(\Gamma)$ is finite;*
(c) *the group Γ is finitely generated;*
(d) *the group Γ has property (FA) (see 3.10) if the following condition is satisfied: for non-archimedean k_α, $\alpha \in A$, any almost k_α-simple factor of the group $\mathbf{G}_\alpha$ has k_α-rank > 1.*

(C) Theorem. *With notation as in Theorem 5.8, we assume the group $\mathbf{G}$ is simply connected and the following condition is satisfied:*

(E_1) *for every almost K-simple factor $\mathbf{G}'$ of $\mathbf{G}$, $\sum_{v \in S} \operatorname{rank}_{K_v} \mathbf{G}' \geq 2$.*

Let Γ_1 and Γ_2 be two subgroups of Γ with $\operatorname{card}(\Gamma_1/(\Gamma_1 \cap \Gamma_2)) \geq 2$, $\operatorname{card}(\Gamma_2/(\Gamma_1 \cap \Gamma_2)) \geq 3$, and $\Gamma = \Gamma_1 * \Gamma_2$. Then there exist a non-archimedean valuation $v \in S$, an almost K_v-simple factor $\tilde{\mathbf{G}}$ of $\mathbf{G}$, and two open subgroups G_1 and G_2 in $\tilde{\mathbf{G}}(K_v)$ such that $\operatorname{rank}_{K_v} \tilde{\mathbf{G}} = 1$, $\Gamma_i = \pi^{-1}(G_i) \cap \Gamma \, (i = 1, 2)$, and $\mathbf{G}(K_v) = \pi^{-1}(G_1) * \pi^{-1}(G_2)$, where $\pi \colon \mathbf{G} \to \tilde{\mathbf{G}}$ is the natural map.

Theorem C can be deduced from Theorem A in essentially the same way as Theorem 5.8 from Theorem 5.7. To this end, it suffices to observe that condition (E_1), the strong approximation theorem (see II.6.8), and Corollary I.2.3.2(b) imply that, for $v \in S$ and any almost K-simple factor $\mathbf{G}'$ of $\mathbf{G}$ with $\operatorname{rank}_{K_v} \mathbf{G}' = 1$, the subgroup $\Gamma \cap \mathbf{G}'(K)$ is dense in $\mathbf{G}'(K_v)$.

(D) Theorem. *With notation as in Theorem 5.8, assume the group $\mathbf{G}$ is almost K-simple and*

(E_2) *there exists $v \in S$ such that at least one of almost K_v-simple factors of the group $\mathbf{G}$ has K_v-rank > 1.*
Then

(a) *the factor group Γ/N of Γ modulo any normal subgroup $N \not\subset Z(\mathbf{G})$ has property (T), and hence (see 2.2) is either non-amenable or finite;*
(b) *the factor group $\Gamma/\mathscr{D}(\Gamma)$ is finite;*
(c) *the group Γ is finitely generated;*
(d) *the group Γ has property (FA), whenever*

(E_3) *for every non-archimedean $v \in S$, the K_v-rank of any connected normal K_v-subgroup of the group* $\mathbf{G}$ *is different from* 1.

In case $\mathbf{G}$ is absolutely almost simple, Theorem D can be deduced from Theorem B in essentially the same manner as Theorem 5.8 from Theorem 5.7 (except that we must consider the diagonal embedding of $\mathbf{G}(K)$ into $\prod_{v\in S_0} \mathbf{G}(K_v)$), where $S_0 = \{v \in S \mid \mathbf{G}$ is isotropic over $K_v\}$, rather than into $\prod_{v\in S_0} \mathbf{G}(K_v)$). In the general case, we consider the adjoint group $\mathbf{G}'$ of G and represent it in the form of $R_{L/K}\hat{\mathbf{G}}$, where L is a finite field extension of K, $\hat{\mathbf{G}}$ is an absolutely almost simple adjoint L-group, and $R_{L/K}$ is the restriction of scalars functor (see I.1.7). Let $\pi\colon \mathbf{G} \to \mathbf{G}'$ be the central K-isogeny, $R = R^0_{L/K}\colon \hat{\mathbf{G}}(L) \to \mathbf{G}'(K)$ the natural isomorphism, and S the set of the extensions of $v \in S$ to L. Then the subgroups $\pi(\mathbf{G}(K(S)))$, $\mathbf{G}'(K(S))$, and $R(\hat{\mathbf{G}}(L(\hat{S})))$ are commensurable (see Corollary I.3.2.9 and Lemma I.3.1.4(ii)). Therefore to reduce the general case to the case of an absolutely almost simple group it suffices to show that the following assertion is valid.

(5.10) Lemma. *Let F be a countable discrete group and let P be one of the following properties of F:*

(a) *property (T);*
(b) *$F/\mathscr{D}(F)$ is finite;*
(c) *F is finitely generated;*
(d) *property (FA).*

Then

(i) *If a subgroup of finite index in F has property P, then so does F; in particular, if F is finite, then F has property P.*
(ii) *If D is a normal subgroup of F and D and F/D have property P, then so does F.*

For the case where P is property (T), see Proposition 2.9 and Corollary 2.15. The lemma is obvious if P is the property that $F/\mathscr{D}(F)$ is finite or the property that F is finitely generated. For property (FA), see 6.3 in [Ser 3].

(5.11) *Remark.* We denote by $F_q[t]$ the ring of polynomials in one variable t over the finite field F_q. The ring $F_q[t]$ coincides with $K(\{v\})$, where $K = F_q(t)$ is the field of rational functions in one variable over F_q and v is a valuation of the field K with $|t|_v > 1$. It then follows from Theorem 5.8(a) that the group $\mathbf{SL}_3(F_q[t])$ has property (T). On the other hand (see [Be 3]), the group $\mathbf{SL}_3(F_q[t])$ is not finitely presentable. Thus, we have constructed a counterexample to Kazhdan's conjecture (see [Kaz 1]) that every countable group with property (T) is finitely presentable.

(5.12) Let X be a measure space with finite normalized measure $\mu(\mu(X) = 1)$ and let M be a topological group of μ-preserving transformations of X. We assume that $\mu(hB \triangle B)$ depends continuously on $h \in M$ for all measurable $B \subset X$. We say that X *has small almost M-invariant sets* if for all $t > 0$, $\varepsilon > 0$, and a finite set $E \subset M$, there is a measurable subset $B \subset X$ such that $0 < \mu(B) < \varepsilon$ and $\mu(hB \triangle B) < t\mu(B)$ for all $h \in E$.

A sequence $\{B_n\}_{n \in \mathbb{N}^+}$ of measurable subsets of the space X is called *asymptotically M-invariant* if $\mu(hB_n \triangle B_n) \to 0$ for all $h \in M$, and is called *trivial* if

$$\mu(B_n)(1 - \mu(B_n)) \to 0.$$

We call an action of the group M on X *strongly ergodic* if any asymptotically M-invariant sequence of subsets of X is trivial.

We denote by $L_2^0(X, \mu)$ the subspace of the space $L_2(X, \mu)$ consisting of the functions with integral 0. Let U be the quasi-regular representation of the group M in $L_2(X, \mu)$, that is

$$(U(h)f)(x) = f(h^{-1}x), \ h \in M, \ f \in L_2(X, \mu), \ x \in X.$$

For $C \subset X$, we denote by $\chi(C)$ the characteristic function of the set C and by $\tilde{\chi}(C)$ the function $\chi(C) - \mu(C)$. Obviously, for C measurable, we have that

(1) $\qquad \tilde{\chi}(C) \in L_2^0(X, \mu);$

(2) $\qquad \|\tilde{\chi}(C)\|^2 = \mu(C)(1 - \mu(C));$

(3) $\qquad \|U(h)\tilde{\chi}(C) - \tilde{\chi}(C)\| = \mu(hC \triangle C), \ h \in M.$

This implies that if X has small almost M-invariant sets or has non-trivial asymptotically M-invariant sequences, with M viewed as a discrete group, then the restriction of U to $L_2^0(X, \mu)$ is close to I_M. On the other hand, if M acts ergodically on X, then the restriction of U to $L_2^0(X, \mu)$ does not contain I_M. Therefore the following two propositions hold.

(1) Proposition. *If M, viewed as a discrete group, has property (T) and acts ergodically on X, then X does not have small almost M-invariant sets.*

(2) Proposition. *If M, viewed as a discrete group, has property (T) and acts ergodically on X, then the action of M on X is strongly ergodic.*

Proposition 2 admits a converse. More precisely, the following assertion holds.

(3) Proposition. (See [Co-W].) *Let Λ be a countable discrete group. If every ergodic measure preserving action of Λ on a probability measure space is strongly ergodic, then Λ has property (T).*

We denote by $L_\infty(X, \mu)$ the space of (essentially) bounded measurable functions on X. A *mean* on $L_\infty(X, \mu)$ is a linear functional l on $L_\infty(X, \mu)$ such that $l(1) = 1$ and $l(f) \geq 0$ for all $f \geq 0$. A mean l is called *M-invariant* if $l(hf) = l(f)$ for all $h \in M$ and $f \in L_\infty(X, \mu)$, where $(hf)(x) = f(h^{-1}x)$, $x \in X$. There exists a natural correspondence between the means on $L_\infty(X,\mu)$ and the finitely additive non-negative normalized functions defined on the ring $\Omega(X,\mu)$ of measurable subsets of X. Indeed, if ν is such a function, then for each $f \in L_\infty(X,\mu)$ one can define the integral $\int f d\nu$. We obtain a mean on $L_\infty(X,\mu)$ by setting $l(f) = \int f d\nu$. Conversely, if l is a mean, then by setting $\nu(B) = l(\chi(B))$, where $\chi(B)$ is the characteristic function of the set B, we obtain a finitely additive set function. Clearly, this correspondence preserves M-invariance. Thus, the following holds.

(4) Proposition. *The integral relative to the measure μ is the unique M-invariant mean if and only if μ is the unique M-invariant finitely additive non-negative normalized set function defined on $\Omega(X,\mu)$.*

It was established in [D-R] that for M countable the integration relative to μ is the only M-invariant mean on $L_\infty(X,\mu)$ if and only if X has no small almost M-invariant sets. On the other hand, since $\mu(hB \triangle B)$ depends continuously on $h \in M$ for all measurable $B \subset X$, every dense subgroup of M acts ergodically on X, if M itself acts ergodically on X. It then follows from Proposition 1 that

(5) Proposition. *If M contains a countable dense subgroup with property (T) as a discrete group and if M acts ergodically on X, then the integral relative to μ is the unique M-invariant mean on $L_\infty(X,\mu)$.*

Proposition 5 implies the following

(6) Proposition. *Suppose the group M is locally compact and acts transitively on X. Assume in addition the action of the group M on X is measurable in the following sense: the map*

$$\varphi\colon\ M \times X \to X,\ \varphi(h,x) = hx,\ h \in M,\ x \in X,$$

is measurable. If M contains a countable dense subgroup with property (T) as a discrete group, then the integral relative to μ is the unique M-invariant mean on $L_\infty(X,\mu)$.

Let us deduce now from Theorem 5.8 and Proposition 6

(7) Proposition. *Let H be a connected simple non-commutative compact Lie group which is not locally isomorphic to $\mathbf{SO}_3(\mathbb{R})$. If H acts transitively and continuously on a smooth manifold Y and preserves the Lebesgue measure μ on Y, then the integral relative to μ is the unique H-invariant mean on $L_\infty(Y,\mu)$.*

Proof. We fix a prime $p \in \mathbb{N}^+$. It is well known that there is an absolutely almost simple $\mathbb{Q}$-group $\mathbf{H}$ such that $\mathbf{H}$ is split over p and the Lie groups $\mathbf{H}(\mathbb{R})^0$ and H are isomorphic (this, for instance, is a consequence of Theorem B in [Bo-Hard]). Since H is not locally isomorphic to $\mathbf{SO}_3(\mathbb{R})$, $\operatorname{rank}_{\mathbb{Q}_p} \mathbf{H} \geq 2$. Let $\mathbb{Z}[1/p]$ be the subring of the ring $\mathbb{Q}$ generated by $1/p$. Since $\operatorname{rank}_{\mathbb{R}} \mathbf{H} = 0$ (because $\mathbf{H}(\mathbb{R}) \simeq H$ and H is compact), and $\mathbf{H}$ is absolutely almost simple, it follows from Theorem

5.8(a) that the group $\mathbf{H}(\mathbb{Z}[1/p])$ has property (T). On the other hand, by the strong approximation theorem (see II.6.8), $\mathbf{H}(\mathbb{Z}[1/p])$ is dense in $\mathbf{H}(\mathbb{R})^0 \simeq H$. Thus, the desired assertion is a consequence of Proposition 6. □

The following assertion is a special case of Proposition 7.

(8) Proposition. (See [Mar 14] or [Sul 2].) *If $n \geq 5$, then the integral relative to the usual Lebesgue measure μ on the sphere S^{n-1} is the unique $\mathbf{SO}_n(\mathbb{R})$-invariant mean on $L_\infty(S^{n-1}, \mu)$.*

(1) *Remark.* Combining Propositions 5–8 with Proposition 4 we obtain assertions concerning the uniqueness of finitely additive invariant set functions defined on the ring of all measurable subsets.

(2) *Remark.* It was shown in [Mar 16] that for $n \geq 3$ the Lebesgue measure μ is the unique (up to a multiple) finitely additive set function, defined on the ring of bounded μ-measurable subsets of $\mathbb{R}^n$, invariant under all rigid motions. The proof of this assertion is technically considerably more complicated than that of Proposition 8, though based on similar ideas.

(3) *Remark.* It was established by Drinfeld in [Dr] that the condition "$n \geq 5$" in Proposition 8 can be replaced by the weaker condition "$n \geq 3$". In fact, it was also shown in that paper that the condition "H is not locally isomorphic to $\mathbf{SO}_3(\mathbb{R})$" in Proposition 7 is superfluous. Drinfeld's proof is based on a theorem of Jacquet-Langlands and the Petersson's conjecture proved by Deligne.

(5.13) Let H be a countable group with property (T), let $h_1, \ldots, h_m$ be the generators of H. Making use of the equalities analogous to (2) and (3) in 5.12, it is not hard to prove the existence of a number $c > 0$ depending only on $h_1, \ldots, h_m$ such that

$$\left| \bigcup_{1 \leq i \leq m} h_i X \right| \geq (1 + c(1 - |X|/|F|)) \cdot |X|$$

for any finite factor group F of H and $X \subset F$ ($h_i X$ denotes the image $\pi(h_i \pi^{-1}(X))$ under the natural epimorphism $\pi\colon H \to F$). This fact has been used in an explicit construction of the so called "expanders". For more details we refer to [A-M] and [Mar 2].

6. Relationship Between the Structure of Closed Subgroups and Property (T) of Normal Subgroups

Let M be a closed normal subgroup of the group H, $\pi\colon H \to H/M$ the natural map, and F a closed subgroup of H. The closure of a subset B of a topological space will as usual be denoted by $\bar{B}$.

In previous sections a series of results on the structure of closed subgroups in a group with property (T) have been proved. In this section we present similar results in case M has property (T) and the subgroup $\overline{\pi(F)}$ is "sufficiently large".

(6.1) Definition. *Let $U \in \tilde{F}$ be a unitary representation of the group F. We say that U is* π-factorable *if there is a (continuous) unitary representation ρ of the group $\overline{\pi(F)}$ on $L(U)$ such that $\rho(\pi(z)) = U(z)$ for all $z \in F$. We say that U is* partially π-factorable *if the restriction of U to a non-zero closed $U(F)$-invariant subspace of $L(U)$ is π-factorable ($L(U)$ denotes, as before, the space of the representation U).*

(6.2) Lemma. *Let $U \in \tilde{F}$. If there is a non-zero vector $x \in L(U)$ such that, for a continuous map $\varphi \colon \overline{\pi(F)} \to L(U)$,*

$$\varphi(\pi(z)) = U(z)x$$

for all $x \in F$, then the representation U is partially π-factorable.

Proof. We set $Y = \{y \in L(U) \mid$ there is a (uniquely determined) continuous map $\varphi_y \colon \overline{\pi(F)} \to L(U)$ with $\varphi_y(\pi(z)) = U(z)y$ for all $z \in F\}$. Since U is unitary,

$$\|\varphi_{y_1}(\pi(z)) - \varphi_{y_2}(\pi(z))\| = \|y_1 - y_2\|$$

for all $y_1, y_2 \in Y$ and $z \in F$. It follows that Y is closed (if $y = \lim_{n\to\infty} y_n$, where $y_n \in Y$, then we set $\varphi_y(g) = \lim_{n\to\infty} \varphi_{y_n}(g)$ and make use of the theorem on the continuity of the limit of a uniformly convergent sequence of continuous maps). Clearly, Y is a linear subspace in $L(U)$. Putting

$$\varphi_{U(z)y}(g) = \varphi_y(g\pi(z))$$

for $z \in F$, $g \in \overline{\pi(F)}$, $y \in Y$, we observe that Y is $U(F)$-invariant. For each $g \in \overline{\pi(F)}$ we define the unitary operator $\rho(g)$ on the space Y by setting $\rho(g)y = \varphi_y(g)$, $y \in Y$. Since $\varphi_y(\pi(z)) = U(z)y$, the maps φ_y are continuous, the set $\pi(F)$ is dense in $\overline{\pi(F)}$, and U is unitary, it follows that ρ is a continuous unitary representation of the group $\overline{\pi(F)}$ on the space Y and $\rho(\pi(z)) = U(z)$ for all $z \in F$. □

(6.3) Theorem. *Suppose*

(A) *M has property (T) and*
(B) *there is a finite H-invariant measure μ on $F \setminus H$.*

Then there exists a neighbourhood W of the representation I_F in $\tilde{F}$ such that

(a) *every $U \in W$ is partially π-factorable and*
(b) *every irreducible $U \in W$ is π-factorable.*

Proof. Since $\mu(F \setminus H) < \infty$, the inducing operation is continuous, and as M has property (T), it follows that there exists an open neighbourhood $W \subset \tilde{F}$ of I_F such that, for every $U \in W$, $\operatorname{Ind}(H,F,U)|_M \geq I_M$. Let $U \in W$, $\rho = \operatorname{Ind}(H,F,U)$, and j be a non-zero $\rho(M)$-invariant vector in the space $L_2(H,F,U)$ of the representation ρ. It then follows from Lemma I.4.1.1(v) and equalities (1) and (2) in 1.8, that after a modification of the vector valued function j on a null set, we have

$$(1) \qquad j(zhM) = U(z)j(h), \; z \in F, \; h \in H.$$

Since $j \neq 0$, there is a function $f \in \mathfrak{A}(H)$ such that $(\rho(f)j)(e) \neq 0$. It follows from 1.8 (2) and Lemma 1.7 that the vector-valued function $\rho(f)j$ is continuous. On the other hand, since the subgroup M is normal, it follows from 1.8 (2) that the set of the vector-valued function j satisfying (1) is invariant relative to $\rho(H)$, and hence, relative to $\rho(f)$. Therefore, replacing j by $\rho(f)j$, we may assume that j is continuous and $j(e) \neq 0$. It follows from (1) that j is constant on $\pi^{-1}(g)$ for each $g \in H/M$. Therefore one can define a map $\varphi\colon \overline{\pi(F)} \to X$ to the space X of the representation U, by setting

$$(2) \qquad \varphi(g) = j(\pi^{-1}(g)), \; g \in \overline{\pi(F)}.$$

Since j is continuous, φ is also continuous. On the other hand, $j(e) \neq 0$ and (1), (2) imply that $\varphi(\pi(z)) = U(z)j(e)$ for all $z \in F$. Thus, according to Lemma 6.2, the representation U is partially π-factorable, hence (a). Property (b) is a consequence of (a) and the definition of irreducibility of a representation. □

The following assertion is a straightforward consequence of the definition of a partially π-factorable representation.

(6.4) Lemma. *If $U \in \tilde{F}$ is partially π-factorable, then there is $x \in L(U)$ with $\|x\| = 1$ such that*

$$\lim_{z \in F, \; \pi(z) \to e} \langle U(z)x, x\rangle = 1.$$

(6.5) Theorem. *Suppose conditions (A) and (B) of Theorem 6.3 are satisfied and the group $\overline{\pi(F)}$ is compactly generated. Then F is compactly generated. In particular, if F is discrete, then F is finitely generated.*

Proof. Since F is σ-compact, there exists a sequence $\{F_n\}_{n \in \mathbb{N}^+}$ of open compactly generated subgroups of F with

$$F = \bigcup_{n \in \mathbb{N}^+} F_n \text{ and } F_n \subset F_{n+1}, \; n \in \mathbb{N}^+.$$

We consider the following two cases:

(i) there exists $n_0 \in \mathbb{N}^+$ such that $e \notin \overline{\pi(F - F_{n_0})}$;

(ii) $e \in \overline{\pi(F - F_n)}$ for all $n \in \mathbb{N}^+$.

(i) Let $K \subset \overline{\pi(F)}$ be a compact set generating the group $\overline{\pi(F)}$ and let W be a neighbourhood of the identity in the group $\overline{\pi(F)}$ with

$$(1) \qquad W \cap \overline{\pi(F - F_{n_0})} = \varnothing.$$

Since K is a compact subset lying in $\overline{\pi(F)}$ and W is open, there exists a finite set $L \subset F$ such that $K \subset \pi(L)W$. Pick a number $n_1 \geq n_0$ with $L \subset F_{n_1}$. It then follows from (1) and the inclusion $F_{n_0} \subset F_{n_1}$ that for each $z \in L$

$$\pi(z)W \cap \overline{\pi(F - F_{n_1})} = \pi(z)(W \cap \overline{\pi(z^{-1}(F - F_{n_1}))} =$$
$$= \pi(z)(W \cap \overline{\pi(F - F_{n_1})}) \subset \pi(z)(W \cap \overline{\pi(F - F_{n_0})}) = \varnothing.$$

But $K \subset \pi(L)W$. Thus

(2) $$K \cap \overline{\pi(F - F_{n_1})} = \varnothing$$

and hence

(3) $$K \subset \overline{\pi(F_{n_1})}.$$

Since F_{n_1} is a subgroup, $\overline{\pi(F - F_{n_1})} \cdot \overline{\pi(F_{n_1})} = \overline{\pi(F - F_{n_1})}$. But $\overline{\pi(F_{n_1})}$ is also a subgroup. Therefore either $\overline{\pi(F - F_{n_1})} \supset \overline{\pi(F_{n_1})}$ or $\overline{\pi(F - F_{n_1})} \cap \overline{\pi(F_{n_1})} = \varnothing$. In view of (2) and (3) this implies that $\overline{\pi(F - F_{n_1})} \cap \overline{\pi(F_{n_1})} = \varnothing$. On the other hand, since K generates $\overline{\pi(F)}$ and $\overline{\pi(F_{n_1})}$ is a subgroup, (3) implies that $\overline{\pi(F)} \supset \overline{\pi(F_{n_1})}$. Thus $\overline{\pi(F - F_{n_1})} = \varnothing$, and hence F coincides with the compactly generated group F_{n_1}.

(ii) In this case there is a sequence $\{z_n\}_{n \in \mathbb{N}^+}$ such that $z_n \in F - F_n$ and

(4) $$\lim_{n \to \infty} \pi(z_n) = e.$$

As in the proof Theorem 2.7 (replacing H by F and H_n by F_n) we define an F-invariant measure μ_n on $F_n \setminus F$ and denote by ρ_n the quasi-regular unitary representation of the group F on the space $L_2(F_n \setminus F, \mu_n)$. Since $F = \bigcup_{n \in \mathbb{N}^+} F_n$, $F_n \subset F_{n+1}$, and $z_n \in F - F_n$, it follows that for each $i \in \mathbb{N}^+$ and every finite $X \subset F_i \setminus F$ there is a number $n(X)$ such that $(Xz_n) \cap X = \varnothing$ for all $n \geq n(X)$. This implies that

(5) $$\lim_{n \to \infty} \langle \rho_i(z_n)f, f \rangle = 0 \text{ for all } i \in \mathbb{N}^+ \text{ and } f \in L_2(F_i \setminus F, \mu_i).$$

As in the proof of Theorem 2.7 we check that the sequence $\{\rho_n\}$ converges to I_F. Thus, by Theorem 6.3, there is $i \in \mathbb{N}^+$ such that ρ_i is partially π-factorable. It then follows from (4) (see Lemma 6.4) that there is a function $f \in L_2(F_i \setminus F, \mu_i)$ such that $\lim_{n \to \infty} \langle \rho_i(z_n)f, f \rangle = 1$. But this contradicts (5). □

(6.6) Theorem. *Suppose conditions (A) and (B) of Theorem 6.3 are satisfied.*

(a) *Let N be a closed normal subgroup of the group F with $\overline{\pi(F)}/\overline{\pi(N)}$ finite. Then the factor group F/N has property (T), and hence (see 2.2) is either non-amenable or compact.*

(b) *If $\overline{\pi(F)}/\overline{\pi(\mathscr{D}(F))}$ is finite, then the factor group $F/\mathscr{D}(F)$ is compact. In particular, if $\overline{\pi(F)}/\overline{\pi(\mathscr{D}(F))}$ is finite and F is discrete, then $F/\mathscr{D}(F)$ is finite.*

Proof. (a) Since $\overline{\pi(F)}/\overline{\pi(N)}$ is finite, it follows from Corollary 2.15 that if $(F \cap \pi^{-1}(\overline{\pi(N)}))/N$ has property (T), then so does F/N. Thus, replacing F by $F \cap \pi^{-1}(\overline{\pi(N)})$, we can assume that

(1) $$\overline{\pi(F)} = \overline{\pi(N)}.$$

Let $\varphi\colon F \to F/N$ be the natural map. Suppose F/N does not have property (T). Let U' be a representation of F/N close to $I_{F/N}$ and not containing $I_{F/N}$. By setting $U(z) = U'(\varphi(z))$, $z \in F$, we define a representation $U \in \tilde{F}$, not containing I_F, which is trivial on N and close to I_F. Since U is close to I_F, it follows from Theorem 6.3 that U is partially π-factorable. Thus, one can find a unitary representation ρ of the group $\overline{\pi(F)}$ and a non-zero vector $x \in L(U)$ with

$$\rho(\pi(z))x = U(z)x \text{ for all } z \in F. \tag{2}$$

Since U is trivial on N and ρ is continuous, (2) implies that $\rho(\overline{\pi(N)})x = x$. On the other hand, (1) and (2) imply that

$$U(F)x = \rho(\pi(F))x \subset \rho(\overline{\pi(F)})x = \rho(\overline{\pi(N)})x.$$

Thus $U(F)x = x$ and, since $x \neq 0$, $U \geq I_F$. This contradiction completes the proof of (a).

(b) is a consequence of (a) and Lemma 2.3. □

(6.7) Theorem. *Suppose conditions (A) and (B) of Theorem 6.3 are satisfied and* $\overline{\pi(F)} = H/M$.

(a) *Let F_1, F_2 be two open subgroups of the group F with $F_1 \neq F \neq F_2$ and, with notation as in 3.8, let $F = F_1 * F_2$. Then there exist two open subgroups G_1 and G_2 of H/M such that $F_i = \pi^{-1}(G_i) \cap F$ $(i = 1, 2)$ and $H/M = G_1 * G_2$.*

(b) *If H/M is not an amalgam, then neither is F.*

(c) *If H/M is connected, then F is not an amalgam.*

Proof. (a) Define as in 3.8 the graph $X = X(F, F_1, F_2)$. Since $F = F_1 * F_2$, the graph X is a tree (see 3.8). Let us consider the standard distance function d on the set $V = (F/F_1) \cup (F/F_2)$ of vertices (see 3.5) of X and the natural action of the group F on X (see 3.8). Let $P_0 = eF_1 \in F/F_1 \subset V$. For $z \in F$, we set $l(z) = d(P_0, zP_0)$. By Proposition 3.6 l is a negative definite function on F. Therefore, by Theorem 3.2, for each $t > 0$, the function $\exp(-tl)$ is positive definite. Thus (see Theorem I.5.4.3) for each $t > 0$ there exist $\rho_t \in \tilde{F}$ and a cyclic vector $x_t \in L(\rho_t)$ for ρ_t such that

$$\langle \rho_t(z)x_t, x_t \rangle = \exp(-tl(z)) \text{ for all } z \in F. \tag{1}$$

Since ρ_t is unitary,

$$\langle \rho_t(z)\rho_t(z_1)x_t,\ \rho_t(z_2)x_t \rangle = \langle \rho_t(z_2^{-1}zz_1)x_t, x_t \rangle$$

for all $t > 0$ and $z, z_1, z_2 \in F$. On the other hand, since $l(z) = d(P_0, zP_0)$ and the distance function d is invariant relative to $\operatorname{Aut} X$, $l(z_2^{-1}zz_1) \geq l(z) - l(z_2^{-1}) - l(z_1)$ for all $z, z_1, z_2 \in F$. Thus, (1) implies that

$$\lim_{z \in F, l(z) \to \infty} \langle \rho_t(z)\rho_t(z_1)x_t, \rho_t(z_2)x_t \rangle = 0$$

for all $t > 0$ and $z_1, z_2 \in F$. But finite linear combinations of the vectors $\rho_t(z)x_t$, $z \in F$, form a dense subset in $L(\rho_t)$ (because x_t is a cyclic vector for ρ_t). Thus,

(2) $$\lim_{z\in F, l(z)\to\infty} \langle \rho_t(z)x, y\rangle = 0 \text{ for all } t > 0 \text{ and } x, y \in L(\rho_t).$$

Since l is continuous, and hence locally bounded, it follows from (1) that ρ_t approaches I_F as $t \to 0$. Therefore by virtue of Theorem 6.3 there is $t > 0$ such that ρ_t is partially π-factorable. It then follows from Lemma 6.4 that there exists $x \in L(\rho_t)$ such that

(3) $$\lim_{z\in F, \pi(z)\to e} \langle \rho_t(z), x, x\rangle = 1.$$

(2) and (3) imply that

(4) $$\lambda \stackrel{\text{def}}{=} \limsup_{z\in F, \pi(z)\to e} l(z) < \infty.$$

We choose a sequence $\{z_i\}_{i\in\mathbb{N}^+}$ in F with

(5) $$\lim_{i\to\infty} \pi(z_i) = e \text{ and } l(z_i) = \lambda \text{ for all } i \in \mathbb{N}^+.$$

We denote by $[w_1, w_2]$ the reduced path joining the vertices w_1 and w_2 of the tree X. Since $F_1 \neq F \neq F_2$, there exist $u_1, u_2 \in F$ such that

(6) $$u_1 P_0 \neq P_0 \neq u_2 P_0 \text{ and } P_0 \in [u_1 P_0, u_2 P_0]$$

(as u_1 and u_2 we can take, respectively, v_1 and $v_2 v_1$, where $v_i \in F_{3-i} - F_i$). We will show that the following assertion is true.

(*) If $z \in F$ and $zP_0 \neq P_0$, then $l(z) < \max\{l(z^2), l(u_1^{-1} z u_1), l(u_2^{-1} z u_2)\}$.

It follows from (6) that P_0 belongs to the union of the paths $[u_1 P_0, zP_0]$ and $[u_2 P_0, zP_0]$. Let $P_0 \in [u_1 P_0, zP_0]$. Then $zP_0 \in [zu_1 P_0, z^2 P_0]$. It follows that zP_0 belongs either to $[P_0, zu_1 P_0]$ or to $[P_0, z^2 P_0]$. If $zP_0 \in [P_0, z^2 P_0]$ and $zP_0 \neq P_0$, then

$$\begin{aligned} l(z^2) = d(P_0, z^2 P_0) &= d(P_0, zP_0) + d(zP_0, z^2 P_0) = \\ &= 2d(P_0, zP_0) = 2l(z) > l(z). \end{aligned}$$

Thus, we can assume that $zP_0 \in [P_0, zu_1 P_0]$. Since $P_0 \in [u_1 P_0, zP_0]$, it then follows that $[P_0, zP_0] \subset [u_1 P_0, zu_1 P_0]$, and hence

$$\begin{aligned} l(u_1^{-1} z u_1) &= d(P_0, u_1^{-1} z u_1 P_0) = d(u_1 P_0, z u_1 P_0) = \\ &= d(u_1 P_0, P_0) + d(P_0, zP_0) + d(zP_0, zu_1 P_0) = l(z) + 2l(u_1). \end{aligned}$$

But $u_1 P_0 \neq P_0$, and so $l(u_1) > 0$. Thus, $l(u_1^{-1} z u_1) > l(z)$, and this proves (*).

Since $\lim_{i\to\infty} \pi(z_i) = e$, it follows that

$$\lim_{i\to\infty} \pi(z_i^2) = \lim_{i\to\infty} \pi(u_1^{-1} z_i u_1) = \lim_{i\to\infty} \pi(u_2^{-1} z_i u_2) = e.$$

But according to (5) $z_i P_0 \neq P_0$, whenever $\lambda > 0$. Thus, (4), (5), and (*) imply that $\lambda = 0$. On the other hand, $\lambda(z) = 0$ if and only if $z \in F_1$. It follows that there is an open neighbourhood $W_1 \subset H/M$ of the identity such that $F \cap \pi^{-1}(W_1) \subset F_1$. According to the assumption of the theorem, $\overline{\pi(F)} = H/M$, and hence

(i) the subgroup $G_1 \stackrel{\text{def}}{=} \overline{\pi(F_1)}$ is open;

(ii) $F \cap \pi^{-1}(G_1) \subset F \cap \pi^{-1}(\pi(F_1)W_1) = F \cap (F_1\pi^{-1}(W_1)) = F_1(F \cap \pi^{-1}(W_1)) = F_1F_1 = F_1$, and so

$$F_1 = F \cap \pi^{-1}(G_1). \tag{7}$$

Similarly we obtain that the subgroup $G_2 \stackrel{\text{def}}{=} \overline{\pi(F_2)}$ is open and

$$F_2 = F \cap \pi^{-1}(G_2). \tag{8}$$

We denote by Y_G (resp. Y_F) the set of elements $g \in G$ (resp. $z \in F$) representable in the form $g = g_n \dots g_1$ (resp. $z = z_n \dots z_1$), where $g_i \in G_{j(i)} - (G_1 \cap G_2)$ (resp. $z_i \in F_{j(i)} - (F_1 \cap F_2)$), $j(i) = 1, 2 (1 \le i \le n)$ and $j(i+1) \ne j(i)$ for all $1 \le i \le n-1$. Since $G_i = \overline{\pi(F_i)}$, it follows that $Y_G \subset \overline{\pi(Y_F)}$. But $F = F_1 * F_2$ implies that $Y_F \cap F_1 \cap F_2 = \varnothing$ (see 3.8). It then follows from (7), (8), and the openness of the subgroups G_1 and G_2 that $Y_G \cap G_1 \cap G_2 = \varnothing$, and hence $G = G_1 * G_2$. This completes the proof of (a).

(b) is a consequence of (a).

(c) is a consequence of (b) and the fact that a connected topological group contains no proper open subgroups. □

(6.8) Lemma. *With A, k_α, $\mathbf{G}_\alpha$, G and Γ as in the statement of Theorem 5.7, for any $B \subset A$ we set*

$$G_B = \prod_{\alpha \in B} \mathbf{G}_\alpha(k_\alpha), \quad G_B^+ = \prod_{\alpha \in B} \mathbf{G}_\alpha(k_\alpha)^+,$$

and denote by pr_B the natural projection $G \to G_B$. Let N be a normal subgroup of Γ. Suppose $N \not\subset \mathscr{Z}(G)$, the lattice Γ is irreducible, and for each $\alpha \in A$ the group $\mathbf{G}_\alpha$ is isotropic and almost simple over k_α. Then

(a) $\overline{\mathrm{pr}_B(N)} \supset G_B^+$ *for any $B \subsetneqq A$,*

(b) *if the groups $\mathbf{G}_\alpha$, $\alpha \in A$, are simply connected, then $\overline{\mathrm{pr}_B(N)} = G_B$ for any $B \subsetneqq A$.*

Proof. (a) Suppose the contrary. Then $\mathbf{G}_\alpha(k_\alpha)^+ \not\subset \overline{\mathrm{pr}_B(N)}$ for some $\alpha \in B$. Since $N \not\subset \mathscr{Z}(G)$ and $\Gamma \cap G_{A-\{\alpha\}} \subset \mathscr{Z}(G_{A-\{\alpha\}})$ (see Theorem II.6.7(d)), $\mathrm{pr}_\alpha(N) \not\subset \mathscr{Z}(\mathbf{G}_\alpha)$ and hence $\mathrm{pr}_\alpha(\overline{\mathrm{pr}_B(N)}) \not\subset \mathscr{Z}(\mathbf{G}_\alpha)$. But the closure of the subgroup $\mathrm{pr}_B(\Gamma)$ contains G_B^+ (see Theorem II.6.7(a)) and N is normal in Γ, so G_B^+ normalizes $\overline{\mathrm{pr}_B(N)}$, and hence $\mathbf{G}_\alpha(k_\alpha)^+$ normalizes $\mathrm{pr}_\alpha(\overline{\mathrm{pr}_B(N)})$. It then follows from Theorem I.1.5.6 that

$$\begin{aligned}\overline{\mathrm{pr}_B(N)} &\supset [\mathbf{G}_\alpha(k_\alpha)^+, \overline{\mathrm{pr}_B(N)}] = [\mathbf{G}_\alpha(k_\alpha)^+, \mathrm{pr}_\alpha(\overline{\mathrm{pr}_B(N)})] \supset \\ &\supset [\mathbf{G}_\alpha(k_\alpha)^+, \mathbf{G}_\alpha(k_\alpha)^+] = \mathbf{G}_\alpha(k_\alpha)^+,\end{aligned}$$

where $[C, D]$ denotes the mutual commutator subgroup of the subgroups C and D. This is a contradiction.

(b) is a consequence of (a) and Theorem I.2.3.1(a). □

(6.9) *Proof of Theorem B in 5.9* Let $\mathbf{G}'_\alpha$ be the adjoint group of $\mathbf{G}_\alpha$ and let $p_\alpha\colon \mathbf{G}_\alpha \to \mathbf{G}'_\alpha$ be the central k_α-isogeny. We set $G' = \prod_{\alpha \in A} \mathbf{G}'_\alpha(k_\alpha)$. The isogenies p_α induce a continuous homomorphism $p\colon G \to G'$. By Proposition I.2.3.4 the subgroup $p(G)$ is closed and normal in G', $G'/p(G)$ is compact and the isomorphism $G/\operatorname{Ker} p \to p(G)$ is an isomorphism of topological groups. Thus $p(\Gamma)$ is a lattice in G'. Since Γ is irreducible, $p(\Gamma)$ is irreducible. By Corollary I.1.4.6 (b) the set $\{\alpha \in A \mid k_\alpha\text{-rank of at least one of almost } k\text{-simple factors of the group } \mathbf{G}'_\alpha \text{ is greater than } 1\}$ coincides with $\hat{A}$, and hence, is not empty. Thus, in view of Lemma 5.10, replacing $\mathbf{G}_\alpha$ and Γ, respectively, by $\mathbf{G}'_\alpha$ and $p(\Gamma)$, we may assume that $\mathbf{G}_\alpha$ is an adjoint group. Then $\mathbf{G}_\alpha$ can be decomposed into a direct product of k_α-simple groups. Thus, we can assume that the groups $\mathbf{G}_\alpha$ are k_α-simple. In that case we have $\hat{A} = \{\alpha \in A \mid \operatorname{rank}_{k_\alpha} \mathbf{G}_\alpha \geq 2\}$. Let G_B, G_B^+, and pr_B be as in the statement of Lemma 6.8. Since Γ is irreducible and $\hat{A} \neq \varnothing$, we have by Theorem II.6.7(a) that

$$(1) \qquad \overline{\mathrm{pr}_{A-\hat{A}}(\Gamma)} \supset G^+_{A-\hat{A}}.$$

It follows from Corollary I.2.3.5 and Theorem I.2.3.1(b) that the group $G^+_{A-\hat{A}}$ is compactly generated and the factor group $G_{A-\hat{A}}/G^+_{A-\hat{A}}$ is compact. It then follows from (1) that the subgroup $\overline{\mathrm{pr}_{A-\hat{A}}(\Gamma)}$ is compactly generated. On the other hand, by Corollary 5.4 the group $G_{\hat{A}}$ has property (T). Thus, by Theorem 6.5, the group Γ is finitely generated, and hence the proof of (c) is complete. Since the groups $\mathbf{G}_\alpha$ are isotropic over k_α, it follows from Theorem I.2.3.1(c) that the factor group $G_{A-\hat{A}}/G^+_{A-\hat{A}}$ is commutative and torsion. Thus the subgroup $\Gamma \cap \mathrm{pr}^{-1}_{A-\hat{A}}(G^+_{A-\hat{A}})$ is of finite index in the finitely generated group Γ. It then follows from Lemma 5.10(i) that we can replace Γ by $\Gamma \cap \mathrm{pr}^{-1}_{A-\hat{A}}(G^+_{A-\hat{A}})$, so we have, according to (1), that

$$(2) \qquad \overline{\mathrm{pr}_{A-\hat{A}}(\Gamma)} = G^+_{A-\hat{A}}.$$

By Corollary 5.4 the group $G_{\hat{A}}$ has property (T). On the other hand, by Theorem I.1.5.6(ii), $\mathscr{D}(G^+_{A-\hat{A}}) = G^+_{A-\hat{A}}$. Thus, assertions (a) and (b) of Theorem B in 5.9 are easy consequences of (2), Theorem 6.6, and Lemma 6.8(a). It remains to prove (d).

We set $A' = \{\alpha \in A \mid k_\alpha \text{ is isomorphic to } \mathbb{R} \text{ or } \mathbb{C}\}$. It follows from Theorem I.2.3.1(c) that $G^+_{A'}$ is a connected open subgroup of $G_{A'}$. Thus, making use of (2) and applying Theorem 6.7(c), we obtain that $\Gamma' \overset{\text{def}}{=} \Gamma \cap \mathrm{pr}^{-1}_{A'}(G^+_{A'})$ is of finite index in Γ and Γ' is not an amalgam, if $A' \supset A - \hat{A}$. It then follows from the assertions (b) and (c) just proved that Γ' has property (FA) whenever $A' \supset A - \hat{A}$ (see 3.10). Thus, by Lemma 5.10(i), the lattice Γ has property (FA) whenever $A' \supset A - \hat{A}$, which proves (d). □

(6.10) *Remark.* Making use of Theorem 6.7(a), we can prove Theorem A of 5.9 by the kind of argument that we used in 6.9 above to prove Theorem B (d) of 5.9, provided that Γ is irreducible and the set $\hat{A}$ is nonempty. Furthermore, the condition $\mathrm{card}(\Gamma_2/(\Gamma_1 \cap \Gamma_2)) \geq 3$ can be replaced by the weaker condition: $\mathrm{card}(\Gamma_2/(\Gamma_1 \cap \Gamma_2)) \geq 2$.

Chapter IV. Factor Groups of Discrete Subgroups

In Section 4 of this chapter we prove a theorem asserting that under certain conditions every normal subgroup N of an irreducible lattice Γ is either central or of finite index. This theorem is deduced from the following two assertions:

(a) if the factor group Γ/N is not amenable, then N is central;
(b) if Γ/N is amenable, then Γ/N is finite.

The proof of (a) is based on the results of Section 2 on Γ-invariant algebras of measurable sets and on Furstenberg's theorem on the existence of equivariant measurable maps to certain spaces of measures. In Section 2 we use generalization of the classical density point theorem. This is proved in Section 1. In the majority of cases assertion (b) is an easy consequence of the results of Chapter III. To handle the rest of the cases, in Section 3 we prove a theorem on the factor groups of lattices lying in direct products.

1. b-metrics, Vitali's Covering Theorem and the Density Point Theorem

Let X be a set and $b \geq 1$. A non-negative function ρ on $X \times X$ is called a *b-distance function* (or a *b-metric*) if the following conditions are satisfied:

$$\rho(x, y) = 0 \text{ if and only if } x = y; \tag{1}$$

$$\rho(x, y) = \rho(y, x); \tag{2}$$

$$\rho(x, z) \leq b[\rho(x, y) + \rho(y, z)], \tag{3}$$

for all $x, y, z \in X$. For $b = 1$, we obtain the usual distance function. For $x \in X$, $F \subset X$, and $\varepsilon > 0$, we set

$$U(x, \varepsilon) = \{y \in X \mid \rho(y, x) \leq \varepsilon\},$$
$$U_0(x, \varepsilon) = \{y \in X \mid \rho(y, x) < \varepsilon\},$$
$$U(F, \varepsilon) = \bigcup_{x \in F} U(x, \varepsilon), \text{ and}$$
$$\delta(F) = \sup_{x, y \in F} \rho(x, y).$$

The latter is called the *diameter* of the set F.

Let X be a topological space endowed with a b-distance function ρ. If the family $\{U_0(x, \varepsilon) \mid x \in X, \varepsilon > 0\}$ forms a base of open sets of the space X, then X is called a *b-metric topological space*. The closure of the set $U(x, \varepsilon)$ will be denoted by $\overline{U(x, \varepsilon)}$. Clearly, $\overline{U(x, \varepsilon)} \subset U(x, b\varepsilon)$.

(1.1) Lemma. *Let X be a compact b-metric space and $B \subset X$. Suppose that a family $\mathscr{F}$ of closed subsets of X has the following property:*

(V) for each $x \in X$ there exists F in $\mathscr{F}$ with $x \in F$ and with arbitrarily small positive diameter $\delta(F)$.

Then either B is contained in a finite union of disjoint members of $\mathscr{F}$ or there is a countable family $\{F_n\}$ of disjoint members of $\mathscr{F}$ such that

$$B \subset F_1 \cup \ldots \cup F_n \cup \bigcup_{k=n+1}^{\infty} U(F_k, 3b\delta(F_k)) \text{ for all } n. \tag{1}$$

Proof. We define the family $\{F_n\}$ by induction. Let F_1 be an arbitrary member of $\mathscr{F}$. Suppose the sets $F_1, \ldots, F_k$ have already been chosen. If $B \subset F_1 \cup \ldots \cup F_k$, then the lemma is obvious. Otherwise, let $\varepsilon_k = \sup \delta(F)$, $F \in \mathscr{F}_k \overset{\text{def}}{=} \{F \in \mathscr{F} \mid F_i \cap U(F, \delta(F)) = \varnothing$ for all i, $1 \leq i \leq k\}$. Since X is a b-metric space and $\bigcup_{i=1}^{k} F_i$ is closed in X, it follows from property (V) that $\mathscr{F}_k$ is not empty and $\varepsilon_k > 0$. We choose F_{k+1} to be an arbitrary member of $\mathscr{F}_k$ with $\delta(F_{k+1}) > 2\varepsilon_k/3$. Now, suppose inclusion (1) is not valid for some $n \geq 1$. Let

$$p \in B - [F_1 \cup \ldots \cup F_n \cup \bigcup_{k=n+1}^{\infty} U(F_k, 3b\delta(F_k))]. \tag{2}$$

It then follows from property (V), since the set $\bigcup_{i=1}^{n} F_i$ is closed, that there exists $F \in \mathscr{F}_n$ such that $p \in F$ and $\delta(F) > 0$. If $k \geq n$ and $F \in \mathscr{F}_k$, then $\delta(F) \leq \varepsilon_k$. Hence $\delta(F_{k+1}) > 2\delta(F)/3$. Since $p \in F$, this implies that

$$U(F, \delta(F)) \subset U(p, 2b\delta(F)) \subset U(p, 3b\delta(F_{k+1})).$$

On the other hand, since $k \geq n$, (2) implies that $p \notin U(F_{k+1}, 3b\delta(F_{k+1}))$. Thus $\mathscr{F}_{k+1} \cap U(F, \delta(F)) = \varnothing$, and we have shown that if $k \geq n$ and $F \in \mathscr{F}_k$, then $F \in \mathscr{F}_{k+1}$. From this and the fact that $F \in \mathscr{F}_n$, we deduce by induction on k that $F \in \mathscr{F}_k$ for all k. Thus $\varepsilon_k \geq \delta(F) > 0$ for all $k > 0$, and hence $\delta(F_k) > 2\delta(F)/3 > 0$ for all $k > 1$. Now consider an arbitrary sequence $\{p_k\}_{k \in \mathbb{N}^+}$ in X with $p_k \in F_k$, $k \in \mathbb{N}^+$. If $i < j$, then $F_i \cap U(F_j, \delta(F_j)) = \varnothing$ and so $\rho(p_i, p_j) \geq \delta(F_j) > 2\delta(F)/3$. It follows that the sequence $\{p_k\}$ does not contain a convergent subsequence, which contradicts the compactness of the space X. Thus inclusion (1) holds for all $n \geq 1$. □

(1.2) Definition. *Let μ be a regular Borel measure on a locally compact b-metric topological space X. We say that a set $B \subset X$ is* covered in the sense of Vitali *by a family $\mathscr{F}$ of closed subsets of X if every $F \in \mathscr{F}$ is of positive μ-measure and there is $\lambda > 0$ such that each point of B belongs to some $F \in \mathscr{F}$ of arbitrarily small diameter with*

$$\mu(U(F, 3b\delta(F)))/\mu(F) \leq \lambda. \tag{1}$$

(1.3) Theorem. (Vitali's covering theorem.) *Let X be a locally compact b-metric topological space and let μ be a (regular Borel) measure on X. If a family $\mathscr{F}$ of closed sets in X covers a set $B \subset X$ in the sense of Vitali, then there is a sequence $\{F_n\}_{n \in \mathbb{N}^+}$ of disjoint members of $\mathscr{F}$ such that $\mu(B - \bigcup_{n=1}^{\infty} F_n) = 0$.*

Proof. First we shall consider the case where X is compact. By virtue of Lemma 1.1 there is a sequence $\{F_n\}$ of disjoint members of $\mathscr{F}$ such that

$$B - \bigcup_{k=1}^{n} F_k \subset \bigcup_{k=n+1}^{\infty} U(F_k, 3b\delta(F_k))$$

for each n. According to inequality (1) in 1.2,

$$\mu(\bigcup_{k=n+1}^{\infty} U(F_k, 3b\delta(F_k)) \leq \sum_{k=n+1}^{\infty} \mu(U(F_k, 3b\delta(F_k)) \leq$$
$$\leq \lambda \sum_{k=n+1}^{\infty} \mu(F_k) \to 0$$

as $n \to \infty$, because $\sum_{k=1}^{\infty} \mu(F_k) \leq \mu(X) < \infty$. Thus, for each $\varepsilon > 0$, there exists a positive integer n_ε such that

$$B - \bigcup_{k=1}^{\infty} F_k \subset B - \bigcup_{k=1}^{n_\varepsilon} F_k \subset \bigcup_{k=n_\varepsilon+1}^{\infty} U(F_k, 3b\delta(F_k))$$

and

$$\mu(\bigcup_{k=n_\varepsilon+1}^{\infty} U(F_k, 3b\delta(F_k))) < \varepsilon.$$

Hence $B - \bigcup_{n=1}^{\infty} F_n$ is a null set.

The general case reduces to the case where X is compact as follows. Since X is σ-compact, there is a sequence $\{X_i\}_{i \in \mathbb{N}^+}$ of disjoint relatively compact open subsets of X such that $\mu(X - \bigcup_{i=1}^{\infty} X_i) = 0$, e.g. $X_i := \{x \in X \mid \lambda_i < f(x) < \lambda_{i+1}\}$, where $f\colon X \to \mathbb{R}$ is a function with $\{x \in X \mid f(x) < c\}$ relatively compact for each $c \in \mathbb{R}$ and $\{\lambda_i\}$ is an increasing sequence in $\mathbb{R}$ with $\lim_{i \to \infty} \lambda_i = \infty$ and where λ_i is not a point of discontinuity of the monotone function $h(c) = \mu\{x \in X \mid f(x) < c\}$. The existence of such a function f is a consequence of the classical Uryhson theorem on the extension of continuous functions on normal spaces. Replacing X by the closure of the set X_i for each $i \in \mathbb{N}^+$, B by $B \cap X_i$, and the family $\mathscr{F}$ by the subfamily $\{F \in \mathscr{F} \mid F \subset X_i\}$ we reduce to the case where X is compact. □

(1.4) With μ and X as in Definition 1.2 we say that a b-distance function is *μ-finite-dimensional* if $\mu(U(x,\varepsilon)) > 0$ for all $x \in X$ and $\varepsilon > 0$, and

(1) $$\sup_{x \in X} \overline{\lim_{\varepsilon \to 0}} (\mu(U(x, c\varepsilon))/\mu(U(x,\varepsilon))) < \infty$$

for all $c > 1$. Note that, if (1) holds for some $c > 1$, then (1) is valid for any $c > 1$.

We say that $x \in X$ is a *density point* of a measurable subset $C \subset X$ if

(2) $$\lim_{\varepsilon \to 0} (\mu(C \cap U(x,\varepsilon))/\mu(U(x,\varepsilon))) = 1.$$

(1.5) Theorem. (Density point theorem.) *With X and μ as in Theorem 1.3, we assume the b-distance function on X is μ-finite-dimensional. Then for every measurable $C \subset X$ almost all (in measure μ) points of C are density points of C.*

Proof. Since μ is regular, there is a sequence $\{W_j\}$ of open subsets of X containing C such that

(1) $$\lim_{j \to \infty} \mu(W_j - C) = 0.$$

For $i \in \mathbb{N}^+$, we set

$$B_i = \{x \in C \mid \liminf_{\varepsilon \to 0} (\mu(C \cap U(x,\varepsilon))/\mu(U(x,\varepsilon))) < 1 - (1/i)\}.$$

Since the sets $W_j \supset C$ are open and the b-distance function on X is μ-finite-dimensional, it follows that for $i, j \in \mathbb{N}^+$ the family

$$\mathscr{F}_{ij} \stackrel{\text{def}}{=} \{U(x,\varepsilon) \mid x \in C,\ \varepsilon > 0,\ U(x,\varepsilon) \subset W_j, \text{ and } (\mu(C \cap U(x,\varepsilon))/\mu(U(x,\varepsilon))) < 1 - (1/i)\}$$

covers B_i in the sense of Vitali. Therefore, by virtue of Theorem 1.3, there is a sequence $\{F_n^{ij}\}_{n \in \mathbb{N}^+}$ of disjoint members of $\mathscr{F}_{ij}$ such that

(2) $$\mu(B_i - \bigcup_{n=1}^{\infty} F_n^{ij}) = 0.$$

Since all the F_n^{ij} belong to the $\mathscr{F}_{ij}$'s and are disjoint,

(3) $$\mu(W_j - C) \geq \sum_{n=1}^{\infty} \mu(F_n^{ij} - C) > \sum_{n=1}^{\infty} (1/i)\mu(F_n^{ij}).$$

It follows from (1), (2), and (3) that $\mu(B_i) = 0$ for all $i \in \mathbb{N}^+$. But

$$B \stackrel{\text{def}}{=} \{x \in C \mid x \text{ is not a density point of } C\} = \bigcup_{i=1}^{\infty} B_i.$$

Thus $\mu(B) = 0$ and the proof is complete. □

(1.6) Corollary. *Let H be a locally compact σ-compact group, $\varphi: H \to H$ a contracting automorphism, and $B \subset H$ a μ_H-measurable subset. For $Y \subset H$, we set*

$$\omega(Y) = \begin{cases} H, & \text{if } e \in Y \\ \varnothing, & \text{if } e \notin Y. \end{cases}$$

Then for almost all (in measure μ_H) $h \in H$ the sequence $\{\varphi^{-n}(hB)\}_{n\in\mathbb{N}^+}$ converges in measure μ_H to $\omega(hB)$ (in the sense of I.4.2).

Proof. Since the group H is σ-compact, it suffices to show that, for every $C \subset H$ and every open relatively compact subset $K \subset H$ containing the identity,

(1) $$\lim_{n\to+\infty} (\mu_H(\varphi^{-n}(hC) \cap K)/\mu_H(K)) = 1 \text{ for almost all } h \in C^{-1}.$$

We then apply (1) to $C = B$ and $C = H - B$. Since φ is an automorphism, the measure μ_H is φ-homogeneous, i.e., $\mu_H(\varphi(W))/\mu_H(W)$ does not depend upon $W \subset H$. Therefore (1) is equivalent to

(2) $$\begin{aligned}&\lim_{n\to+\infty} \mu_H((hC) \cap \varphi^n(K))/\mu_H(\varphi^n(K)) = \\ &= \lim_{n\to+\infty} (\mu_H(C \cap (h^{-1}\varphi^n(K)))/\mu_H(\varphi^n(K))) = 1\end{aligned}$$

for almost all $h \in C^{-1}$.

Since φ is contracting, there exists $N \in \mathbb{N}^+$ such that

(3) $$\varphi^n((K \cup K^{-1}) \cdot K) \subset K \text{ for all } n \geq N.$$

Replacing K by $\bigcup_{0\leq n\leq N-1} \varphi^n(K \cup K^{-1})$ we may assume that

(4) $$K = K^{-1} \text{ and } \varphi(K) \subset K.$$

For $h_1, h_2 \in H$, we set

$$\begin{aligned}&n_K(h_1, h_2) = \max\{n \in \mathbb{Z} \mid h_1^{-1}h_2 \in \varphi^n(K)\} \text{ and} \\ &\rho_K(h_1, h_2) = 2^{-n_K(h_1,h_2)/N}.\end{aligned}$$

It follows from (3) and (4) that ρ_K is a left-invariant 2-distance function on H and

(5) $$h\varphi^n(K) = U(h, 2^{-n/N}) \text{ for all } h \in H \text{ and } n \in \mathbb{Z},$$

where $U(h, 2^{-n/N}) = \{y \in H \mid \rho_K(h, y) \leq 2^{-n/N}\}$. Since φ is contracting, K is open and contains e, and μ_H is φ-homogeneous, it follows from (5) that the group H equipped with the 2-distance function ρ_K is a 2-metric topological space and the distance function ρ_K is μ_H-finite-dimensional. Thus (2) is a consequence of (5) and Theorem 1.5 applied to the distance function ρ_K and the set C. □

2. Invariant Algebras of Measurable Sets

In this section we use the notation introduced in II.3.0. We provide a list of designations we shall actually use below:

A, k_α, $\mathbf{G}_\alpha$, G_α, G, G_B, pr_B, pr_α, $\mathbf{P}_\vartheta$, $\mathbf{P}_\alpha$, $\mathbf{V}_\vartheta^-$, S, P, P^-, V, V^-, Δ, S_ϑ, P_ϑ, P_ϑ^-, V_ϑ, V_ϑ^-, R_ϑ, D_ϑ, and $A(s)$.

We set $L_\vartheta^- = \mathscr{Z}_G(S_\vartheta) \cap V^-$ and $\Psi_\vartheta(C) = V_\vartheta^- \cdot (C \cap L_\vartheta^-)$. Obviously, $L_\varnothing^- = \{e\}$. Since $\mathscr{Z}_G(S_\vartheta) = P_\vartheta \cap P_\vartheta^-$, $P_\vartheta^- = \mathscr{Z}_G(S_\vartheta) \ltimes V_\vartheta^-$, and $V_\vartheta^- \subset V^- \subset P^- \subset P_\vartheta^-$, we have that $L_\vartheta^- = P_\vartheta \cap V^-$ and $V^- = L_\vartheta^- \ltimes V_\vartheta^-$ for any $\vartheta \in \Delta$ (see II.3.0). If $C \subset V_\vartheta^-$, then

$$\Psi_\vartheta(C) = \begin{cases} V_\vartheta^-, & \text{if } e \in C \\ \varnothing, & \text{if } e \notin C. \end{cases}$$

By the *rank* of the group G we mean $\operatorname{rank} G = \sum_{\alpha \in A} \operatorname{rank}_{k_\alpha} \mathbf{G}_\alpha$. We shall also use the notation introduced in I.4.2. In particular, let H be a locally compact group and F, $F' \subset H$ subgroups with $F \subset F'$. Denote by $\mathfrak{M}(H) = \mathfrak{M}(H, \mu_H)$ and $\mathfrak{M}(H/F) = \mathfrak{M}(H/F, \mu_{H/F})$ the algebra of classes of measurable sets in H and H/F and by $\mathfrak{M}(H/F, F') \subset \mathfrak{M}(H/F)$ the preimage of the algebra $\mathfrak{M}(H/F')$ under the natural map $H/F \to H/F'$. We recall that β denotes the map sending every measurable subset to its class.

Let Γ be a lattice in G. The main result in this section (Theorem 2.11(a)) asserts that under certain restrictions on G and Γ every Γ-invariant subalgebra of the algebra $\mathfrak{M}(G/P)$ has the form $\mathfrak{M}(G/P, P_\vartheta)$, where $\vartheta \subset \Delta$. This result is applied in Section 4 to the study of non-amenable factor groups of discrete subgroups.

As in II.7.1 we say that an element $g \in G$ acts ergodically on $\Gamma \setminus G$, if the transformation $x \mapsto xg$ of the measure space $\Gamma \setminus G$ with measure μ_G is ergodic. We recall that the ergodicity of the action of g on $\Gamma \setminus G$ is equivalent to the following: if $B \in \mathfrak{M}(G)$ with $\Gamma \cdot B \cdot g = B$, then either $\mu_G(B) = 0$ or $\mu_G(G - B) = 0$.

(2.1) Lemma. *If $g \in G$ acts ergodically on $\Gamma \setminus G$, then for almost all (in measure μ_G) $g' \in G$ the sets $\{\Gamma g' g^n \mid n \in \mathbb{N}^+\}$ and $\{\Gamma g' g^{-n} \mid n \in \mathbb{N}^+\}$ are dense in G.*

Proof. Since G is second countable, it suffices to show that for every open non-empty $B \subset G$ the sets $\bigcup_{n \in \mathbb{N}^+} \Gamma B g^n$ and $\bigcup_{n \in \mathbb{N}^+} \Gamma B g^{-n}$ are conull. But this is equivalent to the following assertion:

(*) for every open non-empty set $Y \subset \Gamma \setminus G$, the sets $\bigcup_{n \in \mathbb{N}^+} Y g^n$ and $\bigcup_{n \in \mathbb{N}^+} Y g^{-n}$ are conull in $\Gamma \setminus G$.

All that remains now is to observe that, since g acts ergodically on $\Gamma \setminus G$ and $\mu_G(Y) > 0$ for every open non-empty $Y \subset \Gamma \setminus G$, assertion (*) is a consequence of the Birkhoff ergodic theorem applied to the characteristic function of the set Y and the transformation $x \mapsto xg$, $x \in \Gamma \setminus G$. (Application of the Birkhoff ergodic theorem can be replaced by the following elementary argument: If $X = \bigcup_{n \in \mathbb{N}^+} Y g^n$, then $Xg \subset X$ and, since the measure is invariant, $X - gX$ is a null set.) □

Since for each $\alpha \in A$ and every $\vartheta \subset \Delta_\alpha$ the map sending $(v, p) \in \mathbf{V}_\vartheta^- \times \mathbf{P}_\vartheta$ to $vp^{-1} \in \mathbf{G}_\alpha$ is an isomorphism of $\mathbf{V}_\vartheta^- \times \mathbf{P}_\vartheta$ onto a Zariski open subset of $\mathbf{G}_\alpha$ (see I.1.2), we deduce from Propositions I.2.5.3(i) and I.4.3.1 the following

(2.2) Lemma. *Let $\vartheta \subset \Delta$. Then*

(i) the subset $V_\vartheta^- \cdot P_\vartheta$ is open in G and the map $\varepsilon\colon V_\vartheta^- \times P_\vartheta \to V_\vartheta^- \cdot P_\vartheta$, $\varepsilon(v, p) = vp^{-1}$, $v \in V_\vartheta^-$, $p \in P_\vartheta$, is a homeomorphism:

(ii) $\mu_G(G - (V_\vartheta^- \cdot P_\vartheta)) = 0$;

(iii) up to a scalar multiple, the measure μ_G is the image of the measure $\mu_{V_\vartheta^- \times P_\vartheta}$ under the map ε.

(2.3) Lemma. *Let $s \in S$. Suppose $s \in R_\Delta$ and s acts ergodically on $\Gamma \setminus G$. Then for almost all (in measure μ_{V^-}) $v \in V^-$ the set $\{\Gamma v s^{-n} \mid n \in \mathbb{N}^+\}$ is dense in G.*

Proof. We set $W = \{g \in G \mid \text{the set } \{\Gamma g s^{-n} \mid n \in \mathbb{N}^+\} \text{ is dense in } G\}$ and $Y = \{v \in V^- \mid \mu_P(C_v) = 0, \text{ where } C_v = \{p \in P \mid vp^{-1} \notin W\}\}$. Since s acts ergodically on $\Gamma \setminus G$, Lemma 2.1 implies that $\mu_G(G - W) = 0$. It then follows from Lemma 2.2 and Fubini's theorem that

$$\mu_{V^-}(V^- - Y) = 0. \tag{1}$$

Since $s \in R_\Delta$, by virtue of Lemma II.3.1(b) the automorphism $\operatorname{Int} s|_P$ is non-expanding. But $gps^{-n} = gs^{-n}(s^n p s^{-n})$, $g \in G$, $p \in P$. Thus, for each $g \in G$, the set $\{p \in P \mid gp^{-1} \in W\}$ is closed, and hence $V^- \cap W \supset Y$. In view of (1), this completes the proof. □

(2.4) Proposition. *Let X be a locally compact σ-compact group.*

(a) If $\mathfrak{B}$ is a subalgebra of the algebra $\mathfrak{M}(X)$ with $X\mathfrak{B} = \mathfrak{B}$ (i.e. $xB \in \mathfrak{B}$ for all $x \in X$ and $B \in \mathfrak{B}$), then $\mathfrak{B} = \mathfrak{M}(X, H)$ for some closed subgroup H of X.

(b) If F is a closed subgroup of X and $\mathfrak{B}$ is a subalgebra of the algebra $\mathfrak{M}(X/F)$ with $X\mathfrak{B} = \mathfrak{B}$, then $\mathfrak{B} = \mathfrak{M}(X/F, H)$ for some closed subgroup H of X which contains F.

Proof. For every two closed subgroups H and $F (H \supset F)$ of X, the algebra $\mathfrak{M}(X, H)$ is the inverse image of the algebra $\mathfrak{M}(X/F, H)$ under the natural map $X \to X/F$. Therefore (b) is a consequence of (a). We shall prove (a).

Let Ω be the space of real-valued measurable functions f on X with $\{x \in X \mid a < f(x) < b\} \in \mathfrak{B}$ (more precisely, $\beta\{x \in X \mid a < f(x) < b\} \in \mathfrak{B}$) for all $-\infty < a < b < \infty$). We provide Ω with the topology of convergence in measure (for the definition see for example VII.1.2 below). Since $X\mathfrak{B} = \mathfrak{B}$, it follows that $h * f \in \Omega$ for all $h \in L_1(X, \mu_X)$ and bounded $f \in \Omega$, where $*$ denotes the convolution of functions. Making use of Lemma III.1.7 we then deduce that the set $\Omega_0 \stackrel{\text{def}}{=} \{f \in \Omega \mid f \text{ is continuous}\}$ is dense in Ω. We set $H = \{h \in X \mid f(xh) = f(x) \text{ for all } f \in \Omega_0 \text{ and } x \in X\}$. Clearly, H is a closed subgroup in X. We denote by $C(X/H)$ the space of real-valued continuous functions on X/H. Since $f(xH) = f(x)$ for all $f \in \Omega_0$ and $x \in X$, there is a map $q\colon \Omega_0 \to C(X/H)$ such that $q(f) \circ \tilde{q} = f$ for all $f \in \Omega_0$, where $\tilde{q}\colon X \to X/H$

is the natural map. Since $X\mathfrak{B} = \mathfrak{B}$, the set $H_x \stackrel{\text{def}}{=} \{h \in H \mid f(xh) = f(x) \text{ for all } f \in \Omega_0\}$ does not depend on $x \in X$. Hence, $H_x = H$. This implies that $q \in \Omega_0$ separates the points of the space X/H (i.e. for all distinct $y_1, y_2 \in X/H$ there exists a function $f \in q(\Omega_0)$ such that $f(y_1) \neq f(y_2)$). On the other hand, it follows from the definition of the set Ω_0 that $q(\Omega_0)$ is a linear subspace of $C(X/H)$ which is closed with respect to the operations of multiplication and uniform convergence on compact sets. It then follows from the Stone-Weierstrass theorem that $q(\Omega_0) = C(X/H)$. Since Ω_0 is dense in Ω, this implies that $\mathfrak{B} = \mathfrak{M}(X, H)$. □

(2.5) Lemma. *Let $\vartheta \subset \Delta$.*

(a) If $s \in R_\vartheta$ and B is a $\mu_{V_\vartheta^-}$-measurable subset of V_ϑ^-, then for almost all $v \in V_\vartheta^-$ the sequence $\{s^n v B s^{-n}\}_{n \in \mathbb{N}^+}$ converges in measure $\mu_{V_\vartheta^-}$ to $\Psi_\vartheta(vB)$.

(b) If $s \in D_\vartheta$ and B is a μ_{V^-}-measurable subset of V^-, then for almost all $v \in V^-$ the sequence $\{s^n v B s^{-n}\}_{n \in \mathbb{N}^+}$ converges in measure μ_{V^-} to $\Psi_\vartheta(vB)$.

Proof. If $s \in R_\vartheta$, then by Lemma II.3.1(a) the automorphism $\operatorname{Int} s^{-1}|_{V_\vartheta^-}$ is contracting. Thus (a) is a consequence of Corollary 1.6. Since $V^- = L_\vartheta^- \ltimes V_\vartheta^-$ and L_ϑ^- commutes with D_ϑ, (b) is a consequence of (a), Proposition I.4.3.1 and Fubini's theorem. □

(2.6) By virtue of Lemma 2.2, restricting the natural map $G \to G/P$ to V^-, we may identify the measure spaces (V^-, μ_{V^-}) and $(G/P, \mu_{G/P})$ (up to a null set), and hence $\mathfrak{M}(V^-)$ and $\mathfrak{M}(G/P)$. Then the action of the group G by left translations on G/P induces the action of G on V^-; for each $g \in G$ the image $g \circ v$ under the action of g on V^- is defined for almost all $v \in V^-$. Furthermore,

(1) $$v' \circ v = v'v, \quad s \circ v = svs^{-1}, \quad v', \ v \in V^-, \ s \in S.$$

(The first equality is obvious, the second one is a consequence of the inclusion $S \subset P$ and the fact that S normalizes V^-). If $B \in \mathfrak{M}(G/P)$ and $v \in V^-$, then $Bv \in \mathfrak{M}(G/P)$ denotes the right translation by v of the class B viewed as an element of the algebra $\mathfrak{M}(V^-)$. The expressions of the form BY and $\mathfrak{B}Y$, where $B \in \mathfrak{M}(G/P)$, $\mathfrak{B} \subset \mathfrak{M}(G/P)$, and $Y \subset V^-$ make sense as well.

If $v \in V^-$ and $\vartheta \subset \Delta$, then $(vP_\vartheta) \cap V^- = v(P_\vartheta \cap V^-) = vL_\vartheta^-$. Thus, under the above-mentioned identification of the algebras $\mathfrak{M}(V^-)$ and $\mathfrak{M}(G/P)$, we have

(2) $$\mathfrak{M}(G/P, P_\vartheta) = \mathfrak{M}(V^-, L_\vartheta^-) \text{ for every } \vartheta \subset \Delta.$$

(2.7) Lemma. *Let $\mathfrak{B}$ be a subalgebra of the algebra $\mathfrak{M}(G/P) = \mathfrak{M}(V^-)$, $B \in \mathfrak{B}$, $C \in \beta^{-1}(B)$, and $\vartheta \subset \Delta$. Suppose $\Gamma\mathfrak{B} = \mathfrak{B}$ (i.e. $\gamma Y \in \mathfrak{B}$ for all $\gamma \in \Gamma$ and $Y \in \mathfrak{B}$. Further suppose that some $s \in D_\vartheta$ acts ergodically on $\Gamma \setminus G$. Then for almost all $v \in V^-$ and for each $g \in G$ we have that $\beta(g \circ \Psi_\vartheta(vC)) \in \mathfrak{B}$.*

Proof. We set $U_1 = \{v \in V^- \mid \text{the set } \{\Gamma v^{-1} s^{-n} \mid n \in \mathbb{N}^+\} \text{ is dense in } G\}$, $U_2 = \{v \in V^- \mid \text{the sequence } \{s^n v C s^{-n}\}_{n \in \mathbb{N}^+} \text{ converges in measure } \mu_{V^-} \text{ to } \Psi_\vartheta(vC)\}$, and $U = U_1 \cap U_2$. It follows from Lemmas 2.3 and 2.5 that

(1) $$\mu_{V^-}(V^- - U) = 0.$$

Let $v \in U$ and $g \in G$. Since $v \in U_1$, there exist $g_n \in G$ and $\gamma_n \in \Gamma$ $(n \in \mathbb{N}^+)$ such that $\lim_{n\to\infty} g_n = g$ and $g_n = \gamma_n v^{-1} s^{-n}$, i.e. $\gamma_n = g_n s^n v$. But $v \in U_2$, and hence, by equalities (1) in 2.6, the sequence $\{\gamma_n \circ C = (g_n s^n v) \circ C = g_n \circ (s^n v C s^{-n})\}_{n \in \mathbb{N}^+}$ converges in measure $\mu_{V_\vartheta^-}$ to $\lim_{n\to\infty} g_n \circ \Psi_\vartheta(vC) = g \circ \Psi_\vartheta(vC)$. But, since $\mathfrak{B}$ is a subalgebra, $\mathfrak{B}$ is closed with respect to convergence in measure (see I.4.2) and $\beta(\gamma_n \circ C) = \gamma_n \circ B \in \mathfrak{B}$ (because $\Gamma\mathfrak{B} = \mathfrak{B}$). Thus $\beta(g \circ \Psi_\vartheta(vC)) \in \mathfrak{B}$ for all $v \in U$. In view of (1), this completes the proof of the lemma. □

(2.8) Lemma. *If H is a subgroup of G containing P, then $H = P_\vartheta$ for some $\vartheta \subset \Delta$.*

Proof. Let $\alpha \in A$. The subgroups P_ϑ, $\vartheta \subset \Delta_\alpha$, coincide with their normalizers in G_α and are the only subgroups of G_α containing $\mathbf{P}_\alpha(k_\alpha)$ (see Proposition I.1.2.3). Furthermore, $\mathrm{pr}_\alpha(H)$ normalizes the subgroup $H \cap G_\alpha$ which contains $\mathbf{P}_\alpha(k_\alpha)$. Hence, $\mathrm{pr}_\alpha(H) = H \cap G_\alpha = P_{\vartheta_\alpha}$, where $\vartheta_\alpha \subset \Delta_\alpha$, and therefore $H = P_\vartheta$ with $\vartheta = \bigcup_{\alpha \in A} \vartheta_\alpha$. □

(2.9) Lemma. *For every $\vartheta \subset \Delta$ the subgroups $P_{\{b\}}$, $b \in \vartheta$, generate P_δ.*

This lemma is a consequence of the fact that for all $\alpha \in A$ and $\delta \subset \Delta_\alpha$ the subgroups $\mathbf{P}_{\{d\}}(k_\alpha)$, $d \in \delta$, generate $\mathbf{P}_\vartheta(k_\alpha)$ (see Corollary I.1.2.4).

(2.10) Lemma. *Let $\vartheta \subset \Delta$ and $B \in \mathfrak{M}(G/P) = \mathfrak{M}(V^-)$. Assume $B \notin \mathfrak{M}(G/P, P_\vartheta)$. Then $BL^-_{\{b\}} \neq B$ for some $b \in \vartheta$.*

Proof. Since $\mathfrak{M}(G, H) = \{M \in \mathfrak{M}(G) \mid MH = M\}$ for every closed subgroup $H \subset G$ (see I.4.2), Lemma 2.9 implies that $\mathfrak{M}(G, P_\vartheta) = \bigcap_{b \in \vartheta} \mathfrak{M}(G, P_{\{b\}})$. Therefore $\mathfrak{M}(G/P, P_\vartheta) = \bigcap_{b \in \vartheta} \mathfrak{M}(G/P, P_{\{b\}})$. This and equality (2) in 2.6 imply that $B \notin \mathfrak{M}(V^-, L^-_{\{b\}})$ and hence $BL^-_{\{b\}} \neq B$ for some $b \in \vartheta$. □

(2.11) Theorem. *Suppose for each $\alpha \in A$ the group $\mathbf{G}_\alpha$ is simply connected, k_α-isotropic, and almost k_α-simple. Assume in addition that*

(*) *the subgroup $\Gamma \cdot G_{A-\{\alpha\}}$ is dense in G for those $\alpha \in A$ where $\mathrm{rank}_{k_\alpha} \mathbf{G}_\alpha = 1$.*

(a) If $\mathfrak{B}$ is a subalgebra of the algebra $\mathfrak{M}(G/P)$ with $\Gamma\mathfrak{B} = \mathfrak{B}$ (i.e. $\gamma B \in \mathfrak{B}$ for all $\gamma \in \Gamma$ and $B \in \mathfrak{B}$), then $\mathfrak{B} = \mathfrak{M}(G/P, P_\vartheta)$ for some $\vartheta \subset \Delta$.

(b) If $\delta \subset \Delta$ and $\mathfrak{H}$ is a subalgebra of the algebra $\mathfrak{M}(G/P_\vartheta)$ with $\Gamma\mathfrak{H} = \mathfrak{H}$, then $\mathfrak{H} = \mathfrak{M}(G/P_\delta, P_\vartheta)$ for some $\delta \subset \vartheta \subset \Delta$.

(c) If $\mathbf{P}'_\alpha$ is a parabolic k_α-subgroup in $\mathbf{G}'_\alpha$, $P' = \prod_{\alpha \in A} \mathbf{P}'_\alpha(k_\alpha)$, and $\mathfrak{H}$ is a subalgebra of the algebra $\mathfrak{M}(G/P')$ with $\Gamma\mathfrak{H} = \mathfrak{H}$, then there exist parabolic k_α-subgroups $\mathbf{P}''_\alpha$ of $\mathbf{G}_\alpha$ containing $\mathbf{P}'_\alpha$ such that $\mathfrak{H} = \mathfrak{M}(G/P', P'')$, where $P'' = \prod_{\alpha \in A} \mathbf{P}''_\alpha(k_\alpha)$.

Proof. (a) Let ϑ be the minimal subset of Δ with $\mathfrak{B} \supset \mathfrak{M}(G/P, P_\vartheta)$. We shall show that $\mathfrak{B} = \mathfrak{M}(G/P, P_\vartheta)$. Suppose the contrary. Let $B \in \mathfrak{B} - \mathfrak{M}(G/P, P_\vartheta)$ and $C \in \beta^{-1}(B)$. By Lemma 2.10, $BL^-_{\{b\}} \neq B$ for some $b \in \vartheta$. We set $U = \{v \in V^- \mid \mu_{L^-_{\{b\}}}(L^-_{\{b\}} \cap (vC)) > 0$ and $\mu_{L^-_{\{b\}}}(L^-_{\{b\}} - (vC)) > 0\}$. Since $BL^-_{\{b\}} \neq B$, it follows from Fubini's theorem that $\mu_{V^-}(U) > 0$. Observe that the set $D_{\{b\}}$ is not empty (see II.3.0) and pick $s \in D_{\{b\}}$. Since for each $\alpha \in A$ the group $\mathbf{G}_\alpha$ is k_α-isotropic, either

$A(s) = A$ or $A(s) = A - \{\alpha\}$, where $b \in \Delta_\alpha$ and $\operatorname{rank}_{k_\alpha} \mathbf{G}_\alpha = 1$. In both cases the subgroup $\Gamma \cdot G_{A(s)}$ is dense in G (in the first case this is obvious and in the second one this is a consequence of (b)). Thus $s \in D_{\{b\}}$ acts ergodically on $\Gamma \setminus G$ (see Corollary II.7.3(a)). From this, the inequality $\mu_{V^-}(U) > 0$, and Lemma 2.7 we deduce that there exists $v \in U$ such that

$$\beta(g \circ \Psi_{\{b\}}(vC)) \in \mathfrak{B} \text{ for all } g \in G. \tag{1}$$

Let $\mathfrak{B}_1$ denote the subalgebra of $\mathfrak{B}$ which is generated by the classes $\beta(g \circ \Psi_{\{b\}}(vC))$, $g \in G$. We then have that $G\mathfrak{B}_1 = \mathfrak{B}_1$. It follows from Proposition 2.4(b) that $\mathfrak{B}_1 = \mathfrak{M}(G/P, P_{\vartheta'})$ for some $\vartheta' \subset \Delta$. Since $v \in U$, it follows that $\beta(\Psi_{\{b\}}(vC))L^-_{\{b\}} \neq \beta(\Psi_{\{b\}}(vC))$. But $\beta(\Psi_{\{b\}}(vC)) \in \mathfrak{B}_1 = \mathfrak{M}(G/P, P_{\vartheta'}) = \mathfrak{M}(V^-, L^-_{\vartheta'})$. Thus $b \notin \vartheta'$ and so $\vartheta' \cap \vartheta \neq \vartheta$. On the other hand, since $\mathfrak{M}(G/P, P_{\vartheta'}) = \mathfrak{B}_1 \subset \mathfrak{B}$ (see (1)) and $\mathfrak{B} \supset \mathfrak{M}(G/P, P_\vartheta)$, it follows that $\mathfrak{B} \supset \mathfrak{M}(G/P, P_{\vartheta' \cap \vartheta})$. This contradicts the minimality of the set ϑ and completes the proof of (a).

Assertion (b) is a consequence of (a) applied to the inverse image of the subalgebra $\mathfrak{H}$ under the natural map $G/P \to G/P_\delta$. Assertion (c) is a consequence of (b) and the fact that every parabolic k_α-subgroup of the group $\mathbf{G}_\alpha$ is conjugate to a standard parabolic subgroup by an element of $\mathbf{G}_\alpha(k_\alpha)$ (see I.1.2). □

The following is a consequence of Theorem 2.11 and Theorem II.6.7(b).

(2.12) Corollary. *Suppose* rank $G \geq 2$, *the lattice Γ is irreducible (in the sense of Definition II.6.5 or the definition given in III.5.9) and for each $\alpha \in A$ the group $\mathbf{G}_\alpha$ is simply connected, k_α-isotropic, and almost k_α-simple. Then assertions (a), (b), and (c) of Theorem 2.11 hold.*

(2.13) Corollary. *Let $\mathbf{P}'_\alpha$ be a parabolic k_α-subgroup in $\mathbf{G}_\alpha$, $P' = \prod_{\alpha \in A} \mathbf{P}'_\alpha(k_\alpha)$, X a Γ-space with Γ-quasi-invariant σ-finite measure μ, and let $f\colon G/P' \to X$ be a Γ-equivariant measurable map such that the preimage of any null set is a null set. Assume that the conditions of Theorem 2.11 are satisfied. Then there exist a closed subgroup P'' of G which contains P' and a bi-measurable bijection $\varphi\colon G/P'' \to X$ such that $\varphi^{-1} \circ f$ agrees almost everywhere with the natural projection of G/P' onto G/P''.*

(2.14) Corollary 2.13 is a consequence of Theorem 2.11(c) applied to the inverse image of the algebra $\mathfrak{M}(X, \mu)$ under the map f. Notice that the converse is also valid, i.e. assertion (c) of Theorem 2.11 is a consequence of Corollary 2.13.

The following theorem is a topological version of Corollary 2.13.

(2.15) Theorem. (See [Da 5].) *Let $\mathbf{G}$ be a connected semisimple $\mathbb{R}$-group without $\mathbb{R}$-anisotropic almost $\mathbb{R}$-simple factors, Γ an irreducible lattice in $\mathbf{G}(\mathbb{R})$, and let $\mathbf{P}$ be a parabolic $\mathbb{R}$-subgroup in $\mathbf{G}$. Let X be a Hausdorff topological Γ-space and let $f\colon \mathbf{G}(\mathbb{R})/\mathbf{P}(\mathbb{R}) \to X$ be a continuous surjective Γ-equivariant map. Suppose* $\operatorname{rank}_{\mathbb{R}} \mathbf{G} \geq 2$. *Then there exist a parabolic $\mathbb{R}$-subgroup $\mathbf{P}'$ of $\mathbf{G}$ which contains $\mathbf{P}$*

and a Γ-equivariant homomorphism φ: $\mathbf{G}(\mathbb{R})/\mathbf{P}'(\mathbb{R}) \to X$ such that $\varphi^{-1} \circ f$ agrees with the natural projection of $\mathbf{G}(\mathbb{R})/\mathbf{P}(\mathbb{R})$ onto $\mathbf{G}(\mathbb{R})/\mathbf{P}'(\mathbb{R})$.

(2.16) *Remark.* The condition "$\text{rank}_{\mathbb{R}}\, \mathbf{G} \geq 2$" in Theorem 2.15 is essential (see [Spa]). Similarly, condition (*) in Theorem 2.11 can not be omitted (see [Mar 11], Corollary 2.9.1 (II)).

3. Amenable Factor Groups of Lattices Lying in Direct Products

For the definition of an amenable group we refer to I.5.5.1. This section is devoted to the proof of Theorem 3.9 which states that under certain conditions, the amenable factor groups of lattices lying in direct products of locally compact groups G_1 and G_2 which satisfy the following property are finite:

(Q) There exists a compact subgroup $K_i \subset G_i$, $i = 1, 2$, such that $g^{-1} \in K_i g K_i$ for all $g \in G_i$.

(3.1) Notation. By $\bar{A}$ we denote the closure of a subset A in a topological space.

Let H be a locally compact group. As in Chapter III we denote by $\tilde{H}$ (resp. $\hat{H}$) the set of all continuous unitary (resp. irreducible unitary) representations of the group H on separable complex Hilbert spaces. By $I_H \in \hat{H}$ we denote the trivial 1-dimensional representation of the group H. The space of the representation $\rho \in \tilde{H}$ will be denoted by $L(\rho)$. As before, μ_H denotes a left-invariant Haar measure on H. We assume in addition that if $H = H_1 \times H_2$, then $\mu_H = \mu_{H_1} \times \mu_{H_2}$. If $H = H_1 \times H_2$, then for every $\rho \in \tilde{H}$ and $f \in L_1(H_1, \mu_{H_1})$ we set

$$\rho(f) = \int_{H_1} f(h)\rho(h)d\mu_{H_1}(h).$$

For $B \subset H$, $\rho \in \tilde{H}$, and a compact subgroup $K \subset H$ we set $L(\rho)^B = \{x \in L(\rho) \mid \rho(B)x = x\}$ and $\rho^K = (1/\mu_K(K)) \int_K \rho(k) d\mu_K(k)$.

Let $\mathfrak{A}(H)$ denote the set of non-negative continuous functions f on H which have compact support and satisfy $\int_H f d\mu_H = 1$.

As before, hf and fh will denote the left and the right translation of the function f on H.

For $A \subset H$, we set $A^0 = \{e\}$ and $A^i = A^{i-1} \cdot A$, where $i \in \mathbb{N}^+$.

(3.2) Lemma. *Let H be a locally compact group and $f \in \mathfrak{A}(H)$. Then there exists $a(f) > 0$ such that for every $\rho \in \tilde{H}$ and $x \in L(\rho)$*

$$\begin{aligned} (1) \qquad 2(\|x\|^2 - \mathrm{Re}\langle \rho(f), x, x\rangle) &\leq \sup_{h \in \operatorname{supp} f} \|\rho(h)x - x\|^2 \leq \\ &\leq a(f)(\|x\|^2 - \mathrm{Re}\langle \rho(f)x, x\rangle). \end{aligned}$$

Proof. We fix $\rho \in \tilde{H}$ and $x \in L(\rho)$. Since $\|x\|^2 - \mathrm{Re}\langle \rho(h)x, x\rangle = (1/2)\|\rho(h)x - x\|^2$ for all $h \in H$ and $\int_H f d\mu_H = 1$, it follows that

(2) $$\|x\|^2 - \mathrm{Re}\langle \rho(f)x, x\rangle = (1/2)\int_H f(h)\|\rho(h)x - x\|^2 d\mu_H(h).$$

Since $f \in \mathfrak{A}(H)$, the first inequality in (1) is a straightforward consequence of (2). Another consequence of the fact that $f \in \mathfrak{A}(H)$ is that there exist a compact set $K \subset H$ and a relatively compact neighbourhood V of the identity in H such that

(3) $$K \cdot K^{-1} \cdot K \supset \operatorname{supp} f \text{ and}$$

(4) $$a \stackrel{\text{def}}{=} \inf\{f(h) \mid h \in K \cdot V \cdot V\} > 0.$$

Since K is compact, there exists $k \in K$ such that $\|\rho(k)x - x\| = b \stackrel{\text{def}}{=} \sup\{\|\rho(h)x - x\| \mid h \in K\}$. Consider the set $Y = \{v \in V \cdot V \mid \|\rho(kv)x - x\| \geq b/3\}$. Since V is relatively compact, $c \stackrel{\text{def}}{=} \inf\{\Delta_H(h) \mid h \in V\} > 0$, where Δ_H is the modular function of the locally compact group H. We shall show that

(5) $$\mu_H(Y) \geq (c/c+1)\mu_H(V).$$

If $Y \supset V$, then (5) is obvious. Now suppose $V - Y \neq \varnothing$ and $w \in V - Y$. Since $\|\rho(ku)x - x\| < b/3$ for all $u \in V - Y$ and ρ is unitary, for each $v \in V - Y$ it follows that

$$\begin{aligned}\|\rho(kvw)x - x\| &\geq \|\rho(kvw)x - \rho(kv)x\| - \|\rho(kv)x - x\| \geq \\ &\geq \|\rho(kvk^{-1})(\rho(kw)x - \rho(k)x)\| - b/3 = \\ &= \|\rho(kw)x - \rho(k)x\| - b/3 \geq \\ &\geq \|\rho(k)x - x\| - \|\rho(kw)x - x\| - b/3 \geq \\ &\geq b - (b/3) - (b/3) = b/3.\end{aligned}$$

Thus $(V - Y)w \subset Y$. Inequality (5) then follows from $\mu_H((V - Y)w = \Delta_H(w)\mu_H(V - Y) \geq c\mu_H(V - Y)$. Since $f \geq 0$ and μ_H is left-invariant, it follows from (2), (4), and (5) that

(6) $$\begin{aligned}\|x\|^2 - \mathrm{Re}\langle \rho(f)x - x\rangle &\geq (1/2)\int_{kY} f(h)\|\rho(h)x - x\|^2 d\mu_H(h) \geq \\ &\geq (ab^2/9)\mu_H(Y) \geq (ab^2c/9(c+1))\mu_H(V).\end{aligned}$$

Since ρ is unitary, $\|\rho(h_1h_2)x - x\| \leq \|\rho(h_1)x - x\| + \|\rho(h_2)x - x\|$ and $\|\rho(h^{-1})x - x\| = \|\rho(h)x - x\|$ for all h, h_1, $h_2 \in H$. Thus (3) implies that

$$\sup\{\|\rho(h)x - x\| \mid h \in \operatorname{supp} f\} \leq 3\sup\{\|\rho(h)x - x\| \mid h \in K\} = 3b.$$

This and (6) imply that $a(f) = (81(c+1)/ac)\mu_H(V)$ is the desired constant. □

(3.3) Let G_1, G_2 be compactly generated unimodular locally compact second countable groups and let $G = G_1 \times G_2$. We assume the groups G_i, $i = 1, 2$, have the following property:

(Q) There exists a compact subgroup $K_i \subset G_i$ such that $g^{-1} \in K_i g K_i$ for all $g \in G_i$.

We set $K = K_1 \times K_2$. From now until the end of this section G, G_1, G_2, K_1, and K_2 will be defined as above.

(3.4) Lemma. *Let $f_i \in \mathfrak{A}(G_i)$, $i = 1,2$, be bi-invariant relative to K_i (i.e. $f_i(K_i g K_i) = f_i(g)$ for all $g \in G_i$) with $\operatorname{supp} F_i \supset K_i$. Set $f = f_1 \times f_2 (f(g_1, g_2) = f_1(g_1) f_2(g_2)$, $g_1 \in G_1$, $g_2 \in G_2)$. Let $\mathscr{T} \in \hat{G}$. Then*

(i) $\dim L(\mathscr{T})^K \leq 1$;

(ii) *the operators $\mathscr{T}(f)$ and $\mathscr{T}(f_i)$ are Hermitian;*

(iii) *the subspace $L(\mathscr{T})^K$ consists of eigen-vectors both of the operators $\mathscr{T}(f)$ and $\mathscr{T}(f_i)$ which, respectively, correspond to the real eigenvalues $\lambda_f(\mathscr{T})$ and $\lambda_{f_i}(\mathscr{T})$ not exceeding 1 in absolute value. Furthermore, $\lambda_f(\mathscr{T}) = \lambda_{f_1}(\mathscr{T})\lambda_{f_2}(\mathscr{T})$;*

(iv) *the operator $\mathscr{T}(f)$ maps $L(\mathscr{T})$ to $L(\mathscr{T})^K$ and $\|\mathscr{T}(f)\| = |\lambda_f(\mathscr{T})|$;*

(v) *there exists a constant $b(f) > 0$ independent of $\mathscr{T}$ such that*

$$(1) \qquad \|\mathscr{T}^K(\mathscr{T}(g)x - x)\| \leq b(f)\sqrt{1 - \lambda_f(\mathscr{T})}\|\mathscr{T}(g)x - x\|, \quad x \in L(\mathscr{T})^K,\ g \in \operatorname{supp} f;$$

$$(2) \qquad \|\mathscr{T}(g_1)(\mathscr{T}(g_2)x - x) - (\mathscr{T}(g_2)x - x)\| \leq \leq b(f)\sqrt{1 - \lambda_f(\mathscr{T})}\|\mathscr{T}(g_2)x - x\|; \quad x \in L(\mathscr{T})^K,\ g_1 \in \operatorname{supp} f_1,\ g_2 \in G_2.$$

Proof. (i) The following assertion is well known (see [Lan 2], Ch. IV, Theorems 1 and 3):

(A) Let τ be an anti-automorphism of order 2 of the group G with $\tau(g) \in KgK$ for all $g \in G$. Then $\dim L(\rho)^K \leq 1$ for all $\rho \in \hat{G}$.

Since $g^{-1} \in KgK$ for all $g \in G$ and the automorphism $g \mapsto g^{-1}$ is an anti-automorphism of order 2, it follows from (A) that $\dim L(\mathscr{T})^K \leq 1$.

(ii) Since $g^{-1} \in K_i g K_i$ and $f_i(K_i g K_i) = f_i(g)$, we have that $f_i(g^{-1}) = f_i(g)$ for all $g \in G_i$. On the other hand, the representation $\mathscr{T}$ is unitary and the transformation $g \mapsto g^{-1}$, $g \in G_i$, preserves the measure μ_{G_i} (because G_i is unimodular). Thus, $\mathscr{T}(f)$ and $\mathscr{T}(f_i)$ are Hermitian.

(iii) Since $\mathscr{T}(f)$ and $\mathscr{T}(f_i)$ commute with $\mathscr{T}(K)$ (because f_i is bi-invariant relative to K_i), $L(\mathscr{T})^K$ is invariant relative to $\mathscr{T}(f)$ and $\mathscr{T}(f_i)$. But $\dim L(\mathscr{T})^K \leq 1$. Therefore the subspace $L(\mathscr{T})^K$ consists of eigen-vectors of both $\mathscr{T}(f)$ and $\mathscr{T}(f_i)$. We denote the corresponding eigenvalues by $\lambda_f(\mathscr{T})$ and $\lambda_{f_i}(\mathscr{T})$. Since $f = f_1 \times f_2$, $\lambda_f(\mathscr{T}) = \lambda_{f_1}(\mathscr{T})\lambda_{f_2}(\mathscr{T})$. Since the operators $\mathscr{T}(f)$ and $\mathscr{T}(f_i)$ are Hermitian, $\lambda_f(\mathscr{T}) \in \mathbb{R}$ and $\lambda_{f_i}(\mathscr{T}) \in \mathbb{R}$. Finally, since $\mathscr{T}$ is unitary and $f_i \in \mathfrak{A}(G_i)$, we have that $\|\mathscr{T}(f_i)\| \leq 1$, and hence $|\lambda_{f_i}(\mathscr{T})| \leq 1$ and $|\lambda_f(\mathscr{T})| \leq 1$.

(iv) Since $f(Kg) = f(g)$ for all $g \in G$, $\mathscr{T}(K)\mathscr{T}(f) = \mathscr{T}(f)$. Thus $\mathscr{T}(f)$ maps $L(\mathscr{T})$ to $L(\mathscr{T})^K$. Since the operator $\mathscr{T}(f)$ is Hermitian, it follows that $\|\mathscr{T}(f)\| = |\lambda_f(\mathscr{T})|$.

(v) Let $a(f)$ and $a(f_1)$ be the constants defined in Lemma 3.2. We set

$$(3) \qquad b(f) = \max\{\sqrt{a(f)}, \sqrt{a(f)a(f_1)}\}.$$

Let $x \in L(\mathscr{T})^K$. Replacing x by $x/\|x\|$, in case $x \neq 0$, we can assume that $\|x\| = 1$. Since $\mathscr{T}(f)x = \lambda_f(\mathscr{T})x$, it follows from inequality (1) in 3.2 that

(4) $$\|\mathcal{T}(g)x - x\|^2 \leq a(f)(1 - \lambda_f(\mathcal{T})), \ g \in \operatorname{supp} f.$$

Since $\mathcal{T}$ is unitary, $x \in L(\mathcal{T})^K$, and $g^{-1} \in KgK$, it follows that $\langle \mathcal{T}(g)x, x\rangle = \langle \mathcal{T}(g^{-1})x, x\rangle = \overline{\langle \mathcal{T}(g)x, x\rangle}$, where the bar denotes complex conjugation. Thus $\langle \mathcal{T}(g)x, x\rangle \in \mathbb{R}$ for all $g \in G$. Since $\mathcal{T}$ is unitary, $\dim L(\mathcal{T})^K = 1$, $x \in L(\mathcal{T})^K$, $\|x\| = 1$, and $\mathcal{T}^K$ is the orthogonal projector onto $L(\mathcal{T})^K$ (because $\mathcal{T}(K)\mathcal{T}^K = \mathcal{T}^K$ and $\mathcal{T}$ is unitary), it then follows from (4) that

$$\begin{aligned}\|\mathcal{T}^K(\mathcal{T}(g)x - x)\| &= |\langle x, \mathcal{T}(g)x - x\rangle| = 1 - \langle \mathcal{T}(g)x, x\rangle = \\ &= (1/2)\|\mathcal{T}(g)x - x\|^2 = \\ &= (1/2)\|\mathcal{T}(g)x - x\| \cdot \|\mathcal{T}(g)x - x\| \leq \\ &\leq (1/2)(a(f)(1 - \lambda_f(\mathcal{T}))^{1/2}\|\mathcal{T}(g)x - x\|.\end{aligned}$$

From this and (3) we deduce (1). Further, since $\mathcal{T}(f)(x) = \lambda_f(\mathcal{T})x$, $\mathcal{T}(f_1)x = \lambda_{f_1}(\mathcal{T})x$, and $\operatorname{supp} f_1 \supset \operatorname{supp} f$ (because $e \in K_2 \subset \operatorname{supp} f_2$), it follows from inequality (1) in 3.2 that

(5) $$\begin{aligned}2(1 - \lambda_{f_1}(\mathcal{T})) &\leq \sup\{\|\mathcal{T}(g)x - x\|^2 \mid g \in \operatorname{supp} f_1\} \leq \\ &\leq \sup\{\|\mathcal{T}(g)x - x\|^2 \mid g \in \operatorname{supp} f\} \leq a(f)(1 - \lambda_f(\mathcal{T})).\end{aligned}$$

Since $\mathcal{T}(f_1)x = \lambda_{f_1}(\mathcal{T})x$ and the operator $\mathcal{T}(f_1)$ commutes with $\mathcal{T}(G_2)$ (because $f_1 \in \mathfrak{A}(G_1)$), we have that

$$\mathcal{T}(f_1)(\mathcal{T}(g_2)x - x) = \lambda_{f_1}(\mathcal{T})(\mathcal{T}(g_2)x - x) \text{ for all } g_2 \in G_2.$$

This and inequality (1) in 3.2 imply that

$$\|\mathcal{T}(g_1)(\mathcal{T}(g_2)x - x) - (\mathcal{T}(g_2)x - x)\| \leq (a(f_1)(1 - \lambda_{f_1}(\mathcal{T}))^{1/2}\|\mathcal{T}(g_2)x - x\|$$

for all $g_1 \in \operatorname{supp} f_1$, $g_2 \in G_2$. In view of (3) and (5) this proves (2). □

(3.5) Definition. *Let $\rho \in \tilde{G}$ and $x_n \in L(\rho)$, $n \in \mathbb{N}^+$.*

(i) The sequence $\{x_n\}_{n\in\mathbb{N}^-}$ is called asymptotically $\rho(G_1)$-invariant *(resp.* asymptotically $\rho(G)$-invariant*) if $x_n \neq 0$ for all sufficiently large n and*

$$\overline{\lim_{n\to\infty}} \sup_{g\in D} \|\rho(g)x_n - x_n\|/\|x_n\| = 0$$

for any compact $D \subset G_1$ (resp. $D \subset G$).

(ii) A sequence $\{x_n\}$ is said to be $\rho(G)$-uniform, *provided that $x_n \neq 0$ for all sufficiently large n and*

$$\lim_{g\to e} \sup_{n\in\mathbb{N}^+} \|\rho(g)x_n - x_n\|/\|x_n\| = 0.$$

Since ρ is unitary, the $\rho(G)$-uniformity of the sequence $\{x_n\}$ is equivalent to the equicontinuity of the sequence $\{f_n\}$ on compact sets in the sense of III.1.5, where $f_n(g) = \rho(g)x_n/\|x_n\|$, $g \in G$.

(3.6) Lemma. *Let $\rho \in \tilde{G}$ and $x_n \in L(\rho)$, $n \in \mathbb{N}^+$. Suppose the sequence $\{x_n\}_{n\in\mathbb{N}^+}$ is $\rho(G)$-uniform, asymptotically $\rho(G_1)$-invariant, but not asymptotically $\rho(G)$-invariant. Then there is a neighbourhood of the identity $Y \subset G$ such that for $f \in \mathfrak{A}(G)$ with* $\operatorname{supp} f \subset Y$ *the sequence $\{\rho(f)x_n\}$ is $\rho(G)$-uniform, asymptotically $\rho(G_1)$-invariant, but not asymptotically $\rho(G)$-invariant.*

Proof. Since the sequence $\{x_n\}_{n\in\mathbb{N}^+}$ is not asymptotically $\rho(G)$-invariant, there is a compact set $D \subset G_1$ such that

$$r \overset{\text{def}}{=} \varlimsup_{n\to\infty} \sup_{g\in D} \|\rho(g)x_n - x_n\|/\|x_n\| > 0. \tag{1}$$

The sequence $\{x_n\}$ is $\rho(G)$-uniform. Hence one can choose a neighbourhood $Y \subset G$ of the identity such that

$$\|\rho(g)x_n - x_n\| \le r\|x_n\|/3; \; g \in Y, \; n \in \mathbb{N}^+. \tag{2}$$

We shall show that Y is the desired neighbourhood. Let $f \in \mathfrak{A}(G)$ and $\operatorname{supp} f \subset Y$. It then follows from inequality (1) in III.1.1 and (2) above that $\|\rho(f)x_n - x_n\| \le r\|x_n\|/3$ for each $n \in \mathbb{N}^+$. Making use of (1) and the property of ρ being unitary, we deduce that the sequence $\{\rho(f)x_n\}$ is not asymptotically $\rho(G)$-invariant. Since $\|\rho(f)x_n - x_n\| \le r\|x_n\|/3$,

$$\|\rho(f)x_n\| \ge 2r\|x_n\|/3, \; n \in \mathbb{N}^+. \tag{3}$$

Clearly, $\rho(g)\rho(f) - \rho(f) = \rho(gf - f)$ for all $g \in G$. But, since $f \in \mathfrak{A}(G)$, the inequality (1) in III.1.0 implies that $\lim_{g\to e} \|\rho(gf - f)\| = 0$. Thus the $\rho(G)$-uniformity of the sequence $\{\rho(f)x_n\}$ is a consequence of (3). Finally, since

$$\begin{aligned}\rho(g)\rho(f)x - \rho(f)x &= \int_G f(h)(\rho(gh)x - \rho(h)x)d\mu_G(h) = \\ &= \int_G f(h)\rho(h)(\rho(h^{-1}gh)x - x)d\mu_G(h)\end{aligned}$$

for all $g \in G$ and $x \in L(\rho)$, the asymptotic $\rho(G_1)$-invariance of the sequence $\{\rho(f)x_n\}$ is a consequence of (3), the asymptotic $\rho(G_1)$-invariance of the sequence $\{x_n\}_{n\in\mathbb{N}^+}$, and the normality of the subgroup G_1 in G. □

(3.7) Lemma. *Let $\rho \in \tilde{G}$. Suppose $L(\rho)^{G_2} = \{0\}$ and ρ is closed to I_G (in the sense of* III.1.2*). Then there exist $x_n \in L(\rho)$, $n \in \mathbb{N}^+$, such that the sequence $\{x_n\}_{n\in\mathbb{N}^+}$ is $\rho(G)$-uniform, asymptotically $\rho(G_1)$-invariant and not asymptotically $\rho(G)$-invariant.*

Proof. Choose a compact set $D_i \supset K_i$ which generates the group G_i, $i = 1, 2$, and a function $f_i \in \mathfrak{A}(G_i)$ such that $\operatorname{supp} f_i \supset D_i$ and f_i is bi-invariant relative to K_i. We set $D = D_1 \times D_2$ and $f = f_1 \times f_2$. Consider the decomposition

$$\rho = \int_Z \rho_z d\mu(z), \; L(\rho) = \int_Z L(\rho_z)d\mu(z)$$

of ρ into a continuous sum of irreducible representations $\rho_z \in \hat{G}$. For $n \in \mathbb{N}^+$, we set

$$Z_n = \{x \in Z \mid \|\rho_z(f)\| > 1 - 1/n\} \text{ and } W_n = \int_{Z_n} L(\rho_z) d\mu(z) \subset L(\rho).$$

Since ρ is close to I_G, it follows from Proposition III.1.4 that $W_n \neq \{0\}$ for all $n \in \mathbb{N}^+$. On the other hand, Lemma 3.4(iv) implies that $L(\rho_z)^K \neq \{0\}$ for all $z \in Z_1$. Thus, we can choose for each $n \in \mathbb{N}^+$, $y_n \in W_n \cap L(\rho)^K$, $y_n \neq 0$. We set

(1) $$c_n = \sup_{g \in \operatorname{supp} f_2} \|\rho(g) y_n - y_n\|.$$

Since $L(\rho)^{G_2} = \{0\}$, $\operatorname{supp} f_2 \supset D_2$, and D_2 generates G_2, it follows that $c_n \neq 0$. For each $n \in \mathbb{N}^+$ we choose $g_n \in \operatorname{supp} f_2$ with

(2) $$\|\rho(g_n) y_n - y_n\| > c_n/2$$

and set

(3) $$x_n = \rho(g_n)\rho(f) y_n - \rho(f) y_n = \rho(g_n f - f) y_n.$$

Let

(4) $$y_n = \int_Z y_{n,z} d\mu(z)$$

be a decomposition of the vector y_n into a continuous sum of vectors $y_{n,z} \in L(\rho_z)^K$. It follows from Lemma 3.4(iv) that

(5) $$l - \lambda_f(\rho_z) < 1/n \text{ for all } z \in Z_n.$$

Since

$$\begin{aligned}
&\|(\rho_z(g_n)\rho_z(f) y_{n,z} - \rho_z(f) y_{n,z}) - (\rho_z(g_n) y_{n,z} - y_{n,z})\| = \\
&= \|(\rho_z(g_n)\lambda_f(\rho_z) y_{n,z} - \lambda_f(\rho_z) y_{n,z}) - (\rho_z(g_n) y_{n,z} - y_{n,z})\| = \\
&= (1 - \lambda_f(\rho_z)) \|\rho_z(g_n) y_{n,z} - y_{n,z}\|,
\end{aligned}$$

it follows from (3), (4), and (5) that

$$\|x_n - (\rho(g_n) y_n - y_n)\| < \|\rho(g_n) y_n - y_n\|/n.$$

This inequality and (2) imply that

(6) $$\liminf_{n \to \infty} \|x_n\|/c_n > 1/2.$$

For $g \in G$ and $n \in \mathbb{N}^+$, we set $f_{g,n} = g g_n f - g f - g_n f + f$. It then follows from (3) that

(7) $$\begin{aligned}
\rho(g) x_n - x_n &= \rho(g)\rho(g_n f - f) y_n - \rho(g_n f - f) y_n = \\
&= \rho(f_{g,n}) y_n = \int_G f_{g,n}(g') \rho(g') y_n d\mu_G(g') = \\
&= \int_G f_{g,n}(g')(\rho(g') y_n - y_n) d\mu_G(g')
\end{aligned}$$

(the latter equality is a consequence of the equality $\int_G f_{g,n}(g')d\mu_G(g') = 0$). Since $f \in \mathfrak{A}(G)$ and g_n, $e \in \operatorname{supp} f$, we have that

$$\lim_{g \to e} \sup_{n \in \mathbb{N}^+} \int_G |f_{g,n}(g')| d\mu_G(g') = 0 \text{ and}$$

$$\operatorname{supp} f_{g,n} \subset (\operatorname{supp} f)^3, \text{ if } g \in \operatorname{supp} f.$$

But, since $\|\rho(g_1 g_2)x - x\| \leq \|\rho(g_1)x - x\| + \|\rho(g_2)x - x\|$ for all $g_1, g_2 \in G$ and $x \in L(\rho)$ (because ρ is unitary),

$$\sup\{\|\rho(g)y_n - y_n\| \mid g \in \operatorname{supp} f)^3\} \leq$$
$$\leq 3 \sup\{\|\rho(g)y_n - y_n\| \mid g \in \operatorname{supp} f\} = 3c_n.$$

Thus (6) and (7) imply that the sequence $\{x_n\}_{n \in \mathbb{N}^+}$ is $\rho(G)$-uniform. From inequalities (1) and (2) in 3.4 applied to the representations ρ_z and the vectors $\rho_z(f)y_{n,z}$, (3), (4), and (5) we deduce that

$$\lim_{n \to \infty} \|\rho^K x_n - x_n\| / \|x_n\| = 0 \tag{8}$$

and

$$\lim_{n \to \infty} \sup_{g \in \operatorname{supp} f_1} \|\rho(g)x_n - x_n\| / \|x_n\| = 0. \tag{9}$$

Since $\|\rho^K x_n - x_n\| \leq \sup_{g \in K} \|\rho(g)x_n - x_n\|$, it follows from (8) that the sequence $\{x_n\}_{n \in \mathbb{N}^+}$ is not asymptotically $\rho(G)$-invariant. Finally, since $\operatorname{supp} f_1 \supset D_1$, D_1 generates G_1 and, as was observed above, for all $g_1, g_2 \in G$ and $x \in L$, $\|\rho(g_1 g_2)x - x\| \leq \|\rho(g_1)x - x\| + \|\rho(g_2)x - x\|$, it follows from (9) that the sequence $\{x_n\}$ is asymptotically $\rho(G_1)$-invariant. □

(3.8) We begin with some notation and terminology which is needed in stating and proving Theorem 3.9. Let Γ be a lattice in G, N a normal subgroup of Γ, and $\varphi\colon \Gamma \to \Gamma/N$ the natural epimorphism. The family of all left Borel fundamental domains $X \subset G$ for Γ will be denoted by $\mathfrak{f}(G, \Gamma)$. For every domain $X \in \mathfrak{f}(G, \Gamma)$ we define (uniquely) the map $\tau_X\colon G \to \Gamma$ via the inclusion $g \in \tau_X(g)X$, $g \in G$. A domain $X \in \mathfrak{f}(G, \Gamma)$ is called *(Γ, N)-admissible* if for any compact set $M \subset G$ the set $\varphi(\tau_X(X \cdot M))$ is finite. If $\Gamma \setminus G$ is compact, then there exists a (Γ, N)-admissible domain (one can take any relatively compact $X \in \mathfrak{f}(G, \Gamma)$ as such a domain).

(3.9) Theorem. *If there is a (Γ, N)-admissible domain in $\mathfrak{f}(G, \Gamma)$ and $\overline{NG_i} = G_i$, $i = 1, 2$, then the factor group Γ/N has property (T), and hence is either non-amenable or finite (see III.2.2).*

We omit the proof in the general case and restrict our consideration to the special case where $\Gamma \setminus G$ is compact. The general case of the theorem was proved in [Mar 12].

(3.10) *Proof of Theorem 3.9 in case where $\Gamma \setminus G$ is compact.* Suppose Γ/N does not satisfy property (T). Let U' be a unitary representation of the group Γ/N which is close to $I_{\Gamma/N}$ and does not contain $I_{\Gamma/N}$. By setting $U(\gamma) = U'(\varphi(\gamma)), \gamma \in \Gamma$, we define a representation $U \in \tilde{\Gamma}$ which is trivial on N, close to I_Γ, and does not contain I_Γ.

Since $\Gamma \setminus G$ is compact, one can choose a relatively compact domain $X \in \mathfrak{f}(G, \Gamma)$. Let $\rho \in \tilde{G}$ be the representation which is induced in the sense of Mackey from U. We recall that $L(\rho)$ consists of measurable functions $j\colon G \to L(U)$ with

(1) $$j(\gamma g) = U(\gamma) j(g), \gamma \in \Gamma, g \in G,$$

and

(2) $$\|j\|^2 \overset{\text{def}}{=} \int_X \|j(g)\|^2 d\mu_G(g).$$

The representation ρ is defined by the formula

(3) $$(\rho(g)j)(g_1) = j(g_1 g), \qquad g, g_1 \in G.$$

We shall show that

(4) $$L(\rho)^{G_2} = \{0\}.$$

Suppose the contrary. Let $j \in L(\rho)^{G_2}$, $j \neq 0$. Then, from Lemma I.4.1.1(v) and equalities (1) and (3), we deduce, modifying, if necessary, the vector-valued function j on a null set, that

$$j(\gamma G_2 g) = j(\gamma g G_2) = U(\gamma) j(g) \text{ for all } \gamma \in \Gamma \text{ and } g \in G.$$

But U is trivial on N, so $j(NG_2 g) = j(g)$ for all $g \in G$. On the other hand, since $\overline{NG_2} = G$, the action of the group NG_2 by left translations on G is ergodic (see Proposition I.4.5.1). It follows that the vector-valued function j is constant (after a modification on a null set). We then deduce from (3) that ρ contains I_G. But U does not contain I_G. Hence ρ does not contain I_G (see Lemma III.2.11). This contradiction verifies (4).

Since U is close to I_Γ, it follows from Proposition III.1.11 that ρ is close to I_G. But $L(\rho)^{G_2} = \{0\}$. So by Lemma 3.7 there exist $j_n \in L(\rho)$, $n \in \mathbb{N}^+$, such that the sequence $\{j_n\}_{n \in \mathbb{N}^+}$ is $\rho(G)$-uniform, asymptotically $\rho(G_1)$-invariant and not asymptotically $\rho(G)$-invariant. Replacing j_n by $j_n/\|j_n\|$ we may assume that $\|j_n\| = 1$. Since $\{j_n\}_{n \in \mathbb{N}^+}$ is $\rho(G)$-uniform, there is a neighbourhood of the identity $Y_1 \subset G$ such that for every $f \in \mathfrak{A}(G)$ with $\operatorname{supp} f \subset Y_1$ $\|\rho(f) j_n\| > \|j_n\|/2 = 1/2$ for all $n \in \mathbb{N}^+$. Therefore, by Lemma 3.6, there exists a function $f \in \mathfrak{A}(G)$ such that

(5) $$\|\rho(f) j_n\| > 1/2, \; n \in \mathbb{N}^+,$$

and the sequence $\{\rho(f) j_n\}$ is $\rho(G)$-uniform asymptotically $\rho(G_1)$-invariant and not asymptotically $\rho(G)$-invariant. Since $\|j_n\| = 1$, the sequence $\{j_n\}_{n \in \mathbb{N}^+}$ is mean uniformly bounded in the sense of III.1.5. But (3) implies that

$$(\rho(f) j_n)(g_1) = \int_G f(g) j_n(g_1 g) d\mu_G(g), \; n \in \mathbb{N}^+, \; g \in G.$$

Thus the sequence $\rho(f)j_n$ is equicontinuous and uniformly bounded on compact sets (see Lemma III.1.7), and hence the sequence $\{q_n\}_{n\in\mathbb{N}^+}$ is equicontinuous and uniformly bounded on compact sets, where $q_n(g) = \|(\rho(f)j_n)(g) - (\rho(f)j_n)(e)\|$, $g \in G$. It then follows from Lemma III.1.6 that the set $\{q_n \mid n \in \mathbb{N}^+\}$ is relatively compact in the space of continuous functions on G equipped with the topology of uniform convergence on compact sets.

Let q be a cluster point of the set $\{q_n \mid n \in \mathbb{N}^+\}$. Since U is trivial on N, (1) implies that $j(Ng) = j(g)$ for all $j \in L(\rho)$ and $g \in G$. It follows that $q(Ng) = q(g)$ for all $g \in G$. On the other hand, since the sequence $\{\rho(f)j_n\}$ is asymptotically $\rho(G_1)$-invariant, $q(gG_1) = q(g)$ for all $g \in G$. Thus $q(NG_1g) = q(NgG_1) = q(g)$. But q is continuous, $q(e) = 0$, and $\overline{NG_1} = G$. Hence, $q = 0$. Therefore the function $q \equiv 0$ is the unique cluster point of the compact set $\{q_n \mid n \in \mathbb{N}^+\}$. It follows that

$$\lim_{n\to\infty} \sup_{g\in D} \|(\rho(f)j_n)(g) - (\rho(f)j_n)(e)\| = 0$$

for any compact set $D \subset G$. But, taking into consideration (2), (3), (5), and the compactness of the set X, we obtain a contradiction, since $\{\rho(f)j_n\}$ is not asymptotically $\rho(G)$-invariant. □

The following assertion was established in [B-K].

(3.11) Theorem. *Let F_1, F_2 be two locally compact unimodular groups, $F = F_1 \times F_2$, and let Λ be a cocompact lattice in F. Suppose $\overline{\Lambda F_1} = \overline{\Lambda F_2} = F$ and F has the following property:*

(∗) *there is a compact subgroup K in F such that the ring of continuous functions on F (with respect to the operation of convolution) with compact support and bi-invariant relative to K is commutative.*

Then $\mathrm{Hom}(\Lambda, \mathbb{C}) = \mathrm{Hom}_{\mathrm{cont}}(F, \mathbb{C})$, *i.e., every homomorphism of the group Λ to the additive group of the field $\mathbb{C}$ extends to a continuous homomorphism of the group F to $\mathbb{C}$. In particular, if F is compactly generated (or equivalently Λ is finitely generated) and* $\mathrm{Hom}_{\mathrm{cont}}(F, \mathbb{C}) = 0$, *then the factor group $\Lambda/\mathscr{D}(\Lambda)$ is finite.*

Property (∗) is a consequence of property (Q) (see [Lan 2], Ch. IV, §1, Theorem 1). Furthermore, it is easily seen that if F has property (Q), then $\mathrm{Hom}_{\mathrm{cont}}(F, \mathbb{C}) = 0$. Thus, the following is a consequence of Theorem 3.11.

(3.12) Corollary. *With G_1, G_2, G, and Γ as in* 3.3 *and* 3.8, *we assume $\Gamma \setminus G$ is compact and $\overline{\Gamma G_1} = \overline{\Gamma G_2} = G$. Then $\Gamma/\mathscr{D}(\Gamma)$ is finite.*

(3.13) *Remark.* In case $\overline{\mathscr{D}(G)} = G$, Corollary 3.12 is a consequence of Theorem 3.9.

4. Finiteness of Factor Groups of Discrete Subgroups

The main purpose of this section is to prove Theorem 4.9 on the finiteness of factor groups of irreducible lattices modulo non-central normal subgroups.

This theorem is deduced from the results of Sections 2 and 3, Theorem III.6.6, and from Furstenberg's theorem (Theorem 4.5) on the existence of equivariant measurable maps to the spaces of measures, which we also prove in this section.

As before, for every compact space X, we denote by $\mathscr{K}(X)$ the space of continuous functions on X, by $\mathscr{K}(X)^*$ its dual space, by $\mathscr{M}(X)$ the set of regular Borel measures on X, and by $\mathscr{P}(X)$ the set $\{\mu \in \mathscr{M}(X) \mid \mu(X) = 1\}$ identified with the space of positive normalized ($l(1) = 1$) functionals $l \in \mathscr{K}(X)^*$. As usual $\bar{B}$ denotes the closure of a subset B of a topological space.

As in Section 2 we will use here the notation introduced in I.4.2. We begin with the following simple remark.

(4.1) Lemma. *If X is a compact space, $l \in \mathscr{K}(X)^*$ with $l(1) = 1$ and $\|l\| = 1$, then $l \in \mathscr{P}(X)$.*

Proof. Suppose the contrary. Then $l(f) < 0$ for a function $f \in K(X)$ $f \geq 0$, $\|f\| < 1$, and hence $1 = \|l\| \geq \|l(1-f)\|/\|1-f\| > 1/1 = 1$. This is of course a contradiction. □

(4.2) Lemma. *If B is a separable non-amenable locally compact group, then there is a continuous action of B on a metrizable compact space X such that there is no B-invariant measure $\mu \in \mathscr{P}(X)$ on X.*

Proof. Since B is non-amenable, there is a compact B-space Y (not necessarily metrizable) such that there are no B-invariant measures $\mu \in \mathscr{P}(Y)$ on Y (see I.5.5.1). For $f \in \mathscr{K}(Y)$, we set $L(f) = \{\mu \in \mathscr{P}(Y) \mid (b\mu)(f) = \mu(f)$ for all $b \in B\}$. Then the intersection of all $L(f)$, $f \in \mathscr{K}(Y)$ is empty. Therefore, since the space $\mathscr{P}(Y)$ is compact and the sets $L(f)$ are closed in the weak*-topology, there is a finite set $Q \subset \mathscr{K}(Y)$ such that $\bigcap_{q \in Q} L(q) = \emptyset$. We denote by $F \subset \mathscr{K}(Y)$ the C^*-algebra with 1 generated by the set BQ and by X its space of maximal ideals ($F = \mathscr{K}(X)$). Since the group B is separable, it follows that the algebra F is separable, and hence the compact space X is metrizable. Since $B(BQ) = BQ$, the algebra F is B-invariant. Therefore the action of the group B on $\mathscr{K}(Y) \supset F$ induces the action of the group B on X. Suppose there exists a B-invariant measure $\mu \in \mathscr{P}(X)$ on X. By the Hahn-Banach theorem μ can be extended to $l \in \mathscr{K}(Y)^*$ such that $\|l\| = 1$. We then have

1) since $l(1) = \mu(1) = 1$, by Lemma 4.1, $l \in \mathscr{P}(Y)$; and
2) since $F \supset Q$ and μ is B-invariant, $l \in \bigcap_{q \in Q} L(q) = \emptyset$.

This is a contradiction and consequently X is the desired compact space. □

(4.3) As in Section 2, up to the end of this section we will use the notation introduced in II.3.0. In addition, Γ will denote a lattice in G, N a normal subgroup of Γ, $\operatorname{rank} G = \sum_{\alpha \in A} \operatorname{rank}_{k_\alpha} \mathbf{G}_\alpha$ the rank of the group G, and $\mathfrak{f}(G, \Gamma)$ the family of all left Borel fundamental domains $X \subset G$ for Γ.

(4.4) Lemma. *The group P is amenable.*

Proof. Since the groups S and V are solvable, $P = \mathscr{Z}_G(S) \ltimes V$ (see III.3.0), and by virtue of Proposition I.2.3.6 the group $\mathscr{Z}_G(S)/S$ is compact, it follows that P contains the solvable normal subgroup $S \cdot V$ with $P/(S \cdot V)$ compact. Thus the group P is amenable (see I.5.5.2). □

(4.5) Theorem. *Let X be a compact metrizable Γ-space. Then there exists a $\mu_{G/P}$-measurable Γ-equivariant map ω: $G/P \to \mathscr{P}(X)$ (i.e. $\omega(\gamma x) = \gamma\omega(x)$ for all $\gamma \in \Gamma$ and for almost all $x \in G/P$).*

Proof. The group $\Gamma \times G$ acts on $G \times X$ and on G by $(\gamma, g_1)(g, x) = (\gamma g g_1^{-1}, \gamma x)$ and $(\gamma, g_1)g = \gamma g g_1^{-1}$, $g_1, g \in G$, $\gamma \in \Gamma$. Let φ: $G \times X \to G$ be the natural projection. We set

$$Q = \{\mu \in \mathscr{M}(G \times X) \mid \varphi(\mu) = \mu_G \text{ and } (\gamma, e)\mu = \mu \text{ for all } \gamma \in \Gamma\}.$$

Let $F \in \mathfrak{f}(G, \Gamma)$. For $g \in G$, we define $\gamma_g \in \Gamma$ via the inclusion $g \in \gamma_g F$. Pick a point $x_0 \in X$ and set $\varphi'(g) = (g, \gamma_g x_0) \in G \times X$, $g \in G$. We then have that $(\gamma, e)(\varphi'(g)) = \varphi'((\gamma, e)g)$ for all $\gamma \in \Gamma$ and $g \in G$, φ' is a Borel map, and $\varphi \circ \varphi' = \mathrm{Id}$. Thus $\varphi'(\mu_G) \in Q$, and hence Q is not empty. Clearly Q is convex. Since X is compact, Q is a closed (in the weak*-topology) bounded subset in $\mathscr{M}(G \times X)$. Hence Q is compact, convex, and non-empty. On the other hand,

1) since $(\Gamma \times G)\mu_G = \mu_G$ (because G is unimodular) and the map φ is $(\Gamma \times G)$-equivariant, $(\Gamma \times G)Q = Q$;
2) the group P is amenable (see Lemma 4.4).

Therefore there is a measure $\tau \in Q$ such that $(\{e\} \times P)\tau = \tau$ (see I.5.5.1). But $(\gamma, e)\tau = \tau$ for all $\gamma \in \Gamma$. Hence $(\Gamma \times P)\tau = \tau$. Since $\varphi(\tau) = \mu_G$, it follows from the decomposition theorem for measures on a direct product of locally compact spaces (see [Bou 3], Ch. VI, §3, Theorem 1) that

$$\tau = \int_G (\delta_g \times \lambda_g) d\mu_G(g), \text{ where } \lambda_g \in \mathscr{P}(X). \tag{1}$$

Furthermore, the map $g \mapsto \lambda_g$ is measurable and decomposition (1) is unique up to modification of the map $g \mapsto \lambda_g$ on a null set. It follows from the $\Gamma \times P$-invariance of the measures τ and μ_G that for all $\gamma \in \Gamma$ and $p \in P$

$$\begin{aligned} \int_G (\delta_g \times \lambda_g) d\mu_G(g) &= \tau = (\gamma, p)\tau = \\ &= \int_G ((\gamma, p)(\delta_g \times \lambda_g)) d\mu_G(g) = \\ &= \int_G (\delta_{\gamma g p^{-1}} \times \gamma\lambda_g) d\mu_G(g) = \\ &= \int_G (\delta_g \times \gamma\lambda_{\gamma^{-1} g p}) d\mu_G(g). \end{aligned} \tag{2}$$

Therefore, since decomposition (1) is unique, $\lambda_{\gamma^{-1}gp} = \gamma^{-1}\lambda_g$ for almost all $g \in G$. Since for each $p \in P$ we have $\lambda_{gp} = \lambda_g$ for almost all $g \in G$, it follows from

Lemma I.4.1.3(ii) that there is a $\mu_{G/P}$-measurable map $\omega\colon G/P \to \mathscr{P}(X)$ such that $\lambda_g = \omega(\pi(g))$ for almost all $g \in G$, where $\pi\colon G \to G/P$ is the natural map. But for each $\gamma \in \Gamma$ we have $\lambda_{\gamma g} = \gamma\lambda_g$ for almost all $g \in G$. Thus ω is Γ-equivariant. □

(4.6) *Remark 1.* In case $k_\alpha = \mathbb{R}$ for each $\alpha \in A$, Theorem 4.5 is a restatement of Theorem 15.1 in [Fu 5]. Furthermore, the proof presented above is similar to that of given in that paper. Notice that the compact space X in the statement of Theorem 15.1 in [Fu 5] is not assumed to be metrizable. However, the proof presented in that paper is correct only for metrizable X, because, for an arbitrary compact space X, the decomposition (1) in 4.5 is not necessarily unique.

Remark 2. The proof shows that G in Theorem 4.5 can be replaced by any metrizable σ-compact group, Γ by any closed subgroup, and P by any amenable closed subgroup with $\Delta_G(p) = 1$ for all $p \in P$, where Δ_G is the modular function of the group P.

(4.7) Lemma. *If the factor group Γ/N is not amenable, then there is a subalgebra $\mathfrak{B}$ of $\mathfrak{M}(G/P)$ which contains infinitely many elements such that $\Gamma\mathfrak{B} = \mathfrak{B}$ (i.e. $\gamma B \in \mathfrak{B}$ for all $\gamma \in \Gamma$ and $B \in \mathfrak{B}$) and $NB = B$ for all $B \in \mathfrak{B}$.*

Proof. Since the countable group $H \stackrel{\text{def}}{=} \Gamma/N$ is not amenable, it follows from Lemma 4.2 that there exists a continuous action of the group H on a metrizable compact space X such that there is no H-invariant measure $\mu \in \mathscr{P}(X)$ on X. Let $\pi\colon \Gamma \to H$ be the natural epimorphism. We set $\gamma x = \pi(\gamma)x$, $\gamma \in \Gamma$, $x \in X$. By virtue of Theorem 4.5 there exists a $\mu_{G/P}$-measurable Γ-equivariant map $\omega\colon G/P \to \mathscr{P}(X)$. Consider the set $\mathfrak{B} = \{B \in \mathfrak{M}(G/P) \mid$ there is a subset in the class B which is the inverse image under ω of a Borel subset in $\mathscr{P}(X)\}$. Since ω is Γ-equivariant and $Nx = \pi(N)x = ex = x$ for all $x \in X$, we have that $\Gamma\mathfrak{B} = \mathfrak{B}$ and $NB = B$ for all $B \in \mathfrak{B}$. The algebra $\mathfrak{B}$ is infinite, for otherwise the support R of the image of the measure $\mu_{G/P}$ under ω would be finite and the measure $(1/|R|)\sum_{x\in R}\delta_x \in \mathscr{P}(X)$ would be Γ-invariant (because ω is Γ-equivariant), and hence H-invariant. □

(4.8) Theorem. *Suppose* rank $G \geq 2$, *the lattice Γ is irreducible, and for each $\alpha \in A$ the group $\mathbf{G}_\alpha$ is simply connected, k_α-isotropic, and almost k_α-simple. If the factor group Γ/N is non-amenable, then $N \subset \mathscr{Z}(G)$.*

Proof. By virtue of Lemma 4.7 there exists a non-trivial subalgebra $\mathfrak{B}$ of the algebra $\mathfrak{M}(G/P)$ such that

(1) $$\Gamma\mathfrak{B} = \mathfrak{B}$$

and

(2) $$NB = B \text{ for all } B \in \mathfrak{B}.$$

It follows from (1) and Corollary 2.12 that $\mathfrak{B} = \mathfrak{M}(G/P, P_\vartheta)$ for some $\vartheta \subset \Delta$. It then follows from (2) that $Nx = x$ for each $x \in G/P_\vartheta$, and hence

(3)
$$N \subset \bigcap_{g \in G} g P_{\vartheta} g^{-1}.$$

Since every non-central normal subgroup of the group G_α, $\alpha \in A$, coincides with G_α (see Corollary I.2.3.2(a)),

(4)
$$\bigcap_{g \in G} g P_{\vartheta} g^{-1} = \mathscr{Z}(G_L) G_{A-L},$$

where $L = \{\alpha \in A \mid \Delta_\alpha \not\subset \vartheta\}$. Since the subalgebra $\mathfrak{B}$ is non-trivial, $\vartheta \neq \Delta$, and hence $L \neq \varnothing$. Thus $\Gamma \cap \mathscr{Z}(G_L) \cdot G_{A-L} \subset \mathscr{Z}(G)$ (see Theorem II.6.7(d)). It then follows from (3) and (4) that $N \subset \mathscr{Z}(G)$. □

(4.9) Theorem. *Suppose* rank $G \geq 2$, *the lattice* Γ *is irreducible, and for each* $\alpha \in A$ *the group* $\mathbf{G}_\alpha$ *is simply connected,* k_α*-isotropic, and almost* k_α*-simple. Assume in addition that at least one of the following conditions is satisfied:*

(a) the set $\hat{A} = \{\alpha \in A \mid \operatorname{rank}_{k_\alpha} \mathbf{G}_\alpha \geq 2\}$ *is not empty;*

(b) there exists a (Γ, N)*-admissible (in the sense of 3.8) domain* $X \in \mathfrak{f}(G, \Gamma)$*;*

(c) the quotient space $\Gamma \setminus G$ *is compact.*

Then either $N \subset \mathscr{Z}(G)$ *or the factor group* Γ/N *is finite.*

Proof. Assume $N \not\subset \mathscr{Z}(G)$. Then, by Theorem 4.8 Γ/N is amenable. Thus, it remains to show that Γ/N has property (T) (see III.2.2). If $\hat{A} \neq \varnothing$, then by virtue of Theorem B(a) in III.5.9 the factor group Γ/N has property (T). Suppose now that $\hat{A} = \varnothing$ and that (b) is satisfied. It then follows from Corollaries I.2.2.2 and I.2.3.5 that the groups G_α, $\alpha \in A$, have property (Q) and are compactly generated. Furthermore, since rank $G \geq 2$, we have $A = A_1 \cup A_2, A_1 \neq \varnothing$, $A_2 \neq \varnothing$, and by Lemma III.6.8(b) $\overline{NG_{A_i}} = G$, $i = 1, 2$. Thus Theorem 3.9 implies that Γ/N has property (T). All that remains now is to observe that (c) implies (b) (see 3.8). □

(4.10) Theorem 4.9 and the arithmeticity theorems which are proved below imply (see Theorem IX.5.4 and remark (i) in IX.1.2) the following

Theorem. *Suppose* rank $G \geq 2$, *the lattice* Γ *is irreducible, and for each* $\alpha \in A$ *the group* $\mathbf{G}_\alpha$ *has no* k_α*-anisotropic factors. Then either* $N \subset \mathscr{Z}(G)$ *or* Γ/N *is finite.*

(4.11) *Remark.* If rank $G = 1$, then there exists a subgroup Γ' of finite index in Γ containing a free non-commutative normal (in Γ') subgroup of infinite index. This assertion is an easy consequence of Gromov's results on normal subgroups in fundamental groups of manifolds of negative curvature (see Remark IX.7.16 below) and the following observation. If k is a non-archimedean local field and $\mathbf{G}$ is a connected semisimple k-group with $\operatorname{rank}_k \mathbf{G} = 1$, then every discrete torsion free subgroup $\Lambda \subset \mathbf{G}(k)$ is free.

Chapter V. Characteristic Maps

This chapter and the next one are devoted to proofs of the existence theorems for equivariant measurable maps. These will be applied in Chapter VII to the proofs of the superrigidity theorems. In the present chapter we develop a method based on the application of the multiplicative ergodic theorem.

In Section 2 we present the multiplicative ergodic theorem and prove several related assertions. Sections 3–5 are devoted to the following situation: Γ is a lattice in a locally compact group G, T is a representation of the lattice on a finite-dimensional vector space $\mathscr{W}$ over a local field, and s is an element of the group G acting ergodically on $\Gamma \setminus G$. In Section 3 we define the characteristic measurable maps ω_i of the group G into the Grassmann varieties $\mathbf{Gr}_{l_i}(\mathscr{W})$ and study their fundamental properties. These maps are associated in a canonical way to the pair (s, T). To obtain this correspondence we make use of the multiplicative ergodic theorem. In Sections 4 and 5 we consider the pairs (s, T), where the associated characteristic maps have some additional properties. The results of Section 4 (resp. 5) are applied to the case where the Zariski closure $\overline{T(\Gamma)}$ of the subgroup $T(\Gamma)$ is not (resp. is) a semisimple group. It should be noted that we will not be using the results of Section 5 in the sequel. This is due to the fact that whenever $\overline{T(\Gamma)}$ is semisimple it suffices to use the results of Chapter VI. Nevertheless, the author believes that the methods developed in Section 5 are interesting independent of their application to the proof of the superrigidity theorems.

Throughout this chapter k denotes a local field with absolute value $|\ |$, $n \in \mathbb{N}^+$, and $\mathscr{W}$ denotes a vector space over k of dimension n. As in Section 1 of Chapter II, given a coordinate system $y = (y_1, \ldots, y_n)$ we set $\|w\| = \|w\|_y = \sum_{1 \leq j \leq n} |y_j(w)|$, $w \in \mathscr{W}$, and $\|B\| = \|B\|_y = \sup_{w \in \mathscr{W}, w \neq 0} \|Bw\|_y / \|w\|_y$, $B \in \operatorname{End}(\mathscr{W})$, and observe that the function $w \mapsto \|w\|$ is a norm on $\mathscr{W}$ and the function $B \mapsto \|B\|$ is a norm on $\operatorname{End}(\mathscr{W})$. Furthermore, $\|BC\| \leq \|B\| \|C\|$ for all $B, C \in \operatorname{End}(\mathscr{W})$. We equip $\mathscr{W}$, $\mathbf{GL}(\mathscr{W})$, and the Grassmann varieties $\mathbf{Gr}_l(\mathscr{W})$ with the topologies induced from the topology of the field k.

1. Auxiliary Assertions

In this section we shall prove several auxiliary assertions.

(1.1) Lemma. *If $B \in \mathbf{GL}(\mathscr{W})$, then*

$$-n \ln \|B^{-1}\| \le \ln |\det B| \le n \ln \|B\|. \tag{1}$$

Proof. Let $(e_1, \dots, e_n)$ be the basis of the coordinate system y and let b_{ij} be the entries of the matrix of the linear transformation B in this basis. We set $c_j = \sum_{l \le i \le n} |b_{ij}|$. Since

$$\det B = \sum_{d \in D} (-1)^{r(d)} \prod_{1 \le j \le n} b_{d(j),j},$$

where D is the set of permutations of degree n and $r(d) = \pm 1$ (depending on the parity of d), it follows that $\det B \le \prod_{1 \le j \le n} c_j$. But $c_j \le \|Be_j\| \le \|B\|$. Thus $|\det B| \le \|B\|^n$ and taking the logarithm of both sides of this inequality we obtain the right-hand side of (1). The left-hand side of (1) is obtained by applying B^{-1} to its right-hand side. □

(1.2) Lemma. *There exist a Borel map $\vartheta \colon \mathbf{GL}(\mathscr{W}) \to \mathbf{SL}(\mathscr{W})$ and a constant $c > 0$ such that*

(A) $\vartheta(gh) = \vartheta(g)h$ *for all* $g \in \mathbf{GL}(\mathscr{W})$ *and* $h \in \mathbf{SL}(\mathscr{W})$;

(B) $\|\vartheta(g_1)\vartheta(g_2)^{-1}\| \le c\|g_1 g_2^{-1}| \; \| |g_2 g_1^{-1}\|$ *for all* $g_1, g_2 \in \mathbf{GL}(\mathscr{W})$.

Proof. We set $G = \mathbf{GL}(\mathscr{W})$ and $H = \mathbf{SL}(\mathscr{W})$. It follows from the description of local fields that the factor group of the multiplicative group k^* of the field k modulo the subgroup $\{x^n \mid x \in k^*\}$ is compact. Thus, by virtue of Theorem I.4.4.1 (on the existence of Borel sections) there exist Borel maps $\lambda \colon G \to k^*$, $f \colon G \to G$, and $\vartheta \colon G \to H$ such that ϑ has property (A), the set $Z \overset{\text{def}}{=} f(G)$ is relatively compact, and $g = \lambda(g) f(g) \vartheta(g)$ for all $g \in G$. Since Z is relatively compact, there is $r > 0$ such that

$$|\lambda| \|z\| \le r |\det \lambda z|^{l/n}$$

for all $\lambda \in k^*$ and $z \in Z \cup Z^{-1}$. On the other hand, $\det(\lambda(g) f(g)) = \det g$ for each $g \in G$. Thus, for all $g_1, g_2 \in G$, we have

$$\begin{aligned}
\|\vartheta(g_1)\vartheta(g_2)^{-1}\| &= \|f(g_1^{-1})\lambda(g_1)^{-1} g_1 g_2^{-1} \lambda(g_2) f(g_2)\| \le \\
&\le |\lambda(g_1)^{-1}| \, \|f(g_1)^{-1}\| \, |\lambda(g_2)| \, \|f(g_2)\| \, \|g_1 g_2^{-1}\| \le \\
&\le r^2 |\det g_1^{-1}|^{1/n} |\det g_2|^{1/n} \|g_1 g_2^{-1}\| = \\
&= r^2 |\det(g_1 g_2^{-1})|^{-1/n} \|g_1 g_2^{-1}\| \le r^2 \|g_2 g_1^{-1}\| \, \|g_1 g_2^{-1}\|
\end{aligned}$$

(the latter inequality follows from the left-hand side of inequality (1) in 1.1). Hence ϑ and $c = r^2$ satisfy (B). This completes the proof. □

(1.3) Lemma. *Let $H = \mathbf{SL}(\mathscr{W})$. Then there exists $d > 0$ such that the function $b(h) = \|h\|^{-d}$, $h \in H$, belongs to $L_p(H, \mu_H)$ for all $p \ge 1$.*

Proof. We recall that $\mathrm{mod}_k(x)$ denotes the modulus of the element x of the field k and that for each $x \in k$, $\mathrm{mod}_k(x) = |x|^c$ for some $c > 0$ (see I.0.31). We set

$B(r) = \{x \in k \mid \mathrm{mod}_k(x) \leq r\}$, $r \geq 0$. Let k^+ be the additive group of the field k. Since $\mu_{k^+}(xB(1)) = \mathrm{mod}_k(x)\mu_{k^+}(B(1)$ for all $x \in k$ (by the definition of $\mathrm{mod}_k(x)$), $\mu_{k^+}(B(r)) \leq r\mu_{k^+}(B(1)))$ for all $r \geq 0$. Therefore the function

$$\delta(x) = \min\{1, \mathrm{mod}_k(x)^{-3}\} = \min\{1, |x|^{-3c}\}, \ x \in k,$$

is integrable in measure μ_{k^+}.

We set $Q = \mathbf{GL}(\mathscr{W})$ and $P = \{X \in Q \mid \mathrm{mod}_k(\det X) \geq 1\}$. Since the maximum of the absolute values of the entries x_{ij} of the matrix of a linear transformation $X \in \mathbf{GL}(\mathscr{W})$ does not exceed $\|X\|$ and $\|X\| \geq 1$ for all $X \in P$ (see Lemma 1.1), it follows that $\|X\|^{-d} \leq \prod_{i,j} \delta(x_{ij})$ for all $X = (x_{ij}) \in P$, where $d = 3cn^2$. But $d\mu_Q(X) = \mathrm{mod}_k(\det X)^{-n} \bigotimes_{i,j} d\mu_{k^+}(x_{ij})$ $(X = (x_{ij}))$, (see [Bou 3], Ch. VII, §3, paragraph 3, Example 1), the function $\delta(x)$ is μ_{k^+}-integrable, and $\mathrm{mod}_k(\det X)^{-n} \leq 1$ for all $X \in P$. So $\int_P \|X\|^{-d} d\mu_Q(X) < \infty$. On the other hand, by virtue of Proposition I.4.3 the measure μ_Q is, up to a scalar multiple, the image of the measure $\mu_H \times \mu_D$ under the map $(h, d) \mapsto hd^{-1}$ of $H \times D$ onto Q, where $D \subset \mathbf{GL}(\mathscr{W})$ is the subgroup of diagonal matrices of the form

$$\begin{pmatrix} x & & & 0 \\ & 1 & & \\ & & \ddots & \\ 0 & & & 1 \end{pmatrix}, \ x \in k, \ x \neq 0.$$

It then follows from Fubini's theorem applied to the function $f(h, p) = \|hp\|^{-d}$, $h \in H$, $p \in P \cap D$, that $\int_H \|Xp\|^{-d} d\mu_H(X) < \infty$ for some (even for almost all) $p \in P \cap D$. But $\|Xp\| \leq \|X\| \cdot \|p\|$. So the function $b(h) = \|h\|^{-d}$, $h \in H$, is μ_H-integrable. Finally, since $\|h\| \geq 1$ for all $h \in H$, in view of Lemma 1.1 we have that $b \in L_p(H, \mu_H)$ for all $p \geq 1$. □

2. The Multiplicative Ergodic Theorem

(2.0) Let X be a locally compact σ-compact space with finite regular Borel measure μ. By a *Borel automorphism* of the space X we mean a bijection $f: X \to X$ with f and f^{-1} Borel. Let L be a Borel automorphism of X preserving the measure μ. Now consider a *Borel cocycle* u on $\mathbb{Z} \times X$ with respect to the dynamical system $\{L^m\}$ which takes values in the group $\mathbf{GL}(\mathscr{W})$, i.e., $u(m, x)$ is a function of (m, x) which is Borel in x for each m with

(1) $u(m + r, x) = u(m, L^r x)u(r, x)$ for all $m, r \in \mathbb{Z}$ and $x \in X$.

We set $w(x) = u(1, x)$, $b(x) = \ln \|w(x)\|$, and $\tilde{b}(x) = \ln \|w(x)^{-1}\|$. By induction on m, for each $m \in \mathbb{N}^+$ and each $x \in X$, it follows that

(2) $u(m, x) = w(L^{m-1}x) \cdot \ldots \cdot w(x)$ and
$u(-m, x) = u(m, L^{-m}x)^{-1} = w(L^{-m}x)^{-1} \cdot \ldots \cdot w(L^{-1}x)^{-1}$.

Conversely, if $u(m, x)$ and $u(-m, x)$, $m \in \mathbb{N}^+$ are defined via $w(x)$ by (2) and $u(0, x) = E$, where E is the unit matrix, then it can directly be verified that (1) holds. A cocycle u is said to be *integrable* if

(3) $$\int_X \max_{m=\pm 1} \ln^+ \|u(m,x)\| d\mu(x) < \infty.$$

Since L preserves the measure μ, $u(1,x) = w(x)$, and $u(-1,x) = w(L^{-1}x)^{-1}$, it follows that condition (3) is equivalent to

(4) $$b \in L_1(X,\mu) \text{ and } \tilde{b} \in L_1(X,\mu).$$

We recall that an automorphism L of X is said to be ergodic (in measure μ) if the measure of every L-invariant measurable subset of X is either 0 or $\mu(X)$.

Before stating the multiplicative ergodic theorem, we would like to mention the following

Remark. Since the integrability of the cocycle u is equivalent to(4), Lemma 1.1 implies that if u is integrable, then for all $x \in X$ the function $\ln|\det u(1,x)|$ is μ-integrable.

(2.1) Theorem. (Multiplicative ergodic theorem.) *Suppose that the cocycle u is integrable and the automorphism L is ergodic (in measure μ).*

(i) *There exist numbers $r \in \mathbb{N}^+$, $l_i \in \mathbb{N}^+$, $\chi_i \in \mathbb{R}$, and measurable maps $\psi_i \colon X \to \mathbf{GR}_{l_i}(\mathscr{W})$, $1 \le r \le n$, $1 \le i \le r$, with the following properties:*
  (a) *for almost all $x \in X$, the space $\mathscr{W}$ is the direct sum of the subspaces $\psi_i(x)$, $1 \le i \le r$;*
  (b) *for almost all $x \in X$ and for each i, $1 \le i \le r$, the sequence $\{(1/m)\ln\|u(m,x)w\|/\|w\|\}_{m\in\mathbb{N}^+}$ converges to χ_i, uniformly in $w \in \psi_i(x) - \{0\}$ as $m \to \pm\infty$;*
  (c) $\sum_{i=1}^r l_i\chi_i = (1/\mu(X)) \int_X \ln|\det u(1,x)| d\mu(x)$;
  (d) $\chi_i < \chi_j$, *if* $1 \le i < j \le r$.

(ii) *For almost all $x \in X$ and for each $w \in \mathscr{W} - \{0\}$ the following limits exist: $\chi^+(w,x) = \lim_{m\to+\infty} \ln\|u(m,x)w\|/m$ and $\chi^-(w,x) = \lim_{m\to-\infty} \ln\|u(m,x)w\|/m$. Furthermore, for almost all $x \in X$ and for each $a \in \mathbb{R}$*

$$\{0\} \cup \{w \in \mathscr{W} - \{0\} \mid \chi^+(w,x) \le a\} = \bigoplus_{\chi_i \le a} \psi_i(x)$$

and

$$\{0\} \cup \{w \in \mathscr{W} - \{0\} | \chi^-(w,x) \ge a\} = \bigoplus_{\chi_i \ge a} \psi_i(x).$$

We say that the numbers χ_i are the *characteristic exponents* of the cocycle u and the number l_i is the *multiplicity* of the characteristic exponent χ_i. The maps ψ_i are called the *characteristic maps* defined by the cocycle u and the characteristic exponents χ_i.

A cocycle u is called *essential* if it has at least two characteristic exponents or, equivalently, if the multiplicity of any characteristic exponent of u is strictly less than n.

In case $k = \mathbb{R}$, the multiplicative ergodic theorem is due to Oseledec (see [O], Theorem 1 and 4); for similar results, see [Milli]. For an arbitrary field k the theorem will be established by the method of Oseledec in Appendix A. In [Rag 8] and [Ru] another method of proving the multiplicative ergodic theorem was suggested. This is based on the application of the subadditive ergodic theorem.

(1) *Remark.* It follows from assertion (ii) above that the numbers r, l_i, χ_i are uniquely determined and the maps ψ_i are defined uniquely modulo null sets.

(2) *Remark.* The assumption that the automorphism L is ergodic is not essential and an assertion similar to Theorem 2.1 holds for non-ergodic L. But in this case r, l_i, and χ_i are functions rather than numbers and in property (c) above we have to take $\lim_{m\to\pm\infty} \ln|\det u(m,x)|/m$ instead of $(1/\mu(X)) \int_X \ln|\det u(1,x)| d\mu(x)$.

(3) *Remark.* It is well known that every norm $\|\;\|'$ on $\mathscr{W}$ is equivalent to $\|\;\|$, i.e.,

$$0 < \inf_{w\in\mathscr{W}-\{0\}} \|w\|/\|w\|' \leq \sup_{w\in\mathscr{W}-\{0\}} \|w\|/\|w\|' < \infty.$$

Thus r, l_i, χ_i and $\psi_i(x)$ do not depend upon the norm on $\mathscr{W}$.

(4) *Remark.* It follows from assertion (ii) that for almost all $x \in X$ and for each $w \in \mathscr{W} - \{0\}$,

$$\chi^+(w,x) \geq \chi^-(w,x).$$

Equality is attained if and only if $w \in \psi_i(x)$ for some $1 \leq i \leq r$.

(5) *Remark.* In the statement of the multiplicative ergodic theorem we could take L to be an arbitrary automorphism of an arbitrary finite measure space and u be an arbitrary measurable cocycle. However, this generalization would not be essential, because making use of the standard methods of measure theory one can easily reduce the general case to the "Borel" situation.

(6) *Remark.* The method proposed in [Rag 8] and [Ru] is also applicable to the case that k is an arbitrary complete valuation field. In the proof of Oseledec the compactness of the quotient space $\mathbf{GL}_n(k)/B$, where $B \subset \mathbf{GL}_n(k)$ is the group of triangular matrices, is used. Hence this proof does not apply if k is not locally compact.

(2.2) Proposition. *Suppose* $\det u(m,x) \equiv 1$, *the automorphism* L *is ergodic, the cocycle* u *is integrable, and*

$$\text{(1)} \qquad \lim_{m\to+\infty} \sup(1/m) \int_X \ln \|u(m,x)\| d\mu(x) > 0.$$

Then the cocycle u *is essential.*

Proof. Since $\det u(m,x) \equiv 1$, Theorem 2.1(c) implies that $\sum_{i=1}^{r} l_i \chi_i = 0$. Thus, it suffices to show that $\chi_i \neq 0$ for some i. Suppose the contrary. Then, as $m \to +\infty$, the sequence of functions $f_m(x) \stackrel{\text{def}}{=} (1/m) \ln \|u(m,x)\|$ converges almost everywhere to 0. We set

$$d_m(x) = (1/m) \sum_{i=0}^{m-1} b(L^i x), \quad m \in \mathbb{N}^+, \; x \in X,$$

$$d = \int_X b \, d\mu \text{ and } \tilde{f}_m(x) = \min\{f_m(x), d\}.$$

For a function $p \in L_1(X, \mu)$, let $\|p\|_1$ denote its norm. Since $\|BC\| \leq \|B\| \cdot \|C\|$ for all $B, C \in \operatorname{End}(\mathscr{W})$, it follows from equalities (2) in 2.0 that $f_m \leq d_m$. On the other hand, it follows from the statistical ergodic theorem (see I.4.6.4) that $\lim_{m\to\infty} \|d_m - d\|_1 = 0$. Thus, we have

$$\lim_{m \to \infty} \|f_m - \tilde{f}_m\|_1 = 0. \tag{2}$$

Since $f_m \geq 0$, $b \geq 0$ (in view of $\det u(m,x) \equiv 1$ and Lemma 1.1) and $\lim_{m\to\infty} f_m(x) = 0$ for almost all $x \in X$, it follows that $\tilde{f}_m \geq 0$ and $\lim_{m\to\infty} \tilde{f}_m(x) = 0$ for almost all $x \in X$. But $\tilde{f}_m(x) \leq d$, and hence, by virtue of the classical Lebesgue theorem (see [Hal 1], Theorem 4 of Chapter V), $\lim_{m\to\infty} \|\tilde{f}_m\|_1 = 0$. It then follows from (2) that $\lim_{m\to\infty} \|f_m\|_1 = 0$. This contradicts (1) and completes the proof. □

(2.3) Proposition. *Let Q be a linear subspace in $\mathscr{W}$ such that $u(m,x)Q = Q$ for all $m \in \mathbb{Z}$ and $x \in X$. Let $F = \mathscr{W}/Q$ and denote by $p\colon \mathscr{W} \to F$ the natural projection and by $u'\colon \mathbb{Z} \times X \to \mathbf{GL}(F)$ the cocycle induced from the cocycle u and the projection p (more precisely, $u'(m,x)z = p(u(m,x))p^{-1}(z))$, $m \in \mathbb{Z}$, $x \in X$, and $z \in F$). Assume that the cocycle u (and hence the cocycle u') is integrable. Let χ_i, $1 \leq i \leq r$ (resp. χ'_j, $1 \leq j \leq r'$), be the characteristic exponents of the cocycle u (resp. u') and let ψ_i (resp. ψ'_j) be the characteristic maps defined by the cocycle u (resp. u') and the characteristic exponents χ_i (resp. χ'_j). Then*

(i) *for each j, $1 \leq j \leq r$, there is a number $i(j)$ with $1 \leq i(j) \leq r$, such that $\chi'_j = \chi_{i(j)}$ and $\psi'_j(x) = p(\psi_{i(j)}(x))$ for almost all $x \in X$;*
(ii) *if i is a number with $\chi_i \neq \chi'_j$ for all j, then $\psi_i(x) \subset Q$ for almost all $x \in X$;*
(iii) *$Q = \bigoplus_{1 \leq i \leq r} (Q \cap \psi_i(x))$ for almost all $x \in X$.*

Proof. For $a \in \mathbb{R}$ and $x \in X$, we set

$$V_a(x) = \bigoplus_{\chi_i \leq a} \psi_i(x), \quad W_a(x) = \bigoplus_{\chi_i \geq a} \psi_i(x),$$

$$V'_a(x) = \bigoplus_{\chi_j \leq a} \psi_j(x), \text{ and } W'_a(x) = \bigoplus_{\chi'_j \geq a} \psi_j(x).$$

It follows from Theorem 2.1(ii) that for almost all $x \in X$

$$p(V_a(x)) \subset V'_a(x) \text{ and } p(W_a(x)) \subset W'_a(x).$$

On the other hand,

$$\psi_i(x) = V_{\chi_i}(x) \cap W_{\chi_i}(x), \ \psi'_j(x) = V'_{\chi'_j}(x) \cap W'_{\chi'_j}(x), \text{ and}$$
$$V'_a(x) \cap W'_a(x) = \{0\},$$

if $a \notin A \overset{\text{def}}{=} \{\chi'_j \mid 1 \le j \le r'\}$.

Thus, for almost all $x \in X$

(1) $$p(\psi_i(x)) \subset \psi'_j(x), \text{ if } \chi_i = \chi'_j, \text{ and}$$

(2) $$p(\psi_i(x)) = 0, \text{ if } \chi_i \notin A.$$

So (2) implies (ii). Since $\mathscr{W} = \bigoplus_{1 \le i \le r} \psi_i(x)$ and $F = \bigoplus_{1 \le j \le r} \psi'_j(x)$, (i) is a consequence of (1) and (2). Assertion (iii) is a consequence of Theorem 2.1 applied to the restriction $u(m,x)|_Q$ of the cocycle u to Q. □

3. Definition and Fundamental Properties of Characteristic Maps

(3.0) In this section G denotes a unimodular σ-compact locally compact group, Γ a lattice in G, and $\pi\colon G \to \Gamma \setminus G$ the natural map. Let Λ be a subgroup of $\mathrm{Comm}_G(\Gamma)$ containing Γ and let $T\colon \Lambda \to \mathbf{GL}(\mathscr{W})$ be a representation of the group Λ on the space $\mathscr{W}$. We denote by $\Omega = \Omega_{\Gamma,T}$ the set of Borel maps $f\colon G \to \mathbf{GL}(\mathscr{W})$ with

(1) $$f(\gamma g) = f(g)T(\gamma^{-1}), \ \gamma \in \Gamma, \ g \in G.$$

For $f \in \Omega$ and $y, g \in G$, we set

(2) $$B'_f(y,g) = f(y)f(yg)^{-1}.$$

It follows from (1) and (2) that $B'_f(\gamma y, g) = B'_f(y,g)$ for all $f \in \Omega$, $\gamma \in \Gamma$, and $y, g \in G$. So for $f \in \Omega$, one can define

(3) $$B_f(x,g) = B'_f(\pi^{-1}(x), g), \ x \in \Gamma \setminus G, \ g \in G.$$

It follows from (2) that B'_f is a cocycle and hence B_f is a cocycle, i.e.

(4) $$B_f(x, g_1 g_2) = B_f(x, g_1) B_f(x g_1, g_2) \text{ for all } x \in \Gamma \setminus G, \ g_1, g_2 \in G.$$

For a compact subset $M \subset G$ and $x \in \Gamma \setminus G$ we set

(5) $$Q_{M,f}(x) = \sup_{g \in M} (\ln^+ \|B_f(x,g)\| + \ln^+ \|B_f(x,g)^{-1}\|).$$

We further set $\Psi = \Psi_{\Gamma,T} = \{f \in \Omega \mid$ for any compact set $M \subset G$, $Q_{M,f}(x) < \infty$ for almost all $x \in \Gamma \setminus G$ and the function $Q_{M,f}$ belongs to $L_1(\Gamma \setminus G, \mu_G)\}$.

We say that a representation T is *Γ-integrable* if $\Psi_{\Gamma,T} \neq \varnothing$.

As in IV.3.8 denote by $\mathfrak{f}(G,\Gamma)$ the family of all left Borel fundamental domains $X \subset G$ for Γ. For $X \in \mathfrak{f}(G,\Gamma)$, define the Borel map $\tau_X\colon G \to \Gamma$ via the relation $g \in \tau_X(g)X$, $g \in G$. We set

$$f_X(g) = T(\tau_X(g)^{-1}),\ g \in G.$$

Since $\tau_X(\gamma g) = \gamma\tau_X(g)$ for all $g \in G$ and $\gamma \in \Gamma$, we have $f_X \in \Omega_{\Gamma,T}$. A domain $X \in \mathfrak{f}(G,\Gamma)$ is called *(Γ,T)-integrable*, if $f_X \in \Psi_{\Gamma,T}$. If $y \in X$, then $\tau_X(y) = e$, and hence $B'_f(y,g) = T(\tau_X(yg))$ for all $g \in G$. The (Γ,T)-integrability condition for X is equivalent to $Q_{M,X} \in L_1(X,\mu_G)$ for every compact subset M, where

$$Q_{M,X}(y) \overset{\text{def}}{=} \sup_{g\in M}(\ln^+\|T(\tau_X(yg))\| + \ln^+\|T(\tau_X(yg))^{-1}\|),\ y \in X.$$

Remark. If $\Gamma \setminus G$ is compact, then T is Γ-integrable. Indeed, for a compact domain $X \in \mathfrak{f}(G,\Gamma)$ and every compact $M \subset G$ the set $\tau_X(X \cdot M)$ is finite, and hence the function $Q_{M,X}$ is bounded. Thus the domain X is (Γ,T)-integrable which proves that the representation T is Γ-integrable.

As before, we say that $s \in G$ acts ergodically on $\Gamma \setminus G$ if the transformation $x \mapsto xs$ of the space $\Gamma \setminus G$ is ergodic (relative to the measure μ_G). Notice that, since G is unimodular, this transformation preserves the measure μ_G.

We now fix an element $s \in G$ which acts ergodically on $\Gamma \setminus G$.

(3.1) Lemma. *If the representation T is Γ-integrable and $T(\Gamma) \subset \mathbf{SL}(\mathscr{W})$, then there exists a map $f \in \Psi_{\Gamma,T}$ such that $f(G) \subset \mathbf{SL}(\mathscr{W})$.*

Proof. Let $h \in \Psi$ and let $\vartheta\colon \mathbf{GL}(\mathscr{W}) \to \mathbf{SL}(\mathscr{W})$ be the map defined in Lemma 1.2. We set $f = \vartheta \circ h\colon G \to \mathbf{SL}(\mathscr{W})$. Since $T(\Gamma) \subset \mathbf{SL}(\mathscr{W})$, $h \in \Omega$, $\vartheta(gh) = \vartheta(g)h$ for all $g \in \mathbf{GL}(\mathscr{W})$ and $h \in \mathbf{SL}(\mathscr{W})$, it follows that $f \in \Omega$. On the other hand, h is Γ-integrable and the equalities (2), (3) in 3.0, and property (B) of the map ϑ imply that $\ln^+\|B_f(x,g)\| + \ln^+\|B_f(x,g)^{-1}\| \le 2(c + \ln^+\|B_h(x,g)\| + \ln^+\|B_h(x,g)^{-1}\|)$. Thus $f \in \Psi$. □

(3.2) Theorem. *Suppose that the representation T be Γ-integrable and $f \in \Psi_{\Gamma,T}$.*

(i) There exist $r \in \mathbb{N}^+$, $l_i \in \mathbb{N}^+$, $\chi_i \in \mathbb{R}$, and measurable maps ω_i: $G \to \mathbf{Gr}_{l_i}(\mathscr{W})$, $1 \le r \le n$, $1 \le i \le r$, with the following properties:

(a) *for almost all $g \in G$ the space $\mathscr{W}$ is the direct sum of the subspaces $\omega_i(g)$, $1 \le i \le r$;*
(b) *for almost all $g \in G$ and for each i, $1 \le i \le r$, the sequence $(1/m)\ln\|f(gs^m)w\|/\|w\|$ converges to χ_i, uniformly in $w \in \mathscr{W} - \{0\}$ as $m \to \pm\infty$;*
(c) *$\sum_{i=1}^r l_i\chi_i = (1/\mu_G(\Gamma \setminus G)) \cdot \int_G -\ln|\det B_f(x,s)|d\mu_G(x)$;*
(c′) *$\sum_{i=1}^r l_i\chi_i = 0$, if $T(\Gamma) \subset \mathbf{SL}(\mathscr{W})$;*
(d) *$\chi_i < \chi_j$, if $1 \le i < j \le r$.*

(ii) For almost all $g \in G$ and for each $w \in \mathscr{W} - \{0\}$ the following limits exist:

(1) $$\chi^+(w,g) = \lim_{m\to+\infty}(1/m)\ln\|f(gs^m)w\|$$

and

(2) $$\chi^-(w,g) = \lim_{m\to-\infty}(1/m)\ln\|f(gs^m)w\|.$$

Furthermore, for almost all $g \in G$ and for each $a \in \mathbb{R}$, we have

(3) $$\{0\} \cup \{w \in \mathscr{W} - \{0\} \mid \chi^+(w,g) \leq a\} = \bigoplus_{\chi_i \leq a} \omega_i(g)$$

and

(4) $$\{0\} \cup \{w \in \mathscr{W} - \{0\} \mid \chi^-(w,g) \geq a\} = \bigoplus_{\chi_i \geq a} \omega_i(g).$$

(iii) The numbers r, l_i, χ_i are uniquely defined by properties (a), (b), (c), and (d) and the maps ω_i are uniquely defined up to modification on a null set.

(iv) If f is replaced by a map $h \in \Psi$, the numbers r, l_i, and χ_i remain the same and the maps ω_i vary only on a null set.

Proof. We set

(5) $$u(m,x) = B_f(x,s^m)^{-1}, \ m \in \mathbb{Z}, \ x \in \Gamma \setminus G.$$

It then follows from equalities (2) and (3) in 3.0 that for all $g \in G$ and $m \in \mathbb{Z}$

(6) $$f(gs^m) = B'_f(g,s^m)^{-1}f(g) = u(m,\pi(g))f(g).$$

Since B_f is a cocycle, it follows that u is a cocycle with respect to the dynamical system $\{L^m\}$, where L is the automorphism on $\Gamma \setminus G$ which sends $x \in \Gamma \setminus G$ to xs. Since $f \in \Psi$, the cocycle u is integrable. We now apply Theorem 2.1 (Multiplicative ergodic theorem). For this, denote by χ_i, $1 \leq i \leq r$, the characteristic exponents of the cocycle u, by l_i the multiplicity of exponent χ_i, and by ψ_i the characteristic map defined by the cocycle u and the characteristic exponent χ_i. We define the maps $\omega_i\colon G \to \mathbf{Gr}_{l_i}(\mathscr{W})$ by setting

(7) $$\omega_i(g) = f(g)^{-1}\psi_i(\pi(g)), \ g \in G.$$

Since ψ_i, χ_i, and l_i have property (a), (b), (c), and (d) in the statement of Theorem 2.1, it follows from (6) and (7) that ω_i, χ_i, and l_i have property (a), (b), (c), and (d) of the present theorem. Assertion (ii) is a consequence of (6), (7), and (ii) of Theorem 2.1 and (iii) is a consequence of (ii).

We shall prove (iv). Since $f, h \in \Psi$, we have $f(\gamma g)h(\gamma g)^{-1} = f(g)h(g)^{-1}$ for all $\gamma \in \Gamma$ and $g \in G$. It then follows from Corollary I.4.6.3 that for almost all $g \in G$ the lower limits of the sequences

$$\{\|f(gs^m)h(gs^m)^{-1}\|/|m|\} \text{ and } \{\|h(gs^m)f(gs^m)^{-1}\|/|m|\}$$

equal zero both as $m \to +\infty$ and as $m \to -\infty$. Thus, for almost all $g \in G$, the following assertion holds: If for $w \in \mathscr{W} - \{0\}$ the following limits exist

$$\lim_{m\to+\infty}(1/m)\ln\|f(gs^m)w\| \text{ and } \lim_{m\to+\infty}(1/m)\ln\|h(gs^m)w\|,$$

(resp. the limits

$$\lim_{m\to-\infty}(1/m)\ln\|f(gs^m)w\| \text{ and } \lim_{m\to-\infty}(1/m)\ln\|h(gs^m)w\|),$$

then they coincide. This assertion and (ii) imply (iv). If now $f(G) \subset \mathbf{SL}(\mathscr{W})$, then $\det B_f(x,g) \equiv 1$. It then follows from (iv) that (c′) is a consequence of (c) and Lemma 3.1. □

We call the numbers χ_i the *characteristic exponents* of the pair (s, T) and the numbers l_i the *multiplicities* of the characteristic exponents χ_i. The maps ω_i will be called the *characteristic maps* defined by the pair (s, T) and the characteristic exponents χ_i.

(3.3) Theorem. *Suppose the representation T is Γ-integrable. Denote by χ_i, $1 \le i \le r$, the characteristic exponents of the pair (s, T) and by ω_i the characteristic map defined by the pair (s, T) and the characteristic exponent χ_i.*

(i) The map ω_i is Λ-equivariant, i.e., for each $\lambda \in \Lambda$ and for almost all $g \in G$

(1) $$\omega_i(\lambda g) = T(\lambda)\omega_i(g),\ 1 \le i \le r$$

(ii) For $p, p' \in G$ such that

$$\{s^{-m}ps^m \mid m \in \mathbb{N}^+\} \text{ and } \{s^m p' s^{-m} \mid m \in \mathbb{N}^+\}$$

are relatively compact in G, we have for almost all $g \in G$ and for all i, $1 \le i \le r$,

(2) $$\bigoplus_{\chi_j \le \chi_i} \omega_j(gp) = \bigoplus_{\chi_j \le \chi_i} \omega_j(g)$$

and

(3) $$\bigoplus_{\chi_j \ge \chi_i} \omega_j(gp') = \bigoplus_{\chi_j \ge \chi_i} \omega_j(g).$$

(iii) for each $z \in \mathscr{Z}_G(S)$

(4) $$\omega_i(gz) = \omega_i(g)$$

for almost all $g \in G$ and for each i, $1 \le i \le r$.

Proof. Let $f \in \Psi$ and $\lambda \in \Lambda$. We set

(5) $$h_\lambda(g) = f(\lambda g)T(\lambda)f(g)^{-1},\ g \in G.$$

If $\gamma \in \Gamma$ and $\lambda\gamma^{-1}\lambda^{-1} \in \Gamma$, i.e. if $\gamma \in \Gamma_\lambda \stackrel{\text{def}}{=} \Gamma \cap (\lambda^{-1}\Gamma\lambda)$, then, in view of (1) in 3.0, for each $g \in G$ we have

(6) $$\begin{aligned} h_\lambda(\gamma g) &= f(\lambda\gamma g)T(\lambda)f(\gamma g)^{-1} = f(\lambda\gamma\lambda^{-1}\lambda g)T(\lambda)f(\gamma g)^{-1} = \\ &= f(\lambda g)T(\lambda\gamma^{-1}\lambda^{-1})T(\lambda)T(\gamma)f(g)^{-1} = f(\lambda g)T(\lambda)f(g)^{-1} = h_\lambda(g). \end{aligned}$$

Since $\lambda \in \mathrm{Comm}_G(\Gamma)$, the subgroup Γ_λ is of finite index in the lattice Γ, and hence Γ_λ is a lattice in G. It then follows from Corollary I.4.6.3 and (6) that for almost all $g \in G$ the lower limits of the sequences

$$\{\|h_\lambda(gs^m)\|/|m|\} \text{ and } \{\|h_\lambda(gs^m)^{-1}\|/|m|\}$$

equal zero both as $m \to +\infty$ or as $m \to -\infty$. On the other hand, (5) implies that

$$h_\lambda(gs^m)f(gs^m)w = f(\lambda gs^m)T(\lambda)w$$

for all $g \in G$, $w \in W$, and $m \in \mathbb{Z}$. Thus, for almost all $g \in G$ and for each $w \in \mathscr{W} - \{0\}$ we have

$$\chi^+(w,g) = \chi^+(T(\lambda)w, \lambda g) \tag{7}$$

and

$$\chi^-(w,g) = \chi^-(T(\lambda)w, \lambda g), \tag{8}$$

where $\chi^+(w,g)$ and $\chi^-(w,g)$ were defined by the equalities (1) and (2) in 3.2. Clearly,

$$\omega_i(g) = (\bigoplus_{\chi_j \le \chi_i} \omega_j(g)) \cap (\bigoplus_{\chi_j \ge \chi_i} \omega_j(g)). \tag{9}$$

From (7), (8), and (9), making use of equalities (3) and (4) in 3.2, we deduce (i). We set $M = \{s^{-m}ps^m \mid m \in \mathbb{N}^+\}$ and define $Q_{M,f}(x)$, $x \in \Gamma \setminus G$, by equality (5) in 3.0. Since M is relatively compact and $f \in \Psi$, it follows that $Q_{M,f}(x) < \infty$ for almost all $x \in \Gamma \setminus G$. Thus, for almost all $x \in \Gamma \setminus G$ (see Corollary I.4.6.2),

$$\begin{aligned} &\lim_{m \to +\infty} \inf (1/m) \ln^+ \|B_f(xs^m, s^{-m}ps^m)\| \le \\ &\le \lim_{m \to +\infty} \inf (1/m) \ln^+ Q_{M,f}(xs^m) = 0. \end{aligned}$$

Hence, for almost all $g \in G$,

$$\lim_{m \to +\infty} \inf (1/m) \ln^+ \|B_f(gs^m, s^{-m}ps^m)\| = 0.$$

On the other hand, equality (2) in 3.0 implies that

$$f(gs^m) = B'_f(gs^m, s^{-m}ps^m)f(gps^m).$$

Thus $\chi^+(w,g) \le \chi^+(w,gp)$ for almost all $g \in G$ and for each $w \in \mathscr{W} - \{0\}$. Similarly, replacing p by p^{-1} we obtain that $\chi^+(w,g) \ge \chi^+(w,gp)$ for almost all $g \in G$ and for each $w \in \mathscr{W} - \{0\}$. Therefore $\chi^+(w,g) = \chi^+(w,gp)$ for almost all $g \in G$ and for each $w \in \mathscr{W} - \{0\}$. Due to (3) of 3.2, this proves (2). Equality (3) is proved analogously. Since $s^m z s^{-m} = z$ for all $z \in \mathscr{Z}_G(s)$ and $m \in \mathbb{Z}$, assertion (iii) is a consequence of (9) and (ii). □

(3.4) Proposition. *Let $Q \subset \mathscr{W}$ be a $T(\Lambda)$-invariant linear subspace of $\mathscr{W}$. We set $F = \mathscr{W}/Q$ and denote by $p\colon \mathscr{W} \to F$ the natural map. Let $T'\colon \Lambda \to$* $\mathbf{GL}(F)$

be the quotient representation $(T'(\lambda)p(w) = p(T(\lambda)w),\ \lambda \in \Lambda,\ w \in \mathscr{W})$. *Suppose the representation* T *(and hence* T'*) is* Γ*-integrable. Let* χ_i, $1 \le i \le r$ *(resp.* χ'_j, $1 \le j \le r'$*) be the characteristic exponents of the pair* (s, T) *(resp.* (s, T')*) and* ω_i *(resp.* ω'_j*) the characteristic maps defined by the pair* (s, T) *(resp.* (s, T')*) and the characteristic exponents* χ_i *(resp.* χ'_j*). Then*

(i) *for each* j, $1 \le j \le r'$, *there is* $i(j)$, $1 \le i(j) \le r$, *such that* $\chi'_j = \chi_{i(j)}$ *and* $\omega'_j(g) = p(\omega_{i(j)}(g))$ *for almost all* $g \in G$;

(ii) *for each* i *with* χ_i *different from every* χ'_j *we have* $\omega_i(g) \subset Q$ *for almost all* $g \in G$;

(iii) $Q = \bigoplus_{1 \le i \le r}(Q \cap \omega_i(g))$ *for almost all* $g \in G$.

This proposition can be deduced from Theorem 3.2 in the same way as Proposition 2.3 was deduced from the multiplicative ergodic theorem. Furthermore, this proposition is a consequence of Proposition 2.3 applied to the cocycle u which is defined by equality (5) in 3.2.

(3.5) Definition. *We say that a homomorphism* $f\colon D \to H$ *of a subgroup* $D \subset G$ *into a locally compact group* H almost extends *to a continuous homomorphism* $\tilde{f}\colon G \to H$, *if the map* $\rho\colon D \to H$, $\rho(d) = f(d)\tilde{f}(d^{-1})$, $d \in D$, *has the following properties:*

(a) $\rho(D)$ *is relatively compact in* H;

(b) $\rho(D)$ *and* $\tilde{f}(G)$ *commute (i.e.* $\rho(D) \subset \mathscr{Z}_H(\tilde{f}(G))$*).*

It can easily be checked that (b) implies

(c) ρ *is a homomorphism.*

(3.6) Proposition. *Suppose the restriction of the representation* T *to* Γ *almost extends to a continuous representation* $\tilde{T}\colon G \to \mathbf{GL}(\mathscr{W})$. *Denote by* $d_i \in \mathbb{R}$, $1 \le i \le r$, *the characteristic exponents of the transformation* $\tilde{T}(s^{-1})$, *arranged in increasing order. Let* $\mathscr{W}_i \subset \mathscr{W}$ *be the characteristic subspace of the transformation* $\tilde{T}(s^{-1})$ *corresponding to* d_i *(for the definition, see §1 of Chapter II). Then*

(a) *the representation* T *is* Γ*-integrable;*

(b) $\omega_i(g) = \tilde{T}(g)\mathscr{W}_i$ *and* $\chi_i = d_i$ *for almost all* $g \in G$ *and each* i, *where* χ_i *is the characteristic exponent of the pair* (s, T) *and* ω_i *is the characteristic map defined by the pair* (s, T) *and the characteristic exponent* χ_i, $1 \le i \le r$.

Proof. Choose a left Borel fundamental domain $X \subset G$ for Γ and, for $g \in G$, define (uniquely) $\tau(g) \in \Gamma$ via $\tau(g)g \in X$. We set $\rho(\lambda) = T(\lambda)\tilde{T}(\lambda^{-1})$, $\lambda \in \Lambda$, $\vartheta(g) = \rho(\tau(g))$ and $f(g) = \vartheta(g)\tilde{T}(g^{-1})$, $g \in G$. Since $T|_\Gamma$ almost extends to $\tilde{T}$ and $\tau(\gamma g) = \tau(g)\gamma^{-1}$ for all $g \in G$ and $\gamma \in \Gamma$, ϑ and ρ have the following properties:

(A) $\vartheta(\gamma g) = \vartheta(g)\rho(\gamma^{-1})$ for all $g \in G$ and $\gamma \in \Gamma$;

(B) $\vartheta(G) \subset \rho(\Gamma)$ is relatively compact in $\mathbf{GL}(\mathscr{W})$;

(C) $\rho(\Gamma)$ and $\tilde{T}(G)$ commute.

It follows from (A) and (C) that $f(\gamma g) = f(g)T(\gamma^{-1})$ for all $g \in G$ and $\gamma \in \Gamma$, i.e., $f \in \Omega_{\Gamma,T}$. On the other hand, since the representation $\tilde{T}$ is continuous and

$$B'_f(y,g) = f(y)f(yg)^{-1} = \vartheta(y)\tilde{T}(y^{-1})\tilde{T}(yg)\vartheta(yg)^{-1} = \\ = \vartheta(y)\tilde{T}(g)\vartheta(yg)^{-1}$$

for all $y, g \in G$, it follows from (B) that the set $\{B_f(x,g) \mid x \in \Gamma \setminus G,\ g \in M\}$ is relatively compact for every compact set $M \subset G$. Thus $f \in \Psi$, which proves (a). It follows from Lemma II.1.1(d) that for each $a \in \mathbb{R}$

(1) $$\{0\} \cup \{w \in \mathscr{W} - \{0\} \mid \lim_{m \to +\infty} \ln \|\tilde{T}(s^{-m})w\| \leq a\} = \bigoplus_{d_i \leq a} \mathscr{W}_i$$

and

(2) $$\{0\} \cup \{w \in \mathscr{W} - \{0\} \mid \lim_{m \to -\infty} \ln \|\tilde{T}(s^{-m})w\| \geq a\} = \bigoplus_{d_i \geq a} \mathscr{W}_i.$$

On the other hand, for all $g \in G$ and $m \in \mathbb{Z}$

(3) $$f(gs^m) = \vartheta(gs^m)\tilde{T}(s^{-m})\tilde{T}(g^{-1}).$$

Since $\vartheta(G)$ is relatively compact in $\mathbf{GL}(\mathscr{W})$, it follows from (1), (2), and (3) above and Theorem 3.2(ii) that for almost all $g \in G$ and for each $a \in \mathbb{R}$

(4) $$\bigoplus_{\chi_i \leq a} \tilde{T}(g^{-1})\omega_i(g) = \bigoplus_{d_i \leq a} \mathscr{W}_i$$

and

(5) $$\bigoplus_{\chi_i \geq a} \tilde{T}(g^{-1})\omega_i(g) = \bigoplus_{d_i \geq a} \mathscr{W}_i.$$

From (4) and (5), making use of equality (9) in 3.3, we deduce (b). □

4. Effective Pairs

In this section G, Γ, Λ, T, and s will denote the same objects as in §3. The algebraic closure of k will be denoted by K. The unique extension of the absolute value $|\ |$ to the absolute value on K will also be denoted by $|\ |$. We set $\mathscr{W}_K = \mathscr{W} \bigotimes_k K$. The Zariski closure of the subgroup $T(\Lambda)$ in $\mathbf{GL}(\mathscr{W}_K)$ will be denoted by $\overline{T(\Lambda)}$.

In 4.2 we define the concept of effectiveness of the pair (s, T). It turns out that the effectiveness of the pair (s, T) implies the existence of some specific measurable maps of the group G to vector spaces (see 4.3). Later on (see Propositions 4.7 and 4.8 of Chapter VII) we will show that, under certain additional assumptions, the existence of such maps implies that the representation T is extendable to a continuous representation of the group G. The above mentioned Propositions VII.4.7 and VII.4.8 will then be used in proving the superrigidity theorems.

The main result of this section is Theorem 4.12 which asserts that, under certain restrictions on the k-group $\mathbf{F}$ and the homomorphism $\tau\colon \Lambda \to \mathbf{F}(k)$, the pair $(s, \operatorname{Ad} \circ \tau)$ is effective.

(4.1) Given a k-rational action of a k-group $\mathbf{H}$ on a k-variety $\mathbf{M}$, we call a set $X \subset \mathbf{M}(k)$ *strictly* $\mathbf{H}$*-effective* if one of the following equivalent conditions is satisfied:

(i) $\mathbf{H}$ operates effectively on every orbit $\mathbf{H}x$, $x \in X$ (i.e. $x \in X$ and $h'(hx) = hx$ for all $h \in \mathbf{H}$ imply that $h' = e$);

(ii) for each $x \in X$ the stabilizer $\mathbf{H}_x \overset{\text{def}}{=} \{h \in \mathbf{H} \mid hx = x\}$ of the point x contains no non-trivial k-closed normal subgroups of $\mathbf{H}$.

The equivalence of (i) and (ii) is a consequence of the equality $\{h' \in H \mid h'(hx) = hx \text{ for all } h \in \mathbf{H}\} = \bigcap_{h \in H} h\mathbf{H}_x h^{-1}$ and the fact that the subgroup $\bigcap_{h \in \mathbf{H}} h\mathbf{H}_x h^{-1}$ is k-closed and normal in $\mathbf{H}$ for each $x \in \mathbf{M}(k)$.

Let Y be a measure space with measure μ. A measurable map $\Psi\colon Y \to \mathbf{M}(k)$ is said to be *strictly* $\mathbf{H}$*-effective* if there is a null set $Z \subset Y$ such that the set $\Psi(Y - Z)$ is strictly $\mathbf{H}$-effective.

In the sequel we shall use the following obvious observation: if ρ is a rational m-dimensional representation of $\mathbf{H}$ defined over k and there exists a measurable strictly $\mathbf{H}$-effective map of a measure space to k^m, then $\operatorname{Ker} \rho = \{e\}$.

(4.2) The pair (s, T) is called *effective* if the representation T is Γ-integrable and there exists i, $1 \le i \le r$, such that the characteristic map $\omega_i\colon G \to \mathbf{Gr}_{l_i}(\mathscr{W})$ is strictly $\overline{T(\Lambda)}$-effective. In other words, the pair (s, T) is effective if the representation T is Γ-integrable and the following equivalent conditions are satisfied

(a) there exists i, $1 \le i \le r$, such that $\overline{T(\Lambda)}$ acts effectively on the orbit $\overline{T(\Lambda)}\omega_i(g)$ for almost all $g \in G$;

(b) there exists i, $1 \le i \le r$, such that for almost all $g \in G$ the maximal k-closed normal subgroup of the group $\overline{T(\Lambda)}$ which is contained in $\{h \in \overline{T(\Lambda)} \mid h\omega_i(g) = \omega_i(g)\}$ is $\{e\}$.

In the proofs of the superrigidity theorems (Chapter VII) we will use the following

(4.3) Proposition. *Suppose the pair (s, T) is effective. Then there exist $m \in \mathbb{N}^+$, a faithful rational m-dimensional representation ρ of the group $\overline{T(\Lambda)}$ which is defined over k, and two measurable maps $\varphi\colon G \to k^m$ and $\varphi'\colon G/\mathscr{Z}_G(s) \to k^m$ such that*

(i) *the map φ and φ' are strictly $\overline{T(\Lambda)}$-effective with respect to the action of the group $\overline{T(\Lambda)}$ on k^m given by ρ.*

(ii) *for each $\lambda \in \Lambda$, $\varphi(\lambda g z) = \rho(T(\lambda))\varphi(g)$ for almost all $g \in G$ and for each $z \in \mathscr{Z}_G(s)$;*

(iii) *the map φ' is Λ-equivariant, i.e., $\varphi'(\lambda x) = \rho(T(\lambda))\varphi'(x)$ for each $\lambda \in \Lambda$ and almost all $x \in G/\mathscr{Z}_G(s)$.*

Proof. Since the pair (s, T) is effective, it follows from the definition that there exists i, $1 \le i \le r$, such that the characteristic map $\omega_i\colon G \to \mathbf{Gr}_{l_i}(\mathscr{W})$ is strictly $\overline{T(\Lambda)}$-effective. We set $\vartheta(g) = \bigoplus_{1 \le j \le r, j \ne i} \omega_j(g)$. Then $\mathscr{W} = \omega_i(g) \oplus \vartheta(g)$ for almost all $g \in G$. Denote by $\varphi(g) \in \operatorname{End}(\mathscr{W})$ the projector of the space $\mathscr{W}$ onto $\omega_i(g)$

which sends $\vartheta(g)$ to 0. Define a rational representation ρ over k of the group $\overline{T(\Lambda)} \subset \mathbf{GL}(\mathscr{W}_K)$ on $\operatorname{End}(\mathscr{W}_K)$ by setting $\rho(h)x = hxh^{-1}$, $h \in \overline{T(\Lambda)}$, $x \in \operatorname{End}(\mathscr{W}_K)$. According to assertions (i) and (iii) of Theorem 3.3, for all $\lambda \in \Lambda$, $z \in \mathscr{Z}_G(s)$, and j, $1 \leq j \leq r$,

$$\omega_j(\lambda g z) = T(\lambda)\omega_j(g)$$

for almost all $g \in G$. Hence, for each $\lambda \in \Lambda$ and each $z \in \mathscr{Z}_G(s)$,

$$\varphi(\lambda g z) = \rho(T(\lambda))\varphi(g)$$

for almost all $g \in G$. Thus, applying Lemma I.4.1.1(v) and altering φ on a null set, we can assume that φ has property (ii). If $\rho(h)\varphi(g) = \varphi(g)$, then $h\omega_i(g) = \omega_i(g)$, $h \in \overline{T(\Lambda)}$. But the map ω_i is strictly $\overline{T(\Lambda)}$-effective, and hence φ is also strictly $\overline{T(\Lambda)}$-effective. Therefore the map φ is the desired one. It now remains to observe that $\varphi(gz) = \varphi(g)$ for almost all $g \in G$ and for each $z \in \mathscr{Z}_G(s)$, and to set $\varphi'(g\mathscr{Z}_G(s)) = \varphi(g)$, $g \in G$. □

(4.4) Let B be a transformation in $\mathbf{GL}(\mathscr{W})$. As in §1 of Chapter II, we let $\Omega(B)$ denote the set of all eigenvalues of B, $\mathscr{W}_\lambda(B)$ the weight subspace corresponding to $\lambda \in \Omega(B)$, and $\mathscr{W}_d(B)$ the subspace $[\bigoplus_{\ln|\lambda|=d} \mathscr{W}_\lambda(B)]_k$. In particular,

$$\mathscr{W}_0(B) = [\bigoplus_{|\lambda|=1} \mathscr{W}_\lambda(B)]_k.$$

(4.5) Lemma. *Let H be a k-group and $h \in \mathbf{H}(k)$. Let $h_s \in \mathbf{H}(k)$ be the semisimple part of h and $\mathbf{M}$ the Zariski closure of the subgroup $\{h^m \mid m \in \mathbb{Z}\}$. The following conditions are equivalent*

(i) *the subgroup $\{h_s^m \mid m \in \mathbb{Z}\}$ is not relatively compact in $\mathbf{H}(k)$;*
(ii) *there is a (rational) character χ of the group $\mathbf{M}$ such that $|\chi(h)| \neq 1$.*

Proof. (i) ⇒ (ii). We may assume that $\mathbf{H}$ is a k-subgroup of $\mathbf{GL}(\mathscr{W}_K)$. Since the transformation h_s is diagonalizable and the subgroup $\{h_s^m \mid m \in \mathbb{Z}\}$ is not relatively compact, there exists $\lambda \in \Omega(h_s) = \Omega(h)$ such that the subgroup $\{\lambda^m \mid m \in \mathbb{Z}\}$ is not relatively compact in $K - \{0\}$. If $x \in K - \{0\}$ and $|x| = 1$, then from the description of local fields we see that the subgroup $\{x^m \mid m \in \mathbb{Z}\}$ is relatively compact. Thus $|\lambda| \neq 1$. Let z be an eigen-vector of h corresponding to the eigenvalue λ. Since $\mathbf{M}$ is the Zariski closure of the subgroup $\{h^m \mid m \in \mathbb{Z}\}$, z is an eigen-vector of any transformation $y \in \mathbf{M}$. We define the character χ of the group $\mathbf{M}$ by setting $\chi(y)z = y(z)$, $y \in \mathbf{M}$. Then $|\chi(h)| = |\lambda| \neq 1$.

(ii) ⇒ (i). Since every rational algebraic group morphism sends unipotent elements to unipotent ones, $\chi(h_s) = \chi(h)$. Thus the subgroup $\{\chi(h_s^m) \mid m \in \mathbb{Z}\}$ is not relatively compact in $K - \{0\}$, and hence the subgroup $\{h_s^m \mid m \in \mathbb{Z}\}$ is not relatively compact in $\mathbf{H}(k)$. □

(4.6) Definition. *Let H be a k-group. An element h of $\mathbf{H}(k)$ is called* essentially non-compact *if conditions (i) and (ii) in Lemma 4.5 are satisfied.*

(4.7) Lemma. *Let* $\mathbf{M}$ *be a commutative* k*-group and let* $x \in \mathbf{M}(k)$ *be essentially non-compact. Then there is a positive dimensional* k*-subgroup* $\mathbf{M}_x \subset \mathbf{M}$ *which consists of semisimple elements such that* $\rho(\mathbf{M}_x)w = w$ *for every* k*-rational representation* $\rho\colon \mathbf{M} \to \mathbf{GL}(\mathscr{W}_K)$ *defined over* k *and for each* $w \in \mathscr{W}_0(\rho(x))$.

Proof. The commutative k-group $\mathbf{M}$ can be represented in the form of a direct product of k-subgroups (see I.0.20 and I.0.22): $\mathbf{M} = C \times \mathbf{T} \times \mathbf{M}^{(u)}$, where $\mathbf{T}$ is a torus, C is a finite group, and $\mathbf{M}^{(u)} = \{y \in \mathbf{M} \mid y \text{ is unipotent}\}$. Since every rational homomorphism sends unipotent elements to unipotent ones, $\mathbf{M}^{(u)} \subset \operatorname{Ker}\chi$ for each $\chi \in \mathscr{X}(M)$, where as usual $\mathscr{X}(\mathbf{M})$ denotes the character group of the group $\mathbf{M}$. Thus $\mathscr{X}(\mathbf{M}) \cong \mathbb{Z}^r \times C$, where $r = \dim \mathbf{T}$. We set $A = \{\chi \in \mathscr{X}(M) \mid |\chi(x)| = 1\}$ and $\mathbf{M}_x = \{y \in \mathbf{T} \mid \chi(y) = 1 \text{ for all } \chi \in A\}$. Since x is essentially non-compact, the factor group $\mathscr{X}(\mathbf{M})/A$ is infinite. Thus $A \cong \mathbb{Z}^{r'} \times C'$, where $r' < r$ and $C' \subset C$. This implies that $\dim \mathbf{M}_x > 0$. Since A is invariant relative to the automorphisms of the field K over k, the subgroup $\mathbf{M}_x$ is k-closed. But any k-closed subgroup of a k-torus is defined over k. Thus $\mathbf{M}_x$ is a k-subgroup. Now let $\rho\colon \mathbf{M} \to \mathbf{GL}(\mathscr{W}_K)$ be a k-rational representation of the group $\mathbf{M}$. Since $\mathscr{W}_0(\rho(x)) = [\bigoplus_{|\lambda|=1} \mathbf{W}_\lambda(\rho(x))]_k$, any weight of the group $\mathbf{M}$ in the $\rho(\mathbf{M})$-invariant subspace $\mathscr{W}_0(\rho(x))$ is trivial on $\mathbf{M}_x$. But the group $\rho(\mathbf{T})$, as the image of the torus $\mathbf{T}$ under a rational morphism, is diagonalizable. Hence, $\rho(\mathbf{M}_x)w = w$ for all $w \in \mathscr{W}_0(\rho(x))$. □

(4.8) We say that a connected k-group $\mathbf{F}$ has *property (L) over* k, if $\{e\} \neq R_u(\mathbf{F}) \neq \mathbf{F}$, the unipotent radical $R_u(\mathbf{F})$ is defined over k and every non-trivial normal k-closed subgroup of $\mathbf{F}$ contains $R_u(\mathbf{F})$.

From now until the end of this section $\mathbf{F}$ will denote a connected k-group with property (L) over k. We set

$$\mathbf{H} = \mathbf{F}/R_u(\mathbf{F}),\ \mathfrak{F} = \operatorname{Lie}(\mathbf{F}),\ \mathfrak{R} = \operatorname{Lie}(R_u(\mathbf{F})),$$
$$\mathfrak{H} = \operatorname{Lie}(\mathbf{H}) = \mathfrak{F}/\mathfrak{R}.$$

Let $\sigma\colon \mathbf{F} \to \mathbf{H}$ and $\tilde{\sigma}\colon \mathfrak{F} \to \mathfrak{H}$ be the natural epimorphisms. Since $\mathbf{F}$ has property (L) over k and the commutator subgroup $\mathscr{D}(R_u(\mathbf{F}))$ of the nilpotent normal k-subgroup $R_u(\mathbf{F})$ is different from $R_u(\mathbf{F})$ and is a normal k-subgroup of $\mathbf{F}$ it follows that $R_u(\mathbf{F})$, and hence $\mathfrak{R}$, is commutative. Thus one can define the representations $\operatorname{Ad}_{\mathfrak{R}}$ and $\operatorname{ad}_{\mathfrak{R}}$ of the group $\mathbf{H}$ and the Lie algebra $\mathfrak{H}$ on the space $\mathfrak{R}$ by setting

$$\operatorname{Ad}_{\mathfrak{R}} \sigma(f)(r) = \operatorname{Ad} f(r),\ f \in \mathbf{F},\ r \in \mathfrak{R}:$$
$$\operatorname{ad}_{\mathfrak{R}} \tilde{\sigma}(f)(r) = \operatorname{ad} f(r),\ f \in \mathfrak{F},\ r \in \mathfrak{R}.$$

The representations $\operatorname{Ad}_{\mathfrak{R}}$ and $\operatorname{ad}_{\mathfrak{R}}$ should not be confused with the adjoint representations Ad and ad of the group $\mathbf{H}$. For $g \in \mathfrak{H}$ and $r \in \mathfrak{R}$, $\operatorname{ad}_{\mathfrak{R}} g(r)$ will also be denoted by $[g, r]$. For $A \subset \mathfrak{H}$ and $B \subset \mathfrak{R}$, we set $[A, B] = \{[g, r] \mid g \in A, r \in B\}$. We now set

$$\mathscr{Z}_{\mathbf{H}}(\mathfrak{R}) = \operatorname{Ker} \operatorname{Ad}_{\mathfrak{R}} = \{h \in \mathbf{H} \mid \operatorname{Ad}_{\mathfrak{R}} h(r) = r \text{ for all } r \in \mathfrak{R}\} \text{ and}$$
$$\mathscr{Z}_{\mathfrak{H}}(\mathfrak{R}) = \operatorname{Ker} \operatorname{ad}_{\mathfrak{R}} = \{g \in \mathfrak{H} \mid [g, r] = 0 \text{ for all } r \in \mathfrak{R}\}.$$

It is easily seen that

$$\mathscr{Z}_{\mathbf{H}}(\mathfrak{R}) = \sigma(\mathscr{Z}_{\mathbf{F}}(\mathfrak{R})) \text{ and } \mathscr{Z}_{\mathfrak{H}}(\mathfrak{R}) = \tilde{\sigma}(\mathscr{Z}_{\mathfrak{F}}(\mathfrak{R})). \tag{1}$$

As usual, the set of semisimple elements of $\mathfrak{H}$ will be denoted by $\mathfrak{H}^{(s)}$. The symbols $\mathfrak{F}_k$, $\mathfrak{R}_k$, and $\mathfrak{H}_k = \mathfrak{F}_k/\mathfrak{R}_k$ will denote, respectively, the set of k-points in the Lie algebras $\mathfrak{F}$, $\mathfrak{R}$, and $\mathfrak{H}$.

We use that notation until the end of this section.

(4.9) Lemma. *Let $h \in \mathbf{H}(k)$ and $d_1, d_2 \in \mathbb{R}$. Then*

$$[\mathscr{W}_{d_1}(\operatorname{Ad} h),\ \mathscr{W}_{d_2}(\operatorname{Ad}_{\mathfrak{R}} h)] \subset \mathscr{W}_{d_1+d_2}(\operatorname{Ad}_{\mathfrak{R}} h).$$

Proof. Pick $f \in \sigma^{-1}(h)$. Since $\operatorname{Ad} f$ is an automorphism of the Lie algebra $\mathfrak{F}$, it follows from [Bou 7], Chapter VII, §1, Proposition 12] that

$$[\mathscr{W}_{\lambda_1}(\operatorname{Ad} f),\ \mathscr{W}_{\lambda_2}(\operatorname{Ad} f)] \subset \mathscr{W}_{\lambda_1\lambda_2}(\operatorname{Ad} f) \text{ for all } \lambda_1, \lambda_2 \in \Omega(\operatorname{Ad} f).$$

On the other hand, $\sigma(f) = h$. Hence, $\sigma(\mathscr{W}_\lambda(\operatorname{Ad} f)) = \mathscr{W}_\lambda(\operatorname{Ad} h)$ and $\mathfrak{R} \cap \mathscr{W}_\lambda(\operatorname{Ad} f) = \mathscr{W}_\lambda(\operatorname{Ad}_{\mathfrak{R}} h)$. Thus

$$[\mathscr{W}_{\lambda_1}(\operatorname{Ad} h),\ \mathscr{W}_{\lambda_2}(\operatorname{Ad}_{\mathfrak{R}} h)] \subset \mathscr{W}_{\lambda_1\lambda_2}(\operatorname{Ad}_{\mathfrak{R}} h),$$

which, due to the equality $\mathscr{W}_d(B) = [\bigoplus_{\ln|\lambda|=d} \mathscr{W}_\lambda(B)]_k$, proves the lemma. □

(4.10) Lemma. *Let h be an essentially non-compact element of the group $\mathbf{H}(k)$. Assume $\mathfrak{H}^{(s)} \cap \mathscr{Z}_{\mathfrak{H}}(\mathfrak{R}) = \{0\}$. Then*

$$[\mathscr{W}_0(\operatorname{Ad} h), \mathfrak{R}_k] \not\subset \mathscr{W}_0(\operatorname{Ad}_{\mathfrak{R}} h).$$

Proof. Denote by $\mathbf{M}$ the Zariski closure of the subgroup $\{h^m \mid m \in \mathbb{Z}\}$. Since h is essentially non-compact, it follows from Lemma 4.7 that there exists a positive-dimensional k-subgroup $\mathbf{M}_h \subset \mathbf{M}$ consisting of semisimple elements such that $\operatorname{Ad}_{\mathfrak{R}} \mathbf{M}_h(w) = w$ for all $w \in \mathscr{W}_0(\operatorname{Ad}_{\mathfrak{R}} h)$. Then

$$[\mathfrak{M}_k, \mathscr{W}_0(\operatorname{Ad}_{\mathfrak{R}} h)] = \{0\}, \tag{1}$$

where $\mathfrak{M}_k = \operatorname{Lie}(\mathbf{M}_h)_k$. Since $\dim \mathbf{M}_h > 0$ and $\mathbf{M}_h$ consists of semisimple elements, $\mathfrak{M}_k \neq \{0\}$ and $\mathfrak{M}_k \subset \mathfrak{H}^{(s)}$. But $\mathfrak{H}^{(s)} \cap \mathscr{Z}_{\mathfrak{H}}(\mathfrak{R}) = \{0\}$. Hence $[\mathfrak{M}_k, \mathfrak{R}_k] \neq \{0\}$. It then follows from (1) and the equality $\mathfrak{R}_k = \bigoplus_{d\in\mathbb{R}} \mathscr{W}_d(\operatorname{Ad}_{\mathfrak{R}} h)$ that $[\mathfrak{M}_k, \mathscr{W}_d(\operatorname{Ad}_{\mathfrak{R}} h)] \neq \{0\}$ for some non-zero $d \in \mathbb{R}$.

On the other hand, $\mathbf{M}$ is commutative, and hence $\mathfrak{M}_k \subset \mathscr{W}_0(\operatorname{Ad} h)$. Thus $[\mathscr{W}_0(\operatorname{Ad} h), \mathscr{W}_d(\operatorname{Ad}_{\mathfrak{R}} h)] \neq \{0\}$. The lemma follows by applying Lemma 4.9 and the equality $\mathscr{W}_d(\operatorname{Ad}_{\mathfrak{R}} h) \cap \mathscr{W}_0(\operatorname{Ad}_{\mathfrak{R}} h) = \{0\}$. □

(4.11) Lemma. *Suppose* $\operatorname{char} k = 0$. *Then*

(i) $\mathscr{Z}_{\mathbf{F}}(R_u(\mathbf{F})) = R_u(\mathbf{F})$;

(ii) $\mathscr{Z}_{\mathbf{F}}(\mathfrak{R}) = R_u(\mathbf{F})$ *and* $\mathscr{Z}_{\mathfrak{F}}(\mathfrak{R}) = \mathfrak{R}$;

(iii) $\mathscr{Z}_{\mathbf{H}}(\mathfrak{R}) = \{e\}$ *and* $\mathscr{Z}_{\mathfrak{H}}(\mathfrak{R}) = \{0\}$;
(iv) $\mathscr{Z}(\mathbf{F}) = \{e\}$.

Proof. Let $\mathbf{D}$ denote the subgroup $\mathscr{Z}_{\mathbf{F}}(R_u(\mathbf{F}))$. Since the group $R_u(\mathbf{F})$ is commutative (see 4.8), $\mathbf{D} \supset R_u(\mathbf{F})$, and hence $R_u(\mathbf{D}) \supset R_u(\mathbf{F})$. On the other hand, since the subgroup $R_u(\mathbf{F})$ is normal in $\mathbf{F}$, it follows that $\mathbf{D}$, and hence $R_u(\mathbf{D})$, are normal in $\mathbf{F}$. Thus $R_u(\mathbf{D}) = R_u(\mathbf{F})$, and hence $R_u(\mathbf{D}) \subset \mathscr{Z}(\mathbf{D})$. This implies that there is a unique Levi k-subgroup $\mathbf{M}$ in $\mathbf{D}$ and $\mathbf{D} = \mathbf{M} \times R_u(\mathbf{D})$. Since $\mathbf{D}$ is normal in $\mathbf{F}$, the k-subgroup $\mathbf{M}$ is normal in $\mathbf{F}$. But $\mathbf{F}$ has property (L) over k and $\mathbf{M} \cap R_u(\mathbf{F}) = \{e\}$. So $\mathbf{M} = \{e\}$, which implies (i). Since $\operatorname{char} k = 0$, (ii) is a consequence of (i). From (i) and (ii), making use of equality (1) in 4.8, we deduce (iii). Since $R_u(\mathbf{F}) \neq \mathbf{F}$, (iv) is a consequence of (i). □

(4.12) Theorem. *Let $\tau\colon \Lambda \to \mathbf{F}(k)$ be a homomorphism with $\tau(\Lambda)$ Zariski dense in $\mathbf{F}$. Suppose the representation* $\operatorname{Ad} \circ \tau\colon \Lambda \to \mathbf{GL}(\mathfrak{F}_k)$ *is Γ-integrable and the restriction of the homomorphism $\sigma \circ \tau\colon \Lambda \to \mathbf{H}(k)$ to Γ almost extends (in the sense of 3.5) to a continuous homomorphism $\tilde{\tau}\colon G \to \mathbf{H}(k)$. Assume in addition that $\mathfrak{H}^{(s)} \cap \mathscr{Z}_{\mathfrak{H}}(\mathfrak{R}) = \{0\}$ and the element $\tilde{\tau}(s)$ is essentially non-compact. Then the pair* $(s, \operatorname{Ad} \circ \tau)$ *is effective.*

Proof. For $d \in \mathbb{R}$, we set

$$\mathscr{W}_d^{\mathfrak{H}} = \mathscr{W}_d(\operatorname{Ad} \tilde{\tau}(s)) \subset \mathfrak{H}_k \text{ and } \mathscr{W}_d^{\mathfrak{R}} = \mathscr{W}_d(\operatorname{Ad}_{\mathfrak{R}} \tilde{\tau}(s)) \subset \mathfrak{R}_k.$$

Since $\tilde{\tau}(s)$ is essentially non-compact, it follows from Lemma 4.10 that

$$[\mathscr{W}_0^{\mathfrak{H}}, \mathfrak{R}_k] \not\subset \mathscr{W}_0^{\mathfrak{R}}. \tag{1}$$

Since $\sigma \circ \tau|_\Gamma$ almost extends to $\tilde{\tau}$, $\operatorname{Ad} \circ \sigma \circ \tau|_\Gamma$ almost extends to $\operatorname{Ad} \circ \tilde{\tau}\colon G \to \mathbf{GL}(\mathfrak{H}_k)$. Thus zero is a characteristic exponent of the pair $(s, \operatorname{Ad} \circ \sigma \circ \tau)$ (see Proposition 3.6) and for almost all $g \in G$

$$\omega'(g) = \operatorname{Ad} \tilde{\tau}(g)(\mathscr{W}_0^{\mathfrak{H}}), \tag{2}$$

where $\omega'\colon G \to \mathbf{Gr}(\mathfrak{H}_k)$ is the characteristic map defined by the pair $(s, \operatorname{Ad} \circ \sigma \circ \tau)$ and the characteristic exponent 0. It is easily seen that $\operatorname{Ad} \circ \sigma \circ \tau$ is the representation induced from the representation $\operatorname{Ad} \circ \tau$ on $\mathfrak{H}_k = \mathfrak{F}_k/\mathfrak{R}_k$. Thus $\omega'(g) = \tilde{\sigma}(\omega(g))$ for almost all $g \in G$, where $\omega\colon G \to \mathbf{Gr}(\mathfrak{F}_k)$ is the characteristic map defined by the pair $(s, \operatorname{Ad} \circ \tau)$ and the characteristic exponent 0 (see Proposition 3.4(i)). It then follows from (2) that for almost all $g \in G$

$$\tilde{\sigma}(\omega(g)) = \operatorname{Ad} \tilde{\tau}(g)(\mathscr{W}_0^{\mathfrak{H}}). \tag{3}$$

Making use of Proposition 3.6, analogously we obtain from Proposition 3.4 (iii) that

$$\omega(g) \cap \mathfrak{R}_k = \operatorname{Ad}_{\mathfrak{R}} \tilde{\tau}(g)(\mathscr{W}_0^{\mathfrak{R}}) \tag{4}$$

for almost all $g \in G$. (Here we use the fact that $\operatorname{Ad}_{\mathfrak{R}} \circ \sigma \circ \tau$ is the restriction to $\mathfrak{R}$ of the representation $\operatorname{Ad} \circ \tau$). Since $\tilde{\sigma}(\operatorname{Ad} f(\mathfrak{f}) = \operatorname{Ad} \sigma(f)(\tilde{\sigma}(\mathfrak{f}))$ and $\operatorname{Ad} f(r) = \operatorname{Ad}_{\mathfrak{R}} \sigma(f)(r)$ for all $f \in \mathbf{F}$, $\mathfrak{f} \in \mathfrak{F}$, and $r \in \mathfrak{R}$, it follows that

$$[\operatorname{Ad} h(y), \operatorname{Ad}_{\mathfrak{R}} h(r)] = \operatorname{Ad}_{\mathfrak{R}} h([y, r])$$

for all $h \in \mathbf{H}$, $y \in \mathfrak{H}$, and $r \in \mathfrak{R}$. So (1), (3), and (4) imply that

$$(5) \qquad [\tilde{\sigma}(\omega(g)), \mathfrak{R}_k] \not\subset \omega(g) \cap \mathfrak{R}_k.$$

for almost all $g \in G$. Since $[\tilde{\sigma}(\omega(g)), \mathfrak{R}_k] = [\omega(g), \mathfrak{R}_k]$ and $[\mathfrak{H}_k, \mathfrak{R}_k] \subset \mathfrak{R}_k$, it follows from (5) that

$$[\omega(g), \mathfrak{R}_k] \not\subset \omega(g),$$

for almost all $g \in G$ and hence the subgroup $\mathbf{F}_g \overset{\text{def}}{=} \{f \in \mathbf{F} \mid \operatorname{Ad} f(\omega(g)) = \omega(g)\}$ does not contain $R_u(\mathbf{F})$. Denote by $\mathbf{F}'_g$ the maximal normal subgroup of the group $\mathbf{F}$ contained in $\mathbf{F}_g$. Since $\mathbf{F}_g$ is k-closed and does not contain $R_u(\mathbf{F})$, it follows that $\mathbf{F}'_g$ is k-closed. But $\mathbf{F}$ has property (L) over k. So $\mathbf{F}'_g = \{e\}$ for almost all $g \in G$, which proves the theorem. □

(4.13) *Remark.* By virtue of Lemma 4.11 the condition $\mathfrak{H}^{(s)} \cap \mathscr{Z}_{\mathfrak{H}}(\mathfrak{R}) = \{0\}$ is automatically satisfied in case $\operatorname{char} k = 0$.

5. Essential Pairs

In this section we use the notation introduced in 3.0. As in §3, s will denote an element of the group G which acts ergodically on $\Gamma \setminus G$. From now until the end of this section, the representation T is assumed to be Γ-integrable with $T(\Gamma) \subset \mathbf{SL}(\mathscr{W})$. Then one can take a map $f \in \Psi_{\Gamma,T}$ with $f(G) \subset \mathbf{SL}(\mathscr{W})$ (see Lemma 3.1). Let $\overline{T(\Lambda)}$ (resp. $\overline{T(\Gamma)}$) denote the Zariski closure of the subgroup $T(\Lambda)$ (resp. $T(\Gamma)$).

For the most part we will consider representations T with the following property:

(A) the subgroup $T(\Gamma)$ is not relatively compact in $\mathbf{GL}(\mathscr{W})$ (in the topology induced from that of k).

In 5.1 we introduce the concept of an essential pair (s, T) and clarify the relationships between that concept and the concept of an effective pair (s, T) which was studied in §4. The main purpose of this section is to demonstrate results of the following type: if T has property (A) and the group $\overline{T(\Gamma)}$ is semisimple, then, under certain restrictions on G and Γ, the element s can be chosen so that the pair (s, T) is essential.

(5.1) The pair (s, T) is called *essential* if it has at least two characteristic exponents, or equivalently if the multiplicity of any characteristic exponent of the pair (s, T) is strictly less than n. Since $T(\Gamma) \subset \mathbf{SL}(\mathscr{W})$, it follows from Theorem 3.2(i) (c′) that the pair (s, T) is essential if and only if at least one of its characteristic exponents is different from zero. If the pair (s, T) is essential, then it clearly has property (A).

Since for every linear subspace $\mathscr{W}' \subset \mathscr{W}$ the subgroup $\{h \in \overline{T(\Lambda)} \mid h\mathscr{W}' = \mathscr{W}'\}$ is k-closed, the property that the pair (s, T) is essential is equivalent to its effectiveness, provided that the representation T is irreducible over k and the k-group $\overline{T(\Lambda)}$ is k-simple.

(5.2) For $g \in G$, we set

(1) $$\eta(g) = \int_{\Gamma \backslash G} \ln \|B_f(x, g)^{-1}\| d\mu_G(x)$$

and

(2) $$\delta(g) = \lim_{m \to +\infty} \sup (1/m)\eta(g^m).$$

Since the representation T is Γ-integrable, the function η is locally bounded (i.e. bounded on every compact set). Since $f(G) \subset \mathbf{SL}(\mathscr{W})$, equalities (2) and (3) in 3.0 imply that $\det B_f(x, G) \equiv 1$ and hence $\ln \|B_f(x, g)^{-1}\| \geq 0$ for all $x \in \Gamma \setminus G$ and $g \in G$ (see Lemma 1.1). Thus $\eta(g) \geq 0$ and $\delta(g) \geq 0$. Since the characteristic exponents of the pair (s, T) coincide with those of the cocycle u defined by equation (5) in 3.2 (see the proof of Theorem 3.2), the following is a consequence of Proposition 3.2.

(5.3) Proposition. *If $\delta(s) > 0$, then the pair (s, T) is essential.*

(5.4) Definition. *A non-negative function h on a group G is called* semi-additive *if $h(g_1 g_2) \leq h(g_1) + h(g_2)$ for all $g_1, g_2 \in G$.*

(5.5) Lemma.
(i) *The function η is semi-additive.*
(ii) *If two elements g_1 and g_2 of the group G commute, then $\delta(g_1 g_2) \leq \delta(g_1) + \delta(g_2)$.*

Proof. (i) Equation (4) in 3.0 and the unimodularity of the group G imply that

$$\begin{aligned}
\eta(g_1 g_2) &= \int_{\Gamma \backslash G} \ln \|B_f(x, g_1 g_2)^{-1}\| d\mu_G(x) = \\
&= \int_{\Gamma \backslash G} \ln \|B_f(x g_1, g_2)^{-1} B_f(x, g_1)^{-1}\| d\mu_G(x) \leq \\
&\leq \int_{\Gamma \backslash G} \ln \|B_f(x g_1, g_2)^{-1}\| d\mu_G(x) + \int_{\Gamma \backslash G} \ln \|B_f(x, g_1)^{-1}\| d\mu_G(x) = \\
&= \mu(g_1) + \int_{\Gamma \backslash G} \ln \|B_f(x g_1, g_2)^{-1}\| d\mu_G(x g_1) = \eta(g_1) + \eta(g_2).
\end{aligned}$$

ii) If g_1 and g_2 commute, then (i) implies that $\eta((g_1 g_2)^m) \leq \eta(g_1^m) + \eta(g_2^m)$ for all $m \in \mathbb{Z}$, and hence $\delta(g_1 g_2) \leq \delta(g_1) + \delta(g_2)$. □

(5.6) Theorem. (On free discrete subgroups of linear groups.) *Let* **H** *be a semi-simple k-group and let Λ be a finitely generated subgroup of the group* $\mathbf{H}(k)$. *Suppose* *the subgroup Λ is Zariski dense in* $\mathbf{H}$ *and is not relatively compact in* $\mathbf{H}(k)$ *(in the topology induced from that of k). Then Λ contains a noncommutative free discrete subgroup.*

This theorem, was in fact established in [Ti 3], though it was not explicitly stated in that paper. In Appendix B we will show how it can be deduced from the assertions explicitly stated in [Ti 3].

(5.7) Denote by τ the quasi-regular unitary representation of the group G on the space $L_2(\Gamma \setminus G, \mu_G)$, i.e. $(\tau(g)f)(x) = f(xg)$, $x \in \Gamma \setminus G$, $g \in G$, $f \in L_2(\Gamma \setminus G, \mu_G)$. We set

$$L_2^0(\Gamma \setminus G) = \{f \in L_2(\Gamma \setminus G, \mu_G) \mid \int_{\Gamma \setminus G} f(x) d\mu_G(x) = 0\}.$$

A lattice Γ is called *weakly cocompact* if the restriction of the representation τ to $L_2^0(\Gamma \setminus G)$ is isolated from the trivial one-dimensional representation, i.e., if Γ is weakly cocompact in the sense of III.1.8. If the lattice Γ is cocompact, then in view of Corollary III.1.10, Γ is weakly cocompact.

(5.8) From now until the end of this section suppose that the group G is compactly generated. Let M be a compact subset generating G with $M = M^{-1}$ such that M contains a neighbourhood of the identity. We set $M^1 = M$ and $M^i = M^{i-1}M$, where $i \in \mathbb{N}^+$. For $g \in G$, we set

$$c_M(g) = \min\{i \in \mathbb{N} \mid g \in M^i\}.$$

The function c_M is semi-additive, locally bounded, and symmetric (i.e. $c_M(g^{-1}) = c_M(g)$ for all $g \in G$). Furthermore,

$$\sup_{g \in G}[f(g)/c_M(g)] = a(f) < \infty$$

for every locally bounded semi-additive function f on G, where $a(f) = \sup_{g \in M} f(g)$ (we leave the verification of these assertions to the reader). Thus, for any other compact subset $M' \subset G$ with the same properties as M, the functions c_M and $c_{M'}$ are equivalent in the sense that

$$\inf_{g \in G}[c_M(g)/c_{M'}(g)] > 0 \text{ and } \sup_{g \in G}[c_M(g)/c_{M'}(g)] < \infty.$$

(5.9) Proposition. *Suppose the subgroup $T(\Gamma)$ is not relatively compact in* $\mathbf{GL}(\mathscr{W})$, *the group $\overline{T(\Gamma)}$ is semisimple, and the lattice Γ is finitely generated and weakly cocompact. Then*

(1)
$$\limsup_{g \in G, c_M(g) \to \infty} [\eta(g)/c_M(g)] > 0.$$

Proof. We set $H = \mathbf{SL}(\mathscr{W})$. Let E be a regular representation of the group H on the space $L_2(H, \mu_H)$, i.e., $(E(h)f)(h') = f(h'h)$, $h, h' \in H$. Since H is unimodular, E

is unitary (see [Bou 3], Chapter VII, §3, Proposition 6). Since $T(\Gamma) \subset \mathbf{SL}(\mathscr{W})$, one can define a unitary representation U of the group Γ by setting $U(\gamma) = E(T(\gamma))$, $\gamma \in \Gamma$. Let ρ be the representation of the group G on the space $L_2(G, \Gamma, U)$ which is induced in the sense of Mackey from U. We recall that $L_2(G, \Gamma, U)$ consists of measurable vector valued functions j on G which take values in $L_2(H, \mu_H)$ and satisfy

$$j(\gamma g) = U(\gamma) j(g), \gamma \in \Gamma, g \in G, \tag{2}$$

with $\|j\|^2 \overset{\text{def}}{=} \int_X \|j(g)\|^2 d\mu_G(g) < \infty$, where $X \subset G$ is a left Borel fundamental domain for Γ. The inner product on $L_2(G, \Gamma, U)$ and the unitary representation ρ are defined by

$$\langle j_1, j_2 \rangle = \int_X \langle j_1(g), j_2(g) \rangle d\mu_G(x) \tag{3}$$

and

$$(\rho(g) j(g_1) = j(g_1 g), \ g, g_1 \in G. \tag{4}$$

Since the subgroup $\overline{T(\Gamma)}$ is semisimple and the subgroup $T(\Gamma)$ is finitely generated (because Γ is finitely generated) and is not relatively compact in $\mathbf{GL}(\mathscr{W})$, it follows from Theorem 5.6 on free discrete subgroups of linear groups that $T(\Gamma)$ contains a non-commutative free discrete subgroup Φ. Since Φ is discrete, the restriction of the representation E to Φ is a multiple of the regular representation of the group Φ. On the other hand, the regular representation of a non-commutative free group is isolated from the trivial one-dimensional representation (see Proposition III.4.3). But ρ is induced from the representation U, and the lattice Γ is weakly cocompact. So Proposition III.1.11(b) implies that ρ is isolated from the trivial one-dimensional representation, and hence there is a function $r \in \mathfrak{A}(G)$, such that $\|\rho(r)\| < 1$, where $\mathfrak{A}(G)$ is the set of non-negative continuous functions h on G with $\int_G h d\mu_G = 1$ (see Proposition III.1.3).

It follows from Lemma 1.3 that there exists $d > 0$, such that the function $b(h) = \|h\|^{-d}$, $h \in H$, belongs to $L_2(H, \mu_H)$. We set

$$\vartheta(g) = E(f(g)^{-1} b), g \in G, \tag{5}$$

where f is defined at the beginning of this section. Now $f(\gamma g)^{-1} = T(\gamma) f(g)^{-1}$ for all $g \in G$ and $\gamma \in \Gamma$ (because $f \in \Omega_{\Gamma,T}$), $b \in L_2(H, \mu_H)$, and $\mu_G(\Gamma \setminus G) < \infty$. Thus it follows that $\vartheta \in L_2(G, \Gamma, U)$. We set $r_1 = r$, $r_m = r_{m-1} * r$ ($m \in \mathbb{N}^+$), where $*$ denotes the convolution of functions (see I.5.3.1). Then

1) since $r \in \mathfrak{A}(G)$, $r_m \in \mathfrak{A}(G)$ for all $m \in \mathbb{N}^+$;
2) since $\rho(f_1 * f_2) = \rho(f_1)\rho(f_2)$ for all $f_1, f_2 \in L_1(G, \mu_G)$, $\rho(r_m) = \rho(r)^m$ for all $m \in \mathbb{N}^+$.

Thus

$$\inf_{g \in \operatorname{supp} r_m} \langle \rho(g)\vartheta, \vartheta \rangle \leq \|\rho(r)\|^m \|\vartheta\| \text{ for all } m \in \mathbb{N}^+,$$

and since $\|\rho(r)\| < 1$, we have

$$\lim_{m\to\infty} \sup[(1/m) \inf_{g\in \operatorname{supp} r_m} \ln\langle \rho(g)\vartheta, \vartheta\rangle] < 0 \tag{6}$$

(observe that since $b > 0$, (3), and (4) imply that $\langle \rho(g)\vartheta, \vartheta\rangle > 0$ for all $g \in G$). Since r has compact support, $\operatorname{supp} r_m \subset \operatorname{supp} r_{m-i} \operatorname{supp} r$, the function c_M is semi-additive and locally bounded, it follows that

$$\lim_{m\to\infty} \sup[1/m) \sup_{g\in \operatorname{supp} r_m} c_M(g)] < \infty.$$

We then deduce from (6) that

$$\limsup_{g\in G, c_M(g)\to\infty} [\ln\langle \rho(g)\vartheta, \vartheta\rangle / c_M(g)] < 0. \tag{7}$$

Since $b(h) = \|h\|^{-d}$ and $\|h'h\| \leq \|h'\|\,\|h\|$, $\langle E(h)b, b\rangle \geq \langle b, b\rangle \|h\|^{-d}$. It then follows from equation (2) in 3.0 and equation (5) that for all $y, g \in G$

$$\begin{aligned}\langle \vartheta(yg), \vartheta(y)\rangle &= \langle E(f(yg)^{-1})b, E(f(y)^{-1})b\rangle = \\ &= \langle E(f(y))E(f(yg)^{-1})b, b\rangle = \langle E(B'_f(y,g))b, b\rangle \geq \|B'_f(y,g)\|^{-d}\langle b, b\rangle.\end{aligned}$$

Thus, for each $g \in G$,

$$\begin{aligned}\langle \rho(g)\vartheta, \vartheta\rangle &= \int_X \langle \vartheta(yg), \vartheta(y)\rangle d\mu_G(y) \geq \\ &\geq \langle b, b\rangle \int_X \|B'_f(y,g)\|^{-d} d\mu_G(y) = \\ &= \langle b, b\rangle \int_{\Gamma\backslash G} \|B_f(x,g)\|^{-d} d\mu_G(x) = \\ &= \langle b, b\rangle \int_{\Gamma\backslash G} \|B_f(xg^{-1}, g)\|^{-d} \mu_G(xg^{-1}) = \\ &= \langle b, b\rangle \int_{\Gamma\backslash G} \|B_f(x, g^{-1})^{-1}\|^{-d} d\mu_G(x) \geq \\ &\geq \langle b, b\rangle \exp(-d \int_{\Gamma\backslash G} \ln \|B_f(x, g^{-1})^{-1}\| d\mu_G(x)) = \\ &= \langle b, b\rangle \exp(-d\eta(g^{-1})).\end{aligned} \tag{8}$$

The first equality follows from (3) and (4), the second one is a consequence of (3) in 3.0, the third one is obvious, the fourth one follows from (4) in 3.0 and the unimodularity of the group G, and the next one is a consequence of equation (1) in 5.2. The last inequality is a consequence of Jensen's integral inequality:

$$\int_Y \exp h d\mu \geq \exp \int_Y h d\mu,$$

where Y is a measure space with measure μ and h is an integrable function on Y. Taking the logarithm of both sides of inequality (8) we obtain that

$$\eta(g^{-1}) \leq (-1/d)[\ln\langle \rho(g)\vartheta, \vartheta\rangle + \ln\langle b, b\rangle].$$

From this inequality, inequality (7), and the symmetry of the function c_M we deduce (1). This completes the proof. □

(1) *Remark.* It is obvious that if G has property (T), then the lattice Γ is weakly cocompact. On the other hand, if G has property (T), then by Theorem III.2.12 the group Γ has property (T) as well, and hence Γ is finitely generated (see Theorem III.2.7). Therefore in case G has property (T) the condition "Γ is finitely generated and weakly cocompact" is automatically satisfied. Notice that the weak cocompactness of the lattice Γ and Theorem 5.6 have only been used to prove that the representation ρ is isolated from the trivial one-dimensional one. But if G has property (T), this is a consequence of the fact that the representation U (and hence ρ) does not contain the trivial one-dimensional representation (U does not contain the trivial one-dimensional representation, because the subgroup $T(\Gamma)$ is not relatively compact in $\mathbf{GL}(\mathscr{W})$).

(2) *Remark.* We observe that, since G is compactly generated, any cocompact lattice in G is finitely generated (see I.0.40).

(5.10) Definition. *We say that a subgroup $H \subset G$* quasi-generates *a subgroup $F \subset G$ if there exist a compact subset $M \subset G$ which generates G, $i \in \mathbb{N}^+$, and $a > 0$ such that every element $g \in F$ can be represented in the form*

$$g = h_1 m_1 \ldots h_i m_i, \text{ where } h_1, \ldots, h_i \in H,$$
$$m_1, \ldots, m_i \in M, \text{ and } c_M(h_j) < a c_M(g) \text{ for all } j,\ 1 \leq j \leq i.$$

From the semi-additivity and the local boundedness of the function c_M we deduce

(5.11) Lemma.

(a) *If H is a subgroup of the group G and $G = M'HM'$ for a compact set $M' \subset G$, then H quasi-generates G.*

(b) *Suppose the subgroup H quasi-generates the group G, u is a semi-additive function on G, and*

$$\limsup_{h \in H, c_M(h) \to \infty} [u(h)/c_M(h)] = 0.$$

Then

$$\limsup_{g \in G, c_M(g) \to \infty} [u(g)/c_M(g)] = 0.$$

(c) *Let G_1 and G_2 be two compactly generated locally compact groups, $H_1 \subset G_1$ a subgroup which quasi-generates G_1, and $H_2 \subset G_2$ a subgroup which quasi-generates G_2. Then $H_1 \times H_2$ quasi-generates $G_1 \times G_2$;*

(d) *Let F and H be two subgroups of the group G. If F quasi-generates G and H quasi-generates F, then H quasi-generates G. In particular, if F and H are closed, $H \subset F$, and the quotient space F/H is compact, then the condition that F quas-generates G implies that H quasi-generates G.*

(5.12) Proposition. *Let* $\mathbf{G}$ *be a connected semisimple k-group and* $\mathbf{S}$ *a maximal k-split torus in* $\mathbf{G}$*. Then*

(a) *there is a compact subset* $L \subset \mathbf{G}(k)$ *such that* $\mathbf{G}(k) = L \cdot \mathbf{S}(k) \cdot L$*, and*
(b) *the subgroup* $\mathbf{S}(k)$ *quasi-generates* $\mathbf{G}(k)$.

Proof. (a) Let $\tilde{\mathbf{G}}$ be the simply connected covering of the group $\mathbf{G}$, $\tilde{p}\colon \tilde{\mathbf{G}} \to \mathbf{G}$ the central k-isogeny, and $\tilde{\mathbf{S}}$ a maximal k-split torus in $\tilde{\mathbf{G}}$ with $\tilde{p}(\tilde{\mathbf{S}}) = \mathbf{S}$ (see I.1.4). It follows from the Cartan decomposition that $\tilde{\mathbf{G}}(k) = \tilde{L} \cdot \tilde{\mathbf{S}}(k) \cdot \tilde{L}$, where $\tilde{L}$ is a compact subgroup in $\tilde{\mathbf{G}}(k)$ (see Theorem I.2.2.1). On the other hand, the subgroup $\tilde{p}(\tilde{\mathbf{G}}(k))$ is closed and normal in $\mathbf{G}(k)$ and the factor group $\mathbf{G}(k)/\tilde{p}(\tilde{\mathbf{G}}(k))$ is compact (see Proposition I.2.3.4(i)). Hence, $\mathbf{G}(k) = \tilde{p}(\tilde{\mathbf{G}}(k)) \cdot A$ for a compact subset $A \subset \mathbf{G}(k)$. We then have

$$\mathbf{G}(k) = \tilde{p}(\tilde{L}) \cdot (\tilde{p}(\tilde{\mathbf{S}}(k))\tilde{p}(\tilde{L}) \cdot A \subset L \cdot \mathbf{S}(k) \cdot L,$$

where $L = \tilde{p}(\tilde{L}) \cup (\tilde{p}(\tilde{L}) \cdot A)$. This proves (a).
(b) is a consequence of (a) and Lemma 5.11(a). □

Remark. Assertion (b) can be proved without use of the Cartan decomposition. Indeed, choose a minimal parabolic k-subgroup $\mathbf{P}$ of $\mathbf{G}$ which contains $\mathbf{S}$. Making use of the existence of the $\mathbf{S}$-equivariant k-isomorphism $R_u(\mathbf{P}) \to \mathrm{Lie}(R_u(\mathbf{P}))$ (see Proposition I.1.3.3(i)), the cocompactness of the factor group $\mathscr{Z}_{\mathbf{G}}(\mathbf{S})(k)/\mathbf{S}(k)$ (see Proposition I.2.3.6), and the Levi decomposition $\mathbf{P} = R_u(\mathbf{P}) \rtimes \mathscr{Z}_{\mathbf{G}}(\mathbf{S})$ (see I.1.2), it is not hard to prove that $\mathbf{S}(k)$ quasi-generates $\mathbf{P}(k)$. Now, since $\mathbf{G}(k)/\mathbf{P}(k)$ is compact (see Corollary I.2.1.5), it follows at once that $\mathbf{S}(k)$ quasi-generates $\mathbf{G}(k)$.

(5.13) For $i \in \mathbb{N}^+$ and $z = (z_1, \ldots, z_i) \in \mathbb{Z}^i$, we set $\|z\| = \max_{1 \le j \le i} |z_j|$. A homomorphism $\varphi\colon \mathbb{Z}^i \to G$ is called *quasi-geodesic* if

$$\inf_{z \in \mathbb{Z}^i} [c_M(\varphi(z))/\|z\|] > 0.$$

Since the functions c_M and $c_{M'}$ are equivalent for any compact subset $M' \subset G$ with the same properties as M (see 5.8), the question of whether φ is quasi-geodesic does not depend on the choice of the compact set M. A finite set $\{g_1, \ldots, g_i\}$ of mutually commuting elements of the group G will be called *quasi-geodesic*, if the homomorphism $\mathbb{Z}^i \to G$ which sends $(z_1, \ldots, z_i) \in \mathbb{Z}^i$ to $g_1^{z_1} \cdot \ldots \cdot g_i^{z_i}$ is quasi-geodesic.

(5.14) Proposition. *Let* $i \in \mathbb{N}^+$*, and* $g_1, \ldots, g_i$ *be mutually commuting elements of the group* G*. Denote by* H *the commutative subgroup generated by the elements* $g_1, \ldots, g_i$*. Suppose the set* $\{g_1, \ldots, g_i\}$ *is quasi-geodesic and the subgroup* H *quasi-generates the group* G*. Assume in addition that the conditions of Proposition 5.9 are satisfied (i.e. the subgroup* $T(\Gamma)$ *is not relatively compact in* $\mathbf{GL}(\mathscr{W})$*, the group* $T(\Gamma)$ *is semisimple, and the lattice* Γ *is finitely generated and weakly cocompact). Then*

(a) *there exists* j, $1 \le j \le i$*, such that* $\delta(g_j) > 0$*, and*

(b) *if each of the elements* $g_1, \ldots, g_i$ *acts ergodically on* $\Gamma \setminus G$, *then for some* j, $1 \le j \le i$, *the pair* (g_j, T) *is essential.*

Proof. (a) Suppose the contrary. Then $\delta(g_j) = 0$, i.e.

$$(1) \qquad \limsup_{m \to +\infty} [1/m)\eta(g_j^m)] = 0$$

for each j, $1 \le j \le i$. As was observed in 5.2, we recall that $\eta(g) \ge 0$ and $\delta(g) \ge 0$ for all $g \in G$. Since the set $\{g_1, \ldots, g_i\}$ is quasi-geodesic and the function η is semi-additive (see Lemma 5.5), it follows from (1) that

$$(2) \qquad \lim_{h \in H, c_M(h) \to \infty} [\eta(h)/c_M(h)] = 0.$$

From (2) and the semi-additivity of the function η, making use of Lemma 5.11(b), we deduce that

$$(3) \qquad \limsup_{h \in H, c_M(h) \to \infty} [\eta(g)/c_M(g)] = 0.$$

Since the conditions of Proposition 5.9 are satisfied, η satisfies inequality (1) in 5.9. But, this contradicts (3).

(b) is a consequence of (a) and Proposition 5.3. □

(5.15) Theorem. *Let* A *be a finite non-empty set and, for each* $\alpha \in A$, *let* k_α *be a local field and* $\mathbf{G}_\alpha$ *a connected simply connected semisimple* k_α*-group which is isotropic and almost simple over* k_α. *Suppose* $G = \prod_{\alpha \in A} \mathbf{G}_\alpha(k_\alpha)$. *Let* Γ *be a finitely generated weakly cocompact lattice in* G *and let* $T \colon \Gamma \to \mathbf{GL}(\mathscr{W})$ *be a representation. Let* $\mathbf{S}_\alpha$ *be a maximal* k_α*-split torus of the group* $\mathbf{G}_\alpha$. *Set* $S = \prod_{\alpha \in A} \mathbf{S}_\alpha(k_\alpha)$. *Assume that the subgroup* $T(\Gamma)$ *is not relatively compact in* $\mathbf{GL}(\mathscr{W})$ *and that the group* $\overline{T(\Gamma)}$ *is semisimple (in other words, the conditions of Proposition 5.9 are satisfied). Then there exists an element* $s \in S$ *such that* s *acts ergodically on* $\Gamma \setminus G$ *and the pair* (s, T) *is essential.*

Proof. We first observe that, since the groups $\mathbf{G}_\alpha(k_\alpha)$ are compactly generated (see Corollary I.2.3.5), the group G is likewise compactly generated. Let $|\ |_\alpha$ be an absolute value on k_α and let $\mathrm{pr}_\alpha \colon G \to \mathbf{G}_\alpha(k_\alpha)$ be the natural projection. For $s \in S$, let $A(s) = \{\alpha \in A \mid |b(\mathrm{pr}_\alpha(s))|_\alpha \ne 1$ for at least one $b \in \Phi(\mathbf{S}_\alpha, \mathbf{G}_\alpha)\}$. Represent the k_α-split torus $\mathbf{S}_\alpha$ in the form of a direct product of one-dimensional k_α-split tori $\mathbf{S}_{\alpha i}$, $1 \le i \le \dim \mathbf{S}_\alpha$ and consider the tori $\mathbf{S}_{\alpha i}$ as naturally embedded subtori of $\mathbf{S}_\alpha$. Since the group $\mathbf{S}_{\alpha i}(k_\alpha)$ is isomorphic to the multiplicative group of the field k_α, there is an element $s_{\alpha i}$ of $\mathbf{S}_{\alpha i}(k_\alpha)$ such that the subgroup $\tilde{S}_{\alpha i} \stackrel{\text{def}}{=} \{s_{\alpha i}^m \mid m \in \mathbb{Z}\}$ is discrete and the factor group $\mathbf{S}_\alpha(k_\alpha)/\tilde{S}_{\alpha i}$ is compact. Let $\chi_{\alpha i}$ be the character of the torus $\mathbf{S}_\alpha$ which is non-trivial on $\mathbf{S}_{\alpha i}$ and trivial on $\mathbf{S}_{\alpha j}$ for each $j \ne i$. We then have

$$|\chi_{\alpha i}(s_{\alpha i})|_\alpha \ne 1 \text{ and } |\chi_{\alpha i}(s_{\alpha j})|_\alpha = 1 \text{ for } j \ne i.$$

Thus

(1) $$\max_{1\le i\le l_\alpha} |\ln|\chi_{\alpha i}(s_{\alpha 1}^{z_1}\dots s_{\alpha l_\alpha}^{z_{l_\alpha}})|_\alpha| \ge b \max_{1\le i\le l_\alpha}|z_i|$$

for all $z_1,\dots,z_{l_\alpha}\in\mathbb{Z}$, where $l_\alpha=\dim\mathbf{S}_\alpha$ and $b=\min_{1\le i\le l_\alpha}|\ln|\chi_{\alpha i}(s_{\alpha i})|_\alpha|>0$. We regard the k_α-group $\mathbf{G}_\alpha$ as a k_α-subgroup of $\mathbf{GL}_{n_\alpha}$. For $y=(y_1,\dots,y_{n_\alpha})\in k_\alpha^{n_\alpha}$ and $B\in\operatorname{End}(k_\alpha^{n_\alpha})$, define the norms $\|y\|=\sum_{1\le i\le n}|y_i|_\alpha$ and $\|B\|=\sup_{y\in k_\alpha^{n_\alpha}-\{0\}}\|B\|/\|y\|$. Since $\mathbf{G}_\alpha\subset\mathbf{SL}_{n_\alpha}$ (because $\mathbf{G}_\alpha$ is connected and semisimple), $\mathbf{S}_\alpha$ is Zariski closed in $\operatorname{End}(k_\alpha^{n_\alpha})$. Thus $\chi_{\alpha i}(\tilde{s})$ is a polynomial in the entries of the matrix $\tilde{s}\in\mathbf{S}_\alpha$. It follows that there exist positive constants c_1 and c_2 such that $|\ln|\chi_{\alpha i}(\tilde{s})|_\alpha|\le c_1\ln\|\tilde{s}\|+c_2$ for all $\tilde{s}\in S_\alpha(k_\alpha)$ and i, $1\le i\le l_\alpha$. We then obtain the following inequality from (1):

(2) $$\inf_{(z_1,\dots,z_{l_\alpha})\in\mathbb{Z}^{l_\alpha}-\{0\}}[\ln\|s_{\alpha 1}^{z_1}\cdot\ldots\cdot s_{\alpha l_\alpha}^{z_{l_\alpha}}\|/\max_{1\le i\le l_\alpha}|z_i|]>0.$$

For $g\in G$, set $\|g\|=\max_{\alpha\in A}\|\operatorname{pr}_\alpha(g)\|$. The function $g\mapsto\|g\|$ is locally bounded and semi-additive. Hence
$$\sup_{g\in G}[\|g\|/c_M(g)]<\infty$$
(see 5.8). Thus (2) implies that the set $\{s_{\alpha i}|\alpha\in A, 1\le i\le l_\alpha\}$ is quasi-geodesic. Now let $\tilde{S}$ denote the subgroup generated by this set. Since the factor groups $\mathbf{S}_\alpha(k_\alpha)/\tilde{S}_\alpha$ are compact, $S/\tilde{S}$ is compact. It then follows from Proposition 5.12(b) and Lemma 5.11(c) that S quasi-generates G. Therefore $\tilde{S}$ quasi-generates G (see Lemma 5.11(d)). Since the set $\{s_{\alpha i}|\alpha\in A, 1\le i\le l_\alpha\}$ is quasi-geodesic, it follows from Proposition 5.14(a) that $\delta(s_{\alpha_0 i_0})>0$ for some $\alpha_0\in A$ and i_0, $1\le i_0\le l_{\alpha_0}$. Choose an element $\tilde{s}\in\prod_{A-\{\alpha_0\}}\mathbf{S}_\alpha(k_\alpha)$ with $A(\tilde{s})=A-\{\alpha_0\}$ (as $\tilde{s}$ we can take $\prod_{A-\{\alpha_0\}}s_{\alpha 1}$). We then have that
$$A(\tilde{s}^{-1}s_{\alpha_0 i_0}^{-1})=A(\tilde{s}^2 s_{\alpha_0 i_0})=A.$$
But $\delta(s_{\alpha_0 i_0})>0$ and, by Lemma 5.5(ii), $\delta(s_{\alpha_0 i_0})\le\delta(s^{-1}s_{\alpha_0 i_0}^{-1})+\delta(\tilde{s}^2 s_{\alpha_0 i_0})$. Thus, defining $s\in S$ to be either $\tilde{s}^{-1}s_{\alpha_0 i_0}^{-1}$ or $\tilde{s}^2 s_{\alpha_0 i_0}$, we can assume that $A(s)=A$ and $\delta(s)>0$. Then s acts ergodically on $\Gamma\setminus G$ (see Corollary II.7.3(a)) and the pair (s,T) is essential (see Proposition 5.3). □

Remark. If G has property (T), then the condition "the lattice Γ is finitely generated and weakly cocompact" is automatically satisfied (see Remark 1 in 5.9). We recall that

1) $G=\prod_{\alpha\in A}\mathbf{G}_\alpha(k_\alpha)$ satisfies property (T) if and only if $\mathbf{G}_\alpha(k_\alpha)$, satisfies (T) for all $\alpha\in A$ (see Corollary III.2.10);
2) The following are the only cases where $\mathbf{G}_\alpha(k_\alpha)$ does not have property (T):
   (a) k_α is totally disconnected and $\operatorname{rank}_{k_\alpha}\mathbf{G}_\alpha=1$;
   (b) $k_\alpha=\mathbb{C}$ and $\mathbf{G}_\alpha$ is locally isomorphic to $\mathbf{SL}_2$;
   (c) $k_\alpha=\mathbb{R}$ and $\mathbf{G}_\alpha$ is locally isomorphic either to $\mathbf{SO}_{m,l}$ or $\mathbf{SU}_{m,l}$ (see Theorem III.5.6).

Chapter VI. Discrete Subgroups and Boundary Theory

In this chapter we present an approach to the proof of the existence of equivariant measurable maps which is different from that of the previous chapter. That approach is based on boundary theory and can be applied only to the study of irreducible representations, in contrast to the method presented in the previous chapter which is theoretically applicable to the study of arbitrary representations. On the other hand, this new method has a series of advantages, the main of which is that it is not necessary to assume that representations in question are Γ-integrable.

In Sections 1 and 2 we present some fundamental concepts and basic results of Furstenberg's boundary theory. In particular, we introduce the concepts of proximal, strongly proximal, and mean proximal G-spaces as well as the concepts of a boundary and a μ-boundary. In Section 3 we show that, under certain assumptions, projective G-spaces are strongly proximal and mean proximal. In Section 4, making use of the results of Section 3, we establish the existence of equivariant measurable maps.

In this chapter, as before, we denote by $\mathscr{K}(X)$ the space of continuous functions on a compact space X and by $\mathscr{P}(X)$ the set of regular Borel probability measures on X equipped with the weak*-topology. For $\mu \in \mathscr{P}(X)$ and $f \in \mathscr{K}(X)$, we set $\mu(f) = \int_X f(x) d\mu(x)$. As usual $\bar{B}$ denotes the closure of a subset B in a topological space.

1. Proximal G-Spaces and Boundaries

In the present section G denotes a locally compact second countable group and M denotes a compact metric G-space with distance function d. We refer to $\operatorname{diam} F = \sup\{d(x, y) \mid x, y \in F\}$ as the diameter of $F \subset M$. If $\operatorname{diam}(F_n \cup \{x\}) \to 0$ for $x \in M$, $F_n \subset M$, and $n \in \mathbb{N}^+$, we write $F_n \to x$. We set $\delta_M = \{\delta_x \mid x \in M\} \subset \mathscr{P}(M)$, where δ_x denotes as usual the normalized measure with support $\{x\}$. We recall that the G-structure on M induces a G-structure on $\mathscr{P}(M)$.

(1.1) We say that M is *minimal* if M does not contain a proper closed G-invariant subspace. This is obviously equivalent to the density of the orbit Gx in M for each $x \in M$. Clearly, if the space M is homogeneous, then it is minimal.

(1.2) A set $F \subset M$ is called *contractible* if there is a sequence $\{g_n\}_{n \in \mathbb{N}^+}$ in G such that $\operatorname{diam}(g_n F) \to 0$. The space M and the action of the group G on M are called *proximal* if every two-point subset of M is contractible. In other words, M is proximal if for any two points $x, y \in M$ there is a sequence $\{g_n\}_{n \in \mathbb{N}^+}$ in G such that $d(g_n x, g_n y) \to 0$. Since M is compact, M being proximal is equivalent to the following: for all $x, y \in M$, there exist $z \in M$ and a sequence $\{g_n\}_{n \in \mathbb{N}^+}$ in G such that $\lim_{n\to\infty} g_n x = \lim_{n\to\infty} g_n y = z$.

If k is a local field and $m \in \mathbb{N}^+$, then the projective space $\mathbf{P}_{m-1}(k) = \mathbf{P}(k^m)$ with the natural action of the group $G = \mathbf{GL}_m(k)$ is an example of a compact minimal proximal G-spaces.

(1.3) Lemma. *In a proximal space M, for every contractible $F \subset M$ and each $x \in M$, the set $F \cup \{x\}$ is contractible.*

Proof. Let $\{g_n\}_{n \in \mathbb{N}^+}$ be a sequence in G with $\operatorname{diam}(g_n F) \to 0$. Since M is compact, there exist $y, z \in M$ such that, by passing to a subsequence,

$$g_n F \to y \text{ and } g_n x \to z, \text{ as } n \to \infty. \tag{1}$$

Let $\{h_n\}_{n \in \mathbb{N}^+}$ be a sequence in G with $d(h_n y, h_n z) \to 0$. Then, since G acts continuously on M, for each $n \in \mathbb{N}^+$, one can choose neighbourhoods Y_n and Z_n of y and z respectively such that $\operatorname{diam}(h_n Y_n \cup h_n Z_n) \to 0$. On the other hand, in view of (1), we may assume by passing to a subsequence of $\{g_n\}$ that $g_n F \subset Y_n$ and $g_n x \in Z_n$. Thus $\operatorname{diam}(h_n g_n F \cup \{h_n g_n x\}) \to 0$, verifying the lemma. □

From the lemma above, by induction on $\operatorname{card} F$, we deduce

(1.4) Corollary. *If M is proximal and F is a finite subset of M, then F is contractible.*

(1.5) We say that the G-space M is *strongly proximal*, or that the action of G on M is *strongly proximal*, if for any probability measure $\mu \in \mathscr{P}(M)$ there exist a sequence $\{g_n\}_{n \in \mathbb{N}^+}$ in G and $x \in M$ such that $g_n \mu \to \delta_x$. We say that M is a *boundary* of the group G if M is minimal and strongly proximal.

If $x, y, z \in M$, $g_n \in G$, $n \in \mathbb{N}^+$, $\mu = (\delta_y + \delta_z)/2$, and $g_n \mu \to \delta_x$, then $g_n y \to x$ and $g_n z \to x$, and hence $d(g_n y, g_n z) \to 0$. Thus, if the space M is strongly proximal, then M is proximal. Making use of the compactness of M one can easily deduce from Corollary 1.4 that M is proximal if and only if $\overline{G\mu} \cap \delta_M \neq \varnothing$ for any discrete (i.e. supported on a countable set) measure $\mu \in \mathscr{P}(M)$. However, it is not true in general that if M is proximal, then M is strongly proximal. Nevertheless, we have the following

(1.6) Proposition.

(a) *If M is proximal and each point $x \in M$ has a contractible neighbourhood, then M is strongly proximal.*

(b) *If M is minimal, proximal, and contains a non-empty open contractible subset, then M is a boundary of the group G.*

Proof. (a) For $\mu \in \mathscr{P}(M)$, we set $\alpha(\mu) = \sup\{\mu(\{x\}) \mid x \in M\}$ and $\beta(\mu) = \sup\{\alpha(\nu) \mid \nu \in \overline{G\mu}\}$. Since M is compact, $\alpha(\mu) \geq \lim_{n\to\infty} \sup \alpha(\mu_n)$ for every sequence $\{\mu_n\}_{n\in\mathbb{N}^+}$ in $\mathscr{P}(M)$ converging to μ. It then follows from the compactness and metrizability of $\mathscr{P}(M)$ that for any $\mu \in \mathscr{P}(M)$ there is a measure $\nu \in \overline{G\mu}$ such that $\alpha(\nu) = \beta(\mu)$. On the other hand,

1) $\overline{G\nu} \subset \overline{G\mu}$ whenever $\nu \in \overline{G\mu}$;
2) $\alpha(\mu) = 1$ if and only if $\mu \in \delta_M$.

Therefore it suffices to show that

(1) $$\beta(\mu) > \alpha(\mu) \text{ for all } \mu \in \mathscr{P}(M) - \delta_M.$$

Since any point $x \in M$ has a contractible neighbourhood and M is compact, there exists a finite cover of the space M by open contractible subsets U_i, $1 \leq i \leq n$. Let now $\mu \in \mathscr{P}(M) - \delta_M$ and let x be a point in the space M with $\mu(\{x\}) = \alpha(\mu)$. Since $\bigcup_{1\leq i\leq n} U_i = M$, there exists i, $1 \leq i \leq n$, such that

(2) $$\mu(U_i \cup \{x\}) > \alpha(\mu).$$

Since U_i is contractible, it follows from Lemma 1.3 that the set $U_i \cup \{x\}$ is contractible. Let $\{g_n\}_{n\in\mathbb{N}^+}$ be a sequence in G with $\operatorname{diam}(g_n U_i \cup \{g_n x\}) \to 0$. Since the subspaces M and $\mathscr{P}(M)$ are compact and metrizable, by passing to a subsequence we may assume that $g_n\mu \to \nu$, $g_n U_i \to y$, and $g_n x \to y$, where $\nu \in \overline{G\mu}$ and $y \in M$. We then deduce from (2) that

$$\alpha(\nu) \geq \nu(\{y\}) \geq \lim_{n\to\infty} \sup(g_n\mu)(g_n U_i \cup \{g_n x\}) =$$
$$= \mu(U_i \cup \{x\}) > \alpha(\mu),$$

which implies (1).

(b) is a consequence of (a) and the fact that the set of points $x \in M$ having a contractible neighbourhood is open in M and G-invariant. □

(1.7) We shall state without proofs some facts on boundaries which are established in [Fu 5]. Although they will not be used in the sequel, it is helpful to keep them in mind while reading the present chapter.

1) If M' is a minimal G-space and M is a boundary of the group G, then
   (a) $f(M') \subset \delta_M$ for all (continuous) equivariant maps $f\colon M' \to \mathscr{P}(M)$;
   (b) there exists at most one equivariant map of M' to M.
2) If G is amenable, then any boundary of G is a singleton.
3) If G contains a closed amenable subgroup H with G/H compact, then any boundary of the group G is an equivariant image of the G-space G/H.
4) The notion of a boundary of G is not only definable for metrizable compact G-spaces, but also for arbitrary compact ones. Namely, a compact G-space X is said to be a boundary of the group G if $\overline{G\mu}$ contains point measures for all $\mu \in \mathscr{P}(X)$. A compact G-space X is said to be a *universal boundary* of the group G if X is a boundary of G and every boundary of G is an equivariant image of X. Every group G has a (unique up to isomorphism) universal boundary $B(G)$ which is not necessarily metrizable.

If G is a connected Lie group, then G contains a closed subgroup $H(G)$ with $B(G) = G/H(G)$. Furthermore, $H(G)$ is defined uniquely up to conjugacy. If G is a connected semisimple Lie group with finite center and Iwasawa decomposition $G = K \cdot A \cdot N$, then $H(G)$ is the normalizer in G of the solvable subgroup $A \cdot N$.

2. μ-Boundaries

Let G, M, diam F, and δ_M be as in Section 1. Since M is metrizable and compact, one can introduce a distance function on $\mathscr{P}(M)$ which is compatible with the weak*-topology. For distances between points, between points and sets, and between sets in M and $\mathscr{P}(M)$ we use the notation $d(.,.)$.

(2.1) Suppose that for each $n \in \mathbb{N}^+$ we are given a measure space Y_n with probability measure μ_n. We then denote by Ω the cartesian product $\bigoplus_{n=1}^{\infty} Y_n$, i.e., the set of all sequences $\{y_n\}_{n\in\mathbb{N}^+}$ with $y_n \in Y_n$ for all $n \in \mathbb{N}^+$. By Φ_i we mean the cartesian product $\bigoplus_{n=i+1}^{\infty} Y_n$ and by $\mathfrak{S}$ the standard σ-algebra of subsets of Ω generated by the sets of the form $A \times \Omega_i$, where $i \in \mathbb{N}^+$ and $A \subset Y_1 \times \ldots \times Y_i$ is measurable relative to the measure $\mu_1 \times \ldots \times \mu_i$. There is a unique measure τ on Ω which is defined on the σ-algebra $\mathfrak{S}$ such that for every measurable set E of the form $A \times \Omega_i$

$$\tau(E) = (\mu_1 \times \ldots \times \mu_i)(A)$$

(see [Hal 1], §38, Theorem 2). This measure is called the *product* of the measures μ_n and is denoted by $\tau = \bigoplus_{n=1}^{\infty} \mu_n$. With above notation, we now state a special case of the martingale convergence theorem (see [Lo], 29.3).

(2.2) Lemma. *For each $n \in \mathbb{N}^+$ let f_n be a non-negative function on $Y_1 \times \ldots \times Y_n$ which is integrable with respect to the measure $\mu_1 \times \ldots \times \mu_n$. Assume that*

$$f_n(y_1,\ldots,y_n) = \int_{Y_{n+1}} f_{n+1}(y_1,\ldots,y_n,y_{n+1})d\mu_{n+1}(y_{n+1})$$

for all $n \in \mathbb{N}^+$ and for almost all $(y_1,\ldots,y_n) \in Y_1 \times \ldots \times Y_n$. Then for almost all (relative to the measure τ) sequences $(y_1, y_2, \ldots) \in \Omega$, the limit $\lim_{n\to\infty} f_n(y_1,\ldots,y_n)$ *exists.*

(2.3) Let $\mu \in \mathscr{P}(G)$ and $\nu \in \mathscr{P}(M)$. For $g \in G$ and $x \in M$, we set $\alpha(g,x) = gx \in M$. The image of the measure $\mu \times \nu$ under the map α: $G \times M \to M$ is called the *convolution* of the measures μ and ν and is denoted by $\mu * \nu$. Clearly, $\mu * \nu \in \mathscr{P}(M)$. It follows from the definition of $\mu * \nu$ that

$$(\mu * \nu)(f) = \int_G (g\nu)(f)d\mu(g), \tag{1}$$

for every function $f \in \mathscr{K}(M)$. A measure $\nu \in \mathscr{P}(M)$ is called *μ-stationary* if $\mu * \nu = \nu$. The Schauder-Tychonoff fixed point theorem applied to the map

$\nu \mapsto \mu * \nu$, $\nu \in \mathscr{P}(M)$ implies that the set of μ-stationary measures $\nu \in \mathscr{P}(M)$ is not empty.

(2.4) Proposition. *Given $\mu \in \mathscr{P}(G)$ and $\nu \in \mathscr{P}(M)$ with $\mu * \nu = \nu$, set (with notation as in 2.1) $\Omega = \bigoplus_{n=1}^{\infty} G_n$ and $\tau = \bigoplus_{n=1}^{\infty} \mu_n$, where $G_n = G$ and $\mu_n = \mu$ is the measure on G_n. Then for almost all (in measure τ)sequences $(g_1, g_2, \ldots) \in \Omega$, the limit*

$$\lim_{n\to\infty} g_1 g_2 \ldots g_n \nu \tag{1}$$

exists.

Proof. The space $\mathscr{P}(M)$ equipped with weak*-topology and the space $\mathscr{K}(M)$ are separable (because M is metrizable and compact). Thus it suffices to show that if $f \in \mathscr{K}(M)$, then for almost all $(g_1, g_2, \ldots) \in \Omega$, the limit

$$\lim_{n\to\infty} (g_1 \ldots g_n \nu)(f). \tag{2}$$

exists. Since $\mu * \nu = \nu$, it follows from equality (1) in 2.3 that

$$\nu(f) = \int_G (g\nu)(f) d\mu(g),$$

and hence

$$(g_1, \ldots g_n \nu)(f) = \int_G (g_1 \ldots g_n g \nu)(f) d\mu(g).$$

Thus, the existence of limit (2) for almost all $(g_1, g_2, \ldots) \in \Omega$ is a consequence of Lemma 2.2 applied to the functions

$$f_n(g_1, \ldots, g_n) = (g_1 \ldots g_n \nu)(f).$$

□

(2.5) Proposition 2.4 can be restated in terms of probability theory as follows. Suppose $\mu \in \mathscr{P}(G)$ and $\{x_n\}_{n\in\mathbb{N}^+}$ is a sequence of independent identically distributed random variables with values in G and distribution μ. If a measure $\nu \in \mathscr{P}(M)$ is μ-stationary, then with probability 1 the limit $\lim_{n\to\infty} x_1 x_2 \ldots x_n \nu$ exists.

(2.6) Definition. *1) If $\mu \in \mathscr{P}(G)$, $\nu \in \mathscr{P}(M)$, and $\mu * \nu = \nu$, then the pair (M, ν) is called a* (G, μ)-space.

2) A (G, μ)-space (M, ν) is said to be a μ-boundary *if for almost all $(g_1, g_2, \ldots) \in \Omega$, the limit (1) in 2.4 belongs to δ_M. In probability language this definition goes as follows: a (G, μ)-space (M, ν) is called a μ-boundary if, with probability 1, $x_1, x_2 \ldots x_n \nu$ tends to a point measure, where x_n are the same random variables as in 2.5.*

3) Let $\mu \in \mathscr{P}(G)$. We say that a G-space M (and the action of G on M) is μ-proximal *if the (G, μ)-space (M, ν) is a μ-boundary for each μ-stationary measure $\nu \in \mathscr{P}(M)$.*

4) A G-space M (and the action of G on M) is called mean proximal *if it is μ-proximal for each measure $\mu \in \mathscr{P}(G)$ with* $\operatorname{supp} \mu = G$.

(2.7) *Remark.* In [Fu 5] another definition of a proximal G-space was given. In that reference a G-space M is called μ-proximal if for all $x, y \in M$ and $\varepsilon > 0$, $\mu_n\{g \in G \mid d(gx, gy) > \varepsilon\} \to 0$ as $n \to \infty$, where

$$\mu_n = (\mu + \mu^{(2)} + \ldots + \mu^{(n)})/n \text{ and } \mu^{(n)} = \mu * \mu * \ldots * \mu$$

is the n-tuple convolution of the measure μ. The equivalence of this definition of μ-proximality to the previous one was established in the same paper (Theorem 14.1).

(2.8) Lemma. *Let $\sigma \in \mathscr{P}(\mathscr{P}(M))$ be a probability measure on $\mathscr{P}(M)$, $x \in M$, and $\{g_n\}_{n \in \mathbb{N}^+}$ a sequence in G. We set*

$$\nu = \int_{\mathscr{P}(M)} y \, d\sigma(y) \in \mathscr{P}(M) \tag{1}$$

and suppose $g_n \nu \to \delta_x$ as $n \to \infty$. Then

$$\lim_{n \to \infty} \sigma\{y \in \mathscr{P}(M) \mid d(g_n y, \delta_x) > \varepsilon\} = 0 \tag{2}$$

for any $\varepsilon > 0$.

Proof. Since $g_n \nu = \int_{\mathscr{P}(M)} g_n y \, d\sigma(y)$ and $g_n \nu \to \delta_x$, for any $\varepsilon > 0$, $\lim_{n \to \infty} \sigma\{y \in \mathscr{P}(M) \mid (g_n y)(f) > \varepsilon\} = 0$, if $f \in \mathscr{K}(M)$, with $f \geq 0$, $f(x) = 0$. Since the distance function d is compatible with the weak*-topology on $\mathscr{P}(M)$, we obtain (2). □

(2.9) Proposition. *Let $\mu \in \mathscr{P}(G)$, and let $\sigma \in \mathscr{P}(\mathscr{P}(M))$ be a probability measure on $\mathscr{P}(M)$. Suppose σ is μ-stationary and the space M is μ-proximal. Then the support of σ is contained in δ_M.*

Proof. We define $\nu \in \mathscr{P}(M)$ by (1) in 2.8. Since $\mu * \sigma = \sigma$, we have $\mu * \nu = \nu$. But M is μ-proximal. Thus (M, ν) is a μ-boundary, i.e., for almost all (in measure τ) sequences $(g_1, g_2, \ldots) \in \Omega$, limit (1) in 2.4 belongs to δ_M (here and in the remainder of the proof, Ω and τ denote the same objects as in the statement of Proposition 2.4). It then follows from Lemma 2.8 that for any $\varepsilon > 0$ and for almost all $(g_1, g_2, \ldots) \in \Omega$,

$$\lim_{n \to \infty} \sigma\{y \in \mathscr{P}(M) \mid d(g_1 \ldots g_n y, \delta_M) > \varepsilon\} = 0.$$

This implies that for any $\varepsilon > 0$ the $\tau \times \sigma$ measure of the set

$$\{(g_1, g_2, \ldots) \in \Omega,\ y \in \mathscr{P}(M) \mid d(g_1 \ldots g_n y, \delta_M) > \varepsilon\}$$

tends to 0 as $n \to \infty$ (we have used implicitly Fubini's theorem and the Fatou convergence lemma). On the other hand, since $\mu * \sigma = \sigma$, for each $n \in \mathbb{N}^+$ the image of the measure $\tau \times \sigma$ under the map $\alpha_n \colon \Omega \times \mathscr{P}(M) \to \mathscr{P}(M)$, $\alpha_n((g_1, g_2, \ldots), y) =$

$g_1 \dots g_n y$, coincides with σ. Thus, $\sigma(\{y \in \mathscr{P}(M) \mid d(y, \delta_M) > \varepsilon\}) = 0$ for any $\varepsilon > 0$. This completes the proof. □

(2.10) Corollary. *Let M' be a locally compact Hausdorff G-space, $\mu \in \mathscr{P}(G)$, and $\nu' \in \mathscr{P}(M')$. Suppose that $\mu * \nu' = \nu'$ and the space M is μ-proximal.*

(a) *If φ: $M' \to \mathscr{P}(M)$ is a ν'-measurable G-equivariant map, then $\varphi(x) \in \delta_M$ for almost all (in measure ν') $x \in M'$:*

(b) *If φ_1, φ_2: $M' \to M$ are two measurable G-equivariant maps, then $\varphi_1(x) = \varphi_2(x)$ for almost all $x \in M'$.*

Proof. (a) Let $\varphi(\nu') \in \mathscr{P}(\mathscr{P}(M))$ be the image of the measure ν' under the map φ. Since $\mu * \nu' = \nu'$ and φ is G-equivariant, $\mu * \varphi(\nu') = \varphi(\nu')$. Thus (see Proposition 2.9) the support of the measure $\varphi(\nu')$ is contained in δ_M. But this is equivalent to $\varphi(x) \in \delta_M$ for almost all $x \in M'$.

(b) is a straightforward consequence of (a) applied to the map φ: $M' \to \mathscr{P}(M)$, $\varphi(x) = (\delta_{\varphi_1(x)} + \delta_{\varphi_2(x)})/2$, $x \in M'$. □

(2.11) We say that a set $C \subset G$ is *equicontinuous* on $U \subset M$ if, for every $\varepsilon > 0$, there is $\delta > 0$ such that $d(gx, gy) < \varepsilon$ for all $g \in C$ and $x, y \in U$ with $d(x, y) < \delta$.

(2.12) Lemma. *Let C be a subset of G, U an open subset of M, $x \in U$, and let $\{\nu_n\}_{n \in \mathbb{N}^+}$ be a sequence in $\mathscr{P}(M)$. Suppose C is equicontinuous on U and $\nu_n \to \delta_x$ as $n \to \infty$. Then*

(1) $$\lim_{n \to \infty} \sup_{g \in C} d(g\nu_n, \delta_M) = 0.$$

Proof. Since $x \in U$, U is open, and $\nu_n \to \delta_x$, there exist open neighbourhoods $W_n \subset U$ of x, $n \in \mathbb{N}^+$, such that as $n \to \infty$

(2) $$\operatorname{diam} W_n \to 0$$

and

(3) $$\nu_n(W_n) \to 1.$$

Since C is equicontinuous on U and $W_n \subset U$, it follows from (2) that

(4) $$\lim_{n \to \infty} \sup_{g \in C} \operatorname{diam}(gW_n) = 0.$$

From (3), (4), and the equality $g\nu_n(gW_n) = \nu_n(W_n)$ we see that, for every sequence $\{g_n\}_{n \in \mathbb{N}^+}$ in C, all the cluster points of the set $\{g_n \nu_n\}$ belong to δ_M. Since $\mathscr{P}(M)$ is compact, this implies (1).

(2.13) Proposition. *Suppose the following conditions are satisfied:*

(a) *M is strongly proximal;*

(b) *there exist $r \in \mathbb{N}^+$, subsets $C_1, \dots, C_r$ of the group G, and open subsets $U_1, \dots, U_r$ of the space M such that $\bigcup_{1 \le j \le r} C_j = G$, $G \cdot U_j = M$, and for each j, $1 \le j \le r$, the set C_j is equicontinuous on U_j.*

Then the G-space M is mean proximal.

Proof. Let $\mu \in \mathscr{P}(G)$ be a measure with $\operatorname{supp} \mu = G$ and $\nu \in \mathscr{P}(M)$ a μ-stationary measure. With Ω and τ as in Proposition 2.4 and G^l being the direct product of l copies of the group G, we set

$$\Omega_1 = \{\omega = (g_1, g_2, \dots) \in \Omega \mid \lim_{n \to \infty} g_1 g_2 \dots g_n \nu \text{ exists}\}$$

and

$$\Omega_2 = \{(g_1, g_2, \dots) \in \Omega \mid \text{ for each } l \in \mathbb{N}^+$$

the set $\{(g_{n+1}, g_{n+2}, \dots, g_{n+l}) \mid n \in \mathbb{N}^+\}$ is dense in $G^l\}$.

Since $\operatorname{supp} \mu = G$, for each $l \in \mathbb{N}^+$ and any non-empty open set $W \subset G^l$, the following is a consequence of the law of large numbers: for almost all (relative to the measure τ) sequences $(g_1, g_2, \dots) \in \Omega$ the set $\{n \in \mathbb{N}^+ \mid (g_{nl+1}, \dots, g_{nl+l}) \in W\}$ is infinite. But G is second countable. Hence $\tau(\Omega - \Omega_2) = 0$. On the other hand, by Proposition 2.4 $\tau(\Omega - \Omega_1) = 0$. Thus,

$$\tau(\Omega - (\Omega_1 \cap \Omega_2)) = 0. \tag{1}$$

Since $GU_j = M$ and M is compact, for each j, there is a finite subcover of the cover $\{gU_j \mid g \in G\}$ of M. It follows that there exists a finite set $D = \{d_1, \dots, d_1\} \subset G$ such that $DU_j = M$ for each j, $1 \le j \le r$.

Let us consider $\omega = (g_1, g_2, \dots) \in \Omega_1 \cap \Omega_2$. Since M is strongly proximal, there exist $h_k \in G$, $k \in \mathbb{N}^+$, and $x \in M$ such that

$$h_k \nu \to \delta_x \text{ as } k \to \infty. \tag{2}$$

Since $\omega \in \Omega_2$, one can find a subsequence $\{g_{n_k}\}_{k \in \mathbb{N}^+}$ of the sequence $\{g_n\}_{n \in \mathbb{N}^+}$ such that for any i, $1 \le i \le 1$,

$$g_{n_k+1} g_{n_k+2} \dots g_{n_k+i} h_k^{-1} \to d_i^{-1} \text{ as } k \to \infty. \tag{3}$$

By passing to a subsequence of $\{g_{n_k}\}$ we may assume that for some j, $1 \le j \le r$, and all $k \in \mathbb{N}^+$

$$g_1 g_2 \dots g_{n_k} \in C_j. \tag{4}$$

Since $DU_j = M$, there exists an integer i, $1 \le i \le l$, such that $d_i^{-1} x \in U_j$. It follows from (2) and (3) that

$$g_{n_k+1} g_{n_k+2} \dots g_{n_k+i} \nu \to \delta_{d_i^{-1} x} \text{ as } k \to \infty. \tag{5}$$

Since C_j is equicontinuous on U_j and $d_i^{-1} x \in U_j$, it follows from (4), (5), and Lemma 2.12 that

$$\lim_{k \to \infty} d(g_1 g_2 \dots g_{n_k} g_{n_k+1} g_{n_k+2} \dots g_{n_k+i} \nu, \delta_M) = 0. \tag{6}$$

This implies that for each $\omega \in \Omega_1 \cap \Omega_2$ the limit $\lim_{n\to\infty} g_1 g_2 \cdots g_n \nu$ belongs to δ_M. Thus, (1) implies that (M, ν) is a μ-boundary, completing the proof. □

From Propositions 2.13 and 1.6 we deduce

(2.14) Corollary. *Suppose condition (b) in Proposition 2.13 is satisfied and the G-space M is minimal, proximal, and contains a non-empty open contractible subset. Then M is mean proximal.*

3. Projective G-Spaces

In this section, as in Sections 1 and 2, G denotes a locally compact second countable group. Let k be a local field with absolute value $|\ |$, $m \in \mathbb{N}^+$, and $\mathscr{W}$ a vector space over k of dimension m. We equip $\mathscr{W}$, $\mathbf{GL}(\mathscr{W})$, $\operatorname{End}(\mathscr{W})$, and the projective space $\mathbf{P}(\mathscr{W})$ with the topology induced from that of the field k. Denote by π the natural projection of $\mathscr{W} - \{0\}$ onto $\mathbf{P}(\mathscr{W})$. Every $B \in \mathbf{GL}(\mathscr{W})$ induces the transformation $\pi B \pi^{-1}$ of the space $\mathbf{P}(\mathscr{W})$, which will be denoted by B_π. Let $T\colon G \to \mathbf{GL}(\mathscr{W})$ be a continuous representation of the group G on $\mathscr{W}$. Define $\mathbf{P}(\mathscr{W})$ to be a G-space and a $\mathbf{GL}(\mathscr{W})$-space by setting $gx = T(g)_\pi x$ and $Bx = B_\pi x$, $g \in G$, $B \in \mathbf{GL}(\mathscr{W})$, $x \in \mathbf{P}(\mathscr{W})$.

The space $\mathbf{P}(\mathscr{W})$ is compact and metrizable. Let d be a distance function on $\mathbf{P}(\mathscr{W})$ compatible with the topology of $\mathbf{P}(\mathscr{W})$. As in Section 1 we set $\delta_{\mathbf{P}(\mathscr{W})} = \{\delta_x \mid x \in \mathbf{P}(\mathscr{W})\}$ and denote by $\operatorname{diam} F = \sup\{d(x,y) \mid x, y \in F\}$ the diameter of the set $F \subset \mathbf{P}(\mathscr{W})$. We write $F_n \to x$ for $\operatorname{diam}(F_n \cup \{x\}) \to 0$.

Given a system of linear coordinates $y = (y_1, \ldots, y_m)$ in the space $\mathscr{W}$, we define the norms on $\mathscr{W}$ and $\operatorname{End}(\mathscr{W})$ by

$$\|w\| = \sum_{1 \le i \le m} |y_i(w)|,\ w \in \mathscr{W}, \text{ and}$$

$$\|A\| = \sup_{w \in \mathscr{W}, w \ne 0} \|Aw\|/\|w\|,\ A \in \operatorname{End}(\mathscr{W}).$$

(3.1) For $\varepsilon > 0$ we define Ψ_ε to be the family of subsets of the form $\{x \in \mathbf{P}(\mathscr{W}) \mid d(x, \mathbf{P}(\mathscr{W}')) > \varepsilon\}$, where $\mathscr{W}' \in \mathbf{Gr}_{m-1}(\mathscr{W})$ is a linear subspace of codimension 1. Clearly all the members of Ψ_ε are open.

(3.2) Lemma. *For all sufficiently small $\varepsilon > 0$ there exist $r \in \mathbb{N}^+$, $D_1, \ldots, D_r \subset \mathbf{GL}(\mathscr{W})$, and $U_1, \ldots, U_r \in \Psi_\varepsilon$ such that $\mathbf{GL}(\mathscr{W}) = \bigcup_{1\le j\le r} D_j$ and D_j is equicontinuous on U_j (in the sense of 2.11) for all j, $1 \le j \le r$.*

Proof. For each $B \in \operatorname{End}(\mathscr{W})$, $B \ne 0$, we define the transformation $B_\pi = \pi B \pi^{-1}$. This transformation is defined and continuous on $\mathbf{P}(\mathscr{W}) - \mathbf{P}(\operatorname{Ker} B)$. Choose a linear subspace $\mathscr{W}_B$ of codimension 1 containing $\operatorname{Ker} B$ and set $U_B = \{x \in \mathbf{P}(\mathscr{W}) \mid d(x, \mathbf{P}(\mathscr{W}_B)) > \varepsilon\} \in \Psi_\varepsilon$. Since B_π is continuous on $\mathbf{P}(\mathscr{W}) - \mathbf{P}(\operatorname{Ker} B)$ and U_B is relatively compact in $\mathbf{P}(\mathscr{W}) - \mathbf{P}(\operatorname{Ker} B)$, it follows that B_π is uniformly continuous on U_B. Thus, there is a neighbourhood Y_B of the transformation B in the space $\mathrm{End}(\mathscr{W})$ such that Y_B is equicontinuous on U_B. Now observe that $(\lambda B)_\pi = B_\pi$ for all $\lambda \in k - \{0\}$. Finally, denote the natural projection of $\mathrm{End}(\mathscr{W}) - \{0\}$ onto $\mathbf{P}(\mathrm{End}(\mathscr{W}))$ by π_0, choose from the open cover $\{\pi(Y_B) \mid B \in \mathrm{End}(\mathscr{W}), B \neq 0\}$ of the compact space $\mathbf{P}(\mathrm{End}(\mathscr{W}))$ a finite subcover $\{\pi_0(Y_{B_j}) \mid 1 \leq j \leq r\}$, and set $D_j = \mathbf{GL}(\mathscr{W}) \cap \pi_0^{-1}(\pi_0(Y_{B_j}))$ and $U_j = U_{B_j}$. □

(3.3) Lemma. *Let $(e_1, \dots, e_m)$ be a basis of $\mathscr{W}$. Then there exist neighbourhoods $Z_j \subset \mathscr{W}$ of e_j, $1 \leq j \leq m$, such that for every sequence $\{B_n\}_{n \in \mathbb{N}^+}$ in $\mathbf{GL}(\mathscr{W})$ with*

(1) $\mathrm{diam}\{\pi(B_n e_j) \mid 1 \leq j \leq m\} \to 0$ *as* $n \to \infty$,
it follows that

(2) $\inf_{1 \leq j \leq m} \mathrm{diam}\, \pi(B_n Z_j) \to 0$ *as* $n \to \infty$.

Proof. We may assume that $(e_1, \dots, e_m)$ is a basis of the coordinate system $y = (y_1, \dots, y_m)$. We set

$$Z = \{w \in \mathscr{W} \mid |y_i(w)| < 1/2m \text{ for all } i, 1 \leq i \leq m\}$$

and define for each j, $1 \leq j \leq m$, the sets $Z_j = e_j + Z \subset W$ and

$$C_j = \left\{ B \in \mathbf{GL}(\mathscr{W}) \;\middle|\; \|Be_j\| = \sup_{1 \leq i \leq m} \|Be_i\| \right\}.$$

Clearly, $\mathbf{GL}(\mathscr{W}) = \bigcup_{1 \leq j \leq m} C_j$. Thus, by partitioning the sequence $\{B_n\}_{n \in \mathbb{N}^+}$ into a finite number of subsequences, we may assume that the sequence $\{B_n\}$ is the one in C_j for some j, $1 \leq j \leq n$. It then follows from (1) that there exist $\lambda_{n,i} \in k$ $(n \in \mathbb{N}^+, 1 \leq i \leq m)$ such that

$$|\lambda_{n,i}| \leq 1 \tag{3}$$

and

$$\|B_n e_i - \lambda_{n,i} B_n e_j\| / \|B_n e_j\| \to 0 \text{ as } n \to \infty, \tag{4}$$

for all i, $1 \leq i \leq m$. From (4) we deduce that for every relatively compact subset $X \subset \mathscr{W}$

$$\sup_{w \in X} \|B_n w - \sum_{1 \leq i \leq m} \lambda_{n,i} y_i(w) B_n e_j\| / \|B_n e_j\| \to 0 \text{ as } n \to \infty. \tag{5}$$

It follows from (3) and the definition of the set Z that $|\sum_{1 \leq i \leq m} \lambda_{n,i} y_i(w)| < 1/2$ for all $w \in Z$. On the other hand, $Z_j = e_j + Z$. Thus, (5) implies that $\mathrm{diam}\, \pi(B_n Z_j) \to 0$ as $n \to \infty$. □

(3.4) Lemma. *If the representation T is irreducible, then there exists $\varepsilon > 0$ such that $GU = \mathbf{P}(\mathscr{W})$ for all $U \in \Psi_\varepsilon$.*

Proof. For $x \in \mathbf{P}(\mathscr{W})$ and $\mathscr{W}' \in \mathbf{Gr}_{m-1}(\mathscr{W})$, we denote by $f(x, \mathscr{W}')$ the Hausdorff distance between the sets Gx and $\mathbf{P}(\mathscr{W}')$. It is easily seen that the function f is lower semi-continuous. On the other hand, since T is irreducible, f is nowhere vanishing. Thus, the greatest lower bound of the function f on the compact space

$\mathbf{P}(\mathscr{W}) \times \mathbf{Gr}_{m-1}(\mathscr{W})$ is positive, i.e., there exists $\varepsilon > 0$ such that $f(x, \mathscr{W}') > \varepsilon$ for all $x \in \mathbf{P}(\mathscr{W})$ and $\mathscr{W}' \in \mathbf{Gr}_{m-1}(\mathscr{W})$. This is the desired ε. □

(3.5) Let $B \in \mathbf{GL}(\mathscr{W})$. As in Section 1 of Chapter II, we denote by $\mathscr{W}_a(B) \subset \mathscr{W}$ the characteristic subspace of the transformation B corresponding to a characteristic exponent a. Let a_0 be the maximal characteristic exponent of B. We say that B is *proximal* if the subspace $\mathscr{W}_{a_0}(B)$ is 1-dimensional, or equivalently, if $\pi(\mathscr{W}_{a_0}(B))$ is a singleton. From Proposition II.1.2 we deduce

(3.6) Lemma. *If $B \in \mathbf{GL}(\mathscr{W})$ is proximal, then there exist $x \in \mathbf{P}(\mathscr{W})$ and a linear subspace $\mathscr{W}' \subsetneqq \mathscr{W}$ such that B_π attracts $\mathbf{P}(\mathscr{W}) - \mathbf{P}(\mathscr{W}')$ towards x.*

(3.7) Theorem. *Let $\mathbf{H}$ be the Zariski closure of the subgroup $T(G)$ and $\mathbf{H}^0$ the (Zariski) connected component of the identity in $\mathbf{H}$. Suppose the following conditions are satisfied:*

(i) *there exists $g_0 \in G$ such that the transformation $T(g_0)$ is proximal;*
(ii) *the group $\mathbf{H}^0$ acts irreducibly on $\mathscr{W}$, i.e., $\mathscr{W}$ contains no non-trivial $\mathbf{H}^0$-invariant linear subspaces.*

Then

(a) *the action of the group G on $\mathbf{P}(\mathscr{W})$ is strongly proximal;*
(b) *the action of G on $\mathbf{P}(\mathscr{W})$ is mean proximal.*

Proof. (a) Since $T(g_0)$ is proximal, it follows from Lemma 3.6 that there exist $x \in \mathbf{P}(\mathscr{W})$ and a linear subspace $\mathscr{W}' \subsetneqq \mathscr{W}$ such that $T(g_0)$ attracts $\mathbf{P}(\mathscr{W}) - \mathbf{P}(\mathscr{W}')$ towards x. Let $y, z \in \mathbf{P}(\mathscr{W})$. Since $\mathbf{H}^0$ acts irreducibly on $\mathscr{W}$ and $\mathscr{W}' \neq \mathscr{W}$, it follows that

$$H_y \stackrel{\text{def}}{=} \{h \in \mathbf{H}^0 \mid hy \in \mathbf{P}(\mathscr{W}')\} \neq \mathbf{H}^0 \text{ and}$$
$$\mathbf{H}_z \stackrel{\text{def}}{=} \{h \in \mathbf{H}^0 \mid hz \in \mathbf{P}(\mathscr{W}')\} \neq \mathbf{H}^0.$$

But $\mathbf{H}^0$ is connected, $\mathbf{H}_y$ and $\mathbf{H}_z$ are algebraic subvarieties in $\mathbf{H}^0$, and the subgroup $T(G) \cap \mathbf{H}^0$ is Zariski dense in $\mathbf{H}^0$ (because $T(G)$ is Zariski dense in $\mathbf{H}$). Thus, there exists $g \in G$ such that $T(g)y \notin \mathbf{P}(\mathscr{W}')$ and $T(g)z \notin \mathbf{P}(\mathscr{W}')$. Since $T(g_0)$ attracts $\mathbf{P}(\mathscr{W}) - \mathbf{P}(\mathscr{W}')$ towards x, it follows that $T(g_0^n g)y \to x$ and $T(g_0^n g)z \to x$ as $n \to \infty$. Therefore the G-space $\mathbf{P}(\mathscr{W})$ is proximal.

Let $(e_1, \ldots, e_m)$ be a basis of the space $\mathscr{W}$. Since $\mathbf{P}(\mathscr{W})$ is proximal, it follows from Corollary 1.4 that

$$\operatorname{diam}\{\pi(T(g_n)e_i) \mid 1 \leq i \leq n\} \to 0 \text{ as } n \to \infty$$

for some sequence $\{g_n\}_{n \in \mathbf{N}^+}$ in G. It then follows from Lemma 3.3 that there exist open neighbourhoods Z_j of e_j such that

$$\inf_{1 \leq j \leq m} \operatorname{diam} \pi(T(g_n)Z_j) \to 0 \text{ as } n \to \infty.$$

It then follows that at least one of open sets $\pi(Z_j)$, $1 \leq j \leq m$, is contractible. Thus, we have proved that, for every basis $(e_1, \ldots e_m)$ of $\mathscr{W}$, at least one of the points $\pi(e_i)$, $1 \leq i \leq m$, has contractible neighbourhood. Therefore the set

$Y \overset{\text{def}}{=} \{y \in \mathscr{W} - \{0\} \mid \pi(y)$ has no a contractible neighbourhood$\}$ lies in a proper linear subspace of $\mathscr{W}$. But $T(G)$ acts irreducibly on $\mathscr{W}$ and the set Y is $T(G)$-invariant. Thus, $Y = \varnothing$. On the other hand, $\mathbf{P}(\mathscr{W})$ is proximal. Thus, Proposition 1.6(a) implies that the action of G on $\mathbf{P}(\mathscr{W})$ is strongly proximal.

(b) By Lemmas 3.2 and 3.4 there exist $r \in \mathbb{N}^+$, $D_1, \dots, D_r \subset \mathbf{GL}(\mathscr{W})$, and open subsets $U_1, \dots, U_r$ of $\mathbf{P}(\mathscr{W})$ such that $\mathbf{GL}(\mathscr{W}) = \bigcup_{1 \le j \le r} D_j$, $GU_j = \mathbf{P}(\mathscr{W})$, and D_j is equicontinuous on U_j for each j, $1 \le j \le r$. By setting $C_j = T^{-1}(D_j)$, we see that condition (b) of Proposition 2.13 is satisfied. Thus (see Proposition 2.13), (a) above implies that the action of G on $\mathbf{P}(\mathscr{W})$ is mean proximal, completing the proof. □

(3.8) *Remark 1.* Since $\mathbf{H}^0$ coincides with the intersection of all algebraic subgroups of finite index in $\mathbf{H}$ and the subgroup $\{h \in \mathbf{H} \mid h\mathscr{W}' = \mathscr{W}'\}$ is algebraic for any linear subspace $\mathscr{W}' \subset \mathscr{W}$, condition (ii) in Theorem 3.7 is equivalent to the condition that

(ii′) the restriction of the representation T to any subgroup of finite index in G is irreducible.

Remark 2. Let M be a closed G-invariant subset of $\mathbf{P}(\mathscr{W})$. It is not hard to show that if the representation T is irreducible, then conditions (i) and (ii) in Theorem 3.7 are necessary for both the strong and mean proximality of the G-space M. However, this is false for an arbitrary representation T. The action of the group $\{\begin{pmatrix} 1 & x \\ 0 & 1 \end{pmatrix} \mid x \in k\}$ on $\mathbf{P}(k^2)$ provides an elementary example of this.

From Corollary 2.10 and Theorem 3.7(b) we deduce

(3.9) Corollary. *Let $\mu \in \mathscr{P}(G)$ be a measure with* $\operatorname{supp} \mu = G$, *$M'$ a locally compact Hausdorff G-space, and $\nu' \in \mathscr{P}(M')$. Suppose that $\mu * \nu' = \nu'$ and conditions (i) and (ii) in Theorem 3.7 are satisfied.*

(a) *If $\varphi\colon M' \to \mathscr{P}(\mathbf{P}(\mathscr{W}))$ is a ν'-measurable G-equivariant map, then $\varphi(x) \in \delta_{\mathbf{P}(\mathscr{W})}$ for almost all (in measure ν') $x \in M'$;*

(b) *If $\varphi_1, \varphi_2\colon M' \to \mathbf{P}(\mathscr{W})$ are measurable G-equivariant maps, then $\varphi_1(x) = \varphi_2(x)$ for almost all $x \in M'$.*

In the next section we shall need the following

(3.10) Lemma. *Let $\mu \in \mathscr{P}(G)$ be a measure with* $\operatorname{supp} \mu = G$ *and $\nu \in \mathscr{P}(\mathbf{P}(\mathscr{W}))$. Suppose $\mu * \nu = \nu$ and condition (ii) of Theorem 3.7 is satisfied. Then for every proper linear subspace $\mathscr{W}'$ of $\mathscr{W}$, we have $\nu(\mathbf{P}(\mathscr{W}')) = 0$.*

Proof. We set

$$a(s) = \sup\{\nu(\mathbf{P}(\mathscr{W}')) \mid \mathscr{W}' \in \mathbf{Gr}_s(\mathscr{W})\}, 1 \le s \le \dim \mathscr{W}, \text{ and}$$
$$l = \min\{s \in \mathbb{N}^+ \mid a(s) > 0\}.$$

It follows from the definition of the number l that $\nu(\mathbf{P}(\mathscr{W}_1) \cap \mathbf{P}(\mathscr{W}_2)) = 0$ for all distinct $\mathscr{W}_1, \mathscr{W}_2 \in \mathbf{Gr}_l(\mathscr{W})$. Thus, for every $\varepsilon > 0$ the set $\{\mathscr{W}' \in \mathbf{Gr}_l(\mathscr{W}) \mid \nu(\mathbf{P}(\mathscr{W}')) > \varepsilon\}$ is finite, and hence there exists $\mathscr{W}' \in \mathbf{Gr}_l(\mathscr{W})$ such that $\nu(\mathbf{P}(\mathscr{W}')) = a(l)$. Since $\mu * \nu = \nu$, it follows that

$$\nu(\mathbf{P}(\mathscr{W}')) = \int_G \nu(\mathbf{P}(g^{-1}\mathscr{W}'))d\mu(g).$$

But $\nu(\mathbf{P}(\mathscr{W}')) = a(l)$ implies that $\nu(\mathbf{P}(g^{-1}\mathscr{W}')) \le \nu(\mathbf{P}(\mathscr{W}'))$ for all $g \in G$. Thus, $\nu(\mathbf{P}(g^{-1}\mathscr{W}')) = \nu(\mathbf{P}(\mathscr{W}'))$ for almost all (in measure μ) $g \in G$. On the other hand, the support of the measure μ equals G and the set $\{\mathscr{W}'' \in \mathbf{Gr}_l(\mathscr{W}) \mid \mu(\mathbf{P}(\mathscr{W}'')) = \mu(\mathbf{P}(\mathscr{W}')) = a(l) > 0\}$ is finite. Therefore the set $\{g\mathscr{W}' \mid g \in G\} \subset \mathbf{Gr}_l(\mathscr{W})$ is finite. This is equivalent to the finiteness of the index of the subgroup $\{g \in G \mid g\mathscr{W}' = \mathscr{W}'\}$ in G. But condition (ii) of Theorem 3.7 is equivalent to (ii′) in 3.8. Thus, $\mathscr{W}' = \mathscr{W}$, and hence $l = \dim \mathscr{W}$. This completes the proof. □

4. Equivariant Measurable Maps to Algebraic Varieties

As in Section 3 of Chapter II, suppose that we are given a finite non-empty set A. For each $\alpha \in A$, we choose a local field k_α and a connected non-trivial semisimple k_α-group $\mathbf{G}_\alpha$. Denote by G the locally compact group $\prod_{\alpha \in A} \mathbf{G}_\alpha(k_\alpha)$. As in §3 of Chapter II, we remark that G is compactly generated. For $\alpha \in A$, choose a maximal k_α-split torus $\mathbf{S}_\alpha$ in $\mathbf{G}_\alpha$ and the minimal parabolic k_α-subgroup $\mathbf{P}_\alpha$ containing $\mathbf{S}_\alpha$. We set $S = \prod_{\alpha \in A} \mathbf{S}_\alpha(k_\alpha)$ and $P = \prod_{\alpha \in A} \mathbf{P}_\alpha(k_\alpha)$. We remark that G/P is compact, because $\mathbf{G}_\alpha(k_\alpha)/\mathbf{P}_\alpha(k_\alpha)$ is compact for all $\alpha \in A$ (see Corollary I.2.1.5).

Let Γ be a lattice in G and Λ a subgroup of $\mathrm{Comm}_G(\Gamma)$ containing Γ.

(4.1) Proposition. *Let $\mu_0 = \psi_0 \cdot \mu_G \in \mathscr{P}(G)$, where ψ_0 is a continuous function on G with compact support such that the set $\{g \in G \mid \psi_0(g) > 0$ and $\psi_0(g^{-1}) > 0\}$ generates the group G. Let $\nu_0 \in \mathscr{P}(G/P)$ with $\mu_0 * \nu_0 = \nu_0$. Then there is a measure $\mu \in \mathscr{P}(\Gamma)$ with $\mathrm{supp}\,\mu = \Gamma$ such that $\mu * \nu_0 = \nu_0$.*

First we shall prove the following two lemmas.

(A) Lemma. *If $g_1, g_2 \in G$, then there exist $n \in \mathbb{N}^+$ and $\varepsilon > 0$ such that $g_1\mu_0^{(n)} > \varepsilon g_2\mu_0$, where $\mu^{(n)} = \mu * \mu * \ldots * \mu$ is the n-fold convolution of the measure μ.*

Proof. It follows from the conditions imposed on the measure μ_0 that for n sufficiently large, the density of measure $g_1\mu_0^{(n)}$ is positive on the support of the measure $g_2\mu_0$. This implies the assertion of the lemma, because the measure $g_2\mu_0$ has compact support and the densities of the measures $g_1\mu_0^{(n)}$ and $g_2\mu_0$ are continuous. □

(B) Lemma. *For* $g \in G$ *and* $\gamma \in \Gamma$ *there exist* $\varepsilon = \varepsilon(g,\gamma) > 0$ *and* $\omega = \omega(g,\gamma) \in \mathbf{P}(G)$ *such that* $g\nu_0 = \varepsilon\gamma\nu_0 + (1-\varepsilon)\omega * \nu_0$.

Proof. According to Lemma A, $g\mu_0^{(n)} > \varepsilon\gamma\mu_0$ for some $n \in \mathbb{N}^+$ and $\varepsilon > 0$. Since $\mu_0 * \nu_0 = \nu_0$ (and hence $\mu_0^{(n)} * \nu_0 = \nu_0$), it follows that

$$g\nu_0 = g\mu_0^{(n)} * \nu_0 = \varepsilon\gamma\mu_0 * \nu_0 + (g\mu_0^{(n)} - \varepsilon\gamma\mu_0) * \nu_0 = \\ = \varepsilon\gamma\nu_0 + (1-\varepsilon)\omega * \nu_0,$$

where $\omega = (1-\varepsilon)^{-1}(g\mu_0^{(n)} - \varepsilon\gamma\mu_0)$. This completes the proof. □

We now proceed to the proof of Proposition 4.1. For $g \in G$, we set $L(g) = \sup\{l \mid 0 \le l \le 1 \text{ and } g\nu_0 = l\mu' * \nu_0 + (1-l)\mu'' * \nu_0\}$, where $\mu' \in \mathbf{P}(\Gamma)$ with $\operatorname{supp}\mu' = \Gamma$ and $\mu'' \in P(G)$. It follows easily from Lemma B that $L(g) > 0$ for each $g \in G$ (the measure μ' should be selected in such a way that $\mu'(\{\gamma\}) < \varepsilon(g,\gamma)$ for all $\gamma \in \Gamma$). Clearly, $L(\gamma g) = L(g)$ for all $\gamma \in \Gamma$. From the equality

$$g\nu_0 = g\mu_0 * \nu_0 = \int_G g'\nu_0 d\mu_0(g')$$

we easily deduce that

$$L(g) \ge \int_G L(gg')d\mu_0(g').$$

By setting $L'(\Gamma g) = 1 - L(g)$, we obtain a function L' on $\Gamma \setminus G$ such that

$$L'(x) \le \int_G L'(xg)d\mu_0(g), \quad x \in \Gamma \setminus G. \tag{1}$$

Thus, by the Cauchy-Schwarz-Buniakowski inequality,

$$\int_{\Gamma\setminus G} L'(x)^2 d\mu_G(x) \le \int_{\Gamma\setminus G}\int_G L'(xg)^2 d\mu_0(g)d\mu_G(x) = \\ = \int_{\Gamma\setminus G} L'(x)^2\, d\mu_G(x),$$

the equality being attained if and only if $L'(xg) = L'(x)$ for almost all (relative to the measure $\mu_G * \mu_0$) pairs $(x,g) \in (\Gamma \setminus G) \times G$. But $\mu_0 = \psi_0\mu_G$ and $\operatorname{supp}\psi_0$ generates G. Thus the function $L'(x)$ is constant almost everywhere (relative to μ_G). It then follows from (1) and the positivity of the function $L(g)$ that there exists $\varepsilon > 0$ such that $L'(x) \le 1 - \varepsilon$ for all $x \in \Gamma \setminus G$ or, equivalently, $L(g) \ge \varepsilon$ for all $g \in G$. Choose a number $0 < l < \varepsilon$. Then, since $\mu * \nu_0 = \int_G g\nu_0 d\mu(g)$, for each measure $\sigma \in \mathscr{P}(G)$ there exist $\mu' \in \mathscr{P}(\Gamma)$ and $\mu'' \in \mathscr{P}(G)$ such that $\sigma * \nu_0 = l\mu' * \nu_0 + (1-l)\mu'' * \nu_0$ and $\operatorname{supp}\mu' = \Gamma$. Therefore

$$\nu_0 = l\mu_1 * \nu_0 + (1-l)[l\mu_2 * \nu_0 + (1-l)[l\mu_3 * \nu_0 + \dots ,$$

where $\mu_i \in \mathscr{P}(\Gamma)$ with $\operatorname{supp}\mu_i = \Gamma$. Since $l + (1-l)l + (1-l)^2 l + \ldots = 1$, the proof is complete. □

(4.2) Corollary. *There exist $\mu \in \mathscr{P}(\Gamma)$ and $\nu' \in \mathscr{P}(G/P)$ such that $\mu * \nu' = \nu'$, $\operatorname{supp} \mu = \Gamma$, and the measure ν' is quasi-invariant (i.e. ν' is equivalent to $\mu_{G/P}$).*

Proof. Since G/P is compact, there is a continuous function ψ_0 with compact support on G such that the set $Y \stackrel{\text{def}}{=} \{g \in G \mid \psi_0(g) > 0 \text{ and } \psi_0(g^{-1}) > 0\}$ generates the group G and $Yx = G/P$ for all $x \in G/P$. We set $\mu_0 = \psi_0 \mu_G$. Then the measure $\mu_0 * \nu$ is equivalent to $\mu_{G/P}$ for all $\nu \in \mathscr{P}(G/P)$. On the other hand, since G/P is compact, there exists a measure $\nu' \in \mathscr{P}(G/P)$ such that $\mu_0 * \nu' = \nu'$ (see 2.3). Thus, the desired assertion is a consequence of Proposition 4.1. □

(4.3) Theorem. *Let k be a local field, $m \in \mathbb{N}^+$, $\mathscr{W}$ a linear space over k of dimension m, and let $T\colon \Lambda \to \mathbf{GL}(\mathscr{W})$ be a representation of the group Λ on the space $\mathscr{W}$. Denote by $\mathbf{H}$ the Zariski closure of the subgroup $T(\Gamma)$ in $\mathbf{GL}(\mathscr{W})$ and by $\mathbf{H}^0$ the identity component of $\mathbf{H}$. Suppose the following conditions are satisfied:*

(i) *there exists $\gamma_0 \in \Gamma$ such that $T(\gamma_0)$ is proximal (in the sense of 3.5);*
(ii) *$\mathbf{H}^0$ operates irreducibly on $\mathscr{W}$.*
Then the following assertions hold.

(a) *There exists a Γ-equivariant measurable (with respect to $\mu_{G/P}$) map $\psi\colon G/P \to \mathbf{P}(\mathscr{W})$. The map ψ is defined uniquely (up to a modification on a null set) and Λ-equivariant (i.e. for each $\lambda \in \Lambda$ the equality $\psi(\lambda x) = T(\lambda)\psi(x)$ holds for almost all $x \in G/P$).*
(b) *Set $\mathscr{R} = \{x \in \operatorname{End}(\mathscr{W}) \mid x^2 = x \text{ and } \dim \operatorname{Im} x = 1\}$ and define $\mathscr{R}$, equipped with the topology induced from that of k, to be a Λ-space, by setting $\lambda x = T(\lambda)x\, T(\lambda^{-1})$, $\lambda \in \Lambda$, $x \in \mathscr{R}$.*
   1) *There exists a μ_G-measurable map $\varphi\colon G \to \mathscr{R}$ such that for each $\lambda \in \Lambda$ we have $\varphi(\lambda g z) = T(\lambda)\varphi(g)$ for almost all $g \in G$ and for all $z \in \mathscr{Z}_G(S)$.*
   2) *There exists a $\mu_{G/\mathscr{Z}_G(S)}$-measurable Λ-equivariant map $\varphi'\colon G/\mathscr{Z}_G(S) \to \mathscr{R}$.*

Proof. (a) Let $\mu \in \mathscr{P}(\Gamma)$ and $\nu' \in \mathscr{P}(G/P)$ be the measures defined in Corollary 4.2. Since ν' is equivalent to $\mu_{G/P}$, by Theorem IV.4.5 there is a Γ-equivariant ν'-measurable map $\psi_1\colon G/P \to \mathscr{P}(\mathbf{P}(\mathscr{W}))$. It follows from Corollary 3.9(a), that $\psi_1(x) \in \delta_{\mathbf{P}(\mathscr{W})}$ for almost all $x \in G/P$. Then, by setting $\psi_1(x) = \delta_{\psi(x)}$, we define a measurable Γ-equivariant map $\psi\colon G/P \to \mathbf{P}(\mathscr{W})$. The uniqueness of ψ is a consequence of Corollary 4.9(b). Now let $\lambda \in \Lambda$. We set

$$\psi_\lambda(x) = T(\lambda^{-1})\psi(\lambda x), \quad x \in G/P. \tag{1}$$

If $\lambda\gamma\lambda^{-1} \in \Gamma$, then the Γ-equivariance of the map ψ implies that

$$\begin{aligned}\psi_\lambda(\gamma x) &= T(\lambda^{-1})\psi(\lambda\gamma x) = T(\lambda^{-1})\psi(\lambda\gamma\lambda^{-1}\lambda x) = T(\lambda^{-1})T(\lambda\gamma\lambda^{-1})\psi(\lambda x) = \\ &= T(\gamma)T(\lambda^{-1})\psi(\lambda x) = T(\gamma)\psi_\lambda(x)\end{aligned}$$

for almost all $x \in G/P$. Hence, the map ψ_λ is $\lambda^{-1}\Gamma\lambda$-equivariant. We set $\Gamma_\lambda = \Gamma \cap (\lambda^{-1}\Gamma\lambda)$. Since $\lambda \in \operatorname{Comm}_G(\Gamma)$, it follows that Γ_λ is of finite index in the lattice Γ, and hence, Γ_λ is a lattice in G. Thus, from what we have proved above, it follows that any two Γ_λ-equivariant measurable maps of the space G/P into

$\mathbf{P}(\mathscr{W})$ agree almost everywhere. But the maps ψ and ψ_γ are Γ_λ-equivariant. Hence, $\psi(x) = \psi_\lambda(x)$ for almost all $x \in G/P$ and so, in view of (1), ψ is Λ-equivariant.

(b) Let $\mathscr{W}^*$ denote the dual space of $\mathscr{W}$. We define the conjugate (contragradient) representation T^* of T of the group on the space $\mathscr{W}$ by setting $T^*(\lambda) = T(\lambda^{-1})^*$, $\lambda \in \Lambda$, where $*$ denotes the passage to the contragradient transformation. We denote by $(\mathbf{H}^*)^0$ the identity component of the Zariski closure of the group $T^*(\Gamma)$. For $B \in \mathbf{GL}(\mathscr{W})$ the map $B \to (B^{-1})^*$, is regular. Thus $(\mathbf{H}^*)^0 = (\mathbf{H}^0)^* \overset{\text{def}}{=} \{h^* \mid h \in \mathbf{H}^0\}$. But, since $\mathbf{H}^0$ acts irreducibly on $\mathscr{W}$, so does $(\mathbf{H}^0)^*$ on $\mathscr{W}^*$. On the other hand, since the transformation $T(\gamma_0)$ is proximal, so is $T^*(\gamma_0^{-1}) = T(\gamma_0)^*$. Thus, (a) implies the existence of a Λ-equivariant measurable map $\psi^*\colon G/P \to \mathbf{P}(\mathscr{W}^*)$.

Let $q\colon G \to G/P$ be the natural map. We define two measurable maps $\omega\colon G \to \mathbf{P}(\mathscr{W})$ and $\omega^*\colon G \to \mathbf{P}(\mathscr{W}^*)$, by setting $\omega = \psi \circ q$ and $\omega^* = \psi^* \circ q$. Since ψ and ψ^* are Λ-equivariant, for each $\lambda \in \Lambda$ we have

$$\omega(\lambda g p) = T(\lambda)\omega(g) \tag{2}$$

and

$$\omega^*(\lambda g p) = T^*(\lambda)\omega^*(g) \tag{3}$$

for all $p \in P$ and for almost all (relative to μ_G) $g \in G$. We denote by $\pi^*\colon \mathscr{W}^* - \{0\} \to \mathbf{P}(\mathscr{W}^*)$ the natural projection and set $Y = \{(y, y^*) \in \mathbf{P}(\mathscr{W}) \times \mathbf{P}(\mathscr{W}^*) \mid w^*(w) \neq 0$ if $\pi(w) = y$ and $\pi^*(w^*) = y^*\}$. If $w \in \mathscr{W}$, $w^* \in \mathscr{W}^*$, and $w^*(w) \neq 0$, then one can define a 1-dimensional projection $R'(w, w^*) \in \mathscr{R}$, by setting

$$R'(w, w^*)v = \frac{w^*(v)}{w^*(w)} w, \quad v \in \mathscr{W}.$$

Clearly, $R'(w, w^*) = R'(w_1, w_1^*)$, if $\pi(w) = \pi(w_1)$ and $\pi^*(w^*) = \pi^*(w_1^*)$. Thus, one can define a map $R\colon Y \to \mathscr{R}$ by setting

$$R(y, y^*) = R'(\pi^{-1}(y), \pi^{*-1}(y^*)), \quad (y, y^*) \in Y.$$

It is easily seen that the map R is $\mathbf{GL}(\mathscr{W})$-equivariant, i.e.

$$R(By, B^*y^*) = BR(y, y^*)B^{-1} \text{ for all } (y, y^*) \in Y \text{ and } B \in \mathbf{GL}(\mathscr{W}).$$

As before, let us consider the measures $\mu \in \mathscr{P}(\Gamma)$ and $\nu' \in \mathscr{P}(G/P)$ with $\mu * \nu' = \nu'$, $\operatorname{supp} \mu = \Gamma$, and ν' being equivalent to $\mu_{G/P}$. We denote by $\nu \in \mathscr{P}(\mathbf{P}(\mathscr{W}))$ the image of the measure ν' under the Γ-equivariant map ψ. Then $\mu * \nu = \nu$. Thus, $\nu(\mathbf{P}(\mathscr{W})) = 0$ (see Lemma 3.10), and hence $\nu'(\psi^{-1}(\mathbf{P}(\mathscr{W}'))) = 0$ for all $\mathscr{W}' \in \mathbf{Gr}_{m-1}(\mathscr{W})$. But ν' is equivalent to $\mu_{G/P}$, $\omega = \psi \circ q$, and $\mu_G(q^{-1}(A)) = 0$ whenever $\mu_{G/P}(A) = 0$. Therefore $\mu_G(\omega^{-1}(\mathbf{P}(\mathscr{W}'))) = 0$ for all $\mathscr{W}' \in \mathbf{Gr}_{m-1}(\mathscr{W})$. On the other hand, it is clear that $\{u \in \mathbf{P}(\mathscr{W}) \mid (y, \pi(w^*)) \notin Y\} = \operatorname{Ker} w^* \in \mathbf{Gr}_{m-1}(\mathscr{W})$ for each $w^* \in \mathscr{W}^* - \{0\}$. It follows that for each $y^* \in \mathbf{P}(\mathscr{W}^*)$

$$\mu_G\{g \in G \mid (\omega(g), y^*) \notin Y\} = 0. \tag{4}$$

We set

$$Q = \{(g_1, g_2) \in G \times G \mid (\omega(g_1), \omega^*(g_1 g_2)) \notin Y\}.$$

It follows from (4) and Fubini's theorem that $\mu_{G\times G}(Q) = 0$. Thus, there exists $g_0 \in G$ such that $(\omega(g), \omega^*(gg_0)) \in Y$ for almost all $g \in G$. We set

$$\varphi_1(g) = R(\omega(g), \omega^*(gg_0)) \in \mathcal{R}, \; g \in G.$$

The map φ_1 is defined almost everywhere and measurable. It follows from (2), (3), and the $\mathbf{GL}(\mathcal{W})$-equivariance of the map R that for each $\lambda \in \Lambda$

$$\varphi_1(\lambda g h) = T(\lambda)\varphi_1(g)$$

for almost all $g \in G$ and each $h \in P \cap (g_0 P g_0^{-1})$. On the other hand, since the intersection of any two parabolic k_α-subgroups of $\mathbf{G}_\alpha$ contains the centralizer of a maximal k_α-split torus of the group $\mathbf{G}_\alpha$ (see I.0.29) and the maximal k_α-split tori of $\mathbf{G}_\alpha$ are conjugated over k_α (see I.0.24), there exists $u \in G$ such that $u\mathscr{Z}_G(S)u^{-1} \subset P \cap (g_0 P g_0^{-1})$. Thus, by setting $\varphi(g) = \varphi_1(gu^{-1})$, we obtain a measurable map $\varphi\colon G \to \mathcal{R}$ such that for each $\lambda \in \Lambda$,

$$\varphi(\lambda g z) = \varphi_1(\lambda g z u^{-1}) = \varphi_1(\lambda g u^{-1} u z u^{-1}) = T(\lambda)\varphi_1(gu^{-1}) = R(\lambda)\varphi(g)$$

for almost all $g \in G$ and for each $z \in \mathscr{Z}_G(S)$. Finally, by setting $\varphi'(gS) = \varphi(g)$, $g \in G$, we obtain the desired Λ-equivariant measurable map $\varphi'\colon G/\mathscr{Z}_G(S) \to \mathcal{R}$. □

(4.4) Lemma. *Let k be a local field, $m \in \mathbb{N}^+$, and H a subgroup of $\mathbf{GL}_m(k)$. Suppose H is not relatively compact in $\mathbf{GL}_m(k)$ in the topology induced from that of k and* $\operatorname{End}(k^m)$ *is spanned by H. Then there exists $h \in H$ such that at least one of the eigenvalues of h is greater than* 1 *in absolute value (we assume k is equipped with an absolute value $|\;|$ which is extended to every algebraic extension of the field k).*

Proof. Let $\{h_1, \ldots, h_2\} \subset H$ be a basis in $\operatorname{End}(k^m)$. We denote by $\{e_1, \ldots, e_{m^2}\} \subset \operatorname{End}(k^m)$ the dual basis of $\{h_1, \ldots, h_{m^2}\}$ relative to the non-degenerate bilinear form $(x, y) \mapsto \operatorname{Tr} xy$. Then

$$H \subset \sum_{1 \le i \le m^2} \operatorname{Tr}(h_i H) e_i.$$

On the other hand, $h_i H = H$ and the subgroup H is not relatively compact in $\mathbf{GL}_m(k)$, and hence in $\operatorname{End}(k^m)$. Thus, the set of traces $\operatorname{Tr} H$ is not relatively compact in k, and hence there exists $h \in H$ such that $|\operatorname{Tr} h| > m$. Since $|\operatorname{Tr} B| \le m|\lambda(B)|$, where $\lambda(B)$ is an eigenvalue of the transformation $B \in \operatorname{End}(k^m)$ maximal in absolute value, h is the desired element. □

(4.5) Lemma. *Let k be a local field, $\mathbf{H}$ a connected semisimple k-group, and F a subgroup of $\mathbf{H}(k)$ which is Zariski dense in $\mathbf{H}$. Suppose the subgroup F is not relatively compact in the topology induced from that of k. Then there exist $x \in F$, $r > 1$, and a representation ρ of H which is rational, k-irreducible, r-dimensional and defined over k such that the transformation $\rho(x)$ is proximal.*

Proof. Since the k-group $\mathbf{H}$ is connected and semisimple, there exists a faithful completely reducible rational representation ϑ of H which is defined over k (one can choose ϑ as the direct sum of composition factors of any faithful k-rational representation of H). We decompose ϑ into a direct sum of absolutely irreducible m_i-dimensional representations ϑ_i, $1 \leq i \leq t$. The representations ϑ_i are defined over a finite field extension l of k. Since $\operatorname{Ker} \vartheta = \{e\}$, by Corollary I.2.1.3(i), $\vartheta|_{\mathbf{H}(k)}$ is a proper map. But the subgroup F is not relatively compact. Thus, there exists i, $1 \leq i \leq t$, such that the subgroup $\vartheta_i(F)$ is not relatively compact in $\mathbf{GL}_{m_i}(l)$. Since F is Zariski dense in $\mathbf{H}$ and ϑ_i is rational and absolutely irreducible, the restriction of ϑ_i to F is absolutely irreducible. Therefore by Burnside's theorem (see [Wae 2], §III) $\operatorname{End}(l^{m_i})$ is spanned by $\vartheta_i(F)$. Now by Lemma 4.4 one can find an element $x \in F$ such that at least one eigenvalue of the transformation $\vartheta_i(x)$ and hence $\vartheta(x)$ is greater than 1 in absolute value. Let us denote by q the dimension of the characteristic subspace of the transformation $\vartheta(x)$ corresponding to a maximal characteristic exponent (for the definition, see §1 of Chapter II). By replacing the representation ϑ by an appropriate composition factor of its q-th exterior power, we can assume that $q = 1$, i.e., that the representation $\vartheta(x)$ is proximal, and that the maximum of absolute values of the eigenvalues of the transformation $\vartheta(x)$ are, as before, greater than 1. All that remains now is to set $\rho = \vartheta$ and to observe that any 1-dimensional representation of the group $\mathbf{H}$ is trivial, because $\mathbf{H}$ is semisimple. □

(4.6) Given a k-rational action of the k-group $\mathbf{H}$ on a k-variety $\mathbf{M}$, we recall that the notions of a strictly $\mathbf{H}$-effective set and a strictly $\mathbf{H}$-effective measurable map were introduced in V.4.1. Since for each $x \in \mathbf{M}(k)$ the stabilizer $\{h \in \mathbf{H} \mid hx = x\}$ of the point x is k-closed, the following assertion is true:

(*) if the group $\mathbf{H}$ is k-simple and a set $X \subset \mathbf{M}(k)$ does not contain $\mathbf{H}$-invariant points, then X is strictly $\mathbf{H}$-effective.

(4.7) Theorem. *Let k be a local field, $\mathbf{H}$ a connected k-simple k-group, and τ: $\Lambda \to \mathbf{H}(k)$ a homomorphism. Suppose the subgroup $\tau(\Gamma)$ is Zariski dense in $\mathbf{H}$ and is not relatively compact in the topology induced from that of k.*

(a) *There exist a k-rational action of the group $\mathbf{H}$ on a k-variety $\mathbf{M}$ and a measurable (with respect to $\mu_{G/P}$) strictly $\mathbf{H}$-effective Λ-equivariant map ψ: $G/P \to \mathbf{M}(k)$ (ψ is Λ-equivariant means that for each $\lambda \in \Lambda$ the equality $\psi(\lambda x) = \tau(\lambda)\psi(x)$ holds for almost all $x \in G/P$).*

(b) *There exist $m \in \mathbb{N}^+$, a faithful rational m-dimensional representation ρ of H which is defined over k and two measurable maps φ: $G \to k^m$ and φ': $G/\mathscr{Z}_G(S) \to k^m$ such that*
   (i) *the maps φ and φ' are strictly $\mathbf{H}$-effective relative to the action of the group $\mathbf{H}$ defined by the equality $hx = \rho(h)x$, $h \in H$, $x \in k^m$;*
   (ii) *for each $\lambda \in \Lambda$, $\varphi(\lambda g z) = \rho(\tau(\lambda))\varphi(g)$ for almost all (with respect to μ_G) $g \in G$ and for all $z \in \mathscr{Z}_G(S)$;*
   (iii) *the map φ' is Λ-equivariant, i.e., for each $\lambda \in \Lambda$ the equality $\varphi'(\lambda x) = \rho(\tau(\lambda))\varphi'(x)$ holds for almost all (with respect to $\mu_{G/\mathscr{Z}_G(S)}$) $x \in G/\mathscr{Z}_G(S)$.*

Proof. By Lemma 4.5 there exist $\lambda \in \Lambda$, $r > 1$, and a rational k-irreducible r-dimensional representation ρ' of H which is defined over k such that the transformation $\rho'(\tau(\lambda))$ is proximal. We define a representation ρ of the group $\mathbf{H}$ on the space End_r by setting $\rho(h)x = \rho'(h)x\rho'(h^{-1})$, $h \in \mathbf{H}$, $x \in \mathrm{End}_r$. Since the group $\mathbf{H}$ is k-simple, $r > 1$, and the representation ρ is irreducible over k, it follows from (*) in 4.6 that the set $\mathbf{P}(k^r)$ and the set $\mathscr{R} \subset \mathrm{End}(k^r)$ of 1-dimensional projections are strictly $\mathbf{H}$-effective. Thus, assertions (a) and (b) follow from the corresponding assertions of Theorem 4.3. □

(4.8) Suppose the following conditions are satisfied:

(A) for each $\alpha \in A$ the group $\mathbf{G}_\alpha$ is simply connected, k_α-isotropic, and k_α-simple;

(B) the lattice Γ is finitely generated and weakly cocompact (in the sense of V.5.7);

(C) there is a faithful irreducible rational representation ρ of the group $\mathbf{H}$ which is defined over k such that the representation $T \overset{\text{def}}{=} \rho \circ \tau$ is Γ-integrable.

Then Theorem 4.7 is a consequence of Theorem V.5.15. Indeed, since the subgroup $\tau(\Gamma)$ is not relatively compact and by Corollary I.2.1.3(i) the restriction $\rho|_{\mathbf{H}(k)}$ is proper, it follows that the subgroup $T(\Gamma)$ is not relatively compact. Thus, if conditions (A), (B), and (C) above are satisfied, then Theorem V.5.15 implies the existence of an element $s \in S$ such that s acts ergodically on $\Gamma \setminus G$ and the pair (s, T) is essential. Since the representation T is irreducible over k and the k-group $\overline{T(\Lambda)} = \rho(H)$ is k-simple, the pair (s, T) is effective (see V.5.1). By Proposition V.4.3, this proves assertion (b) of Theorem 4.7. Let ω be the characteristic map defined by the pair (s, T) and a maximal characteristic exponent χ. Since any two minimal parabolic k_α-subgroups in $\mathbf{G}_\alpha$ are conjugate by an element of $\mathbf{G}_\alpha(k_\alpha)$, it follows from Lemma II.3.1(b) that, after replacing the subgroups $\mathbf{P}_\alpha$ by their conjugates, we can assume that the set $\{s^m p s^{-m} \mid m \in \mathbb{N}^+\}$ is relatively compact in G for each $p \in P$. Since the characteristic exponent χ is maximal, it then follows from Theorem V.3.3(ii), that for each $p \in P$ the equality $\omega(gp) = \omega(g)$ holds for almost all $g \in G$. Now, making use of Lemma I.4.1.1(v), we can modify ω on a null set so that the equality $\omega(gp) = \omega(g)$ holds for almost all $g \in G$ and for all $p \in P$. Let $\mathscr{W}$ be the space of the representation ρ and l is the multiplicity of the characteristic exponent χ. Define a map $\psi\colon G/P \to \mathbf{Gr}_l(\mathscr{W})$, by setting $\psi(gP) = \omega(g)$. Since by Theorem V.3.3(i) for each $\lambda \in \Lambda$ the equality $\omega(\lambda g) = T(\lambda)\omega(g)$ holds for almost all $g \in G$, the map ψ is Λ-equivariant. Now, to complete the proof of assertion (a) of Theorem 4.7 in the case under consideration, it suffices to observe that the strict $\mathbf{H}$-effectiveness of the map ψ is a consequence of three facts: (1) the pair (s, T) being essential, (2) the irreducibility of the representation T, and (3) the k-simplicity of the group $\overline{T(\Lambda)} = \rho(\mathbf{H})$.

Chapter VII. Rigidity

The main section in this chapter is Section 5, where the superrigidity theorems for discrete subgroups are established. The proofs are based on the study of equivariant measurable maps to linear spaces. The existence of such maps with required properties was proved in Chapters V and VI. The reduction to the results of these chapters is realized with the help of the results of Sections 1–4. In Sections 6 and 7 certain consequences of the superrigidity theorems are obtained. In Section 8 some results on the rigidity of ergodic actions of semisimple groups are stated.

We now provide some notions and notation that will be used in Section 1–4. If X is a measure space with measure μ and Y is a topological space, then a subset $M \subset X \times Y$ is called *measurable* if M belongs to the σ-algebra generated by the sets of the form $A \times B$, where $A \subset X$ is μ-measurable and $B \subset Y$ is Borel. A map of the space $X \times Y$ into a topological space (resp. measure space) is called *measurable* if the inverse image of every open (resp. measurable) set is measurable in the sense indicated above.

For a measure space X with measure μ and a topological space Y, we denote by $F_0(X, \mu, Y)$ or simply by $F_0(X, Y)$ the set of measurable maps (functions) $f\colon X \to Y$. We consider the equivalence relation on $F_0(X, Y)$ by identifying maps which differ on a null set. The quotient set modulo this relation will be denoted by $F(X, \mu, Y)$ or simply by $F(X, Y)$. The natural map of $F_0(X, Y)$ onto $F(X, Y)$, which assigns to every measurable map its equivalence class, will be denoted by β. If $\Omega_1 \subset F_0(X_1, Y_1)$, $\Omega_2 \subset F_0(X_2, Y_2)$, and $A\colon \Omega_1 \to \Omega_2$ is a map sending equivalent functions to equivalent ones, then A induces the map of the set $\beta(\Omega_1)$ to $\beta(\Omega_2)$ that will be denoted by A_β (more precisely, $A_\beta(\beta(f)) = \beta(A(f))$, $f \in \Omega_1$). If Y is a topological vector space over a topological field K, then $F_0(X, Y)$ and $F(X, Y)$, with usual operations of pointwise addition and multiplication by elements of K, are vector spaces over K as well.

1. Auxiliary Assertions

In paragraphs 1.1–1.3 we denote by X a measure space with σ-finite measure μ and by Y a separable metric space with distance function d.

(1.1) Lemma. *For all $f, h \in F_0(X, Y)$ and $\alpha > 0$ the set $B_\alpha = \{x \in X \mid d(f(x), h(x)) > \alpha\}$ is measurable.*

Proof. Let Z be a countable dense subset of Y. Then the assertion of the lemma is a consequence of the following equality

$$B_\alpha = \bigcup_{z \in Z, n \in \mathbb{N}^+} [\{x \in X \mid d(f(x), z) < 1/n\} \cap \\ \cap \{x \in X \mid d(h(x), z) > \alpha + 1/n\}]$$

□

(1.2) We define on $F(X, Y)$ the *topology of convergence in measure* by considering the sets of the form

$$\{f \in F(X, Y) \mid \mu(S_{k,\alpha}(f_0, h_0)) < \varepsilon \text{ for } f_0 \in f\}$$

as a base of open sets, where $\varepsilon > 0$, $\alpha > 0$, $h_0 \in F_0(X, Y)$, and $K \subset X$ is a set of finite measure, and

$$S_{K,\alpha}(f_0, h_0) = \{x \in K \mid d(f_0(x), h_0(x)) > \alpha\}.$$

Since the measure μ is σ-finite, the space $F(X, Y)$ with this topology is Hausdorff. If Y is a topological vector space over a topological field l and the topology on Y is compatible with the distance function d, then the topology of convergence in measure with natural operations of addition and multiplication by elements of l defines on $F(X, Y)$ the structure of a topological vector space over l.

(1.3) Lemma. *Let Z be a measure space with σ-finite measure τ and $q\colon X \times Z \to Y$ a measurable (relative to the measure $\mu \times \tau$) map. We set $q_z(x) = q(x, z)$, $x \in X$, $z \in Z$, and $\vartheta_q(z) = \beta(q_z)$, $z \in Z$. Then the map $\vartheta_q\colon Z \to F(X, Y)$ is measurable (with respect to the measure τ).*

Proof. It suffices to prove that for any subset $K \subset X$ of finite measure and for all $h \in F_0(X, Y)$, $\varepsilon > 0$, and $\alpha > 0$, the set

$$D_{K,\alpha,h,\varepsilon} \stackrel{\text{def}}{=} \{z \in Z \mid \mu(S_{K,\alpha,h,z}) < \varepsilon\}$$

is measurable, where

$$S_{K,\alpha,h,z} = \{x \in K \mid d(q_z(x), h(x)) > \alpha\}.$$

But $S_{K,\alpha,h,z}$ is the z-section (i.e. the intersection of $X \times \{z\}$ and the measurable set

$$\{x \in K, z \in Z \mid d(q(x, z), h(x)) > \alpha\} \subset X \times Z),$$

and by Fubini's theorem the measure of the z-section of any measurable set $B \subset X \times Z$ is a measurable function of $z \in Z$. Thus, $D_{K,\alpha,h,\varepsilon}$ is measurable. □

(1.4) Lemma. *Let G and H be locally compact σ-compact groups. Then any measurable (relative to μ_G) homomorphism $f\colon G \to H$ is continuous.*

Proof. Let W be a neighbourhood of the identity in H. We choose a neighbourhood $V \subset H$ of the identity such that $V^{-1} \cdot V \subset W$. Since H is σ-compact, there is a countable cover of H by the sets of the form hV, $h \in H$. Thus, there exists $h_0 \in H$ such that $\mu_G(f^{-1}(h_0V)) > 0$. On the other hand, for every subset B of finite Haar measure in the group G the measure $\mu_G((Bg)\triangle B)$ is a continuous function of $g \in G$ (see [Hal 1], Theorem 1 in §61). Hence, for every subset $C \subset G$ of positive measure the set $\{g \in G \mid (Cg) \cap C \neq \varnothing\}$ contains a neighbourhood of the identity of the group G. Thus, there is a neighbourhood R of the identity in G such that for each $r \in R$ there exists $g_r \in G$ with $f(g_r) \in h_0V$ and $f(g_r r) \in h_0V$. Since f is a homomorphism,

$$f(r) = f(g_r)^{-1}f(g_r r) \in (h_0V)^{-1} \cdot (h_0V) = V^{-1} \cdot V \subset W$$

for all $r \in R$, which proves the continuity of the homomorphism f at e, and hence the continuity of f on G. □

(1.5) A representation U of a locally compact group G on a topological vector space $\mathscr{W}$ is called *measurable* if for all $w \in \mathscr{W}$ the map $g \mapsto U(g)w$, $g \in G$, is measurable with respect to the measure μ_G, and is called *continuous* if the map $(g, w) \mapsto U(g)w$ is continuous in both variables.

(1.6) Lemma. *Let G be a locally compact σ-compact group, k a local field, and $\mathscr{W}$ a Hausdorff finite-dimensional topological vector space over k. Then any measurable representation U of the group G on the space $\mathscr{W}$ is continuous.*

Proof. By assigning to every linear transformation $h \in \mathbf{GL}(\mathscr{W})$ its matrix in a fixed basis, we obtain an isomorphism $\lambda\colon \mathbf{GL}(\mathscr{W}) \to \mathbf{GL}_n(k)$, $n = \dim \mathscr{W}$. Since $\mathscr{W}$ as a topological vector space is isomorphic to k^n (see I.0.31), it follows that the homomorphism $\lambda \circ U\colon G \to \mathbf{GL}_n(k)$ is continuous (resp. measurable) if and only if the representation U is continuous (resp. measurable). All that remains now is to apply Lemma 1.4 to the homomorphism $\lambda \circ U$. □

2. Cocycles on G-Spaces

(2.0) Let G be a locally compact σ-compact group and X a right G-space with σ-finite measure μ. We assume that the map $X \times G \to X$, $(x, g) \mapsto xg$, is measurable and the measure μ is quasi-invariant (i.e. $\mu(Ag) = 0$ if and only if $\mu(A) = 0$). Let M be a separable metrizable locally compact group and Y a separable metric left M-space. A map $\sigma\colon X \times G \to M$ is called a *cocycle* on the G-space X if σ is measurable and

$$\sigma(x, g_1g_2) = \sigma(x, g_1)\sigma(xg_1, g_2) \tag{1}$$

for all $x \in X$ and $g_1, g_2 \in G$. For the cocycle σ and $g \in G$, we define the transformation $\rho_0^\sigma(g)$ on $F_0(X, Y)$ by

(2) $$(\rho_0^\sigma(g)f)(x) = \sigma(x,g)f(xg), \text{ for } f \in F_0(X,Y) \text{ and } x \in X.$$

It follows directly from (1) that ρ_0^σ is a homomorphism of the group G into the transformation group of $F_0(X,Y)$. Since μ is quasi-invariant, the transformations $\rho_0^\sigma(g)$ send equivalent maps to equivalent ones. Thus, by setting $\rho^\sigma(g) = \rho_0^\sigma(g)_\beta$, we obtain the homomorphism ρ^σ of the group G to the transformation group of $F(X,Y)$. If Y is a vector space and H acts by linear transformations on Y, then ρ_0^σ and ρ^σ define (linear) representations of the group G on $F_0(X,Y)$ and $F(X,Y)$ respectively.

(2.1) Lemma. *For each $f \in F(X,Y)$ the map $\vartheta\colon G \to F(X,Y)$, $g \mapsto \rho^\sigma(g)(f)$, is measurable.*

To prove the lemma it suffices to apply Lemma 1.3 to the map $q\colon X \times G \to Y$, $q(x,g) = \sigma(xg)f(xg), x \in X, g \in G$.

Remark. In fact, under some weak assumptions on G and X, the map ϑ is continuous and moreover the map $(g,f) \mapsto \rho^\sigma(g)f$ is continuous in both variables. In particular, this is valid if G is metrizable and μ is a regular Borel measure on a separable complete metric space X. We omit the routine proof of this fact.

(2.2) Lemma. *Let $f \in F(X,Y)$ be $\rho^\sigma(G)$-invariant. Then*

(a) *there is a map f_0 in the class f such that for almost all $x \in X$ the equality*

(1) $$\sigma(x,g)f_0(xg) = f_0(x)$$

holds for all $g \in G$;

(b) *if Y is a vector space and H acts by linear transformations on Y, then there exists a map $f_0 \in f$ such that (1) holds for all $x \in X$ and $g \in G$ (i.e. f_0 is $\rho_0^\sigma(G)$-invariant).*

Proof. Choose an arbitrary $f_1 \in f$ and denote by Ψ the set of those $x \in X$ for which there is an element $q(x) \in Y$ such that

(2) $$\sigma(x,g)f_1(xg) = q(x) \text{ for almost all } g \in G.$$

Clearly, $q(x)$ is uniquely determined by condition (2). Since f is $\rho^\sigma(G)$-invariant, it follows from Fubini's theorem that $\mu(X - \Psi) = 0$ and $q(x) = f_1(x)$ for almost all $x \in X$. On the other hand, equality (1) in 2.0 implies that $\Psi G = \Psi$ and $\sigma(x,g)q(xg) = q(x)$ for all $x \in \Psi$ and $g \in G$. Thus, by setting $f_0(x) = q(x)$, $x \in \Psi$, we obtain (a). To show (b) it suffices to extend f_0 to $X - \Psi$, by setting $f_0(x) = 0$ for all $x \in X - \Psi$. □

3. Finite-Dimensional Invariant Subspaces

(3.0) Let G, X, and μ be as in 2.0, k a local field, $n \in \mathbb{N}^+$, $\sigma\colon X \times G \to \mathbf{GL}(k^n)$ a cocycle on the G-space X, and U a continuous representation of G on a finite-dimensional vector space $\mathscr{W}$ over k. We denote by Ψ the vector space of linear maps of $\mathscr{W}$ into k^n and define a cocycle $\vartheta\colon X \times G \to \mathbf{GL}(\Psi)$, by setting for $x \in X$, $g \in G$, $\psi \in \Psi$,

$$(1) \qquad \vartheta(x,g)(\psi) = \sigma(x,g)\psi U(g^{-1}).$$

Finally, as in Section 2 we assign to the cocycle σ the representations ρ_0^σ and ρ^σ of the group G on the spaces $F_0(X,k^n)$ and $F(X,k^n)$ respectively, and to the cocycle ϑ the representations ρ_0^ϑ and ρ^ϑ of the group G on the spaces $F_0(X,\Psi)$ and $F(X,\Psi)$ respectively.

We recall the definition of intertwining operators. Given two representations T_1 and T_2 of a group H on the vector spaces V_1 and V_2 over k, we call an operator $A\colon V_1 \to V_2$ *intertwining* if $AT_1(h) = T_2(h)A$ for all $h \in H$. The set of intertwining operators forms a vector space over k which will be denoted by $\mathscr{L}(T_1,T_2)$. The dimension of the space $\mathscr{L}(T_1,T_2)$ is called the *intertwining number* of the representations T_1 and T_2 and is denoted by $c(T_1,T_2)$.

(3.1) Lemma. *(a) For each $A \in \mathscr{L}(U,\rho^\sigma)$ there exists $A_0 \in \mathscr{L}(U,\rho_0^\sigma)$ such that $A = \beta \circ A_0$.*

(b) The intertwining number $c(U,\rho_0^\sigma)$ (resp. $c(U,\rho^\sigma)$) coincides with the dimension of the space of $\rho_0^\vartheta(G)$-invariant (resp. $\rho^\vartheta(G)$-invariant) elements of the space $F_0(X,\Psi)$ (resp. $F(X,\Psi)$).

Proof. We identify $F_0(X,\Psi)$ with the space of linear maps of $\mathscr{W}$ into $F_0(X,k^n)$, by setting

$$(f_0 w)(x) = f_0(x)w, \quad f_0 \in F_0(X,\Psi), \ x \in X, \ w \in \mathscr{W}.$$

If $f_0, h_0 \in F_0(X,\Psi)$, $\beta(f_0) = \beta(h_0)$ and $w \in W$, then $\beta(f_0 w) = \beta(h_0 w)$. Thus, we can identify $F(X,\Psi)$ with the space of linear maps of the space $\mathscr{W}$ into $F(X,k^n)$, by setting

$$\beta(f_0)w = \beta(f_0 w), \ f_0 \in F_0(X,\Psi), \ w \in \mathscr{W}.$$

For all $g \in G$, $f_0 \in F_0(X,\Psi)$, $w \in \mathscr{W}$, and $x \in X$ we have

$$(\rho_0^\sigma(g)f_0 w)(x) = \sigma(x,g)(f_0 w)(xg) = \sigma(x,g)f_0(xg)w$$

and

$$(f_0 U(g)w)(x) = f_0(x)U(g)w.$$

On the other hand, from equality (1) in 3.0 and the definition of the representation ρ_0^ϑ we obtain that for all $g \in G$, $f_0 \in F_0(X,\Psi)$, $x \in X$

$$(\rho_0^\vartheta(g)f_0)(x) = \sigma(x,g)f_0(xg)U(g^{-1}).$$

Therefore $f_0 \in \mathscr{L}(U,\rho_0^\sigma)$ (resp. $f \in \mathscr{L}(U,\rho^\sigma)$) if and only if $\rho_0^\vartheta(G)f = f$ (resp. $\rho^\vartheta(G)f = f$). Now (b) is a straightforward consequence of this observation and (a) follows from Lemma 2.2(b). □

(3.2) Proposition. *Let Z be the space of $\rho^\sigma(G)$-invariant elements of the space $F(X,k^n)$. Suppose G acts ergodically on X. Then $\dim Z \leq n$.*

Proof. For every finite collection $f_1,\ldots,f_i \in \beta^{-1}(Z)$, the dimension $m(x)$ of the linear span of the vectors $f_1(x),\ldots,f_i(x)$ is a measurable function of $x \in X$. If $g \in G$, then $f_j(x) = \sigma(x,g)f_j(xg)$ for almost all $x \in X$, $1 \le j \le i$, and hence, $m(x) = m(xg)$ for almost all $x \in X$. But G acts ergodically on X. Thus, $m(x)$ is almost everywhere equal to an integer which will be called the *rank* of the collection $f_1,\ldots,f_i$. Clearly, this rank $\le n$. Suppose i is an integer such that there is a collection $f_1,\ldots,f_i \in \beta^{-1}(Z)$ of rank i and every collection of $i+1$ maps $f'_1,\ldots,f'_{i+1} \in \beta^{-1}(Z)$ is of rank $\le i$. Then for each $f \in \beta^{-1}(Z)$ the vector $f(x)$ can be represented uniquely in the form of a linear combination

$$f(x) = \sum_{j=1}^{i} c_{j,f}(x)f_j(x) \text{ for almost all } x \in X.$$

If $f \in \beta^{-1}(Z)$, then for each $g \in G$

$$f(x) = \sigma(x,g)\cdot f(xg) = \sigma(x,g)\cdot \sum_{j=1}^{i} c_{j,f}(xg)f_j(xg) =$$
$$= \sum_{j=1}^{i} c_{j,f}(xg)\sigma(x,g)f_j(xg) = \sum_{j=1}^{i} c_{j,f}(xg)f_j(x),$$

and hence $c_{j,f}(xg) = c_{j,f}(x)$ for almost all $x \in X$. But G acts ergodically on X. Thus, there exist constants c'_{j,f_i} such that $c_{j,f}(x) = c'_{j,f}$ for almost all $x \in X$. Then $\beta(f) = \sum_{j=1}^{i} c'_{j,f}\beta(f_j)$ for all $f \in \beta^{-1}(Z)$. It follows that Z is spanned by $\beta(f_1),\ldots,\beta(f_i)$. But $1 \le n$, and hence $\dim Z \le n$.

(3.3) Corollary. *Suppose G acts ergodically on X. Then the intertwining number $c(U,\rho)$ is finite and does not exceed $n\cdot \dim \mathscr{W}$.*

Proof. We denote by Z_ϑ the space of $\rho^\vartheta(G)$-invariant elements of the space $F(X,\Psi)$. By Lemma 3.1(b) we have $c(U,\rho) = \dim Z_\vartheta$. But in view of Proposition 3.2 $\dim Z_\vartheta \le \dim \Psi = n \dim \mathscr{W}$. Thus, $c(U,\rho) \le n \dim \mathscr{W}$. □

(3.4) In 3.5–3.7 B will denote a linear finite-dimensional $\rho^\sigma(G)$-invariant subspace of $F(X,k^n)$ and ρ_B the restriction of the representation ρ^σ to B (i.e. $\rho_B(g) = \rho^\sigma(g)|_B$). From Lemmas 1.6 and 2.1 we deduce

(3.5) Lemma. *The representation ρ_B is continuous.*

(3.6) Proposition. *Let Φ be the set of linear transformations of the space $F(X,k^n)$ which commute with $\rho^\sigma(G)$. Suppose G acts ergodically on X. Then the dimension of the linear subspace Δ spanned by the set $\Phi B \overset{\text{def}}{=} \cup_{\varphi\in\Phi}\varphi(B)$ is finite and does not exceed $n(\dim B)^2$.*

Proof. Let us denote by Φ_B the set of the restrictions of the transformations $\varphi \in \Phi$ to B. Choose a basis $b_1,\ldots,b_i$ ($i = \dim B$) of the space B. Since Φ

commutes with $\rho^\sigma(G)$, it follows that $\Phi_B \subset \mathscr{L} \stackrel{\text{def}}{=} \mathscr{L}(\rho_B, \rho^\sigma)$. On the other hand, $\mathscr{L}B \subset \mathscr{L}b_1 + \ldots + \mathscr{L}b_i$ and by Corollary 3.3 and Lemma 3.5 $\dim \mathscr{L} \leq n \dim B$. Hence $n{\cdot}\dim(\mathscr{L}b_j) \leq n{\cdot}\dim B$ for all j, $1 \leq j \leq i$, and therefore $\dim \Delta \leq n(\dim B)^2$. □

(3.7) Lemma. *There is a map* $\alpha\colon B \to F_0(X, k^n)$ *such that*

(1) $\beta(\alpha(b)) = b$ *for all* $b \in B$;

(2) α *is equivariant, i.e.,* $\alpha(\rho^\sigma(g)b) = \rho_0^\sigma(g)\alpha(b)$ *for all* $g \in G$, $b \in B$.

To prove the lemma it suffices, making use of Lemma 3.5, to apply Lemma 3.1(a) to the representation ρ_B and the identity operator $A\colon B \to F(X, k^n)$, $A(b) = b$, $b \in B$.

4. Equivariant Measurable Maps and Continuous Extensions of Representations

(4.0) Let G be a locally compact σ-compact group, $\Lambda \subset G$ a countable subgroup, $H \subset G$ a closed subgroup, k a local field, $n \in \mathbb{N}^+$, and T a representation of the group Λ on the space k^n. We introduce the discrete topology on Λ and define G to be a right $(\Lambda \times G)$-space by setting $x(\lambda, g) = \lambda^{-1}xg$ for $\lambda \in \Lambda$, and $x, g \in G$. We say that the pair (Λ, H) is *ergodic* if $\Lambda \times H$ acts ergodically on G, i.e., if the following condition is satisfied: for a μ_G-measurable subset $Y \subset G$ with $\Lambda Y H = Y$, either $\mu_G(Y) = 0$ or $\mu_G(G - Y) = 0$. If the subgroup Λ is discrete, then the pair (Λ, H) is ergodic if and only if the action of the group H by right translation on $\Lambda \setminus G$ is ergodic.

We now define the representations ρ_0^T and ρ^T of the group G induced from the representation T. The space Ω_0^T of the representation ρ_0^T consists of $\omega \in F_0(G, k^n) = F_0(G, \mu_G, k^n)$ with

(1) $$\omega(\lambda g) = T(\lambda)\omega(g) \text{ for all } g \in G \text{ and } \lambda \in \Lambda,$$

and the representation ρ_0^T of the group G on Ω_0^T is defined by

(2) $$(\rho_0^T(g)\omega)(x) = \omega(xg),\ \omega \in \Omega_0^T,\ g \in G.$$

The space Ω^T of the representation ρ^T is $\beta(\Omega_0^T)$ and $\rho^T(g) = \rho_0^T(g)_\beta$. According to the terminology accepted before, we call a measurable (relative to μ_G) map $\omega\colon G \to k^n$ *Λ-equivariant* if for each $\lambda \in \Lambda$ equality (1) holds for almost all $g \in G$. If a measurable map $\omega\colon G \to k^n$ is Λ-equivariant, then by Lemma 2.2, $\beta(\omega) \in \Omega^T$. It is easily seen that Ω_0^T (resp. Ω^T) is the space of $\rho_0^\sigma(\Lambda \times \{e\})$-invariant (resp. $\rho^\sigma(\Lambda \times \{e\})$-invariant) elements in $F_0(G, k^n)$ (resp. $F(G, k^n)$) and $\rho_0^T(g) = \rho_0^\sigma(e, g)\,|_{\Omega_0^T}$ (resp. $\rho^T(g) = \rho^\sigma(e, g)|_{\Omega^T}$) for all $g \in G$, where $\sigma\colon G \times (\Lambda \times G) \to \mathbf{GL}(k^n)$ is the cocycle on the $(\Lambda \times G)$-space G defined by the equality

$$\sigma(x, (\lambda, g)) = T(\lambda^{-1}),\ \lambda \in \Lambda,\ x, g \in G.$$

Thus, from Proposition 3.6 and Lemmas 3.5 and 3.7 we deduce Proposition 4.2 and Lemmas 4.1 and 4.3 below.

(4.1) Lemma. *The restriction of the representation ρ^T to any linear finite-dimensional $\rho^T(G)$-invariant subspace $B \subset \Omega^T$ is continuous.*

Remark. It can easily be shown that in fact the representation ρ^T is itself continuous.

(4.2) Proposition. *Let $B \subset \Omega^T$ be a linear finite-dimensional $\rho^T(H)$-invariant subspace. Suppose the pair (Λ, H) is ergodic. Then the dimension of the linear subspace spanned by the set $\rho^T(\mathscr{Z}_G(H))B$ is finite and does not exceed $n(\dim B)^2$, where as usual $\mathscr{Z}_G(H)$ denotes the centralizer of H in G.*

(4.3) Lemma. *For a linear finite-dimensional $\rho^T(G)$-invariant subspace $B \subset \Omega^T$ there is a map $\alpha\colon B \to \Omega_0^T$ such that*

1) $\beta(\alpha(b)) = b$ *for all* $b \in B$;
2) α *is equivariant, i.e.,* $\alpha(\rho^T(g)b) = \rho_0^T(g)\alpha(b)$ *for all* $g \in G$ *and* $b \in B$.

(4.4) We say that a subgroup F of G is *strongly k-dense* in G if, for every finite-dimensional vector space $\mathscr{W}$ over k and every continuous representation $\rho\colon G \to \mathbf{GL}(\mathscr{W})$, any $\rho(F)$-invariant linear subspace $\mathscr{V} \subset \mathscr{W}$ is $\rho(G)$-invariant.

Clearly, if F is dense in G, then F is strongly k-dense in G. Let $g \in G$. Since $\rho(g)$ transforms $\rho(F)$-invariant subspaces to $\rho(gFg^{-1})$-invariant ones, the subgroup gFg^{-1} is strongly k-dense in G if and only if F is strongly k-dense in G.

Remark. Let F be strongly k-dense in G. Then the following two assertions hold

(a) Let $\mathbf{H}$ be a k-group and let $\varphi\colon G \to \mathbf{H}(k)$ be a continuous homomorphism. Then the Zariski closure of $\varphi(F)$ contains $\varphi(G)$.
(b) Let l be a finite field extension of k, $\mathbf{H}$ an algebraic k-group and let $\varphi\colon G \to \mathbf{H}(l)$ be a continuous homomorphism. If $\varphi(F) \subset \mathbf{H}(k)$ then $\varphi(G) \subset \mathbf{H}(l)$.

To prove (a) and (b), it suffices to use the same argument as in the reductions of Corollaries II.2.6 and II.2.8 to Theorem II.2.5.

(4.5) We denote by $\overline{T(\Lambda)}$ the Zariski closure of the subgroup $T(\Lambda)$ in $\mathbf{GL}_n$. According to the definition in V.4.1 we call a set $X \subset k^n$ *strictly $\overline{T(\Lambda)}$-effective* if the following equivalent conditions are satisfied:

(i) $\overline{T(\Lambda)}$ operates effectively on any orbit $\overline{T(\Lambda)}x$, $x \in X$;
(ii) for each $x \in X$ the stabilizer $\{h \in \overline{T(\Lambda)} \mid hx = x\}$ of the point x contains no non-trivial k-closed normal subgroups of $\overline{T(\Lambda)}$.

A measurable map $\varphi\colon G \to k^n$ is called *strictly $\overline{T(\Lambda)}$-effective* if there is a null set $Y \subset G$ such that $\omega(G - Y)$ is strictly $\overline{T(\Lambda)}$-effective.

(4.6) Proposition. *Suppose the subgroup Λ is strongly k-dense in G and there exist a Λ-equivariant strictly $\overline{T(\Lambda)}$-effective measurable map $\varphi\colon G \to k^n$ and a linear subspace $M \subset k^n$ such that*

(a) *the linear subspace $B \subset \Omega^T$ spanned by $\rho^T(G)\beta(\varphi)$ is finite-dimensional;*

(b) *for almost all $g \in G$, the linear subspace of k^n spanned by $T(\Lambda)\varphi(g)$ coincides with M.*

Then T extends to a continuous representation of the group G on k^n.

Proof. It follows from (b) that the subspace M is $T(\Lambda)$-invariant and hence is $\overline{T(\Lambda)}$-invariant. Thus, we can define a rational representation f defined over k of the group $\overline{T(\Lambda)}$ on the space M by setting $f(h)x = hx$, $h \in \overline{T(\Lambda)}$, $x \in M$. Since the map φ is strictly $\overline{T(\Lambda)}$-effective and M contains $T(\Lambda)\varphi(g)$ for almost all $g \in G$, it follows that $\operatorname{Ker} f = \{e\}$. Thus, according to Corollary I.2.1.3(i), the subgroup $f(\overline{T(\Lambda)}(k))$ is closed in $\mathbf{GL}(M)$ and $f\colon \overline{T(\Lambda)}(k) \to f(\overline{T(\Lambda)}(k))$ is a topological group isomorphism and there exists a finite field extension l of k such that $f^{-1}(f(\overline{T(\Lambda)})(k)) \subset \overline{T(\Lambda)}(l)$. If $f \circ T$ extends to a continuous representation δ of G on M then, in view of Remark in 4.4, $\delta(G) \subset f(\overline{T(\Lambda)})$ and $f^{-1} \circ \delta$ is a continuous representation of G on k^n which extends T. Therefore, replacing k^n by M and T by $f \circ T$, we may assume that the following modification of (b) is satisfied:

(b′) for almost all $g \in G$, the linear subspace of k^n spanned by $T(\Lambda)\varphi(g)$ coincides with k^n.

The space B is $\rho^T(G)$-invariant. By Lemma 4.3, there is a linear equivariant map $\alpha\colon B \to \Omega_0^T$ such that $\beta(\alpha(b)) = b$ for each $b \in B$. We set $\varphi_0 = \alpha(\beta(\varphi))$. Since $\varphi_0 \in \beta(\varphi)$, the maps φ and φ_0 agree almost everywhere. Thus, (b′) implies that there exists $g_0 \in G$ such that the linear span of $T(\Lambda)\varphi_0(g_0)$ coincides with k^n. We define a linear operator $Q\colon B \to k^n$, by setting

$$Q(b) = (\alpha(b))(g_0), \quad b \in B. \tag{1}$$

Since $\alpha(B) \subset \Omega_0^T$ and α is equivariant, equalities (1) and (2) in 4.0 imply that for all $\lambda \in \Lambda$ and $b \in B$

$$\begin{aligned} Q(\rho^T(g_0^{-1}\lambda g_0)b) &= (\alpha(\rho^T(g_0^{-1}\lambda g_0)b))(g_0) = \\ &= (\rho^T(g_0^{-1}\lambda g_0)\alpha(b))(g_0) = (\alpha(b))(g_0 g_0^{-1}\lambda g_0) = \\ &= (\alpha(b))(\lambda g_0) = T(\lambda)(\alpha(b)(g_0)) = T(\lambda)Q(b). \end{aligned} \tag{2}$$

By Lemma 4.1 the restriction of ρ^T to B is continuous. On the other hand, Λ (and hence $g_0^{-1}\Lambda g_0$) is strongly k-dense in G and (2) implies that $\operatorname{Ker} Q$ is $\rho^T(g_0^{-1}\Lambda g_0)$-invariant. Thus, $\operatorname{Ker} Q$ is $\rho^T(G)$-invariant, and hence one can define a continuous representation T' of the group G on the space $Q(B)$ by setting

$$T'(g)Q(b) = Q(\rho(g_0^{-1} g g_0)(b) \text{ for } g \in G \text{ and } b \in B. \tag{3}$$

It follows from (2) that $Q(B)$ is $T(\Lambda)$-invariant. But $\varphi_0(g_0) = Q(\varphi_0)$ and k^n is spanned by $T(\Lambda)\varphi_0(g_0)$. Thus, $Q(B) = k^n$. On the other hand, (2) and (3) imply that $T'(\lambda)Q(b) = T(\lambda)Q(b)$ for all $\lambda \in \Lambda$ and $b \in B$. Thus, T' is the desired extension of T. □

(4.7) Proposition. *Suppose the subgroup Λ is dense in G and there is a Λ-equivariant $\overline{T(\Lambda)}$-effective measurable map φ: $G \to k^n$. Then T extends to a continuous representation of the group G on k^n.*

Proof. Since Λ is dense in G, by Proposition I.4.5.1 the action of Λ by left translation on G is ergodic. In other words, the pair $(\Lambda, \{e\})$ is ergodic. It then follows from Proposition 4.2 that the linear subspace spanned by $\rho^T(G)\beta(\varphi)$ is finite-dimensional. Denote by M_g the linear subspace of k^n spanned by $T(\Lambda)\varphi(g)$. Since φ is Λ-equivariant, $M_{\lambda g} = M_g$ for all $\lambda \in \Lambda$ and for almost all $g \in G$. Since the pair $(\Lambda, \{e\})$ is ergodic, this implies that there exists a linear subspace $M \subset k^n$ such that $M = M_g$ for almost all $g \in G$. All that remains now is to observe that Λ is strongly k-dense in G (because Λ is dense in G) and to apply Proposition 4.6. □

(4.8) Proposition. *Suppose the subgroup Λ is strongly k-dense in G and there are $i \in \mathbb{N}^+$, closed subgroups $H_1, \ldots, H_i$ of G, and a Λ-equivariant strictly $\overline{T(\Lambda)}$-effective measurable map φ: $G \to k^n$ such that*

(a) *$H_{j+1} \subset \mathscr{Z}_G(H_j)$ for all j, $1 \leq j \leq i-1$;*
(b) *$\mathscr{Z}_G(H_i) \cdot \ldots \cdot \mathscr{Z}_G(H_1) = G$;*
(c) *the pair (Λ, H_j) is ergodic for all j, $1 \leq j \leq i$;*
(d) *$\varphi(gh) = \varphi(g)$ for almost all $g \in G$ and for all $h \in H_1$.*

Then T extends to a continuous representation of the group G on k^n.

Proof. We denote by L_j, $1 \leq j \leq i$, the linear subspace spanned by

$$\rho^T(\mathscr{Z}_G(H_j) \cdot \ldots \cdot \mathscr{Z}_G(H_1))\beta(\varphi)$$

and we shall show by induction on j that

$$\dim L_j < \infty. \tag{1}$$

Since $\rho^T(H_1)\beta(\varphi) = \beta(\varphi)$ (see (d)) and the pair (Λ, H_1) is ergodic, in case $j = 1$ inequality (1) is a consequence of Proposition 4.2. Since $\mathscr{Z}_G(H_j)$ is a subgroup, L_j is $\mathscr{Z}_G(H_j)$-invariant and hence is invariant relative to $H_{j+1} \subset \mathscr{Z}_G(H_j)$. But the pair (Λ, H_{j+1}) is ergodic. Thus, by Proposition 4.2, the finiteness of the dimension of L_j implies the finiteness of the dimension of L_{j+1}. This completes the proof of (1). From (1), making use of (b), we deduce that the linear span of $\rho^T(G)\beta(\varphi)$ is finite-dimensional. As in the proof of Proposition 4.7, for $g \in G$ we denote by M_g the linear subspace spanned by $T(\Lambda)\varphi(g)$. It follows from (d) and the Λ-equivariance of the map φ that $M_{\lambda g h} = M_g$ for almost all $g \in G$ and for all $\lambda \in \Lambda$ and $h \in H_1$. But the pair (Λ, H_1) is ergodic. Thus, there is a linear subspace $M \subset K^n$ such that $M = M_g$ for almost all $g \in G$. All that remains now is to recall that B is finite-dimensional and to apply Proposition 4.6. □

5. Superrigidity (Continuous Extensions of Homomorphisms of Discrete Subgroups to Algebraic Groups Over Local Fields)

Given a finite non-empty set A and, for each $\alpha \in A$, a local field k_α and a connected non-trivial semisimple k_α-group $\mathbf{G}_\alpha$ without k_α-anisotropic factors, we denote by G the locally compact group $\prod_{\alpha \in A} \mathbf{G}_\alpha(k_\alpha)$. For each $\alpha \in A$ choose a maximal k_α-split torus $\mathbf{S}_\alpha$ in $\mathbf{G}_\alpha$ and set $S = \prod_{\alpha \in A} \mathbf{S}_\alpha(k_\alpha) \subset G$. Let $\operatorname{rank} G = \sum_{\alpha \in A} \operatorname{rank}_{k_\alpha} \mathbf{G}_\alpha$ denote the rank of the group G. As in Section 3 of Chapter II we set $G^+ = \prod_{\alpha \in A} \mathbf{G}_\alpha(k_\alpha)^+$ and denote by pr_α the natural projection $G \to \mathbf{G}_\alpha(k_\alpha)$.

Let Γ be a lattice in G and Λ a countable subgroup of $\operatorname{Comm}_G(\Gamma)$ containing Γ (the assumption that Λ is countable is superfluous, because the group $\operatorname{Comm}_G(\Gamma)$ is countable, see 6.3 below).

If G' is a closed subgroup of G containing Γ, then $\Gamma \setminus G'$ will be viewed as a right G'-space. We say, as before, that $H \subset G'$ acts ergodically on $\Gamma \setminus G'$ if the following condition is satisfied: for any $\mu_{G'}$-measurable subset $Y \subset \Gamma \setminus G'$ with $YH = Y$, either $\mu_{G'}(Y) = 0$ or $\mu_{G'}((\Gamma \setminus G') - Y) = 0$.

(5.1) Lemma. *(a) Let $p\colon F \to F'$ and $q\colon F \to F'$ be two group homomorphisms. For $x \in F$, define $\omega(x) \in F'$ via the equality $p(x) = \omega(x) \cdot q(x)$. Then the map $\omega\colon F \to F'$ is a homomorphism if and only if $\omega(F)$ commutes with $q(F)$.*

(b) Let F be a group, $\mathbf{H}$ an algebraic group, f and f' two homomorphisms of the group F into $\mathbf{H}$, and B a subgroup of F with $\operatorname{Comm}_F(B) = F$. Denote by $\mathbf{D}$ the Zariski closure of the subgroup $f(F)$ in $\mathbf{H}$. Suppose the group $\mathbf{D}$ is connected, the restrictions of the homomorphisms f and f' to B agree, and the subgroup $f(B) = f'(B)$ is Zariski dense in $\mathbf{D}$. Then there exists a homomorphism $\nu\colon F \to \mathscr{Z}_{\mathbf{H}}(D)$ such that $f'(x) = \nu(x) f(x)$ for all $x \in F$.

Proof. (a) Since

$$\omega(x_1x_2)q(x_1x_2) = p(x_1x_2) = p(x_1)p(x_2) = \omega(x_1)q(x_1)\omega(x_2)q(x_2)$$

and

$$\omega(x_1)\omega(x_2)q(x_1x_2) = \omega(x_1)\omega(x_2)q(x_1)q(x_2),$$

it follows that ω is a homomorphism if and only if $q(x_1)$ commutes with $\omega(x_2)$ for all $x_1, x_2 \in F$.

(b) For $x \in F$, we define $\nu(x) \in \mathbf{H}$ via the equality $f'(x) = \nu(x)f(x)$. Let $x \in F$ and $y \in B_x \overset{\text{def}}{=} B \cap (x^{-1}Bx)$. Then

$$f'(x)f(y)f'(x)^{-1} = f'(xyx^{-1}) = f(xyx^{-1}) = f(x)f(y)f(x)^{-1}.$$

This implies that $\nu(x)$ commutes with $f(B_x)$. On the other hand, since the subgroup $f(B)$ is Zariski dense in $\mathbf{D}$, the subgroup B_x is of finite index in B (because $F = \operatorname{Comm}_F(B)$), and the group $\mathbf{D}$ is connected, and hence contains no algebraic subgroups of finite index, it follows that the subgroup $f(B_x)$ is Zariski dense in $\mathbf{D}$. Thus, $\nu(F) \subset \mathscr{Z}_{\mathbf{H}}(\mathbf{D})$. All that remains now is to make use of assertion (a). □

(5.2) We recall that if ϑ is a homomorphism of the field K to a field K' and $\mathbf{M}$ is a K-variety, then ${}^{\vartheta}\mathbf{M}$ denotes the K'-variety resulting from $\mathbf{M}$ via ϑ and $\vartheta^0\colon \mathbf{M}(k) \to {}^{\vartheta}\mathbf{M}(K')$ denotes the map induced by ϑ (see I.1.7). If $\mathbf{M}$ is a K-group, then ϑ^0 is a group homomorphism.

We recall that the definition of a special epimorphism of semisimple algebraic groups was given in I.1.4.14.

(5.3) Proposition. *Let G' be a closed subgroup of G containing G^+, k a local field, $\mathbf{H}$ a connected absolutely almost simple k-group, and let $f\colon G' \to \mathbf{H}(k)$ be a continuous homomorphism with $f(G')$ Zariski dense in $\mathbf{H}$. Suppose that either the group $\mathbf{H}$ is adjoint or for each $\alpha \in A$ the group $\mathbf{G}_\alpha$ is simply connected. Then*

(a) *there exist (uniquely determined) $\alpha \in A$, a continuous homomorphism $\vartheta\colon k_\alpha \to k$, and a special k-epimorphism $\eta\colon {}^{\vartheta}\mathbf{G}_\alpha \to \mathbf{H}$ such that $f(g) = \eta(\vartheta^0(\mathrm{pr}_\alpha(g)))$ for all $g \in G$;*

(b) *f extends uniquely to a continuous homomorphism $\tilde{f}\colon G \to \mathbf{H}(k)$.*

Proof. (a) We decompose the group $\mathbf{G}_\alpha$ into the almost direct product of almost k_α-simple k_α-groups $\mathbf{G}_j$, where j runs through a finite set J_α. For $j \in J_\alpha$, we set $G_j^+ = \mathbf{G}_j(k_\alpha)^+$ and denote by $\mathbf{H}_j$ the Zariski closure of the subgroup $f(G_j^+)$ in $\mathbf{H}$. Since the subgroup G_j^+ is normal in G and the subgroup $f(G')$ is Zariski dense in $\mathbf{H}$, it follows that the subgroup $\mathbf{H}_j$ is normal in the connected almost simple group $\mathbf{H}$, and hence either $\mathbf{H}_j = \mathbf{H}$ or $\mathbf{H}_j \subset \mathscr{Z}(\mathbf{H})$. On the other hand, since $\mathscr{D}(G_j^+) = G_j^+$, it follows that $\mathscr{D}(\mathbf{H}_j) = \mathbf{H}_j$ (see Theorem I.1.5.6(ii)). Thus, either $\mathbf{H}_j = \mathbf{H}$ or $\mathbf{H}_j = \{e\}$.

Let us set $J = \cup_{\alpha \in A} J_\alpha$. If $j_1, j_2 \in J$, $j_1 \neq j_2$, then the subgroups $G_{j_1}^+$ and $G_{j_2}^+$ commute. Hence, $\mathbf{H}_{j_1}$ and $\mathbf{H}_{j_2}$ commute. Since $\mathbf{H}$ is not commutative, this implies that there exists $j_0 \in J$ such that $\mathbf{H}_j \neq \mathbf{H}$, and hence $\mathbf{H}_j = \{e\}$ for all $j \in J - \{j_0\}$. Assume $\mathbf{H}_{j_0} \neq \mathbf{H}$. Then $\mathbf{H}_{j_0} = \{e\}$ and therefore $f(G_j^+) \subset \mathbf{H}_j = \{e\}$ for all $j \in J$. But by, Proposition I.1.5.4(iv), $\mathbf{G}_\alpha(k_\alpha)^+$ is an almost direct product of the subgroups G_j^+, $j \in J$. Thus, $f(G^+) = \{e\}$. On the other hand, since for each $\alpha \in A$, the factor group $\mathbf{G}_\alpha(k_\alpha)/\mathbf{G}_\alpha(k_\alpha)^+$ is commutative, so is the factor group G/G^+ (see Theorem I.2.3.1(c)). Hence, the subgroup $f(G')$ is commutative. This is a contradiction, because $\mathbf{H}$ is non-commutative and $f(G')$ is Zariski dense in $\mathbf{H}$. Therefore we have proved that $\mathbf{H}_{j_0} = \mathbf{H}$. Let $\alpha \in A$ with $j_0 \in J_\alpha$. Then by Theorem I.1.8.1 and Remark I.1.8.2 (III) there exist a uniquely determined continuous homomorphism $\vartheta\colon k_\alpha \to k$, a special k-epimorphism $\eta\colon {}^{\vartheta}\mathbf{G}_\alpha \to \mathbf{H}$, and a homomorphism $\mu\colon \mathbf{G}_\alpha(k_\alpha)^+ \to \mathscr{Z}(\mathbf{H}(k))$ such that $f(g) = \mu(g)\eta(\vartheta^0(\mathrm{pr}_\alpha(g)))$ for each $g \in G^+$. Since the group $\mathscr{Z}(\mathbf{H}(k)) \subset \mathscr{Z}(\mathbf{H})$ is finite and any subgroup of finite index in $\mathbf{G}_\alpha(k_\alpha)^+$ coincides with $\mathbf{G}_\alpha(k_\alpha)^+$ (see Corollary I.1.5.7), the homomorphism μ is trivial, i.e., $f(g) = \eta(\vartheta^0(\mathrm{pr}_\alpha(g)))$ for all $g \in G^+$. By setting $\tilde{f}(g) = \eta(\vartheta^0(\mathrm{pr}_\alpha(g)))$, $g \in G'$, we define a homomorphism $\tilde{f}\colon G' \to \mathbf{H}(k)$. Since $\mathbf{H}_{j_0} = \mathbf{H}$, it follows that $f(G^+)$ is Zariski dense in $\mathbf{H}$. On the other hand, the homomorphisms f and $\tilde{f}$ agree on G^+ and the subgroup G^+ is normal in G'. Thus, there is a homomorphism $\nu\colon G' \to \mathscr{Z}(\mathbf{H})$ such that $\tilde{f}(g) = \nu(g)f(g)$ for each $g \in G$ (see Lemma 5.1(b)). It remains to show that the homomorphism ν is trivial. If $\mathbf{H}$ is adjoint, then $\mathscr{Z}(\mathbf{H}) = \{e\}$. Thus, we may assume that the groups $\mathbf{G}_\alpha$ are simply connected. It then follows from Theorem I.2.3.1 (a') and Corollary I.2.3.2(b) that $G = G^+$ and G contains no proper subgroups of finite index. Since $\mathscr{Z}(\mathbf{H})$ is finite, this implies, that v is trivial.

(b) is a consequence of (a). □

(5.4) Theorem. *Let k be a local field, $\mathbf{H}$ a connected adjoint k-simple k-group, and $\tau\colon \Lambda \to \mathbf{H}(k)$ a homomorphism. Suppose the following conditions are satisfied:*

(i) *the closure of Λ contains G^+;*

(ii) *the subgroup $\tau(\Gamma)$ is Zariski dense in $\mathbf{H}$ but is not relatively compact in $\mathbf{H}(k)$ (in the topology induced from that of the field k).*

Then

(a) *if $\mathbf{H}$ is absolutely simple, then there exist (uniquely determined) $\alpha \in A$, a continuous homomorphism $\vartheta\colon k_\alpha \to k$, and a special k-epimorphism $\eta\colon {}^{\vartheta}\mathbf{G}_\alpha \to \mathbf{H}$ such that $\tau(\lambda) = \eta(\vartheta^0(\mathrm{pr}_\alpha(\lambda)))$ for all $\lambda \in \Lambda$;*

(b) *τ extends uniquely to a continuous homomorphism $\tilde{\tau}\colon G \to \mathbf{H}(k)$.*

Proof. The group $\mathbf{H}$ can be represented in the form $\mathbf{H} = R_{k'/k}\mathbf{H}'$, where $\mathbf{H}'$ is an absolutely simple adjoint k'-group and $R^0_{k'/k}\colon \mathbf{H}'(k') \to \mathbf{H}(k)$ is an isomorphism of topological groups. Thus, replacing $\mathbf{H}$ by $\mathbf{H}'$ and τ by $(R^0_{k'/k})^{-1} \circ \tau$, we may assume that $\mathbf{H}$ is absolutely simple.

By Theorem VI.4.7(b) there exist $m \in \mathbb{N}^+$, a faithful k-rational m-dimensional representation ρ of the group $\mathbf{H}$, and a strictly $\mathbf{H}$-effective measurable map $\varphi\colon G \to k^m$ such that for each $\lambda \in \Lambda$

$$\varphi(\lambda g z) = \rho(\tau(\lambda))\varphi(g) \tag{1}$$

for almost all (with respect to μ_G) $g \in G$ and all $z \in \mathscr{Z}_G(S)$.

As in Section 4, we define G to be a $(\Lambda \times G)$-space, by setting $x(\lambda, g) = \lambda^{-1}xg$, $\lambda \in \Lambda$, $x, g \in G$, and consider the cocycle $\sigma\colon G \times (\Lambda \times G) \to \mathbf{GL}(k^m)$ defined by $\sigma(x, (\lambda, g)) = \rho(\tau(\lambda^{-1}))$, $\lambda \in \Lambda$, $g \in G$. Applying Lemma 2.2(b) to the cocycle σ, one can adjust φ on a null set so that (1) holds for all $\lambda \in \Lambda$, $g \in G$, and $z \in \mathscr{Z}_G(S)$ (leaving φ $\mathbf{H}$-effective).

We denote by G' the closure of the subgroup Λ in G. According to condition (ii), $G' \supset G^+$. On the other hand, since $\mathbf{G}_\alpha(k_\alpha) = \mathbf{G}_\alpha(k_\alpha)^+ \cdot \mathscr{Z}_{\mathbf{G}_\alpha(k_\alpha)}(\mathbf{S}_\alpha(k_\alpha))$ for all $\alpha \in A$ (see Proposition I.1.5.4(vi)), $G = G^+ \cdot \mathscr{Z}_G(S)$. Thus, $G = G' \cdot \mathscr{Z}_G(S)$. We denote by φ' the restriction of φ to G'. Since $G = G' \cdot \mathscr{Z}_G(S)$, φ is measurable and strictly $\mathbf{H}$-effective, and $\varphi(gz) = \varphi(g)$ for all $g \in G$ and $z \in \mathscr{Z}_G(S)$, we obtain, making use of Corollary I.4.4.3, that the map φ' is measurable and strictly $\mathbf{H}$-effective. Let us set $T = \rho \circ \tau$. Since φ is strictly $\mathbf{H}$-effective and the subgroup $\tau(\Lambda)$ is Zariski dense in $\mathbf{H}$, it follows that φ' is strictly $\overline{T(\Lambda)}$-effective in the sense of 4.5. On the other hand, since $\varphi(\lambda g) = \rho(\tau(\lambda))\varphi(g)$ for all $\lambda \in \Lambda$ and $g \in G$, the map φ' is Λ-equivariant. Thus, by Proposition 4.7, T extends to a continuous representation $\tilde{T}$ of the group G' on k^m.

Since $\operatorname{Ker}\rho = \{e\}$, it follows from Corollary I.2.1.3(i) that the subgroup $\rho(\mathbf{H}(k))$ is closed in $\mathbf{GL}(k^m)$ and the homomorphism $\rho\colon \mathbf{H}(k) \to \rho(\mathbf{H}(k))$ is a topological group isomorphism. Observe that, since Λ is dense in G', $\tilde{T}(G') \subset \rho(\mathbf{H}(k))$.

Now set $\tau' = \rho^{-1} \circ \tilde{T}$. Then the homomorphism $\tau'\colon G' \to \mathbf{H}(k)$ is a continuous extension of τ. Furthermore, this extension is unique, because Λ is dense in G'. Therefore assertion (b) of the theorem is a consequence of Proposition 5.3(b). All that remains now is to apply Proposition 5.3(a). □

(5.5) Lemma. *Let G' be the closure of the subgroup $\Gamma \cdot G^+$. Suppose for each $\alpha \in A$ the group $\mathbf{G}_\alpha$ is k_α-isotropic and almost k_α-simple. Assume in addition that* rank $G \geq 2$ *and the lattice Γ is irreducible (in the sense of Definition II.6.5). Then there exist $i \in \mathbb{N}^+$ and cyclic subgroups $S_1, \ldots, S_i$ of $S \cap G^+$ such that*

(a) $\mathscr{Z}_{G'}(S_i) \cdot \ldots \cdot \mathscr{Z}_{G'}(S_1) = G'$;

(b) *S_j acts ergodically on $\Gamma \setminus G'$ for each j, $1 \leq j \leq i$.*

Proof. As in Section 3 of Chapter II, for $s \in S$, we set $A(s) = \{\alpha \in A \mid |b(\mathrm{pr}_\alpha(s))|_\alpha \neq 1$ for at least one $b \in \Phi(\mathbf{S}_\alpha, \mathbf{G}_\alpha)\}$, where $|\ |_\alpha$ is the absolute value on k_α. Further, we set

$$S' = \{s \in S \mid A(s) \neq \varnothing\}.$$

Since by Theorem II.7.2(b) each $s \in S' \cap G^+$ acts ergodically on $\Gamma \setminus G'$, it suffices to find $s_1, \ldots, s_i \in S' \cap G^+$ such that

$$\mathscr{Z}_{G'}(s_i) \cdot \ldots \cdot \mathscr{Z}_{G'}(s_1) = G'.$$

Let $\tilde{\mathbf{G}}_\alpha$ be the simply connected covering of the group $\mathbf{G}_\alpha$, $p_\alpha\colon \tilde{\mathbf{G}}_\alpha \to \mathbf{G}_\alpha$ the central k_α-isogeny, and $\tilde{\mathbf{S}}_\alpha$ a maximal k_α-split torus in $\tilde{\mathbf{G}}_\alpha$ with $p_\alpha(\tilde{\mathbf{S}}_\alpha) = \mathbf{S}_\alpha$. We set

$$\tilde{G} = \prod_{\alpha \in A} \tilde{\mathbf{G}}_\alpha(k_\alpha),\ \tilde{S} = \prod_{\alpha \in A} \tilde{\mathbf{S}}_\alpha(k_\alpha),$$

and denote by $p\colon \tilde{G} \to G$ the continuous homomorphism induced from the isogenies p_α. In complete analogy with the definition of $S' \subset G$, we define $\tilde{S}' \subset \tilde{S}$. Suppose we have found $\tilde{s}_1, \ldots, \tilde{s}_i \in \tilde{S}'$ such that

$$\mathscr{Z}_{\tilde{G}}(\tilde{s}_i) \cdot \ldots \cdot \mathscr{Z}_{\tilde{G}}(\tilde{s}_1) = \tilde{G}.$$

Since $\tilde{\mathbf{G}}_\alpha(k_\alpha)^+ = \tilde{\mathbf{G}}_\alpha(k_\alpha)$ and $p_\alpha(\tilde{\mathbf{G}}_\alpha(k_\alpha)^+) = \mathbf{G}_\alpha(k_\alpha)^+$ (see Theorem I.2.3.1(a) and Proposition I.1.5.5), it then follows that $p(\tilde{G}) = G^+$ and

$$\mathscr{Z}_{G^+}(p(\tilde{s}_i)) \cdot \ldots \cdot \mathscr{Z}_{G^+}(p(\tilde{s}_1)) = G^+.$$

On the other hand, since $G = G^+ \cdot \mathscr{Z}_G(S)$ and $G' \supset G^+$ (see the proof of Theorem 5.4), it follows that $G' = \mathscr{Z}_{G'}(S \cap G') \cdot G^+$, and hence $G' = \mathscr{Z}_{G'}(p(\tilde{s}_i)) \cdot G^+$. Thus,

$$\begin{aligned}
&\mathscr{Z}_{G'}(p(\tilde{s}_i)) \cdot \ldots \cdot \mathscr{Z}_{G'}(p(\tilde{s}_1)) \supset \\
&\supset \mathscr{Z}_{G'}(p(\tilde{s}_i)) \cdot \mathscr{Z}_{G^+}(p(\tilde{s}_i)) \cdot \ldots \cdot \mathscr{Z}_{G^+}(p(\tilde{s}_1)) = \\
&= \mathscr{Z}_{G'}(p(\tilde{s}_i)) \cdot G^+ = G'.
\end{aligned}$$

But, since for all $\alpha \in A$ and $s \in \tilde{S}_\alpha$, the eigenvalues of the transformations $\mathrm{Ad}\, s$ and $\mathrm{Ad}\, p_\alpha(s)$ coincide (because the isogeny p_α is central), $p(\tilde{S}') \subset S'$. Hence, replacing $\tilde{\mathbf{G}}_\alpha$ by $\mathbf{G}_\alpha$, we see that it suffices to find $s_1, \ldots, s_i \in S'$ such that

$$\mathscr{Z}_G(s_i) \cdot \ldots \cdot \mathscr{Z}_G(s_1) = G.$$

We consider two cases:

(i) $\operatorname{card} A \geq 2$ and
(ii) $\operatorname{card} A = 1$.

(i) Since the group $\mathbf{G}_\alpha$ is isotropic over k_α, by Proposition I.2.4.1 the set $\mathbf{S}_\alpha(k_\alpha) \cap S'$ is not empty for all $\alpha \in A$. Let $s_\alpha \in \mathbf{S}_\alpha(k_\alpha) \cap S'$. Then

$$\prod_{\alpha' \in A - \{a\}} \mathbf{G}_{\alpha'}(k_{\alpha'}) \subset \mathscr{Z}_G(s_\alpha)$$

and, since $\operatorname{card} A \geq 2$, the elements s_α, $\alpha \in A$, are the required ones.

(ii) Let $A = \{\alpha\}$. Since $\operatorname{rank}_{k_\alpha} \mathbf{G}_\alpha = \operatorname{rank} G \geq 2$, by Proposition I.1.2.2 there exist $i \in \mathbb{N}^+$ and one-dimensional subtori $\mathbf{S}_1, \ldots, \mathbf{S}_i$ of the torus $\mathbf{S}_\alpha$ such that

$$\mathscr{Z}_G(\mathbf{S}_i(k_\alpha)) \cdot \ldots \cdot \mathscr{Z}_G(\mathbf{S}_1(k_\alpha)) = G. \tag{1}$$

Since $\dim \mathbf{S}_j > 0$, it follows from Proposition I.2.4.1 that for each j, $1 \leq j \leq i$, the set $\mathbf{S}_j(k_\alpha) \cap S'$ is not empty. Let $s_j \in \mathbf{S}_j(k_\alpha) \cap S'$. Then (1) implies that the elements s_j, $1 \leq j \leq i$, are the required ones. □

(5.6) Theorem. *Let k be a local field, $\mathbf{H}$ a connected adjoint k-simple k-group, and $\tau\colon \Lambda \to \mathbf{H}(k)$ a homomorphism. Suppose the following conditions are satisfied:*

(i) $\operatorname{rank} G \geq 2$;
(ii) *the lattice Γ is irreducible (in the sense of III.5.9);*
(iii) *the subgroup $\tau(\Gamma)$ is Zariski dense in $\mathbf{H}$ and is not relatively compact in $\mathbf{H}(k)$.*

Then

(a) *if $\mathbf{H}$ is absolutely simple, then there exist (uniquely determined) $\alpha \in A$, a continuous homomorphism $\vartheta\colon k_\alpha \to k$, and a special k-epimorphism $\eta\colon {}^{\vartheta}\mathbf{G}_\alpha \to \mathbf{H}$ such that $\tau(\lambda) = \eta(\vartheta^0(\operatorname{pr}_\alpha(\lambda)))$ for all $\lambda \in \Lambda$;*
(b) *τ extends uniquely to a continuous homomorphism $\tilde{\tau}\colon G \to \mathbf{H}(k)$.*

Proof. As in the proof of Theorem 5.4, we may assume that the group $\mathbf{H}$ is absolutely simple. If τ_1 and τ_2 are continuous homomorphisms of the group G into $\mathbf{H}(k)$ with $\tau_1(\gamma) = \tau_2(\gamma)$ for all $\gamma \in \Gamma$, then, by Lemma II.2.3 and Proposition II.4.6, $\tau_1(g) = \tau_2(g)$ for all $g \in G^+$. Thus, in view of Proposition 5.3 it suffices to prove that τ extends to a continuous homomorphism $\tilde{\tau}\colon G \to \mathbf{H}(k)$.

Let $\mathbf{G}'_\alpha$ be the adjoint group of $\mathbf{G}_\alpha$ and $p_\alpha\colon \mathbf{G}_\alpha \to \mathbf{G}'_\alpha$ the central k_α-isogeny. We set $G'' = \prod_{\alpha \in A} \mathbf{G}'_\alpha(k_\alpha)$. The isogenies p_α induce the continuous homomorphism $p\colon G \to G''$. We observe as in the proof of Theorem B in III.5.9 (see III.6.9) that $p(\Gamma)$ is an irreducible lattice in G'' and $\operatorname{rank}_{k_\alpha} \mathbf{G}'_\alpha = \operatorname{rank}_{k_\alpha} \mathbf{G}_\alpha$. Since $\mathbf{H}$ is absolutely simple and $\tau(\Lambda)$ is Zariski dense in $\mathbf{H}$, it follows that $\Lambda \cap \operatorname{Ker} p \subset \Lambda \cap \mathscr{Z}(G) \subset \operatorname{Ker} \tau$. As was observed in I.1.4.14, the composition of a special epimorphism and a central isogeny is a special epimorphism. Thus, replacing $\mathbf{G}_\alpha$ by $\mathbf{G}'_\alpha$, Γ by $p(\Gamma)$, Λ by $p(\Lambda)$, and τ by $\tau \circ p^{-1}$, we may assume that $\mathbf{G}_\alpha$ is an adjoint group. Then $\mathbf{G}_\alpha$ can be decomposed into a direct product of almost k_α-simple groups. Therefore we can assume that the groups $\mathbf{G}_\alpha$ are k_α-simple.

We denote by G' the closure of the subgroup $\Gamma \cdot G^+$ in G. Making use of Theorem VI.4.7(b) and arguing as in the proof of Theorem 5.4, we can prove that there exist $m \in \mathbb{N}^+$, a faithful k-rational m-dimensional representation ρ of the group $\mathbf{H}$, and a measurable Λ-equivariant map $\varphi'\colon G' \to k^m$ such that φ' is strictly $\overline{T(\Lambda)}$-effective in the sense of 4.5, where $T = \rho \circ \tau$, and $\varphi'(gz) = \varphi'(g)$ for all $g \in G'$ and $z \in G' \cap \mathscr{Z}_G(S)$. By Lemma 5.5 there exist subgroups $S_1, \ldots, S_i$ of the group $S \cap G^+ \subset S \cap G'$ such that

1) $\mathscr{Z}_{G'}(S_i) \cdot \ldots \cdot \mathscr{Z}_{G'}(S_1) = G'$"
2) S_j acts ergodically on $\Gamma \setminus G'$, and hence the pair (Λ, S_j) is ergodic (in the sense of Section 4) for each j, $1 \leq j \leq i$.

It follows from Lemma II.2.3 and Theorem II.4.2(a) that the subgroup Γ is strongly k-dense in G' (in the sense of 4.4). Now, observing that $S_{j+1} \subset \mathscr{Z}_{G'}(S_j)$, we can apply Proposition 4.8 and obtain that the representation $T = \rho \circ \tau$ extends to a continuous representation $\tilde{T}$ of the group G' on k^m.

By Lemma II.2.3 and Proposition II.4.3 the Zariski closure of the subgroup $T(\Gamma)$ contains $\tilde{T}(G^+)$. But $T(\Gamma) \subset \rho(\mathbf{H})$ and $\rho(\mathbf{H})$ is an algebraic subgroup in $\mathbf{GL}_m$. Thus, $\tilde{T}(G') \subset \rho(\mathbf{H})(k)$. Since $\operatorname{Ker} \rho = \{e\}$, one can define the homomorphism $\tau' = \rho^{-1} \circ \tilde{T}$ of the group G' into $\mathbf{H}$. Since $\tilde{T}$ is an extension of $T = \rho \circ \tau$, τ' is an extension of τ. Further, since $\operatorname{Ker} \rho = \{e\}$, by Corollary I.2.1.3, $\rho^{-1}(\rho(\mathbf{H})(k)) \subset \mathbf{H}(l)$, where l is a finite field extension of k and the homomorphism $\rho^{-1}\colon \rho(\mathbf{H})(k) \to \rho^{-1}(\rho(\mathbf{H})(k))$ is a topological group isomorphism. Thus, τ' is a continuous homomorphism of the group G' into $\mathbf{H}(l)$. We then have by Lemma II.2.3 and Proposition II.4.7 that $\tau'(G') \subset \mathbf{H}(k)$. Extending τ' by Proposition 5.3(b) to a continuous homomorphism $\tilde{\tau}\colon G \to \mathbf{H}(k)$, we obtain the desired extension of τ. □

(5.7) Theorem. *Let $\mathscr{B} = \{p \in \mathbb{N}^+ \mid p \text{ is prime}\} \cup \{\infty\}$, A' a finite subset of $\mathscr{B}$. Given, for each $p \in A'$, a connected non-trivial semisimple $\mathbb{Q}_p$-group $\mathbf{G}'_p$ without $\mathbb{Q}_p$-anisotropic factors, define $G' = \prod_{p \in A'} \mathbf{G}'_p(\mathbb{Q}_p)$ and denote by pr_p the natural projection $G' \to \mathbf{G}'_p(\mathbb{Q}_p)$. Let Γ' be a lattice in G' and Λ' a countable subgroup of $\operatorname{Comm}_{G'}(\Gamma')$ containing Γ', let k be a local field, $\mathbf{H}$ a connected adjoint k-simple k-group, and let $\tau\colon \Lambda' \to \mathbf{H}(k)$ be a homomorphism with $\tau(\Gamma')$ dense in $\mathbf{H}$ in the Zariski topology. Suppose that either $\operatorname{rank} G' \overset{\text{def}}{=} \sum_{p \in A'} \operatorname{rank}_{\mathbb{Q}_p} \mathbf{G}'_p \geq 2$ and the lattice Γ' is irreducible or the closure of the subgroup Λ' in G' contains $\prod_{p \in A'} \mathbf{G}'_p(\mathbb{Q}_p)^+$.*

(a) *If k is isomorphic to a finite field extension of $\mathbb{Q}_p$ for no $p \in A'$, then the subgroup $\tau(\Gamma')$ is relatively compact in $\mathbf{H}(k)$.*
(b) *If the subgroup $\tau(\Gamma')$ is not relatively compact in $\mathbf{H}(k)$ and k is a finite field extension of $\mathbb{Q}_p$ for some $p \in A'$, then there is a unique k-epimorphism $\eta\colon \mathbf{G}_p \to H$ such that $\tau(\lambda) = \eta(\operatorname{pr}_p(\lambda))$ for all $\lambda \in \Lambda$.*
(c) *If the subgroup $\tau(\Gamma')$ is not relatively compact in $\mathbf{H}(k)$, then τ extends uniquely to a continuous homomorphism $\tilde{\tau}\colon G' \to \mathbf{H}(k)$.*

Proof. Since the group $\mathbf{H}$ can be represented in the form $\mathbf{H} = R_{k'/k}\mathbf{H}'$, where $\mathbf{H}'$ is an absolutely simple adjoint k-group, the universality property allows us to restrict to the case where $\mathbf{H}$ is absolutely simple (see I.1.7). Now it suffices to make use of Theorems 5.4 and 5.6 and to observe that the following assertions are valid:

(A) if $p \in \mathscr{B}$ and k is not isomorphic to a finite field extension of $\mathbb{Q}_p$, then there exist no continuous homomorphisms of the field $\mathbb{Q}_p$ into k;

(B) if $p \in \mathscr{B}$ and k is a finite field extension of $\mathbb{Q}_p$, then every continuous homomorphism of the field $\mathbb{Q}_p$ into k is the identity map. □

(5.8) In case $\operatorname{char} k_\alpha = 0$ for each $\alpha \in A$, Theorem 5.7 is in fact a restatement of Theorems 5.4 and 5.6. This can easily be shown by applying the restriction of scalars and the classification of local fields. The following assertion is a special case of Theorem 5.7.

(5.9) Theorem. *Let $\mathbf{G}$ be a connected semisimple $\mathbb{R}$-group with no $\mathbb{R}$-anisotropic almost $\mathbb{R}$-simple factors, Γ' a lattice in $\mathbf{G}(\mathbb{R})$, and Λ' a countable subgroup of* $\operatorname{Comm}_{\mathbf{G}(\mathbb{R})}(\Gamma')$ *containing Γ'. Let k be a local field, $\mathbf{H}$ a connected adjoint k-simple k-group, and $\tau\colon \Lambda' \to \mathbf{H}(k)$ a homomorphism with $\tau(\Gamma')$ Zariski dense in $\mathbf{H}$. Suppose that either* $\operatorname{rank}_{\mathbb{R}} \mathbf{G} \geq 2$ *and the lattice Γ' is irreducible or the closure of the subgroup Λ' in $\mathbf{G}(\mathbb{R})$ contains the identity component of the Lie group $\mathbf{G}(\mathbb{R})$.*

(a) *If k is not isomorphic to $\mathbb{R}$ or $\mathbb{C}$, i.e., if k is non-archimedean, then the subgroup $\tau(\Gamma')$ is relatively compact in $\mathbf{H}(k)$.*

(b) *If k is $\mathbb{R}$ or $\mathbb{C}$ and the subgroup $\tau(\Gamma')$ is not relatively compact in $\mathbf{H}(k)$, then τ extends uniquely to a k-epimorphism $\eta\colon \mathbf{G} \to \mathbf{H}$ and to a continuous homomorphism $\tilde{\tau}\colon \mathbf{G}(\mathbb{R}) \to \mathbf{H}(k)$.*

(5.10) In the statements of Theorems 5.4 and 5.6 and assertions (b) and (c) in Theorem 5.7 and also Theorem 5.9(b) the condition "the subgroup $\tau(\Gamma)$ (or $\tau(\Gamma')$) is not relatively compact in $\mathbf{H}(k)$" is essential. This is shown by the following example.

Let $d \neq 1$ be a positive square-free integer, $K = \mathrm{Q}(\sqrt{d})$ a quadratic field extension of $\mathbb{Q}$ obtained by adjoining of $\sqrt{d}$, $L \subset K$ the ring of integers over $\mathbb{Z}$ in the field K, and $\sigma\colon K \to K$ a unique non-trivial automorphism of the field K, $\sigma(a + b\sqrt{d}) = a - b\sqrt{d}$, $a, b \in \mathrm{Q}$. Let us consider the quadratic form

$$f = x_1^2 + x_2^2 + x_3^2 - \sqrt{d}x_4^2 - \sqrt{d}x_5^2$$

and the K-group $\mathbf{G} = \mathbf{SO}(f)$ of unimodular linear transformations preserving the form f. The group $\mathbf{G}$ is connected and absolutely almost simple. Since $-\sqrt{d} < 0$, the form f reduces over $\mathbb{R}$ to the form $x_1^2 + x_2x_3 + x_4x_5$. Thus, $\operatorname{rank}_{\mathbb{R}} G = 2$. We set $\Gamma' = \mathbf{G}(L)$, $\Lambda' = \mathbf{G}(K)$, and observe that, by Lemma I.3.1.1(v), Λ' consists of Γ'-rational elements. We denote by ${}^\sigma\! f$ the quadratic form obtained by applying σ to the coefficients of the form f. Since $\sigma(-\sqrt{d}) = \sqrt{d} > 0$, the form ${}^\sigma\! f$ is positive definite. But ${}^\sigma\mathbf{G} = \mathbf{SO}({}^\sigma\! f)$. Thus, the group ${}^\sigma\mathbf{G}(\mathbb{R})$ is compact. Applying Theorem I.3.2.7, we see that Γ' is a lattice in $\mathbf{G}(\mathbb{R})$. We set $\tau = \sigma^0\colon \Gamma' \to {}^\sigma\mathbf{G}(\mathbb{R})$. Since

Γ' is a lattice in $\mathbf{G}(\mathbb{R})$, by Corollary II.4.4, Γ' is Zariski dense in $\mathbf{G}$, and hence $\tau(\Gamma) = \sigma^0(\Gamma')$ is Zariski dense in ${}^{\sigma}\mathbf{G}$. Since $\tau(\Gamma') \neq \{e\}$ and every continuous homomorphism of the connected non-compact almost simple Lie group $\mathbf{G}(\mathbb{R})$ to the compact Lie group ${}^{\sigma}\mathbf{G}(\mathbb{R})$ is trivial, τ can not be extended to a continuous homomorphism $\tilde{\tau}$: $\mathbf{G}(\mathbb{R}) \to {}^{\sigma}\mathbf{G}(\mathbb{R})$. All that remains is now to observe that the subgroup $\Lambda = \mathbf{G}(K)$ is dense in $\mathbf{G}(\mathbb{R})$ (this fact can easily be deduced both from the weak approximation theorem and from the connectedness and almost simplicity of the Lie group $\mathbf{G}(\mathbb{R})$, non-discreteness of Λ' in $\mathbf{G}(\mathbb{R})$, and Zariski density of Γ' in $\mathbf{G}$).

(5.11) Let k be a local field, $\mathbf{H}$ a connected non-commutative almost k-simple k-group, ν a non-trivial homomorphism of the group Λ into $\mathscr{Z}(\mathbf{H})(k)$, and τ': $G \to \mathbf{H}(k)$ a continuous homomorphism with $\tau'(G)$ Zariski dense in $\mathbf{H}$. For $\lambda \in \Lambda$, we set $\tau(\lambda) = \nu(\lambda) \cdot \tau'(\lambda)$. Since ν and τ are homomorphisms and $\nu(\Lambda)$ commutes with $\tau'(\Lambda)$, by Lemma 5.1(a) τ is a homomorphism. Assume that for each $\alpha \in A$ the group $\mathbf{G}_\alpha$ is simply connected. Then τ can not be extended to a continuous homomorphism of the group G into $\mathbf{H}(k)$. Indeed, if $\tilde{\tau}$ is such an extension, then $\tilde{\tau}$ agrees with τ' on $\operatorname{Ker}\nu$. But, since the group $\mathscr{Z}(\mathbf{H})(k)$ is finite, the subgroup $\operatorname{Ker}\nu$ is of finite index in Λ, and hence $\Gamma \cap \operatorname{Ker}\nu$ is a lattice in G. Thus, by Lemma II.2.3 and Corollary II.4.6, $\tilde{\tau} = \tau'$. This contradicts the assumption that ν is not trivial. Notice that, since the subgroup $\tau'(G)$ is Zariski dense in $\mathbf{H}$, $\Gamma \cap \operatorname{Ker}\nu$ is a lattice in G, and τ agrees with τ' on $\operatorname{Ker}\nu$. Lemma II.2.3 and Proposition II.4.3 imply that the subgroup $\tau(\Gamma)$ is Zariski dense in $\mathbf{H}$.

We present an example of the situation described above. Denote by $\mathbb{F}_3$ the residue field modulo 3 and let ψ: $\mathbf{SL}_3(\mathbb{Z}[1/2]) \to \mathbf{SL}_3(\mathbb{F}_3)$ be the homomorphism induced from the natural epimorphism $\mathbb{Z} \to \mathbb{F}_3$, where $\mathbb{Z}[1/2]$ denotes the ring obtained by adjoining of $1/2$ to $\mathbb{Z}$. We set

$$\Delta = \left\{ \begin{pmatrix} 1 & 0 & x \\ 0 & 1 & 0 \\ 0 & 0 & 1 \end{pmatrix} \middle| x \in \mathbb{F}_3 \right\} \subset \mathbf{SL}_3(\mathbb{F}_3), \ \Gamma = \pi^{-1}(\Delta) \cap \mathbf{SL}_3(\mathbb{Z}),$$

and $G = \mathbf{SL}_3(\mathbb{R})$. We take as Λ either $\mathbf{SL}_3(\mathbb{Z}[1/2])$ or Γ. It follows from Theorem I.3.2.8(a) that Γ is a lattice in G. We set $k = \mathbb{C}$ and $\mathbf{H} = \mathbf{SL}_3$. Since the groups Δ and $\mathscr{Z}(\mathbf{H})$ are isomorphic to the cyclic group of order 3, there is a non-trivial homomorphism τ: $\Lambda \to \mathscr{Z}(\mathbf{H})(k)$. Let τ': $G \to \mathbf{H}(k)$ be the identity homomorphism ($\tau'(g) = g$). All that remains now is to observe that the group $\mathbf{SL}_3$ is absolutely almost simple and simply connected.

In the example above $\operatorname{rank} G = 2$ and the subgroup $\Lambda = \mathbf{SL}_3(\mathbb{Z}[1/2])$ is dense in G. Furthermore, in this example the subgroup $\tau(\Gamma)$ is not relatively compact in $\mathbf{H}(k)$, where as before $\tau(\lambda) = \nu(\lambda) \cdot \tau'(\lambda)$. Thus, in Theorems 5.4, 5.6, 5.7(b) and (c), and 5.9(b) the condition "$\mathbf{H}$ is a connected adjoint k-simple k-group" can not be replaced by the condition "$\mathbf{H}$ is a connected almost k-simple k-group". However, we shall show below that if for each $\alpha \in A$ the group $\mathbf{G}_\alpha$ is simply connected and we have made the replacement indicated above in the statements of Theorems 5.4 and 5.6, then there exist a continuous homomorphism $\tilde{\tau}$: $G \to \mathbf{H}(k)$ and a homomorphism $v\colon \Lambda \to \mathscr{Z}(\mathbf{H})$ such that $\tau(\lambda) = v(\lambda) \cdot \tilde{\tau}(\lambda)$ for all $\lambda \in \Lambda$ (see 5.13). For proving this and similar assertions we shall need the following

(5.12) Lemma. *Let k be a local field, $p\colon \mathbf{H} \to \mathbf{H}_1$ a central k-isogeny of connected semisimple k-groups, and $f_1\colon G \to \mathbf{H}_1(k)$ a continuous homomorphism with $f_1(G)$ Zariski dense in $\mathbf{H}_1$. Suppose for each $\alpha \in A$ the group $\mathbf{G}_\alpha$ is simply connected. Then there exists a unique continuous homomorphism $f\colon G \to \mathbf{H}(k)$ such that $f_1 = p \circ f$.*

Proof. We first show the uniqueness of the homomorphism f. Let $f, f'\colon G \to \mathbf{H}(k)$ be two homomorphisms with $f_1 = p \circ f = p \circ f'$. We set $v(g) = f'(g)f(g)^{-1}$. Since $p \circ f = p \circ f'$ and $\operatorname{Ker} p \subset \mathscr{Z}(\mathbf{H})$, it follows that $v(G) \subset \mathscr{Z}(\mathbf{H})$. It follows from Lemma 5.1(a) that v is a homomorphism. Since the group $\mathscr{Z}(\mathbf{H})$ is finite and $v(G) \subset \mathscr{Z}(\mathbf{H})$, the kernel of v is of finite index in G. On the other hand, by Corollary I.2.3.2(b) any subgroup of finite index in G coincides with G. Thus, $\operatorname{Ker} v = G$ and hence $f = f'$.

Now we shall prove the existence of f. We first consider the case in which $\mathbf{H}$ is simply connected and $\mathbf{H}_1$ is adjoint. Then, by decomposing the groups $\mathbf{H}$ and $\mathbf{H}_1$ into a direct product of almost k-simple k-groups (see Proposition I.1.4.10), we can assume that $\mathbf{H}$ and $\mathbf{H}_1$ are almost k-simple. Further, by representing the groups $\mathbf{H}$ and $\mathbf{H}_1$ in the form $R_{k'/k}\mathbf{H}'$ and $R_{k'/k}\mathbf{H}'_1$, where $\mathbf{H}'$ and $\mathbf{H}'_1$ are absolutely almost simple (see I.1.7), we may assume that $\mathbf{H}$ and $\mathbf{H}_1$ are absolutely almost simple. Then by Proposition 5.3 there exist $\alpha \in A$, a continuous homomorphism $\vartheta\colon k_\alpha \to k$, and a k-epimorphism $\eta\colon {}^{\vartheta}\mathbf{G}_\alpha \to \mathbf{H}_1$ such that $f_1(g) = \eta(\vartheta^0(\mathrm{pr}_\alpha(g)))$ for all $g \in G$. Since the group $\mathbf{G}_\alpha$ is simply connected, by Proposition I.1.4.11(ii) there exists a k-morphism $\eta'\colon {}^{\vartheta}\mathbf{G}_\alpha \to \mathbf{H}$ such that $\eta = p \circ \eta'$. We set $f(g) = \eta'(\vartheta^0(\mathrm{pr}_\alpha(g))$, $g \in G$. Then f is the desired homomorphism.

We now turn to the general case. Let $\operatorname{Ad}\mathbf{H}_1$ be the adjoint group of $\mathbf{H}_1$, $\tilde{\mathbf{H}}$ the simply connected covering of the group $\mathbf{H}$, and let $\tilde{p}\colon \tilde{\mathbf{H}} \to \mathbf{H}$ be the central k-isogeny. It follows from Proposition I.1.4.7 that the composition of central isogenies of reductive groups is a central isogeny. On the other hand, according to Proposition I.1.4.11(iv) the isogeny $\operatorname{Ad}\colon \mathbf{H}_1 \to \operatorname{Ad}\mathbf{H}_1$ is central. Thus, $\operatorname{Ad} \circ p \circ \tilde{p}\colon \tilde{\mathbf{H}} \to \operatorname{Ad}\mathbf{H}_1$ is a central isogeny. But the case in which $\mathbf{H}$ is simply connected and $\mathbf{H}_1$ is adjoint has already been examined. Hence, there is a continuous homomorphism $\tilde{f}\colon G \to \tilde{\mathbf{H}}(k)$ such that $\operatorname{Ad} \circ f_1 = \operatorname{Ad} \circ p \circ \tilde{p} \circ \tilde{f}$. It then follows from the uniqueness of f established above that $f_1 = p \circ \tilde{p} \circ \tilde{f}$. Thus, $f = \tilde{p} \circ \tilde{f}$ is the desired homomorphism. □

(5.13) Theorem. *Let k be a local field, $\mathbf{H}$ a connected semisimple k-group, $\tau\colon \Lambda \to \mathbf{H}(k)$ a homomorphism. Suppose the following conditions are satisfied:*

(i) *either* $\operatorname{rank} G \geq 2$ *and the lattice Γ is irreducible or the closure of the subgroup Λ contains G^+;*

(ii) *the subgroup $\tau(\Gamma)$ is Zariski dense in $\mathbf{H}$;*

(iii) *either the subgroup $\mathbf{H}$ is adjoint or for each $\alpha \in A$ the group $\mathbf{G}_\alpha$ is simply connected.*

Then the following assertions hold.

(a) *There exist two connected normal k-subgroups* $\mathbf{H}'$ *and* $\mathbf{H}''$ *of* $\mathbf{H}$, *a continuous homomorphism* $\tilde{\tau}\colon G \to \mathbf{H}'(k)$, *and a homomorphism* $\omega\colon \Lambda \to \mathbf{H}''(k) \cdot \mathscr{Z}(\mathbf{H})(k)$ *such that* $\mathbf{H}$ *is the almost direct product of the subgroups* $\mathbf{H}'$ *and* $\mathbf{H}''$, *the subgroup* $\tilde{\tau}(G)$ *is Zariski dense in* $\mathbf{H}'$, *the subgroup* $\omega(\Gamma)$ *is relatively compact in* $\mathbf{H}(k)$, *and* $\tau(\lambda) = \omega(\lambda) \cdot \tilde{\tau}(\lambda)$ *for all* $\lambda \in \Lambda$.

(b) *The restriction of the homomorphism* τ *to* Γ *almost extends (in the sense of Definition V.3.5) to a continuous homomorphism* $\tilde{\tau}\colon G \to \mathbf{H}(k)$.

(c) *If the subgroup* $\tau(\Gamma)$ *is not relatively compact in* $\mathbf{H}(k)$ *and* $\mathbf{H}$ *is almost k-simple, then there exist uniquely determined homomorphisms* $\tilde{\tau}\colon G \to \mathbf{H}(k)$ *and* $\nu\colon \Lambda \to \mathscr{Z}(\mathbf{H})$, $\tilde{\tau}$ *being continuous, such that* $\tau(\lambda) = \nu(\lambda) \cdot \tilde{\tau}(\lambda)$ *for all* $\lambda \in \Lambda$.

(d) *If the subgroup* $\tau(\Gamma)$ *is not relatively compact in* $\mathbf{H}(k)$ *and* $\mathbf{H}$ *is absolutely almost simple, then there exist (uniquely determined)* $\alpha \in A$, *a continuous homomorphism* $\vartheta\colon k_\alpha \to k$, *a special k-epimorphism* $\eta\colon {}^{\vartheta}\mathbf{G}_\alpha \to \mathbf{H}$, *and a homomorphism* $\nu\colon \Lambda \to \mathscr{Z}(\mathbf{H})$ *such that* $\tau(\lambda) = \nu(\lambda) \cdot \eta(\vartheta^0(\mathrm{pr}_\alpha(\lambda)))$ *for all* $\lambda \in \Lambda$.

Proof. (a) We decompose the group $\mathbf{H}$ into an almost direct product $\prod_{i \in J} \mathbf{H}_i$ of almost k-simple k-groups $\mathbf{H}_i$. Then $\operatorname{Ad}\mathbf{H}$ is the direct product of the adjoint k-simple k-groups $\operatorname{Ad}\mathbf{H}_i$. We denote by π_i the natural projection $\operatorname{Ad}\mathbf{H} \to \operatorname{Ad}\mathbf{H}_i$. Let $J' = \{i \in J \mid$ the subgroup $(\pi_i \circ \operatorname{Ad} \circ \tau)(\Gamma)$ is not relatively compact in $(\operatorname{Ad}\mathbf{H}_i)(k)\}$ and let $J'' = J - J'$. We set $\mathbf{H}' = \prod_{i \in J'} \mathbf{H}_i$, $\mathbf{H}'' = \prod_{i \in J''} \mathbf{H}_i$, and denote the natural projection $\prod_{i \in J'} \pi_i$ of the group $\operatorname{Ad}\mathbf{H}$ onto $\operatorname{Ad}\mathbf{H}'$ by π'. If $i \in J'$, then, by Theorems 5.4(b) and 5.6(b), the homomorphism $\pi_i \circ \operatorname{Ad} \circ \tau$ extends to a continuous homomorphism $\tau_i\colon G \to (\operatorname{Ad}\mathbf{H}_i)(k)$. Thus, the homomorphism $\pi' \circ \operatorname{Ad} \circ \tau$ extends to a continuous homomorphism $\tau'\colon G \to (\operatorname{Ad}\mathbf{H}')(k)$. Since the subgroup $\tau(\Gamma)$ is Zariski dense in $\mathbf{H}$, the subgroup $\tau'(G)$ which contains $(\pi' \circ \operatorname{Ad} \circ \tau)(\Gamma)$ is Zariski dense in $\mathbf{H}'$. The isogeny $\mathbf{H}' \to \operatorname{Ad}\mathbf{H}'$ is central and is a k-isomorphism if $\mathbf{H}$ is adjoint (see I.1.4). Thus there exists a continuous homomorphism $\tilde{\tau}\colon G \to \mathbf{H}'(k)$ such that $\tau' = \operatorname{Ad} \circ \tilde{\tau}$ (see Lemma 5.12). For $\lambda \in \Lambda$, we define $\omega(\lambda) \in \mathbf{H}(k)$ via the equality $\tau(\lambda) = \omega(\lambda) \cdot \tilde{\tau}(\lambda)$. It then follows from the definitions of the set J' and the homomorphisms τ' and $\tilde{\tau}$ that $(\operatorname{Ad} \circ \omega)(\Lambda) \subset (\operatorname{Ad}\mathbf{H}'')(k)$ and the set $(\operatorname{Ad} \circ \omega)(\Gamma)$ is relatively compact in $(\operatorname{Ad}\mathbf{H})(k)$. But $\operatorname{Ker}\operatorname{Ad} \subset \mathscr{Z}(\mathbf{H})$ and by Proposition I.2.3.4(ii) the map $\operatorname{Ad}\colon \mathbf{H}(k) \to (\operatorname{Ad}\mathbf{H})(k)$ is proper. Thus, $\omega(\Lambda) \subset \mathbf{H}''(k) \cdot \mathscr{Z}(\mathbf{H})(k)$ and the set $\omega(\Gamma)$ is relatively compact in $\mathbf{H}(k)$. Now it remains to observe that, since the set $\omega(\Lambda) \subset \mathbf{H}'' \cdot \mathscr{Z}(\mathbf{H})$ commutes with $\mathbf{H}'$, Lemma 5.1(a) implies that ω is a homomorphism.

(b) is a weak version of (a).

(c) In view of (a), all that remains is to prove the uniqueness of the homomorphisms $\tilde{\tau}$ and ν. Let τ_1 and τ_2 be continuous homomorphisms of the group G into $\mathbf{H}(k)$ and ν_1, ν_2 two homomorphisms of the group Λ into $\mathscr{Z}(\mathbf{H})$ such that $\tau(\lambda) = \nu_1(\lambda)\tau_1(\lambda) = \nu_2(\lambda)\tau_2(\lambda)$ for all $\lambda \in \Lambda$. We set $\Lambda_0 = \operatorname{Ker}\nu_1 \cap \operatorname{Ker}\nu_2$. Since $\mathscr{Z}(\mathbf{H})$ is finite, Λ_0 is of finite index in Λ, and hence $\Lambda_0 \cap \Gamma$ is a lattice in G. But $\tau_1(\lambda) = \tau_2(\lambda)$ for all $\lambda \in \Lambda_0$. Thus $\tau_1(g) = \tau_2(g)$ for all $g \in G^+$ (see Lemma II.2.3 and Corollary II.4.6). From this, making use of Proposition 5.3(b), we obtain that $\tau_1 = \tau_2$ and so $\nu_1 = \nu_2$.

(d) is a consequence of (c) and Proposition 5.3(a). □

Remark. Making use of the uniqueness of the homomorphisms $\tilde{\tau}$ and ν in (c) and Lemma 5.18 below, it is not hard to show that $\mathbf{H}'$, $\mathbf{H}''$, $\tilde{\tau}$, and ω in (a) are uniquely determined.

(5.14) Theorem. *With $\mathscr{B}$, A', $\mathbf{G}'_p$, G', pr_p, Γ', and Λ' as in the statement of Theorem 5.7, we let $\mathbf{H}$ be a connected non-commutative almost k-simple k-group, $\tau\colon \Lambda' \to \mathbf{H}(k)$ a homomorphism with $\tau(\Gamma')$ Zariski dense in $\mathbf{H}$, and suppose that either* $\operatorname{rank} G' \overset{\text{def}}{=} \sum_{p\in A'} \operatorname{rank}_{\mathbb{Q}_p} \mathbf{G}'_p \geq 2$ *and the lattice Γ' is irreducible or the closure of the subgroup Λ' in G' contains $\prod_{p\in A'} \mathbf{G}'_p(\mathbb{Q}_p)^+$. Assume in addition that either $\mathbf{H}$ is adjoint or for each $p \in A'$ the group $\mathbf{G}'_p$ is simply connected.*

(a) *If k is not isomorphic to a finite field extension of $\mathbb{Q}_p$ for all $p \in A'$, then the subgroup $\tau(\Gamma')$ is relatively compact in $\mathbf{H}(k)$.*
(b) *If the subgroup $\tau(\Gamma')$ is not relatively compact in $\mathbf{H}(k)$ and k is a finite field extension of $\mathbb{Q}_p$ for some $p \in A'$, then there exist (uniquely determined) a k-epimorphism $\eta\colon \mathbf{G}'_p \to \mathbf{H}$ and a homomorphism $\nu\colon \Lambda' \to \mathscr{Z}(\mathbf{H})$ such that $\tau(\lambda) = \nu(\lambda)\cdot\eta(\mathrm{pr}_p(\lambda))$ for all $\lambda \in \Lambda$.*
(c) *If the subgroup $\tau(\Gamma')$ is not relatively compact in $\mathbf{H}(k)$, then there exist a uniquely determined continuous homomorphism $\tilde{\tau}\colon G' \to \mathbf{H}(k)$ and a uniquely determined homomorphism $\nu\colon \Lambda' \to \mathscr{Z}(\mathbf{H})$ such that $\tau(\lambda) = \nu(\lambda)\cdot\tilde{\tau}(\lambda)$ for all $\lambda \in \Lambda'$.*

This theorem can be deduced from assertions (c) and (d) of Theorem 5.13 in the same way as Theorem 5.7 was deduced from Theorems 5.4 and 5.6.

The following is a special case of Theorem 5.14.

(5.15) Theorem. *Let $\mathbf{G}$, Γ', Λ' and k be as above in Theorem 5.9, $\mathbf{H}$ a connected non-commutative almost k-simple k-group, and $\tau\colon \Lambda' \to \mathbf{H}(k)$ a homomorphism with $\tau(\Gamma')$ Zariski dense in $\mathbf{H}$. Suppose that either* $\operatorname{rank}_{\mathbb{R}} \mathbf{G} \geq 2$ *and the lattice Γ' is irreducible or the closure of the subgroup Λ' in $\mathbf{G}(\mathbb{R})$ contains the identity component of the Lie group $\mathbf{G}(\mathbb{R})$. Assume in addition that either $\mathbf{H}$ is adjoint or $\mathbf{G}$ is simply connected.*

(a) *If k is not isomorphic to $\mathbb{R}$ or $\mathbb{C}$, i.e., if k is totally disconnected, then the subgroup $\tau(\Gamma')$ is relatively compact in $\mathbf{H}(k)$.*
(b) *If k is isomorphic to $\mathbb{R}$ or $\mathbb{C}$ and the subgroup $\tau(\Gamma')$ is not relatively compact in $\mathbf{H}(k)$, then there exist (uniquely determined) a k-epimorphism $\eta\colon \mathbf{G} \to \mathbf{H}$ and a homomorphism $\nu\colon \Lambda' \to \mathscr{Z}(\mathbf{H})$ such that $\tau(\lambda) = \nu(\lambda)\cdot\eta(\lambda)$ for all $\lambda \in \Lambda'$.*
(c) *If k is isomorphic to $\mathbb{R}$ or $\mathbb{C}$ and the subgroup $\tau(\Gamma')$ is not relatively compact in $\mathbf{H}(k)$, then there exist (uniquely determined) a continuous homomorphism $\tilde{\tau}\colon \mathbf{G}(\mathbb{R}) \to \mathbf{H}(k)$ and a homomorphism $\nu\colon \Lambda' \to \mathscr{Z}(\mathbf{H})$ such that $\tau(\lambda) = \nu(\lambda)\cdot\tilde{\tau}(\lambda)$ for all $\lambda \in \Lambda'$.*

(5.16) Let k be a local field, $\mathbf{H}$ a connected semisimple k-group. We represent the group $\operatorname{Ad}\mathbf{H}$ in the form

$$\operatorname{Ad}\mathbf{H} = \prod_{1\leq i\leq n} R_{k_i/k}\mathbf{H}_i,$$

where k_i is a finite separable field extension of k and $\mathbf{H}_i$ is an absolutely simple adjoint k_i-group (see I.1.4 and I.1.7). We denote by π_i the natural projection $\mathrm{Ad}\,\mathbf{H} \to R_{k_i/k}\mathbf{H}_i$ and identify $(R_{k_i/k}\mathbf{H}_i)(k)$ with $\mathbf{H}(k_i)$ by $R^0_{k_i/k}$. Since the map $\mathrm{Ad}\colon \mathbf{H}(k) \to (\mathrm{Ad}\,\mathbf{H})(k)$ is proper (see Proposition I.1.4.11(iv) and Proposition I.2.3.4(ii)), the set $X \subset \mathbf{H}(k)$ is relatively compact if and only if the set $\pi_i(\mathrm{Ad}\,X)$ is relatively compact in $\mathbf{H}_i(k_i)$ for all i, $1 \le i \le n$. Thus, from Theorem 5.4 and 5.6 we deduce the following

Corollary. *Let k be a local field, $\mathbf{H}$ a connected semisimple k-group, and $\tau\colon \Lambda \to \mathbf{H}(k)$ a homomorphism with $\tau(\Gamma)$ Zariski dense in $\mathbf{H}(k)$. Suppose that either the closure of Λ contains G^+ or $\operatorname{rank} G \ge 2$ and the lattice Γ is irreducible. We assume in addition that for each $\alpha \in A$ the fields k_α and k are of different type in the sense of I.0.31 (i.e. there are no continuous homomorphisms of the field k_α to a finite field extension of k). Then the subgroup $\tau(\Gamma)$ is relatively compact in $\mathbf{H}(k)$.*

This corollary shows that assertion (a) of Theorem 5.14 (resp. (a) of Theorem 5.15) remains valid if in the statement of the theorem we omit the assumption that "either $\mathbf{H}$ is adjoint or for each $p \in A'$ the group $\mathbf{G}'_p$ is simply connected" (resp. "either $\mathbf{H}$ is adjoint or $\mathbf{G}$ is simply connected") and replace the condition "$\mathbf{H}$ is a connected non-commutative almost k-simple k-group" by the condition "$\mathbf{H}$ is a semisimple k-group".

(5.17) Let k be a local field and τ a homomorphism of the group Λ into the set of k-rational points of a k-group. So far in this section we have restricted ourselves to the case where the Zariski closure of the subgroup $\tau(\Lambda)$ is semisimple. We now consider the case where the k-group $\overline{T(\Lambda)}$ has property (L) (for the definition of property (L) see V.4.8). The corresponding theorem will be stated and proved in 5.19. Next, as a consequence we shall obtain some results on 1-dimensional cohomology of the group Γ. In Section 6, Theorem 5.19 will be used in the proof of the fact that under certain restrictions on G, Λ, and Γ, the Zariski closure of the subgroup $\rho(\Lambda)$ is semisimple for every homomorphism ρ of Λ to an algebraic group over a field of characteristic 0. For proving Theorem 5.19 we shall need the following

(5.18) Lemma. *Let k be a local field, $\mathbf{H}$ an algebraic k-group, and $\varphi\colon G \to \mathbf{H}(k)$ a continuous homomorphism.*

(i) *If $\varphi(G^+) \ne \{e\}$, then the group $\varphi(G)$ is not relatively compact in $\mathbf{H}(k)$. Furthermore, there exists $s \in S$ such that the element $\varphi(s)$ is essentially non-compact (in the sense of Definition V.4.6).*

(ii) *If $\varphi(G) \ne \{e\}$ and for each $\alpha \in A$ the group $\mathbf{G}_\alpha$ is simply connected, then there exists $s \in S$ such that s acts ergodically on $\Gamma \setminus G$ and the element $\varphi(s)$ is essentially non-compact.*

Proof. We shall use the notation introduced in II.3.0. In particular, let $D_\varnothing = \{s \in S \mid |b(\mathrm{pr}_\alpha(s))|_\alpha < 1 \text{ for all } \alpha \in A \text{ and } b \in \Delta\}$. As was remarked in II.3.0, $D_\varnothing \ne \varnothing$. Let $s \in D_\varnothing$. According to Lemma II.3.1(a), the automorphisms

$\operatorname{Int} s|_{V_\varnothing}$ and $\operatorname{Int} s^{-1}|_{V_\varnothing^-}$ are contracting. From this and the continuity of the homomorphism φ we deduce that for each $v \in V_\varnothing \cup V_\varnothing^-$ the closure of the set $\{\varphi(s)^i\varphi(v)\varphi(s)^{-i} \mid i \in \mathbb{Z}\}$ contains e. So by Lemma II.1.4 at least one of the following two assertions holds:

(a) at least one of the eigenvalues of the transformation $\varphi(s)$ is different from 1 in absolute value, and hence the element $\varphi(s)$ is essentially non-compact;

(b) $\varphi(V_\varnothing \cup V_\varnothing^-) = \{e\}$.

According to Proposition I.1.5.4(iii), for each $\alpha \in A$ the subgroups $\mathbf{V}_\varnothing(k_\alpha)$ and $\mathbf{V}_\varnothing^-(k_\alpha)$ generate $\mathbf{G}_\alpha(k_\alpha)^+$. Thus, if $\varphi(G^+) \neq \{e\}$, then $\varphi(V_\varnothing \cup V_\varnothing^-) \neq \{e\}$, and hence (a) holds. Now, to complete the proof of the lemma it remains to observe that if for each $\alpha \in A$ the group $\mathbf{G}_\alpha$ is simply connected, then $G = G^+$ (see Theorem I.2.3.1(a)) and every element of $D_\varnothing$ acts ergodically on $\Gamma \setminus G$ (see Corollary II.7.3(a) and Proposition I.1.4.10). □

(5.19) In this paragraph we use the notation accepted in V.4.8. In particular, k denotes a local field, $\mathbf{F}$ denotes a connected k-group with property (L), $\mathbf{H} = \mathbf{F}/R_u(\mathbf{F})$, $\mathfrak{F} = \operatorname{Lie}(\mathbf{F})$, $\mathfrak{R} = \operatorname{Lie}(R_u(\mathbf{F}))$, $\mathfrak{H} = \operatorname{Lie}(\mathbf{H}) = \mathfrak{F}/\mathfrak{R}$, $\mathfrak{H}^{(s)}$ the set of semisimple elements in $\mathfrak{H}$, $\sigma\colon \mathbf{F} \to \mathbf{H}$ denotes the natural epimorphism, and $\mathscr{Z}_{\mathfrak{H}}(\mathfrak{R}) = \operatorname{Ker} \operatorname{ad}_{\mathfrak{R}} \subset \mathfrak{H}$, where $\operatorname{ad}_{\mathfrak{R}}$ is the representation of the Lie algebra $\mathfrak{H}$ on the space $\mathfrak{R}$ defined in V.4.5.

We shall also be using the concept of Γ-integrable representation introduced in V.3.0.

Theorem. *Let $\tau\colon \Lambda \to \mathbf{F}(k)$ be a homomorphism with $\tau(\Lambda)$ Zariski dense in $\mathbf{F}$. Suppose the following conditions are satisfied:*

(i) *for each $\alpha \in A$ the group $\mathbf{G}_\alpha$ is simply connected;*

(ii) *either* rank $G \geq 2$ *and the lattice Γ is irreducible or the subgroup Λ is dense in G;*

(iii) *the restriction of the homomorphism $\sigma \circ \tau\colon \Lambda \to \mathbf{H}(k)$ to Γ almost extends (in the sense of Definition V.3.5) to a continuous homomorphism $\rho\colon G \to \mathbf{H}(k)$ such that $\rho(G) \neq \{e\}$;*

(iv) *the representation* $\operatorname{Ad} \circ \tau\colon \Lambda \to \mathbf{GL}(\mathfrak{F}_k)$ *is Γ-integrable;*

(v) $\mathfrak{H}^{(s)} \cap \mathscr{Z}_{\mathfrak{H}}(\mathfrak{R}) = \{0\}$.

Then τ extends uniquely to a continuous homomorphism $\tilde{\tau}\colon G \to \mathbf{F}(k)$.

Proof. It follows from Lemma 5.18(a) that there exists $s \in S$ acting ergodically on $\Gamma \setminus G$ with $\varphi(s)$ essentially non-compact. Now, making use of conditions (iii), (iv), and (v) and applying Theorem V.4.12, we see that the pair $(s, \operatorname{Ad} \circ \tau)$ is effective. Thus, there exist $m \in \mathbb{N}^+$, a faithful rational m-dimensional representation ρ of the group $\operatorname{Ad} \mathbf{F}$ which is defined over k and a measurable strictly $(\operatorname{Ad} \mathbf{F})$-effective map $\varphi\colon G \to k^m$ such that for each $\lambda \in \Lambda$

$$\varphi(\lambda g z) = \rho(\operatorname{Ad}(\tau(\lambda)))\varphi(g)$$

for almost all $g \in G$ and each $z \in \mathscr{Z}_G(S) \subset \mathscr{Z}_G(s)$ (see Proposition V.4.3). Then, arguing as in the proof of Theorem 5.4 (if Λ is dense) or as in the proof of Theorem

5.6 (if $\operatorname{rank} G \geq 2$ and Γ is irreducible), we obtain that the representation $\operatorname{Ad} \circ \tau$ extends to a continuous representation $\tilde{T}\colon G \to \mathbf{GL}(\mathfrak{F}_k)$. The reduction given in the proof of Theorem 5.6 to the case that $\mathbf{G}_\alpha$ is almost k_α-simple is replaced by the following argument: "Since $\mathbf{G}_\alpha$ is simply connected, $\mathbf{G}_\alpha$ can be decomposed into a direct product of almost k_α-simple groups. Thus, we may assume that the groups $\mathbf{G}_\alpha$ are almost k_α-simple".

Since $\mathbf{F}$ has property (L), Ker Ad is either trivial or contains $R_u(\mathbf{F})$, where Ad is the adjoint representation of the group $\mathbf{F}$. If $\operatorname{Ker} \operatorname{Ad} \supset R_u(\mathbf{F})$, then $\mathscr{Z}_{\mathfrak{H}}(\mathfrak{R}) = \mathfrak{H}$. This contradicts (v). Thus, $\operatorname{Ker} \operatorname{Ad} = \{E\}$. Now set $\tilde{\tau} = \operatorname{Ad}^{-1} \circ \tilde{T}$. An argument similar to the one completing the proof of Theorem 5.6 shows that the homomorphism $\tilde{\tau}$ is a continuous extension of τ mapping G to $\mathbf{F}(k)$. Finally, as in the proof of Theorem 5.6, the uniqueness of $\tilde{\tau}$ is deduced from Lemma II.2.3 and Proposition II.4.6. □

(5.20) *Remark.* It follows from Theorem 5.13(b) that condition (iii) of Theorem 5.19 is satisfied if the subgroup $(\sigma \circ \tau)(\Gamma)$ is Zariski dense in $\mathbf{H}$ and is not relatively compact in $\mathbf{H}(k)$.

In paragraphs 5.21–5.25 below we will be using the notation introduced in I.0.41.

(5.21) Corollary. *Let k be a local field, $n \in \mathbb{N}^+$, and ρ a continuous non-trivial absolutely irreducible representation of the group G on the space k^n. Suppose the following conditions are satisfied:*

(i) *for each $\alpha \in A$ the group $\mathbf{G}_\alpha$ is simply connected;*

(ii) *either* $\operatorname{rank} G \geq 2$ *and the lattice Γ is irreducible or the subgroup Λ is dense in G;*

(iii) *every representation of the group Λ on a finite-dimensional space over k is Γ-integrable.*

Then the restriction map of cohomology groups $H^1_{\text{cont}}(G, \rho) \to H^1(\Lambda, \rho)$ is an isomorphism.

Proof. Denote by K the algebraic closure of k, by $\mathbf{H}$ the Zariski closure of the subgroup $\rho(G)$ in $\mathbf{GL}_n$, and by $\mathbf{F}$ the semidirect product $K^n \rtimes \mathbf{H}$. It is easily seen that the k-group $\mathbf{F}$ can be realized as the group of the matrices of order $n+1$ of the form

$$\left(\begin{array}{c|c} h & x \\ \hline 0 & 1 \end{array}\right), \quad h \in \mathbf{H},\ x \in K^n.$$

Under this realization $\mathbf{H}$ and K^n are groups of the matrices of the form

$$\left(\begin{array}{c|c} h & 0 \\ \hline 0 & 1 \end{array}\right) \text{ and } \left(\begin{array}{c|c} E & x \\ \hline 0 & 1 \end{array}\right),$$

where E is a unit matrix of order n.

It follows from Corollary I.2.3.2(b) that G does not contain proper subgroups of finite index. Thus, the k-group $\mathbf{H}$ does not contain proper algebraic subgroups of finite index, and hence is connected. If a connected algebraic group admits a faithful rational absolutely irreducible representation, then it is reductive (see I.0.25). But the representation ρ is absolutely irreducible and $\mathbf{H}$ is the Zariski closure of the subgroup $\rho(G)$. Thus, the connected k-group $\mathbf{H}$ is reductive. From this and the above matrix realization of $\mathbf{F}$ we deduce that $R_u(\mathbf{F}) = K^n$ and $\mathbf{H}$ is a Levi subgroup in $\mathbf{F}$.

Now we shall prove the following assertion.

(∗) Every non-trivial algebraic subgroup $\mathbf{P}$ of the group $K^n = R_u(\mathbf{F})$ normalized by $\mathbf{H}$ coincides with K^n.

First, we observe that it is straightforward to check that

$$(x,h)(y,e)(x,h)^{-1}(y,e)^{-1} = (hy - y, e) \tag{1}$$

for all $x, y \in K^n$ and $h \in \mathbf{H}$. Assume further that $y = (y,e) \in \mathbf{P}$, $y \neq 0$. Since the representation ρ is non-trivial and absolutely irreducible, $\mathbf{H} \supset \rho(G)$, and $y \neq 0$, it follows that $\mathbf{H}y \neq \{y\}$. But the algebraic group $\mathbf{H}$ is connected and normalizes $\mathbf{P}$. So (1) implies that $\dim \mathbf{P} > 0$, and hence,

$$\mathrm{Lie}(\mathbf{P}) \neq 0. \tag{2}$$

It is easily seen that, under the above matrix realization of the group $\mathbf{F}$,

$$\mathrm{Lie}(K^n) = \left\{ \left(\begin{array}{c|c} 0 & x \\ \hline 0 & 0 \end{array} \right) \middle| \, x \in K^n \right\} \tag{3}$$

and

$$\mathrm{Ad}\, h \left(\begin{array}{c|c} 0 & x \\ \hline 0 & 0 \end{array} \right) = \left(\begin{array}{c|c} 0 & hx \\ \hline 0 & 0 \end{array} \right). \tag{4}$$

Since ρ is absolutely irreducible and $\mathbf{H} \supset \rho(G)$, (2), (3), and (4) imply that $\mathrm{Lie}(\mathbf{P}) = \mathrm{Lie}(K^n)$. But the algebraic group K^n is connected, so $\mathbf{P} = K^n$, which proves (∗).

Let $\mathbf{M}$ be a non-trivial algebraic normal subgroup of $\mathbf{F}$. Since $\mathbf{M} \neq \{e\}$, K^n normalizes $\mathbf{M}$, and $\mathbf{H}$ acts effectively on K^n, it follows from (1) that $\mathbf{M} \cap K^n \neq \{e\}$. But $\mathbf{H}$ normalizes the algebraic subgroup $\mathbf{M} \cap K^n$. So (∗) implies that $\mathbf{M} \supset K^n$. On the other hand, $\{e\} \neq R_u(\mathbf{F}) = K^n \neq \mathbf{F}$. Thus, $\mathbf{F}$ has property (L).

Let $\sigma\colon \mathbf{F} = K^n \rtimes \mathbf{H} \to \mathbf{H}$ be the natural epimorphism. As was remarked in I.0.41 there exists the natural one-to-one correspondence between $H^1(\Lambda, \rho)$ (resp. $H^1_{\mathrm{cont}}(G, \rho)$) and a set of equivalence classes of all (resp. continuous) homomorphisms of the group Λ (resp. G) to $\mathbf{F}(k) = k^n \rtimes \mathbf{H}(k)$ covering ρ. Thus the assertion which we want to prove is equivalent to the following one:

(A) *Let $\tau\colon \Lambda \to \mathbf{F}(k)$ be a homomorphism with $\sigma \circ \tau = \rho|_\Lambda$. Then τ extends uniquely to a continuous homomorphism $\tilde{\tau}\colon G \to \mathbf{F}(k)$ such that $\sigma \circ \tilde{\tau} = \rho$.*

We shall prove (A). Let $\overline{\tau(\Lambda)}$ be the Zariski closure of the subgroup $\tau(\Lambda)$ in $\mathbf{F}$. Since Γ is a lattice in G, it follows from Lemma II.2.3 and Proposition

II.4.3(b) that the subgroup $\rho(\Gamma) \subset \rho(\Lambda)$ is Zariski dense in $\mathbf{H}$. But $\sigma \circ \tau = \rho|_\Lambda$ and the image of an algebraic subgroup under an algebraic group morphism is an algebraic subgroup. Thus

$$(5) \qquad \sigma(\overline{\tau(\Lambda)}) = \mathbf{H}.$$

We will now show that τ extends to a continuous homomorphism $\tilde{\tau}\colon G \to \mathbf{F}(k)$. We consider two cases (a) $\overline{\tau(\Lambda)} \neq \mathbf{F}$ and (b) $\overline{\tau(\Lambda)} = \mathbf{F}$.

(a) Since the subgroup $K^n \subset \mathbf{F}$ is commutative and normal, $K^n \cdot \overline{\tau(\Lambda)}$ normalizes $K^n \cap \overline{\tau(\Lambda)}$. On the other hand, since $\mathbf{F} = K^n \rtimes \mathbf{H}$, $\operatorname{Ker}\sigma = K^n$, and $\sigma(\overline{\tau(\Lambda)}) = \mathbf{H}$ (see (5)), we have that $K^n \cdot \overline{\tau(\Lambda)} = \mathbf{F}$. From this and (*) we deduce that in the case under consideration $K^n \cap \overline{\tau(\Lambda)} = \{e\}$, and hence $\operatorname{Ker}\sigma_0 = \{e\}$, where $\sigma_0 = \sigma|_{\overline{\tau(\Lambda)}}$. It then follows from Corollary I.2.1.3 that $\sigma_0^{-1}(\mathbf{H}(k)) \subset \overline{\tau(\Lambda)}(l)$, where l is a finite field extension of k and the homomorphism $\sigma_0^{-1}\colon \mathbf{H}(k) \to \sigma_0^{-1}(\mathbf{H}(k))$ is a topological group isomorphism. We set $\tilde{\tau} = \sigma_0^{-1} \circ \rho$. Since $\sigma \circ \tau = \rho|_\Lambda$, $\sigma_0 = \sigma|_{\overline{\tau(\Lambda)}}$, and the homomorphisms σ_0^{-1} and ρ are continuous, it follows that the homomorphism $\tilde{\tau}\colon G \to \overline{\tau(\Lambda)}(l)$ is a continuous extension of τ. Since Γ is a lattice in G and $\tilde{\tau}(\Gamma) = \tau(\Gamma) \subset \overline{\tau(\Lambda)}(k)$, it follows from Lemma II.2.3 and Proposition II.4.7 that $\tilde{\tau}(G) \subset \mathbf{F}(k)$.

(b) As we have shown above, $\mathbf{F}$ has property (L). We clearly have

$$\operatorname{Lie}(\mathbf{H}) \subset \left\{ \begin{pmatrix} A & 0 \\ 0 & 0 \end{pmatrix} \middle| A \in \operatorname{End}_n \right\} \text{ and}$$

$$\left[\begin{pmatrix} A & 0 \\ 0 & 0 \end{pmatrix}, \begin{pmatrix} 0 & x \\ 0 & 0 \end{pmatrix} \right] = \begin{pmatrix} 0 & Ax \\ 0 & 0 \end{pmatrix}$$

for all $A \in \operatorname{End} K^n$ and $x \in K^n$. From this and (3) we deduce that $[h, \operatorname{Lie}(K^n)] \neq 0$ for each $h \in \operatorname{Lie}(\mathbf{H})$, $h \neq 0$. Thus, the centralizer $\mathscr{Z}_{\mathbf{F}}(\operatorname{Lie}(K^n))$ coincides with $\operatorname{Lie}(K^n)$ and, since $K^n = R_u(\mathbf{F})$, condition (v) of Theorem 5.19 is satisfied. So we can apply Theorem 5.19 and obtain the existence of the desired extension.

As in the proofs of Theorems 5.6 and 5.19, the uniqueness of the extension $\tilde{\tau}$ is a consequence of Lemma II.2.3 and Proposition II.4.6. Finally, since Γ is a lattice in $\mathbf{G}$ and $(\sigma \circ \tilde{\tau})(\gamma) = (\sigma \circ \tau)(\gamma) = \rho(\gamma)$ for all $\gamma \in \Gamma$, once again applying Lemma II.2.3 and Proposition II.4.6, we obtain that $\sigma \circ \tilde{\tau} = \rho$. □

(5.22) Lemma. *Let D be a topological group, l a topological field of characteristic 0, $n \in \mathbb{N}^+$, and ρ a continuous representation of D on the space l^n. Suppose the following condition is satisfied:*

(U) for every l-group $\mathbf{B}$ and every continuous homomorphism $f\colon D \to \mathbf{B}(l)$, the Zariski closure of the subgroup $f(D)$ in $\mathbf{B}$ is reductive.

Then $H^1_{\text{cont}}(D, \rho) = 0$.

Proof. Let K be the algebraic closure of l, $\mathbf{H}$ the Zariski closure of the subgroup $\rho(D)$ in $\mathbf{GL}_n$, and let $\mathbf{F} = K^n \rtimes \mathbf{H}$. Since the l-group $\mathbf{H}$ is reductive (see (U)), as in the proof of Corollary 5.21 we check that $R_u(\mathbf{F}) = K^n$ and $\mathbf{H}$ is a Levi subgroup in $\mathbf{F}$. Let $\sigma\colon \mathbf{F} = K^n \rtimes \mathbf{H} \to \mathbf{H}$ be the natural epimorphism and $\tau\colon D \to \mathbf{F}(l)$ a continuous homomorphism with $\sigma \circ \tau = \rho$. We denote by $\overline{\tau(D)}$ the Zariski closure of the subgroup $\tau(D)$ in $\mathbf{F}$. By condition (U) the l-group $\overline{\tau(D)}$ is reductive. But $R_u(\mathbf{F}) = K^n$, $\mathbf{H}$ is a Levi l-subgroup in $\mathbf{F}$, and $\operatorname{char} l = 0$. Thus, there exists $x \in K^n$ such that $x \cdot \overline{\tau(D)} \cdot x^{-1} \subset \mathbf{H}$ (see I.0.28). Since $\sigma \circ \tau = \rho$ it then follows that $x\tau(d)x^{-1} = \rho(d)$ for all $d \in D$. Therefore we have shown that any continuous homomorphism $\tau\colon D \to \mathbf{F}(k)$ covering ρ is equivalent to ρ. In view of I.0.41, this proves the lemma. □

(5.23) Corollary. *Let k be a local field, $n \in \mathbb{N}^+$, and ρ a non-trivial continuous absolutely irreducible representation of the group G on the space k^n. Suppose* $\operatorname{char} k = 0$ *and conditions (i), (ii), and (iii) of Corollary 5.21 are satisfied. Then* $H^1(\Lambda, \rho) = 0$.

Proof. According to Proposition I.2.6.4(b), for any k-group $\mathbf{H}$ and any continuous homomorphism $f\colon G \to \mathbf{H}(k)$, the Zariski closure of the subgroup $f(G)$ in $\mathbf{H}$ is semisimple. From this and Lemma 5.22 we deduce that $H^1_{\text{cont}}(G, \rho) = 0$. All that remains now is to make use of Corollary 5.21. □

(5.24) *Remark.* (i) As was remarked in V.3.0, if $\Gamma \setminus G$ is compact, then any representation of the group Γ on a finite-dimensional vector space over a local field k is Γ-integrable. Thus, condition (iv) of Theorem 5.19 and condition (iii) of Corollary 5.21 are automatically satisfied if $\Gamma \setminus G$ is compact.

(ii) We shall show in the sequel (see Corollary IX.5.9) that if $\operatorname{rank} G \geq 2$, the lattice Γ is irreducible, and $\operatorname{char} k_\alpha = 0$ for some $\alpha \in A$, then $H^1(\Gamma, \rho) = 0$ for any representation ρ of Γ on a finite-dimensional vector space over a field of characteristic 0.

(iii) In the statements of all the results of this section the condition "$\operatorname{rank} G \geq 2$ and the lattice Γ is irreducible" can be replaced by the following one:

(∗) the subgroup $\Gamma \cap \mathbf{G}'(k_\alpha)$ is not a lattice in $\mathbf{G}'(k_\alpha)$ for each $\alpha \in A$ and every almost k_α-simple factor $\mathbf{G}'$ of the group $\mathbf{G}_\alpha$ such that $\operatorname{rank}_{k_\alpha} \mathbf{G}' = 1$.

In case $\mathbf{G}_\alpha$ is almost k_α-simple for each $\alpha \in A$, it is not hard to show that condition (∗) is equivalent both to the condition "the subgroup $\operatorname{pr}_\alpha(\Gamma)$ is not discrete in $\mathbf{G}_\alpha(k_\alpha)$, whenever $\alpha \in A$ and $\operatorname{rank}_{k_\alpha} \mathbf{G}_\alpha = 1$" and the condition

(∗∗) the closure of the subgroup $\operatorname{pr}_\alpha(\Gamma)$ in $\mathbf{G}_\alpha(k_\alpha)$ contains $\mathbf{G}_\alpha(k_\alpha)^+$, whenever $\alpha \in A$ and $\operatorname{rank}_{k_\alpha} \mathbf{G}_\alpha = 1$.

To convince oneself that the condition "$\operatorname{rank} G \geq 2$ and the lattice Γ is irreducible" may be replaced by (∗) it suffices to show that the statement of Lemma 5.5 remains valid if condition (∗∗) is satisfied instead of the condition "$\operatorname{rank} G \geq 2$ and the lattice Γ is irreducible."

(5.25) Let H be a locally compact group and Φ a countable subgroup of H. Let us denote by $\mathscr{A}(\Phi, H)$ the space of all homomorphisms of the group Φ to H. We equip this with the topology of point-wise convergence. There is the natural action of H on $\mathscr{A}(\Phi, H)$ by inner automorphisms:

$$(hf)(x) = hf(x)h^{-1}, \; h \in H, \; x \in \Phi, \; f \in \mathscr{A}(\Phi, H).$$

Let id: $\Phi \to H$, $\mathrm{id}(x) = x$, be the identity embedding. The subgroup Φ is called *locally* (or *infinitesimally*) *rigid* if the orbit of the homomorphism id under H is open in $\mathscr{A}(\Phi, H)$.

If k is a local field, $\mathbf{G}$ a k-group, $H = \mathbf{G}(k)$, and Φ is finitely generated, then by a result of Weil (see [Weil 3] and also [Rag 5], Theorem 6.7) the triviality of the cohomology group $H^1(\Phi, \mathrm{Ad})$ implies the local rigidity of the subgroup Φ. From this and Remark (ii) in 5.24 we deduce

(A) Theorem. *Suppose* $\operatorname{rank} G \geq 2$, *the lattice* Γ *is irreducible, and* $\operatorname{char} k_\alpha = 0$ *for some* $\alpha \in A$. *Then the subgroup* Γ *is locally rigid.*

In Theorem A the condition "$\operatorname{char} k_\alpha = 0$ for some $\alpha \in A$" is essential. This is shown in the following

(B) Theorem. (See [Pr 4].) *Let* F *be a finite field,* $k = F((t))$ *the field of formal power series in one variable over* F, $\mathbf{G}$ *a connected non-trivial semisimple* F*-group with trivial centre and* Γ' *a finitely generated lattice in* $\mathbf{G}(k)$. *Then the subgroup* Γ' *is not locally rigid and hence* $H^1(\Gamma', \mathrm{Ad}) \neq 0$.

Thus, condition "$\operatorname{char} k = 0$" in Corollary 5.23 is essential. We present a sketch of the proof of Theorem B. For each $i > 1$ there is a unique continuous automorphism ϑ_i of the field $F((t))$ over F sending t to $t + t^i$. Since the group $\mathbf{G}$ is defined over F, ϑ_i induces a continuous automorphism $\vartheta_i^0 \colon \mathbf{G}(k) \to \mathbf{G}(k)$. Making use of Lang's theorem according to which any non-trivial connected semisimple F-group is isotropic over F, it is not hard to show that the automorphisms ϑ_i^0 are not inner. But by Lemma II.2.3 and Proposition II.4.8, if $\varphi \colon G \to G$ is a continuous automorphism and $\varphi(\gamma) = \gamma$ for all $\gamma \in \Gamma'$, then $\varphi(g) = g$ for all $g \in G$. Therefore the homomorphisms $\vartheta_i^0|_{\Gamma'} \in \mathscr{A}(\Gamma', \mathbf{G}(k))$ do not belong to the $\mathbf{G}(k)$-orbit of the identity embedding id: $\Gamma \to \mathbf{G}(k)$. All that remains now is to observe that for each $g \in \mathbf{G}(k)$ the sequence $\{\vartheta_i^0(g)\}$ converges to g as $i \to \infty$. □

Let us denote by $\operatorname{Aut} H$ (resp. $\operatorname{Int} H$) the group of continuous (resp. inner) automorphisms of the group H. The above proof of Theorem B was based on the fact that the subgroup $\operatorname{Int} \mathbf{G}(k)$ is not open in $\operatorname{Aut} \mathbf{G}(k)$. It turns out that in the majority of cases this is the only reason for the lattice Γ' to be locally non-rigid. Namely, the following is true.

(C) Theorem. *Suppose* $\operatorname{rank} G \geq 2$ *and the lattice* Γ *is irreducible. Then the orbit of the identity embedding* id: $\Gamma \to G$ *under the natural action of the group* $\operatorname{Aut} G$ *on* $\mathscr{A}(\Gamma, G)$ *is open in* $\mathscr{A}(\Gamma, G)$.

We omit the proof of Theorem C. Note however that this theorem can easily be deduced from Theorem 5.6 with the help of the Borel-Wang density theorem.

(5.26) There is a different approach to the proof of the superrigidity theorems (see [Fu 8], [Gu], [Mar 7], [Mar 9], [Ti 4], [Zi 1], [Zi 4], [Zi 8]) which is based on the study of equivariant measurable maps not only to vector spaces but also to more general algebraic varieties. We shall illustrate this approach by considering the proof of Theorem 5.9. It is not hard to show that this theorem is a consequence of the following two propositions.

(1) Proposition. *With* $\mathbf{G}$, Γ', Λ', k, $\mathbf{H}$, *and* τ *as in the statement of Theorem 5.9, let* $\mathbf{P}$ *be a minimal parabolic* $\mathbb{R}$*-subgroup in* $\mathbf{G}$ *and suppose the subgroup* $\tau(\Gamma')$ *is not relatively compact in* $\mathbf{H}(k)$. *Then there exist a k-rational action of the group* $\mathbf{H}$ *on a k-variety* $\mathbf{M}$ *and a measurable strictly* $\mathbf{H}$*-effective (in the sense of V.4.1)* Λ'*-equivariant map* $\psi\colon \mathbf{G}(\mathbb{R})/\mathbf{P}(\mathbb{R}) \to \mathbf{M}(k)$.

(2) Proposition. *Suppose that a k-group* $\mathbf{H}$ *acts k-rationally on a k-variety* $\mathbf{M}$. *Let* $\psi\colon \mathbf{G}(\mathbb{R})/\mathbf{P}(\mathbb{R}) \to \mathbf{M}(k)$ *be a measurable* Λ'*-equivariant map and assume that* $\operatorname{rank}_{\mathbb{R}} \mathbf{G} \geq 2$ *and the lattice* Γ' *is irreducible.*

(a) *If k is* $\mathbb{R}$ *or* $\mathbb{C}$, *then* ψ *agrees almost everywhere (with respect to the measure* $\mu_{\mathbf{G}(\mathbb{R})/\mathbf{P}(\mathbb{R})}$*) with a rational map.*

(b) *If k is neither isomorphic to* $\mathbb{R}$ *nor to* $\mathbb{C}$, *i.e., if k is non-archimedean, then* ψ *agrees almost everywhere with a constant map to a point.*

Proposition 1 is a special case of Theorem VI.4.7(a). At present there are various proofs of this proposition. The original proof, which was presented in Section 5 of Chapter V, is based on the multiplicative ergodic theorem (see [Mar 7] and [Ti 4]). In other proofs the first step consists of the use of Furstenberg's theorem on the existence of Γ'-equivariant measurable maps to the space $\mathscr{P}(X)$ of measures, where X is a compact metric Γ'-space (see Theorem IV.4.5). Next, as in Section 4 of Chapter VI, either probability arguments are applied (see [Fu 6], [Fu 8], [Gu]) or one considers equivariant measurable maps of the space of measures on a projective space to certain algebraic varieties (see [Mar 9]) or one uses the smoothness of certain actions of algebraic groups on suitable spaces of measures (see [Zi 1], [Zi 4], [Zi 8]). Notice that in [Mar 9], [Zi 1], and [Zi 4] Γ'-equivariant measurable maps have been constructed. However their Λ'-equivariance was not proved.

The proof of Proposition 2 is based upon the study of actions of algebraic groups on suitable spaces of measures and the theorem on rationality of a function which is rational in each of its variable.

The approach described above can be directly applied in case $\operatorname{char} k_\alpha = 0$ for all $\alpha \in A$. If $\operatorname{char} k_\alpha \neq 0$ for some $\alpha \in A$, then one encounters technical difficulties. These were surmounted in [Ve]. Notice also that in the case where the subgroup Λ contains G^+, all the results of the present section can be deduced from appropriate generalizations of Proposition 1 without using Proposition 2.

(5.27) In the statements of the results of the first part of this section (until 5.17), an analysis of the proofs shows that it is not necessary to assume that the discrete subgroup is a lattice. It suffices only to assume that the following conditions are satisfied:

(a) there exist $i \in \mathbb{N}^+$ and $s_1, \ldots, s_i \in S$ such that $\mathscr{Z}_G(s_1) \cdot \ldots \cdot \mathscr{Z}_G(s_i) = G$ and s_j acts ergodically on $\Gamma \setminus G$ for all j, $1 \leq j \leq i$;

(b) there is a function f on G such that
  1) f is continuous, has compact support, and $\int_G f \, d\mu_G = 1$;
  2) if L is a bounded measurable function on $\Gamma \setminus G$ and $L(x) \geq \int_G L(xg) f(g) \, d\mu_G(g)$

for all $x \in \Gamma \setminus G$, then L is almost everywhere (in μ_G) constant.

Furthermore, if the closure of the subgroup Λ contains G^+, then condition (a) is superfluous. It seems likely that if (a) is satisfied (b) is satisfied as well. We also remark that the following two assertions hold.

(i) *Suppose the group G has property (T). Then condition* (b) *is equivalent to the statement that Γ is a lattice in G.*

(ii) *Suppose the group $\mathbf{G}_{\alpha i}(k_\alpha)$ has property (T) for each $\alpha \in A$ and every almost k_α-simple factor $\mathbf{G}_{\alpha i}$ of the group $\mathbf{G}_\alpha$. Then every discrete subgroup Γ in G, such that some $s \in S$ acts ergodically on $\Gamma \setminus G$, is a lattice in G.*

It should be noted that the problem of ergodicity of the action of an element $s \in S$ on $\Gamma \setminus G$ in the case where G is locally isomorphic to the group of motions of Lobachevsky space and $\Gamma \setminus G$ is of infinite measure was considered (though in a somewhat different way) in [Shir], [Sul 1], and in some other papers.

6. Homomorphisms of Discrete Subgroups to Algebraic Groups Over Arbitrary Fields

In this section we let A, k_α, $\mathbf{G}_\alpha$, G, G^+, pr_α, $\operatorname{rank} G$, Γ, and Λ be as in Section 5, $\bar{M}$ will denote the Zariski closure of a subset M in an algebraic variety, and $\mathbf{H}^0$ will denote the identity component of an algebraic group $\mathbf{H}$. Given an l-group $\mathbf{H}$ and a homomorphism σ of the field l to a field l', we define as in I.1.7 and I.1.8 the l'-group ${}^\sigma\mathbf{H}$ and the homomorphism

$$\sigma^0\colon \mathbf{H}(l) \to {}^\sigma\mathbf{H}(l').$$

In the preceding section we studied the homomorphisms of the form $\tau\colon \Lambda \to \mathbf{H}(k)$, where k is a local field and $\mathbf{H}$ is a k-group. Now we shall consider homomorphisms of the group Λ to the sets of l-rational points of algebraic groups defined over an arbitrary field l. The proofs of the theorems of this section are based on a reduction to the results of Section 5. In this reduction the following plays a main role.

(6.1) Lemma. *Let l be a finitely generated field, $\mathbf{H}$ a connected non-trivial reductive l-group, and Φ a subgroup of $\mathbf{H}(l)$ which is Zariski dense in $\mathbf{H}$. Then there exist a semisimple element $g \in \Phi$, a local field k, and a homomorphism $\sigma\colon l \to k$ such* *that $\sigma^0(g) \in {}^{\sigma}\mathbf{H}(k)$ is essentially non-compact (in the sense of Definition V.4.6), and hence the subgroup $\sigma^0(\Phi)$is not relatively compact in ${}^{\sigma}\mathbf{H}(k)$.*

This lemma will be proved in 6.22. We shall also need the following

(6.2) Lemma. *Let l be a field, $\mathbf{H}$ a connected l-group, and let Φ be a subgroup of $\mathbf{H}(l)$ which is Zariski dense in $\mathbf{H}$. Denote by $\tilde{\Phi}$ the commensurability subgroup* $\mathrm{Comm}_{\mathbf{H}}(\Phi)$.

(i) *If $\mathrm{Ad}\colon \mathbf{H} \to \mathrm{Ad}\,\mathbf{H}$ is an l-group isomorphism, then $\tilde{\Phi} \subset \mathbf{H}(l)$.*
(ii) *If the group $\mathbf{H}$ is semisimple and adjoint, then $\tilde{\Phi} \subset \mathbf{H}(l)$.*
(iii) *If $\mathrm{char}\, l = 0$ and $\mathscr{Z}(\mathbf{H}) = \{e\}$, then $\tilde{\Phi} \subset \mathbf{H}(l)$.*

Proof. (i) Let $h \in \tilde{\Phi}$. We set $B = \Phi \cap h^{-1}\Phi h$. Since Φ is Zariski dense in $\mathbf{H}$, the subgroups Φ and $h^{-1}\Phi h$ are commensurable and the connected group $\mathbf{H}$ does not contain algebraic subgroups of finite index, it follows that B is Zariski dense in $\mathbf{H}$. But $B \subset \mathbf{H}(l)$ and $(\mathrm{Int}\, h)(B) \subset \mathbf{H}(l)$. Thus the automorphism $\mathrm{Int}\, h\colon \mathbf{H} \to \mathbf{H}$ is defined over l (see I.0.11 (II)), and hence $\mathrm{Lie}(\mathbf{H})_l$ is invariant relative to the transformation $\mathrm{Ad}\, h$ which is the differential of the automorphism $\mathrm{Int}\, h$ (see I.0.15). Therefore $\mathrm{Ad}\, h \in (\mathrm{Ad}\,\mathbf{H})(l)$. But $\mathrm{Ad}\colon \mathbf{H} \to \mathrm{Ad}\,\mathbf{H}$ and $\mathrm{Ad}^{-1}\colon \mathrm{Ad}\,\mathbf{H} \to \mathbf{H}$ are both l-group l-isomorphisms. So $h \in \mathbf{H}(l)$.

Assertion (ii) is a consequence of (i) and Proposition I.1.4.11(iv). If $\mathrm{char}\, l = 0$, then any bijective l-group l-morphism is an l-isomorphism and $\mathrm{Ker}\,\mathrm{Ad} = \mathscr{Z}(\mathbf{H})$ (see I.0.15). Thus, (iii) is a consequence of (i) as well. □

It is good to keep the following in mind.

(6.3) Corollary. *The commensurability subgroup* $\mathrm{Comm}_G(\Gamma)$ *is countable.*

Proof. Take $\alpha \in A$. Since the group Γ is countable, for some countable subfield $l_\alpha \subset k_\alpha$ there is an l_α-structure of the Lie algebra $\mathrm{Lie}(\mathbf{G}_\alpha)$ which is invariant under $\mathrm{Ad}(\mathrm{pr}_\alpha(\Gamma))$. By Lemma II.2.3, Lemma II.2.4, and Corollary II.4., the subgroup $\mathrm{Ad}(\mathrm{pr}_\alpha(\Gamma))$ is Zariski dense in $\mathrm{Ad}\,\mathbf{G}_\alpha$. Therefore, by Lemma 6.2(ii), the above-mentioned l_α-structure is invariant under $\mathrm{Ad}(\mathrm{pr}_\alpha(\mathrm{Comm}_G(\Gamma)))$, and hence the group $\mathrm{Ad}(\mathrm{pr}_\alpha(\mathrm{Comm}_G(\Gamma)))$ is countable. From this and the finiteness of the kernels of the morphisms $\mathrm{Ad}\colon \mathbf{G}_\alpha \to \mathrm{Ad}\,\mathbf{G}_\alpha$ we deduce the countability of the group $\mathrm{Comm}_G(\Gamma)$. □

(6.4) If Φ is a finitely generated subgroup of $\mathbf{GL}_n$, then $\Phi \subset \mathbf{GL}_n(l)$, where l is a finitely generated field (l coincides with the field generated by the matrix coefficients of generators of the group Φ). Thus, in the study of homomorphisms of an arbitrary finitely generated group Ω into linear algebraic groups it suffices to consider the homomorphisms $f\colon \Omega \to \mathbf{H}(l)$, where l is a finitely generated field and $\mathbf{H}$ is an l-group.

(6.5) Theorem. *Let l' be a field, $l \subset l'$ a finitely generated subfield, $\mathbf{H}$ a connected non-commutative absolutely almost simple l-group, and $\delta\colon \Lambda \to \mathbf{H}(l')$ a homomorphism with $\delta(\Gamma) \subset \mathbf{H}(l)$. Suppose the following conditions are satisfied:*

(i) *either* rank $G \geq 2$ *and the lattice* Γ *is irreducible or the closure of the subgroup* Λ *contains* G^+;
(ii) *the subgroup* $\delta(\Gamma)$ *is Zariski dense in* $\mathbf{H}$;
(iii) *either the group* $\mathbf{H}$ *is adjoint or the group* $\mathbf{G}_\alpha$ *is simply connected for each* $\alpha \in A$;
(iv) *the lattice* Γ *is finitely generated.*

Then there exist $\alpha \in A$, *a finite field extension* k *of* k_α, *a field extension* k' *of* k, *a homomorphism* $\sigma\colon l' \to k'$ *which maps* l *into* k, *a special (in the sense of I.1.4.14) k-epimorphism* $\eta\colon \mathbf{G}_\alpha \to {}^\sigma\mathbf{H}$, *and a homomorphism* $\nu\colon \Lambda \to \mathscr{Z}({}^\sigma\mathbf{H})$ *such that* $\sigma^0(\delta(\lambda)) = \nu(\lambda) \cdot \eta(\mathrm{pr}_\alpha(\lambda))$ *for all* $\lambda \in \Lambda$.

Proof. Since the l-group $\mathbf{H}$ is connected, non-trivial, and reductive, and the subgroup $\delta(\Gamma) \subset \mathbf{H}(l)$ is Zariski dense in $\mathbf{H}$ and finitely generated, it follows from Lemma 6.1 that there exist a local field k and a homomorphism $\sigma\colon l \to k$ such that the subgroup $\sigma^0(\delta(\Gamma))$ is not relatively compact in ${}^\sigma\mathbf{H}(k)$. An arbitrary extension of the homomorphism σ to a homomorphism of the field l' into an extension k' of k will also be denoted by σ.

We first consider the case where $\mathbf{H}$ is adjoint. It then follows from Lemma 6.2(ii) that $\sigma^0(\delta(\Lambda)) \subset {}^\sigma\mathbf{H}(k)$. By applying Theorem 5.13(d) or Theorems 5.4(a) and 5.6(a) to the homomorphism $\sigma^0 \circ \delta\colon \Lambda \to {}^\sigma\mathbf{H}(k)$, we see that there exist $\alpha \in A$, a continuous homomorphism $\vartheta\colon k_\alpha \to k$, and a special k-epimorphism $\eta\colon {}^\vartheta\mathbf{G}_\alpha \to {}^\sigma\mathbf{H}$ such that $\sigma^0(\delta(\lambda)) = \nu(\lambda) \cdot \eta(\vartheta^0(\mathrm{pr}_\alpha(\lambda)))$ for all $\lambda \in \Lambda$. By identifying k_α and $\sigma(k_\alpha)$ via the homomorphism ϑ, we may assume that $\vartheta^0 = \mathrm{Id}$ and k is a field extension of k_α. This extension is finite, because the fields k and k_α are local. Thus, for $\mathbf{H}$ adjoint, the theorem is true. Suppose now that the groups $\mathbf{G}_\alpha$ are simply connected. Since the theorem is established for $\mathbf{H}$ adjoint, there exist $\alpha \in A$, a finite field extension k of k_α, a field extension k' of k, a homomorphism $\sigma\colon l' \to k'$ mapping l to k, and a special k-epimorphism $\eta'\colon \mathbf{G}_\alpha \to {}^\sigma\mathrm{Ad}\,\mathbf{H}$ such that

(1) $$(\mathrm{Ad} \circ \sigma^0 \circ \delta)(\lambda) = (\sigma^0 \circ \mathrm{Ad} \circ \delta)(\lambda) = \eta'(\mathrm{pr}_\alpha(\lambda)) \text{ for all } \lambda \in \Lambda.$$

The group $\mathbf{G}_\alpha$ is simply connected. Therefore there is a k-epimorphism $\eta\colon \mathbf{G}_\alpha \to {}^\sigma\mathbf{H}$ such that $\eta' = \mathrm{Ad} \circ \eta$ (see Proposition I.1.4.11(ii) and (iv)). Since the epimorphism η' is special and by Proposition I.1.4.11(iv) the isogeny $\mathrm{Ad}\colon \mathbf{H} \to \mathrm{Ad}\,\mathbf{H}$ is central, it follows from I.1.4.14 that the epimorphism η is also special. For $\lambda \in \Lambda$, we define $\nu(\lambda) \in {}^\sigma\mathbf{H}$ via the equality $\sigma^0(\delta(\lambda)) = \nu(\lambda) \cdot \eta(\mathrm{pr}_\alpha(\lambda))$. Since $\eta' = \mathrm{Ad} \circ \eta$, (1) implies that $\nu(\lambda) \in \mathrm{Ker}\,\mathrm{Ad} = \mathscr{Z}({}^\sigma\mathbf{H})$. Thus $\nu\colon \Lambda \to \mathscr{Z}({}^\sigma\mathbf{H})$ is a homomorphism (see Lemma 5.1(a)). □

(6.6) Theorem. *Let* $\mathbf{G}$ *be a connected semisimple* $\mathbb{R}$*-group without* $\mathbb{R}$*-anisotropic factors,* Γ' *a lattice in* $\mathbf{G}(\mathbb{R})$, *and* Λ' *a countable subgroup of* $\mathrm{Comm}_{\mathbf{G}(\mathbb{R})}(\Gamma')$ *containing* Γ'. *Given a finitely generated field* l, *its algebraic closure* $\bar{l}$, *a connected non-commutative absolutely almost simple l-group* $\mathbf{H}$, *and a homomorphism* $\delta\colon \Lambda' \to \mathbf{H}(\bar{l})$ *with* $\delta(\Gamma') \subset \mathbf{H}(l)$, *we assume the following conditions are satisfied:*

(i) *either* $\mathrm{rank}_{\mathbb{R}}\,\mathbf{G} \geq 2$ *and the lattice* Γ' *is irreducible or the closure of the subgroup* Λ' *in* $\mathbf{G}(\mathbb{R})$ *contains the identity component of the Lie group* $\mathbf{G}(\mathbb{R})$;

(ii) *the subgroup* $\delta(\Gamma')$ *is Zariski dense in* $\mathbf{H}$;
(iii) *either the group* $\mathbf{H}$ *is adjoint or the group* $\mathbf{G}$ *is simply connected;*
(iv) *the lattice* Γ' *is finitely generated.*

Then there exist a homomorphism $\sigma\colon \bar{l} \to \mathbb{C}$, *a* $\mathbb{C}$-*epimorphism* $\eta\colon \mathbf{G} \to {}^{\sigma}\mathbf{H}$, *and a homomorphism* $\nu\colon \Lambda \to \mathscr{Z}({}^{\sigma}\mathbf{H})$ *such that* $\sigma^0(\delta(\lambda)) = \nu(\lambda)\cdot\eta(\lambda)$ *for all* $\lambda \in \Lambda$.

(6.7) *Remark.* The conclusions of Theorems 6.5 and 6.6 can be restated, respectively, as follows: "Then there exist a field extension k' of l', a subfield k of k' containing l, $\alpha \in A$, a homomorphism $\sigma\colon k_\alpha \to k$, a special k-epimorphism $\eta\colon {}^{\sigma}\mathbf{G}_\alpha \to \mathbf{H}$, and a homomorphism $\nu\colon \Lambda \to \mathscr{Z}(\mathbf{H})$ such that the degree of the field k over $\sigma(k_\alpha)$ is finite and $\delta(\lambda) = \nu(\lambda)\cdot\eta(\sigma^0(\mathrm{pr}_\alpha(\lambda)))$ for all $\lambda \in \Lambda$" and "Then there exist an isomorphism σ of the field $\mathbb{C}$ onto a field extension k of $\bar{l}$, a k-epimorphism $\eta\colon {}^{\sigma}\mathbf{G} \to \mathbf{H}$, and a homomorphism $\nu\colon \Lambda \to \mathscr{Z}(\mathbf{H})$ such that $\delta(\lambda) = \nu(\lambda)\cdot\eta(\sigma^0(\lambda))$ for all $\lambda \in \Lambda$".

(6.8) Lemma. *Let* $\mathbf{H}$ *be a connected non-solvable algebraic group. Then* $\mathbf{H}$ *contains a normal algebraic subgroup* $\mathbf{F}$ *such that the factor group* $\mathbf{H}/\mathbf{F}$ *is non-commutative and absolutely simple.*

Proof. Passing to the factor group of $\mathbf{H}$ modulo its radical we can assume that $\mathbf{H}$ is semisimple. Then $\mathbf{H}$ can be decomposed into an almost direct product of absolutely almost simple subgroups $\mathbf{H}_i$, $1 \le i \le n$, and $\mathbf{F} = \mathscr{Z}(\mathbf{H})\cdot\mathbf{H}_1\cdot\ldots\cdot\mathbf{H}_{n-1}$ is the desired subgroup. □

(6.9) Lemma. *Let* F *be a group,* $\mathbf{H}$ *an algebraic group,* $\delta\colon F \to \mathbf{H}$ *a homomorphism, and* Φ *a subgroup of* F *with* $F = \mathrm{Comm}_F(\Phi)$ *(i.e. for each* $x \in F$ *the subgroups* $x\Phi x^{-1}$ *and* Φ *are commensurable).*

(a) *The group* $\overline{\delta(F)}$ *normalizes the subgroup* $(\overline{\delta(\Phi)})^0$.
(b) *Suppose for any subgroup* Φ_0 *of finite index in* Φ *the subgroup generated by* $\bigcup_{x\in F} x\Phi_0 x^{-1}$ *is of finite index in* F. *Then* $(\overline{\delta(F)})^0 = (\overline{\delta(\Phi)})^0$.

Proof. If $\mathbf{D}$ is an algebraic group, then $\mathbf{D}/\mathbf{D}^0$ is finite and $\mathbf{D}^0$ is contained in every algebraic subgroup of finite index of $\mathbf{D}$. But, since $F = \mathrm{Comm}_F(\Phi)$, the subgroups $\overline{\delta(\Phi)}$ and $\overline{\delta(x\Phi x^{-1})}$ are commensurable for each $x \in F$. Thus,

$$\delta(x)(\overline{\delta(\Phi_0)})^0\delta(x)^{-1} = (\overline{\delta(x\Phi_0 x^{-1})})^0 = (\overline{\delta(\Phi)})^0$$

for each $x \in F$ and every subgroup Φ_0 of finite index in Φ. This easily implies both assertions (a) and (b). □

(6.10) Corollary. *Let* $\bar{l}$ *be a field with* $\mathrm{char}\,\bar{l} \ne \mathrm{char}\,k_\alpha$ *for all* $\alpha \in A$. *Suppose the following conditions are satisfied.:*

(i) *either* rank $G \ge 2$ *and the lattice* Γ *is irreducible or the closure of the subgroup* Λ *contains* G^+;
(ii) *the lattice* Γ *is finitely generated;*
(iii) *for every subgroup* Γ_0 *of finite index in* Γ, *the subgroup generated by* $\bigcup\limits_{\lambda\in\Lambda} \lambda\Gamma_0\lambda^{-1}$ *is of finite index in* Λ.

Then for an $\bar{l}$-group $\mathbf{H}$ *and a homomorphism* $\delta\colon \Lambda \to \mathbf{H}(\bar{l})$, *the group* $\delta(\Lambda)$ *contains a solvable subgroup of finite index.*

Proof. Suppose the contrary, i.e., $\delta(\Lambda)$ does not contain a solvable subgroup of finite index. We may assume that $\mathbf{H} = \overline{\delta(\Lambda)}$ and the field $\bar{l}$ is algebraically closed. The group $\mathbf{H}/\mathbf{H}^0$ is finite, any subgroup of finite index in a finitely generated group is finitely generated, and, by Corollary I.1.5.7, G^+ does not contain proper subgroups of finite index. Therefore we can replace Λ by $\delta^{-1}(\mathbf{H}^0)$ and Γ by $\Gamma \cap \delta^{-1}(\mathbf{H}^0)$ and assume that $\mathbf{H}$ is connected. Since $\mathbf{H}$ is connected and non-solvable and the field $\bar{l}$ is algebraically closed, it follows from Lemma 6.8 that $\mathbf{H}$ contains a normal $\bar{l}$-subgroup $\mathbf{F}$ such that $\mathbf{H}/\mathbf{F}$ is non-commutative and absolutely simple. Let $\pi\colon \mathbf{H} \to \mathbf{H}/\mathbf{F}$ be the natural epimorphism. Replacing δ by $\pi \circ \delta$, we can then assume that $\mathbf{H}$ is non-commutative and absolutely simple. The group $\mathbf{H}$ can be viewed as an $\bar{l}$-subgroup of $\mathbf{GL}_n$. Since Γ is finitely generated, there exists a finitely generated subfield $l \subset \bar{l}$ such that $\delta(\Gamma) \subset \mathbf{GL}_n(l)$ (see 6.4). Then the group $\overline{\delta(\Gamma)} \subset \mathbf{GL}_n$ is defined over l. Since $\Lambda \subset \mathrm{Comm}_G(\Gamma)$, the group $\mathbf{H} = \overline{\delta(\Lambda)}$ is connected, and condition (iii) is satisfied, it follows from Lemma 6.9(b) that $\overline{\delta(\Gamma)} = \mathbf{H}$. Thus, we have a finitely generated field $l \subset \bar{l}$, a connected non-commutative absolutely simple l-group $\mathbf{H}$, and a homomorphism $\delta\colon \Lambda \to \mathbf{H}(\bar{l})$ with $\delta(\Gamma) \subset \mathbf{H}(l)$ and $\overline{\delta(\Gamma)} = \mathbf{H}$. By Theorem 6.5 this contradicts the condition that $\mathrm{char}\, l \neq \mathrm{char}\, k_\alpha$ for all $\alpha \in A$. □

The following assertion is a special case of Corollary 6.10.

(6.11) Corollary. *With* $\mathbf{G}$, Γ', *and* Λ' *as in the statement of Theorem 6.6 we assume the following conditions are satisfied:*

(i) *either* $\mathrm{rank}_{\mathbb{R}}\, \mathbf{G} \geq 2$ *and the lattice* Γ' *is irreducible or the closure of the subgroup* Λ' *in* $\mathbf{G}(\mathbb{R})$ *contains the identity component of the Lie group* $\mathbf{G}(\mathbb{R})$;

(ii) *the lattice* Γ' *is finitely generated;*

(iii) *for every subgroup* Γ_0 *of finite index in* Γ', *the subgroup generated by* $\bigcup_{\lambda \in \Lambda} \lambda \Gamma_0 \lambda^{-1}$ *is of finite index in* Λ'.

Then for every homomorphism δ *of the group* Λ' *into an algebraic group over a field of positive characteristic, the group* $\delta(\Lambda')$ *contains a solvable subgroup of finite index.*

(6.12) *Remark.* As follows from one of the assertions below on finite generation of lattices (see IX.3.1(vi)), condition (iv) of Theorem 6.6 and condition (ii) of Corollary 6.11 are actually automatically satisfied.

(6.13) Lemma. *If* $f\colon \mathbf{H} \to \mathbf{H}'$ *is an algebraic group epimorphism, then* $f(R_u(\mathbf{H})) = R_u(\mathbf{H}')$.

Proof. The epimorphism f induces a rational epimorphism $\tilde{f}\colon \mathbf{H}/R_u(\mathbf{H}) \to \mathbf{H}'/f(R_u(\mathbf{H}))$. The group $\mathbf{H}/R_u(\mathbf{H})$ is reductive and the image of a reductive group under an algebraic group morphism is reductive. Therefore the group $\mathbf{H}'/f(R_u(\mathbf{H})) = \tilde{f}(\mathbf{H}/R_u(\mathbf{H}))$ is reductive, and hence $f(R_u(\mathbf{H})) \supset R_u(\mathbf{H}')$. Finally, since $f(\mathbf{H}^{(u)}) \subset \mathbf{H}'^{(u)}$ (see I.0.20), it follows that $f(R_u(\mathbf{H})) \subset R_u(\mathbf{H}')$. □

(6.14) In V.4.8 we defined property (L) over k for k-groups. According to that definition we say that a connected algebraic group $\mathbf{F}$ has *absolute property (L)* if $\{e\} \neq R_u(\mathbf{F}) \neq \mathbf{F}$ and any non-trivial algebraic normal subgroup of $\mathbf{F}$ contains $R_u(\mathbf{F})$. If a connected k-group $\mathbf{F}$ has absolute property (L) and the unipotent radical $R_u(\mathbf{F})$ is defined over k, then $\mathbf{F}$ has property (L) over k. In particular, if a connected k-group $\mathbf{F}$ has absolute property (L) and $\operatorname{char} k = 0$, then $\mathbf{F}$ has property (L) over k.

(6.15) Lemma. *Let l be a field of characteristic 0 and $\mathbf{F}$ a connected l-group. Suppose $\mathscr{D}(\mathbf{F}) = \mathbf{F}$ and $R_u(\mathbf{F}) \neq \{e\}$. Then $\mathbf{F}$ contains a normal algebraic subgroup $\mathbf{B}$ such that the factor group $\mathbf{F}/\mathbf{B}$ has absolute property (L).*

Proof. From the results presented in I.0.24 we obtain that

(a) if an algebraic group $\mathbf{H}$ is connected, reductive, and coincides with its commutator subgroup, then $\mathbf{H}$ is semisimple;

(b) a semisimple algebraic group contains only finitely many normal algebraic subgroups.

Since $\mathbf{F}$ is connected and $\mathscr{D}(\mathbf{F}) = \mathbf{F}$, it follows from (a) that the reductive group $\mathbf{F}/R_u(\mathbf{F})$ is semisimple, and hence $\mathbf{F}/R_u(\mathbf{F})$ contains only a finite numbers of algebraic normal subgroups (see (b)). On the other hand, since $\operatorname{char} l = 0$, every algebraic subgroup of the unipotent l-group $R_u(\mathbf{F})$ is connected (see I.0.20), and hence, any ascending chain of algebraic subgroups in $R_u(\mathbf{F})$ stabilizes. Therefore every ascending chain of normal algebraic subgroups of $\mathbf{F}$ stabilizes. This implies the existence of a normal algebraic subgroup $\mathbf{B}$ in $\mathbf{F}$ such that $R_u(\mathbf{F}) \not\subset \mathbf{B}$ and $\mathbf{B}$ has the following maximality property: if $\mathbf{B}'$ is an normal algebraic subgroup of $\mathbf{F}$ and $\mathbf{B}' \supsetneqq \mathbf{B}$, then $\mathbf{B}' \supset R_u(\mathbf{F})$. We shall show that $\mathbf{F}/\mathbf{B}$ has absolute property (L). Let $\mathbf{H}$ be a non-trivial normal algebraic subgroup of $\mathbf{F}/\mathbf{B}$. We set $\mathbf{H}' = \pi^{-1}(\mathbf{H})$, where $\pi\colon \mathbf{F} \to \mathbf{F}/\mathbf{B}$ is the natural epimorphism. Since $\mathbf{H} \neq \{e\}$, it follows that $\mathbf{H}' \supsetneqq \mathbf{B}$ and, by maximality of the subgroup $\mathbf{B}$ we have that $\mathbf{H}' \supset R_u(\mathbf{F})$. But $\pi(R_u(\mathbf{F})) = R_u(\mathbf{F}/\mathbf{B})$ (see Lemma 6.13), so $\mathbf{H} = \pi(\mathbf{H}') \supset R_u(\mathbf{F}/\mathbf{B})$. Since $R_u(\mathbf{F}) \not\subset \mathbf{B}$, it follows that $\{e\} \neq R_u(\mathbf{F}/\mathbf{B})$. Finally, since $\mathscr{D}(\mathbf{F}) = \mathbf{F}$, and hence $\mathscr{D}(\mathbf{F}/\mathbf{B}) = \mathbf{F}/\mathbf{B}$, and the unipotent subgroup $R_u(\mathbf{F}/\mathbf{B})$ is nilpotent, $R_u(\mathbf{F}/\mathbf{B}) \neq \mathbf{F}/\mathbf{B}$. □

(6.16) Theorem. *Suppose the following conditions are satisfied:*

(i) *for each $\alpha \in A$ the group $\mathbf{G}_\alpha$ is simply connected;*

(ii) *either* $\operatorname{rank} G \geq 2$ *and the lattice Γ is irreducible or the subgroup Λ is dense in G;*

(iii) *any representation of the group Λ on a finite-dimensional vector space over a local field of characteristic 0 is Γ-integrable;*

(iv) *the lattice Γ is finitely generated;*

(v) *for every subgroup Γ_0 of finite index in Γ, the subgroup generated by $\bigcup_{\lambda \in \Lambda} \lambda \Gamma_0 \lambda^{-1}$ is of finite index in Λ;*

(vi) *the factor group $\Gamma_0/\mathscr{D}(\Gamma_0)$ is finite for every subgroup Γ_0 of finite index in Γ.*

Then for a field $\bar{l}$ of characteristic 0, an $\bar{l}$-group $\mathbf{F}$, and a homomorphism $\delta\colon \Lambda \to \mathbf{F}(\bar{l})$, the $\bar{l}$-group $\overline{\delta(\Lambda)}$ is semisimple.

Proof. Suppose the contrary, i.e., the group $\overline{\delta(\Lambda)}$ is not semisimple. We may assume that $\mathbf{F} = \overline{\delta(\Lambda)}$ and the field $\bar{l}$ is algebraically closed. The group $\mathbf{F}/\mathbf{F}^0$ is finite, any subgroup of finite index in a finitely generated group is finitely generated, and, in view of Corollary I.2.3.2(b), G does not contain proper subgroups of finite index. Therefore we can replace Λ by $\delta^{-1}(\mathbf{F}^0)$ and Γ by $\Gamma \cap \delta^{-1}(\mathbf{F}^0)$ and assume that $\mathbf{F}$ is connected. Since $\Lambda \subset \mathrm{Comm}_G(\Gamma)$, the group $\mathbf{F} = \overline{\delta(\Lambda)}$ is connected, and condition (v) is satisfied, it follows from Lemma 6.9(b) that $\overline{\delta(\Gamma)} = \mathbf{F}$. But the factor group $\Gamma/\mathscr{D}(\Gamma)$ is finite. Therefore $\mathbf{F}/\mathscr{D}(\mathbf{F})$ is finite and, since $\mathbf{F}$ is connected, $\mathbf{F} = \mathscr{D}(\mathbf{F})$. Since the group $\mathbf{F} = \overline{\delta(\Lambda)}$ is non-semisimple, connected, and perfect (i.e. coincides with its commutator subgroup), it follows from assertion (a) in the proof of Lemma 6.15 that $R_u(\mathbf{F}) \neq \{e\}$. Now, since $\bar{l}$ is algebraically closed, by Lemma 6.15 we can choose a normal $\bar{l}$-subgroup $\mathbf{B}$ in $\mathbf{F}$ such that $\mathbf{F}/\mathbf{B}$ has absolute property (L). Let $\pi\colon \mathbf{F} \to \mathbf{F}/\mathbf{B}$ be the natural epimorphism. Replacing δ by $\pi \circ \delta$ we can assume that $\mathbf{F}$ has absolute property (L). We realize $\mathbf{F}$ as an $\bar{l}$-subgroup of $\mathbf{GL}_n$. Since Γ is finitely generated, there is a finitely generated field $l \subset \bar{l}$ such that $\delta(\Gamma) \subset \mathbf{GL}_n(l)$ (see 6.4). Therefore the group $\mathbf{F} = \overline{\delta(\Gamma)}$ is defined over l. Since the l-group $\mathbf{F}$ has absolute property (L) and $\operatorname{char} l = 0$, it follows from Lemma V.4.11(iv) that $\mathscr{Z}(\mathbf{F}) = \{e\}$. But $\delta(\Gamma) \subset \mathbf{F}(l)$, $\mathbf{F} = \overline{\delta(\Gamma)}$, and $\Lambda \subset \mathrm{Comm}_G(\Gamma)$. Thus, Lemma 6.2(iii) implies that $\delta(\Lambda) \subset \mathbf{F}(l)$.

So we have a finitely generated field l of characteristic 0, a connected l-group $\mathbf{F}$ with absolute property (L), and a homomorphism $\delta\colon \Lambda \to \mathbf{F}(l)$ with $\overline{\delta(\Gamma)} = \mathbf{F}$. We set $\mathbf{H} = \mathbf{F}/R_u(\mathbf{F})$ and denote by σ the natural epimorphism $\mathbf{F} \to \mathbf{H}$. Since $R_u(\mathbf{F}) \neq \mathbf{F}$ and the group $\mathbf{F} = \overline{\delta(\Gamma)}$ is connected, the subgroup $\sigma(\delta(\Gamma))$ is infinite. But the group Γ (and hence the group $\sigma(\delta(\Gamma))$) is finitely generated. Therefore there is a local field k containing l such that the subgroup $\sigma(\delta(\Gamma))$ is not relatively compact in $\mathbf{H}(k)$ (see Lemma 6.1). Thus, by Theorem 5.19, Remark V.4.13, and Remark 5.20, δ extends to a continuous homomorphism $\delta_0\colon G \to \mathbf{F}(k)$. Since $\operatorname{char} k = 0$, it follows from Proposition I.2.6.4(b) that the group $\overline{\delta_0(G)} \subset \mathbf{F}$ is semisimple. But this contradicts the fact that $\delta(\Lambda) \subset \delta_0(G)$ and the assumption that $\mathbf{F} = \overline{\delta(\Lambda)}$ is not semisimple. □

From Theorem 6.16 and Lemma 5.22 we deduce

(6.17) Corollary. *Suppose conditions (i)–(vi) of Theorem 6.16 are satisfied. Then* $H^1(\Lambda, \rho) = 0$ *for every representation* ρ *of the group* Λ *on a finite-dimensional vector space over a field of characteristic* 0.

(6.18) Corollary. *Suppose the following conditions are satisfied:*

(i) *the quotient space* $\Gamma \backslash G$ *is compact;*

(ii) *either* $\operatorname{rank} G \geq 2$ *and the lattice* Γ *is irreducible or the subgroup* $\Lambda \cap G^+$ *is dense in* G^+*;*

(iii) *for every subgroup* Γ_0 *of finite index in* Γ, *the subgroup generated by* $\bigcup_{\lambda \in \Lambda} \lambda \Gamma_0 \lambda^{-1}$ *is of finite index in* Λ.

Then

(a) *If* $\bar{l}$ *is a field of characteristic* 0, $\mathbf{F}$ *is an* $\bar{l}$*-group, and* $\delta\colon \Lambda \to \mathbf{F}(\bar{l})$ *is a homomorphism, then the* $\bar{l}$*-group* $\overline{\delta(\Lambda)}$ *is semisimple;*

(b) $H^1(\Lambda, \rho) = 0$ *for every representation* ρ *of the group* Λ *on a finite-dimensional vector space over a field of characteristic* 0.

Proof. (a) Since the groups $\mathbf{G}_\alpha(k_\alpha)$ are compactly generated (see Corollary I.2.3.5), so is G. But $\Gamma \setminus G$ is compact and any uniform lattice in a compactly generated group is finitely generated (see I.0.40). Therefore Γ is finitely generated. But, since the factor groups $\mathbf{G}_\alpha(k_\alpha)/\mathbf{G}_\alpha(k_\alpha)^+$ are commutative and torsion (see Corollary I.2.3.1(c)), the factor group $\mathbf{G}/G^+$ is commutative and torsion. Thus, the subgroup $\Gamma \cap G^+$ is of finite index in Γ, and hence we can substitute $\Gamma \cap G^+$ for Γ and $\Lambda \cap G^+$ for Λ and assume that $\Lambda \subset G^+$. Let $\tilde{\mathbf{G}}_\alpha$ be the simply connected covering of the group $\mathbf{G}_\alpha$ and $p_\alpha\colon \tilde{\mathbf{G}}_\alpha \to \mathbf{G}_\alpha$ the central k-isogeny. We set $\tilde{G} = \prod_{\alpha \in A} \tilde{\mathbf{G}}_\alpha(k_\alpha)$ and $\operatorname{rank} \tilde{G} = \sum_{\alpha \in A} \operatorname{rank}_{k_\alpha} \tilde{\mathbf{G}}_\alpha$. The isogenies p_α induce a continuous homomorphism $p\colon \tilde{G} \to G$. By Corollary I.1.4.6, Proposition I.1.5.5, Theorem I.2.3.1, and Proposition I.2.3.4, for each $\alpha \in A$ the group $\tilde{\mathbf{G}}_\alpha$ has no k_α-anisotropic almost k_α-simple factors, $\operatorname{rank}_{k_\alpha} \tilde{\mathbf{G}}_\alpha = \operatorname{rank}_{k_\alpha} \mathbf{G}_\alpha$, $\tilde{\mathbf{G}}_\alpha(k_\alpha)^+ = \tilde{\mathbf{G}}_\alpha(k_\alpha)$, the subgroup $\mathbf{G}_\alpha(k_\alpha)^+$ is closed in $\mathbf{G}_\alpha(k_\alpha)$, $p_\alpha(\tilde{\mathbf{G}}_\alpha(k_\alpha)) = \mathbf{G}_\alpha(k_\alpha)^+$, and the isomorphism $\tilde{\mathbf{G}}_\alpha(k_\alpha)/(\operatorname{Ker} p_\alpha)(k_\alpha) \to \mathbf{G}_\alpha(k_\alpha)^+$ is a topological group isomorphism. Thus, $\operatorname{rank} \tilde{G} = \operatorname{rank} G$, the subgroup G^+ is closed in G, $p(\tilde{G}) = G^+$ and the isomorphism $\tilde{G}/\operatorname{Ker} p \to G^+$ is a topological group isomorphism. Notice that $\Lambda \subset G^+$ and, since Γ is an irreducible cocompact lattice in $\tilde{G}$, it follows that $p^{-1}(\Gamma)$ is an irreducible uniform lattice in G. Thus, we replace $\mathbf{G}_\alpha$ by $\tilde{\mathbf{G}}_\alpha$, Γ by $p^{-1}(\Gamma)$, and Λ by $p^{-1}(\Lambda)$, and assume that conditions (i) and (ii) of Theorem 6.16 are satisfied.

As was observed in V.3.0, since Γ/G is compact, any representation of the group Λ on a finite-dimensional vector space over a local field is Γ-integrable. If Γ_0 is a subgroup of finite index in Γ, then by Theorem IV.4.9 and Lemma II.6.3(iii), the factor group $\Gamma_0/\mathscr{D}(\Gamma_0)$ is finite. Thus, all conditions of Theorem 6.16 are satisfied and all that remains is only to apply this theorem.

(b) is a consequence of Lemma 5.22(a). □

(6.19) We now turn to the proof of Lemma 6.1. We first present two lemmas from field theory.

(6.20) Lemma. (See [Ti 3], Lemma 2.3.) *Let* l *be a finitely generated field and let* $m \in \mathbb{N}^+$. *Then there exist only finitely many roots of unity satisfying an equation of degree* m *over* l.

(6.21) Lemma. (See [Ti 3], Lemma 4.1.) *Let* l *be a finitely generated field,* l^* *the multiplicative group of* l*, and* $t \in l^*$ *an element of infinite order. Then there exist a local field* k *with absolute value* $|\ |$ *and a homomorphism* $\sigma\colon l \to k$ *such that* $|\sigma(t)| \neq 1$.

Proof. We realize $\mathbf{H}$ as an l-subgroup of $\mathbf{GL}_n$. Let us denote by Δ the set of roots of unity satisfying an equation of degree n over l and set $\Omega = \{h \in \mathbf{H} \mid$ all the eigenvalues of the matrix h belong to $\Delta\}$. According to Lemma 6.20 the set Δ is finite. Therefore the set Ω is Zariski closed in $\mathbf{H}$. Since $\mathbf{H}$ is connected, reductive, and different from $\{e\}$, $\mathbf{H}$ contains a torus of positive dimension, which implies that $\Omega \neq \mathbf{H}$. Thus, the set $\mathbf{H} - \Omega$ is Zariski open and non-empty. On the other hand, the subgroup Φ is Zariski dense in $\mathbf{H}$ and in a connected reductive group the set of semisimple elements contains a Zariski open and dense subset (see I.0.24). Therefore there is a semisimple element g of Φ belonging to $\mathbf{H} - \Omega$. Since $g \in \mathbf{H} - \Omega$ and all the eigenvalues of the matrix $g \in \mathbf{GL}_n(l)$ satisfy an equation of degree n over l, at least one of the eigenvalues of g, say λ, is not a root of unity. By extending the field l we may assume that $\lambda \in l$. By Lemma 6.21 there exist a local field k with absolute value $|\ |$ and a homomorphism $\sigma\colon l \to k$ such that $|\sigma(\lambda)| \neq 1$. Then the element $\sigma^0(g) \in {}^{\sigma}\mathbf{H}(k)$ is essentially non-compact. □

(6.23) *Remark.* (i) Lemma 6.1 was in fact proved in [Ti 3]. Furthermore, in [Ti 3] it was in fact established that if the group $\mathbf{H}$ in Lemma 6.1 is semisimple, then there exist a local field k and a homomorphism $\sigma\colon l \to k$ such that the group $\sigma^0(\Phi) \subset {}^{\sigma}\mathbf{H}(k)$ contains a non-commutative free discrete subgroup generated by semisimple elements (in this connection, see Appendix B).

(ii) The closure of any infinite subfield of a local field is a local field. Therefore, by passing from k to the closure of the subfield $\sigma(l)$, we may assume in Lemma 6.1 that $\sigma(l)$ is dense in k.

7. Strong Rigidity (Continuous Extensions of Isomorphisms of Discrete Subgroups)

As in Section 6 we recall that for an l-group $\mathbf{H}$ and a homomorphism σ of the field l to a field l' the l'-group ${}^{\sigma}\mathbf{H}$ and the homomorphism $\sigma^0\colon \mathbf{H}(l) \to {}^{\sigma}\mathbf{H}(l')$ are defined as in I.1.7 and I.1.8.

(7.1) Theorem. *Let A and A' be finite sets. For each $\alpha \in A$ (resp. $\alpha' \in A'$) let k_α (resp. $k'_{\alpha'}$) be a local field and $\mathbf{G}_\alpha$ (resp. $\mathbf{G}'_{\alpha'}$) a connected semisimple adjoint algebraic group which is defined, isotropic, and simple over k_α (resp. $k'_{\alpha'}$). Set $G = \prod_{\alpha \in A} \mathbf{G}_\alpha(k_\alpha)$ and $G' = \prod_{\alpha' \in A'} \mathbf{G}'_{\alpha'}(k'_{\alpha'})$. Let Γ be a lattice in G, Γ' a lattice in G', and $f\colon \Gamma \to \Gamma'$ an isomorphism. Suppose* $\operatorname{rank} G \stackrel{\text{def}}{=} \sum_{\alpha \in A} \operatorname{rank}_{k_\alpha} \mathbf{G}_\alpha \geq 2$ *and the lattice Γ is irreducible.*

(a) *The isomorphism f extends uniquely to a continuous group homomorphism $\tilde{f}\colon G \to G'$ which is a topological group isomorphism.*
(b) *If for all $\alpha \in A$ and $\alpha' \in A'$ the groups $\mathbf{G}_\alpha$ and $\mathbf{G}'_{\alpha'}$ are absolutely simple, then there exist (uniquely determined) a bijection $i\colon A \to A'$, continuous homomorphisms $\vartheta_\alpha\colon k_\alpha \to k'_{i(\alpha)}$, and special (in the sense of I.1.4.14) $k'_{i(\alpha)}$-epimorphisms $\eta_\alpha\colon {}^{\vartheta_\alpha}\mathbf{G}_\alpha \to \mathbf{G}'_{i(\alpha)}$ such that the product $\tilde{f} = \prod_{\alpha \in A}(\eta_\alpha \circ \vartheta^0_\alpha)\colon G \to G'$ of the homomorphisms $\eta_\alpha \circ \vartheta^0_\alpha\colon \mathbf{G}_\alpha(k_\alpha) \to \mathbf{G}'_{i(\alpha)}(k'_{i(\alpha)})$ is an extension of the isomorphism f. Furthermore, the ϑ_α's are locally compact field isomorphisms and the η_α's are algebraic group isomorphisms.*

Proof. As was observed in I.1.7, if $\mathbf{G}$ is an adjoint k-simple k-group, then there exist a finite separable field extension k' of k and a connected adjoint absolutely simple k-group $\mathbf{G}'$ such that $\mathbf{G} = R_{k'/k}\mathbf{G}'$. Furthermore, $\operatorname{rank}_k G = \operatorname{rank}_{k'} \mathbf{G}'$ and if k is a local field, then $R^0_{k'/k}\colon \mathbf{G}'(k') \to \mathbf{G}(k)$ is a topological group isomorphism. Thus we can assume that the groups $\mathbf{G}_\alpha$ and $\mathbf{G}'_{\alpha'}$ are absolutely simple. We denote, respectively, by pr_α, $\mathrm{pr}_{\alpha'}$, and $\overline{\mathrm{pr}_{\alpha'}(\Gamma')}$ the natural projections $G \to \mathbf{G}_\alpha(k_\alpha)$ and $G' \to \mathbf{G}'_{\alpha'}(k'_{\alpha'})$ and the closure of the subgroup $\mathrm{pr}_{\alpha'}(\Gamma')$ in $\mathbf{G}'_{\alpha'}(k'_{\alpha'})$ (in the topology induced from the topology of the field $k'_{\alpha'}$). Since Γ' is a lattice in G', by Lemma II.6.1 $\overline{\mathrm{pr}_{\alpha'}(\Gamma)} \setminus \mathbf{G}'_{\alpha'}(k'_{\alpha'})$ has a finite invariant measure, and hence the subgroup $\overline{\mathrm{pr}_{\alpha'}(\Gamma')}$ is Zariski dense in $\mathbf{G}'_{\alpha'}$ (see Lemma II.2.3 and Corollary II.4.4). On the other hand, by Proposition I.2.3.6 the group $\mathbf{G}'_{\alpha'}(k'_{\alpha'})$ is non-compact. So the subgroup $\overline{\mathrm{pr}_{\alpha'}(\Gamma')}$ (and hence the subgroup $\mathrm{pr}_{\alpha'}(\Gamma')$) is Zariski dense in $\mathbf{G}'_{\alpha'}$ and is not relatively compact in $\mathbf{G}'_{\alpha'}(k'_{\alpha'})$, $\alpha' \in A'$. Now we can apply Theorem 5.6 to the homomorphisms $\mathrm{pr}_{\alpha'} \circ f\colon \Gamma \to \mathbf{G}'_{\alpha'}(k'_{\alpha'})$ and obtain the following assertions.

(i) For each $\alpha' \in A'$ there exist (uniquely determined) $j(\alpha') \in A$, a continuous homomorphism $\tilde{\vartheta}_{\alpha'}\colon k_{j(\alpha')} \to k'_{\alpha'}$, and a special $k'_{\alpha'}$-epimorphism $\tilde{\eta}_{\alpha'}\colon {}^{\tilde{\vartheta}_{\alpha'}}\mathbf{G}_{j(\alpha')} \to \mathbf{G}'_{\alpha'}$ such that

$$\mathrm{pr}_{\alpha'}(f(\gamma)) = \eta_{\alpha'}(\tilde{\vartheta}^0_{\alpha'}(\mathrm{pr}_{j(\alpha')}(\gamma))) \text{ for each } \gamma \in \Gamma. \tag{1}$$

(ii) For each $\alpha' \in A'$ the homomorphism $\mathrm{pr}_{\alpha'} \circ f\colon \Gamma \to \mathbf{G}'_{\alpha'}(k'_{\alpha'})$ extends uniquely to a continuous homomorphism $f_{\alpha'}\colon G \to \mathbf{G}'_{\alpha'}(k'_{\alpha'})$. Furthermore, $f_{\alpha'} = \tilde{\eta}_{\alpha'} \circ \tilde{\vartheta}^0_{\alpha'} \circ \mathrm{pr}_{j(\alpha')}$.

Assertion (ii) implies the first part of assertion (a), i.e., the existence and the uniqueness of the extension $\tilde{f}\colon G \to G'$; furthermore, $\tilde{f} = \prod_{\alpha' \in A'} f_{\alpha'}$. We set $H_{\alpha'} = (\tilde{\eta}_{\alpha'} \circ \tilde{\vartheta}^0_{\alpha'})(\mathbf{G}_{j(\alpha')}(k_{j(\alpha')})) \subset \mathbf{G}'_{\alpha'}(k'_{\alpha'})$. Since the groups $\mathbf{G}_\alpha$ are absolutely simple and $\mathbf{G}'_{\alpha'} \neq \{e\}$, the kernels of the epimorphisms $\eta_{\alpha'}$ are trivial. It then follows from Corollary I.2.1.3 that

(iii) for each $\alpha' \in A'$ the subgroup $H_{\alpha'}$ is closed in $\mathbf{G}'_{\alpha'}(k'_{\alpha'})$ and the homomorphism $\tilde{\eta}_{\alpha'} \circ \vartheta^0_{\alpha'}\colon \mathbf{G}_{j(\alpha')}(k_{j(\alpha')}) \to H_{\alpha'}$ is a topological group isomorphism.

Since $\overline{\mathrm{pr}_{\alpha'}(\Gamma)} \setminus \mathbf{G}_{\alpha'}(k_{\alpha'})$ has a finite invariant measure and the subgroup $H_{\alpha'}$ is closed, non-discrete and in view of (1) contains $\mathrm{pr}_{\alpha'}(\Gamma')$, it follows from Theorem II.5.1(a) that $H_{\alpha'} \supset \mathbf{G}'_{\alpha'}(k'_{\alpha'})^+$. Thus the following assertion holds (see Remark I.1.8.2(iv)).

(iv) For each $\alpha' \in A'$, $\tilde{\vartheta}_{\alpha'}$ is an isomorphism of locally compact fields and $\tilde{\eta}_{\alpha'}$ is an algebraic group isomorphism.

As in Section 3 of Chapter II, for $B \subset A$ we set $G_B = \prod_{\alpha \in B} \mathbf{G}_\alpha(k_\alpha)$ and denote by pr_B the natural projection $G \to G_B$. We similarly define $G'_{B'}$ and $\mathrm{pr}_{B'}\colon G' \to G_{B'}$, where $B' \subset A'$. Since $\tilde{f} = \prod_{\alpha' \in A'} f_{\alpha'}$, $f_{\alpha'} = \tilde{\eta}_{\alpha'} \circ \vartheta^0_{\alpha'} \circ \mathrm{pr}_{j(\alpha')}$, and $\operatorname{Ker} \tilde{\eta}_{\alpha'} = \{e\}$, it follows that $\operatorname{Ker} \tilde{f} = G_{A - j(A')}$. But the homomorphism $\tilde{f}$ is continuous and the subgroup $\tilde{f}(\Gamma) = f(\Gamma) = \Gamma'$ is discrete in G'. Therefore the subgroup $\mathrm{pr}_{j(A')}(\Gamma)$ is discrete in $G_{j(A')}$. On the other hand, since the lattice Γ is irreducible, whenever $\varnothing \neq B \neq A$, by Lemma II.6.4 the subgroup $\mathrm{pr}_B(\Gamma)$ is not discrete in G_B. Hence,

(2) $$A = j(A').$$

For $\alpha' \in A'$, we set $J_{\alpha'} = \{\alpha'' \in A' \mid j(\alpha'') = j(\alpha')\}$, denote by $h_{\alpha'}$ the homomorphism $\prod_{\alpha'' \in J_{\alpha'}} (\eta_{\alpha''} \circ \vartheta^0_{\alpha''})$ of the group $\mathbf{G}_{j(\alpha')}(k_{j(\alpha')})$ into $G'_{J_{\alpha'}}$, and set $F_{\alpha'} = h_{\alpha'}(\mathbf{G}_{j(\alpha')}(k_{j(\alpha')}))$. It follows from (iii) that the subgroup $F_{\alpha'}$ is closed in $G_{J_{\alpha'}}$ and the multiplication map defines a homeomorphism of the topological space $G_{J_{\alpha'}-\{\alpha'\}} \times F_{\alpha'}$ onto $G_{J_{\alpha'}}$. Thus, there is a $G_{J_{\alpha'}-\{\alpha'\}}$-equivariant homeomorphism of the space $F_{\alpha'} \setminus G_{J_{\alpha'}}$ onto $G_{J_{\alpha'}-\{\alpha'\}}$. But, since the groups $\mathbf{G}'_{\alpha''}(k'_{\alpha''})$, $\alpha'' \in A'$, are non-compact, $G_{J_{\alpha'}-\{\alpha'\}}$ does not carry a finite invariant measure if $J_{\alpha'} \neq \{\alpha'\}$. Hence, if $J_{\alpha'} \neq \{\alpha'\}$, then $F_{\alpha'} \setminus G_{J_{\alpha'}}$ does no carry a finite invariant measure. On the other hand, since Γ' is a lattice in G' and, according to (1), $F_{\alpha'} \supset \mathrm{pr}_{J_{\alpha'}}(\Gamma')$, it follows from Lemma II.6.1 and I.0.36 that $F_{\alpha'} \setminus G_{J_{\alpha'}}$ has a finite invariant measure. Thus, $J_{\alpha'} = \{\alpha'\}$. From this and (2) we deduce that the map $j\colon A' \to A$ is bijective. We set $i = j^{-1}$, $\eta_\alpha = \tilde{\eta}_{i(\alpha)}$, and $\vartheta_\alpha = \vartheta_{i(\alpha)}$, $\alpha \in A$. Now, to complete the proof it remains to make use of equality (1), assertion (iv), and to observe that the uniqueness of the map i and the homomorphisms ϑ_α and η_α is a consequence of the uniqueness of $\tilde{f}$, the equality $\tilde{f}(\mathbf{G}_\alpha(k_\alpha)) = \mathbf{G}'_{i(\alpha)}(k'_{i(\alpha)})$, and the uniqueness of the homomorphisms $\tilde{\vartheta}_{\alpha'}$ and $\tilde{\eta}_{\alpha'}$. □

(7.2) *Remark.* Applying Theorem 7.1 it is not hard to show that the condition "rank $G \geq 2$ and the lattice Γ is irreducible" in the statement of this theorem can be replaced by the condition "the subgroup $\mathrm{pr}_{\alpha(\Gamma)}$ is not discrete in $\mathbf{G}_\alpha(k_\alpha)$ for every $\alpha \in A$ with $\mathrm{rank}_{k_\alpha} \mathbf{G}_\alpha = 1$".

(7.3) If k is a field, K is a finite field extension of k, $f\colon \mathbf{W} \to \mathbf{W}'$ is a K-group K-isomorphism, $\pi\colon R_{K/k}\mathbf{W} \to \mathbf{W}$ and $\pi'\colon R_{K/k}\mathbf{W}' \to \mathbf{W}'$ are the natural projections, then the universality property of the maps π and π' given in I.1.7 implies the existence and the uniqueness of a k-group k-morphism $h\colon R_{K/k}\mathbf{W} \to R_{K/k}\mathbf{W}'$ such that $f \circ \pi = \pi' \circ h$. On the other hand, any connected semisimple adjoint k-group $\mathbf{F}$ can be represented in the form $\mathbf{F} = \prod_{1 \leq i \leq n} R_{k_i/k}\mathbf{F}_i$, where $\mathbf{F}_i$ is an absolutely simple adjoint k_i-group (see I.1.4 and I.1.7). Thus, assertions (A) and (B) in the proof of Theorem 5.7 and Theorem 7.1 imply the following

(7.4) Theorem. *Let $\mathscr{B} = \{p \in \mathbb{N}^+ \mid p \text{ is prime}\} \cup \{\infty\}$, A' and A'' two finite subsets of B, and for each $p \in A'$ (resp. $p \in A''$), let $\mathbf{G}'_p$ (resp. $\mathbf{G}''_p$) be a connected semisimple adjoint $\mathbb{Q}_p$-group without $\mathbb{Q}_p$-anisotropic $\mathbb{Q}_p$-simple factors. We set $G' = \prod_{p \in A'} \mathbf{G}'_p(\mathbb{Q}_p)$ and $G'' = \prod_{p \in A''} \mathbf{G}''_p(\mathbb{Q}_p)$. Let $\Gamma' \subset G'$ and $\Gamma'' \subset G''$ be lattices and $f\colon \Gamma' \to \Gamma''$ an isomorphism. Suppose* rank $G' \overset{\text{def}}{=} \sum_{p \in A'} \mathrm{rank}_{\mathbb{Q}_p} \mathbf{G}'_p \geq 2$ *and the lattice Γ' is irreducible. Then*

(a) *the isomorphism f extends uniquely to a continuous homomorphism $\tilde{f}$ of the group G' to G'' which is a topological group isomorphism;*
(b) *$A' = A''$ and there exist uniquely determined $\mathbb{Q}_p$-epimorphisms $\eta_p\colon \mathbf{G}'_p \to \mathbf{G}''_p$, $p \in A'$ such that $\prod_{p \in A'} \eta_p\colon G' \to G''$ is an extension of the isomorphism f. Furthermore, all of the η_p's are algebraic group isomorphisms.*

The following assertion is a special case of Theorem 7.4.

(7.5) Theorem. *Let* $\mathbf{G}'$ *and* $\mathbf{G}''$ *be connected semisimple adjoint* $\mathbb{R}$*-groups without* $\mathbb{R}$*-anisotropic almost* $\mathbb{R}$*-simple factors, let* $\Gamma' \subset \mathbf{G}'(\mathbb{R})$ *and* $\Gamma'' \subset \mathbf{G}''(\mathbb{R})$ *be lattices, and* $f\colon \Gamma' \to \Gamma''$ *an isomorphism. Suppose* $\operatorname{rank}_{\mathbb{R}} \mathbf{G}' \geq 2$ *and the lattice* Γ' *is irreducible. Then*

(a) *The isomorphism* f *extends uniquely to a Lie group isomorphism* $\tilde{f}\colon \mathbf{G}'(\mathbb{R}) \to \mathbf{G}''(\mathbb{R})$.

(b) *The isomorphism* f *extends uniquely to an* $\mathbb{R}$*-isomorphism* $\eta\colon \mathbf{G}' \to \mathbf{G}''$.

(7.6) *Remark.* In the statement of Theorem 7.5 the condition "$\operatorname{rank}_{\mathbb{R}} \mathbf{G}' \geq 2$" can be replaced by the condition "the group $\mathbf{G}'$ is not isomorphic over $\mathbb{R}$ to the group $\mathbf{PSL}_2 = \mathbf{SL}_2/\{\pm E\}$ (see [Most 7] and [Pr 1]). It should be noted that the similar replacement is impossible in the statement of Theorem 5.9 (for further details we refer to 2.5 in Appendix C).

8. Rigidity of Ergodic Actions of Semisimple Groups

In this section we present some of Zimmer's results on rigidity and orbit equivalence of ergodic actions of semisimple groups. Theorems 8.2 and 8.3 stated below on superrigidity for cocycles can be viewed as generalizations of Theorems 5.7 and 5.9.

All G-spaces X in the present section are considered to be standard Borel i.e. there is a Borel isomorphism of X onto a Borel subset of a complete separable metric space. Furthermore, we consider only Borel measures on X and assume that the map $X \times G \to X$, $(x, g) \mapsto xg$, is Borel.

(8.1) Let G, X, μ, and M be as in 2.0. A cocycle on a G-space X is a Borel map $\sigma\colon X \times G \to M$ which satisfies (i) in 2.0. Two cocycles $\sigma_1, \sigma_2\colon X \times G \to M$ are said to be *equivalent* or *cohomologous* if there is a Borel map $\eta\colon X \to M$ such that for each $g \in G$

$$\sigma_2(x, g) = \eta(g)\sigma_1(x, g)\eta(xg)^{-1}$$

for almost all (with respect to the measure μ) $x \in X$. We then write $\sigma_1 \sim \sigma_2$.

If $\pi\colon G \to M$ is a continuous homomorphism, then the map $\sigma_\pi\colon X \times G \to M$, $\sigma_\pi(x, g) = \pi(g)$, is a cocycle. The cocycles of the form σ_π are precisely those which are independent of $x \in X$.

A cocycle $\sigma\colon X \times G \to M$ is called *relatively compact* if σ is equivalent to a cocycle which takes values in a compact subgroup of M. Let k be a local field and $\mathbf{H}$ an algebraic k-group. We say that a cocycle $\sigma\colon X \times G \to \mathbf{H}(k)$ is *Zariski dense* in $\mathbf{H}$ if for every proper k-subgroup $\mathbf{L} \subset \mathbf{H}$ the cocycle σ is not equivalent to a cocycle with values in $\mathbf{L}(k)$.

Before stating the superrigidity theorem for cocycles, we give one more definition.

With $G = \prod_{\alpha \in A} \mathbf{G}_\alpha(k_\alpha)$ as in Section 5, we call a $\mathbf{G}$-space (X, μ) *irreducible* if any infinite standard (in the sense of III.5.9) normal subgroup of G acts ergodically on X.

(8.2) Theorem. (See [Zi 8], Theorem 10.1.6.) *Set* $\mathscr{B} = \{p \in \mathbb{N}^+ \mid p$ *is prime*$\} \cup \{\infty\}$. *Let* A *be a finite subset of* $\mathscr{B}$ *and for each* $p \in A$ *let* $\mathbf{G}_p$ *be a connected simply connected non-trivial semisimple* $\mathbb{Q}_p$*-group without* $\mathbb{Q}_p$*-anisotropic factors. Let* $G = \prod_{p \in A} \mathbf{G}_p(\mathbb{Q}_p)$ *and let* pr_p *be the natural projection* $G \to \mathbf{G}_p(\mathbb{Q}_p)$. *Further let* $p(k) \in \mathscr{B}$, $k = \mathbb{Q}_{p(k)}$, $\mathbf{H}$ *be a connected adjoint* k*-simple* k*-group,* X *an irreducible* G*-space with finite invariant measure, and* $\vartheta \colon X \times G \to \mathbf{H}(k)$ *a cocycle. Suppose*

$$\operatorname{rank} G \overset{\text{def}}{=} \sum_{p \in A} \operatorname{rank}_{\mathbb{Q}_p} \mathbf{G}_p \geq 2.$$

Assume in addition that the cocycle ϑ *is Zariski dense in* $\mathbf{H}$ *and is not relatively compact. Then*

(i) $p(k) \in A$ *and there exists a* $\mathbb{Q}_{p(k)}$*-epimorphism* $\eta \colon \mathbf{G}_{p(k)} \to \mathbf{H}$ *such that* $\vartheta \sim \sigma_{\eta \circ \mathrm{pr}_{p(k)}}$, *where* $\sim$ *is the equivalence relation for cocycles defined in 8.1 and* $\sigma_{\eta \circ \mathrm{pr}_{p(k)}}(x, g) = \eta(\mathrm{pr}_{p(k)}(g))$, $x \in X$, $g \in G$;

(ii) *there is a continuous homomorphism* $\pi \colon G \to \mathbf{H}(k)$ *such that* $\vartheta \sim \sigma_\pi$.

(8.3) Theorem. (See [Zi 8], Theorem 5.2.5.) *Let* $\mathbf{G}$ *be a connected semisimple* $\mathbb{R}$*-group without* $\mathbb{R}$*-anisotropic factors. Let* $\mathbf{G}(\mathbb{R})^0$ *be the identity component of the Lie group* $\mathbf{G}(\mathbb{R})$, k *a local field of characteristic* 0, $\mathbf{H}$ *a connected almost* k*-simple* k*-group,* X *an irreducible* $\mathbf{G}(\mathbb{R})^0$*space with finite invariant measure and* $\vartheta \colon X \times \mathbf{G}(\mathbb{R})^0 \to \mathbf{H}(k)$ *a cocycle. Suppose* $\operatorname{rank}_{\mathbb{R}} \mathbf{G} \geq 2$ *and the cocycle* ϑ *is Zariski dense in* $\mathbf{H}$. *Then*

(i) *if* k *is non-archimedean, then the cocycle* ϑ *is relatively compact;*

(ii) *if* k *is* $\mathbb{R}$ *or* $\mathbb{C}$ *and the group* $\mathbf{H}$ *is simple over* k *(or equivalently* $\mathscr{Z}(\mathbf{H}) = \{e\}$*), then either the cocycle* ϑ *is relatively compact or there is a* k*-epimorphism* $\eta \colon \mathbf{G} \to \mathbf{H}$ *such that* $\vartheta \sim \sigma_{\eta|_{\mathbf{G}(\mathbb{R})^0}}$.

Notice that for $\mathbf{G}$ simply connected, Theorem 8.3 is a particular case of Theorem 8.2.

(8.4) With G, k, and $\mathbf{H}$ as in Theorem 8.2, we let Γ be a lattice in G and $\tau \colon \Gamma \to \mathbf{H}(k)$ a homomorphism. As in V.3.0 we denote by $\Omega = \Omega_{\Gamma,\tau}$ the set of Borel maps $f \colon G \to \mathbf{H}(k)$ satisfying the condition $f(\gamma g) = f(g)\tau(\gamma^{-1})$, $\gamma \in \Gamma$, $g \in G$, and associate with each $f \in \Omega$ the cocycle $B_f \colon (\Gamma \setminus G) \times G \to \mathbf{H}(k)$ (recall that $B_f(\pi(y), g) = f(y)f(yg)^{-1}$, where $y \in G$, and $\pi \colon G \to \Gamma \setminus G$ is the natural projection). The equivalence class of the cocycle B_f does not depend upon the choice of $f \in \Omega$. It is easy to check that the cocycle B_f is relatively compact (resp. Zariski dense) in $\mathbf{H}$ if and only if the subgroup $\pi(\Gamma)$ is relatively compact (resp. Zariski dense) in $\mathbf{H}$. It can easily be verified that for a continuous homomorphism $\pi \colon G \to \mathbf{H}(k)$ the following conditions are satisfied:

(a) there exists $h \in \mathbf{H}(k)$ such that $\pi(\gamma) = h\tau(\gamma)h^{-1}$ for all $\gamma \in \Gamma$;

(b) the cocycles B_f are equivalent to the cocycle σ_π for all $f \in \Omega_{\Gamma,\tau}$.

It follows from Theorem II.6.7(b) that the irreducibility of the lattice Γ is equivalent to that of the G-space $\Gamma \setminus G$. So Theorem 5.7 is a consequence of

Theorem 8.2 (at any rate for $\mathbf{G}'_p$ simply connected). Similarly, Theorem 5.9 is a consequence of Theorem 8.3.

(8.5) Theorems 8.2 and 8.3 can also be applied to prove results which at first sight have nothing in common with the superrigidity theorems for discrete subgroups. These results concern the problem of orbit equivalence of ergodic actions. We recall the necessary definitions.

Let G and G' be two locally compact groups and suppose that X is a (right) G-space with quasi-invariant measure μ and X' a (right) G'-space with quasi-invariant measure μ'. The actions of G on X and G' on X' are called *orbit equivalent* if there are conull Borel subsets $X_0 \subset X$, $X'_0 \subset X'$ and a Borel isomorphism $\varphi\colon X_0 \to X'_0$ such that:

(a) if $x, y \in X_0$, then x and y are in the same G-orbit if and only if $\varphi(x)$ and $\varphi(y)$ are in the same G'-orbit and

(b) φ transfers μ to a measure equivalent to μ' (i.e. $\mu(Y) = 0$ if and only if $\mu'(\varphi(Y)) = 0$).

If $G = G'$ and φ is G-equivariant, then the actions of the group G on X and X' are called *conjugate*.

Suppose a Borel isomorphism $\varphi\colon X \to X'$ realizes the orbit equivalence and the action of the group G' on X' is essentially free (i.e. for almost all $x \in X'$ the stabilizer of the point x is trivial). Then, possibly after discarding null sets from X and X', for all $x \in X$ and $g \in G$, via the equality $\varphi(x)\vartheta(x,g) = \varphi(xg)$, one can define uniquely $\vartheta(x,g) \in G'$. It is straightforward that ϑ is a cocycle. It is not hard to show that if $\vartheta \sim \sigma_\pi$, where $\pi\colon G \to G'$ is a continuous isomorphism, then, after identification of the groups G and G' via π the actions of the group G on X and X' are conjugate.

Making use of the just mentioned relation between the orbit equivalence and cocycles, we deduce Theorem 8.6 and 8.8 from Theorems 8.2 and 8.3.

(8.6) Theorem. (See [Zi 8], Theorem 10.1.8.) *Given two finite sets A and A', for each $\alpha \in A$ (resp. $\alpha' \in A'$) let k_α (resp. $k'_{\alpha'}$) be a local field of characteristic* 0 *and let $\mathbf{G}_\alpha$ (resp. $\mathbf{G}'_{\alpha'}$) be a connected semisimple adjoint algebraic group, which defined, isotropic, and simple over k_α (resp. $k'_{\alpha'}$). Set $G = \prod_{\alpha \in A} \mathbf{G}_\alpha(k_\alpha)$ and $G' = \prod_{\alpha' \in A'} \mathbf{G}'_{\alpha'}(k'_{\alpha'})$. Let X (resp. X') be an essentially free irreducible G-space (resp. G'-space) with finite invariant measure. Suppose* $\operatorname{rank} G \overset{\text{def}}{=} \sum_{\alpha \in A} \operatorname{rank}_{k_\alpha} \mathbf{G}_\alpha \geq 2$. *If the action of the group G on X is orbit equivalent to that of the group G' on X', then the groups G and G' are isomorphic (as topological groups) and these two actions are conjugate modulo an isomorphism $\pi\colon G \to G'$.*

(8.7) For every connected semisimple Lie group G we set $\operatorname{rank} G = \operatorname{rank}_{\mathbb{R}} \overline{\operatorname{Ad} G}$, where $\overline{\operatorname{Ad} G}$ is the Zariski closure of the group $\operatorname{Ad} G$ (for further details see the beginning of Section 6 in Chapter IX).

(8.8) Theorem. (See [Zi 8], Theorem 5.2.1.) *Let G and G' be connected semisimple Lie groups with finite center and no non-trivial compact factor groups, X (resp.*

X') an essentially free ergodic irreducible G (resp. G')-space with finite invariant measure, and assume that the actions of G on X and G' on X' are orbit equivalent. Assume in addition that $\operatorname{rank} G \geq 2$. *Then*

(i) *G and G' are locally isomorphic.*
(ii) *If $\mathscr{Z}(G) = \mathscr{Z}(G') = \{e\}$, then the groups G and G' are isomorphic and identifying G and G' via this isomorphism $\pi\colon G \to G'$, the actions of G on X and X' are conjugate.*

(8.9) The following assertion seems to be true.

(*) *With A, k_α, $\mathbf{G}_\alpha$, $G = \prod_{\alpha \in A} \mathbf{G}_\alpha(k_\alpha)$, and* $\operatorname{rank} G$ *as in Section 5, we let k be a local field, $\mathbf{H}$ a connected adjoint k-simple k-group, X an irreducible G-space with finite invariant measure, and $\vartheta\colon X \times G \to \mathbf{H}(k)$ a cocycle. Suppose* $\operatorname{rank} G \geq 2$ *and the cocycle ϑ is Zariski dense in $\mathbf{H}$ and is not relatively compact. Then there is a continuous homomorphism $\pi\colon G \to \mathbf{H}(k)$ such that $\vartheta \sim \sigma_\pi$.*

In case $\operatorname{char} k = 0$ and $\operatorname{char} k_\alpha = 0$ for all $\alpha \in A$, assertion (*) can easily be deduced from Theorem 8.2(ii).

Theorems 8.2 and 8.3 were proved in [Zi 9] by a method analogous in many respects to the method developed for proving the superrigidity theorems that were outlined in 5.26. It seems likely that this method, in the form presented in [Ve], is also applicable to the proof of (*). On the other hand, it is very likely that (*) can be proved by modifying the proof of Theorem 5.6. It is also very likely that one can obtain results for ergodic actions similar to those in the second part of Section 5 (after 5.17).

(8.10) *Remark.* From the superrigidity theorems for cocycles one can deduce a series of other consequences (we refer the reader to[Zi 8], Ch. 9 and other original papers of Robert Zimmer, in particular, to [Zi 6]).

Chapter VIII. Normal Subgroups and "Abstract" Homomorphisms of Semisimple Algebraic Groups Over Global Fields

In this chapter K denotes a global field, $\mathscr{R}$ the set of all (inequivalent) valuations of the field K, and $\mathscr{R}_\infty \subset \mathscr{R}$ the (finite) set of archimedean valuations of K. The value of a valuation $v \in \mathscr{R}$ at $x \in K$ will be denoted by $|x|_v$. As before, K_v denotes the completion of the field K corresponding to the valuation $v \in \mathscr{R}$. If $S \subset \mathscr{R}$ and $\mathbf{H}$ is a (reductive) K-group, we set

$$\operatorname{rank}_S \mathbf{H} = \sum_{v \in S} \operatorname{rank}_{K_v} \mathbf{H} \in \mathbb{N} \cup \{\infty\}.$$

As in I.0.32, for any $S \subset \mathscr{R}$, we denote by $K(S)$ the ring of S-integers of the field K. As in I.3.1, for every K-subgroup $\mathbf{H}$ of $\mathbf{GL}_n$, we set

$$\mathbf{H}(K(S)) = \mathbf{H} \cap \mathbf{GL}_n(K(S)) \text{ and } \mathbf{H}(\mathfrak{a}) = \mathbf{H} \cap \mathbf{GL}_n(\mathfrak{a}),$$

where $\mathfrak{a}$ is a non-zero ideal of $K(S)$. As in I.3.1.2 a subgroup of $\mathbf{H}$ is called *S-arithmetic* if it is commensurable with $\mathbf{H}(K(S))$.

As in Section 6 of Chapter VII, we denote by $\bar{M}$ the Zariski closure of a set M and by $\mathbf{H}^0$ the connected component of the identity in an algebraic group $\mathbf{H}$.

For an l-group $\mathbf{H}$ and a homomorphism σ of the field l into a field l', the l'-group ${}^\sigma\mathbf{H}$ and the homomorphism $\sigma^0\colon \mathbf{H}(l) \to {}^\sigma\mathbf{H}(l')$ were defined in I.1.7 and I.1.8.

Throughout this chapter $\mathbf{G} \subset \mathbf{GL}_n$ denotes a connected non-commutative almost K-simple K-group. We set $\mathscr{T} = \mathscr{T}(G) = \{v \in \mathscr{R} \mid$ the group $\mathbf{G}(K_v)$ is compact, or equivalently, $\mathbf{G}$ is anisotropic over $K_v\}$ and pick a subset $S \subset \mathscr{R}$ containing $\mathscr{R}_\infty - \mathscr{T}$.

In this chapter a series of results will be established on normal subgroups and homomorphisms of S-arithmetic subgroups of $\mathbf{G}$ to algebraic groups. In particular, the following theorems will be proved.

(A) *Let Λ be an S-arithmetic subgroup of $\mathbf{G}$ and let N be a normal subgroup of Λ. Assume* $\operatorname{rank}_S \mathbf{G} \geq 2$ *and that either $\mathbf{G}$ is connected or S is finite. Then either $N \subset \mathscr{Z}(\mathbf{G})$ or Λ/N is finite.*

(B) *Let Λ be an S-arithmetic subgroup of $\mathbf{G}$, l a field, $\mathbf{H}$ an algebraic l-group, and $\delta\colon \Lambda \to \mathbf{H}(l)$ a homomorphism. Suppose* $\operatorname{char} K = 0$ *and* $\operatorname{rank}_S \mathbf{G} \geq 2$. *Then*

(i) *if* $\operatorname{char} l \neq 0$ *and $\mathbf{G}$ is simply connected, then the group $\delta(\Lambda)$ is finite;*
(ii) *if* $\operatorname{char} l = 0$, *then the group $\overline{\delta(\Lambda)}$ is semisimple;*

(iii) *if* $\operatorname{char} l = 0$ *(i.e.* $l \supset \mathbb{Q}$*) and* $\mathbf{G}$ *is simply connected, then there exist (uniquely determined) an l-morphism* $\varphi\colon R_{K/\mathbb{Q}}\mathbf{G} \to \mathbf{H}$ *and a homomorphism* $\nu\colon \Lambda \to \mathbf{H}$ *such that the subgroup* $\nu(\Lambda)$ *is finite and commutes with* $\varphi(R_{K/\mathbb{Q}}\mathbf{G})$ *and* $\delta(\lambda) = \nu(\lambda)\cdot\varphi(R^0_{K/\mathbb{Q}})$ *for all* $\lambda \in \Lambda$.

The scheme of the proof of Theorems (A) and (B) is the following. Choose a finite subset $\mathscr{E} \subset S$ with $\mathscr{E} \supset \mathscr{R}_\infty - \mathscr{R}$ and $\operatorname{rank}_{\mathscr{E}} \mathbf{G} \geq 2$. We set $\Gamma = \Lambda \cap \mathbf{G}(K(\mathscr{E}))$ and identify the group Γ with its image under the diagonal embedding in $G_{\mathscr{E}} \overset{\text{def}}{=} \prod_{v\in\mathscr{E}} \mathbf{G}(K_v)$. According to the Borel-Harish-Chandra-Behr-Harder reduction theorem (see Theorem I.3.2.5) Γ is a lattice in $G_{\mathscr{E}}$. We can then apply Theorem IV.4.9 to prove Theorem (A) and Theorems VII.6.5 and VII.6.16 to prove Theorem (B). In case S is infinite it is necessary to make use of the strong approximation theorem. The realization of this scheme encounters no essential obstructions in any of the following cases:

(a) the group $\mathbf{G}$ is anisotropic over K;
(b) in proving Theorem (A) $\operatorname{rank}_{K_v} \mathbf{G} \geq 2$ for some $v \in S$;
(c) in proving Theorem (B) the group $\overline{\delta(\Lambda)}$ is semisimple.

In the other cases we have to use certain properties of fundamental domains for S-arithmetic subgroups. We consider these properties in Section 1. For the above mentioned fundamental domains we choose as usual the subsets of finite unions of Siegel's domains.

If we discard in Theorem (B) the assumption "$\operatorname{char} K = 0$", then according to Theorem I.3.2.4(b), Theorem B of VII.5.25, and Lemma VII.5.22 the group $\overline{\delta(\Lambda)}$ is in general not semisimple. Nevertheless, without any restriction on the characteristic of the field K, one can prove the following theorem (see 3.4):

(C) *Let* Λ *be an S-arithmetic subgroup of* $\mathbf{G}$, l *a field,* $\mathbf{H}$ *a connected noncommutative absolutely almost simple l-group, and* $\delta\colon \Lambda \to \mathbf{H}(l)$ *a homomorphism with* $\overline{\delta(\Lambda)} = \mathbf{H}$. *Suppose that* $\operatorname{rank}_S \mathbf{G} \geq 2$ *and either* $\mathbf{G}$ *is simply connected or* $\mathbf{H}$ *is adjoint. Then there exist (uniquely determined) a homomorphism* $\sigma\colon K \to l$, *a special (in the sense of I.1.4.14) l-epimorphism* $\eta\colon {}^\sigma\mathbf{G} \to \mathbf{H}$, *and a homomorphism* $\nu\colon \Lambda \to \mathscr{Z}(\mathbf{H})$ *such that* $\delta(\lambda) = \nu(\lambda)\cdot\eta(\sigma^0(\lambda))$ *for all* $\lambda \in \Lambda$.

We would like to mention a few words about the proof of Theorem (C). Making use of Theorem VII.6.5 and the Borel-Harish-Chandra-Behr-Harder reduction theorem mentioned above, one can easily establish the existence of the homomorphisms σ, η, and ν after replacing l by a field extension of l. It then remains to prove the inclusion $\sigma(K) \subset l$ and the uniqueness of the homomorphisms σ, η, and ν. If $K = \mathbb{Q}$, the inclusion $\sigma(K) \subset l$ is evident and the uniqueness is a consequence of the fact that any subgroup of finite index in Λ is Zariski dense in $\mathbf{G}$ (see Proposition I.3.2.10). The case where $\operatorname{char} K = 0$ can easily be reduced to the case where $K = \mathbb{Q}$ with the help of the restriction of scalars functor. However, if $\operatorname{char} K \neq 0$, a more detailed argument is needed. We also remark that in our proof of Theorem (C) we shall not be using the results of Section 1.

1. Some Properties of Fundamental Domains for S-Arithmetic Subgroups

In this section we assume the set S is finite. Let Γ be an S-arithmetic subgroup of $\mathbf{G}(K)$. We set $G_S = \prod_{v \in S} \mathbf{G}(K_v)$ and identify the group Γ with its image under the diagonal embedding in G_S. The set of left Borel fundamental domains $X \subset G_S$ for Γ will be denoted by $\mathfrak{f}(G_S, \Gamma)$. For each $X \in \mathfrak{f}(G_S, \Gamma)$ we define (uniquely) the Borel maps $\tau_X\colon G_S \to \Gamma$ and $\omega_X\colon G_S \to X$ via the equation

$$g = \tau_X(g) \cdot \omega_X(g), \; g \in G, \tau_X(g) \in \Gamma, \; \omega_X(g) \in X.$$

(1.1) We recall that G is viewed as a K-subgroup of the group $\mathbf{GL}_n$, $n \in \mathbb{N}^+$. For $v \in S$, $y = (y_1, \ldots, y_n) \in K_v^n$, $B \in \operatorname{End}(K_v^n)$, and $G \in G_S$, we define the norms

$$\|y\|_v = \sum_{1 \le i \le n} |y_i|_v, \; \|B\|_v = \sup_{y \in K_v^n, y \neq 0} \|By\|/\|y\|, \text{ and}$$

$$\|g\| = \max_{v \in S} \|\operatorname{pr}_v(g)\|,$$

where $\operatorname{pr}_v\colon G_S \to \mathbf{G}(K_v)$ is the natural projection.

A Borel set $X \subset G_S$ is called *quasi-bounded* if the following condition is satisfied

$$\text{(1)} \qquad \int_X \ln^+(\|y\| + \|y^{-1}\|) d\mu_{G_S}(y) < \infty.$$

Remark. Since $\mathscr{D}(\mathbf{G}) = \mathbf{G}$, it follows that $\mathbf{G} \subset \mathbf{SL}_n$. For $A \in \mathbf{SL}_n$, the entries of A^{-1} are polynomials in the entries of the matrix A. Therefore there exist positive constants c and d such that $\|g^{-1}\| < c\|g\|^d$ for all $g \in G_S$. Furthermore, since $G \subset \mathbf{SL}_n$, by Lemma IV.1.1 $\ln\|g\| \ge 0$ for all $g \in G_S$. Thus, condition (1) is equivalent to the condition

$$\text{(1')} \qquad \int_X \ln\|y\| d\mu_G(y) < \infty.$$

(1.2) Proposition. *There exist a quasi-bounded fundamental domain $X \in \mathfrak{f}(G_s, \Gamma)$, $i \in \mathbb{N}^+$, and minimal parabolic K-subgroups $\mathbf{P}_1, \ldots, \mathbf{P}_i$ of $\mathbf{G}$ such that*

(a) *for any compact set $M \subset G_S$ there is a finite set $F = F(M) \subset \Gamma$ such that $\tau_X(X \cdot M) \subset \bigcup_{1 \le j \le i} (R_u(\mathbf{P}_j) \cap \Gamma) \cdot F$;*
(b) *for any compact set $M \subset G_S$ the set $\{\omega_X(yg)^{-1} y \mid y \in X, \; g \in M\}$ is relatively compact in G_S.*

We shall only sketch the proof of this proposition. Suppose we have found a Borel set $\Omega \subset G_S$ such that

(i) Ω is a fundamental set for Γ, i.e., $G_S = \Gamma \cdot \Omega$;
(ii) the set Ω is quasi-bounded;

(iii) there exist $i \in \mathbb{N}^+$ and minimal parabolic K-subgroups $\mathbf{P}_1, \dots, \mathbf{P}_i$ of $\mathbf{G}$ such that for any compact set $M \subset G_S$ the set $\{\gamma \in \Gamma \mid \gamma\Omega \cap \Omega M \neq \varnothing\}$ is contained in a finite union of sets of the form $(R_u(\mathbf{P}_j) \cap \Gamma)\gamma$, $1 \leq j \leq i$, $\gamma \in \Gamma$;

(iv) for any compact set $M \subset G_S$ the set $\{z^{-1}y \mid y, z \in \Omega,\ z \in \Gamma y M\}$ is relatively compact in G_S.

Then as the desired fundamental domain we can take any domain $X \in \mathfrak{f}(G_S, \Gamma)$ with $X \subset \Omega$. The construction of the domain Ω is carried out in [Bo 5] in case $K = \mathbb{Q}$ and $S = \{\infty\}$, and in [Be 2] and [Har 2] in the general case. One can take Ω to be finite unions of (generalized) Siegel's domains.

We recall the definition of a Siegel domain in case $K = \mathbb{Q}$ and $S = \{\infty\}$ (i.e. $G_S = \mathbf{G}(\mathbb{R})$ and Γ is an arithmetic subgroup in $\mathbf{G}(\mathbb{Q})$). Let $\mathbf{P}$ be a minimal parabolic $\mathbb{Q}$-subgroup of $\mathbf{G}$, $\mathbf{A}$ the maximal $\mathbb{Q}$-split torus of $\mathbf{G}$ that is contained in $\mathbf{P}$, L the maximal compact subgroup in $\mathbf{G}(\mathbb{R})$ whose Lie algebra is orthogonal relative to the Killing form to the Lie algebra of the subgroup $\mathbf{A}(\mathbb{R})$. We define on the root system $\Phi(\mathbf{A}, \mathbf{G})$ the ordering so that $\mathbf{P}$ is the group $\mathbf{P}_\varnothing$ (with notation as in I.1.2). Let Δ be the set of simple roots relative to the ordering. We denote by ${}_FA$ the identity component of the Lie group $\mathbf{A}(\mathbb{R})$ and for any $t \in \mathbb{R}^+$ we set ${}_FA_t = \{a \in {}_FA \mid \alpha(a) \leq t \text{ for all } \alpha \in \Delta\}$. Let $\mathbf{U} = R_u(\mathbf{P})$. We recall that $\mathbf{P} = \mathscr{Z}_\mathbf{G}(\mathbf{A}) \ltimes \mathbf{U}$. Furthermore, $\mathscr{Z}_\mathbf{G}(\mathbf{A})$ can be represented in the form $\mathbf{A} \cdot \mathbf{F}$, where $\mathbf{F}$ is the largest connected $\mathbb{Q}$-anisotropic $\mathbb{Q}$-subgroup of $\mathscr{Z}_\mathbf{G}(\mathbf{A})$. On the other hand, from the Iwasawa decomposition we have $\mathbf{P}(\mathbb{R}) \cdot L = \mathbf{G}(\mathbb{R})$. This yields the following decomposition:

$$\mathbf{G}(\mathbb{R}) = \mathbf{U}(R) \cdot {}_FA \cdot \mathbf{F}(R) \cdot L.$$

A *Siegel domain* in $\mathbf{G}(\mathbb{R})$ associated to L, $\mathbf{P}$, and $\mathbf{A}$ is any set of the form

$$\sigma_{t,\eta,\omega} = \omega \cdot {}_FA_t \cdot \eta \cdot L,$$

where η (resp. ω) is a compact set in $\mathbf{F}(\mathbb{R})$ (resp. $\mathbf{U}(\mathbb{R})$).

As we have noted above we choose Ω to be a finite union of Siegel domains. More precisely there are a Siegel domain $\sigma = \sigma_{t,\eta,\omega}$ and a finite set $C \subset \mathbf{G}(\mathbb{Q})$ such that $\Omega = C \cdot \sigma$ satisfies conditions (i)–(iv). The verification of (iii)–(iv) was in fact carried out in [Be 2] while proving that $\mathbf{G}(K(S))$ is finitely generated. The quasi-boundedness of the set σ (and hence Ω) can be proved in exactly the same way as the finiteness of volume of these sets (see [Bo 5], §12, Proposition 3). For (i) in the case where $K = Q$ and $S = \{\infty\}$, see Theorem 2 in §14 of [Bo 5].

(1.3) Proposition. *Let $n \in \mathbb{N}^+$, k a local field, and let $T\colon \Gamma \to \mathbf{GL}_n(k)$ be a representation of the group Γ on the space k^n. Suppose* $\operatorname{char} k = 0$, $S \subset \mathscr{R}_\infty$, *and the following condition is satisfied:*

($*$) *there is a subgroup Γ_1 of finite index in Γ such that $T(\Gamma_1^{(u)}) \subset \mathbf{GL}_n^{(u)}$ (i.e. the matrix $T(\gamma)$ is unipotent whenever γ is a unipotent element of Γ_1).*

Then the representation T is Γ-integrable (in the sense of §3 of Chapter V).

Proof. According to Lemma I.3.1.4 we can replace $\mathbf{G}$ by $R_{K/\mathbb{Q}}\mathbf{G}$ and Γ by $R^0_{K/\mathbb{Q}}(\Gamma)$ and assume $K = \mathbb{Q}$ and $S = \{\infty\}$. Then $\mathbf{G}$ is a $\mathbb{Q}$-group, Γ is an arithmetic subgroup in $\mathbf{G}(\mathbb{Q})$, and $G_S = \mathbf{G}(\mathbb{R})$. Let fundamental domain $X \in \mathfrak{f}(G_S, \Gamma)$ and K-subgroups $\mathbf{P}_1, \ldots, \mathbf{P}_i$ be as in Proposition 1.2. We shall show that X is (Γ, T)-integrable, i.e., for any compact set $M \subset G_S$, $Q_{M,X}(y) \overset{\text{def}}{=} \sup_{g\in M}(\ln^+ \|T(\tau_X(yg))\| + \ln^+ \|T(\tau_X(yg))^{-1}\|) < \infty$ for almost all $y \in X$ and $Q_{M,X} \in L_1(X, \mu_{G_S})$. Since

$$\tau_X(yg) = yg\omega_X(yg)^{-1} = yg\left(\omega_X(yg)^{-1}y\right)y^{-1},$$

it follows from the quasi-boundedness of X and (b) of Proposition 1.2 that for any compact set $M \subset G_S$ the function

$$\mathfrak{f}_M(y) \overset{\text{def}}{=} \sup_{g\in M}(\ln^+ \|\tau_X(yg)\| + \ln^+ \|\tau_X(yg)^{-1}\|), \; y \in X,$$

belongs to $L_1(X, \mu_{G_S})$, where $\|B\|$, $B \in \operatorname{End}(k^n)$, is defined as in Chapter II or in Chapter V. Further, since Γ_1 is of finite index in Γ and

$$\tau_X(X \cdot M) \subset \bigcup_{1\le j\le i} (R_u(\mathbf{P}_j) \cap \Gamma) \cdot F,$$

where $F \subset \Gamma$ is finite, there exists a finite set $F_1 \subset \Gamma$ such that

$$(1) \qquad \tau_X(X \cdot M) \subset \bigcup_{1\le j\le i} (R_u(\mathbf{P}_j) \cap \Gamma_1) \cdot F_1,$$

We first consider the case in which k is isomorphic to $\mathbb{R}$ or $\mathbb{C}$. Since the restriction of the homomorphism T to the arithmetic subgroup $R_u(\mathbf{P}_j) \cap \Gamma_1$ of $R_u(\mathbf{P})(\mathbb{Q})$ is unipotent, it can be extended by Lemma I.3.3.4(a) to a rational homomorphism $T_j\colon R_u(\mathbf{P}_j) \to \mathbf{GL}_n$. Therefore the entries of the matrix $T(u)$ are polynomials in the entries of the matrix $u \in R_u(\mathbf{P}_j) \cap \Gamma_1$, and hence there exist constants c_1 and c_2 such that

$$(2) \qquad \ln^+ \|T(u)\| \le c_1 \ln^+ \|u\| + c_2, \; u \in R_u(\mathbf{P}_j) \cap \Gamma_1, \; 1 \le j \le i.$$

Since $f_M \in L_1(X, \mu_{G_S})$ and the set F_1 is finite, it follows from (1), (2), and the inequality $\|BC\| \le \|B\|\,\|C\|$ $(B, C \in \operatorname{End} k^n)$ that $Q_{M,X} \in L_1(X, \mu_{G_S})$. Suppose now that k is non-archimedean. Then Lemma I.3.3.4(b) implies that the subgroup $T(R_u(\mathbf{P}_j) \cap \Gamma)$ is relatively compact in $\mathbf{GL}_n(k)$ for each j, $1 \le j \le i$. From this, (1), and the finiteness of F_1 we deduce that the set $\tau_X(X \cdot M)$ is relatively compact in $\mathbf{GL}_n(k)$. Therefore the function $Q_{M,X}$ is bounded, and hence belongs to $L_1(X, \mu_{G_S})$. Thus, we have proved that the domain X is (Γ, T)-integrable, and hence the representation T is Γ-integrable. □

(1.4) *Remark.* In Proposition 1.3 the condition "$S \subset R_\infty$" was imposed just to simplify the proof and it can be dropped. However the condition "$\operatorname{char} K = 0$" is not in general superfluous. Indeed, let $K = F(t)$ be the field of rational functions in one variable over a finite field F, $S = \{v\}$ consist of a single valuation, and

$\Gamma = \mathbf{SL}_2(K(S))$ be viewed as a discrete subgroup in $\mathbf{SL}_2(K_v)$. It can then be shown that there is a finite-dimensional representation T of the group Γ over K_v such that T is not Γ-integrable and transfers unipotent elements of Γ to unipotent matrices. Nevertheless, there are good reasons also to believe that in case $\operatorname{rank}_S \mathbf{G} \geq 2$ any representation $T\colon \Gamma \to \mathbf{GL}_n(k)$ is Γ-integrable for fields K of positive characteristic.

2. Finiteness of Factor Groups of S-Arithmetic Subgroups

We define $\Psi = \Psi(\mathbf{G})$ to be the subgroup $\operatorname{Ad}^{-1}((\operatorname{Ad}\mathbf{G})(K))$ of $\mathbf{G}$. In other words, $\Psi = \{g \in \mathbf{G} \mid \text{the } K\text{-structure } \operatorname{Lie}(\mathbf{G})_K \text{ is } (\operatorname{Ad} g)\text{-invariant}\}$. We remark that $\mathbf{G}(K) \subset \Psi$.

(2.1) Lemma. *(i) The subgroup $\mathbf{G}(K)$ is normal in Ψ and the factor group $\Psi/\mathbf{G}(k)$ is commutative and torsion, and the orders of its elements are uniformly bounded.*

(ii) Let $\pi\colon \mathbf{G} \to \mathbf{G}'$ be a central K-isogeny of K-groups. Then $\pi(\Psi(\mathbf{G})) = \Psi(\mathbf{G}')$.

Proof. (i) According to Proposition I.1.4.11(iv) the isogeny $\operatorname{Ad}\colon \mathbf{G} \to \operatorname{Ad}\mathbf{G}$ is central. All that remains now is to apply Remark 2 of 1.4.2 in Chapter I.

(ii) It follows from Proposition I.1.4.7 that the composition of central isogenies of reductive groups is a central isogeny. But the isogenies $\operatorname{Ad}_{\mathbf{G}} = \operatorname{Ad}\colon \mathbf{G} \to \operatorname{Ad}\mathbf{G}$ and $\operatorname{Ad}_{\mathbf{G}'} = \operatorname{Ad}\colon \mathbf{G}' \to \operatorname{Ad}\mathbf{G}'$ are central and $\operatorname{Ad}\mathbf{G}$ and $\operatorname{Ad}\mathbf{G}'$ are adjoint groups. Thus, we have two central K-isogenies $\operatorname{Ad}_{\mathbf{G}'} \circ \pi\colon \mathbf{G} \to \operatorname{Ad}\mathbf{G}'$ and $\operatorname{Ad}_{\mathbf{G}}\colon \mathbf{G} \to \operatorname{Ad}\mathbf{G}$ of the group $\mathbf{G}$ into adjoint K-groups (see Proposition I.1.4.11(iv)). Then by virtue of Proposition I.1.4.11(iii) there is a K-isomorphism $f\colon \operatorname{Ad}\mathbf{G}' \to \operatorname{Ad}\mathbf{G}$ such that $f \circ \operatorname{Ad}_{\mathbf{G}'} \circ \pi = \operatorname{Ad}_{\mathbf{G}}$. Identifying $\operatorname{Ad}\mathbf{G}'$ and $\operatorname{Ad}\mathbf{G}$ by f we have that

$$\pi(\Psi(\mathbf{G})) = \pi(\operatorname{Ad}_{\mathbf{G}}^{-1}(\operatorname{Ad}\mathbf{G})(K)) = (\pi \circ \pi^{-1} \circ \operatorname{Ad}_{\mathbf{G}'}^{-1})(\operatorname{Ad}\mathbf{G})(K) = \Psi(\mathbf{G}').$$

□

(2.2) Lemma. *Let Λ be an S-arithmetic subgroup of $\mathbf{G}$. Suppose $S \not\subset \mathcal{T}$ or, equivalently, that* $\operatorname{rank}_S \mathbf{G} > 0$. *Then*

(i) *$\Lambda \subset \Psi$;*

(ii) *if N is a subgroup of Ψ normalized by Λ and $N \not\subset \mathscr{Z}(\mathbf{G})$, then $\overline{N \cap \mathbf{G}(K)} = \mathbf{G}$, and hence $N \cap \mathbf{G}(K) \not\subset \mathscr{Z}(\mathbf{G})$.*

Proof. (i) By Proposition I.3.2.10 the subgroup $\Lambda_0 \overset{\text{def}}{=} \Lambda \cap \mathbf{G}(K(S))$ is Zariski dense in $\mathbf{G}$, and hence the subgroup $\operatorname{Ad}\Lambda_0$ of the group $(\operatorname{Ad}\mathbf{G})(K)$ is Zariski dense in the semisimple adjoint group $\operatorname{Ad}\mathbf{G}$. On the other hand, since Λ_0 is of finite index in Λ, it follows that $\operatorname{Ad}\Lambda \subset \operatorname{Comm}_{\operatorname{Ad}\mathbf{G}}(\operatorname{Ad}\Lambda_0)$. Thus, by Lemma VII.6.2(ii) $\operatorname{Ad}\Lambda \subset (\operatorname{Ad}\mathbf{G})(K)$, and hence $\Lambda \subset \Psi$.

(ii) According to Proposition I.3.2.10 $\bar{\Lambda} = \mathbf{G}$, and hence $\overline{\mathrm{Ad}\,\Lambda} = \mathrm{Ad}\,\mathbf{G}$. Since $\mathrm{Ker}\,\mathrm{Ad} = \mathscr{Z}(\mathbf{G})$ (see I.0.24), $N \not\subset \mathscr{Z}(\mathbf{G})$, $\mathrm{Ad}\,N \subset (\mathrm{Ad}\,\mathbf{G})(K)$, and $\mathrm{Ad}\,\Lambda$ normalizes $\mathrm{Ad}\,N$, it follows that the subgroup $\overline{\mathrm{Ad}\,N}$ is different from $\{e\}$, defined over K, and normal in $\overline{\mathrm{Ad}\,\Lambda} = \mathrm{Ad}\,\mathbf{G}$. But since $\mathbf{G}$ is connected and almost K-simple, the group $\mathrm{Ad}\,\mathbf{G}$ is K-simple. Therefore $\mathrm{Ad}\,\bar{N} = \overline{\mathrm{Ad}\,N} = \mathrm{Ad}\,\mathbf{G}$. From this, the connectedness of $\mathbf{G}$, and the finiteness of $\mathrm{Ker}\,\mathrm{Ad}$ we deduce that $\bar{N} = G$, and hence $\overline{\mathscr{D}(N)} = \mathbf{G}$. But by Lemma 2.1(i), $\mathscr{D}(N) \subset \mathscr{D}(\Psi) \subset \mathbf{G}(K)$. Thus, $\overline{N \cap \mathbf{G}(K)} = \mathbf{G}$. □

(2.3) Lemma. *Let Λ be an S-arithmetic subgroup of $\mathbf{G}$, N a subgroup of Ψ normalized by Λ, and $\mathscr{E}$ a subset of S with $\mathscr{E} \not\subset \mathscr{T}$. We set $\Lambda_{\mathscr{E}} = \Lambda \cap \mathbf{G}(K(\mathscr{E}))$.*

(a) *If $N \not\subset \mathscr{Z}(\mathbf{G})$, then $N \cap \Lambda_{\mathscr{E}} \not\subset \mathscr{Z}(\mathbf{G})$.*
(b) *Suppose $\mathbf{G}$ is connected and absolutely almost simple, $\mathscr{E} \supset S \cap \mathscr{T}$, $\Lambda \subset \mathbf{G}(K)$, $N \not\subset \mathscr{Z}(\mathbf{G})$. Then $\Lambda = (N \cap \Lambda) \cdot \Lambda_{\mathscr{E}}$, and hence the factor group $\Lambda/(N \cap \Lambda)$ is isomorphic to $\Lambda_{\mathscr{E}}/(N \cap \Lambda_{\mathscr{E}})$.*
(c) *Suppose $\mathbf{G}$ is simply connected. Then $N \cap \Lambda$ is of finite index in Λ whenever $N \cap \Lambda_{\mathscr{E}}$ is of finite index in $\Lambda_{\mathscr{E}}$.*

Proof. According to Lemma 2.2(ii), $N \cap \mathbf{G}(K) \not\subset \mathscr{Z}(\mathbf{G})$. Let $g \in (N \cap \mathbf{G}(K)) - \mathscr{Z}(\mathbf{G})$. We set $\Lambda_g = \Lambda_{\mathscr{E}} \cap (g^{-1}\Lambda_{\mathscr{E}} g)$. Since $g \in \mathbf{G}(K)$, by Lemma I.3.1.3(b) Λ_g is of finite index in the $\mathscr{E}$-arithmetic subgroup $\Lambda_{\mathscr{E}}$, and hence $\bar{\Lambda}_g = \mathbf{G}$ (see Proposition I.3.2.10). But $g \notin \mathscr{Z}(\mathbf{G})$, $\mathbf{G}$ is connected, and the center $\mathscr{Z}(\mathbf{G})$ is finite. Therefore there is $\lambda \in \Lambda_g$ such that $g\lambda g^{-1}\lambda^{-1} \notin \mathscr{Z}(\mathbf{G})$. Since $g \in N_1$, $\lambda \in \Lambda_g$, and $\Lambda_{\mathscr{E}}$ normalizes N, it follows that $g\lambda g^{-1}\lambda^{-1} \in N \cap \Lambda_{\mathscr{E}}$. Thus, $N \cap \Lambda_{\mathscr{E}} \not\subset \mathscr{Z}(\mathbf{G})$.

(b) Since $\mathbf{G}(K(S))$ is the union of the subgroups $\mathbf{G}(K(S'))$, where S' runs through all finite subsets of S, we can assume that S is finite. Then by induction arguments we may restrict ourselves to the case in which $S = \mathscr{E} \cup \{v\}$, where $v \in \mathscr{R} - \mathscr{R}_\infty - \mathscr{E} \subset \mathscr{R} - \mathscr{R}_\infty - \mathscr{T}$. Denote by Λ_v and N_v the closures of the subgroups Λ and $N \cap \mathbf{G}(K)$ in $\mathbf{G}(K_v)$. Since $\mathscr{E} \not\subset \mathscr{T}$ and $v \in S - \mathscr{E}$, by the strong approximation theorem the subgroup $\mathbf{G}(K(S))$ is dense in $\mathbf{G}(K_v)$ (see Remark 2 of II.6.8). But the subgroup Λ is commensurable with $\mathbf{G}(K(S))$ and normalizes $N \cap \mathbf{G}(K)$, and, by Lemma 2.2(ii), $N \cap \mathbf{G}(K) \not\subset \mathscr{Z}(\mathbf{G})$. Thus, the subgroup Λ_v is of finite index in $\mathbf{G}(K_v)$ and normalizes $N_v \not\subset \mathscr{Z}(\mathbf{G})$. On the other hand, since $\mathbf{G}$ is simply connected, absolutely almost simple, and K_v-isotropic, by Corollary I.2.3.2 the group $\mathbf{G}(K_v)$ does not contain proper subgroups of finite index and every normal subgroup of $\mathbf{G}(K_v)$ is either contained in $\mathscr{Z}(\mathbf{G})$ or equal to $\mathbf{G}(K_v)$. Thus, $N_v = \mathbf{G}(K_v)$, i.e., the subgroup $N \cap \mathbf{G}(K)$ is dense in $\mathbf{G}(K_v)$. Denoting by $\mathcal{O}_v$ the ring of integers of the field K_v, it remains to make use of the fact that $\Lambda_{\mathscr{E}} = \Lambda \cap \mathbf{G}(\mathcal{O}_v)$ and that the subgroup $\mathbf{G}(\mathcal{O}_v)$ is open in $\mathbf{G}(K_v)$.

(c) Replacing Λ by $\Lambda \cap \mathbf{G}(K)$ and N by $N \cap \mathbf{G}(K)$ we may assume that $\Lambda \subset \mathbf{G}(K)$ and $N \subset \mathbf{G}(K)$. Let us represent $\mathbf{G}$ in the form $\mathbf{G} = R_{K'/K}\mathbf{G}'$, where K' is a finite separable field extension of K and $\mathbf{G}'$ is a simply connected absolutely almost simple K-group (see I.1.7). Then by Lemma I.3.1.4, we can replace $\mathbf{G}$, Λ, and N by $\mathbf{G}'$, $(R^0_{K'/K})^{-1}(\Lambda)$, and $(R^0_{K'/K})^{-1}(N)$. Thus we may assume that $\mathbf{G}$ is absolutely almost simple. Replacing $\mathscr{E}$ by $\mathscr{E} \cup (S \cap \mathscr{T})$, since the subgroups $\mathbf{G}(K(\mathscr{E}))$ and $\mathbf{G}(K(\mathscr{E} \cup \mathscr{T}))$ are commensurable, we can further assume, that

$\mathscr{E} \supset S \cap \mathscr{T}$ (see I.3.2.3). After all these reductions, making use of (b) we can claim that the factor group $\Lambda/(N \cap \Lambda)$ is isomorphic to the finite group $\Lambda_{\mathscr{E}}/(N \cap \Lambda_{\mathscr{E}})$. □

(2.4) Lemma. *Let Λ be an S-arithmetic subgroup of $\mathbf{G}$, N a normal subgroup of Λ, $\mathbf{P}$ a minimal parabolic K-subgroup of the group $\mathbf{G}$, and $\mathbf{V} = R_u(\mathbf{P})$. Suppose $N \not\subset \mathscr{Z}(\mathbf{G})$, $\mathbf{G}$ is simply connected, S is finite, and* $\operatorname{rank}_S \mathbf{G} \geq 2$. *Then*

(i) *There exists a non-zero ideal $\mathfrak{a}$ in $K(S)$ such that $N \supset (g\mathbf{V}g^{-1})(\mathfrak{a})$ for all $g \in \mathbf{G}(K)$ (see [Rag 7], Theorem 2.1).*
(ii) *The subgroup $N \cap \mathbf{V}$ is of finite index in $\Lambda \cap \mathbf{V}$, or in other words, the image of the subgroup $\Lambda \cap \mathbf{V}$ under the natural epimorphism $\varphi\colon \Lambda \to \Lambda/N$ is finite* (this follows from (i) and Lemma I.3.1.1(iii)).

From Proposition 1.2(a) and Lemma 2.4(ii) we deduce

(2.5) Corollary. *Let Λ be an S-arithmetic subgroup of the group $\mathbf{G}(K)$ and let N be a normal subgroup of Λ. Suppose $N \not\subset \mathscr{Z}(\mathbf{G})$, $\mathbf{G}$ is simply connected, S is finite, and* $\operatorname{rank}_S \mathbf{G} \geq 2$. *We set $G_S = \prod_{v \in S} \mathbf{G}(K_v)$, identify the group Λ with its (discrete) image under the diagonal embedding into G_S, and denote by $\mathfrak{f}(G_S, \Lambda)$ the family of left Borel fundamental domains $X \subset G_S$ for Λ. Then there is a (Λ, N)-admissible domain $X \in \mathfrak{f}(G_S, \Lambda)$.* (As in IV.3.8 a domain $X \in \mathfrak{f}(G_S, \Lambda)$ is called (Λ, N)-admissible if for any compact set $L \subset G_S$ the set $\varphi(\tau_X(X \cdot L))$ is finite, where $\varphi\colon \Lambda \to \Lambda/N$ is the natural epimorphism and the map $\tau_X\colon G_S \to \Lambda$ is defined via the inclusion $g \in \tau_X(g)X$, $g \in G$).

(2.6) Theorem. *Let Λ be an S-arithmetic subgroup of G, N a subgroup of Ψ normalized by Λ.*

(i) *If S is finite and* $\operatorname{rank}_S \mathbf{G} \geq 2$, *then either $N \subset \mathscr{Z}(\mathbf{G})$ or $N \cap \Lambda$ is of finite index in Λ.*
(ii) *If S is infinite and G is simply connected, then either $N \subset \mathscr{Z}(\mathbf{G})$ or $N \cap \Lambda$ is of finite index in Λ.*

Proof. (i) Let $\tilde{\mathbf{G}}$ be the simply connected covering of $\mathbf{G}$ and let $\pi\colon \tilde{\mathbf{G}} \to \mathbf{G}$ be the central K-isogeny (see Proposition I.1.4.11(i)). Since S is finite, it follows from Corollary I.3.2.9 that the subgroups $\pi(\tilde{\mathbf{G}}(K(S))$ and $\mathbf{G}(K(S))$ are commensurable. On the other hand, for each $v \in \mathscr{R}$ the groups $\mathbf{G}$ and $\tilde{\mathbf{G}}$ have the same K_v-rank (see Corollary I.1.4.6) and so by Lemma 2.1(ii) $\pi^{-1}(\Psi(\mathbf{G})) = \Psi(\tilde{\mathbf{G}})$. Therefore we can replace $\mathbf{G}$ by $\tilde{\mathbf{G}}$, Λ by $\pi^{-1}(\Lambda)$, and N by $\pi^{-1}(N)$ and assume that $\mathbf{G}$ is simply connected. Furthermore, according to Lemma 2.3(a) and the fact that the subgroups Λ and $\Lambda \cap \mathbf{G}(K)$ are commensurable, one can replace Λ by $\Lambda \cap \mathbf{G}(K)$ and N by $N \cap \Lambda$ and assume that $\Lambda \subset \mathbf{G}(K)$ and $N \subset \Lambda$. Then in the same way as in the proof of Lemma 2.3(c), making use of the restriction of scalars functor, we reduce to the case in which $\mathbf{G}$ is absolutely almost simple. As was observed in I.3.2.3, for any $S' \subset \mathscr{R}$ the subgroup $\mathbf{G}(K(S'))$ is of finite index in $\mathbf{G}(K(S' \cup \mathscr{T}))$. Therefore, replacing S by $S - \mathscr{T}$, we may assume that $\mathbf{G}$ is isotropic over K_v for all $v \in S$. We set $G_S = \prod_{v \in S} \mathbf{G}(K_v)$ and identify the group Λ with its image under the diagonal embedding in G_S. By the Borel-Harish-Chandra-Behr-Harder reduction theorem (see Theorem I.3.2.5) Λ is a lattice in G_S. Obviously, $\Lambda \cap \prod_{v \in S'} \mathbf{G}(K_v) = \{e\}$ for all $S' \subsetneqq S$. Since G is absolutely almost simple, this implies that the lattice $\Lambda \subset G_S$ is irreducible. According to Corollary 2.5 there is a (Λ, N)-admissible domain $X \in \mathfrak{f}(G_S, \Lambda)$. Now we can apply Theorem IV.4.9 and conclude that either $N \subset \mathscr{Z}(\mathbf{G}(K)) = \mathscr{Z}(\mathbf{G})(K)$ or Λ/N is finite.

(ii) Since S is infinite and $\mathscr{T}$ is finite, there is a finite subset $\mathscr{E} \subset S$ such that $\mathscr{E} \supset R_\infty - \mathscr{T}$ and $\operatorname{rank}_{\mathscr{E}} \mathbf{G} \geq 2$. We set $\Lambda_{\mathscr{E}} = \Lambda \cap \mathbf{G}(K(\mathscr{E}))$. Assume $N \not\subset \mathscr{Z}(\mathbf{G})$. It then follows from Lemma 2.3(a) that $N \cap \Lambda_{\mathscr{E}} \not\subset \mathscr{Z}(\mathbf{G})$. From this and (i) we deduce that $N \cap \Lambda_{\mathscr{E}}$ is of finite index in $\Lambda_{\mathscr{E}}$, and hence $N \cap \Lambda$ is of finite index in Λ (see Lemma 2.3(a)). □

(2.7) *Remarks on Theorem 2.6.* (i) For $\mathbf{G}$ K-anisotropic the quotient space $\mathbf{G}(K(S)) \setminus G_S$ is compact (see Theorem I.3.2.4(b)). Thus, if $\mathbf{G}$ is anisotropic over K, then, in the proof of Theorem 2.6, while reducing to Theorem IV.4.9 it is not necessary to make use of Corollary 2.5. This is also superfluous when $\operatorname{rank}_{K_v} \mathbf{G} \geq 2$ for some $v \in S$.

(ii) Theorem (A) stated at the beginning of this chapter is a particular case $(N \subset \Lambda)$ of Theorem 2.6.

(iii) If $\operatorname{rank}_S \mathbf{G} = 1$, then there is an S-arithmetic subgroup Λ of $\mathbf{G}$ containing a non-commutative free normal (in Λ) subgroup of infinite index. The assertion is a consequence of Remark IV.4.11.

(iv) Consider $d \in \mathbb{N}^+$, $k = \mathbb{Q}(\sqrt{-d})$ the imaginary quadratic field extension of $\mathbb{Q}$, and B the ring of integers over $\mathbb{Z}$ of the field k. Then (see [G-S 2]) there is a subgroup Γ of finite index in $\mathbf{SL}_2(B)$ such that Γ has a free non-commutative factor group.

(2.8) Corollary. *Let Λ be an S-arithmetic subgroup of $\mathbf{G}$. Then the factor group $\Lambda/\mathscr{D}(\Lambda)$ is finite in either of the following two cases:*

(a) *S is finite and* $\operatorname{rank}_S \mathbf{G} \geq 2$;

(b) *S is infinite and $\mathbf{G}$ is simply connected.*

Proof. According to Proposition I.3.2.10, $\bar{\Lambda} = \mathbf{G}$ and hence $\overline{\mathscr{D}(\Lambda)} = \mathscr{D}(\mathbf{G}) = \mathbf{G}$. Thus, $\mathscr{D}(\Lambda) \not\subset \mathscr{Z}(\mathbf{G})$ and it remains to make use of Theorem 2.6. □

(2.9) *Remark.* If $\mathbf{G}$ is anisotropic over K or if $\operatorname{rank}_{K_v} \mathbf{G} \geq 2$ for some $v \in S$, then Corollary 2.8 can also be deduced from Theorem D (b) of III.5.9 and Corollary IV.3.12 in exactly the same way as Theorem 2.6 is deduced from Theorem IV.4.9.

The following theorem is a particular case $(S = \mathscr{R}$ and $N \subset \Lambda)$ of Theorem 2.6.

(2.10) Theorem. *Let Λ be a subgroup of finite index in $\mathbf{G}(K)$ and let N be a normal subgroup of Λ. Suppose $\mathbf{G}$ is simply connected. Then either $N \subset \mathscr{Z}(\mathbf{G})$ or Λ/N is finite.*

(2.11) *Remark.* If $\mathbf{G}$ is isotropic over K and $\operatorname{rank}_{K_v} \mathbf{G} = 1$ for all $v \in \mathscr{R}$, then $\mathbf{G}$ is quasi-split over K. But it is well known that if k is an infinite field and $\mathbf{H}$ a connected simply connected quasi-split semisimple k-group, then $\mathbf{H}(k) = \mathbf{H}(k)^+$. So $\mathbf{H}(k)$ does not contain proper subgroups of finite index and any normal subgroup of $\mathbf{H}(k)$ which is not contained in $\mathscr{Z}(\mathbf{H})$ coincides with $\mathbf{H}(k)$ (see Theorem I.1.5.6 and Corollary I.1.5.7). Therefore, as was remarked in [Pr 5], we may assume in proving Theorem 2.10 that either $\mathbf{G}$ is anisotropic over K or $\operatorname{rank}_{K_v} \mathbf{G} \geq 2$ for some $v \in \mathscr{R}$. From this and Remark 2.7(i) we see that in proving Theorem 2.10 it is not necessary to make use of Corollary 2.5.

Theorem 2.6 admits the following generalization.

(2.12) Theorem. *Let $\tilde{\mathbf{G}}$ be the simply connected covering of $\mathbf{G}$, $\pi\colon \tilde{\mathbf{G}} \to \mathbf{G}$ the central K-isogeny, $\tilde{\Lambda}$ an S-arithmetic subgroup of $\tilde{\mathbf{G}}(K)$, and N a subgroup of Ψ normalized by $\pi(\tilde{\Lambda})$. Suppose* $\operatorname{rank}_S \mathbf{G} \geq 2$. *Then either $N \subset \mathscr{Z}(\mathbf{G})$ or the factor group $\pi(\tilde{\Lambda})/(N \cap \pi(\tilde{\Lambda}))$ is finite.*

Proof. By Corollary I.1.4.6, for each $v \in \mathscr{R}$ the groups $\mathbf{G}$ and $\tilde{\mathbf{G}}$ have the same K_v-rank. It follows that $\operatorname{rank}_S \tilde{\mathbf{G}} \geq 2$. On the other hand, $\tilde{\Lambda}$ normalizes $\pi^{-1}(N)$ and by Lemma 2.1(ii) we have $\pi^{-1}(N) \subset \pi^{-1}(\Psi(G)) = \Psi(\tilde{G})$. Thus, by Theorem 2.6, either $\pi^{-1}(N) \subset \mathscr{Z}(\tilde{\mathbf{G}})$ or $\pi^{-1}(N) \cap \tilde{\Lambda}$ is of finite index in $\tilde{\Lambda}$. □

(2.13) Corollary. *Let Λ be an S-arithmetic subgroup of $\mathbf{G}$ and N a normal subgroup of Λ. Suppose* $\operatorname{rank}_S \mathbf{G} \geq 2$. *Then either $N \subset \mathscr{Z}(\mathbf{G})$ or the orders of all elements of the factor group Λ/N are bounded and Λ/N contains a normal subgroup F of finite index such that the commutator subgroup $\mathscr{D}(F)$ is finite.*

To prove this corollary we need the following

(2.14) Lemma. *Let $\varphi\colon \mathbf{F}' \to \mathbf{F}$ be a central K-isogeny of connected reductive K-groups. Then*

(i) *there exists an S-congruence subgroup $\mathbf{F}(\mathfrak{a})$ (of finite index) in $\mathbf{F}(K(S))$ such that $\mathscr{D}(\mathbf{F}(\mathfrak{a})) \subset \varphi(\mathbf{F}'(K(S)))$;*

(ii) *there exists $r \in \mathbb{N}^+$ such that $g^r \in \varphi(\mathbf{F}'(K(S)))$ for all $g \in \mathbf{F}(K(S))$.*

Proof. Since φ is a central K-isogeny of connected reductive K-groups, there exist $d \in \mathbb{N}^+$ and regular maps $\chi\colon \mathbf{F} \times \mathbf{F} \to \mathbf{F}'$ and $\mu\colon \mathbf{F} \to \mathbf{F}'$ defined over K such that $\chi(\varphi(u_0, v_0)) = u_0 v_0 u_0^{-1} v_0^{-1}$ and $\mu(\varphi(u_0)) = u_0^d$ for all $u_0, v_0 \in \mathbf{F}$ (see I.1.4.1). Clearly $\chi(e, e) = e$ and $\mu(e) = e$. Therefore there is a non-zero ideal $\mathfrak{a}$ of the ring $K(S)$ such that $\chi(\mathbf{F}(\mathfrak{a}) \times \mathbf{F}(\mathfrak{a})) \subset \mathbf{F}'(K(S))$ and $\mu(\mathbf{F}(\mathfrak{a})) \subset \mathbf{F}'(K(S))$ (see Lemma I.3.1.1(i)). Then $uvu^{-1}v^{-1} = \varphi(\chi(u, v)) \in \varphi(\mathbf{F}'(K(S)))$ for all $u, v \in \mathbf{F}(\mathfrak{a})$. This proves (i). Furthermore, $u^d \in \varphi(\mathbf{F}(K(S)))$ for each $u \in \mathbf{F}(\mathfrak{a})$. This proves (ii), because $g^t \in \mathbf{F}(\mathfrak{a})$ for all $g \in \mathbf{F}(K(S))$, where $t = \operatorname{card}(\mathbf{F}(K(S))/\mathbf{F}(\mathfrak{a}))$. □

(2.15) *Proof of Corollary 2.13.* Let $\tilde{\mathbf{G}}$ be the simply connected covering of $\mathbf{G}$, $\pi\colon \tilde{\mathbf{G}} \to \mathbf{G}$ the central K-isogeny. We set $\tilde{\Lambda} = \pi^{-1}(\Lambda) \cap \tilde{\mathbf{G}}(K(S))$. Since Λ and

$\mathbf{G}(K(S))$ are commensurable, it follows from Lemma 2.14 that there exist $t \in \mathbb{N}^+$ and a subgroup Λ_0 of finite index in Λ such that $g^t \in \pi(\tilde{\Lambda})$ for each $g \in \Lambda$ and $\mathscr{D}(\Lambda_0) \subset \pi(\tilde{\Lambda})$. All that remains now is to set $F = \Lambda_0/(N \cap \Lambda_0)$ and to observe that if $N \not\subset \mathscr{Z}(\mathbf{G})$, then, by Theorem 2.12, $\pi(\tilde{\Lambda})/(N \cap \pi(\tilde{\Lambda}))$ is finite. □

(2.16) For $\mathbf{G}$ K-isotropic, various cases of Theorem 2.6 were proved earlier by algebraic methods. In particular, in case $\operatorname{rank}_K \mathbf{G} \geq 2$, this was proved by Vasserstein [Va 2] for all classical groups and by Raghunathan [Rag 7] for all $\mathbf{G}$. In [Va 2] some results on normal subgroups were also announced, e.g., in case $\mathbf{G}$ is a K-rank 1 classical group, Theorem 2.6. By algebraic methods it is possible to handle completely the case of K-isotropic groups (see [Rag 2]. The proofs suggested in [Rag 7] and [Va 2] are divided into two parts. The first part is assertion (i) of Lemma 2.4 above. The second part is devoted to a demonstration of the fact that if $\mathbf{G}$ is isotropic over K and $\operatorname{rank}_S \mathbf{G} \geq 2$, then, for any non-zero ideal $\mathfrak{a}$ of the ring $K(S)$, the subgroup $E_{\mathfrak{a}}$ generated by $\bigcup_{g \in \mathbf{G}(K)} (g\mathbf{V}g^{-1})(\mathfrak{a})$ is of finite index in $\mathbf{G}(K(S))$, where $\mathbf{V} = R_u(\mathbf{P})$ and $\mathbf{P}$ is a maximal parabolic K-subgroup of $\mathbf{G}$. If $\mathbf{G}$ is anisotropic over K, then $E_{\mathfrak{a}} = \{e\}$. Furthermore, for any arbitrary K-anisotropic $\mathbf{G}$, even in case $\operatorname{rank}_S \mathbf{G} \geq 2$, it seems unikely that one can explicitly find a class Ω of subsets of the group $\mathbf{G}(K(S))$ with the following properties:

(a) if N is a normal subgroup of an S-arithmetic subgroup of $\mathbf{G}$ and $N \not\subset \mathscr{Z}(\mathbf{G})$, then $N \supset X$ for some $X \in \Omega$,

(b) any subset $X \in \Omega$ generates a subgroup of finite index in $\mathbf{G}(K(S))$.

Therefore it seems to be extremely difficult to apply algebraic methods in case $\operatorname{rank}_K \mathbf{G} = 0$. Nevertheless, Kneser (see [Kn 1], [Kn 4], and [Kn 5]) succeeded in giving an algebraic proof of Theorem 2.6 for spinor groups $\mathbf{G} = \mathbf{Spin}(f)$, where f is a degenerate quadratic form over K in sufficiently large (≥ 5) number of variables. Kneser's methods are probably applicable to numerous other classical groups. But it seems impossible to prove Theorem 2.6 in the general case by these methods, e.g. in case $\mathbf{G}(K) = \mathbf{SL}_1(D) = \{x \in D \mid \operatorname{Nrd}_{D/K}(x) = 1\}$, where D is a central division algebra and $\operatorname{Nrd}_{D/K}$ is the reduced norm.

It should be noted that the main purpose of the above-mentioned papers [Va 2], [Rag 7], [Kn 1], [Kn 4], and [Kn 5] was the solution of the congruence subgroup problem rather than the proof of the finiteness of the factor groups of S-arithmetic subgroups modulo their non-central normal subgroups. In the form due to Serre, the congruence subgroup problem is stated as follows. We set $\Gamma = \mathbf{G}(K(S))$ and denote by Ω the set of normal subgroups of finite index of Γ and by $\Omega_0 \subset \Omega$ the set of all congruence subgroups of Γ. The factor groups Γ/H, $H \in \Omega$, form an inverse system and one can consider its projective limit $\hat{\Gamma} = \varprojlim \Gamma/H$, $H \in \Omega$. Similarly, given Ω_0 instead of Ω, one can define the projective limit $\tilde{\Gamma} = \varprojlim \Gamma/H$, $H \in \Omega_0$. There is the natural epimorphism $\hat{\Gamma} \to \tilde{\Gamma}$, whose kernel is denoted by $C(S, \mathbf{G})$ and is called the *congruence kernel.* The congruence subgroup problem consists of describing of this kernel. We remark that the congruence kernel is a profinite (and hence) compact group and it is trivial if and only if any subgroup of finite index in Γ contains an S-congruence subgroup.

If $\mathbf{G}$ is not simply connected and $\operatorname{rank}_S \mathbf{G} > 0$, then the subgroup $C(S, \mathbf{G})$ is infinite. Furthermore, it can easily be verified that the compact group $\prod_{v \in \mathcal{T} \cap S} \mathbf{G}(K_v)$ is a factor group of the group $C(S, \mathbf{G})$. So $\mathbf{G}$ is usually assumed to be simply connected and $\mathcal{T} \cap S$ is assumed to be empty. In this case the general scheme of the solution of the congruence subgroup problem is the following (see [Pr-R 2]):

First, prove that the subgroup $C(S, \mathbf{G})$ is central in $\hat{\Gamma}$. From this, with the help of the theory of central extensions, one can prove that if $\mathcal{D}(\mathbf{G}(K)) = \mathbf{G}(K)$ and $\mathcal{R} = \emptyset$, then the group $C(S, \mathbf{G})$ is Pontrjagin dual to the so called *metaplectic kernel* $M(S, \mathbf{G})$. The group $M(S, \mathbf{G})$ is defined as the kernel of the restriction homomorphism

$$H^2(G_{\mathcal{R}-S}, \mathbb{R}/\mathbb{Z}) \to H^2(\mathbf{G}(K), \mathbb{R}/\mathbb{Z}),$$

where $G_{\mathcal{R}-S}$ denotes the subgroup of $\mathbf{G}(\mathbb{A}_K)$ consisting of all those adèles whose v-components equal e ($v \in S$), $H^2(G_{\mathcal{R}-S}, \mathbb{R}/\mathbb{Z})$ is the second cohomology group of the group $G_{\mathcal{R}-S}$ with coefficients in $\mathbb{R}/\mathbb{Z}$ defined by Borel cocycles and $H^2(\mathbf{G}(K), \mathbb{R}/\mathbb{Z})$ is the usual second cohomology group with coefficients in $\mathbb{R}/\mathbb{Z}$ of the abstract group $\mathbf{G}(K)$.

It is known that $C(S, \mathbf{G})$ is central in $\hat{\Gamma}$ if the following condition is satisfied: $\mathbf{G}$ is isotropic over K and $\operatorname{rank}_S \mathbf{G} \geq 2$.

This was established in [B-M-S] for $\mathbf{SL}_n$, $n \geq 3$, and $\mathbf{SP}_{2n}$, in [Ser 1] for $\mathbf{SL}_2$, in [Mat] for all K-split $\mathbf{G}$, in [Va 2] for classical groups of K-rank ≥ 2, and in [Rag 7] for all groups of K-rank ≥ 2 (as was noted above, the methods of the papers [Va 2] and [Rag 7] cover the case $\operatorname{rank}_K \mathbf{G} = 1$ as well, see [Rag 12]).

Kneser proved in 1979 (see [Kn 5]) that the group $C(S, \mathbf{G})$ is central for spinor groups $\mathbf{G} = \mathbf{Spin}(f)$, where f is a degenerate quadratic form over K in sufficiently large (≥ 5) number of variables (under the assumption $\operatorname{rank}_S \mathbf{G} \geq 2$). Later under the same assumption $\operatorname{rank}_S \mathbf{G} \geq 2$, the centrality of $C(S, \mathbf{G})$ was proved by Rapinchuk for groups of type $\mathbb{G}_2$ (see [Rap 2]) and by Tomanov for some groups of type $\mathbb{A}_n$, for groups of type $\mathbb{C}_n$ ($n \geq 4$) and, under some conditions in the case where $S \subset \mathcal{R}_\infty$, for groups of type $\mathbb{D}_n$ ($n \geq 5$) (see [To 4] and [To 5]) and also for groups of type $\mathbb{G}_2$ (see [To 2]). In just mentioned results, the group $\mathbf{G}$ is not supposed to be K-isotropic.

If $\operatorname{rank}_S \mathbf{G} = 1$, then the congruence kernel is infinite (see [Lu 1], [Lu 2] and [Ser 1]).

In [Pr-R 2] Prasad and Raghunathan established that, for any K-isotropic $\mathbf{G}$, the group $M(S, \mathbf{G})$ is trivial if S contains non-archimedean valuations. In general, they showed that it is isomorphic to a subgroup of the group $\mu(K)$ of roots of unity of the field K. For many K-anisotropic groups, the metaplectic kernel was essentially described by Rapinchuk (see [Rap 1]).

Summarizing the above discussion of the case that $\mathbf{G}$ is isotropic over K we obtain the following

Theorem. *Suppose* $\mathbf{G}$ *is simply connected and* K*-isotropic,* $\operatorname{rank}_S \mathbf{G} \geq 2$ *and* $\mathcal{D}(\mathbf{G}(K)) = \mathbf{G}(K)$. *Then the group* $C(S, \mathbf{G})$ *is trivial if* S *contains non-archimedean*

valuations. In general it is isomorphic to a subgroup of the group of roots of unity of K.

Remark. If $\mathbf{G}$ is simply connected, K-isotropic, and is not isomorphic to some groups of type $\mathbb{E}_6$ or $\mathbb{D}_4$, then $\mathbf{G}(K)$ does not contain non-central normal subgroups, and hence $\mathscr{D}(\mathbf{G}(K)) = \mathbf{G}(K)$ (see [Pl 4], 9.1). This assertion seems to be true for all simply connected K-isotropic groups $\mathbf{G}$ (for the case where $\mathbf{G}$ is anisotropic over K, see 2.17).

(2.17) Let f be a nondegenerate quadratic form over K in n variables. We denote the spinor group $\mathbf{Spin}(f)$ by $\mathbf{G}_f$. In [Kn 1] Kneser proved that for $n \geq 5$ the group $\mathbf{G}_f(K)/\mathscr{Z}(\mathbf{G}_f(K))$ is abstractly simple. If $n = 3, 4$, then in general $\mathscr{T}(\mathbf{G}_f) \not\subset \mathscr{R}_\infty$ and so there exist infinitely many normal subgroups in $\mathbf{G}_f(K)$ of the form $\mathbf{G}_f(K) \cap W$, where W is an open normal subgroup of the compact totally disconnected group $\mathbf{G}(K_v)$, $v \in \mathscr{T}(\mathbf{G}_f) - \mathscr{R}_\infty$. Therefore if $n = 3$ or 4 the group $\mathbf{G}_f(K)/\mathscr{Z}(\mathbf{G}_f(K))$ is not necessarily abstractly simple. However, if $\mathscr{T}(\mathbf{G}_f) \subset \mathscr{R}_\infty$ the group $\mathbf{G}_f(K)/\mathscr{Z}(\mathbf{G}_f(K))$ is abstractly simple even for $n = 3, 4$. This assertion was stated as a conjecture in [Kn 1] mentioned above and proved in [Mar 13] with the help of Theorem 2.5 stated above and the methods of [Pl-R 1]. It is good to keep in mind that for $n = 3, 4$ the group $\mathbf{G}_f(K)$ is isomorphic to $\mathbf{SL}_1(D)$, where D is the central quaternion algebra over K, or over a quadratic extension of K ($n = 4$). Note also that any simply connected K-group of type $\mathbb{A}_1$ is isomorphic over K to a group $\mathbf{G}_f$, where f is a nondegenerate form over K in 3 variables.

In [Pl 3] Platonov conjectured that the group $\mathbf{G}(K)/\mathscr{Z}(\mathbf{G}(K))$ is abstractly simple whenever $\mathbf{G}$ is simply connected and $\mathscr{T}(\mathbf{G}) \subset \mathscr{R}_\infty$. It seems natural to generalize Platonov's conjecture as follows.

(∗) *Let Λ be a subgroup of finite index in $\mathbf{G}(K)$ and let N be a normal subgroup of Λ. Denote by $G_{\mathscr{T} \cap (\mathscr{R} - \mathscr{R}_\infty)}$ the non-compact totally disconnected group $\prod_{v \in \mathscr{T} \cap (\mathscr{R} - \mathscr{R}_\infty)} \mathbf{G}(K_v)$ and identify the group $\mathbf{G}(K)$ with its image under the diagonal embedding in $G_{\mathscr{T} \cap (\mathscr{R} - \mathscr{R}_\infty)}$. Suppose $\mathbf{G}$ is simply connected. Then either $N \subset \mathscr{Z}(\mathbf{G})$ or N is representable in the form $N = \mathbf{G}(K) \cap W$, where W is an open subgroup of the group $G_{\mathscr{T} \cap (\mathscr{R} - \mathscr{R}_\infty)}$.*

According to Theorem 2.6 conjecture (∗) can be viewed as a particular case ($S = \mathscr{R} - \mathscr{T}$) of the congruence subgroup problem. For the groups $\mathbf{G}$ of type $\mathbb{A}_1$ this conjecture was proved in [Mar 13] under the condition $\operatorname{char} K \neq 2$ and in [Rag 11] in case $\operatorname{char} K = 2$. Platonov's conjecture was proved by Borovoj for all groups of type $\mathbb{C}_n$ ($n \geq 3$), $\mathbb{F}_4$, $\mathbb{G}_2$, and some groups of type $\mathbb{A}_n$ ($n \geq 2$) (see [Boro 1]), by Borovoj and Tomanov for groups of type $\mathbb{D}_n$ ($n \geq 5$) and for classical groups of type $\mathbb{D}_4$ (see [Boro 2] and [To 1]) and by Chernousov for groups split over a quadratic extension of K and, in particular, for groups of type $\mathbb{B}_n$, $\mathbb{C}_n$, $\mathbb{G}_2$, $\mathbb{F}_4$, $\mathbb{E}_7$ and $\mathbb{E}_8$ (see [Ch 1] and [Ch 3]). Another proof for types $\mathbb{C}_n$, $\mathbb{F}_4$ and $\mathbb{G}_2$ was given in [To 3]. Recently Tomanov proved Platonov's conjecture for all groups of type $\mathbb{A}_3$ as well as conjecture (∗) for many groups of type $\mathbb{A}_3$ and, in particular, in case $\mathbf{G}(K) = \mathbf{SL}_1(D)$, where D is a central division algebra over K of index 4 (see [To 5]). It should be noted that, in the above-mentioned paper [To 1], conjecture (∗) was in fact proved for groups of type $\mathbb{D}_n$ ($n \geq 5$)

and for classical groups of type $\mathbb{D}_4$. The general case has not yet been treated. In particular, conjecture (*) remains unproved in case $\mathbf{G}(K) = \mathbf{SL}_1(D)$, where D is a central division algebra over K of index $n \geq 3$, $n \neq 4$. In this case we have the following partial result.

Theorem. *Let D be a finite-dimensional central division algebra over K of index n. Denote by $\mathcal{T}_1$ the set of those non-archimedean valuations of the field K for which $D_v = D \otimes_K K_v$ is a skew field. We identify the group D with its image under the diagonal embedding in $\prod_{v\in\mathcal{T}_1} \mathbf{SL}_1(D_v)$. Let W be an open subgroup of the group $\prod_{v\in\mathcal{T}_1} \mathbf{SL}_1(D_v)$, and set $\Lambda_1 = W \cap \mathbf{SL}_1(D)$. Then $\mathcal{D}(\Lambda) = \mathcal{D}(W) \cap \mathbf{SL}_1(D)$. In particular, if $\mathcal{T}_1 = \emptyset$, then the group $\mathbf{SL}_1(D)$ coincides with its commutator subgroup.*

This theorem was proved by Platonov and Rapinchuk (see [Pl-R 2]) in case $\operatorname{char} K = 0$ and $v((n,2)) = 1$ for each (multiplicative) valuation $v \in \mathcal{T}_1$ ($(n,2)$ denotes the greatest common divisor of the numbers n and 2). In the general case the theorem was proved by Raghunathan (see [Rag 11]).

The following assertion can be viewed as a step in proving (*).

(2.18) Proposition. *Let Λ_1 be a subgroup of finite index in $\mathbf{G}(K)$, N_1 a normal subgroup of Λ_1, and C the automorphism group of $\mathbf{G}(K)$. Suppose $\mathbf{G}$ is simply connected. Then N_1 contains a C-invariant subgroup N' of finite index.*

Proof. Choose a set $\mathcal{E} \supset \mathcal{T}$ with $\operatorname{rank}_{\mathcal{E}} \mathbf{G} \geq 2$. Then the group $\mathbf{G}(K(\mathcal{E}))$ is finitely generated (see Theorem 3.3 below). But in any finitely generated group the number of subgroups of fixed finite index is finite and, by virtue of Lemma 2.3(b), every non-central normal subgroup of the group $\mathbf{G}(K)$ is defined uniquely by its intersection with $\mathbf{G}(K(\mathcal{E}))$. Thus, the number of normal subgroups of fixed finite index in $\mathbf{G}(K)$ is finite. All that remains now is to set $N' = \cap_{c\in C} c(N)$ and to apply Theorem 2.10. □

(2.19) *Remark.* According to Proposition 2.18, it suffices to prove conjecture (*) only for C-invariant N_1. In particular, if $\mathbf{G}(K) = \mathbf{SL}_1(D)$ it suffices to prove (*) for subgroups N_1 normalized by the group of invertible elements of D.

3. Homomorphisms of S-Arithmetic Subgroups to Algebraic Groups

As in Section 2 we set $\Psi = \Psi(G) = \operatorname{Ad}^{-1}((\operatorname{Ad}\mathbf{G})(K))$. Denote by $\Omega_S = \Omega_S(\mathbf{G})$ the class of subgroups Λ in $\mathbf{G}$ with $\Lambda \subset \Psi$ and $\Lambda \cap (\mathbf{G}(K(S))$ being of finite index in $\mathbf{G}(K(S))$. If $S \not\subset \mathcal{T}$ and $S \subset S' \subset \mathcal{R}$, then by Lemma 2.2(i) every S'-arithmetic subgroup of $\mathbf{G}$ belongs to Ω_S.

Given a field homomorphism $\sigma\colon k \to l$ and a field extension k' of k, σ is assumed to be extended to a homomorphism of the field k' into a field extension

l' of l. Thus, we assume that for a k-variety $\mathbf{M}$ the map σ^0: $\mathbf{M}(k) \to {}^{\sigma}\mathbf{M}$ is defined not only on $\mathbf{M}(k)$ but also on $\mathbf{M}$ (more precisely on $\mathbf{M}(k_1)$, where k_1 is an arbitrary field extension of k).

(3.1) Lemma. *Let π: $\tilde{\mathbf{G}} \to \mathbf{G}$ be a K-group K-isogeny and let $\Lambda \in \Omega_S(\mathbf{G})$. Then $\pi^{-1}(\Lambda) \in \Omega_S(\tilde{\mathbf{G}})$.*

Proof. Since $\Lambda \cap \mathbf{G}(K(S))$ is of finite index in $\mathbf{G}(K(S))$, by Lemma I.3.13 (a), $\pi^{-1}(\Lambda) \cap \tilde{\mathbf{G}}(K(S))$ is of finite index in $\tilde{\mathbf{G}}(K(S))$. On the other hand, since $\Lambda \subset \Psi(\mathbf{G})$, by Lemma 2.1(ii), $\pi^{-1}(\Lambda) \subset \Psi(\tilde{\mathbf{G}})$. □

(3.2) Lemma. *Let $\Lambda \in \Omega_S$, $\mathscr{E} \subset S$, and Γ be a subgroup of finite index of the group $\Lambda \cap \mathbf{G}(K(\mathscr{E}))$. Suppose $\Lambda \subset \mathbf{G}(K)$ and $\mathscr{E} \not\subset \mathscr{T}$.*

(i) *If $\mathbf{G}$ is simply connected, then the subgroup F of Λ which is generated by $\bigcup_{\lambda \in \Lambda} \lambda \Gamma \lambda^{-1}$ is of finite index in Λ.*
(ii) *Let $\mathbf{H}$ be an algebraic group, δ: $\Lambda \to \mathbf{H}$ a homomorphism with $\overline{\delta(\Lambda)} = \mathbf{H}$. Suppose either $\mathbf{G}$ is simply connected or $\mathscr{D}(\mathbf{H}^0) = \mathbf{H}^0$. Then $\overline{\delta(\Gamma)} \supset \mathbf{H}^0$.*

Proof. Since F is a normal subgroup of Λ, (i) is a consequence of Lemma 2.3(c). Furthermore, by Lemma I.3.1.3(b), it follows that $\Lambda \subset \mathbf{G}(K) \subset \mathrm{Comm}_{\mathbf{G}}(\Gamma)$. Therefore in case $\mathbf{G}$ is simply connected, (ii) is a consequence of (i) and Lemma VII.6.9(b). Suppose now that $\mathscr{D}(\mathbf{H}^0) = \mathbf{H}^0$. The group $\mathbf{H}/\mathbf{H}^0$ is finite. Therefore we can replace Λ by $\delta^{-1}(\mathbf{H}^0)$ and assume that $\mathbf{H}$ is connected. Let $\tilde{\mathbf{G}}$ be the simply connected covering of $\mathbf{G}$ and let π: $\tilde{\mathbf{G}} \to \mathbf{G}$ be the central K-isogeny (see Proposition I.1.4.11(i)). We set $\Lambda_0 = \pi^{-1}(\Lambda) \cap \tilde{\mathbf{G}}(K)$ and $\Gamma_0 = \pi^{-1}(\Gamma) \cap \tilde{\mathbf{G}}(K(\mathscr{E}))$. Since Γ is of finite index in $\Lambda \cap \mathbf{G}(K(\mathscr{E}))$ and, by virtue of Lemma I.3.1.3(a), the subgroup $\tilde{\mathbf{G}}(K(\mathscr{E})) \cap \pi^{-1}(\mathbf{G}(K(\mathscr{E})))$ is of finite index in $\tilde{\mathbf{G}}(K(\mathscr{E}))$, it follows that Γ_0 is of finite index in $\Lambda_0 \cap \tilde{\mathbf{G}}(K(\mathscr{E}))$. On the other hand, since $\mathscr{E} \not\subset \mathscr{T}(\mathbf{G})$ and for each $v \in \mathscr{R}$ the groups $\mathbf{G}$ and $\tilde{\mathbf{G}}$ have the same K_v-rank, it follows that $\mathscr{E} \not\subset \mathscr{T}(\tilde{\mathbf{G}})$ (see Corollary I.1.4.6.). This implies, since the case in which $\mathbf{G}$ is simply connected was already examined, that $(\overline{\delta(\pi(\Lambda_0))})^0 = (\overline{\delta(\pi(\Gamma_0))})^0$. All that remains now is to observe that, since $\mathscr{D}(\mathbf{G}(K)) \subset \pi(\tilde{\mathbf{G}}(K))$, we have $\mathscr{D}(\Lambda) \subset \pi(\Lambda_0)$ (see Remark I.1.4.2(i)). Hence,

$$(\overline{\delta(\pi(\Lambda_0))})^0 \supset (\overline{\delta(\mathscr{D}(\Lambda))})^0 = \mathscr{D}(\overline{\delta(\Lambda)})^0 = (\mathscr{D}(\mathbf{H}))^0 = \mathbf{H}$$

(the latter equality is true because $\mathbf{H} = \mathbf{H}^0 = \mathscr{D}(\mathbf{H}^0)$). □

In this section we use the results of Section 6 in Chapter VII where we essentially only considered finitely generated lattices. In this connection we shall need the following

(3.3) Theorem. *Suppose S is finite and at least one of the following conditions is satisfied:*

(a) $\operatorname{char} K = 0$;
(b) *G is anisotropic over K;*

(c) $\operatorname{rank}_S \mathbf{G} \geq 2$.

Then any S-arithmetic subgroup Γ of $\mathbf{G}$ *is finitely generated.*

For the case "char $K = 0$", see [Bo-Hari] and [Bo 3]. Let $\mathbf{G}$ be anisotropic over K. It then follows from Theorem I.3.2.4(b) that under the diagonal embedding in $G_S \overset{\text{def}}{=} \prod_{v \in S} \mathbf{G}(K_v)$ the group Γ is a cocompact lattice in G_S. But according to Corollary I.2.3.5, the group G_S is compactly generated. So Γ is finitely generated (see I.0.40). Finally, for the case "$\operatorname{rank}_S \mathbf{G} \geq 2$", we refer to [Be 2]. We remark that in [Be 2] certain restrictions are imposed on the characteristic of the field K. But these restrictions are not necessary for proving the results of [Har 2]. Notice also that the finite generation of Γ in "almost all" cases was proved in Chapter III (see Theorem D (c) of III.5.9).

Now we proceed to the proof of Theorem (C) mentioned at the beginning of the present chapter. We shall state this theorem in a slightly different form considering the larger class Ω_S instead of the class of S-arithmetic subgroups.

(3.4) Theorem. *Let l be a field, $\Lambda \in \Omega_S$,* $\mathbf{H}$ *a connected non-commutative absolutely almost simple l-group, and $\delta\colon \Lambda \to \mathbf{H}(l)$ a homomorphism with $\overline{\delta(\Lambda)} = \mathbf{H}$.*

(a) *Suppose* $\operatorname{rank}_S \mathbf{G} \geq 2$ *and either* $\mathbf{G}$ *is simply connected or* $\mathbf{H}$ *is adjoint. Then there exist a field extension l' of l, a homomorphism $\sigma\colon K \to l'$, a special l'-epimorphism $\eta\colon {}^{\sigma}\mathbf{G} \to \mathbf{H}$, and a homomorphism $\nu\colon \Lambda \to \mathscr{Z}(\mathbf{H})$ such that*

$$(*) \qquad \delta(\Lambda) = \nu(\lambda) \cdot \eta(\sigma^0(\lambda)) \text{ for all } \lambda \in \Lambda.$$

(b) *Let there be given a field extension l' of l, a homomorphism $\sigma\colon K \to l'$, a special l'-epimorphism $\eta\colon {}^{\sigma}\mathbf{G} \to \mathbf{H}$, and a homomorphism $\nu\colon \Lambda \to \mathscr{Z}(\mathbf{H})$ such that $(*)$ holds. Suppose* $\operatorname{rank}_S \mathbf{G} > 0$*, or equivalently, $S \not\subset \mathscr{I}(\mathbf{G})$. Then $\sigma(K) \subset l$ and η is defined over l. Furthermore, σ, η, and $\nu|_{\Lambda \cap \mathbf{G}(K)}$ are defined uniquely via $(*)$.*

Proof. (a) Let $\tilde{\mathbf{G}}$ be the simply connected covering of $\mathbf{G}$ and let $\pi\colon \tilde{\mathbf{G}} \to \mathbf{G}$ be the central K-isogeny (see Proposition I.1.4.11(i)). If the group $\mathbf{H}$ is adjoint, then, for a homomorphism $\sigma\colon K \to l'$ and a special l'-epimorphism $\eta\colon {}^{\sigma}\tilde{\mathbf{G}} \to \mathbf{H}$, there is a special l'-epimorphism $\eta\colon {}^{\sigma}\mathbf{G} \to \mathbf{H}$ such that $\tilde{\eta} = {}^{\sigma}\pi \circ \eta$ (see Proposition I.1.4.11(iii) and I.1.4.14). Since $\Lambda \in \Omega_S(\mathbf{G})$, it follows from Lemma 3.1 that $\pi^{-1}(\Lambda) \in \Omega_S(\tilde{\mathbf{G}})$. Since for each $v \in \mathscr{R}$ the groups $\mathbf{G}$ and $\tilde{\mathbf{G}}$ have the same K_v-rank, $\operatorname{rank}_S \tilde{\mathbf{G}} = \operatorname{rank}_S \mathbf{G}$ (see Corollary I.1.4.6). Thus, since the center of any adjoint group is trivial, replacing $\mathbf{G}$ by $\tilde{\mathbf{G}}$, Λ by $\pi^{-1}(\Lambda)$, and δ by $\pi \circ \delta$, we may assume that $\mathbf{G}$ is simply connected even if $\mathbf{H}$ is adjoint. Then the group $\mathbf{G}$ can be represented in the form $\mathbf{G} = R_{K'/K}\mathbf{G}'$, where K' is a finite separable field extension of K and $\mathbf{G}'$ is a simply connected absolutely almost simple K'-group (see I.1.7). Let $p\colon \mathbf{G} \to \mathbf{G}'$ be the natural K-epimorphism. By Lemma I.3.1.4 we can replace $\mathbf{G}$, Λ, and δ by $\mathbf{G}'$, $p(\Lambda)$, and $\delta \circ p^{-1}$ and assume that $\mathbf{G}$ is absolutely almost simple.

Now choose a finite set $\mathscr{E} \subset S$ with $\operatorname{rank}_{\mathscr{E}} \mathbf{G} \geq 2$, $\mathscr{E} \supset \mathscr{R}_\infty - \mathscr{T}$, and $\mathscr{E} \cap \mathscr{T} = \varnothing$. We set

$$\Lambda_0 = \Lambda \cap \mathbf{G}(K), \quad \Gamma = \Lambda \cap \mathbf{G}(K(\mathscr{E})) \subset \Lambda_0, \text{ and}$$
$$G_{\mathscr{E}} = \prod_{v \in \mathscr{E}} \mathbf{G}(K_v).$$

We identify the group $\mathbf{G}(K)$ with its image under the diagonal embedding in $G_{\mathscr{E}}$. According to Theorem I.3.2.5, Γ is a lattice in $G_{\mathscr{E}}$. It is easily seen that the lattice $\Gamma \subset G_{\mathscr{E}}$ is irreducible (see the proof of Theorem 2.6). According to Theorem 3.3, Γ is finitely generated. By virtue of Lemma 2.1, $\mathscr{D}(\Lambda) \subset \Lambda_0$. But $\overline{\delta(\Lambda)} = \mathbf{H}$ and $\mathscr{D}(\mathbf{H}) = \mathbf{H}$. Therefore $\overline{\delta(\Lambda_0)} = \mathbf{H}$ and, since $\Lambda_0 \in \Omega_S$, $\Lambda_0 \subset \mathbf{G}(K)$, and $\mathbf{H}$ is connected, Lemma 3.2(ii) implies that $\overline{\delta(\Gamma)} = \mathbf{H}$. We realize $\mathbf{H}$ as an l-subgroup of $\mathbf{GL}_n$. Since Γ is finitely generated, by VII.6.4 there is a finitely generated field $l_0 \subset l$ such that $\delta(\Gamma) \subset \mathbf{GL}_n(l_0)$. Then the group $\mathbf{H} = \overline{\delta(\Gamma)}$ is defined over l_0. Note that, by Lemma I.3.1.3, $\Lambda_0 \subset \mathbf{G}(K) \subset \mathrm{Comm}_{\mathbf{G}}(\Gamma)$. Thus, we have the following:

1) a finite set $\mathscr{E} \subset \mathscr{R}$,
2) an absolutely almost simple simply connected K-group $\mathbf{G}$ with $\mathrm{rank}_{\mathscr{E}} \mathbf{G} \geq 2$ which is isotropic over K_v for all $v \in \mathscr{E}$,
3) a finitely generated irreducible lattice Γ in $G_{\mathscr{E}} = \prod_{v \in \mathscr{E}} \mathbf{G}(K_v)$,
4) a subgroup Λ_0 of $\mathrm{Comm}_{\mathbf{G}}(\Gamma)$,
5) a finitely generated field l_0 and a field extension l of l_0,
6) a non-commutative absolutely simple l_0-group $\mathbf{H}$ and a homomorphism $\delta\colon \Lambda_0 \to \mathbf{H}(l)$ with $\delta(\Gamma) \subset \mathbf{H}(l_0)$ and $\delta(\Gamma)$ Zariski dense in $\mathbf{H}$.

Theorem VII.6.5 and Remark VII.6.7 then imply that there exist $v \in \mathscr{E}$, a homomorphism σ of the field K_v into a field extension l' of l, a special l'-epimorphism $\eta\colon {}^{\sigma}\mathbf{G} \to \mathbf{H}$, and a homomorphism $\nu_0\colon \Lambda_0 \to \mathscr{Z}(\mathbf{H})$ such that

$$(1) \qquad \delta(\lambda) = \nu_0(\lambda) \cdot \eta(\sigma^0(\lambda)) \text{ for all } \lambda \in \Lambda_0$$

(with Λ_0 viewed as a subgroup of $\mathbf{G}(K_v)$). We set $\Lambda_1 = \mathrm{Ker}\, \nu_0$. Since the center $\mathscr{Z}(\mathbf{H})$ is finite, Λ_1 is of finite index in Λ_0. But Lemma 2.1(i) implies that the subgroup Λ_0 is normal in Λ. Therefore $\Lambda \subset \mathrm{Comm}_{\mathbf{G}}(\Lambda_1)$. It follows from (1) that the homomorphisms δ and $\eta \circ \sigma^0$ agree on Λ_1. Since the subgroup $\delta(\Lambda_0) \supset \delta(\Gamma)$ is Zariski dense in $\mathbf{H}$, Λ_1 is of finite index in Λ_0, and $\mathbf{H}$ is connected, and hence does not contain algebraic subgroups of finite index, it follows that $\overline{\delta(\Lambda_1)} = \mathbf{H}$. Thus, Lemma VII.5.1(b) implies that there is a homomorphism $\nu\colon \Lambda \to \mathscr{Z}(\mathbf{H})$ such that $\delta(\lambda) = \nu(\lambda) \cdot \eta(\sigma^0(\lambda))$ for all $\lambda \in \Lambda$. This completes the proof of (a).

(b) The proof in case $\mathrm{char}\, K \neq 0$ will be presented in 3.14–3.27.

For the time being assume $\mathrm{char}\, K = 0$. Then K is a finite extension of $\mathbb{Q}$ and we may assume l' contains the algebraic closure $\overline{\mathbb{Q}}$ of $\mathbb{Q}$. Let $\{\sigma_1 = \mathrm{Id}, \sigma_2, \ldots, \sigma_d\}$ be the set of all distinct embeddings of the field K in $\overline{\mathbb{Q}}$. We represent the group $R_{K/\mathbb{Q}}\mathbf{G}$ in the form $R_{K/\mathbb{Q}}\mathbf{G} = {}^{\sigma_1}\mathbf{G} \times \ldots \times {}^{\sigma_d}\mathbf{G}$. The natural projection $R_{K/\mathbb{Q}}\mathbf{G} \to {}^{\sigma_i}\mathbf{G}$, $1 \leq i \leq d$, will be denoted by p_i. To each pair (σ, η), where $\sigma = \sigma_i\colon K \to \overline{\mathbb{Q}}$ is an embedding, we associate the l'-epimorphism $\alpha_{\sigma,\eta} = p_i \circ \eta\colon R_{K/\mathbb{Q}}\mathbf{G} \to \mathbf{H}$. Clearly, if $(\sigma, \eta) \neq (\tilde{\sigma}, \tilde{\eta})$, then $\alpha_{\sigma,\eta} \neq \alpha_{\tilde{\sigma},\tilde{\eta}}$. Suppose we are given two homomorphisms σ, $\tilde{\sigma}\colon K \to l'$, l'-epimorphisms $\eta\colon {}^{\sigma}\mathbf{G} \to \mathbf{H}$ and $\tilde{\eta}\colon {}^{\tilde{\sigma}}\mathbf{G} \to \mathbf{H}$, and two homomorphisms ν, $\tilde{\nu}\colon \Lambda \to \mathscr{Z}(\mathbf{H})$ such that

$$\nu(\lambda)\cdot\eta(\sigma^0(\lambda)) = \tilde{\nu}(\lambda)\cdot\tilde{\eta}(\tilde{\sigma}^0(\lambda)) \text{ for all } \lambda \in \Lambda.$$

We set $\Lambda_1 = \mathbf{G}(K) \cap \operatorname{Ker}\nu \cap \operatorname{Ker}\tilde{\nu}$. Then

$$(2) \qquad \eta(\sigma^0(\lambda)) = \tilde{\eta}(\tilde{\sigma}^0(\lambda)) \text{ for all } \lambda \in \Lambda_1,$$

Since $\Lambda \in \Omega_S$ and the group $\mathscr{Z}(\mathbf{H})$ is finite, the subgroup $\Lambda_1 \cap \mathbf{G}(K(S))$ is of finite index in $\mathbf{G}(K(S))$. Therefore the subgroup $R^0_{K/\mathbb{Q}}(\Lambda_1)$ is Zariski dense in $R_{K/\mathbb{Q}}\mathbf{G}$ (see Proposition I.3.2.10). It follows from the definitions of the homomorphisms $R^0_{K/\mathbb{Q}}$, $\alpha_{\sigma,\eta}$, and $\alpha_{\tilde{\sigma},\tilde{\eta}}$ that $\eta \circ \sigma^0 = \alpha_{\sigma,\eta} \circ R^0_{K/\mathbb{Q}}$ and $\tilde{\eta} \circ \tilde{\sigma}^0 = \alpha_{\tilde{\alpha},\tilde{\eta}} \circ R^0_{K/\mathbb{Q}}$. From this and (2) above we deduce that the rational epimorphisms $\alpha_{\alpha,\eta}$ and $\alpha_{\tilde{\sigma},\tilde{\eta}}$ agree on $R^0_{K/\mathbb{Q}}(\Lambda_1)$ which is Zariski dense in $R_{K/\mathbb{Q}}\mathbf{G}$. Hence $\alpha_{\sigma,\eta} = \alpha_{\tilde{\sigma},\tilde{\eta}}$. This implies that $\sigma = \tilde{\sigma}$ and $\eta = \tilde{\eta}$, and hence $\nu(\lambda) = \tilde{\nu}(\lambda)$ for all $\lambda \in \Lambda \cap \mathbf{G}(K)$. Thus, σ, η, and $\nu|_{\Lambda\cap\mathbf{G}(K)}$ are determined uniquely via (*).

Now suppose (*) holds for σ, η, and ν and $\sigma(K) \not\subset l$. Since the subgroup $\mathscr{Z}(\mathbf{H})$ is finite, $\operatorname{Ker}\nu$ is of finite index in Λ. Therefore, replacing Λ by $\operatorname{Ker}\nu$, we may assume that $\delta = \eta \circ \sigma^0|_\Lambda$. Since $\sigma(K) \not\subset l$, there is an automorphism ϑ of the algebraic closure of the field l' which is the identity on l and not the identity on $\sigma(K)$. We set $\tilde{\sigma} = \vartheta \circ \sigma$ and $\tilde{\eta} = {}^\vartheta\eta$. Clearly, $\tilde{\eta} \circ \tilde{\sigma}^0 = \vartheta^0 \circ \eta \circ \sigma^0$. But since $(\eta \circ \sigma^0)(\Lambda) = \delta(\Lambda) \subset \mathbf{H}(l)$ and the automorphism ϑ is the identity on l, we have $(\vartheta^0 \circ \eta \circ \sigma^0)(\lambda) = (\eta \circ \sigma^0)(\lambda)$ for all $\lambda \in \Lambda$. Therefore $\tilde{\eta} \circ \tilde{\sigma}^0(\lambda) = \eta(\sigma(\lambda))$ for all $\lambda \in \Lambda$. Since ϑ is not the identity on $\sigma(K)$, it follows that $\tilde{\sigma} \neq \sigma$. Since the homomorphisms σ, η, and ν are unique we obtain a contradiction. Hence, $\sigma(K) \subset l$.

It now remains to show that if (*) holds for σ, η, and ν, then η is defined over l. As above we can replace Λ by $\operatorname{Ker}\nu$ and assume that $\delta = \eta \circ \sigma^0|_\Lambda$. Then $\eta(\sigma^0(\Lambda)) \subset \mathbf{H}(l)$. On the other hand, $\sigma^0(\Lambda \cap \mathbf{G}(K)) \subset {}^\sigma\mathbf{G}(l)$ (because, as was shown above, $\sigma(K) \subset l$) and by Proposition I.3.2.10 $\overline{\Lambda \cap \mathbf{G}(K)} = \mathbf{G}$. Hence, $\overline{\sigma^0(\Lambda \cap \mathbf{G}(K))} = {}^\sigma\mathbf{G}$. Thus, η is defined over l (see I.0.11(ii)). □

(3.5) Let l be a field, $\mathbf{H}$ an algebraic l-group, $\Lambda \in \Omega_S$, and K' a field extension of K with $\Lambda \subset \mathbf{G}(K')$. A homomorphism $\tau\colon \Lambda \to \mathbf{H}$ is called *quasi-regular* if there exist $i \in \mathbb{N}^+$, homomorphisms $\sigma_1, \ldots, \sigma_i$ of the field K into a field extension l' of l, and an l'-morphism $\varphi\colon {}^{\sigma_1}\mathbf{G} \times \ldots \times {}^{\sigma_i}\mathbf{G} \to \mathbf{H}$ of l'-groups such that

$$\tau(\lambda) = \varphi(\sigma_1^0(\lambda), \ldots, \sigma_i^0(\lambda)) \text{ for all } \lambda \in \Lambda.$$

If $\operatorname{char} K = \operatorname{char} l = 0$ (i.e. $K \supset \mathbb{Q}$ and $l \supset \mathbb{Q}$) and $\Lambda \subset \mathbf{G}(K)$, then, by the definition of the functor $R_{K/\mathbb{Q}}$ and the homomorphism $R^0_{K/\mathbb{Q}}$, the homomorphism $\tau\colon \Lambda \to \mathbf{H}$ is quasi-regular if and only if there is a morphism $\varphi\colon R_{K/\mathbb{Q}}\mathbf{G} \to \mathbf{H}$ of algebraic groups such that

$$\tau(\lambda) = \varphi(R^0_{K/\mathbb{Q}}(\lambda)) \text{ for all } \lambda \in \Lambda.$$

□

(3.6) Theorem. *Let l be a field, $\Lambda \in \Omega_S$, $\mathbf{H}$ a connected semisimple l-group, and $\delta\colon \Lambda \to \mathbf{H}(l)$ a homomorphism with $\overline{\delta(\Lambda)} = \mathbf{H}$. Suppose* $\operatorname{rank}_S \mathbf{G} \geq 2$ *and either* $\mathbf{G}$ *is simply connected or* $\mathbf{H}$ *is adjoint.*

(i) *There exist a quasi-regular homomorphism* $\tau\colon \Lambda \to \mathbf{H}$ *and a homomorphism* $\nu\colon \Lambda \to \mathscr{Z}(\mathbf{H})$ *such that* $\delta(\lambda) = \nu(\lambda)\cdot\tau(\lambda)$ *for all* $\lambda \in \Lambda$.

(ii) *Suppose* $\operatorname{char} K = \operatorname{char} l = 0$ *(i.e.* $K \supset \mathbb{Q}$ *and* $l \supset \mathbb{Q}$*) and let* $\Lambda \subset \mathbf{G}(K)$. *Then*

(a) *there are a rational epimorphism* $\varphi\colon R^0_{K/\mathbb{Q}}\mathbf{G} \to \mathbf{H}$ *of algebraic groups and a homomorphism* $\nu\colon \Lambda \to \mathscr{Z}(\mathbf{H})$ *such that*

$$(*) \qquad \delta(\lambda) = \nu(\lambda)\cdot\varphi(R^0_{K/\mathbb{Q}}(\lambda)) \text{ for all } \lambda \in \Lambda;$$

(b) *if* (*) *holds for* φ *and* ν, *then* φ *is defined over* l. *Furthermore,* φ *and* ν *are uniquely determined via* (*).

Proof. (i) In case $\mathbf{H}$ is adjoint it suffices to decompose $\mathbf{H}$ into a direct product of absolutely almost simple groups $\mathbf{H}_1, \ldots, \mathbf{H}_i$ and to apply Theorem 3.4(a) to the homomorphisms $\pi_j \circ \delta$, $1 \le j \le i$, where $\pi_j\colon \mathbf{H} \to \mathbf{H}_j$ is the natural projection. Assume now that $\mathbf{G}$ is simply connected. From what we have proved we deduce that the homomorphism $\operatorname{Ad}\circ\delta\colon \Lambda \to \operatorname{Ad}\mathbf{H}$ is quasi-regular. On the other hand, since $\mathbf{G}$ is simply connected, by Proposition I.1.4.11, for any quasi-regular homomorphism $\tau'\colon \Lambda \to \operatorname{Ad}\mathbf{H}$, there is a quasi-regular homomorphism $\tau''\colon \Lambda \to \mathbf{H}$ such that $\operatorname{Ad}\circ\tau'' = \tau'$. Therefore there is a quasi-regular homomorphism $\tau\colon \Lambda \to \mathbf{H}$ such that $\operatorname{Ad}\circ\tau = \operatorname{Ad}\circ\delta$. For $\lambda \in \Lambda$, we define $\nu(\lambda)$ via the equality $\delta(\lambda) = \nu(\lambda)\cdot\tau(\lambda)$. Since $\operatorname{Ad}\circ\tau = \operatorname{Ad}\circ\delta$ and $\operatorname{Ker}\operatorname{Ad} \subset \mathscr{Z}(\mathbf{H})$, it follows that $\nu(\Lambda) \subset \mathscr{Z}(\mathbf{H})$. All that remains now is to make use of Lemma VII.5.1(a).

(ii) Assertion (a) is a consequence of (i) and the observations in 3.5. We will now prove (b). Since the group $\mathscr{Z}(\mathbf{H})$ is finite, $\operatorname{Ker}\nu$ is of finite index in Λ. Therefore we can replace Λ by $\operatorname{Ker}\nu$ and assume that $\delta = \varphi \circ R^0_{K/\mathbb{Q}}|_\Lambda$. By virtue of Proposition I.3.2.10 the subgroup $R^0_{K/\mathbb{Q}}(\Lambda)$ is Zariski dense in $R_{K/\mathbb{Q}}\mathbf{G}$. It now remains to make use of I.0.11(ii) and to observe that an algebraic group morphism $\alpha\colon \mathbf{F} \to \mathbf{F}'$ is uniquely determined by its values on a Zariski dense subgroup $A \subset \mathbf{F}$. □

(3.7) Lemma. *Let* l *be a field,* $\mathbf{H}$ *an algebraic* l*-group, and* F *a subgroup of* $\mathbf{H}(l)$. *Suppose* F *is torsion and the orders of all elements of* F *are uniformly bounded. Then*

(a) *F contains a unipotent subgroup of finite index;*

(b) *if* $\operatorname{char} l = 0$, *then* F *is finite.*

Proof. (a) We may assume that $\mathbf{H} = \bar{F}$. The group $\mathbf{H}/\mathbf{H}^0$ is finite. Therefore, replacing F by $F \cap \mathbf{H}^0$, we can assume that $\bar{F} = \mathbf{H} = \mathbf{H}^0$. Choose $d \in \mathbb{N}^+$ with $x^d = e$ for all $x \in F$. Then, since $\bar{F} = \mathbf{H}$ and the elements of order d form an algebraic subvariety in $\mathbf{H}$, it follows that $h^d = e$ for all $h \in \mathbf{H}$. Assume now that the connected group $\mathbf{H}$ is not unipotent. Then $\mathbf{H}$ contains a torus $\mathbf{T}$ of positive dimension (see I.0.22). But in any torus the number of elements of order d is finite (see I.0.22). This contradicts $h^d = e$ for all $h \in \mathbf{H}$. Thus, $\mathbf{H} = \mathbf{H}^{(u)}$, and hence F is unipotent.

Finally, (b) is a consequence of (a) and the fact that if $\operatorname{char} l = 0$ and $g \in \mathbf{H}(l)^{(u)} - \{e\}$, then g is of infinite order (see I.0.20). □

(3.8) Theorem. *Let l be a field, $\Lambda \in \Omega_S$, $\mathbf{H}$ an algebraic l-group, and $\delta\colon \Lambda \to \mathbf{H}(l)$ a homomorphism. Suppose* $\operatorname{rank}_S \mathbf{G} \geq 2$ *and* $\operatorname{char} l \neq \operatorname{char} K$.

(i) *The group $\delta(\Lambda)$ is torsion, the orders of all elements of $\delta(\Lambda)$ are uniformly bounded, and $\delta(\Lambda)$ contains a unipotent subgroup of finite index.*

(ii) *(a) If $\mathbf{G}$ is simply connected and $\Lambda \cap \mathbf{G}(K)$ is of finite index in Λ, then $\delta(\Lambda)$ is finite.*
(b) If S is finite and Λ is an S-arithmetic subgroup of $\mathbf{G}$, then $\delta(\Lambda)$ is finite.
(c) If $\operatorname{char} l = 0$, *then $\delta(\Lambda)$ is finite.*

Proof. First we shall show (ii) (a). As in the proof of Theorem 3.4, making use of the restriction of scalars functor, we reduce to the case in which $\mathbf{G}$ is absolutely almost simple. Choose a finite subset $\mathscr{E} \subset S$ with $\operatorname{rank}_{\mathscr{E}} \mathbf{G} \geq 2$, $\mathscr{E} \supset \mathscr{R}_\infty - \mathscr{T}$, and $\mathscr{E} \cap \mathscr{T} = \varnothing$. We set $\Lambda_0 = \Lambda \cap \mathbf{G}(K)$, $\Gamma = \Lambda \cap \mathbf{G}(K(\mathscr{E})) \subset \Lambda_0$, and $G_{\mathscr{E}} = \prod_{v \in \mathscr{E}} \mathbf{G}(K_v)$.

Now, identify the group Λ_0 with its image under the diagonal embedding in $G_{\mathscr{E}}$. As in the proof of Theorem 3.4, we verify that Γ is a finitely generated irreducible lattice in $G_{\mathscr{E}}$ and $\Lambda_0 \subset \operatorname{Comm}_{G_{\mathscr{E}}}(\Gamma)$. It then follows from Corollary VII.6.10 and Lemma 3.2(i) that the group $\delta(\Lambda_0)$ contains a solvable subgroup of finite index. In other words, replacing Λ by a subgroup of finite index, we can assume that the group $\delta(\Lambda)$, and hence the group $\mathbf{F} \stackrel{\text{def}}{=} \overline{\delta(\Lambda)}$, are solvable (see I.0.13). The group $\mathbf{F}/\mathbf{F}^0$ is finite. Therefore we may replace Λ by $\delta^{-1}(\mathbf{F}^0)$ and assume that $\mathbf{F}$ is connected. It then follows from Lemma 3.2(ii) that $\mathbf{F} = \overline{\delta(\Gamma)}$. On the other hand, by virtue of Corollary 2.8 the factor group $\Gamma/\mathscr{D}(\Gamma)$ and hence $\overline{\delta(\Gamma)}/\mathscr{D}(\overline{\delta(\Gamma)})$ is finite. Therefore $\mathscr{D}(\mathbf{F}) = \overline{\mathscr{D}(\mathbf{F})}$ is of finite index in $\mathbf{F}$. But $\mathbf{F}$ is connected and solvable, and a connected algebraic group does not contain proper algebraic subgroups of finite index. Hence, $\mathbf{F} = \{e\}$, which proves (ii) (a).

(i) Let $\tilde{\mathbf{G}}$ be the simply connected covering of $\mathbf{G}$ and $\pi\colon \tilde{\mathbf{G}} \to \mathbf{G}$ the central K-isogeny. Since $\Lambda \in \Omega_S(\mathbf{G})$, it follows from Lemma 3.1 that $\pi^{-1}(\Lambda) \in \Omega_S(\tilde{\mathbf{G}})$. On the other hand, since for each $v \in \mathscr{R}$ the groups $\mathbf{G}$ and $\tilde{\mathbf{G}}$ have the same K_v-rank, we have $\operatorname{rank}_S \tilde{\mathbf{G}} = \operatorname{rank}_S \mathbf{G}$ (see Corollary I.1.4.6). Therefore we can replace $\mathbf{G}$ by $\tilde{\mathbf{G}}$, Λ by $\pi^{-1}(\Lambda)$, and δ by $\delta \circ \pi$ and assume that $\mathbf{G}$ is simply connected. It then follows from (ii) (a) that the group $\delta(\Lambda \cap \mathbf{G}(K))$ is finite. But by Lemma 2.1(i) the orders of all elements of the factor group $\Lambda/(\Lambda \cap \mathbf{G}(K))$ are uniformly bounded. So the orders of all elements of the group $\delta(\Lambda)$ are uniformly bounded. All that remains now is to make use of Lemma 3.7(a).

(ii) (b) Arguing as at the beginning of the proof of Theorem 2.6(i), we reduce to the case in which $\mathbf{G}$ is simply connected. But in this case the finiteness of the group $\delta(\Lambda)$ is a consequence of (ii) (a) and the fact that the subgroup $\Lambda \cap \mathbf{G}(K) \supset \Lambda \cap \mathbf{G}(K(S))$ is of finite index in Λ.

(ii) (c) is a consequence of (i) and Lemma 3.7(b). □

(3.9) Lemma. *Let Γ be an S-arithmetic subgroup of $\mathbf{G}$, l a field, $\mathbf{H}$ an algebraic l-group, and $\delta\colon \Gamma \to \mathbf{H}(l)$ a homomorphism. Suppose* $\operatorname{rank}_S \mathbf{G} \geq 2$ *and $\mathbf{G}$ is simply connected. Then there is a subgroup Γ_1 of finite index in Γ such that $\delta(\Gamma_1^{(u)}) \subset \mathbf{H}^{(u)}$.*

Proof. We may assume that $\overline{\delta(\Gamma)} = \mathbf{H}$. The group $\mathbf{H}/\mathbf{H}^0$ is finite. Therefore, replacing Γ by $\delta^{-1}(\mathbf{H}^0)$, we may assume that $\mathbf{H}$ is connected. Let $\pi\colon \mathbf{H} \to \mathbf{H}/R_u(\mathbf{H})$

be the natural epimorphism. Since $\operatorname{Ker}\pi$ is is unipotent, it follows from I.0.20(v) that $\pi^{-1}((\mathbf{H}/R_u(\mathbf{H}))^{(u)}) = \mathbf{H}^{(u)}$. Therefore, replacing $\mathbf{H}$ by $\mathbf{H}/R_u(\mathbf{H})$ and δ by $\pi \circ \delta$, we may assume that $\mathbf{H}$ is reductive. According to Corollary 2.8 the factor group $\Gamma/D(\Gamma)$, and hence $\overline{\delta(\Gamma)}/\mathscr{D}(\overline{\delta(\Gamma)})$, are finite. But $\overline{\delta(\Gamma)} = \mathbf{H}$, the commutator subgroup $\mathscr{D}(\mathbf{H})$ is an algebraic subgroup of $\mathbf{H}$, and the connected group $\mathbf{H}$ does not contain proper algebraic subgroups of finite index. Thus, $\mathbf{H} = \mathscr{D}(\mathbf{H})$. From this and the connectedness and the reductivity of $\mathbf{H}$, we deduce that $\mathbf{H}$ is semisimple. It then follows from Theorem 3.6(i) that there exist a quasi-regular homomorphism $\tau\colon \Gamma \to \mathbf{H}$ and a homomorphism $\nu\colon \Gamma \to Z(\mathbf{H})$ such that $\delta(\gamma) = \nu(\gamma)\cdot\tau(\gamma)$ for each $\gamma \in \Gamma$. We set $\Gamma_1 = \operatorname{Ker}\nu$. Since the group $\mathscr{Z}(\mathbf{H})$ is finite, Γ_1 is of finite index in Γ. On the other hand, $\delta(\gamma) = \tau(\gamma)$ for all $\gamma \in \Gamma_1$ and by I.0.20(ii) the image of a unipotent element under a quasi-regular homomorphism is unipotent. Thus, Γ_1 is the desired subgroup. □

(3.10) Theorem. *Let l be a field, $\Lambda \in \Omega_S$, $\mathbf{H}$ an algebraic l-group, and $\delta\colon \Lambda \to \mathbf{H}(l)$ a homomorphism. Suppose* $\operatorname{rank}_S \mathbf{G} \geq 2$ *and* $\operatorname{char} l = 0$. *Then the l-group $\overline{\delta(\Lambda)}$ is semisimple.*

Proof. If $\operatorname{char} K \neq 0$, then by virtue of Theorem 3.8(ii)(c) the group $\delta(\Lambda)$ is finite. Therefore we may assume that $\operatorname{char} K = 0$. Let $\tilde{\mathbf{G}}$ be the simply connected covering of $\mathbf{G}$ and let $\pi\colon \tilde{\mathbf{G}} \to \mathbf{G}$ be the central K-isogeny. As in proving Theorem 3.4(a), replacing $\mathbf{G}$ by $\tilde{\mathbf{G}}$, Λ by $\pi^{-1}(\Lambda)$, and δ by $\delta \circ \pi$, we can assume that $\mathbf{G}$ is semisimple. Further, as in the proof of Theorem 3.4, with the help of the restriction of scalars functor we come to the case in which $\mathbf{G}$ is absolutely almost simple.

Since $\operatorname{char} K = 0$, we have $\mathscr{R}_\infty \neq \emptyset$. On the other hand, $\operatorname{rank}_S \mathbf{G} \geq 2$. Therefore we can choose a finite set $\mathscr{E} \subset S$ such that

(a) $\mathscr{E} \not\subset \mathscr{T}$, $\mathscr{E} \supset \mathscr{R}_\infty - \mathscr{R}$, and $\mathscr{E} \cap \mathscr{T} = \emptyset$;
(b) if $\mathbf{G}$ is isotropic over K, then $\mathscr{E} \supset \mathscr{R}_\infty$.

We set

$$\Lambda_0 = \Lambda \cap \mathbf{G}(K),\ \Gamma = \Lambda \cap \mathbf{G}(K(\mathscr{E})) \subset \Lambda_0, \text{ and}$$
$$G_{\mathscr{E}} = \prod_{v \in \mathscr{E}} \mathbf{G}(K_v).$$

Identifying the group Λ_0 with its image under the diagonal embedding in $G_{\mathscr{E}}$ we verify as in the proof of Theorem 3.4 that Γ is a finitely generated irreducible lattice in $G_{\mathscr{E}}$ and $\Lambda_0 \subset \operatorname{Comm}_{G_{\mathscr{E}}}(\Gamma)$. We shall prove that

$$(*)\qquad \text{either } \operatorname{rank}_{\mathscr{E}} G \geq 2 \text{ or } \Lambda_0 \text{ is dense in } G_{\mathscr{E}}.$$

Assume $\operatorname{rank}_{\mathscr{E}} \mathbf{G} < 2$. Then, since $\operatorname{rank}_S \mathbf{G} \geq 2$, we have $S - \mathscr{E} \not\subset \mathscr{T}$. Therefore according to the strong approximation theorem (see Remark II.6.8.2) the subgroup $\mathbf{G}(K(S))$ is dense in $G_{\mathscr{E}}$. We denote by F the closure of the subgroup Λ_0 in $G_{\mathscr{E}}$. Since $\mathbf{G}(K(S))$ is dense in $G_{\mathscr{E}}$ and $\Lambda_0 \cap \mathbf{G}(K(S))$ is of finite index in $\mathbf{G}(K(S))$ (because $\Lambda \in \Omega_S$), it follows that F is of finite index in $G_{\mathscr{E}}$. On the other hand, since $\mathscr{E} \cap \mathscr{T} = \emptyset$ and $\mathbf{G}$ is absolutely almost simple and simply connected, by virtue of Corollary I.2.3.2(b) $G_{\mathscr{E}}$ does not contain proper subgroups of finite index. Therefore $F = G_{\mathscr{E}}$, which proves (*). Now let us verify that

(**) every representation $T: \Lambda_0 \to \mathbf{GL}_n(k)$ of the group Λ_0 on a finite-dimensional vector space over an arbitrary local field k is Γ-integrable.

If $\mathbf{G}$ is anisotropic over K, then by Theorem I.2.3.4(b) the quotient space $\Gamma \setminus G_{\mathscr{E}}$ is compact and (**) is a consequence of the remark at the beginning of Section 3 in Chapter V. Let $\mathbf{G}$ be isotropic over K. Then condition (b) implies that $\mathscr{E} \subset \mathscr{R}_\infty$. On the other hand, since $\operatorname{rank}_S \mathbf{G} \geq 2$, $\mathbf{G}$ is simply connected, and $\Lambda \in \Omega_S$, it follows from Lemma 3.9 that the subgroup $\Lambda \cap \mathbf{G}(K(S))$ containing Γ contains a subgroup Λ_1 of finite index such that $T(\Lambda_1^{(u)}) \subset \mathbf{GL}_n^{(u)}$. Therefore the representation T is Γ-integrable (see Proposition 1.3). Thus, we have proved (*) and (**). We further observe that for any subgroup Γ_0 of finite index in Γ the factor group $\Gamma_0/\mathscr{D}(\Gamma_0)$ is finite (see Corollary 2.8) and the subgroup generated by $\bigcup_{\lambda \in \Lambda_0} \lambda \Gamma_0 \lambda^{-1}$ is of finite index in Λ_0 (see Lemma 3.2(i)). Furthermore, we recall that Γ is a finitely generated irreducible lattice in $G_{\mathscr{E}}$ and the group $\mathbf{G}$ is isotropic over K_v for all $v \in \mathscr{E}$, absolutely almost simple, and simply connected. Now we can apply Theorem VII.6.16 and deduce that the group $\overline{\delta(\Lambda_0)}$ is semisimple. According to Lemma 2.1(i) the orders of all elements of the factor group Λ/Λ_0 (and hence of $\overline{\delta(\Lambda)}/\overline{\delta(\Lambda_0)}$) are uniformly bounded. Since $\operatorname{char} l = 0$ and $\overline{\delta(\Lambda)}/\overline{\delta(\Lambda_0)}$ is an l-group, it then follows from Lemma 3.7, that the group $\overline{\delta(\Lambda)}/\overline{\delta(\Lambda_0)}$ is finite. Thus, the group $\overline{\delta(\Lambda)}$ contains a semisimple algebraic subgroup $\overline{\delta(\Lambda_0)}$ of finite index, and hence is itself semisimple. □

(3.11) We turn now to Theorem (B) stated at the beginning of the present chapter. Assertion (i) of this theorem is a particular case of Theorem 3.8 (ii) (a) and assertion (ii) is a special case of Theorem 3.10. We shall state and prove an insignificant generalization of Theorem (B) (iii).

(3.12) Theorem. *Let l be a field, $\Lambda \in \Omega_S$, $\mathbf{H}$ an algebraic l-group, and $\delta: \Lambda \to \mathbf{H}(l)$ a homomorphism. Suppose* $\operatorname{char} K = \operatorname{char} l = 0$ *(i.e. $K \supset \mathbb{Q}$ and $l \supset \mathbb{Q}$), $\Lambda \subset G(K)$, $\mathbf{G}$ is simply connected and* $\operatorname{rank}_S \mathbf{G} \geq 2$. *Then*

(a) *there exist (uniquely determined) a morphism $\varphi: R_{K/\mathbb{Q}}\mathbf{G} \to \mathbf{H}$ of algebraic groups and a homomorphism $\nu: \Lambda \to H$ such that the subgroup $\nu(\Lambda)$ is finite and commutes with $\varphi(R_{K/\mathbb{Q}}\mathbf{G})$, and*

$$(*) \qquad \delta(\lambda) = \nu(\lambda) \cdot \varphi(R^0_{K/\mathbb{Q}}(\lambda)) \text{ for all } \lambda \in \Lambda;$$

(b) *the morphism φ in (a) above is defined over l.*

Proof. We may assume that $\overline{\delta(\Lambda)} = \mathbf{H}$. Then by Theorem 3.10 the group $\mathbf{H}$ is semisimple. We set $\Lambda' = \delta^{-1}(\mathbf{H}^0)$. Since $\mathbf{H}/\mathbf{H}^0$ is finite and $\Lambda \in \Omega_S$, it follows that $\Lambda' \in \Omega_S$. Therefore there exist (uniquely determined) an algebraic group morphism $\varphi: R_{K/\mathbb{Q}}\mathbf{G} \to \mathbf{H}$ and a homomorphism $\nu': \Lambda' \to \mathscr{Z}(\mathbf{H})$ such that $\delta(\lambda) = \nu'(\lambda) \cdot \varphi(R^0_{K/\mathbb{Q}}(\lambda))$ for all $\lambda \in \Lambda'$ (see Theorem 3.6(ii)). Furthermore, the morphism φ is defined over l. We set $\Lambda'' = \operatorname{Ker} \nu'$. Then the homomorphisms δ and $\varphi \circ R^0_{K/\mathbb{Q}}$ agree on Λ''. On the other hand, since $\overline{\delta(\Lambda)} = \mathbf{H}$, $\delta(\Lambda'') \subset \mathbf{H}^0$, and

Λ'' is of finite index in Λ (because the groups $\mathbf{H}/\mathbf{H}^0$ and $\mathscr{Z}(\mathbf{H}^0)$ are finite), we have $\overline{\delta(\Lambda'')} = \mathbf{H}^0$ and $\Lambda \subset \mathrm{Comm}_{\mathbf{G}}(\Lambda'')$. Therefore Lemma VII.5.1(b) implies that there is a homomorphism $v\colon \Lambda \to \mathscr{Z}_{\mathbf{H}}(\mathbf{H}^0)$ such that (*) holds. All that remains now is to observe that, since $\mathbf{H}$ is semisimple, the centralizer $\mathscr{Z}_{\mathbf{H}}(\mathbf{H}^0)$ is finite. □

(3.13) *Remarks.* (i) Since the set $\mathscr{T}(\mathbf{G})$ is finite, the condition "$\mathrm{rank}_S\, \mathbf{G} \geq 2$" frequently mentioned above is satisfied automatically whenever S is infinite.

(ii) Since the set $\mathscr{R}$ is infinite and $K = K(\mathscr{R})$, it follows from Remark (i) that $\Lambda := \mathbf{G}(K)$ satisfies the assumptions of all theorems in this section.

(iii) Let $\Lambda \in \Omega_S$. If $\mathrm{rank}_S\, \mathbf{G} > 0$, then, by virtue of Proposition I.3.2.10, $\bar{\Lambda} = \mathbf{G}$. Thus, Theorems 3.4 and 3.6 yield a description of all automorphisms of the group Λ in case $\mathrm{rank}_S\, \mathbf{G} \geq 2$.

(iv) There are numerous papers devoted to the study of "abstract homomorphisms" (for the bibliography until 1979, see [Me 1] and [Me 2], for recent papers, see [J-W-W] and [Weis 2]). In these papers, as a rule, purely algebraic methods are used. Roughly speaking, these are based on the study of the "geometry" of "abstract" algebraic groups. These methods yield a more or less complete investigation of the case of isotropic groups. However, for anisotropic groups it is extremely difficult to apply these methods. Furthermore, for groups with poor "geometry" (in particular, for groups of units of many skew fields) they seem to be completely inapplicable. We remark that

1) for automorphisms of the group $\Lambda = \mathbf{G}(K)$, Theorem 3.4 and 3.6 were proved earlier by the algebraic methods for the majority of the classical K-isotropic groups as well as for many K-anisotropic groups;
2) for $\mathbf{G}$ K-isotropic and the subgroups Λ containing $\mathbf{G}(K)^+$, Theorems 3.4 and 3.6 were essentially proved in [Bo-T 4];
3) in case $\mathrm{rank}_K\, \mathbf{G} \geq 2$ and $\mathbf{G}$ is almost K-simple, the theorems of this section were essentially proved in [Rag 7];
4) the results stated in this section can be deduced by the well known reduction from the finiteness of the congruence-kernel, which has been established for all K-isotropic and simply connected $\mathbf{G}$, and for some K-anisotropic $\mathbf{G}$ (see 2.13);
5) in case δ is an automorphism, S is finite, and Λ is an S-arithmetic subgroup, Theorem 3.4 can be deduced from the strong rigidity theorem (see §7 of Chapter VII).

(3.14) The rest of the present section will be devoted to the proof of Theorem 3.4(b) in the case where $\mathrm{char}\, K \neq 0$. First we shall make several reductions. Since $\Lambda \subset \Psi$ and $\mathscr{D}(\Psi) \subset \mathbf{G}(K)$ (see Lemma 2.1(i)), it follows that $\mathscr{D}(\Lambda) \subset \mathbf{G}(K)$. But $\overline{\delta(\mathscr{D}(\Lambda))} = \mathscr{D}(\overline{\delta(\Lambda)}) = \mathscr{D}(\mathbf{H}) = \mathbf{H}$. Therefore, replacing Λ by $\Lambda \cap \mathbf{G}(K)$, we may assume $\Lambda \subset \mathbf{G}(K)$. Since $\mathscr{Z}(\mathbf{H})$ is finite, $\mathrm{Ker}\, v$ is of finite index in Λ. So, replacing Λ by $\mathrm{Ker}\, v$, we may assume that $\delta = \eta \circ \sigma^0|_\Lambda$.

Suppose we are given two morphisms $\varphi, \varphi'\colon {}^\sigma\mathbf{G} \to \mathbf{H}$ with $\mathrm{Ad} \circ \varphi = \mathrm{Ad} \circ \varphi'$. Then $\varphi = \varphi'$, because $\mathbf{G}$ is connected and the kernel of the morphism $\mathrm{Ad}\colon \mathbf{H} \to \mathrm{Ad}\, \mathbf{H}$ is finite. On the other hand, the composition of a central isogeny and a special epimorphism is a special epimorphism (see I.1.4.14). Therefore, by Proposition I.1.4.11 ((iii) and (iv)), we can replace $\mathbf{H}$ by $\operatorname{Ad}\mathbf{H}$ and δ by $\operatorname{Ad}\circ\delta$ and assume that $\mathbf{H}$ is an adjoint group.

We may assume that S is finite. It then follows from Corollary I.3.2.9 and Proposition I.1.4.11(iv) that the subgroups $\operatorname{Ad}(\mathbf{G}(K(S)))$ and $(\operatorname{Ad}\mathbf{G})(K(S))$ are commensurable, and hence $\operatorname{Ad}\Lambda \in \Omega_S(\operatorname{Ad}\mathbf{G})$. The group $\mathbf{H}$ is adjoint. Therefore for a special l'-epimorphism $\eta\colon {}^\sigma\mathbf{G} \to \mathbf{H}$ there is a special l'-epimorphism $\eta'\colon {}^\sigma\operatorname{Ad}\mathbf{G} = \operatorname{Ad}({}^\sigma\mathbf{G}) \to \mathbf{H}$ such that $\eta = \operatorname{Ad}\circ\eta'$ (see Proposition I.1.4.11 and Section I.1.4.14). Furthermore, since for each $v \in \mathscr{R}$ the groups $\mathbf{G}$ and $\operatorname{Ad}\mathbf{G}$ have the same K_v-rank, we have $\operatorname{rank}_S \operatorname{Ad}\mathbf{G} = \operatorname{rank}_S \mathbf{G}$ (see Corollary I.1.4.6 and Proposition I.14.11 (iv)). Thus, replacing $\mathbf{G}$ by $\operatorname{Ad}\mathbf{G}$, Λ by $\operatorname{Ad}\Lambda$, and δ by $\delta\circ\operatorname{Ad}^{-1}$, we can assume that $\mathbf{G}$ is an adjoint group. Then $\mathbf{G} = R_{K'/K}\mathbf{G}'$, where K' is a finite separable field extension of K and $\mathbf{G}'$ is a connected adjoint absolutely almost simple K'-group (see I.1.7). Let $\{\sigma_1 = \operatorname{Id}, \sigma_2, \dots, \sigma_d\}$ be the set of all distinct embeddings of the field K' into the algebraic closure $\bar{K}$ of the field K which agree with the identity on K. We represent the group $\mathbf{G} = R_{K'/K}\mathbf{G}'$ in the form

$$\mathbf{G} = {}^{\sigma_1}\mathbf{G} \times \ldots \times {}^{\sigma_d}\mathbf{G}'$$

(see I.1.7). Let $p_i\colon \mathbf{G} \to {}^{\sigma_i}\mathbf{G}'$, $1 \le i \le d$, be the natural projection. Since $\mathbf{H}$ is absolutely almost simple, for any rational epimorphism $\varphi\colon {}^\sigma\mathbf{G} \to \mathbf{H}$ there exist i, $1 \le i \le d$, and a rational epimorphism $\varphi_i\colon {}^{\sigma\circ\sigma_i}\mathbf{G}' \to \mathbf{H}$ such that $\varphi = \varphi_i \circ {}^\sigma p_i$. Therefore, replacing $\mathbf{G}$ by $\mathbf{G}'$, Λ by $(R^0_{K'/K})^{-1}(\Lambda)$, and δ by $\delta\circ R^0_{K'/K}$, we may assume that $\mathbf{G}$ is absolutely almost simple, and hence η is an isogeny.

We set $\Gamma = \Lambda \cap \mathbf{G}(K(S))$. Since $\Lambda \subset \mathbf{G}(K)$, $S \not\subset \mathscr{T}$, and $\overline{\delta(\Lambda)} = \mathbf{H} = \mathbf{H}^0 = \mathscr{D}(\mathbf{H}^0)$, it follows from Lemma 3.2(ii) that $\overline{\delta(\Gamma)} = \mathbf{H}$. But, since $\Gamma \subset \Lambda$ and $\delta = \eta\circ\sigma^0|_\Lambda$, we have $\delta(\Gamma) = \eta(\sigma^0(\Gamma))$. Thus,

$$\overline{\eta(\sigma^0(\Gamma))} = \mathbf{H}. \tag{1}$$

It follows from Lemma I.3.1.3(b) that

$$\mathbf{G}(K) \subset \operatorname{Comm}_{\mathbf{G}}(\Gamma). \tag{2}$$

Since $\delta(\Gamma) \subset \delta(\Lambda) \subset \mathbf{H}(l)$ and $\mathbf{H}$ is an adjoint group, in view of Lemma VII.6.2(ii), (1) and (2) above imply that $\eta(\sigma^0(\mathbf{G}(K))) \subset \mathbf{H}(l)$. Suppose we are given a homomorphism $\bar\sigma\colon K \to l'$ and an l'-morphism $\bar\eta\colon {}^{\bar\sigma}\mathbf{G} \to \mathbf{H}$ such that $\eta(\sigma^0(\gamma)) = \bar\eta(\bar\sigma^0(\gamma))$ for all $\gamma \in \Gamma$. Since $\mathscr{Z}(\mathbf{H}) = \{e\}$, it then follows from (1), (2), and Lemma VII.5.1(b) that $\eta(\sigma^0(g)) = \bar\eta(\bar\sigma^0(g))$ for all $g \in \mathbf{G}(K)$. Thus, by setting $\delta(g) = \eta(\sigma^0(g))$, $g \in \mathbf{G}(K)$, with $\mathbf{G}(K)$ in place of Λ, we can assume that $\Lambda = \mathbf{G}(K)$.

Thus, we have reduced to the case in which $\mathbf{G}$ and $\mathbf{H}$ are absolutely almost simple and adjoint and $\Lambda = \mathbf{G}(K)$. This will be assumed below.

According to Proposition I.3.2.10, $\bar\Lambda = \mathbf{G}$, and hence $\overline{\sigma^0(\Lambda)} = {}^\sigma\mathbf{G}$. But $\delta(\Lambda) = \eta(\sigma^0(\Lambda)) \subset \mathbf{H}(l)$. Therefore if $\sigma(K) \subset l$, then η is defined over l (see Proposition I.0.11(ii)). Furthermore, since $\overline{\sigma^0(\Lambda)} = {}^\sigma\mathbf{G}$, for a fixed σ the isogeny η is uniquely determined via (*). Thus, it suffices to show that if, for a homomorphism $\sigma\colon K \to l'$ and an l'-isogeny $\eta\colon {}^\sigma\mathbf{G} \to \mathbf{H}$,

(**) $$\delta(g) = \eta(\sigma^0(g)) \text{ for all } g \in \Lambda = \mathbf{G}(K),$$

then $\sigma(K) \subset l$, and to show that σ is uniquely determined via (**).

(3.15) Let $\mathbf{F}$ be a connected non-commutative absolutely almost simple l-group and let $\mathbf{T}$ be a maximal torus in $\mathbf{F}$. As in I.1.1, for each $b \in \Phi(\mathbf{T}, \mathbf{F})$ we let $\mathbf{U}_b$ denote the one-parameter root subgroup associated with b and set $\mathfrak{U}_b = \mathrm{Lie}(\mathbf{U}_b) = \{v \in \mathrm{Lie}(\mathbf{F}) \mid \mathrm{Ad}\, t(v) = b(t)v \text{ for all } t \in \mathbf{T}\}$. It follows from the well known results on semisimple algebraic groups that the following conditions are equivalent (see [Bo-T 4, 3.1–3.8], [Bo-T 3, Proposition 2.13], and [Ste, Lemma 15]):

(i) any special isogeny $\beta\colon \mathbf{F} \to \mathbf{F}'$ is central;
(ii) any $\mathrm{Ad}\,\mathbf{F}$-invariant linear subspace of the Lie algebra $\mathrm{Lie}\,\mathbf{F}$ either contains $\sum_{b \in \Phi(\mathbf{T},\mathbf{F})} \mathfrak{U}_b$ or is contained in $\mathscr{Z}(\mathrm{Lie}(\mathbf{F})) \subset \mathrm{Lie}(\mathbf{T})$;
(iii) if $a, b \in \Phi(\mathbf{T}, \mathbf{F})$ and $a + b \in \Phi(\mathbf{T}, \mathbf{F})$, then $[\mathfrak{U}_a, \mathfrak{U}_b] = \mathfrak{U}_{a+b}$;
(iv) either $\mathrm{char}\, l \notin \{2; 3\}$ or $\mathrm{char}\, l = 2$ and $\mathbf{F}$ is not a group of type $\mathbb{B}_n$, $\mathbb{C}_n$, nor $\mathbb{F}_4$, or $\mathrm{char}\, l = 3$ and $\mathbf{F}$ is not of type $\mathbb{G}_2$.

The group $\mathbf{F}$ is said to be *standard* if conditions (i)–(iv) are satisfied; otherwise $\mathbf{F}$ is *non-standard*.

(3.16) Since $\Lambda = \mathbf{G}(K)$, for $\mathbf{G}$ K-isotropic, Theorem 3.4(b) is a straightforward consequence of Theorem I.1.8.1. But if $\mathrm{char}\, K > 0$ and $\mathbf{G}$ is anisotropic over K, then $\mathbf{G}$ is standard (see [Har 5], Corollary 1 to Theorem C). Therefore is suffices to consider the case where $\mathbf{G}$ is standard.

Since $\mathbf{H}$ is connected, non-commutative, and almost simple and $\overline{\delta(\Lambda)} = \mathbf{H}$, the subgroup $\delta(\Lambda) = \Lambda / \mathrm{Ker}\, \delta$ does not contain a normal subgroup of finite index whose commutator subgroup is finite. Therefore $\mathrm{Ker}\, \delta \subset \mathscr{Z}(\mathbf{G}) = \{e\}$ (see Corollary 2.13), i.e., δ is a monomorphism. But in case $\Lambda = \mathbf{G}(K)$, δ is a monomorphism, $\mathbf{G}$ and $\mathbf{H}$ are adjoint, and $\mathbf{G}$ is of type $\mathbb{A}_1$, Theorem 3.4 has been proved by Weisfeiler (see [Weis 1], Theorem 4.1) and is true not only for global fields K but also for arbitrary infinite fields K. Thus, we may assume that $\mathbf{G}$ is not of type $\mathbb{A}_1$ (this assumption will in fact only be used in the case where $\mathrm{char}\, K = 2$).

(3.17) Let $\mathbf{G}$ be standard. Then the special isogeny $\eta\colon {}^\sigma\mathbf{G} \to \mathbf{H}$ is central. Therefore $\mathrm{Tr}\,\mathrm{Ad}\, g = \mathrm{Tr}\,\mathrm{Ad}(\eta(g))$ for all $g \in {}^\sigma\mathbf{G}$ (see Corollary I.1.4.8). On the other hand, $\mathrm{Tr}\,\mathrm{Ad}(\sigma^0(g)) = \sigma^0(\mathrm{Tr}\,\mathrm{Ad}\,\mathbf{g})$ for all $g \in \mathbf{G}(K)$. Now from (**) we deduce that

(1) $$\mathrm{Tr}\,\mathrm{Ad}(\delta(g)) = \sigma^0(\mathrm{Tr}\,\mathrm{Ad}\, g) \text{ for all } g \in \Lambda = \mathbf{G}(K).$$

We denote by $K_{\mathbf{G}}$ the subfield of K generated by the set $\mathrm{Tr}\,\mathrm{Ad}\,\Lambda = \{\mathrm{Tr}\,\mathrm{Ad}\, g \mid g \in \mathbf{G}(K)\}$. It follows from (1) that $\sigma|_{K_{\mathbf{G}}}$ is uniquely defined via (**). Since the l-structure $\mathrm{Lie}(\mathbf{H})_l$ is $\mathrm{Ad}(\mathbf{H}(l))$-invariant and $\delta(\Lambda) \subset \mathbf{H}(l)$, it follows that $\mathrm{Tr}\,\mathrm{Ad}(\delta(\Lambda)) \subset l$ (see I.0.15). Therefore (1) implies that $\sigma(K_{\mathbf{G}}) \subset l$. Thus, to complete the proof of Theorem 3.4(b) it suffices to prove the following

(3.18) Lemma. *Let k be an infinite field, $\mathbf{F}$ a connected non-commutative absolutely almost simple k-group. Denote by $k_{\mathbf{F}}$ the subfield of k generated by the set $\{\operatorname{Tr} \operatorname{Ad} g \mid g \in \mathbf{F}(k)\}$. Suppose $\mathbf{F}$ is standard and adjoint. Assume in addition that either $\mathbf{F}$ is not of type $\mathbb{A}_1$ or $\operatorname{char} k \neq 2$. Then $k_{\mathbf{F}} = k$.*

(3.19) We shall prove Lemma 3.18 in 3.23. For proving this lemma we shall need some results on definition fields of linear groups. These results will also be used in Chapter IX. We first present the necessary definitions.

Let V be a finite-dimensional vector space over a field k, Δ a family of linear transformations of V. A subfield k_0 of k is called a *definition field* of the family Δ is there is a Δ-invariant k_0-structure on V, i.e., if there is a basis in V in which any transformation from Δ can be written as a matrix with entries in k_0. In this case we also say that Δ is *definable* over k_0.

A rational representation ρ of an algebraic group $\mathbf{F}$ is said to be *almost faithful* if $\operatorname{Ker} \rho$ is finite and is said to be *separable* if the morphism $\mathbf{F} \to \rho(\mathbf{F})$ is separable.

(3.20) Proposition. *Let k be an algebraically closed field, k_0 a subfield of k, $\mathbf{F}$ a reductive k-group, and Δ a Zariski dense subgroup of $\mathbf{F}$. Suppose that for some rational almost faithful separable absolutely irreducible representation $\rho\colon \mathbf{F} \to \mathbf{GL}(W)$ of the group $\mathbf{F}$ on a finite-dimensional vector space W over k we have $\operatorname{Tr} \rho(\Delta) \subset k_0$. Then the group $\operatorname{Ad} \Delta$ of linear transformations of the space $\operatorname{Lie}(\mathbf{F})$ is definable over k_0.*

Proof. We denote by $M \subset \operatorname{End}(W)$ (resp. $M_0 \subset \operatorname{End}(W)$) the k-linear (resp. k_0-linear) span of the set $\rho(\Delta)$. In other words, M (resp. M_0) is the set of linear combinations of vectors in $\rho(\Delta)$ with coefficients in k (resp. k_0). Since the representation ρ is absolutely irreducible and the subgroup Δ is Zariski dense in $\mathbf{F}$, the restriction of ρ to Δ is absolutely irreducible as well. Therefore by Burnside's theorem (see [Wae 2], §III) we have $M = \operatorname{End}(W)$. Let $m = \dim W$ and let $(h_1, \dots, h_{m^2}) \subset \rho(\Delta)$ be a basis of $M = \operatorname{End}(W)$. We denote by $(e_1, \dots, e_{m^2})$ the dual basis to $(h_1, \dots, h_{m^2})$ in $M = \operatorname{End}(W)$ relative to the bilinear form $(x, y) \to \operatorname{Tr} xy$. Since the basis $(h_1, \dots, h_{m^2})$ is contained in $\rho(\Delta)$ and $\operatorname{Tr} h_i h_j \in \operatorname{Tr} \rho(\Delta) \subset k_0$, it follows that

$$(1) \qquad (e_1, \dots, e_{m^2}) \subset M_0.$$

Clearly,

$$(2) \qquad M_0 \subset \sum_{1 \le i \le m^2} \operatorname{Tr}(h_i M_0) e_i.$$

Since $(h_1, \dots, h_{m^2}) \subset \rho(\Delta)$, $\rho(\Delta)$ is a subgroup, and $\operatorname{Tr} \rho(\Delta) \subset k_0$, it follows that $\operatorname{Tr}(h_i \rho(\Delta)) \subset k_0$, and hence $\operatorname{Tr}(h_i M_0) \subset k_0$ for all i, $1 \le i \le m^2$. Therefore (1) and (2) imply that M_0 is a k_0-structure on $M = \operatorname{End}(W)$.

We define the homomorphism f of the algebra M into the algebra $\operatorname{End}(M)$ by setting $f(x)y = xy$, $x, y \in M$. Since M_0 is the k_0-linear span of the subgroup

$\rho(\Delta)$, it follows that $xy \in M_0$ for all $x \in \rho(\Delta)$ and $y \in M_0$. On the other hand, M_0 is a k_0-structure on M. Therefore the group $\overline{f(\rho(\Delta))}$ of linear transformations of the space M is definable over k_0. We set $\mathbf{F}_0 = \overline{f(\rho(\Delta))}$ and observe that, since $\bar{\Delta} = \mathbf{F}$, we have $\mathbf{F}_0 = f(\rho(\mathbf{F}))$. Since $f(\rho(\Delta))$ is definable over k_0, according to I.0.11(i) we may assume that $\mathbf{F}_0$ is a k_0-subgroup in $\mathbf{GL}(M)$ and $f(\rho(\Delta)) \subset \mathbf{F}_0(k_0)$. But the k_0-structure $\mathrm{Lie}(\mathbf{F}_0)_{k_0}$ is $\mathrm{Ad}(\mathbf{F}_0(k_0))$-invariant (see I.o.15). Thus, the group $(\mathrm{Ad} \circ f \circ \rho)(\Delta)$ is definable over k_0.

Since $M = \mathrm{End}(W)$, for each $x \in M$, $x \neq 0$, there is $y \in M$ such that $xy \neq 0$. Therefore $\mathrm{Ker} f = \{0\}$. This implies that $f\colon M \to f(M)$ is an algebra isomorphism, and hence $f\colon \rho(\mathbf{F}) \to f(\rho(\mathbf{F})) = \mathbf{F}_0$ is an algebraic group isomorphism. But the isogeny $\rho\colon \mathbf{F} \to \rho(\mathbf{F})$ is separable. Therefore the isogeny $f \circ \rho\colon \mathbf{F} \to \mathbf{F}_0$ is separable, i.e., the differential $d(f \circ \rho)\colon \mathrm{Lie}(\mathbf{F}) \to \mathrm{Lie}(\mathbf{F}_0)$ is a Lie algebra isomorphism. From this and the definability of the group $\mathrm{Ad} \circ f \circ \rho)(\Delta)$ over k_0 we deduce that the group $\mathrm{Ad}\,\Delta$ is definable over k_0. □

(3.21) Lemma. *With* $\mathbf{F}$, $\mathbf{T}$, *and* $\mathfrak{U}_b$ *as in 3.15, we denote by* L *a minimal* $\mathrm{Ad}\,\mathbf{F}$*-invariant linear subspace of the Lie algebra* $\mathrm{Lie}(\mathbf{F})$ *containing* $\sum_{b \in \Phi(\mathbf{T},\mathbf{F})} \mathfrak{U}_b$ *and by* ρ *the restriction of the representation* Ad *to* L *(i.e.* $\rho(h)x = \mathrm{Ad}\,h(x)$, $h \in \mathbf{F}$, $x \in L$*). Suppose the group* $\mathbf{F}$ *is standard and adjoint. Then*

(i) *the representation* ρ *is almost faithful and absolutely irreducible;*
(ii) *if* $\mathbf{F}$ *is not of type* $\mathbb{A}_1$ *or if* $\mathrm{char}\, l \neq 2$, *then* $\rho\colon \mathbf{F} \to \rho(\mathbf{F})$ *is an algebraic group isomorphism, and hence the representation* ρ *is separable.*

Proof. (i) Pick $b \in \Phi(\mathbf{T},\mathbf{F})$. Since $\mathfrak{U}_b \subset L$, $\rho(t)x = b(t)x$ for all $t \in \mathbf{T}$ and $x \in \mathfrak{U}_b$, and b is a non-trivial character of the torus $\mathbf{T}$, it follows that $\mathbf{T} \not\subset \mathrm{Ker}\,\rho$. From this and the semisimplicity of $\mathbf{F}$ we deduce that ρ is almost faithful. Further, since $\mathbf{F}$ is adjoint (see [Bo-T 3], 2.13 and 2.23(a)), it follows that $\mathscr{Z}(\mathrm{Lie}(\mathbf{F})) = \{0\}$. But $\mathbf{F}$ is standard. Therefore any $\rho(F)$-invariant linear subspace of L is either $\{0\}$ or contains $\sum_{b \in \Phi(\mathbf{T},\mathbf{F})} \mathfrak{U}_b$ and hence L (see condition (ii) in 3.15). Thus, ρ is absolutely irreducible.

(ii) Pick $b \in \Phi(\mathbf{T},\mathbf{F})$. If $\mathbf{F}$ is not of type $\mathbb{A}_1$, then there is a root $a \in \Phi(\mathbf{T},\mathbf{F})$ such that $a+b \in \Phi(\mathbf{T},\mathbf{F})$, and hence $[\mathfrak{U}_b, \mathfrak{U}_a] \neq \{0\}$. If $\mathrm{char}\, l \neq 2$, then $[\mathfrak{U}_b, \mathfrak{U}_{-b}] \neq \{0\}$ because, as follows from the results of §13 of Chapter IV in [Bo 6], the group generated by $\mathfrak{U}_b \cup \mathfrak{U}_{-b}$ is isomorphic either to $\mathbf{SL}_2$ or $\mathbf{PGL}_2 \overset{\mathrm{def}}{=} \mathbf{GL}_2/\mathscr{Z}(\mathbf{GL}_2)$. In both cases $(d\rho)(\mathfrak{U}_b) \neq 0$, and hence the isogeny $\rho\colon \mathbf{F} \to \rho(\mathbf{F})$ is special. But the group $\mathbf{F}$ is standard and adjoint. Thus, $\rho\colon \mathbf{F} \to \rho(\mathbf{F})$ is an algebraic group isomorphism. □

(3.22) Proposition. *Let* k *be an algebraically closed field,* k_0 *a subfield of* k, $\mathbf{F}$ *a connected non-commutative absolutely almost simple* k*-group, and* Δ *a Zariski dense subgroup of* $\mathbf{F}$. *Suppose* $\mathbf{F}$ *is standard and adjoint. Assume in addition that either* $\mathbf{F}$ *is not of type* $\mathbb{A}_1$ *or* $\mathrm{char}\, k \neq 2$. *Then* k_0 *is a definition field of the group* $\mathrm{Ad}\,\Delta$ *if and only if* $k_0 \supset \mathrm{Tr}\,\mathrm{Ad}\,\Delta$.

Proof. With $\mathbf{T}$, $\mathfrak{U}_b$, L, and ρ as in the statement of Lemma 3.21, we have that $\mathrm{Tr}\,\mathrm{Ad}\,t = \mathrm{Tr}\,\rho(t)$ for all $t \in \mathbf{T}$, because $L \supset \sum_{b \in \Phi(\mathbf{T},\mathbf{F})} \mathfrak{U}_b$. But the set of semisimple elements of the group $\mathbf{F}$ is Zariski dense in $\mathbf{F}$ and coincides with $\bigcup_{g \in \mathbf{F}} g\mathbf{T}g^{-1}$ (see I.0.22 and I.0.24). Therefore $\operatorname{Tr} \operatorname{Ad} g = \operatorname{Tr} \rho(g)$ for all $g \in F$, and hence $\operatorname{Tr} \operatorname{Ad} \Delta = \operatorname{Tr} \rho(\Delta)$. According to Lemma 3.21 the representation ρ is almost faithful and separable. Thus, if $k_0 \supset \operatorname{Tr} \operatorname{Ad} \Delta = \operatorname{Tr} \rho(\Delta)$, then the group $\operatorname{Ad} \Delta$ is definable over k_0 (see Proposition 3.20). On the other hand, if k_0 is a definition field of $\operatorname{Ad} \Delta$, then $k_0 \supset \operatorname{Tr} \operatorname{Ad} \Delta$. □

(3.23) *Proof of Lemma 3.18.* Since $\mathbf{F}$ is adjoint, it follows from Proposition I.1.4.11(iv) that $\operatorname{Ad}\colon \mathbf{F} \to \operatorname{Ad} \mathbf{F}$ is a k-isomorphism. We identify $\mathbf{F}$ and $\operatorname{Ad} \mathbf{F}$ by Ad. It follows from Proposition 3.22 that the group $\mathbf{F}(k)$ of linear transformations of the space $\operatorname{Lie}(\mathbf{F})$ is definable over $k_{\mathbf{F}}$. But the subgroup $\mathbf{F}(k)$ is Zariski dense in $\mathbf{F}$ (see I.0.24). Therefore we can assume that $\mathbf{F}$ is a $k_{\mathbf{F}}$-subgroup in $\mathbf{GL}(\operatorname{Lie}(\mathbf{F}))$ and $\mathbf{F}(k) = \mathbf{F}(k_{\mathbf{F}})$ (see I.0.11(i)). Hence, it suffices to prove the following

(3.24) Proposition. *Let k be an infinite field, k_0 a subfield of k, $\mathbf{F}$ a connected non-unipotent k_0-group. If $k_0) \neq k$, then $\mathbf{F}(k_0 \neq \mathbf{F}(k)$.*

For proving this proposition we shall need Lemma 3.26. Before stating this lemma we make the following

(3.25) Definition. *A rational function $R \in k(x_1, \ldots, x_n)$ in n variables with coefficients in the field k is called* separable *if the following equivalent conditions hold:*

(a) $R \notin k(x_1^p, \ldots, x_n^p)$, *where* $p = \operatorname{char} k$;
(b) *there is i, $1 \leq i \leq n$, such that the rational function $\frac{\partial R}{\partial x_i}$ does not vanish.*

The equivalence of conditions (a) and (b) is a consequence of the following remarks.

(i) Let L be a field of characteristic p, L_0 a subfield of L, and Ω the set of all derivations D of the field L with $D(L_0) = 0$ (we recall that a map $D\colon L \to L$ is called a derivation of L if $D(x+y) = D(x)+D(y)$ and $D(xy) = xD(y)+yD(x)$ for all $x, y \in L$). Then the set $\{x \in L \mid D(x) = 0 \text{ for all } D \in \Omega\}$ coincides with the subfield of L generated by L_0 and $\{x^p \mid x \in L\}$ (see [Z-S], Chapter II, §17, the remark following Corollary 5).
(ii) The subfield of the field $k(x_1, \ldots, x_n)$ generated by k and $\{Q^p \mid Q \in k(x_1, \ldots, x_n)\}$ coincides with $k(x_1^p, \ldots, x_n^p)$.
(iii) Any derivation D of the field $k(x_1, \ldots, x_n)$ which maps k to 0 can be represented in the form

$$D = \sum_{1 \leq i \leq n} R_i \frac{\partial}{\partial x_i}, \text{ where } R_i \in k(x_1, \ldots, x_n).$$

We also remark that in case $\operatorname{char} k = 0$ the separability of the function R is equivalent to R not being constant.

(3.26) Lemma. *Let k be an infinite field, $n \in \mathbb{N}^+$, $\bar{k}$ its algebraic closure, $\mathbf{V}$ a proper algebraic subvariety in $\bar{k}^n$, k_0 a subfield of k different from k, and $R \in k_0(x_1, \ldots, x_n)$. Suppose R is separable. Then $R(k^n - \mathbf{V}) \not\subset k_0$.*

Proof. If k_0 is finite, then $R(k^n - \mathbf{V}) \not\subset k_0$, because k is infinite. Therefore we may assume that k_0 is infinite. Let i be an integer with $\partial R/\partial x_i \neq 0$. Then there exist $x_1, \dots, x_{i-1}, x_{i+1}, \dots, x_n \in k_0$ such that the rational function $Q(x) \stackrel{\text{def}}{=} R(x_1, \dots, x_{i-1}, x, x_{i+1}, \dots, x_n)$ is separable and the set $\{x \in k \mid (x_1, \dots, x_{i-1}, x, x_{i+1}, \dots, x_n) \in \mathbf{V}\}$ is finite. Therefore we can restrict ourselves to the case that $n = 1$, i.e., that R is a rational function in one variable y and $\mathbf{V}$ is finite. If k contains an element t transcendental over k_0, then $R(t+z) \notin k_0 \cup \{\infty\}$ for all $z \in k_0$, and hence there is an element z of the infinite field k_0 such that $t+z \notin \mathbf{V}$, $R(t+z) \neq \infty$, and $R(t+z) \notin k_0$. Thus, replacing k by a subextension of k_0, we may assume that k is a finite field extension of k_0. Let d be the degree of this extension and $\{\omega_1 = 1, \omega_2, \dots, \omega_d\}$ a basis of the field k over k_0. We represent $y \in k$ in the form $y = \sum_{1 \le j \le d} y_j \omega_j$, $y_j \in k_0$, and identify y with $(y_1, \dots, y_d) \in k_0^d$. Then the rational function R can be viewed as a rational map of the space k_0^d into itself, i.e.,

$$R(y) = \sum_{1 \le j \le d} R_j(y)\omega_j, \quad y = (y_1, \dots, y_d) \in k_0^d,$$

where R_j is a rational function on k_0^d defined over k_0. Let R' be the derivative of R and $y = (y_1, \dots, y_d)$ and $t = (t_1, \dots, t_d)$ two elements of $k = k_0^d$. Then

$$\begin{aligned} R(y+t) &= R(y) + R'(y)t + Q_y(t)t^2 = \\ &= R(y) + \sum_{1 \le j \le d} R'(y)\omega_j t_j + Q_y(t) \sum_{1 \le j,r \le d} \omega_j \omega_r t_j t_r, \end{aligned}$$

where $Q_y(t)$ is a rational function of t and $Q_y(0) \neq \infty$. It follows that $\partial R_j/\partial y_j = R'(y)$ for all j, $1 \le j \le d$. But $R' \neq 0$, because R is separable. Therefore $\partial R_j/\partial y_j \neq 0$, and hence $R_j \neq 0$ for all j, $1 \le j \le d$. From this, making use of the facts that k_0 is infinite and $\mathbf{V}$ is finite, we deduce the existence of an element $y = (y_1, \dots, y_d) \in k - \mathbf{V} = k_0^d - \mathbf{V}$ such that $R_d(y) \neq \infty$ and $R_d(y) \neq 0$. Since $d > 1$ (because $k \neq k_0$) and $\omega_1 = 1$, we then have that $R(y) \notin k_0$ which proves the lemma. □

(3.27) *Proof of Proposition 3.24.* Since the k_0-group $\mathbf{F}$ is not unipotent, it follows from I.0.22 that $\mathbf{F}$ contains a torus of positive dimension defined over k_0. It is known that $\mathbf{T}$ is unirational over k_0 (see [Bo 6], Chapter III, 8.13), i.e, the field $k_0(\mathbf{T})$ of rational functions on $\mathbf{T}$ defined over k_0 can be embedded for some $n \in \mathbb{N}^+$ in $k_0(x_1, \dots, x_n)$. Let $f_1, \dots, f_i$ be a system of generators of the algebra $k_0[\mathbf{T}] \subset k_0(\mathbf{T}) \subset k_0(x_1, \dots, x_n)$. We may assume that at least one of the rational functions $f_1, \dots, f_i$ is separable (because otherwise after the change of variables $x_j \to x_j^p$, $1 \le j \le n$, $p = \operatorname{char} k$, we can decrease p times the power of the numerator and denominator of each of the functions $f_1, \dots, f_i$). Let f_1 be separable and let $\mathbf{V}$ be the algebraic subvariety of points at which at least one of the functions $f_1, \dots, f_i$ is not defined or equals ∞. It then follows from Lemma 3.26 that there is $x \in k^n - \mathbf{V}$ such that $f_1(x) \notin k_0$. Now, viewing $f_1(t), \dots, f_i(t)$ as the coordinates of the point t of the affine variety $\mathbf{T}$, we see that the point $(f_1(x), \dots, f_i(x))$ belongs to $\mathbf{T}(k) - \mathbf{T}(k_0)$. Thus, $\mathbf{T}(k_0) \neq \mathbf{T}(k)$, and hence $\mathbf{F}(k_0) \neq \mathbf{F}(k)$. □

(3.28) *Remarks.*

(i) If k is a finite separable field extension of k_0, then the proof of Proposition 3.24 can be simplified by the observation that $R^0_{k/k_0}(\mathbf{T}(k)) = (R_{k/k_0}\mathbf{T})(k_0)$.

(ii) Slightly modifying the proof, one can show that under the assumptions of Proposition 3.24 the set $\mathbf{F}(k) - \mathbf{F}(k_0)$ is Zariski dense in $\mathbf{F}$.

Chapter IX. Arithmeticity

In this chapter the arithmeticity theorems will be stated and proved (see Sections 1 and 2). A series of consequences will be deduced from these theorems and, in particular, we shall show how they can be applied to strengthen some of the results of Chapters IV and VII (Sections 4 and 5). In Section 3 we prove some results on finite generation of lattices which are used in Section 1. Sections 6 and 7 are largely devoted to restating some results obtained in the present and previous chapters in terms of Lie group theory, the theory of symmetric spaces, and complex manifold theory.

As at the beginning of Chapter VIII we recall that for an l-group $\mathbf{F}$, a field homomorphism $\sigma\colon l \to l'$, and an l-morphism $p\colon \mathbf{F} \to \mathbf{F}'$ we have the associated l'-group ${}^{\sigma}\mathbf{F}$, the homomorphism $\sigma^0\colon \mathbf{F}(l) \to {}^{\sigma}\mathbf{F}(l')$, and the l'-morphism ${}^{\sigma}p\colon {}^{\sigma}\mathbf{F} \to {}^{\sigma}\mathbf{F}'$ (see I.1.7 and I.1.8).

1. Statement of the Arithmeticity Theorems

Let A be a finite non-empty set. For each $\alpha \in A$ choose a local field k_α and a connected nontrivial semisimple k_α-group $\mathbf{G}_\alpha$. We denote by $\mathbf{G}_\alpha^{\mathrm{is}} \subset \mathbf{G}_\alpha$ the subgroup which is the (almost direct) product of k_α-isotropic factors of $\mathbf{G}_\alpha$. We set $G = \prod_{\alpha\in A} \mathbf{G}_\alpha(k_\alpha)$ and $G^{\mathrm{is}} = \prod_{\alpha\in A} \mathbf{G}_\alpha^{\mathrm{is}}(k_\alpha) \subset G$. We remark that the factor groups $\mathbf{G}_\alpha(k_\alpha)/\mathbf{G}_\alpha^{\mathrm{is}}(k_\alpha)$ are compact (see Proposition I.2.3.8), and hence the factor group G/G^{is} is compact.

As in Section 5 of Chapter VII, $\operatorname{rank} G = \sum_{\alpha\in A} \operatorname{rank}_{k_\alpha} \mathbf{G}_\alpha$ denotes the rank of the group G.

(1.1) Let k be a local field, $\mathbf{F}$ a connected semisimple k-group, $\tilde{\mathbf{F}}$ the simply connected covering of $\mathbf{F}$, and $\pi\colon \tilde{\mathbf{F}} \to \mathbf{F}$ the central k-isogeny. We set $\mathbf{F}(k)^0 = \pi(\tilde{\mathbf{F}}(k))$. It follows from Theorem I.2.3.1 (a′), Proposition I.1.5.4(iv), and Proposition I.1.5.5 that $\mathbf{F}(k)^0 \cap \mathbf{F}^{\mathrm{is}}(k) = \mathbf{F}(k)^+$, where $\mathbf{F}^{\mathrm{is}} \subset \mathbf{F}$ denotes the product of the k-isotropic factors of $\mathbf{F}$. In particular, if $\mathbf{F}$ has no k-anisotropic factors, then $\mathbf{F}(k)^0 = \mathbf{F}(k)^+$.

Since the group of $\mathbb{C}$-rational points of any connected $\mathbb{C}$-group is connected, it follows that $\mathbf{F}(k)^0 = \mathbf{F}(k)$ for $k = \mathbb{C}$. Since the group of $\mathbb{R}$-rational points of any connected simply connected semisimple $\mathbb{R}$-group is connected (see Remark

I.2.3.1.2), for $k = \mathbb{R}$ the subgroup $\mathbf{F}(k)^0$ coincides with the identity component of the Lie group $\mathbf{F}(k)$.

It follows from Proposition I.2.3.4(i) that $\mathbf{F}(k)^0$ is a closed normal subgroup in $\mathbf{F}(k)$ and the factor group $\mathbf{F}(k)/\mathbf{F}(k)^0$ is compact, commutative and torsion. Furthermore, if $\operatorname{char} k = 0$, then the subgroup $\mathbf{F}(k)^0$ is open and of finite index in $\mathbf{F}(k)$.

Since π is a K-isogeny and the subgroup $\tilde{\mathbf{F}}(k)$ is Zariski dense in $\tilde{\mathbf{F}}$ (see Proposition I.2.5.3(ii)), the subgroup $\mathbf{F}(k)^0$ is Zariski dense in $\mathbf{F}(k)$. Therefore if $\mathbf{F} \neq \{e\}$, or equivalently, if $\dim \mathbf{F} > 0$, then the subgroup $\mathbf{F}(k)^0$ is infinite.

If k is a finite separable field extension of a local field l, then the standard properties of the restriction of scalars functor imply that $R^0_{k/l}(\mathbf{F}(k)^0) = (R_{k/l}\mathbf{F})(l)^0$ (see I.1.7).

(1.2) We set

$$G^0 = \prod_{\alpha \in A} \mathbf{G}_\alpha(k_\alpha)^0 \subset G.$$

From what we have mentioned in 1.1 above we deduce the following assertions:

(a) Let G^+ denote the subgroup $\prod_{\alpha \in A} \mathbf{G}_\alpha(k_\alpha)^+$ of G. Then $G^0 \cap G^{\text{is}} = G^+$. In particular, if for each $\alpha \in A$ the group $\mathbf{G}_\alpha$ has no k_α-anisotropic factors, then $G^0 = G^+$.

(b) If k_α is $\mathbb{R}$ or $\mathbb{C}$ for each $\alpha \in A$, then G^0 coincides with the identity component of the Lie group G.

(c) G^0 is a closed normal subgroup of G and G/G^0 is compact.

(d) If $\operatorname{char} k_\alpha = 0$ for each $\alpha \in A$, then the subgroup G^0 is open and of finite index in G.

(e) Suppose, that for each $\alpha \in A$ there is a central k_α-isogeny $p_\alpha : \mathbf{G}_\alpha \to \mathbf{G}'_\alpha$ of semisimple k_α-groups and $p = \prod_{\alpha \in A} p_\alpha$. Then $p(G^0) = G'^0 \overset{\text{def}}{=} \prod_{\alpha \in A} \mathbf{G}'_\alpha(k_\alpha)^0$.

(f) The subgroup G^0 is infinite.

(g) Suppose that for each $\alpha \in A$ the field k_α is a finite separable field extension of a local field l_α. We set $\mathbf{D}_\alpha = R_{k_\alpha/l_\alpha}\mathbf{G}_\alpha$. The isomorphisms $R^0_{k_\alpha/l_\alpha}$: $\mathbf{G}_\alpha(k_\alpha) \to \mathbf{D}_\alpha(l_\alpha)$ induce the isomorphism R^0: $G \to D \overset{\text{def}}{=} \prod_{\alpha \in A} \mathbf{D}_\alpha(l_\alpha)$. Then $R^0(G^0) = D^0 \overset{\text{def}}{=} \prod_{\alpha \in A} \mathbf{D}_\alpha(l_\alpha)^0$. Thus, roughly speaking, the subgroup G^0 is invariant under the application of the restriction of scalars functor to the groups $\mathbf{G}_\alpha$.

Definition. *Let F be a subgroup of G. We denote by $\overline{G^{\text{is}} \cdot F}$ the closure of the subgroup $G^{\text{is}} \cdot F$ in G. We shall say that F has* property (QD) *in G if the subgroup $G^0 \cap \overline{G^{\text{is}} \cdot F}$ is of finite index in G^0.*

Remarks.

(i) Obviously, if for each $\alpha \in A$ the group $\mathbf{G}_\alpha$ has no k_α-anisotropic factors, then any subgroup of G has property (QD) in G.

(ii) As was observed at the beginning of the present section, the factor group G/G^{is} is compact. Therefore if $\overline{G^{\text{is}} \cdot F}$ is open in G, then $\overline{G^{\text{is}} \cdot F}$ is of finite index in G, and hence F has property (QD) in G.

(iii) Clearly, if F_1 and F_2 are two commensurable subgroups of G, then the subgroups $\overline{G^{\mathrm{is}} \cdot F_1}$ and $\overline{G^{\mathrm{is}} \cdot F_2}$ are commensurable. Hence F_1 has property (QD) in G if and only if F_2 does.

(iv) Suppose that for each $\alpha \in A$, there is a k_α-group $\mathbf{G}'_\alpha$ and a central k_α-isogeny $p_\alpha\colon \mathbf{G}_\alpha \to \mathbf{G}'_\alpha$. We set $\mathbf{G}' = \prod_{\alpha \in A} \mathbf{G}'_\alpha(k_\alpha)$. The isogenies p_α induce the continuous homomorphism $p\colon G \to G'$. Then a subgroup F of G has property (QD) in G' if and only if $p(F)$ has property (QD) in G'. This is a consequence of assertion (e), the finiteness of the kernel $\operatorname{Ker} p$, and the fact that 1) in view of Proposition I.2.3.4(ii), p is a proper map, and hence, p maps closed sets to closed sets and that 2) in view of Corollary I.1.4.6(c), $p_\alpha(\mathbf{G}_\alpha^{\mathrm{is}})$ coincides with the product of k_α-isotropic factors of the group $\mathbf{G}'_\alpha$.

(v) Since $G^0 \cap G^{\mathrm{is}} = G^+$, the subgroup G^0 is closed in G. By virtue of Proposition I.2.3.1(b), the factor group G/G^+ is compact. It follows that the image of G^0 under the natural epimorphism $G \to G/G^{\mathrm{is}}$ is compact. But in a compact group the class of closed subgroups of finite index coincides with the one of open subgroups. Therefore for any closed subgroup H in G containing G^{is} the following are equivalent:
 (1) $G^0 \cap H$ is of finite index in G^0;
 (2) $G^0 \cap H$ is open in G^0.
 Hence, a subgroup F of G has property (QD) in G if and only if the subgroup $G^0 \cap \overline{G^{\mathrm{is}} \cdot F}$ is open in G^0.

(vi) It follows from assertion (d) and remark (v) above that, in case $\operatorname{char} k_\alpha = 0$ for each $\alpha \in A$, a subgroup F in G has property (QD) in G if and only if the closure of the subgroup $G^{\mathrm{is}} \cdot F$ in G is open in G.

(vii) Making use of the theorem on the algebraicity of a compact real linear group (see I.0.31) one can easily show that in case k_α is $\mathbb{R}$ or $\mathbb{C}$ for each $\alpha \in A$, the factor group G/G^{is} is connected. It then follows from (b) that in this case the following are equivalent:
 (1) a subgroup F in G has property (QD) in G;
 (2) the subgroup $G^{\mathrm{is}} \cdot F$ is dense in G.

(viii) It follows from (f) that if $\operatorname{rank} G = 0$, or equivalently if $G^{\mathrm{is}} = \{e\}$, then any subgroup of G with property (QD) is infinite.

(ix) Property (QD) is invariant under the application of the restriction of scalars functor to the groups $\mathbf{G}_\alpha$. More precisely, with D and $R^0\colon G \to D$ as in assertion (g), a subgroup F of G has property (QD) if and only if the subgroup $R^0(F)$ has property (QD) in D.

(1.3) As in Chapter VIII, let K be a global field, $\mathscr{R}$ the set of all (inequivalent) valuations of K, and $\mathscr{R}_\infty \subset \mathscr{R}$ the set of archimedean valuations. As before, the value of the valuation $v \in \mathscr{R}$ at $x \in K$ will be denoted by $|x|_v$ and K_v will denote the completion of the field K corresponding to the valuation $v \in \mathscr{R}$.

Decompose the group $\mathbf{G}_\alpha$ into an almost direct product of connected non-commutative almost k_α-simple subgroups $\mathbf{G}_{\alpha i}$, $i \in I_\alpha$.

Let $\mathbf{H}$ be a connected non-commutative absolutely almost simple K-group and suppose we are given $v \in \mathscr{R}$, $\alpha \in A$, $i \in I_\alpha$, and a continuous homomorphism $f\colon \mathbf{H}(K_v) \to \mathbf{G}_{\alpha i}(k_\alpha)$. We say that the homomorphism f belongs to the *class* Ψ_0, if there exist a closed subfield K'_v of the field K_v, a bicontinuous isomorphism $\omega\colon K'_v \to k_\alpha$, and, after identification of the fields K'_v and k_α by ω, a central k_α-isogeny $\tau\colon R_{K_v/K'_v}\mathbf{H} \to \mathbf{G}_{\alpha i}$ such that K_v is a finite separable field extension of K'_v and $f = \tau \circ R^0_{K_v/K'_v}$. We remark that if the group $\mathbf{G}_{\alpha i}$ is absolutely almost simple, then $K'_v = K_v$. Thus, in the case where $\mathbf{G}_{\alpha i}$ is absolutely almost simple, f belongs to the class Ψ_0 if there exist a continuous isomorphism $\omega\colon K_v \to k_\alpha$ and a central k_α-isogeny $\tau\colon {}^\omega\mathbf{H} \to \mathbf{G}_{\alpha i}$ such that $f(h) = \tau(\omega^0(h))$ for all $h \in \mathbf{H}(K_v)$.

Suppose B is finite set in $\mathscr{R}$. We shall say that the continuous homomorphism $\varphi\colon H_B \overset{\text{def}}{=} \prod_{v\in B} \mathbf{H}(K_v) \to G$ belongs to the *class* Ψ if there exist a map $\delta\colon B \to A$, a bijection $\nu\colon B \to I \overset{\text{def}}{=} \bigcup_{\alpha\in A} I_\alpha$, and continuous homomorphisms $f_v\colon H(K_v) \to G_{\delta(v)\nu(v)}(k_{\delta(v)})$ such that $\nu(v) \in I_{\delta(v)}$, f_v belongs to the class Ψ_0 and $\varphi = \prod_{v\in S} f_v$.

Remarks.

(i) In the case in which the groups $\mathbf{G}_\alpha$ are absolutely almost simple the homomorphism $\varphi\colon H_B \to G$ belongs to the class Ψ if there exist a bijective map $\delta\colon B \to A$, continuous isomorphisms $\omega_v\colon K_v \to k_{\delta(v)}$, and central $k_{\delta(v)}$-isogenies $\tau_v\colon {}^{\omega_v}\mathbf{H} \to \mathbf{G}_{\delta(v)}$ such that $\varphi = \prod_{v\in B}(\tau_v \circ \omega_v^0)$.

(ii) It follows from the definition of the class Ψ and Proposition I.2.3.4 that for any continuous homomorphism $\varphi\colon H_B \to G$ belonging to the class Ψ, the kernel $\operatorname{Ker}\varphi$ is finite, the subgroup $\varphi(H_B)$ is closed and normal in G, the factor group $G/\varphi(H_B)$ is compact, and $H_B/\operatorname{Ker}\varphi \to H_B$ is an isomorphism of topological groups. Making use of the results in 1.8, 1.5, and 2.3 of Chapter I, one could show that inverse assertion is true in the case in which the groups $\mathbf{G}_\alpha$ have no k_α-anisotropic factors for all $\alpha \in A$, i.e., in this case the class Ψ coincides with the set of continuous homomorphisms $\varphi\colon H_B \to G$ with finite kernel, the subgroup $\varphi(H_B)$ is closed and normal in G, and the factor group $G/\varphi(H_B)$ is compact. If for each $\alpha \in A$, the field k_α is $\mathbb{R}$ or $\mathbb{C}$, then the continuous homomorphism $\varphi\colon H_B \to G$ belongs to Ψ if and only if $\operatorname{Ker}\varphi$ is finite and $\varphi(H_B)$ contains the identity component of the Lie group G. (This assertion is an easy consequence of the regularity of continuous homomorphisms of connected semisimple real Lie groups, and can also easily be deduced from Proposition I.1.4.13.)

(iii) Suppose that either the group $\mathbf{H}$ is adjoint or for each $\alpha \in A$ the group $\mathbf{G}_\alpha$ is simply connected. Then the following assertions are true:

(a) if $\varphi\colon H_B \to G$ belongs to Ψ, then φ is a topological group isomorphism (this follows immediately from the definitions);

(b) if for each $\alpha \in A$ the group $\mathbf{G}_\alpha$ has no k_α-anisotropic factors, then a homomorphism $\varphi\colon H_B \to G$ belongs to Ψ if and only if φ is a topological group isomorphism (this follows from the description of the class Ψ given in remark (ii)).

(iv) With $\mathbf{G}'_\alpha$, p_α, G', and p as in remark (v) of 1.2 the following two assertions hold:

(a) if a homomorphism $\varphi\colon H_B \to G$ belongs to Ψ, then the homomorphism $p \circ \varphi\colon H_B \to G'$ belongs to Ψ;

(b) if a homomorphism $\varphi'\colon H_B \to G'$ belongs to Ψ and the group H is simply connected, then there is a homomorphism $\varphi\colon H_B \to G$ in the class Ψ such that $\varphi' = p \circ \varphi$.

Assertion (a) follows easily from the fact that the composition of central isogenies of reductive groups is central (see I.1.4.7). For proving assertion (b) we have to use Proposition I.1.4.11(ii). We leave further details of the proofs of (a) and (b) as elementary exercises.

(v) Let $\pi\colon \tilde{\mathbf{H}} \to \mathbf{H}$ be a central K-isogeny of connected K-groups. The isogeny π induces the continuous homomorphism $\pi_B\colon \tilde{H}_B \stackrel{\text{def}}{=} \prod_{v\in B} \tilde{\mathbf{H}}(K_v) \to H_B$. If a homomorphism $\varphi\colon H_B \to G$ belongs to Ψ, then so does the homomorphism $\varphi \circ \pi_B\colon \tilde{H}_B \to G$.

(vi) With D and $R^0\colon G \to D$ as in (g) of 1.2, an isomorphism $\varphi\colon H_B \to G$ belongs to Ψ if and only if $R^0 \circ \varphi\colon H_B \to D$ belongs to Ψ.

(vii) If there is a continuous homomorphism $\varphi\colon H_B \to G$ in the class Ψ, then $\operatorname{rank} G = \sum_{v\in B} \operatorname{rank}_{K_v} \mathbf{H}$. This is a consequence of the fact that

$$\operatorname{rank} G = \sum_{\alpha\in A} \operatorname{rank}_{k_\alpha} \mathbf{G}_\alpha = \sum_{\alpha\in A}\sum_{i\in I_\alpha} \operatorname{rank}_{k_\alpha} \mathbf{G}_{\alpha i},$$

and

$$\operatorname{rank}_{K_v} \mathbf{H} = \operatorname{rank}_{K'_v} R_{K_v/K'_v}\mathbf{H} = \operatorname{rank}_{k_{\delta(v)}} \mathbf{G}_{\delta(v)\nu(v)}$$

(see Corollary I.1.4.6(a) and I.1.7).

(1.4) With K, $\mathscr{R}$, $\mathscr{R}_\infty$, $|x|_v$, K_v, and $\mathbf{H}$ as in 1.3, we set $\mathscr{T} = \mathscr{T}(\mathbf{H}) = \{v \in \mathscr{R} \mid$ the group $\mathbf{H}$ is anisotropic over K_v, or equivalently, $\mathbf{H}(K_v)$ is compact$\}$ and consider a finite set $S \subset \mathscr{R}$ with $S \supset \mathscr{R}_\infty - \mathscr{T}$. As in I.0.32, for $S \subset \mathscr{R}$ we set $K(S) = \{x \in K \mid |x|_v \le 1 \mid \text{for all } v \in \mathscr{R} - \mathscr{R}_\infty - S\}$, i.e. $K(S)$ is the ring of S-integers of the field K.

With $\mathbf{H}(K(S))$ as in I.3.1, we set $H_S = \prod_{v\in S} \mathbf{H}(K_v)$ and identify the group $\mathbf{H}(K(S))$ with its image under the diagonal embedding in H_S. It then follows from Theorem I.3.2.5 that $\mathbf{H}(K(S))$ is a lattice in H_S. Since $\mathbf{H}$ is absolutely almost simple, the lattice $\mathbf{H}(K(S))$ is irreducible.

Suppose we are given a homomorphism $\varphi\colon H_S \to G$ belonging to the class Ψ. Since $\mathbf{H}(K(S))$ is an irreducible lattice in H_S and $\operatorname{Ker}\varphi$ is finite (see remark (ii) in 1.3), the subgroup $\varphi(H_S)$ is closed and normal in G, the factor group $G/\varphi(H_S)$ is compact, and $H_S/\operatorname{Ker}\varphi \to \varphi(H_S)$ is an isomorphism of topological groups, it follows that $\varphi(\mathbf{H}(K(S)))$ is an irreducible lattice in G. We now provide the definition of an arithmetic lattice.

Definition. *We call an irreducible lattice $\Gamma \subset G$* arithmetic *if there exist K, $\mathbf{H}$, S, as above, and a continuous homomorphism $\varphi\colon \prod_{v\in S} \mathbf{H}(K_v) \to G$ belonging to the class Ψ such that the subgroup $\varphi(\mathbf{H}(K(S)))$ is commensurable with Γ.*

We remark that arithmetic lattices are sometimes called S-arithmetic.

(1.5) For the case of $\mathbb{R}$-groups we provide one other definition of an arithmetic lattice (for the equivalence of this definition and the definition above in 1.4, see remark (ii) in 1.6).

Definition. *Let* $\mathbf{G}$ *be a connected semisimple* $\mathbb{R}$*-group,* $\mathbf{G}^{\text{is}} \subset \mathbf{G}$ *the (almost direct) product of* $\mathbb{R}$*-isotropic factors of the group* $\mathbf{G}$*, and* Γ_0 *an irreducible lattice in* $\mathbf{G}(\mathbb{R})$ *with* $\mathbf{G}^{\text{is}} \cdot \Gamma_0$ *dense in* $\mathbf{G}(\mathbb{R})$*. We call the lattice* Γ_0 arithmetic *if there exist a connected non-commutative almost* $\mathbb{Q}$*-simple* $\mathbb{Q}$*-group* $\mathbf{F}$ *and an* $\mathbb{R}$*-epimorphism* $\tau\colon \mathbf{F} \to \mathbf{G}$ *such that the Lie group* $(\operatorname{Ker}\tau)(\mathbb{R})$ *is compact and the subgroups* $\tau(\mathbf{F}(\mathbb{Z}))$ *and* Γ_0 *are commensurable.*

In the case where all the fields k_α are archimedean, we mention another description of arithmetic lattices. Let $K \subset \mathbb{C}$ be a finite field extension of $\mathbb{Q}$. An embedding $\sigma\colon K \to \mathbb{C}$ is called *real* if $\sigma(K) \subset \mathbb{R}$; otherwise it is called *imaginary*. Two embeddings σ and σ' of the field K in $\mathbb{C}$ are called *equivalent* if $\sigma'(x) = \overline{\sigma(x)}$, where the bar denotes complex conjugation. Choose a representative in each equivalence class of embeddings of K in $\mathbb{C}$. Let $\mathbb{R}'_\infty = \{\sigma_1 = \mathrm{Id}, \ldots, \sigma_i\}$ be the set of these representatives. For each $\sigma \in \mathscr{R}'_\infty$, we set $k_\sigma = \mathbb{R}$ if σ is real and $k_\sigma = \mathbb{C}$ if σ is imaginary. Then (σ, k_σ) is the completion of the field K. As was observed in I.0.32, the set of equivalence classes of valuations of K can be naturally identified with the set of equivalence classes of completions of K. Therefore we can identify $\mathscr{R}'_\infty$ with the set $\mathscr{R}_\infty$ of archimedean valuations of the field K. Furthermore, if $v \in \mathscr{R}_\infty$ is a valuation corresponding to an embedding $\sigma \in \mathscr{R}'_\infty$, then σ extends to a continuous isomorphism of the field K_v onto the field k_σ.

Now let $\mathbf{H}$ be a connected non-commutative almost K-simple K-group. We set $\mathscr{T}' = \{\sigma \in \mathscr{R}'_\infty \mid \sigma \text{ is real and the group } {}^\sigma\mathbf{H}(\mathbb{R}) \text{ is compact}\}$. Suppose we are given a subset $S \subset \mathscr{R}'_\infty$ with $S \supset \mathscr{R}'_\infty - \mathscr{T}'$. We set $H_S = \prod_{\sigma \in S} {}^\sigma\mathbf{H}(k_\sigma)$. Denote by J the ring of integers of the field K and identify the group $\mathbf{H}(J)$ with its image under the embedding $\prod_{\sigma \in S} \sigma^0\colon \mathbf{H}(J) \to H_S$. Then the following assertion is true.

(*) *Suppose that for each* $\alpha \in A$ *the field* k_α *is archimedean, let* $\varphi\colon H_S \to G$ *be a continuous homomorphism such that* $\operatorname{Ker}\varphi$ *is finite and* $\varphi(H_S)$ *contains the identity component of the Lie group* G*. Then*

(a) $\varphi(\mathbf{H}(J))$ *is an irreducible lattice in* G*;*

(b) *for any irreducible arithmetic lattice* $\Gamma \subset G$ *there exist* $\mathbf{H}$*,* S*, and* φ *as above such that* Γ *is commensurable with* $\varphi(\mathbf{H}(J))$*. Furthermore, we may assume that* $\mathbf{H}$ *is absolutely almost simple.*

Assertion (*) can easily be deduced from the equality $J = K(\mathscr{R}_\infty)$ and the natural identification of $\mathscr{R}'_\infty$ with $\mathscr{R}_\infty$ which was mentioned above. We remark that the transition in (a) to the case where $\mathbf{H}$ is absolutely almost simple is realized with the help of the restriction of scalars functor. In this context we have to use Lemma I.3.1.4 and the arguments of remarks (i) and (ii) in 1.6 below. In addition, in (b) we have to use the description of the class Ψ given in remark (ii) of 1.3.

The following assertion is a particular case of (*).

(**) *Let* k *be* $\mathbb{R}$ *or* $\mathbb{C}$*,* $\mathbf{G}$ *a connected absolutely almost simple* k*-isotropic* k*-group, and* Γ_0 *a lattice in* $\mathbf{G}(k)$*. Then the following are equivalent*

(a) The lattice Γ_0 *is arithmetic.*

(b) There exist a finite field extension $K \subset k$ *of* $\mathbb{Q}$, *a connected noncommutative absolutely almost simple* K*-group* $\mathbf{H}$, *and a continuous homomorphism* $\varphi\colon \mathbf{H}(k) \to \mathbf{G}(k)$ *such that*

1) K *is dense in* k;
2) $\sigma(K) \subset \mathbb{R}$ *and the group* ${}^\sigma\mathbf{H}(\mathbb{R})$ *is compact for all embeddings* $\sigma\colon K \to \mathbb{C}$ *inequivalent to the identity.*
3) $\operatorname{Ker}\sigma$ *is finite and* $\varphi(\mathbf{H}(k))$ *contains the identity component of the Lie group* G;
4) *the subgroups* $\varphi(\mathbf{H}(J))$ *and* Γ_0 *are commensurable, where* J *is the ring of integers of the field* K.

(1.6) *Remarks.*

(i) We may assume that the group $\mathbf{H}$ in Definition 1.4 is simply connected. Indeed, let $\tilde{\mathbf{H}}$ be the simply connected covering of $\mathbf{H}$, $\pi\colon \tilde{\mathbf{H}} \to \mathbf{H}$ the central K-isogeny, and $\varphi\colon \prod_{v\in S} \mathbf{H}(K_v) \to G$ a homomorphism in the class Ψ such that the subgroup $\varphi(H(K(S)))$ is commensurable with Γ. Then 1) the homomorphism $\varphi \circ \pi_S\colon \prod_{v\in S} \tilde{\mathbf{H}}(K_v) \to G$ belongs to Ψ, where $\pi_S\colon \prod_{v\in S} \tilde{\mathbf{H}}(K_v) \to \prod_{v\in S} \mathbf{H}(K_v)$ is induced from π (see remark (v) in 1.3); 2) since the subgroups $\pi(\tilde{\mathbf{H}}(K(S)))$ and $\mathbf{H}(K(S))$ are commensurable (see Corollary I.3.2.9), it follows that the subgroup $(\varphi \circ \pi_S)(\tilde{\mathbf{H}}(K(S)))$ is commensurable with Γ. Also notice that we can similarly assume that the group $\mathbf{F}$ in Definition 1.5 is simply-connected.

(ii) Definitions 1.4 and 1.5 are equivalent in the following sense. Suppose that k_α is $\mathbb{R}$ or $\mathbb{C}$ for each $\alpha \in A$. We set $\hat{\mathbf{G}}_\alpha = \mathbf{G}_\alpha$ if $k_\alpha = \mathbb{R}$ and $\hat{\mathbf{G}}_\alpha = R_{\mathbb{C}/\mathbb{R}}\mathbf{G}_\alpha$ if $k_\alpha = \mathbb{C}$. Let $\mathbf{G}$ by the $\mathbb{R}$-group $\prod_{\alpha\in A} \hat{\mathbf{G}}_\alpha$ and identify the groups G and $\mathbf{G}(\mathbb{R})$ by $R^0_{\mathbb{C}/\mathbb{R}}$. Then an irreducible lattice $\Gamma \subset G$ is arithmetic in the sense of Definition 1.4 if and only if Γ is arithmetic in the sense of Definition 1.5. This equivalence can easily be deduced from Lemma I.3.1.4 if, assuming that the $\mathbb{Q}$-group $\mathbf{F}$ is simply connected, we represent $\mathbf{F}$ in the form $R_{K/\mathbb{Q}}\mathbf{H}$, where K is a finite field extension of $\mathbb{Q}$ and $\mathbf{H}$ is a connected non-commutative absolutely almost simple K-group.

(iii) If in Definition 1.5 the lattice Γ_0 is not cocompact, then $\mathbf{G}$ has no $\mathbb{R}$-anisotropic factors (i.e. $\mathbf{G} = \mathbf{G}^{\text{is}}$) and the epimorphism τ is an isogeny. Indeed, let Γ_0 be non-cocompact. Then, since $\tau(\mathbf{F}(\mathbb{Z}))$ and Γ_0 are commensurable, the quotient space $\mathbf{F}(\mathbb{Z}) \setminus \mathbf{F}(\mathbb{R})$ is non-compact. Therefore $\mathbf{F}$ is isotropic over $\mathbb{Q}$ (see Theorem I.3.2.8(b)). From this and the almost $\mathbb{Q}$-simplicity of $\mathbf{F}$ we easily deduce that

1) $\mathbf{F}$, and hence $\mathbf{G}$, have no $\mathbb{R}$-anisotropic factors;
2) every infinite normal $\mathbb{R}$-subgroup of $\mathbf{F}$ is isotropic over $\mathbb{R}$.

On the other hand, the subgroup $\operatorname{Ker}\tau$ is defined over $\mathbb{R}$ and is normal in $\mathbf{F}$. Therefore $\operatorname{Ker}\tau$ is either finite or isotropic over $\mathbb{R}$, and hence the group $(\operatorname{Ker}\tau)(\mathbb{R})$ is non-compact. But $(\operatorname{Ker}\tau)(\mathbb{R})$ is compact. Thus, $\operatorname{Ker}\tau$ is finite, i.e., τ is an isogeny.

(iv) If $\operatorname{rank} G > 0$ and an irreducible lattice $\Gamma \subset G$ is arithmetic, then Γ has property (QD) in G. Indeed, according to remark (i) we may assume that the group $\mathbf{H}$ in the definition of 1.4 is simply connected. It then follows from the strong approximation theorem (see II.6.8) that the subgroup $\mathbf{H}(K(S))$ under the diagonal embedding is dense in $\prod_{v \in \mathscr{T}} \mathbf{H}(K_v)$ whenever $S \not\subset \mathscr{T} \overset{\text{def}}{=} \{v \in \mathscr{R} \mid \mathbf{H}$ is anisotropic over $K_v\}$. Thus, in view of remark (ii) in 1.2, the subgroup $\mathbf{H}(K(S))$ has property (QD) in $\prod_{v \in S} \mathbf{H}(K_v)$. All that remains now is to make use of remarks (iii) and (iv) in 1.2.

(v) With $\mathbf{G}'_\alpha$, p_α, G', and p as in remark (iv) of 1.2, the following conditions are equivalent:

(1) Γ is an arithmetic lattice in G;
(2) $p(\Gamma)$ is an arithmetic lattice in G'.

The implication (1) $\Rightarrow$ (2) is a straightforward consequence of remark (iv)(a) in 1.3; (2) $\Rightarrow$ (1) is a consequence of remark (iv)(b) in 1.2 and the fact that in the definition of 1.4 we can assume that the group $\mathbf{H}$ is simply-connected (see remark (i)).

(vi) The arithmeticity of a lattice Γ is invariant relative to the application of the restriction of scalars functor to the groups $\mathbf{G}_\alpha$. More precisely, with D and $R^0\colon G \to D$ as in (g) of 1.2, Γ is an arithmetic lattice in G if and only if $R^0(\Gamma)$ is an arithmetic lattice in D.

(vii) It follows from remark (ii) in 1.3 that any homomorphism in the class Ψ transfers cocompact lattices to cocompact ones and non-cocompact lattices to non-cocompact ones. But, with notation of 1.4, the lattice $\mathbf{H}(K(S)) \subset H_S$ is cocompact if and only if $\mathbf{H}$ is anisotropic over K (see Theorem I.3.2.4(b)). Thus, the lattice Γ in the definition of 1.4 is cocompact if and only if $\mathbf{H}$ is anisotropic over K.

(viii) It is known (see [Har 5], §3, Korollar 1) that only groups of type $\mathbb{A}$ have anisotropic forms over global fields of positive characteristic. From this and remark (vii) we deduce that if the set $\{\alpha \in A \mid \operatorname{char} k_\alpha > 0$ and at least one of almost simple factors of the group $\mathbf{G}_\alpha$ is not of type $\mathbb{A}\}$ is not empty, then G does not contain cocompact irreducible arithmetic lattices.

(1.7) *Examples of arithmetic lattices.*

(i) The group $\mathbf{SL}_n(\mathbb{Z})$ is a non-cocompact arithmetic lattice in $\mathbf{SL}_n(\mathbb{R})$.

(ii) Let p be a prime and $n \in \mathbb{N}^+$, $n \geq 2$. We denote by $\mathbb{Z}[1/p]$ the subring of the field $\mathbb{Q}$ generated by $1/p$. Identify the group $\Gamma = \mathbf{SL}_n(\mathbb{Z}[1/p])$ with its image under the diagonal embedding in $G = \mathbf{SL}_n(\mathbb{R}) \times \mathbf{SL}_n(\mathbb{Q}_p)$. Then Γ is an irreducible non-cocompact arithmetic lattice in G.

(iii) We provide a generalization of example (ii). Let $B = \{p_1, \ldots, p_i\}$ be a finite set of primes and $n \in \mathbb{N}^+$, $n \geq 2$. We denote by $\mathbb{Q}(B)$ the subring of the field $\mathbb{Q}$ generated by the elements $1/p_1, \ldots, 1/p_i$. Identify the group $\Gamma = \mathbf{SL}_n(\mathbb{Q}(B))$ with its image under the diagonal embedding in $G = \mathbf{SL}_n(\mathbb{R}) \times \mathbf{SL}_n(\mathbb{Q}_{p_1}) \times \ldots \times \mathbf{SL}_n(\mathbb{Q}_{p_i})$. Then Γ is an irreducible non-cocompact arithmetic lattice in G.

(iv) Let us denote by $F_q((t))$ the field of formal power series in one variable t over the finite field F_q and by $F_q[t^{-1}] \subset F_q(t))$ the polynomial ring in one variable t^{-1}. Then $\mathbf{SL}_n(F_q[t^{-1}])$ is a non-cocompact arithmetic lattice in $\mathbf{SL}_n(F_q((t)))$.

(v) Let K be a finite field extension of $\mathbb{Q}$, J the ring of integers of the field K, and $n \in \mathbb{N}^+$, $n \geq 2$. With $\mathscr{R}'_\infty$ and k_σ as in 1.5, we identify the group $\Gamma = \mathbf{SL}_n(J)$ with its image under the diagonal embedding in $G = \prod_{\sigma \in \mathscr{R}'_\infty} \mathbf{SL}_n(k_\sigma)$. Then Γ is an irreducible non-cocompact arithmetic lattice in G. In particular,
   (a) if $K = \mathbb{Q}(\sqrt{-d})$, $d \in \mathbb{N}^+$, is an imaginary quadratic field extension of $\mathbb{Q}$, then $\mathbf{SL}_n(J)$ is a non-cocompact arithmetic lattice in $\mathbf{SL}_n(\mathbb{C})$;
   (b) if $K = \mathbb{Q}(\sqrt{d})$, $d \in \mathbb{N}^+$, is a real quadratic field extension of $\mathbb{Q}$, then $\mathbf{SL}_n(J)$ is an irreducible non-cocompact arithmetic lattice in $\mathbf{SL}_n(\mathbb{R}) \times \mathbf{SL}_n(\mathbb{R})$;
   (c) if K is a cubic field extension of $\mathbb{Q}$ with one real and one imaginary embedding in $\mathbb{C}$, e.g., if $K = \mathbb{Q}(\sqrt[3]{2})$, then $\mathbf{SL}_n(J)$ is an irreducible non-cocompact arithmetic lattice in $\mathbf{SL}_n(\mathbb{R}) \times \mathbf{SL}_n(\mathbb{C})$.

(vi) Let $n \geq 3$, $K \subset \mathbb{C}$ a finite field extension of $\mathbb{Q}$, and $\Phi = \sum_{1 \leq i,j \leq n} a_{ij} x_i x_j$ a nondegenerate quadratic form in n variables with coefficients in K (as usual, $a_{ij} = a_{ji}$). With $\mathscr{R}'_\infty$ and k_σ as in 1.5, we set for $\sigma \in \mathscr{R}'_\infty$, ${}^\sigma\Phi = \sum_{1 \leq i,j \leq n} \sigma(a_i) x_i x_j$. We further set $\mathscr{T}' = \{\sigma \in \mathscr{R}'_\infty \mid k_\sigma = \mathbb{R}$ and the form ${}^\sigma\Phi$ is either positive definite or negative definite$\}$ and choose a subset $S \subset \mathscr{R}'_\infty$ with $S \supset \mathscr{R}'_\infty - \mathscr{T}'$. We denote by $\mathbf{SO}_\Phi$ the group of unimodular matrices preserving the form Φ. This group is a K-subgroup of $\mathbf{SL}_n$. Denote by J the ring of integers of the field K and identify the group $\mathbf{SO}_\Phi(J)$ with its image under the embedding $\prod_{\sigma \in S} \sigma^0 \colon \mathbf{SO}_\Phi(J) \to \mathbf{SO}_{\Phi,S} \stackrel{\text{def}}{=} \prod_{\sigma \in S} \mathbf{SO}_{\sigma_\Phi}(k_\sigma)$. Suppose
   1) $\mathscr{T}' \neq \mathscr{R}'_\infty$ and 2) the group $\mathbf{SO}_\Phi$ is almost K-simple

   (it is well known that this condition is equivalent to that of either $n \neq 4$ or $n = 4$ and the discriminant of the form Φ is not a square in K.) Then $\mathbf{SO}_\Phi(J)$ is an irreducible arithmetic lattice in $\mathbf{SO}_{\Phi,S}$. This lattice is non-cocompact if Φ represents 0 over K and otherwise is cocompact.

   We now provide some examples which arise as particular cases of the constructions described above. In all these examples we assume $n \geq 3$ and denote by $\mathbb{Q}(\beta)$ the field extension of $\mathbb{Q}$ generated by $\mathbb{Q}$ and an additional number β.
   (1) Let $\Phi = x_1^2 + \ldots + x_{n-1}^2 + x_n^2$. Then $\mathbf{SO}_\Phi(\mathbb{Z})$ is a non-cocompact arithmetic lattice in $\mathbf{SO}_\Phi(\mathbb{R})$.
   (2) Let $\Phi = x_1^2 + \ldots + x_{n-1}^2 + 2x_n^2$, $n \geq 4$, and let $k = \mathbb{Q}(\sqrt{-d})$, $d \in \mathbb{N}^+$, be an imaginary quadratic field extension of $\mathbb{Q}$. Then $\mathbf{SO}_\Phi(J)$ is an irreducible non-cocompact arithmetic lattice in $\mathbf{SO}_\Phi(\mathbb{C})$.
   (3) Let $\Phi = x_1^2 + \ldots + x_{n-1}^2 - x_n^2$ and let $K = \mathbb{Q}(\sqrt{d})$, $d \in \mathbb{N}^+$, be a real quadratic field extension of $\mathbb{Q}$. Then $\mathbf{SO}_\Phi(J)$ is an irreducible non-cocompact arithmetic lattice in $\mathbf{SO}_\Phi(\mathbb{R}) \times \mathbf{SO}_\Phi(\mathbb{R})$.
   (4) Let $\Phi = x_1^2 + \ldots + x_{n-2}^2 - \sqrt{2}x_n^2$ and $K = \mathbb{Q}(\sqrt{2})$. Then $\mathbf{SO}_\Phi(J)$ is a cocompact arithmetic lattice in $\mathbf{SO}_\Phi(\mathbb{R})$.
   (5) Let P be an irreducible polynomial of degree 3 over $\mathbb{Q}$ having one negative and two positive roots δ and δ', $K = \mathbb{Q}(\delta)$ the cubic field

extension of $\mathbb{Q}$, and $\sigma: K \to \mathbb{R}$ a homomorphism with $\sigma(\delta) = \delta'$. Let us consider the form $\Phi = x_1^2 + \ldots + x_{n-1}^2 - \delta x_n^2$. Since δ and $\sigma(\delta)$ are of same sign, the forms Φ and ${}^{\sigma}\Phi$ are equivalent over $\mathbb{R}$. Therefore we can identify $\mathbf{SO}_{\Phi}(\mathbb{R})$ and $\mathbf{SO}_{\sigma_\Phi}(\mathbb{R})$. Hence $\mathbf{SO}_{\Phi}(J)$ is an irreducible cocompact arithmetic lattice in $\mathbf{SO}_{\Phi}(\mathbb{R}) \times \mathbf{SO}_{\Phi}(\mathbb{R})$.

(6) Let $\Phi = x_1^2 + \ldots + x_{n-1}^2 - x_n^2$ and K a cubic field extension of $\mathbb{Q}$ with one real and one imaginary embedding in $\mathbb{C}$ (for example, $K = \mathbb{Q}(\sqrt[3]{2})$. Then $\mathbf{SO}_{\Phi}(J)$ is an irreducible non-cocompact arithmetic lattice in $\mathbf{SO}_{\Phi}(\mathbb{R}) \times \mathbf{SO}_{\Phi}(\mathbb{C})$.

(vii) Let $n \geq 3$ and let Φ be a positive definite quadratic form in n variables with coefficients in $\mathbb{Q}$. We further consider as in example (iii) a finite set $B = \{p_1, \ldots, p_i\}$ of primes and the subring $\mathbb{Q}(B)$ of the field $\mathbb{Q}$ generated by the elements $1/p_1, \ldots, 1/p_i$. We identify the group $\Gamma = \mathbf{SO}_{\Phi}(\mathbb{Q}(B))$ with its image under the diagonal embedding in $G = \mathbf{SO}_{\Phi}(\mathbb{Q}_{p_1}) \times \ldots \times \mathbf{SO}_{\Phi}(\mathbb{Q}_{p_i})$. Suppose

1) for at least one j, $1 \leq j \leq i$, the form Φ represents 0 over $\mathbb{Q}_{p_j}$;
2) the group $\mathbf{SO}_{\Phi}$ is almost $\mathbb{Q}$-simple.

Then Γ is an irreducible cocompact arithmetic lattice in G. In particular, if $\Phi = x_1^2 + \ldots + x_{n-1}^2 + 2x_n^2$, then $\mathbf{SO}_{\Phi}(\mathbb{Z}[1/3])$ is a cocompact arithmetic lattice in $\mathbf{SO}_{\Phi}(\mathbb{Q}_3)$.

(viii) With notation as in example (iv), we set $k = F_q((t))$ and $J = F_q[t^{-1}]$. Let $K \subset k$ be the field of rational functions in one variable t; K is the field of fractions of the ring J. Let $m \geq 5$ and Φ be a nondegenerate quadratic form in m variables with coefficients in J. Suppose $\operatorname{char} k \neq 2$ or, in other words, q is not a power of 2. Then the group $\mathbf{SO}_{\Phi} \subset \mathbf{GL}_n$ is defined over K and absolutely almost simple, and $\mathbf{SO}_{\Phi}(J)$ is a non-cocompact arithmetic lattice in $\mathbf{SO}_{\Phi}(k)$.

(1.8) Let $\mathbf{F}$ be a connected non-commutative absolutely almost simple l-group. We recall (see VIII.3.15) that the group $\mathbf{F}$ is called standard if any special isogeny $\beta: \mathbf{F} \to \mathbf{F}'$ is central. We shall say that $\mathbf{F}$ is *admissible* if $\mathbf{F}$ is standard and either $\mathbf{F}$ is not of type $\mathbb{A}_1$ or $\operatorname{char} l \neq 2$. In view of the description of standard groups given in VIII.3.15 the admissibility of the group $\mathbf{F}$ is equivalent to the following condition

(*) either $\operatorname{char} l \notin \{2; 3\}$ or $\operatorname{char} l = 2$ and $\mathbf{F}$ is not of type $\mathbb{A}_1$, $\mathbb{B}_n$, $\mathbb{C}_n$, and $\mathbb{F}_4$, or $\operatorname{char} l = 3$ and $\mathbf{F}$ is not of type $\mathbb{G}_2$.

We can restate Proposition VIII.3.22 as follows

(a) *Let Δ be a Zariski dense subgroup of* $\mathbf{F}$. *Suppose* $\mathbf{F}$ *is admissible and adjoint. Then the group* $\operatorname{Ad} \Delta$ *is definable over the field generated by the set* $\operatorname{Tr} \operatorname{Ad} \Delta$.

We call a connected semisimple algebraic group $\mathbf{G}$ *admissible* if every almost simple factor of the group $\operatorname{Ad} \mathbf{G}$ is admissible. Since the admissibility of the group $\mathbf{F}$ is equivalent to (*), it follows from (a) that

(b) *if* $\mathbf{G}$ *is a connected semisimple l-group and* $\operatorname{char} l \notin \{2; 3\}$, *then* $\mathbf{G}$ *is admissible.*

(1.9) Now we state two theorems on arithmeticity of lattices.

(A) Theorem. *Let Γ be an irreducible lattice in G. Suppose that* $\operatorname{rank} G \geq 2$, *and that the group Γ is finitely generated and has property* (QD) *in G. Then the lattice Γ is arithmetic.*

(B) Theorem. *Let Γ be an irreducible lattice in G. Suppose that Γ is of infinite index in* $\mathrm{Comm}_G(\Gamma)$, *and that the group Γ is finitely generated and has property* (QD) *in G. Then the lattice Γ is arithmetic.*

The proofs of Theorems A and B are based on the superrigidity theorems. In the situation where for each $\alpha \in A$ the group $\mathbf{G}_\alpha$ is admissible, these will be given in Section 2 (more precisely, in 2.11). For the case of non-admissable groups, see [Ve].

Remark 1. If $\operatorname{rank} G = 0$, in view of Proposition I.2.3.6 the group G is compact, and hence any lattice $\Gamma \subset G$ is finite. It then follows from remark (viii) in 1.2 that if $\operatorname{rank} G = 0$, then the group G does not contain lattices with property (QD). So we can assume in the statement of Theorem B that $\operatorname{rank} G > 0$. But the case "$\operatorname{rank} G \geq 2$" has already been treated in Theorem A. Thus, compared with Theorem A, Theorem B is interesting only in the case where $\operatorname{rank} G = 1$.

Remark 2. As we remark below in 3.10, if $\operatorname{rank} G \geq 2$, then any irreducible lattice $\Gamma \subset G$ with property (QD) is finitely generated. So in Theorem A, the condition "Γ is finitely generated" is superfluous.

(1.10) According to remark (ii) in 1.2, if the closure of the subgroup $G^{is} \cdot F$ in G is open in G, then F has property (QD) in G. So from Theorems A and B and Remark 2 in 1.9 we deduce

Theorem. *Let $\Gamma \subset G$ be an irreducible lattice such that the closure of the subgroup $G^{is} \cdot \Gamma$ in G is open in G. Suppose either* $\operatorname{rank} G \geq 2$ *or Γ is finitely generated and is of infinite index in* $\mathrm{Comm}_G(\Gamma)$. *Then the lattice Γ is arithmetic.*

Theorems 1.11 and 1.12 below are special cases of the above theorem.

(1.11) Theorem. *Suppose* $\operatorname{rank} G \geq 2$ *and for each $\alpha \in A$ the group $\mathbf{G}_\alpha$ has no k_α-anisotropic factors. Then any irreducible lattice $\Gamma \subset G$ is arithmetic.*

(1.12) Theorem. *Suppose for each $\alpha \in A$ the group $\mathbf{G}_\alpha$ has no k_α-anisotropic factors. Then any irreducible finitely generated lattice $\Gamma \subset G$, which is of infinite index in* $\mathrm{Comm}_G(\Gamma)$, *is arithmetic.*

(1.13) It is very likely that in Theorem B of 1.9, and hence in Theorems 1.10 and 1.12, the condition "Γ is finitely generated" is superfluous. However, in the general case we have been unable to eliminate this condition. Nevertheless it is possible to do that in almost all particular cases. Namely, from the results given below on finite generation of lattices (see 3.1(v), Theorem 3.2 and 3.10 (*)) and Theorem B of 1.9 we deduce in a straightforward fashion the following result.

Theorem. *Let $\Gamma \subset G$ be an irreducible lattice with property (QD) in G. Suppose that Γ is of infinite index in* $\mathrm{Comm}_G(\Gamma)$. *Assume in addition that at least one of the following equivalent conditions is satisfied:*

(a) the lattice Γ is cocompact;

(b) there is $\alpha \in A$ such that $\operatorname{char} k_\alpha = 0$;

(c) $\operatorname{rank} G \geq 2$.

Then the lattice Γ is arithmetic.

(1.14) As was observed in 1.2 and 1.10, if the closure in G of the subgroup $G^{\mathrm{is}} \cdot F$ is open in G, then F has property (QD) in G. Therefore Theorem 1.13 has the following consequence.

Theorem. *Let $\Gamma \subset G$ be an irreducible lattice such that the closure of the subgroup $G^{\mathrm{is}} \cdot \Gamma$ in G is open in G. Suppose that Γ is of infinite index in* $\mathrm{Comm}_G(\Gamma)$. *Assume in addition that at least one of conditions (a), (b), (c) in the statement of Theorem 1.13 is satisfied. Then the lattice Γ is arithmetic.*

(1.15) The following is a particular case of Theorems 1.10 and 1.14.

Theorem. *Let $\Gamma \subset G$ be an irreducible lattice such that the closure in G of the subgroup $G^{\mathrm{is}} \cdot \Gamma$ is open in G. Suppose that either* $\operatorname{rank} G \geq 2$ *or Γ is of infinite index in* $\mathrm{Comm}_G(\Gamma)$. *If* $\operatorname{char} k_\alpha = 0$ *for all $\alpha \in A$, then the lattice Γ is arithmetic.*

(1.16) The following theorem is a particular case of Theorem 1.15.

Theorem. *Let $\mathbf{G}$ be a connected semisimple $\mathbb{R}$-group, $\mathbf{G}^{\mathrm{is}} \subset \mathbf{G}$ the almost direct product of $\mathbb{R}$-isotropic factors of $\mathbf{G}$, and Γ_0 an irreducible lattice in $\mathbf{G}(\mathbb{R})$ with $\mathbf{G}^{\mathrm{is}}(\mathbb{R}) \cdot \Gamma_0$ dense in $\mathbf{G}(\mathbb{R})$. Suppose that either* $\operatorname{rank}_{\mathbb{R}} \mathbf{G} \geq 2$ *or Γ_0 is of infinite index in* $\mathrm{Comm}_{\mathbf{G}(\mathbb{R})}(\Gamma_0)$. *Then the lattice Γ_0 is arithmetic.*

2. Proof of the Arithmeticity Theorems

In this section we shall prove Theorems A and B of 1.9 in the situation of admissible groups.

With A, k_α, $\mathbf{G}_\alpha$, $\mathbf{G}_\alpha^{\mathrm{is}}$, G, G^{is}, and $\operatorname{rank} G$ as in Section 1, we set as in Section 3 of Chapter II

$$G_B = \prod_{\alpha \in B} \mathbf{G}_\alpha(k_\alpha) \text{ and } G_B^+ = \prod_{\alpha \in B} \mathbf{G}_\alpha(k_\alpha)^+,$$

where $B \subset A$ and $G^+ = G_A^+$. The natural projections $G \to G_B$ and $G \to \mathbf{G}_\alpha(k_\alpha)$, where $B \subset A$ and $\alpha \in A$, will be denoted, respectively, by pr_B and pr_α. As in

Section 6 of Chapter II we set $A_0 = \{\alpha \in A \mid$ the group $\mathbf{G}_\alpha$ is isotropic over k_α or, equivalently, the group $\mathbf{G}_\alpha(k_\alpha)$ is non-compact$\}$ and observe that the group G_{A-A_0} is compact.

If k is a local field and $\mathbf{F}$ is a connected semisimple k-group, then $\mathbf{F}(k)^0$ will denote the same subgroup as in 1.1.

The Zariski closure of a subset M of an algebraic variety will be denoted by $\bar{M}$.

Throughout this section Γ denotes a lattice in G.

The proofs of Theorems A and B of 1.9 will be provided in 2.11. First we prove some auxiliary assertions.

(2.1) Lemma. *Suppose Γ has property (QD) in G. Then*

(i) for each $\alpha \in A$ the subgroup $\mathrm{pr}_\alpha(\Gamma)$ is Zariski dense in $\mathbf{G}_\alpha$;

(ii) if k is a local field, $B \subset A$, and for each $\alpha \in B$, $\vartheta_\alpha\colon k_\alpha \to k$ is a continuous isomorphism, then the subgroup $(\prod_{\alpha \in B} \vartheta_\alpha^0)(\mathrm{pr}_B(\Gamma))$ is Zariski dense in the k-group $\prod_{\alpha\in B} {}^{\vartheta_\alpha}\mathbf{G}_\alpha$;

(iii) if for $\alpha \in A$ the field k_α is a finite separable field extension of a local field k'_α, then the subgroup $\mathrm{pr}_\alpha(\Gamma)$ is Zariski dense in $R_{k_\alpha/k'_\alpha}\mathbf{G}_\alpha$. (Here the group $\mathbf{G}_\alpha(k_\alpha)$ is identified with $((R_{k_\alpha/k'_\alpha}\mathbf{G}_\alpha)(k'_\alpha)$ by $R^0_{k_\alpha/k'_\alpha}$.)

Proof. (i) Since Γ is a lattice in G, in view of Lemmas II.2.3 and II.2.4 the subgroup $\mathrm{pr}_\alpha(\Gamma)$ has property (S) in $\mathbf{G}_\alpha(k_\alpha)$. So by the Borel-Wang density theorem (see Corollary II.4.4) we have

$$\overline{\mathrm{pr}_\alpha(\Gamma)} \supset \mathbf{G}_\alpha^{\mathrm{is}}. \tag{1}$$

We denote by T_α the closure of the subgroup $\mathrm{pr}_\alpha(\Gamma)\cdot \mathbf{G}_\alpha^{\mathrm{is}}(k_\alpha)$ in $\mathbf{G}_\alpha(k_\alpha)$. Since Γ has property (QD) in G, the subgroup $\mathbf{G}_\alpha(k_\alpha)^0 \cap T_\alpha$ is of finite index in $\mathbf{G}_\alpha(k_\alpha)^0$. But the subgroup $\mathbf{G}_\alpha(k_\alpha)^0$ is Zariski dense in $\mathbf{G}_\alpha$ (see 1.1) and the group $\mathbf{G}_\alpha$ is connected, and hence contains no algebraic subgroups of finite index. Therefore the subgroup T_α, and hence the subgroup $\mathrm{pr}_\alpha(\Gamma)\cdot \mathbf{G}_\alpha^{\mathrm{is}}(k_\alpha)$, are Zariski dense in $\mathbf{G}_\alpha$. From this and (1) one readily deduces (i). To prove (ii) we have to identify k_α, $\alpha \in B$, with k by ϑ_α, to set $k_\beta = k$, to denote the k_β-group $\prod_{\alpha\in B}\mathbf{G}_\alpha$ by $\mathbf{G}_\beta$, and replacing A by $(A-B)\cup\{\beta\}$ to apply assertion (i) to $\mathrm{pr}_\beta(\Gamma) = \mathrm{pr}_B(\Gamma)$. Finally, to deduce (iii) from (i) it suffices to replace k_α by k'_α and $\mathbf{G}_\alpha$ by $R_{k_\alpha/k'_\alpha}\mathbf{G}_\alpha$ (taking into consideration remark (ix) in 1.2). □

(2.2) Let F be a group. We shall say that F is *decomposable* if there exist two infinite normal subgroups F_1 and F_2 in F such that the subgroup $F_1 \cap F_2$ is finite and the subgroup $F_1 \cdot F_2$ is of finite index in F; otherwise F is called *indecomposable.* It is easily seen that the following assertions hold.

(a) If F' is a subgroup of finite index in F, then F is decomposable if and only if F' is decomposable.

(b) If N is a finite normal subgroup of F, then F is decomposable if and only if the factor group F/N is decomposable.

(2.3) Lemma. *Suppose that for each $\alpha \in A$ the group $\mathbf{G}_\alpha$ is almost k_α-simple. Then*

(i) if the lattice Γ is irreducible and has property (QD) in G, then Γ is indecomposable;

(ii) if $A = A_0$, then the lattice Γ is irreducible if and only if Γ is indecomposable.

Proof. (i) Suppose Γ is decomposable. Then, replacing Γ by a subgroup of Γ of finite index we can assume that $\Gamma = \Gamma_1 \cdot \Gamma_2$, where Γ_1 and Γ_2 are infinite normal subgroups in Γ such that $\Gamma_1 \cap \Gamma_2$ is finite. For $i = 1, 2$, we set

$$A_i = \{\alpha \in A \mid \mathrm{pr}_\alpha(\Gamma_i) \subset \mathscr{Z}(\mathbf{G}_\alpha)\}.$$

Since the subgroups $\mathscr{Z}(\mathbf{G}_\alpha)$ are finite and the groups Γ_i are infinite, it follows that for $i = 1, 2$

(1) $$A_i \neq A \text{ or, equivalently, } A - A_i \neq \varnothing.$$

Let $\alpha \in A$ and $i = 1, 2$. Since the subgroup Γ_i is normal in Γ and by Lemma 2.1(i) $\overline{\mathrm{pr}_\alpha(\Gamma)} = \mathbf{G}_\alpha$, the subgroup $\overline{\mathrm{pr}_\alpha(\Gamma_i)}$ is normal in $\mathbf{G}_\alpha$. But the subgroup $\overline{\mathrm{pr}_\alpha(\Gamma_i)}$ is defined over k_α (because $\mathrm{pr}_\alpha(\Gamma) \subset \mathbf{G}_\alpha(k_\alpha)$) and $\mathbf{G}_\alpha$ is almost k_α-simple. Therefore

(2) $$\text{either } \overline{\mathrm{pr}_\alpha(\Gamma_i)} = \mathbf{G}_\alpha \text{ or } \overline{\mathrm{pr}_\alpha(\Gamma_i)} \subset \mathscr{Z}(\mathbf{G}_\alpha).$$

Since the subgroups Γ_1 and Γ_2 are normal in Γ and $|\Gamma_1 \cap \Gamma_2| < \infty$, the mutual commutator subgroup of Γ_1 and Γ_2 is finite. Therefore the mutual commutator subgroup of the subgroups $\mathrm{pr}_\alpha(\Gamma_1)$ and $\mathrm{pr}_\alpha(\Gamma_2)$ is finite. The same is therefore true for the mutual commutator subgroup of the subgroups $\overline{\mathrm{pr}_\alpha(\Gamma_1)}$ and $\overline{\mathrm{pr}_\alpha(\Gamma_2)}$. From this and (2) and $\mathscr{D}(\mathbf{G}_\alpha) = \mathbf{G}_\alpha$, we deduce that either $\mathrm{pr}_\alpha(\Gamma_1) \subset \mathscr{Z}(\mathbf{G}_\alpha)$ or $\mathrm{pr}_\alpha(\Gamma_2) \subset \mathscr{Z}(\mathbf{G}_\alpha)$. Hence $A = A_1 \cup A_2$ or, in other words,

(3) $$(A - A_1) \cap (A - A_2) = \varnothing.$$

Since the subgroups $\mathscr{Z}(\mathbf{G}_\alpha)$ are finite, $G_{A-A_i} \cap \Gamma_i$ is of finite index in Γ_i, and hence the subgroup $(G_{A-A_1} \cap \Gamma) \cdot (G_{A-A_2} \cap \Gamma)$ is of finite index in $\Gamma = \Gamma_1 \cdot \Gamma_2$. But in view of (1) and (3) this contradicts the irreducibility of Γ.

(ii) If Γ is irreducible, then in view of assertion (i) Γ is indecomposable. Suppose now Γ is reducible, i.e., the subgroup $(G_B \cap \Gamma) \cdot (G_{A-B} \cap \Gamma)$ is of finite index in Γ for some $B \subset A$, $B \neq \varnothing$, $B \neq A$. Since Γ is a lattice in G, it then follows that $G_B \cap \Gamma$ is a lattice in G_B and $G_{A-B} \cap \Gamma$ is a lattice in G_{A-B}. But since $A = A_0$, $B \neq \varnothing$, and $B \neq A$, the subgroup G_B and G_{A-B} are non-compact, and hence do not contain finite lattices. So the groups $G_B \cap \Gamma$ and $G_{A-B} \cap \Gamma$ are infinite. Summarizing the above discussion we obtain that Γ contains the subgroup $(G_B \cap \Gamma) \cdot (G_{A-B} \cap \Gamma)$ of finite index representable in the form of a direct product of the infinite subgroups $G_B \cap \Gamma$ and $G_{A-B} \cap \Gamma$. Thus, if the lattice Γ is reducible, then Γ is decomposable. □

(2.4) Lemma.

(i) If $A \supset B \supset A_0$, then the subgroup $\Gamma \cap G_{A-B}$ is finite and $\mathrm{pr}_B(\Gamma)$ is a lattice in G_B.

(ii) Assume that for each $\alpha \in A$ the group $\mathbf{G}_\alpha$ is almost k_α-simple. If the lattice Γ is irreducible and has property (QD) in G, then $\mathrm{pr}_{A_0}(\Gamma)$ is an irreducible lattice in G_{A_0}.

Proof. Assertion (i) is a consequence of the compactness of the group $G_{A-B} \subset G_{A-A_0}$, because Γ is a lattice in G. For (ii) note that, according to Lemma 2.3(i) the group Γ is indecomposable. But $\mathrm{pr}_{A_0}(\Gamma) = \Gamma/(\Gamma \cap G_{A-A_0})$ and the subgroup $\Gamma \cap G_{A-A_0}$ is finite (see assertion (i)). Therefore the group $\mathrm{pr}_{A_0}(\Gamma)$ is indecomposable (see 2.2(b)), and hence the lattice $\mathrm{pr}_{A_0}(\Gamma) \subset G_{A_0}$ is irreducible (see Lemma 2.3(ii)). □

(2.5) Lemma. *Suppose Γ has property (QD) in G. Then*
(i) if $C \subset A$ and the subgroup $\Gamma \cap G_C$ is finite, then $\mathrm{Comm}_G(\Gamma) \cap G_C \subset \mathscr{Z}(G)$;
(ii) the subgroup $\mathrm{Comm}_G(\Gamma) \cap G_{A-A_0}$ is contained in $\mathscr{Z}(G)$, and therefore is finite;
(iii) if Γ is of infinite index in $\mathrm{Comm}_G(\Gamma)$, then $\mathrm{pr}_{A_0}(\Gamma)$ is of infinite index in $\mathrm{pr}_{A_0}(\mathrm{Comm}_G(\Gamma))$.

Proof. (i) Pick $g \in \mathrm{Comm}_G(\Gamma) \cap G_C$ and $\alpha \in A$. We set $g_\alpha = \mathrm{pr}_\alpha(g) \in \mathbf{G}_\alpha(k_\alpha)$. Since $g \in \mathrm{Comm}_G(\Gamma) \cap G_C$ and the subgroup G_C is normal in G, there exists a subgroup Γ_g of finite index in Γ such that $\{g\gamma g^{-1}\gamma^{-1} \mid \gamma \in \Gamma_g\} \subset \Gamma \cap G_C$. Hence,

$$\{g_\alpha h g_\alpha^{-1} h^{-1} \mid h \in \mathrm{pr}_\alpha(\Gamma_g)\} \subset \mathrm{pr}_\alpha(\Gamma \cap G_C). \tag{1}$$

For $h \in \mathbf{G}_\alpha$, we set $\varphi(h) = g_\alpha h g_\alpha^{-1} h^{-1}$. Since the map $\varphi\colon \mathbf{G}_\alpha \to \mathbf{G}_\alpha$ is regular, it follows that $\varphi(\overline{\mathrm{pr}_\alpha(\Gamma_g)}) \subset \overline{\varphi(\mathrm{pr}_\alpha(\Gamma_g))}$. On the other hand,
(a) since the subgroup $\Gamma \cap G_C$ is finite, (1) implies that the set $\varphi(\mathrm{pr}_\alpha(\Gamma_g))$, and hence the set $\overline{\varphi(\mathrm{pr}_\alpha(\Gamma_g))}$, are finite;
(b) since $\overline{\mathrm{pr}_\alpha(\Gamma)} = \mathbf{G}_\alpha$ (see Lemma 2.1), Γ_g is of finite index in Γ, and $\mathbf{G}_\alpha$ is connected, it follows that $\overline{\mathrm{pr}_\alpha(\Gamma_g)} = \mathbf{G}_\alpha$.

So the set $\varphi(\mathbf{G}_\alpha)$ is finite and, since φ is regular, $\mathbf{G}_\alpha$ is connected, and $\varphi(e) = e$, it follows that $\varphi(\mathbf{G}_\alpha) = \{e\}$. But this is equivalent to $\mathrm{pr}_\alpha(g) = g_\alpha \in \mathscr{Z}(\mathbf{G}_\alpha)$. Hence, $\mathrm{pr}_\alpha(\mathrm{Comm}_G(\Gamma) \cap G_C \subset \mathscr{Z}(\mathbf{G}_\alpha)$ for all $\alpha \in A$ which implies that $\mathrm{Comm}_G(\Gamma) \cap G_C \subset \mathscr{Z}(G)$.

Since the subgroup $\Gamma \cap G_{A-A_0}$ if finite, (ii) is a consequence of (i) (see Lemma 2.4(i)). Finally, since G_{A-A_0} coincides with the kernel of the homomorphism pr_{A_0}, (iii) is a consequence of (ii). □

(2.6) Lemma. *Let $B \subset A$, $B \neq \varnothing$ and assume that for each $\alpha \in A$ the group $\mathbf{G}_\alpha$ is almost k_α-simple. Suppose the lattice Γ is irreducible and has property (QD) in G. Then*

(i) $\mathrm{Comm}_G(\Gamma) \cap G_{A-B} \subset \mathscr{Z}(G)$;
(ii) if for each $\alpha \in A$ the group $\mathbf{G}_\alpha$ is adjoint, then $\mathrm{Comm}_G(\Gamma) \cap G_{A-B} = \{e\}$, and hence $\mathrm{pr}_B\colon \mathrm{Comm}_G(\Gamma) \to \mathrm{pr}_B(\mathrm{Comm}_G(\Gamma))$ is an isomorphism.

Proof. If for each $\alpha \in A$ the group $\mathbf{G}_\alpha$ is adjoint, then $\mathscr{Z}(G) = \{e\}$. Thus, (ii) is a consequence of (i). For (i), recall that $\mathrm{pr}_{A_0}(\Gamma)$ is an irreducible lattice in G_{A_0} (see Lemma 2.4(ii)). It then follows from Theorem II.6.7(c) that

$\Gamma \cap G_{A-B} \subset G_{A-A_0} \cdot \mathscr{Z}(G)$. But since the subgroup $\mathscr{Z}(G)$ and the subgroup $\Gamma \cap G_{A-A_0}$ are finite (see Lemma 2.4(i)), the subgroup $\Gamma \cap (G_{A-A_0} \cdot \mathscr{Z}(G))$ is finite. So the subgroup $\Gamma \cap G_{A-B}$ is finite. All that remains now is to make use of Lemma 2.5(i). □

(2.7) Lemma. *Let D denote the closure of* $\mathrm{Comm}_G(\Gamma)$ *in G. Suppose that the lattice Γ is irreducible and has property (QD) in G, and that for each $\alpha \in A$ the group $\mathbf{G}_\alpha$ is almost k_α-simple. If Γ is of infinite index in* $\mathrm{Comm}_G(\Gamma)$, *then $D \supset G^+$.*

Proof. Since there is a finite G-invariant measure on $\Gamma \backslash G$ and the closed subgroup D contains Γ, it follows that $D \setminus G$ also has a finite G-invariant measure (see I.0.36). Therefore in view of Theorem II.6.2(a) there is a subset $B \subset A_0$ such that $D \supset G_B^+$ and $\mathrm{pr}_{A_0-B}(D)$ is a lattice in G_{A_0-B}. Now the subgroup D is not discrete in G, because Γ is a lattice in G and Γ is of infinite index in D. Furthermore, the subgroup $\mathrm{Comm}_G(\Gamma)$ is dense in D and, according to Lemma 2.5(ii), the subgroup $\mathrm{Comm}_G(\Gamma) \cap G_{A-A_0}$ is finite. Thus the subgroup $\mathrm{pr}_{A_0}(D)$ is not discrete in G_{A_0}, and hence $B \neq \varnothing$. On the other hand, since Γ is irreducible and has property (QD) in G, it follows from Lemma 2.4(ii) that $\mathrm{pr}_{A_0}(\Gamma)$ is an irreducible lattice in G_{A_0}. Now applying Theorem II.6.4(a) we obtain that the closure of the subgroup $\Gamma \cdot G_B^+$ in G contains G^+. But the subgroup D is closed and contains both Γ and G_B^+. Thus, $D \supset G^+$. □

(2.8) Let k be a local field. We define on k a natural absolute value that will be denoted by $|\ |_k$. If k is archimedean, i.e., if k is isomorphic to $\mathbb{R}$ or $\mathbb{C}$, we choose as $|\ |_k$ the usual absolute value. If k is non-archimedean, we set $|\ |_k = \mathrm{mod}_k$, where as in I.0.31 $\mathrm{mod}_k(x)$ denotes the modulus of the automorphism $y \mapsto yx$ of the additive group of the field k. It is easily seen that the following assertions hold:

(a) If $\varphi\colon k \to k'$ is a continuous isomorphism of local fields, then $|\varphi(x)|_{k'} = |x|_k$ for all $x \in k$.
(b) Let k' be a finite field extension of k and $x \in k$. Then $|x|_{k'} = |x|_k^n$, where $n = 1$ if k is archimedean, and n is the degree of the extension k' over k if k is non-archimedean. In particular, $|x|_{k'} \geq |x|_k$, if $|x|_k \geq 1$.

(2.9) Lemma. *Let l be an infinite finitely generated field. We denote by l_0 the prime subfield of l and by n the transcendence degree of the field l over l_0. Suppose l is not a global field, i.e. $n > 0$ if* $\mathrm{char}\, l = 0$ *and $n > 1$ if* $\mathrm{char}\, l > 0$. *Then, for each element u of l which is transcendental over l_0 and any $c > 0$, there exist a local field k and a homomorphism $\sigma\colon l \to k$ such that $\sigma(l)$ is dense in k and $|\sigma(u)|_k > c$.*

Proof. We consider two cases: (i) $\mathrm{char}\, l = 0$ and (ii) $\mathrm{char}\, l > 0$.

(i) Since x is transcendental over $l_0 = \mathbb{Q}$, the field l is finitely generated, and $\mathbb{C}$ is algebraically closed and is of infinite transcendence degree over $\mathbb{Q}$, it follows that for each transcendental number $y \in \mathbb{C} - \mathbb{R}$ there is a homomorphism $\sigma_y\colon l \to \mathbb{C}$ such that $\sigma_y(l)$ is dense in $\mathbb{C}$ and $\sigma_y(u) = y$. But the set $\{y \in \mathbb{C} - \mathbb{R} \mid y \text{ is transcendental}\}$ is dense in $\mathbb{C}$. So there is a homomorphism $\sigma\colon l \to \mathbb{C}$ such that $\sigma(l)$ is dense in $\mathbb{C}$ and $|\sigma(u)| > c$.

(ii) Choose a transcendence basis B of the field l over l_0 with $u \in B$. Since $n = \operatorname{card} B \geq 2$, one can choose $v \in B$, $v \neq u$. We denote by $l' \subset l$ the subfield generated by the set B, by $k' = l_0((x))$ the (local) field of formal power series in one variable x over the finite field l_0, and by Ω the set $\{y \in k' \mid$ the elements y and x are algebraically independent over $l_0\}$. Since k' is of infinite transcendence degree over l_0 and the set B is finite, for each $y \in \Omega$ there is a homomorphism $\sigma_y \colon l' \to k'$ such that $\sigma_y(v) = x$ and $\sigma_y(u) = y$. But the set $k' - \Omega$ is countable, and hence the set Ω is dense in k'. Therefore there is a homomorphism $\sigma' \colon l' \to k'$ such that $|\sigma'(u)|_{k'} > c+1$ and $\sigma'(v) = x$. Since x generates a dense subfield in k', it follows that $\sigma'(l')$ is dense in k'. Since l is finitely generated, the degree of the field l over l' is finite, and hence the homomorphism $\sigma' \colon l' \to k'$ extends to a homomorphism $\sigma \colon l \to k$, where k is a local finite field extension of k'. All that remains now is to observe that

1) since $\sigma'(l')$ is dense in k', we can assume that $\sigma(l)$ is dense in k

and

2) since $|\sigma'(u)|_k > c + 1$ and σ is an extension of σ', it follows from 2.8(b) that $|\sigma(u)|_k > c + 1$. □

(2.10) The following lemma plays a basic role in the proof of Theorems A and B of 1.9. It will be deduced from the superrigidity theorems of Chapter VII.

Lemma. *Set* $\Gamma_0 = \operatorname{pr}_{A_0}(\Gamma)$ *and* $\Lambda_0 = \operatorname{pr}_{A_0}(\operatorname{Comm}_G(\Gamma))$. *Let k be a local field,* $\mathbf{H}$ *a connected non-trivial adjoint absolutely simple k-group, and* $\tau \colon \Lambda_0 \to \mathbf{H}(k)$ *a homomorphism with* $\tau(\Gamma_0)$ *Zariski dense in* $\mathbf{H}$. *Suppose that for each* $\alpha \in A$ *the group* $\mathbf{G}_\alpha$ *is absolutely almost simple and the lattice* Γ *is irreducible and has property (QD) in G. Assume in addition that either* $\operatorname{rank} G \geq 2$ *or* Γ *is of infinite index in* $\operatorname{Comm}_G(\Gamma)$. *Then either the subgroup* $\tau(\Gamma_0)$ *is relatively compact in* $\mathbf{H}(k)$ *or there exist* $\alpha \in A_0$, *a continuous homomorphism* $\vartheta \colon k_\alpha \to k$ *and a special k-isogeny* $\eta \colon {}^{\vartheta}\mathbf{G}_\alpha \to \mathbf{H}$ *such that* $\tau(\lambda) = \eta(\vartheta^0(\operatorname{pr}_\alpha(\lambda)))$ *for all* $\lambda \in \Lambda_0$.

Proof. According to Lemma 2.4(ii), Γ_0 is an irreducible lattice in G_{A_0}. Since Γ_0 is a lattice in G_{A_0} and for each $\alpha \in A_0$ the group $\mathbf{G}_\alpha$ is almost k_α-simple and k_α-isotropic, it follows from Corollary VII.6.3 that $\operatorname{Comm}_{G_{A_0}}(\Gamma_0)$ is countable. Hence the subgroup Λ_0 of this commensurability subgroup is likewise countable. If Γ is of infinite index in $\operatorname{Comm}_G(\Gamma)$, then by Lemma 2.7 the closure in G of $\operatorname{Comm}_G(\Gamma)$ contains G^+, and hence the closure in G_{A_0} of the subgroup Λ_0 contains $G^+ = G^+_{A_0}$. Further, since for each $\alpha \in A - A_0$ the group $\mathbf{G}_\alpha$ is anisotropic over k_α, we have

$$\operatorname{rank} G_{A_0} \overset{\text{def}}{=} \sum_{\alpha \in A_0} \operatorname{rank}_{k_\alpha} \mathbf{G}_\alpha = \operatorname{rank} G.$$

All that remains now is to make use of Theorem VII.5.4(a) (in the case where Γ is of infinite index in $\operatorname{Comm}_G(\Gamma)$) and Theorem VII.5.6(a) (in case $\operatorname{rank} G \geq 2$) and to observe that any rational epimorphism of an absolutely almost simple group onto a connected non-trivial algebraic group is an isogeny.

(2.11) We now turn to the proof of Theorems A and B of 1.9 in the case where the groups $\mathbf{G}_\alpha$ are admissible. We shall divide the proof into several steps, both theorems being proved simultaneously.

Step 1. *Reduction to the case in which the groups* $\mathbf{G}_\alpha$ *are adjoint and absolutely simple.*

Let $\mathbf{G}'_\alpha$ be the adjoint group of $\mathbf{G}_\alpha$ and let $p_\alpha\colon \mathbf{G}_\alpha \to \mathbf{G}'_\alpha$ be the central k_α-isogeny. We set $G' = \prod_{\alpha\in A} \mathbf{G}'_\alpha(k_\alpha)$. The isogenies p_α induce the continuous homomorphism $p\colon G \to G'$. As in the proof of Theorem III.5.9 (see III.6.9) we observe that $p(\Gamma)$ is an irreducible lattice in G. According to Corollary I.1.4.6(a) $\operatorname{rank}_{k_\alpha} \mathbf{G}'_\alpha = \operatorname{rank}_{k_\alpha} \mathbf{G}_\alpha$. So if $\operatorname{rank} G \geq 2$, then $\operatorname{rank} G' \overset{\text{def}}{=} \sum_{\alpha\in A} \operatorname{rank}_{k_\alpha} \mathbf{G}'_\alpha \geq 2$. Since $p(\operatorname{Comm}_G(\Gamma)) \subset \operatorname{Comm}_{G'}(p(\Gamma))$ and the kernel of the homomorphism p is finite, it follows that $p(\Gamma)$ is of infinite index in $\operatorname{Comm}_{G'}(p(\Gamma))$ whenever Γ is of infinite index in $\operatorname{Comm}_G(\Gamma)$. It follows from the description of admissible groups presented in 1.7 that the admissibility of the group $\mathbf{G}_\alpha$ is equivalent to that of $\mathbf{G}'_\alpha$. Since Γ is finitely generated and has property (QD) in G, it follows that $p(\Gamma)$ is finitely generated and has property (QD) in G' (see remark 1.2(iv)). According to remark 1.6(v) Γ is an S-arithmetic lattice in G if and only if $p(\Gamma)$ is an S-arithmetic lattice in G'. Thus, we can replace $\mathbf{G}_\alpha$ by $\mathbf{G}'_\alpha$ and Γ by $p(\Gamma)$ and assume that $\mathbf{G}_\alpha$ is adjoint. Then the group $\mathbf{G}_\alpha$ can be represented in the form

$$\mathbf{G}_\alpha = \prod_{i\in I_\alpha} R_{k_{\alpha i}/k_\alpha}(\mathbf{F}_{\alpha i}),$$

where $k_{\alpha i}$ is a finite separable field extension of k_α and $\mathbf{F}_{\alpha i}$ is an adjoint absolutely simple $k_{\alpha i}$-group (see Proposition I.1.4.10 and I.1.7). So we can (and do) assume that for each $\alpha \in A$ the group $\mathbf{G}_\alpha$ is adjoint and absolutely simple (we apply here 1.2(d) and remark 1.6(vi)).

Step 2. We set $\Gamma_\alpha = \operatorname{pr}_\alpha(\Gamma)$ and denote by $L_\alpha \subset k_\alpha$ the field generated by the set $\operatorname{Tr}\operatorname{Ad}\Gamma_\alpha$. We reduce to the case in which $\mathbf{G}_\alpha$ is an L_α-group and $\Gamma_\alpha \subset \mathbf{G}_\alpha(L_\alpha)$ for all $\alpha \in A$. Now the group $\mathbf{G}_\alpha$ is adjoint, absolutely simple, and admissible. Furthermore, the subgroup Γ_α is Zariski dense in $\mathbf{G}_\alpha$ (see Lemma 2.1(i)). Thus, it follows from 1.7(a) that the subgroup $\operatorname{Ad}\Gamma_\alpha$ is definable over L_α. So we can choose a basis $x^\alpha = (x_1^\alpha, \ldots, x_{m_\alpha}^\alpha)$, $m_\alpha = \dim \mathbf{G}_\alpha$, in the Lie algebra $\operatorname{Lie}(G_\alpha)$ in which transformations from $\operatorname{Ad}\Gamma_\alpha$ are written as matrices with entries in L_α. We define the rational homomorphism $\varphi_\alpha\colon \mathbf{G}_\alpha \to \mathbf{GL}_{m_\alpha}$ by assigning to $g \in \mathbf{G}_\alpha$ the matrix $\varphi_\alpha(g)$ of the transformation $\operatorname{Ad} g$ in the basis x^α. We set $\mathbf{H}_\alpha = \varphi_\alpha(\mathbf{G}_\alpha)$. Since $\bar{\Gamma}_\alpha = \mathbf{G}_\alpha$, it follows that $\overline{\varphi_\alpha(\Gamma_\alpha)} = \mathbf{H}_\alpha$. But $\Gamma_\alpha \subset \mathbf{G}_\alpha(k_\alpha)$ and by definition of the basis x^α and the homomorphism φ_α we have $\varphi_\alpha(\Gamma_\alpha) \subset \mathbf{GL}_{m_\alpha}(L_\alpha)$. So $\mathbf{H}_\alpha$ is an L_α-subgroup in $\mathbf{GL}_{m_\alpha}$ and the morphism $\varphi_\alpha\colon \mathbf{G}_\alpha \to \mathbf{H}_\alpha$ is defined over k_α (see I.0.11). Since the group $\mathbf{G}_\alpha$ is adjoint and the isogeny $\operatorname{Ad}\colon \mathbf{G}_\alpha \to \operatorname{Ad}\mathbf{G}_\alpha$ is central (see Proposition I.1.4.11(iv)), $\varphi_\alpha\colon \mathbf{G}_\alpha \to \mathbf{H}_\alpha$ is an algebraic group isomorphism. Thus, identifying $\mathbf{G}_\alpha$ and $\mathbf{H}_\alpha$ by φ_α, we may assume that $\mathbf{G}_\alpha$ is defined over L_α and $\Gamma_\alpha = \operatorname{pr}_\alpha(\Gamma) \subset \mathbf{G}_\alpha(L_\alpha)$.

Step 3. *The homomorphisms* $\sigma^0 \circ \mathrm{pr}_\alpha$.

Since $\mathrm{pr}_\alpha(\Gamma) \subset \mathbf{G}_\alpha(L_\alpha)$, for a homomorphism σ of the field L_α into a field l, one can define the homomorphism $\sigma^0 \circ \mathrm{pr}_\alpha\colon \Gamma \to {}^\sigma\mathbf{G}_\alpha(l)$.

(1) Lemma. *Suppose that* $\alpha_0 \in A$, *k is a local field, and* $\sigma\colon L_{\alpha_0} \to k$ *is a homomorphism with* $\sigma(L_{\alpha_0})$ *dense in k. Then either the subgroup* $\sigma^0(\mathrm{pr}_{\alpha_0}(\Gamma))$ *is relatively compact in* ${}^\sigma\mathbf{G}_{\alpha_0}(k)$ *or there exist* $\alpha_\sigma \in A_0$, *a continuous isomorphism* $\vartheta_\sigma\colon k_{\alpha_\sigma} \to k$, *and a k-isomorphism* $\eta_\sigma\colon {}^{\vartheta_\sigma}\mathbf{G}_{\alpha_\sigma} \to {}^\sigma\mathbf{G}_{\alpha_0}$ *such that*

(1) $$\sigma^0(\mathrm{pr}_{\alpha_0}(\gamma)) = \eta_\sigma(\vartheta_\sigma^0(\mathrm{pr}_{\alpha_\sigma}(\gamma))) \text{ for all } \gamma \in \Gamma.$$

Proof. The group $\mathbf{G}_{\alpha_0}$ (and hence the group ${}^\sigma\mathbf{G}_{\alpha_0}$) is connected, adjoint, and absolutely simple. Since the group $\mathbf{G}_{\alpha_0}$ is adjoint, $\mathrm{pr}_\alpha(\mathrm{Comm}_G(\Gamma)) \subset \mathrm{Comm}_{G_{\alpha_0}}(\mathrm{pr}_{\alpha_0}(\Gamma))$, the subgroup $\mathrm{pr}_{\alpha_0}(\Gamma)$ is contained in $\mathbf{G}_{\alpha_0}(L_{\alpha_0})$ and is Zariski dense in $\mathbf{G}_{\alpha_0}$ (see Lemma 2.1), it follows from Lemma VII.6.2(ii) that $\mathrm{pr}_\alpha(\mathrm{Comm}_G(\Gamma)) \subset \mathbf{G}_{\alpha_0}(L_{\alpha_0})$. So the homomorphism $\sigma^0 \circ \mathrm{pr}_{\alpha_0}$ is defined on $\mathrm{Comm}_G(\Gamma)$ as well. According to Lemma 2.6(ii), $\mathrm{pr}_{A_0}\colon \mathrm{Comm}_G(\Gamma) \to \mathrm{pr}_{A_0}(\mathrm{Comm}_G(\Gamma))$ is an isomorphism. Therefore there is a homomorphism $\tau\colon \mathrm{pr}_{A_0}(\mathrm{Comm}_G(\Gamma)) \to {}^\sigma G_{\alpha_0}(k)$ such that $\sigma^0 \circ \mathrm{pr}_{\alpha_0} = \tau \circ \mathrm{pr}_{A_0}$. Since the subgroup $\mathrm{pr}_{\alpha_0}(\Gamma)$ is Zariski dense in $\mathbf{G}_{\alpha_0}$ (see Lemma 2.1), it follows that the subgroup $\tau(\Gamma_0) = \sigma^0(\mathrm{pr}_{\alpha_0}(\Gamma))$ is Zariski dense in ${}^\sigma\mathbf{G}_{\alpha_0}$. Suppose the subgroup $\sigma^0(\mathrm{pr}_{\alpha_0}(\Gamma)) = \tau(\mathrm{pr}_{A_0}(\Gamma))$ is not relatively compact in ${}^\sigma\mathbf{G}_{\alpha_0}(k)$. Then in view of Lemma 2.10 there exist $\alpha_\sigma \in A_0$, a continuous homomorphism $\vartheta_\sigma\colon k_{\alpha_\sigma} \to k$, and a special k-isogeny $\eta_\sigma\colon {}^{\vartheta_\sigma}\mathbf{G}_{\alpha_\sigma} \to {}^\sigma\mathbf{G}_{\alpha_0}$ such that

$$\tau(h) = \eta_\sigma(\vartheta_\sigma^0(\mathrm{pr}_{\alpha_\sigma}(h)))$$

for all $h \in \mathrm{pr}_{A_0}(\Gamma)$. But $\sigma^0 \circ \mathrm{pr}_{\alpha_0} = \tau \circ \mathrm{pr}_{A_0}$ and $\mathrm{pr}_{\alpha_\sigma} \circ \mathrm{pr}_{A_0} = \mathrm{pr}_{\alpha_\sigma}$. Thus, for α_σ, ϑ_σ, η_σ, and all $\gamma \in \Gamma$ (1) holds. Since the group $\mathbf{G}_{\alpha_\sigma}$ (hence the group ${}^{\vartheta_\sigma}\mathbf{G}_{\alpha_\sigma}$) is absolutely simple, adjoint, and admissible, the special k-isogeny η_σ is a k-isomorphism. From this and (1) we deduce that

(2) $$\sigma(\mathrm{Tr}\,\mathrm{Ad}\,\mathrm{pr}_{\alpha_0}(\gamma)) = \vartheta_\sigma(\mathrm{Tr}\,\mathrm{Ad}\,\mathrm{pr}_{\alpha_\sigma}(\gamma)) \text{ for all } \gamma \in \Gamma.$$

Since $\mathrm{Tr}\,\mathrm{Ad}\,\mathrm{pr}_{\alpha_0}(\Gamma)$ generates L_{α_0}, (2) implies that $\vartheta_\sigma(k_{\alpha_\sigma}) \supset \sigma(L_{\alpha_0})$. But $\sigma(L_{\alpha_0})$ is dense in k. Thus, the continuous homomorphism $\vartheta_\sigma\colon k_{\alpha_\sigma} \to k$ is an isomorphism (see I.0.31) and this completes the proof of the lemma. □

As was observed above, (1) implies (2). So from Lemma 1 we deduce

(2) Lemma. *With* α_0, k, *and* σ *as in Lemma 1, either the subgroup* $\sigma^0(\mathrm{pr}_{\alpha_0}(\Gamma))$ *is relatively compact in* ${}^\sigma\mathbf{G}_{\alpha_0}(k)$ *or there exist* $\alpha_\sigma \in A_0$ *and a continuous isomorphism* $\vartheta_\sigma\colon k_{\alpha_\sigma} \to k$ *such that (2) holds for each* $\gamma \in \Gamma$.

It is easily seen that if $n \in \mathbb{N}^+$, k is a local field, $h \in \mathbf{GL}_n(k)$, and the subgroup $\{h^m \mid m \in \mathbb{Z}\}$ is relatively compact in $\mathbf{GL}_n(k)$, then the absolute value of any eigenvalue of the matrix h equals 1, and hence $|\mathrm{Tr}\, h|_k \leq n$ (the absolute value $|\ |_k$ on k extended to any field extension of k was defined in 2.8). From Lemma 2 and 2.8(a) we then deduce

(3) Lemma. *Pick $\gamma \in \Gamma$ and set*

$$c(\gamma) = \max\{\max_{\alpha \in A} \dim \mathbf{G}_\alpha, \max_{\alpha \in A} |\operatorname{Tr} \operatorname{Ad} \operatorname{pr}_\alpha(\gamma)|_k\}.$$

Then, with α_0, k, and σ as in Lemma 1,

$$|\sigma(\operatorname{Tr} \operatorname{Ad} \operatorname{pr}_{\alpha_0}(\gamma))|_k \leq c(\gamma). \tag{3}$$

Step 4. *The fact that fields L_α are global.*

As was observed above (see Step 2) one can choose a basis x^α in the Lie algebra $\operatorname{Lie}(\mathbf{G}_\alpha)$ so that transformations from the group $\operatorname{Ad} \operatorname{pr}_\alpha(\Gamma)$ are written as matrices with entries in L_α. Denote by L'_α the field generated by the entries of these matrices. Clearly, $L'_\alpha \subset L_\alpha$. But since $\operatorname{Tr} \operatorname{Ad} \operatorname{pr}_\alpha(\Gamma) \subset L'_\alpha$ and L_α is generated by the set $\operatorname{Tr} \operatorname{Ad} \operatorname{pr}_\alpha(\Gamma)$, we have $L_\alpha \subset L'_\alpha$. Thus, $L_\alpha = L'_\alpha$. Since the group Γ is finitely generated and the entries of a product of matrices are polynomials with integer coefficients in the entries of the factors, the field $L_\alpha = L'_\alpha$ is finitely generated.

We shall show that L_α is a global field. Suppose the contrary. Since $\operatorname{Tr} \operatorname{Ad} \operatorname{pr}_\alpha(\Gamma)$ generates L_α and the field L_α is finitely generated and is not global, one can choose $\gamma \in \Gamma$ such that the element $\operatorname{Tr} \operatorname{Ad} \operatorname{pr}_\alpha(\gamma)$ is transcendental over the prime subfield of L_α. It then follows from Lemma 2.9 that there exist a local field k and a homomorphism $\sigma\colon L_\alpha \to k$ such that $\sigma(L_\alpha)$ is dense in k and $|\sigma(\operatorname{Tr} \operatorname{Ad} \operatorname{pr}_\alpha(\gamma))|_k > c(\gamma)$, where $c(\gamma)$ was defined in Lemma 3. But this contradicts inequality (3).

Step 5. *The field K, the group $\mathbf{H}$, and the set S.*

Choose $\alpha_0 \in A$, set $K = L_{\alpha_0}$, denote the K-group $\mathbf{G}_{\alpha_0}$ by $\mathbf{H}$, and observe that $\operatorname{pr}_{\alpha_0}(\Gamma) \subset \mathbf{H}(K)$. Let $\mathscr{R}$ denote the set of all (inequivalent) valuations of the global field K and $\mathscr{R}_\infty \subset \mathscr{R}$ the set of archimedean valuations of K. We shall use the standard notation $|x|_v$, K_v, and $K(S)$ defined in 1.3 and 1.4. We set

$S = \{v \in \mathscr{R} \mid$ there exist $\delta(v) \in A$, a continuous isomorphism $\omega_v\colon K_v \to k_{\delta(v)}$, and a $k_{\delta(v)}$-isomorphism $\tau_v\colon {}^{\omega_v}\mathbf{H} \to \mathbf{G}_{\delta(v)}$ such that

$$\operatorname{pr}_{\delta(v)}(\gamma) = \tau_v(\omega_v^0(\operatorname{pr}_{\alpha_0}(\gamma))) \tag{4}$$

for all $\gamma \in \Gamma\}$.

By setting $\vartheta_v = \omega_v^{-1}$ and $\eta_v = {}^{\vartheta_v}\tau_v^{-1}$, we have

(*) $S = \{v \in \mathscr{R} \mid$ *there exist $\delta(v) \in A$, a continuous isomorphism $\vartheta_v\colon k_{\delta(v)} \to K_v$, and a K_v-isomorphism $\eta_v\colon {}^{\vartheta_v}\mathbf{G}_{\delta(v)} \to \mathbf{H}$ such that*

$$\eta_v(\vartheta_v^0(\operatorname{pr}_{\delta(v)}(\gamma))) = \operatorname{pr}_{\alpha_0}(\gamma) \tag{5}$$

for all $\gamma \in \Gamma\}$.

Since η_v is a K_v-isomorphism, (4) implies that

$$\operatorname{Tr}\operatorname{Ad}\operatorname{pr}_{\delta(v)}(\gamma) = \omega_v(\operatorname{Tr}\operatorname{Ad}\operatorname{pr}_{\alpha_0}(\gamma)) \tag{6}$$

for all $\gamma \in \Gamma$. Let $v, v' \in S$ and $\delta(v) = \delta(v')$. We then have from (6) that $\omega_v(x) = \omega_{v'}(x)$ for each $x \in \operatorname{Tr}\operatorname{Ad}\operatorname{pr}_{\alpha_0}(\Gamma)$. But the set $\operatorname{Tr}\operatorname{Ad}\operatorname{pr}_{\alpha_0}(\Gamma)$ generates $K = L_{\alpha_0}$. So the continuous isomorphism $\omega_v^{-1} \circ \omega_{v'}\colon K_{v'} \to K_v$ is the identity on K. Hence the valuations v and v' are equivalent, i.e., $v = v'$. Thus, we have shown that the map $\delta\colon S \to \delta(S)$ is one-to-one.

If $(\delta(v), \vartheta_v, \tau_v)$ and $(\tilde{\delta}(v), \tilde{\vartheta}_v, \tilde{\tau}_v)$ are two collections for which (5) holds and $\delta(v) \neq \tilde{\delta}(v)$, then by setting $\Phi = \operatorname{pr}_{\{\delta(v), \tilde{\delta}(v)\}}(\Gamma)$ and $\Phi' = (\vartheta_v^0 \times \tilde{\vartheta}_v^0)(\Phi) \subset {}^{\vartheta_v}\mathbf{G}_{\delta(v)} \times {}^{\tilde{\vartheta}_v}\mathbf{G}_{\tilde{\delta}(v)}$, we obtain that

1) $(\eta_v \times \tilde{\eta}_v)(\Phi') \subset \{(h,h) \mid h \in \mathbf{H}\}$, and hence the subgroup Φ' is not Zariski dense in the K_v-group ${}^{\vartheta_v}\mathbf{G}_{\delta(v)} \times {}^{\tilde{\vartheta}_v}\mathbf{G}_{\tilde{\vartheta}(v)}$;
2) in view of Lemma 2.1(ii) the subgroup Φ' is Zariski dense in ${}^{\vartheta_v}\mathbf{G}_{\delta(v)} \times {}^{\tilde{\vartheta}_v}\mathbf{G}_{\tilde{\vartheta}(v)}$.
This contradiction shows that $\delta(v)$ is defined uniquely by $v \in S$.

As in 1.4 we set $H_S = \prod_{v \in S} \mathbf{H}(K_v)$ and $\mathscr{T} = \{v \in \mathscr{R} \mid$ the group $\mathbf{H}$ is anisotropic over K_v or, equivalently, the group $\mathbf{H}(K_v)$ is compact$\}$. Since the map $\delta\colon S \to \delta(S)$ is one-to-one, one can define the isomorphism

$$\varphi = \prod_{v \in S} (\tau_v \circ \omega_v^0)\colon H_S \to G_{\delta(S)}.$$

We denote by diag_S the diagonal embedding of the group $\mathbf{H}(K)$ in H_S. Then in view of (4) we have

$$\operatorname{pr}_{\delta(S)}(\gamma) = \varphi(\operatorname{diag}_S(\operatorname{pr}_{\alpha_0}(\gamma)))$$

for all $\gamma \in \Gamma$, and hence

$$\operatorname{pr}_{\delta(S)}(\Gamma) = \varphi(\operatorname{diag}_S(\operatorname{pr}_{\alpha_0}(\Gamma))). \tag{7}$$

On the other hand, remark (i) in 1.3 shows that the homomorphism φ belongs to the class Ψ. So to prove the arithmeticity of the lattice Γ it suffices to establish the following three assertions:

(a) $S \supset \mathscr{R}_\infty - \mathscr{T}$.
(b) the subgroup $\operatorname{pr}_{\alpha_0}(\Gamma)$ is commensurable with $\mathbf{H}(K(S))$.
(c) $\delta(S) = A$.

Step 6. *The set* S_0.

Put $S_0 = \{v \in \mathscr{R} \mid$ the subgroup $\operatorname{pr}_{\alpha_0}(\Gamma)$ is not relatively compact in $\mathbf{H}(K_v)\}$. If $v \in S_0$, Lemma 1 applied to the inclusion map $\sigma\colon L_{\alpha_0} = K \to K_v$ and (*) imply that $v \in S$. Hence,

$$S_0 \subset S. \tag{8}$$

We realize $\mathbf{H}$ as a K-subgroup of $\mathbf{GL}_n$ and for $v \in \mathscr{R} - \mathscr{R}_\infty$ consider $\mathbf{H}(\mathcal{O}_v) = \mathbf{H} \cap \mathbf{GL}_n(\mathcal{O}_v)$, where $\mathcal{O}_v$ denotes the ring of integers of the field K_v. Since the subgroup $\mathbf{H}(\mathcal{O}_v)$ is open in $\mathbf{H}(K_v)$, the subgroup $\mathrm{pr}_{\alpha_0}(\Gamma) \cap \mathbf{H}(\mathcal{O}_v)$ is of finite index in $\mathrm{pr}_{\alpha_0}(\Gamma)$ for each $v \in \mathcal{R} - \mathcal{R}_\infty - S_0$. But $\mathbf{H}(K(\tilde{S})) = \{h \in \mathbf{H}(K) \mid h \in \mathbf{H}(\mathcal{O}_v)$ for all $v \in \mathcal{R} - \mathcal{R}_\infty - \tilde{S}\}$ for any $\tilde{S} \subset \mathcal{R}$. Therefore if $\tilde{S} \subset \mathcal{R}$ is finite, then the subgroup $\mathrm{pr}_{\alpha_0}(\Gamma) \cap \mathbf{H}(K(S_0))$ is of finite index in $\mathrm{pr}_{\alpha_0}(\Gamma) \cap \mathbf{H}(K(S_0 \cup \tilde{S}))$. On the other hand, since the group Γ (and hence the group $\mathrm{pr}_{\alpha_0}(\Gamma) \subset \mathbf{H}(K)$) is finitely generated and $\mathbf{H}(K)$ coincides with the union of the subgroups $\mathbf{H}(K(\tilde{S}))$, where $\tilde{S}$ runs through all finite subsets of $\mathcal{R}$, we find that there exists a finite subset $S' \subset \mathcal{R}$ such that $\mathrm{pr}_{\alpha_0}(\Gamma) \subset \mathbf{H}(K(S'))$. Now, replacing Γ by a subgroup of finite index in Γ (the S-arithmeticity is invariant under such a replacement), we obtain that $\mathrm{pr}_{\alpha_0}(\Gamma) \subset \mathbf{H}(K(S_0))$. In view of (8) we then have that

$$\text{(9)} \qquad \mathrm{pr}_{\alpha_0}(\Gamma) \subset \mathbf{H}(K(S)).$$

Step 7. *Proof of assertions (a) and (b).*

Let $w \in \mathcal{R}_\infty - \mathcal{T}$. Then the group $R^0_{K_w/\mathbb{R}}(\mathrm{pr}_{\alpha_0}(\Gamma))$ is Zariski dense in the $\mathbb{R}$-group $R_{K_w/\mathbb{R}}\mathbf{H}$. So, by the theorem on the algebraicity of a compact real linear group, $w \in S_0$ (see I.0.31). Thus $S_0 \supset \mathcal{R}_\infty - \mathcal{T}$ which, in view of (8), proves (a).

According to assertion (a), $S \supset \mathcal{R}_\infty - \mathcal{T}$. Therefore $\mathrm{diag}_S(\mathbf{H}(K(S)))$ is a lattice in H_S (see Theorem I.3.2.5). Since $\varphi\colon H_S \to G_{\delta(S)}$ is a topological group isomorphism, $\varphi(\mathrm{diag}_S(\mathbf{H}(K(S))))$ is a lattice in $G_{\delta(S)}$. From this, (7) and (9) we deduce that the subgroup $\mathrm{pr}_{\delta(S)}(\Gamma)$ is discrete in $G_{\delta(S)}$. On the other hand, since $\mathrm{pr}_{A_0}(\Gamma)$ is an irreducible lattice in G_{A_0} (see Lemma 2.4(ii)), Lemma II.6.4 implies that $\mathrm{pr}_B(\Gamma)$ is not discrete in G_B for any non-empty $B \subset A$, $B \not\supset A_0$. Therefore

$$\text{(10)} \qquad \delta(S) \supset A_0.$$

It follows from (7), (10), and Lemma 2.4(i) that the subgroup $\varphi(\mathrm{diag}_S(\mathrm{pr}_{\alpha_0(\Gamma)}))$ is a lattice in $G_{\delta(S)}$. But

1) in view of (9), $\varphi(\mathrm{diag}_S(\mathrm{pr}_{\alpha_0}(\Gamma)))$ is contained in the subgroup $\varphi(\mathrm{diag}_S(\mathbf{H}(K(S))))$ which is, as was observed above, a lattice in $G_{\delta(S)}$;
2) any lattice is of finite index in any ambient discrete subgroup.

So $\varphi(\mathrm{diag}_S(\mathrm{pr}_{\alpha_0}(\Gamma)))$ is of finite index in $\varphi(\mathrm{diag}_S(H(K(S))))$ and, since φ is an isomorphism, $\mathrm{pr}_{\alpha_0}(\Gamma)$ is of finite index in $\mathbf{H}(K(S))$. This completes the proof of (b).

Now to complete the proof of Theorems A and B of 1.9 it remains to prove assertion (c). We remark that in the case where for each $\alpha \in A$ the group $\mathbf{G}_\alpha$ has no k_α-anisotropic factors, (c) is a consequence of (10).

Step 8. *Proof of assertion (c).*

In view of (10) it suffices to show that $\delta(S) \supset A - A_0$. Let $\alpha \in A - A_0$. Since the field L_α is finitely generated and the subgroup $\mathrm{pr}_\alpha(\Gamma)$ of $\mathbf{G}_\alpha(L_\alpha)$ is Zariski dense in $\mathbf{G}_\alpha$ (see Lemma 2.1(i)), it follows from Lemma VII.6.1 and remark (ii) in VII.6.23 that there is a local field k containing L_α as a dense subfield such that the subgroup $\mathrm{pr}_\alpha(\Gamma)$ is not relatively compact in $\mathbf{G}_\alpha(k_\alpha)$. It then follows from Lemma 1 applied to the identity embedding $\sigma\colon L_\alpha \to k$ that there exist $\alpha' \in A_0$, a continuous isomorphism $\vartheta\colon k_{\alpha'} \to k$, and a k-isomorphism $\eta\colon {}^{\vartheta}\mathbf{G}_{\alpha'} \to \mathbf{G}_\alpha$ such that

$$\mathrm{pr}_\alpha(\gamma) = \eta(\vartheta^0(\mathrm{pr}_{\alpha'}(\gamma))) \tag{11}$$

for all $\gamma \in \Gamma$. Since $\alpha' \in A_0$, it follows from (10) that $\alpha' = \delta(v')$ for some $v' \in S$. Now (4) and (11) imply that

$$\mathrm{pr}_\alpha(\gamma) = (\eta \circ \vartheta^0 \circ \tau_{v'} \circ \omega_{v'}^0)(\mathrm{pr}_{\alpha_0}(\gamma)) \text{ for all } \gamma \in \Gamma.$$

Then, since $\vartheta^0 \circ \tau_{v'} = {}^{\vartheta}\tau_{v'} \circ \vartheta^0$

$$\mathrm{pr}_\alpha(\gamma) = (\tau \circ \omega_\alpha^0)(\mathrm{pr}_{\alpha_0}(\gamma)) \tag{12}$$

for all $\gamma \in \Gamma$, where the isomorphism $\omega_\alpha\colon K_{v'} \to k$ and the k-isomorphism $\tau_\alpha\colon {}^{\omega_\alpha}\mathbf{H} \to \mathbf{G}_\alpha$ are defined by the equalities $\omega_\alpha = \tau \circ \omega_{v'}$ and $\tau_\alpha = \eta \circ {}^{\vartheta}\tau_{v'}$. Since τ_α is a k-isomorphism, (12) implies that

$$\mathrm{Tr}\,\mathrm{Ad}\,\mathrm{pr}_\alpha(\gamma) = \omega_\alpha(\mathrm{Tr}\,\mathrm{Ad}\,\mathrm{pr}_{\alpha_0}(\gamma))$$

for all $\gamma \in \Gamma$. On the other hand, the sets $\mathrm{Tr}\,\mathrm{Ad}\,\mathrm{pr}_\alpha(\Gamma)$ and $\mathrm{Tr}\,\mathrm{Ad}\,\mathrm{pr}_{\alpha_0}(\Gamma)$ generate L_α and $K = L_{\alpha_0}$ respectively. Thus,

$$\omega_\alpha(K) = L_\alpha. \tag{13}$$

Let us denote by k'_α the closure in k_α of the subfield L_α. Then in view of (13) $(\omega_\alpha|_K, k'_\alpha)$ is a completion of the field K, and hence there is a valuation $v \in \mathcal{R}$ such that $\omega_{\alpha|K}$ extends to a continuous isomorphism $\omega_v\colon K_v \to k'_\alpha$. We shall show that $\delta(v) = \alpha$ (and hence (c)). Since $\omega_{\alpha|K} = \omega_{v|K}$, we can rewrite (12) in the form

$$\mathrm{pr}_\alpha(\gamma) = \tau_\alpha(\omega_v^0(\mathrm{pr}_{\alpha_0}(\gamma)), \ \gamma \in \Gamma. \tag{14}$$

Recalling the definition of the map δ, we see that the equality $\delta(v) = \alpha$ is a consequence of (14) and the following two assertions:

(U) the morphism τ_α is defined over L_α, and hence over k_α;
(V) $k'_\alpha = k_\alpha$.

We first prove (U). Since $\omega_\alpha(K) = L_\alpha$ and the subgroup $\mathrm{pr}_{\alpha_0}(\Gamma)$ is contained in $\mathbf{H}(K)$ and is Zariski dense in $\mathbf{H} = \mathbf{G}_{\alpha_0}$ (see Lemma 2.1(i)), the subgroup $\omega_\alpha^0(\mathrm{pr}_{\alpha_0}(\Gamma))$ is contained in $({}^{\omega_\alpha}\mathbf{H})(L_\alpha)$ and is Zariski dense in ${}^{\omega_\alpha}\mathbf{H}$. On the other hand,

1) in view of (12)
$$\tau_\alpha(\omega_\alpha^0(\mathrm{pr}_{\alpha_0}(\Gamma))) = \mathrm{pr}_\alpha(\Gamma) \subset \mathbf{G}_\alpha(L_\alpha);$$
2) if $f\colon \mathbf{M} \to \mathbf{M}'$ is an l-variety morphism, $B \subset \mathbf{M}(l)$, $\bar{B} = \mathbf{M}$, and $f(B) \subset \mathbf{M}'(l)$, then the morphism f is defined over l (see I.0.11(ii)).

Thus, τ_α is defined over L_α.

We now turn to the proof of (V). Let $\tilde{\mathbf{G}}_\alpha$ be the simply connected covering of $\mathbf{G}_\alpha$ and $\pi\colon \tilde{\mathbf{G}}_\alpha \to \mathbf{G}_\alpha$ the central L_α-isogeny. Now $\mathrm{pr}_\alpha(\Gamma) \subset \mathbf{G}_\alpha(L_\alpha) \subset \mathbf{G}_\alpha(k'_\alpha)$, the subgroup Γ (and hence the subgroup $\mathrm{pr}_\alpha(\Gamma)$) is finitely generated, and the subgroup $\pi(\tilde{\mathbf{G}}_\alpha(k'_\alpha))$ is closed and normal in $\mathbf{G}_\alpha(k'_\alpha)$ (see Proposition I.2.3.4(i)). Furthermore, the factor group $\mathbf{G}_\alpha(k'_\alpha)/\pi(\tilde{\mathbf{G}}_\alpha(k'_\alpha))$ is commutative and torsion. Thus the subgroup $F_\alpha \cap \pi(\tilde{\mathbf{G}}_\alpha(k'_\alpha))$ is of finite index in F_α, where F_α denotes the closure of the subgroup $\mathrm{pr}_\alpha(\Gamma)$ in $\mathbf{G}_\alpha(k_\alpha)$. But since $\alpha \in A - A_0$ and Γ has property (QD) in G, it follows that the subgroup $F_\alpha \cap \pi(\tilde{\mathbf{G}}_\alpha(k_\alpha))$ is of finite index in $\pi(\tilde{\mathbf{G}}_\alpha(k_\alpha))$. Therefore $\pi(\tilde{\mathbf{G}}_\alpha(k'_\alpha))$ is of finite index in $\pi(\tilde{\mathbf{G}}_\alpha(k_\alpha))$ and, since $\operatorname{Ker}\pi$ is finite, $\tilde{\mathbf{G}}_\alpha(k'_\alpha)$ is of finite index in $\tilde{\mathbf{G}}_\alpha(k_\alpha)$. From this and the closeness of the subfield $k'_\alpha \subset k_\alpha$ we deduce that $k'_\alpha = k_\alpha$ (see Proposition I.2.5.5).

(2.12) *Remark.* As we mentioned earlier, Theorems A and B of 1.9 were proved in full generality in [Ve]. We note which changes should be made in the above proofs in order to handle the case where "the set $A_{\mathrm{adm}} \overset{\mathrm{def}}{=} \{\alpha \in A \mid$ at least one of almost k_α-simple factors of the group $\mathbf{G}_\alpha$ is admissible$\}$ is not empty." These changes are the following:

1) in Steps 2 and 4 we must assume that $\alpha \in A_{\mathrm{adm}}$;
2) in Lemmas 1, 2, and 3 in Step 3 we have to impose the additional condition that either $\alpha_0 \in A_{\mathrm{adm}}$ or $A_0 \subset A_{\mathrm{adm}}$;
3) in Step 5 we have to choose as α_0 an element of A_{adm};
4) in Step 8 it should be noted that in view of (10) $A_0 \subset A_{\mathrm{adm}}$, and hence the existence of $\alpha' \in A_0 \subset A_{\mathrm{adm}}$ and a k-isomorphism $\eta\colon {}^{\vartheta}\mathbf{G}_{\alpha'} \to \mathbf{G}_\alpha$ implies that $\alpha \in A_{\mathrm{adm}}$.

Furthermore, in Step 8, before proving the inclusion $\alpha \in A_{\mathrm{adm}}$ (i.e. before the equality (ii)), we have to consider the field generated by the entries of the matrices representing the elements of $\mathrm{Ad}\,\mathrm{pr}_\alpha(\Gamma)$ in some basis of the Lie algebra $\mathrm{Lie}(\mathbf{G}_\alpha)_{k_\alpha}$ instead of L_α.

3. Finite Generation of Lattices

In this section A, k_α, $\mathbf{G}_\alpha$, G, $\operatorname{rank} G$, G_B, G_B^+, pr_B, pr_α, and A_0 will denote the same objects as in Sections 1 and 2.

(3.1) We recall some results on finite generation of lattices:

(i) *If H is a compactly generated locally compact group and Λ is a cocompact lattice in H, then Λ is finitely generated* (see I.0.40).

(ii) *If H is a connected Lie group and Λ is an arbitrary lattice in H, then Λ is finitely generated* (see [Rag 5], 6.18).

(iii) *Suppose for each $\alpha \in A$ that the k_α-rank of any almost k_α-simple factor of the group $\mathbf{G}_\alpha$ is not equal to 1. Then any lattice $\Gamma \subset G$ is finitely generated* (see Theorem III.5.7(c)).

(iv) *Suppose the set $\{\alpha \in A \mid$ the k_α-rank of at least one of the almost k_α-simple factors of the group $\mathbf{G}_\alpha$ is greater than $1\}$ is not empty, and for each $\alpha \in A$, the group $\mathbf{G}_\alpha$ has no k_α-anisotropic factors. Then any irreducible lattice $\Gamma \subset G$ is finitely generated* (see Theorem B (c) of III.5.9).

Since the groups $\mathbf{G}_\alpha(k_\alpha)$ are compactly generated (see Corollary I.2.3.5), the group G is compactly generated. So (i) implies

(v) *If Γ is a cocompact lattice in G, then Γ is finitely generated.*

If for each $\alpha \in A$, the field k_α is isomorphic either to $\mathbb{R}$ or $\mathbb{C}$, then the Lie groups $\mathbf{G}_\alpha(k_\alpha)$ have only finitely many connected component (see I.0.31). Hence, the Lie group G has only finitely many connected components. On the other hand, if H is a Lie group with finitely many connected components and Λ is a lattice in H, then (ii) implies that Λ is finitely generated. So the following holds.

(vi) *Suppose for each $\alpha \in A$, the field k_α is isomorphic either to $\mathbb{R}$ or $\mathbb{C}$. Then any lattice $\Gamma \subset G$ is finitely generated.*

(3.2) In Section 1, when reducing Theorem 1.13 to Theorems A and B of 1.9, we made use of the following

Theorem. *Let $\Gamma \subset G$ be an irreducible lattice with property (QD) in G. Suppose at least one of the following two conditions holds.*

(i) There is $\alpha \in A$ such that $\operatorname{char} k_\alpha = 0$.

(ii) There is $\alpha \in A$ such that the k_α-rank of at least one of almost k_α-simple factors of the group $\mathbf{G}_\alpha$ is greater than 1.

Then Γ is finitely generated.

The proof of this theorem will be presented in 3.8 below. Before proceeding we establish some auxiliary assertions.

(3.3) Lemma. *Let H be a topological group, $\Lambda \subset H$ a dense subgroup, and $W \subset H$ an open subset. If W generates H, then $W \cap \Lambda$ generates Λ.*

Proof. Let $h \in \Lambda$. Since W is open and generates H and Λ is dense in H, it follows that $W \cap \Lambda$ generates a dense subgroup of H. Therefore there exist $w_1,\dots,w_m \in (W \cap \Lambda) \cup (W \cap \Lambda)^{-1}$ such that $w_1 \cdot \ldots \cdot w_m \in hW$. In other words, $w_1 \cdot \ldots \cdot w_m = hw$, where $w \in W$. But $w_i \in \Lambda$ and $h \in \Lambda$. So $w = h^{-1} \cdot w_1 \cdot \ldots \cdot w_m \in \Lambda$. Thus, $w \in W \cap \Lambda^{-1}$, and hence the element $h = w_1 \cdot \ldots \cdot w_m \cdot w^{-1}$ belongs to the subgroup generated by $W \cap \Lambda$. □

(3.4) Proposition. *Suppose H and F are locally compact groups and Ω is a subgroup of $H \times F$. Denote by π_H the natural projection $H \times F \to H$ and by $\overline{\pi_H(\Omega)}$ the closure of the subgroup $\pi_H(\Omega)$ in H. Assume the following conditions are satisfied:*

(i) the group $\overline{\pi_H(\Omega)}$ is compactly generated;

(ii) there is an open subgroup U of H such that the subgroup $(U \times F) \cap \Omega$ is finitely generated.

Then Ω is finitely generated.

Proof. We first remark that replacing H by $\overline{\pi_H(\Omega)}$ we can restrict ourselves to the case where the subgroup $\pi_H(\Omega)$ is dense in H. Since U is open in H and the group $H = \overline{\pi_H(\Omega)}$ is compactly generated, there is a finite set $Y \subset H$ such that

$W \overset{\text{def}}{=} U \cdot Y$ generates H. The subset W is open in H and generates H and the subgroup $\pi_H(\Omega)$ is dense in H. It then follows from Lemma 3.3 that $W \cap \pi_H(\Omega)$ generates $\pi_H(\Omega)$, and hence $(W \times F) \cap \Omega$ generates Ω. Now to complete the proof it remains to observe that, since W consists of finitely many cosets modulo U, $(W \times F) \cap \Omega$ consists of finitely many cosets modulo $(U \times F) \cap \Omega$. □

(3.5) Lemma. *Let k be a local field of characteristic* 0 *and H an algebraic k-group. Then*

(i) there is a neighbourhood W of the identity e in $\mathbf{H}(k)$ containing no finite subgroups of $\mathbf{H}(k)$ apart from $\{e\}$;

(ii) if the field k is non-archimedean, then there is an open neighbourhood $U \subset \mathbf{H}(k)$ of the identity such that each element $h \in U - \{e\}$ generates a non-discrete subgroup.

Assertion (i) is a particular case of Corollary 1 of §4 of Chapter III in [Bou 5]. We shall show (ii). Since k is a non-archimedean local field, k is totally disconnected. Hence, the group $\mathbf{H}(k)$ is totally disconnected. So W contains an open compact subgroup U, where W is the neighbourhood in (i) above. Let $h \in U - \{e\}$. Then the subgroup $\{h^m \mid m \in \mathbb{Z}\} \subset U$ is infinite (because $U \subset W$) and relatively compact, and hence non-discrete. Thus, U is the desired neighbourhood. □

(3.6) Lemma. (See [Rag 5], Theorem 1.12.) *Let H be a locally compact second countable group and Λ a non-cocompact lattice in H. Then there exist two sequences $\{h_n\}_{n \in \mathbb{N}^+}$ in H and $\{\gamma_n\}_{n \in \mathbb{N}^+}$ in $\Lambda - \{e\}$ such that $\lim_{n \to \infty} h_n \gamma_n h_n^{-1} = e$.*

(3.7) Proposition. *Suppose for each $\alpha \in A$, the field k_α is non-archimedean and* $\operatorname{char} k_\alpha = 0$. *Then any lattice $\Gamma \subset G$ is cocompact.*

Proof. It follows from Lemma 3.5(ii) that there is a neighbourhood $U \subset G$ of the identity such that each $g \in G - \{e\}$ generates a non-discrete subgroup in G. Since the subgroup $g\Gamma g^{-1}$ is discrete, $U \cap (g\Gamma g^{-1}) = \{e\}$ for each $g \in G$. From this and Lemma 3.6 we deduce that the lattice Γ is cocompact. □

(3.8) *Proof of Theorem 3.2.* Arguing as in the proof of Theorem III.5.9 (see III.6.9) and making use of remark (iv) of 1.2 we come to the case in which the groups $\mathbf{G}_\alpha$, $\alpha \in A$, are almost k_α-simple. Since the lattice Γ is irreducible and has property (QD) in G, it follows from Lemma 2.4(ii) that $\operatorname{pr}_{A_0}(\Gamma)$ is an irreducible lattice in G_{A_0}. On the other hand, since the subgroup $\Gamma \cap G_{A - A_0}$ is finite (see Lemma 2.4(i)), the finite generation of Γ is equivalent to that of $\operatorname{pr}_{A_0}(\Gamma)$. Therefore, replacing G and Γ by G_{A_0} and $\operatorname{pr}_{A_0}(\Gamma)$, we may assume that for each $\alpha \in A$ the almost k_α-simple group $\mathbf{G}_\alpha$ is isotropic over k_α. Thus, if condition (ii) is satisfied, then, according to (iv) in 3.1, Γ is finitely generated.

Suppose now that condition (i) is satisfied. We set $B = A_{\mathbb{R}} \overset{\text{def}}{=} \{\alpha \in A \mid k_\alpha$ is isomorphic to $\mathbb{R}$ or $\mathbb{C}\}$ if $A_{\mathbb{R}}$ is not empty and $B = \{\alpha \in A \mid \operatorname{char} k_\alpha = 0\}$ if $A_{\mathbb{R}}$

is empty. For each $\alpha \in A - B$ the field k_α is totally disconnected. Therefore the group G_{A-B} is totally disconnected, and hence G_{A-B} contains an open compact subgroup F. We set $\Gamma_F = \Gamma \cap (G_B \cdot F)$. Then, since Γ is a lattice in G, $\mathrm{pr}_B(\Gamma_F)$ is a lattice in G_B and the subgroup $\Gamma_F \cap \mathrm{Ker}\,\mathrm{pr}_B = \Gamma \cap F$ is finite. On the other hand, assertions (v) and (vi) of 3.1 and Proposition 3.7 imply that any lattice in G_B is finitely generated. So the group $\Gamma_F = \Gamma \cap (G_B \cdot F)$ is finitely generated. Now in view of Proposition 3.4 it remains to show that the group $\overline{\mathrm{pr}_{A-B}(\Gamma)}$ is compactly generated, where $\overline{\mathrm{pr}_{A-B}(\Gamma)}$ is the closure of the subgroup $\mathrm{pr}_{A-B}(\Gamma)$ in G_{A-B}. But this is a consequence of the following two observations.

(i) Since the set B is not empty (because condition (i) is satisfied, the lattice Γ is irreducible, and for each $\alpha \in A$ the group $\mathbf{G}_\alpha$ is almost k_α-simple and k_α-isotropic), it follows from Theorem II.6.7(a) that $\overline{\mathrm{pr}_{A-B}(\Gamma)}$ contains G^+_{A-B}.

(ii) Since G^+_C is compactly generated (in view of Corollary I.2.3.5) and the factor group G_C/G^+_C is compact (in view of Theorem I.2.3.1(b)), for every $C \subset A$ any closed subgroup of G_C containing G^+_C is compactly generated. □

(3.9) Theorem. *Suppose* $\mathrm{char}\, k_\alpha = 0$ *for each* $\alpha \in A$. *Then any lattice* $\Gamma \subset G$ *is finitely generated.*

Proof. As in the proof of Theorem 3.2 we may assume that for each $\alpha \in A$ the group $\mathbf{G}_\alpha$ is almost k_α-simple and k_α-isotropic. Partition the set A into subsets $A_1, \dots, A_n$ so that $G_{A_i} \cap \Gamma$ is an irreducible lattice in G_{A_i}, $1 \leq i \leq n$, and the subgroup $(G_{A_1} \cap \Gamma) \cdot \ldots \cdot (G_{A_n} \cap \Gamma)$ is of finite index in Γ. In view of Theorem 3.2 and remark (i) in 1.2, the groups $G_{A_i} \cap \Gamma$ are finitely generated. Thus, the finitely generated subgroup $(G_{A_1} \cap \Gamma) \cdot \ldots \cdot (G_{A_n} \cap \Gamma)$ is of finite index in Γ. Hence, Γ is finitely generated. □

(3.10) Let us fix a finite field F_q and set, with notation of 1.7(iv), $k = F_q((t))$ and $M = F_q[t^{-1}]$. The group $\mathbf{SL}_2(M)$ is not finitely generated and furthermore, the factor group of this group modulo its commutator subgroup contains an infinite-dimensional vector space over the field in $p = \mathrm{char}\, F_q$ elements (see [Ser 1], Theorem 6). But, as was observed in 1.7(iv), $\mathbf{SL}_2(M)$ is a non-cocompact lattice in $\mathbf{SL}_2(k)$. Thus, we have constructed an example of a non-cocompact lattice in $\mathbf{SL}_2(k)$ which is not a finitely generated group. Analogous examples can be constructed for an arbitrary connected non-commutative almost k-simple k-group of k-rank 1 (see [Be 4]). Moreover, if $\mathbf{G}$ is such a k-group then any non-cocompact lattice in $\mathbf{G}(k)$ is not finitely generated (see [Lu 2]). On the other hand, the following assertion is true.

(*) *if* $\mathrm{rank}\, G \geq 2$, *then any irreducible lattice* $\Gamma \subset G$ *with property (QD) is finitely generated.*

In view of Theorem 3.2 and assertion (v) of 3.1, it remains to verify this assertion in the case where $\mathrm{char}\, k_\alpha \neq 0$, $\mathrm{rank}_{k_\alpha} \mathbf{G}_\alpha \leq 1$ for all $\alpha \in A$, and the lattice Γ is non-cocompact. Then, arguing as in the proof of Theorem 3.2 (see 3.8), we come to the case in which the groups $\mathbf{G}_\alpha$, $\alpha \in A$, are almost k_α-simple and of k_α-rank 1. But in this case, the assertion (*) was recently proved by Raghunathan (see [Rag 13]).

4. Consequences of the Arithmeticity Theorems I

In this section A, k_α, $\mathbf{G}_\alpha$, G, and $\operatorname{rank} G$ will denote the same objects as in Section 1.

(4.1) Let $\mathbf{G}_{\alpha i}$, $i \in I_\alpha$, be the (absolutely) almost simple factors of the group $\mathbf{G}_\alpha$. We shall say that the group G is *typewise homogeneous* if the following conditions are satisfied:

(i) $\operatorname{char} k_\alpha = \operatorname{char} k_\beta$ for all $\alpha, \beta \in A$;

(ii) for all $\alpha, \beta \in A$, $i \in I_\alpha$, $j \in I_\beta$, the groups $\mathbf{G}_{\alpha i}$ and $\mathbf{G}_{\beta j}$ are strictly isogeneous or, equivalently, have common type (in the sense of I.0.27).

With K, $\mathbf{H}$, and S as in 1.4, if there is a continuous homomorphism $\varphi: \prod_{v \in S} \mathbf{H}(K_v) \to G$ belonging to the class Ψ, the group G is typewise homogeneous. This follows easily from the definition of the class Ψ. So from Theorem A of 1.9, the assertion (*) in 3.10 and the definition of an arithmetic lattice we deduce

(4.2) Corollary. *Suppose that* $\operatorname{rank} G \geq 2$ *and the group* G *is not typewise homogeneous. Then* G *does not contain irreducible lattices with property (QD) in* G.

(4.3) If $\mathbf{G}$ is a connected non-commutative almost k-simple k-group, then all absolutely almost simple factors of $\mathbf{G}$ have common type. Therefore if the group G is not typewise homogeneous and for each $\alpha \in A$ the group $\mathbf{G}_\alpha$ has no k_α-anisotropic factors, then $\operatorname{rank} G \geq 2$. From this, remark 1.2(i), and Corollary 4.2 we deduce the following.

(4.4) Corollary. *Suppose the group* G *is not typewise homogeneous and for each* $\alpha \in A$ *the group* $\mathbf{G}_\alpha$ *has no* k_α*-anisotropic factors. Then* G *does not contain irreducible lattices.*

(4.5) The following is a particular case of Corollary 4.4.

Corollary. *Let* k *be a local field and* $\mathbf{G}$ *be a connected semisimple* k*-group without* k*-anisotropic factors. Assume* $\mathbf{G}$ *has two non-isogeneous absolutely almost simple factors. Then* $\mathbf{G}(k)$ *does not contain irreducible lattices.*

(4.6) Assertions (i)–(iii) below are special cases of Corollary 4.4.

(ii) *Pick* $n, m \in \mathbb{N}^+$, $n > m \geq 2$. *Then for any local fields* k *and* k' *the group* $\mathbf{SL}_n(k) \times \mathbf{SL}_m(k')$ *does not contain irreducible lattices. In particular, the groups* $\mathbf{SL}_n(\mathbb{R}) \times \mathbf{SL}_m(\mathbb{R})$, $\mathbf{SL}_n(\mathbb{R}) \times \mathbf{SL}_m(\mathbb{C})$, *and* $\mathbf{SL}_n(\mathbb{R}) \times \mathbf{SL}_m(\mathbb{Q}_p)$ *do not contain irreducible lattices.*

We remark that for each $n \geq 2$ the groups $\mathbf{SL}_n(\mathbb{R}) \times \mathbf{SL}_n(\mathbb{R})$, $\mathbf{SL}_n(\mathbb{R}) \times \mathbf{SL}_n(\mathbb{C})$, and $\mathbf{SL}_n(\mathbb{R}) \times \mathbf{SL}_n(\mathbb{Q}_p)$ contain irreducible lattices (see examples (ii) and (v) (b) in 1.7).

(ii) *Let k and k' be two local fields and $n \in \mathbb{N}^+$, $n \geq 2$. Assume that* $\operatorname{char} k \neq \operatorname{char} k'$. *Then the group* $\mathbf{SL}_n(k) \times \mathbf{SL}_n(k')$ *does not contain irreducible lattices.*

(iii) *Denote by* $\mathbf{SO}_{n,1}$ *the group of linear transformations of determinant* 1 *leaving the quadratic form* $x_1^2 + \ldots + x_n^2 - x_{n+1}^2$ *invariant. Pick* $n, m \in \mathbb{N}^+$, $n > m \geq 2$. *Assume that if* $m = 2$, *then* $n \geq 4$. *Then the group* $\mathbf{SO}_{n,1}(\mathbb{R}) \times \mathbf{SO}_{m,1}(\mathbb{R})$ *does not contain irreducible lattices.*

Remark. In the light of examples (vi) (3) and (vi) (5) in 1.7, for each $n \in \mathbb{N}^+$, the group $\mathbf{SO}_{n,1}(\mathbb{R}) \times \mathbf{SO}_{n,1}(\mathbb{R})$ contains irreducible lattices (both cocompact and non-cocompact). The group $\mathbf{SO}_{2,1}(\mathbb{R}) \times \mathbf{SO}_{3,1}(\mathbb{R})$ contains irreducible lattices as well, because the group $\mathbf{SO}_{3,1}$ is R-isogenous to the group $R_{\mathbb{C}/\mathbb{R}}\mathbf{SO}_{2,1}$, and according to example (vi) (6) in 1.7 the group $\mathbf{SO}_{2,1}(\mathbb{R}) \times \mathbf{SO}_{2,1}(\mathbb{C})$ contains irreducible lattices.

(4.7) In [Bo-Hard] the following was established.

(A) Theorem. *Let k be a local field of characteristic* 0, $\mathbf{F}$ *a reductive k-group. Then* $\mathbf{F}(k)$ *contains cocompact lattices.*

The following plays a basic role in the proof of Theorem A.

(B) Theorem. *Let K be a global field of characteristic* 0, *S a finite set of inequivalent valuations of K, and for each $v \in S$ let there be given a connected non-commutative absolutely almost simple K_v-group $\mathbf{H}_v$. Suppose that for each $v \in S$ the group $\mathbf{H}_v$ is either simply connected or adjoint. Assume in addition that for all $v, v' \in S$ the algebraic groups $\mathbf{H}_v$ and $\mathbf{H}_{v'}$ are isomorphic or, equivalently, have common type. Then there is a K-group $\mathbf{H}$ such that $\mathbf{H}$ is K_v-isomorphic to $\mathbf{H}_v$ for all $v \in S$.*

It is not hard to deduce the following results from Theorem B.

(C) Theorem. *Suppose the group G is typewise homogeneous and* $\operatorname{char} k_\alpha = 0$ *for all $\alpha \in A$. Then G contains cocompact irreducible arithmetic lattices.*

It is very likely that by analogous methods one can prove that

(i) *if G is typewise homogeneous, then also in the case where* $\operatorname{char} k_\alpha > 0$, *$G$ contains an irreducible arithmetic lattice;*

(ii) *if G is typewise homogeneous and all almost simple factors of the groups $\mathbf{G}_\alpha$ are of type $\mathbb{A}$, then G contains cocompact irreducible arithmetic lattices.*

We remark that if the almost simple factors of the groups $\mathbf{G}_\alpha$ are not all of type $\mathbb{A}$, then in assertion (i) the lattice Γ is not cocompact (see remark (viii) in 1.6). In connection with assertion (ii) we state one other theorem established in the above-mentioned paper [Bo-Hard].

(D) Theorem. *Let k be a local field (not necessarily of characteristic* 0*) and let $\mathbf{F}$ be a connected reductive k-group. Suppose that the almost simple factors of the* *commutator subgroup* $\mathscr{D}(\mathbf{F})$ *of* $\mathbf{F}$ *are of type* $\mathbb{A}$. *Then* $\mathbf{F}(k)$ *contains cocompact lattices.*

(4.8) We now present a corollary which shows that if $\operatorname{char} k_\alpha > 0$, then as a rule the group G does not contain cocompact lattices.

Corollary.

(i) Suppose that $\operatorname{rank} G \geq 2$ *and that the set* $\{\alpha \in A \mid \operatorname{char} k_\alpha > 0$ *and at least one of almost simple factors of* $\mathbf{G}_\alpha$ *is not of type* $\mathbb{A}\}$ *is not empty. Then* G *does not contain irreducible cocompact lattices with property (QD) in* G.

(ii) Suppose $\operatorname{rank} G \geq 2$, *for each* $\alpha \in A$ *the group* $\mathbf{G}_\alpha$ *is admissible and has no* k_α*-anisotropic factors, and the set* $\{\alpha \in A \mid \operatorname{char} k_\alpha > 0$ *and at least one of almost simple factors of* $\mathbf{G}_\alpha$ *is not of type* $\mathbb{A}\}$ *is not empty. Then* G *does not contain irreducible cocompact lattices.*

(iii) Suppose that there exist $\alpha \in A$ *and an almost* k_α*-simple factor* $\mathbf{G}_{\alpha i}$ *of* $\mathbf{G}_\alpha$ *such that* $\operatorname{char} k_\alpha > 0$, $\operatorname{rank}_{k_\alpha} \mathbf{G}_{\alpha i} \geq 2$, *and* $\mathbf{G}_{\alpha i}$ *is not of type* $\mathbb{A}$. *Then* G *does not contain cocompact lattices.*

(iv) Let k *be a local field of positive characteristic and let* $\mathbf{F}$ *be a connected non-commutative almost* k*-simple* k*-group. Suppose* $\operatorname{rank}_k \mathbf{F} \geq 2$ *and* $\mathbf{F}$ *is not of type* $\mathbb{A}$. *Then* $\mathbf{F}(k)$ *contains no cocompact lattices.*

Assertion (i) is a consequence of Theorem A in 1.9 and remark (viii) in 1.6. In view of remark (i) of 1.2, assertions (ii) and (iv) are special case of (i). If Γ is a cocompact lattice in G, then A can be partitioned into subsets $A_1, \dots, A_n$ so that $G_{A_i} \cap \Gamma$ is an irreducible cocompact lattice in G_{A_i}, $1 \leq i \leq n$. Therefore as in the proof of Theorem 3.2, to deduce (iii) from (ii) it suffices to make a reduction to the case where for each $\alpha \in A$ the group $\mathbf{G}_\alpha$ is almost k_α-simple and k_α-isotropic.

(4.9) *Remarks.*

(i) Corollary 4.3 shows that in case $\operatorname{card} A \geq 2$ the group G does not as a rule contain irreducible lattices.

(ii) Suppose given $n \geq 3$, k a local field of characteristic > 2, and Φ the quadratic form $x_1x_2 + x_3x_4 + \ldots + x_{2n-1}x_{2n}$ over k. As before we denote by $\mathbf{SO}_\Phi$ the k-subgroup of $\mathbf{GL}_{2n}$ consisting of unimodular matrices leaving Φ invariant. The group $\mathbf{SO}_\Phi$ is absolutely almost simple and of type $\mathbb{D}_n$. On the other hand, since the Witt index of Φ equals n, $\operatorname{rank}_k \mathbf{SO}_\Phi = n \geq 3$. So by Corollary 4.8(iv) the group $\mathbf{SO}_\Phi(k)$ does not contain cocompact lattices. But $\mathbf{SO}_\Phi(k)$ contains non-cocompact lattices (see example (viii) in 1.7). Thus, we have constructed a locally compact group containing non-cocompact lattices but not containing cocompact ones. This example seems to be the first of that kind.

(4.10) Before proceeding the next series of consequences we introduce some notation and one definition.

As in I.2.6, we denote by $\mathscr{P}$ the set of primes supplemented by ∞ and for any local field k of characteristic 0, define $p(k) \in \mathscr{P}$ via the following condition:

k is isomorphic to a finite field extension of $\mathbb{Q}_{p(k)}$ (as before, $\mathbb{Q}_\infty = \mathbb{R}$). The set $\mathscr{B}$ is naturally identified with the set of valuations of $\mathbb{Q}$ and for every $S \subset \mathscr{B}$, $\mathbb{Q}(S)$ denotes the ring of S-integers over the field $\mathbb{Q}$ (see I.0.32). If $S \subset B$, K is an algebraic field extension of $\mathbb{Q}$, and $x \in K$, then it is an easy exercise to check that the following are equivalent:

(a) an element x is an integer over the ring $\mathbb{Q}(S)$;
(b) $x = tz$, where $t \in \mathbb{Q}(S)$ and z is an algebraic integer.

If conditions (a) and (b) are satisfied for $x \in K$, then x is called an *S-integer*. The set of S-integers of K form a subring of K that will be denoted by $K(S)$. It can easily be verified that if the extension K is finite, then $K(S) = K(S')$, where S' is the set of valuations of K which are extensions of valuations in S.

(4.11) Corollary. *Let Γ be a lattice in G, $\alpha_0 \in A$, K an algebraic field extension of $\mathbb{Q}$ which is contained in k_{α_0}, $\mathbf{H}$ an algebraic K-group, and $\varphi\colon \mathbf{G}_{\alpha_0} \to \mathbf{H}$ a k_{α_0}-morphism of k_{α_0}-groups with $\varphi(\mathrm{pr}_{\alpha_0}(\Gamma)) \subset \mathbf{H}(K)$. We set $S_A = \{p(k_\alpha) \mid \alpha \in A$ and $\operatorname{char} k_\alpha = 0\}$. Suppose $\operatorname{rank} G \geq 2$ and the lattice Γ is irreducible and has property (QD) in G. Then the subgroup $\varphi(\mathrm{pr}_{\alpha_0}(\Gamma)) \cap \mathbf{H}(K(S_A))$ is of finite index in $\varphi(\mathrm{pr}_{\alpha_0}(\Gamma))$ or, in other words, there is a subgroup Γ_1 of finite index in Γ such that $\varphi(\mathrm{pr}_{\alpha_0}(\Gamma_1)) \subset \mathbf{H}(K(S_A))$.*

It is not hard to deduce this corollary from Theorem 1.13 making use of Remark 2.12. But it can also be proved directly by applying the superrigidity theorems and therefore eliminating the lengthy reduction of the arithmeticity theorems to the results on superrigidity. We present this proof for the case in which the group $\mathbf{G}_\alpha$ has no k_α-anisotropic factors for any $\alpha \in A$ (for the general case we refer to remark 4.12(ii)).

Since the set $\{\alpha \in A \mid \operatorname{char} k_\alpha = 0\}$ includes α_0, and hence is not empty, by Theorem 3.2 the group Γ (and hence the group $\varphi(\mathrm{pr}_{\alpha_0}(\Gamma)) \subset \mathbf{H}(K)$) is finitely generated. On the other hand, since K is an algebraic field extension of $\mathbb{Q}$, K is representable as a union of finite field extensions of $\mathbb{Q}$. So we can assume that K is a finite field extension of $\mathbb{Q}$. Let us denote by $\mathscr{R}$ the set of all (inequivalent) valuations of K and by $\mathscr{R}_\infty \subset \mathscr{R}$ the set of archimedean valuations of K. Since the group $\varphi(\mathrm{pr}_{\alpha_0}(\Gamma)) \subset \mathbf{H}(K)$ is finitely generated and $\mathbf{H}(K)$ coincides with the union of the subgroups $\mathbf{H}(K(\tilde{S}))$, where $\tilde{S} \subset \mathscr{R}$ runs through all finite subsets of $\mathscr{R}$, there is a finite subset $S_0 \subset \mathscr{R}$ such that

$$\varphi(\mathrm{pr}_{\alpha_0}(\Gamma)) \subset \mathbf{H}(K(S_0)). \tag{1}$$

Since the subgroup $\mathrm{pr}_{\alpha_0}(\Gamma)$ is Zariski dense in $\mathbf{G}_{\alpha_0}$ (see Lemma 2.1(i)), replacing $\mathbf{H}$ by $\varphi(\mathbf{G}_{\alpha_0})$ we may assume that $\mathbf{H}$ is semisimple and the subgroup $\varphi(\mathrm{pr}_{\alpha_0}(\Gamma))$ is Zariski dense in $\mathbf{H}$.

We denote by S' the set of those valuations of K which are extensions of valuations from S_A. Let $v \in \mathscr{R} - \mathscr{R}_\infty - S'$. Then for each $\alpha \in A$, the fields k_α and K_v are of different type (in the sense of I.0.31). So by Corollary VII.5.16 the subgroup $\varphi(\mathrm{pr}_{\alpha_0}(\Gamma))$ is relatively compact in $\mathbf{H}(K_v)$. With $\mathbf{H}$ viewed as a K-subgroup of $\mathbf{GL}_n$, we set $\mathbf{H}(\mathscr{O}_v) = \mathbf{H} \cap \mathbf{GL}_n(\mathscr{O}_v)$, where O_v is the ring of integers of the field

K_v. Since the subgroup $\mathbf{H}(\mathcal{O}_v)$ is open in $\mathbf{H}(K_v)$ and the subgroup $\mathrm{pr}_{\alpha_0}(\Gamma)$ is relatively compact in $\mathbf{H}(K_v)$, the subgroup $\varphi(\mathrm{pr}_{\alpha_0}(\Gamma) \cap \mathbf{H}(\mathcal{O}_v)$ is of finite index in $\varphi(\mathrm{pr}_{\alpha_0}(\Gamma))$ for each $v \in \mathcal{R} - \mathcal{R}_\infty - S'$. But $\mathbf{H}(K(\tilde{S})) = \{h \in \mathbf{H}(K) \mid h \in \mathbf{H}(\mathcal{O}_v)$ for all $v \in \mathcal{R} - \mathcal{R}_\infty - \tilde{S}\}$ for any $\tilde{S} \subset \mathcal{R}$. Therefore (i) and the finiteness of the set S_0 imply that $\varphi(\mathrm{pr}_{\alpha_0}(\Gamma)) \cap \mathbf{H}(K(S'))$ is of finite index in $\mathrm{pr}_{\alpha_0}(\Gamma)$. All that remains now is to observe that $K(S_A) = K(S')$ (see 4.10). □

(4.12) *Remarks.*

(i) During the course of the proof of Corollary 4.11 we in fact repeated some of the arguments of Step 6 of the proof of Theorem A and B in 1.9 (see 2.11).

(ii) Making use of Lemma 2.4(ii), it is not hard to show that in the statement of Corollary VII.5.16 as well as of some other results in §5 of Chapter VII the assumption "for each $\alpha \in A$ the group $\mathbf{G}_\alpha$ has no k_α-anisotropic factors" can be replaced by the assumption "Γ has property (QD) in G". Having done this we can directly extend the proof presented in 4.11 to the general case.

(4.13) The following assertion is a particular case of Corollary 4.11.

Corollary. *Let $\mathbf{G}$ be a connected semisimple $\mathbb{R}$-group, Γ_0 a lattice in $\mathbf{G}(\mathbb{R})$, K an algebraic field extension of $\mathbb{Q}$ with $K \subset \mathbb{R}$, $\mathbf{H}$ an algebraic K-group, and $\varphi\colon \mathbf{G} \to \mathbf{H}$ an $\mathbb{R}$-group $\mathbb{R}$-morphism. Denote by L the ring of algebraic integers of K and by $\mathbf{G}^{\mathrm{is}} \subset \mathbf{G}$ the almost direct product of the $\mathbb{R}$-isotropic factors of $\mathbf{G}$. Suppose* $\mathrm{rank}_{\mathbb{R}}\, \mathbf{G} \geq 2$, *the lattice Γ_0 is irreducible, and the subgroup $\mathbf{G}^{\mathrm{is}}(\mathbb{R}) \cdot \Gamma_0$ is dense in $\mathbf{G}(\mathbb{R})$. If $\varphi(\Gamma_0) \subset \mathbf{H}(K)$, then the subgroup $\varphi(\Gamma_0) \cap \mathbf{H}(L)$ is of finite index in $\varphi(\Gamma_0)$. In particular, if $\mathbf{G}$ is defined over $\mathbb{Q}$ and $\Gamma_0 \subset \mathbf{G}(\mathbb{Q})$, then the subgroup $\Gamma_0 \cap \mathbf{G}(\mathbb{Z})$ is of finite index in Γ_0.*

(4.14) The group $\mathbf{SL}_n(\mathbb{Z})$ is a lattice in $\mathbf{SL}_n(\mathbb{R})$ (see Theorem I.3.2.8). Therefore if Γ is a lattice in $\mathbf{SL}_n(\mathbb{R})$ and the subgroup $\Gamma \cap \mathbf{SL}_n(\mathbb{Z})$ is of finite index in Γ, then the subgroups Γ and $\mathbf{SL}_n(\mathbb{Z})$ are commensurable. From this and Corollary 4.13 we deduce the following:

(i) *If $n \geq 3$, then any lattice in $\mathbf{SL}_n(\mathbb{R})$ which is contained in $\mathbf{SL}_n(\mathbb{Q})$ is commensurable with $\mathbf{SL}_n(\mathbb{Z})$.*

Similarly, from Theorem I.3.2.7 and Corollary 4.11 we deduce the following:

(ii) *Let p be a prime and $n \in \mathbb{N}^+$, $n \geq 2$. If a subgroup Γ of $\mathbf{SL}_n(\mathbb{Q})$ under the diagonal embedding in $H \overset{\mathrm{def}}{=} \mathbf{SL}_n(\mathbb{R}) \times \mathbf{SL}_n(\mathbb{Q}_p)$ is a lattice in H, then the subgroups Γ and $\mathbf{SL}_n(\mathbb{Z}[1/p])$ are commensurable, where $\mathbb{Z}[1/p]$ as usual denotes the subring in $\mathbb{Q}$ generated by $1/p$.*

The following assertion (iii) is a particular case of Corollary 4.15 stated below. Let us note that it is not difficult to deduce assertion (iii) from Corollary 4.11, Corollary VII.5.16, Corollary II.4.4 (Borel-Wang density theorem), and Lemma II.2.3.

(iii) *If p is a prime and $n \geq 3$, then no lattice of $\mathbf{SL}_n(\mathbb{Q}_p)$ is contained in $\mathbf{SL}_n(\mathbb{Q})$.*

(4.15) We now state a corollary which is a generalization of assertions (i)–(iii) above.

Corollary. *Let K be a global field and let $\mathbf{H}$ be a connected non-commutative almost K-simple K-group. With $\mathscr{R}$, $\mathscr{R}_\infty$, $\mathscr{T} = \mathscr{T}(\mathbf{H})$, and K_v as in 1.3 and 1.4, we choose a finite $S \subset \mathscr{R}$ and assume that*

$\sum_{v \in S} \operatorname{rank}_{K_v} \mathbf{H} \geq 2.$

Then if a subgroup Γ of $\mathbf{H}(K)$ is a lattice in $H_S \overset{\text{def}}{=} \prod_{v \in S} \mathbf{H}(K_v)$ under the diagonal embedding in H_S, then $S \supset \mathscr{R}_\infty - \mathscr{T}$ and the subgroups Γ and $\mathbf{H}(K(S))$ are commensurable.

We shall not give a detailed proof of this. Note however that it is not hard to deduce the corollary either from Theorem 1.11 or, for admissible $\mathbf{H}$, from Theorem VII.5.6 and Proposition VIII.3.22 (when doing this we need in both proofs to make first the reduction to the case where $\mathbf{H}$ is absolutely almost simple).

(4.16) Corollary. *Suppose that for each $\alpha \in A$ the field k_α is a finite field extension of $\mathbb{Q}_{p(k_\alpha)}$ and $p(k_\alpha) \in \mathscr{B}$. We set $S_A = \{p(k_\alpha) \mid \alpha \in A\}$. Suppose* rank $G \geq 2$. *Then if $\Gamma \subset G$ is an irreducible lattice with property (QD) in G, then*

(i) for each $\alpha \in A$, the group $\operatorname{Ad}\operatorname{pr}_\alpha(\Gamma)$ of linear transformations of the Lie algebra $\operatorname{Lie}(\mathbf{G}_\alpha)_{k_\alpha}$ is definable (in the sense of VIII.3.19) over a finite field extension of $\mathbb{Q}$ which is contained in k_α;

(ii) there is a subgroup Γ_1 of finite index in Γ such that for each $\alpha \in A$ the group $\operatorname{Ad}\operatorname{pr}_\alpha(\Gamma_1)$ is written in some basis of the Lie algebra $\operatorname{Lie}(\mathbf{G}_\alpha)_{k_\alpha}$ by matrices with algebraic S_A-integer entries;

(iii) for all $\alpha \in A$ and $\gamma \in \Gamma$, the eigenvalues of the transformation $\operatorname{Ad}\operatorname{pr}_\alpha(\gamma)$ are algebraic S_A-integers.

Proof. Let $\alpha \in A$ and K be a finite field extension of $\mathbb{Q}$. Let $\mathbf{H}$, $\mathscr{R}$, I_α, and $\mathbf{G}_{\alpha i}$ be as in 1.3. Then, for $v \in \mathscr{R}$, $i \in I_\alpha$, and a continuous homomorphism $f: \mathbf{H}(K_v) \to \mathbf{G}_{\alpha i}(k_\alpha)$ from the class Ψ_0 defined in 1.3, we have that the group $\operatorname{Ad} f(\mathbf{H}(K))$ is defined over some finite field extension $K_{\alpha i} \subset k_\alpha$ of $\mathbb{Q}$. But since $\operatorname{char} k_\alpha = 0$, it follows that $\operatorname{Lie}(\mathbf{G}_\alpha)$ is the direct sum of the Lie algebras $\operatorname{Lie}(\mathbf{G}_{\alpha i})$, $i \in I_\alpha$. Thus, given a finite set $B \subset \mathscr{R}$ and a continuous homomorphism $\varphi: \prod_{v \in B} \mathbf{H}(K_v) \to G$ belonging to Ψ, we have that the group $\operatorname{Ad}\operatorname{pr}_\alpha(\varphi(\mathbf{H}(K)))$ is definable over some finite field extension of $\mathbb{Q}$. Now to complete the proof of (i) it suffices to make the following observations:

(a) In view of Theorem A of 1.9 the lattice Γ is arithmetic, and hence there exist a finite field extension K of $\mathbb{Q}$, a connected non-commutative absolutely almost simple K-group $\mathbf{H}$, a finite set S of valuations of K, and a continuous homomorphism $\varphi: \prod_{v \in S} \mathbf{H}(K_v) \to G$ belonging to Ψ such that the subgroup $\varphi(\mathbf{H}(K(S)))$ is commensurable with Γ.

(b) In view of Lemma 2.1(i) the subgroup $\operatorname{pr}_\alpha(\Gamma)$ is Zariski dense in $\mathbf{G}_\alpha$.

(c) If $\mathbf{F}$ is a connected semisimple algebraic group, l a field, $\Lambda \subset \mathbf{F}$ a Zariski dense subgroup, $\Lambda_1 \subset \Lambda$ a subgroup of finite index, and the group $\operatorname{Ad}\Lambda_1$ is written in some basis of the Lie algebra Lie(**F**) by matrices with entries in l, then, as easily follows from Lemma VII.6.2(ii), the group $\operatorname{Ad} \Lambda$ is written in the same basis by matrices with entries in l.

Assertion (ii) is a consequence of (i) and Corollary 4.11. Assertion (iii) is a consequence of (ii) and the following observations. :

1) Since Γ_1 is of finite index in Γ, there exists $n \in \mathbb{N}^+$ such that $\gamma^n \in \Gamma_1$ for all $\gamma \in \Gamma_1$.

2) The eigenvalues of the transformation $\operatorname{Ad} \operatorname{pr}_\alpha(\gamma)$ are n-th roots of the eigenvalues of $\operatorname{Ad} \operatorname{pr}_\alpha(\gamma^n)$.

We also remark that it is not hard to deduce (ii) directly from Theorem A of 1.9. □

(4.17) For the case of $\mathbb{R}$-groups Corollary 4.16 can be stated as follows.

Corollary. *With* $\mathbf{G}$, Γ_0, *and* $\mathbf{G}^{\mathrm{is}}$ *as in Corollary 4.13, we assume* $\operatorname{rank}_{\mathbb{R}} \mathbf{G} \geq 2$, *the lattice* Γ_0 *is irreducible, and the subgroup* $\mathbf{G}^{\mathrm{is}}(\mathbb{R}) \cdot \Gamma_0$ *is dense in* $\mathbf{G}(\mathbb{R})$. *Then*

(i) the group $\operatorname{Ad} \Gamma_0$ *of linear transformations of the Lie algebra* $\operatorname{Lie}(\mathbf{G})_{\mathbb{R}}$ *is definable over some field extension of* $\mathbb{Q}$ *which is contained in* $\mathbb{R}$;

(ii) there is a subgroup Γ_1 *in* Γ_0 *of finite index such that the group* $\operatorname{Ad} \Gamma_1$ *is written in some basis of the Lie algebra* $\operatorname{Lie}(\mathbf{G})_{\mathbb{R}}$ *by matrices whose entries are algebraic integers;*

(iii) for each $\gamma \in \Gamma$ *the eigenvalues of* $\operatorname{Ad} \gamma$ *are algebraic integers.*

(4.18) As an example we mention the following consequence of Corollary 4.17. The matrix $\begin{pmatrix} 1/2 & 0 & 0 \\ 0 & 1 & 0 \\ 0 & 0 & 2 \end{pmatrix}$ belongs to no lattice of $\mathbf{SL}_3(\mathbb{R})$.

(4.19) By a *$\mathbb{Z}$-lattice* in a finite-dimensional vector space W over a field of characteristic 0 we mean a $\mathbb{Z}$-module $L \subset W$ generated by some basis of W. We say that a subgroup $\Lambda \subset \mathbf{GL}(W)$ is *definable* over $\mathbb{Z}$ if Λ leaves some $\mathbb{Z}$-lattice invariant. The following lemma is a particular case of Theorem 3 in [Vi 4].

Lemma. *Let* l *be a field of characteristic* 0, $\mathbf{H}$ *a semisimple* l*-group (not necessarily connected),* $\Lambda \subset \mathbf{H}$ *a Zariski dense subgroup, and* $\Lambda_1 \subset \Lambda$ *a subgroup of finite index. If the group* $\operatorname{Ad} \Lambda_1$ *is definable over* $\mathbb{Z}$, *then so is* $\operatorname{Ad} \Lambda$.

(4.20) We now deduce from Theorem 1.16 the following

Corollary. *Let* $\mathbf{G}$ *be a connected semisimple* $\mathbb{R}$*-group without* $\mathbb{R}$*-anisotropic factors and let* Γ_0 *be an irreducible non-cocompact lattice in* $\mathbf{G}(\mathbb{R})$. *Suppose that either* $\operatorname{rank}_{\mathbb{R}} \mathbf{G} \geq 2$ *or* Γ_0 *is of infinite index in* $\operatorname{Comm}_{\mathbf{G}(\mathbb{R})}(\Gamma_0)$. *Then the group* $\operatorname{Ad} \Gamma_0$ *of linear transformations of the space* $\operatorname{Lie}(\mathbf{G})$ *is definable over* $\mathbb{Z}$, *and hence for each* $\gamma \in \Gamma_0$ *the coefficients of the characteristic polynomial of the transformation* $\operatorname{Ad} \gamma_0$ *are integers.*

Proof. According to Theorem 1.16 the lattice Γ_0 is arithmetic. Therefore there exist a connected non-commutative almost $\mathbb{Q}$-simple $\mathbb{Q}$-group $\mathbf{F}$ and an $\mathbb{R}$-isogeny $\tau\colon \mathbf{F} \to \mathbf{G}$ such that the subgroups $\tau(\mathbf{F}(\mathbb{Z}))$ and Γ_0 are commensurable (see Definition 1.5 and remark (iii) in 1.6). In view of Lemma I.3.1.1 there is a subgroup D of finite index in $\mathbf{F}(\mathbb{Z})$ such that $\operatorname{Ad} D \subset (\operatorname{Ad}\mathbf{H})(\mathbb{Z})$, and hence the group $\operatorname{Ad} D$ is definable over $\mathbb{Z}$. On the other hand, since τ is an isogeny of algebraic groups over a field of characteristic 0, the differential $d\tau\colon \operatorname{Lie}(F) \to \operatorname{Lie}(G)$ is an isomorphism. So Γ_0 contains the subgroup $\tau(D)$ of finite index with $\operatorname{Ad}\tau(D)$ definable over $\mathbb{Z}$. It remains to make use of Lemma 4.19 and to observe that the lattice Γ_0 is Zariski dense in G (see Lemma 2.1).

(4.21) The following assertion is an easy consequence of Corollary 4.20.

(A) *Let* $\mathbf{G}$ *be a connected semisimple* $\mathbb{R}$*-group without* $\mathbb{R}$*-anisotropic factors. Suppose* $\operatorname{rank}_{\mathbb{R}} \mathbf{G} \geq 2$. *Then there is a neighbourhood* $W \subset \mathbf{G}(\mathbb{R})$ *of the identity such that for any irreducible non-cocompact lattice* $\Gamma_0 \subset \mathbf{G}(\mathbb{R})$, *the intersection* $W \cap \Gamma_0$ *consists of unipotent elements.*

Assertion (A) admits the following generalization which, in fact, can easily be deduced from (A).

(A′) *Suppose* $\operatorname{rank} G \geq 2$ *and for each* $\alpha \in A$, $\operatorname{char} k_\alpha = 0$ *and* $\mathbf{G}_\alpha$ *has no* k_α*-anisotropic factors. Then there is a neighbourhood* $W \subset G$ *of the identity such that if* $\Gamma \subset G$ *is an irreducible non-cocompact lattice and* $\gamma \in W \cap \Gamma$, *then the element* $\operatorname{pr}_\alpha(\gamma)$ *is unipotent for each* $\alpha \in A$.

It is very likely that for cocompact lattices an assertion analogous to (A) is true. Namely the following conjecture is plausible.

(B) *Let* $\mathbf{G}$ *be a connected semisimple* $\mathbb{R}$*-group. Suppose* $\operatorname{rank}_{\mathbb{R}} G \geq 2$. *Then there exists a neighbourhood* $U \subset \mathbf{G}(\mathbb{R})$ *of the identity such that for any irreducible cocompact lattice* $\Gamma_0 \subset \mathbf{G}(\mathbb{R})$ *the intersection* $U \cap \Gamma_0$ *consists of elements of finite order.*

Some interesting number-theoretic questions are related to this conjecture. In particular, (B) could be easily deduced from Theorem 1.16 if one succeeded in proving the following assertion.

(*) *Let* $P = x^n + a_{n-1}x^{n-1} + \ldots + a_0$ *be an irreducible monic polynomial with integral coefficients. Denote by* $\beta_1(P), \ldots, \beta_n(P)$ *the roots of* P *and by* $m(P)$ *the number of those* i *with* $1 \leq i \leq n$ *and* $|\beta_i(P)| \neq 1$. *Then*

$$M(P) \stackrel{\text{def}}{=} \prod_{1 \leq i \leq n} \max\{1, |\beta_i(P)|\} > d,$$

where the constant $d > 1$ *depends only on* $m(P)$ *(and does not depend upon* n*).*

J.-P. Serre communicated to the author the following result of M. Laurent.

$$M(P) \geq c^{r^2/n \log(1+n/r)},$$

where $c > 1$ is an absolute constant and r is the number of real roots of P.

(4.22) The following assertion is in fact a restatement of Theorem B in 1.9.

Proposition. *Let $\Gamma \subset G$ be an irreducible non-arithmetic lattice with property (QD) in G. Suppose Γ is finitely generated. Then there is a discrete subgroup $\Lambda \subset G$ containing Γ such that any subgroup of G commensurable with Γ is contained in Λ.*

Proof. By Theorem *B* of 1.9 Γ is of finite index in $\mathrm{Comm}_G(\Gamma)$. So the subgroup $\mathrm{Comm}_G(\Gamma)$ is discrete. On the other hand, $\mathrm{Comm}_G(\Gamma)$ clearly contains any subgroup of G commensurable with Γ. Thus, $\Lambda = \mathrm{Comm}_G(\Gamma)$ is the desired subgroup. □

(4.23) From Proposition 4.22, making use of Theorem 3.2(b) in 1.8 and remark (i) of 1.2, we deduce the following

Proposition. *Let $\Gamma \subset G$ be an irreducible non-arithmetic lattice. Suppose for each $\alpha \in A$, $\mathrm{char}\, k_\alpha = 0$ and $\mathbf{G}_\alpha$ has no k_α-anisotropic factors. Then there is a discrete subgroup $\Lambda \subset G$ containing Γ such that any subgroup of G commensurable with Γ is contained in Λ.*

(4.24) Theorem. *Let Γ be a lattice in G. Suppose that*

(a) $\mathrm{char}\, k_\alpha = 0$ *for all* $\alpha \in A$;
(b) for each $\alpha \in A$ *the group* $\mathbf{G}_\alpha$ *has no* k_α*-anisotropic factors.*

Then there exists $c > 0$ such that $c\mu_G(\Gamma_1 \setminus G)$ is an integer for any subgroup $\Gamma_1 \subset G$ commensurable with Γ, where μ_G is Haar measure on G.

This theorem was established in [Bo 12] under the assumption that all the groups $\mathbf{G}_\alpha$ are isogenous to $\mathbf{SL}_2$ and in the general case in [Ma-R]. The proof goes along the following lines. First we make the reduction to the case where the lattice Γ is irreducible. Then if Γ is not arithmetic it suffices to make use of Proposition 4.22 ($c^{-1} = \mu_G(\Lambda \setminus G)$). If Γ is arithmetic, the description of maximal arithmetic subgroups is used.

(4.25) *Remark.* Condition (a) in Theorem 4.24 seems to be superfluous in contrast with condition (b) which is essential. Indeed, if $\mathbf{G}$ is a connected $\mathbb{R}$-anisotropic semisimple $\mathbb{R}$-group, then any finite subgroup $\Lambda \subset \mathbf{G}(\mathbb{R})$ is a lattice in the compact Lie group $\mathbf{G}(\mathbb{R})$. But $\mathbf{G}(\mathbb{R})$ contains finite subgroups of arbitrarily large order.

5. Consequences of the Arithmeticity Theorems II

In this section, making use of the arithmeticity theorems, we strengthen certain results of Chapters IV and VII.

Let A, k_α, G_α, G, and $\operatorname{rank} G$ be as in Section I. The Zariski closure of a subset M of an algebraic variety will be denoted by $\bar{M}$.

We shall say that a group Λ is *quasi-simple* if any normal subgroup of Λ is either finite or of finite index in Λ.

(5.1) Lemma. *Let $f\colon \Lambda \to \Lambda'$ be a group homomorphism. Then*

(i) if Λ is quasi-simple and $f(\Lambda)$ is of finite index in Λ', then Λ' is quasi-simple;
(ii) if Λ' is quasi-simple, $f(\Lambda) = \Lambda'$, and $\operatorname{Ker} f$ *is finite, then Λ is quasi-simple.*

Proof. (i) Let N' be an infinite normal subgroup of Λ'. Then since $f(\Lambda)$ is of finite index in Λ', the normal subgroup $f^{-1}(N')$ of Λ is infinite. But Λ is quasi-simple. So $f^{-1}(N')$ is of finite index in Λ and, since $f(\Lambda)$ is of finite index in Λ', it follows that N' is of finite index in Λ'.

(ii) Let N be an infinite normal subgroup of Λ. Since $\operatorname{Ker} f$ is finite, the normal subgroup $f(N)$ of $f(\Lambda) = \Lambda'$ is infinite. But Λ' is quasi-simple. Therefore $f(N)$ is of finite index in Λ', and, since $\operatorname{Ker} f$ is finite, N is of finite index in Λ. □

(5.2) Lemma. *Let $\Gamma \subset G$ be a lattice with property (QD) in G.*

(i) If N is a finite normal subgroup of Γ, then $N \subset \mathscr{Z}(G)$.
(ii) The following are equivalent
  (a) Γ is quasi-simple;
  (b) If N is a normal subgroup of Γ, then either $N \subset \mathscr{Z}(G)$ or Γ/N is finite.
(iii) If Γ is quasi-simple, then the factor group $\Gamma/\mathscr{D}(\Gamma)$ of Γ by its commutator subgroup is finite.

Proof. (i) Since the subgroup N is finite and normal in Γ and, $\overline{\mathrm{pr}_\alpha(\Gamma)} = \mathbf{G}_\alpha$ (see Lemma 2.1(i)), it follows that for each $\alpha \in A$ the algebraic subgroup $\mathscr{Z}_{\mathbf{G}_\alpha}(\mathrm{pr}_\alpha(N))$ is of finite index in $\mathbf{G}_\alpha$. But $\mathbf{G}_\alpha$ is connected. So $\mathscr{Z}_{\mathbf{G}_\alpha}(\mathrm{pr}_\alpha(N)) = \mathbf{G}_\alpha$, and hence $N \subset \mathscr{Z}(G)$.

Assertion (ii) is a consequence of (i) and the finiteness of the center $\mathscr{Z}(G)$. Since $\mathbf{G}_\alpha$ is semisimple, $\mathscr{D}(\mathbf{G}_\alpha) = \mathbf{G}_\alpha$. On the other hand, $\overline{\mathrm{pr}_\alpha(\Gamma)} = \mathbf{G}_\alpha$ (see Lemma 2.1(i)). Thus, the commutator subgroup $\mathscr{D}(\Gamma)$ is infinite, which proves (iii). □

(5.3) Proposition. *Let Γ be an irreducible arithmetic lattice in G. Suppose* $\operatorname{rank} G \geq 2$. *Then*

(i) if N is a normal subgroup of Γ, then either $N \subset \mathscr{Z}(G)$ or Γ/N is finite;
(ii) the factor group $\Gamma/\mathscr{D}(\Gamma)$ of Γ by its commutator subgroup is finite.

Proof. Since Γ is arithmetic, by Definition 1.4 there exist a global field K, a connected non-commutative absolutely almost simple K-group $\mathbf{H}$, a finite set S of valuations of K, and a continuous homomorphism $\varphi\colon \prod_{v \in S} \mathbf{H}(K_v) \to G$ belonging to the class Ψ such that the subgroup $\varphi(\mathbf{H}(K(S)))$ is commensurable with Γ. We set $\Lambda = \varphi^{-1}(\Gamma) \cap \mathbf{H}(K(S))$. Since the subgroups $\varphi(\mathbf{H}(K(S)))$ and Γ are commensurable, it follows that Λ is an S-arithmetic subgroup in $\mathbf{H}(K)$ and $\varphi(\Lambda)$ is of finite index in Γ. In view of remark (vii) of 1.3 we have

$$\operatorname{rank}_S \mathbf{H} \stackrel{\text{def}}{=} \sum_{v \in S} \operatorname{rank}_{K_v} \mathbf{H} = \operatorname{rank} G \geq 2.$$

Making use of Theorem (A) stated at the beginning of Chapter VIII, we can now claim that the group Λ is quasi-simple. But $\varphi(\Lambda)$ is of finite index in Γ. Thus, Γ is quasi-simple (see Lemma 5.1) and in view of Lemma 5.2(i) and remark (iv) in 1.6 this completes the proof. □

In view of remark 2 in 1.9, Theorem A in 1.9 and Proposition 5.3 imply the following

(5.4) Theorem. *Let $\Gamma \subset G$ be an irreducible lattice with property (QD) in G. Suppose that* $\operatorname{rank} G \geq 2$.

Then

(i) if N is a normal subgroup of Γ, then either $N \subset \mathscr{Z}(G)$ or Γ/N is finite;
(ii) $\Gamma/\mathscr{D}(\Gamma)$ is finite.

(5.5) We present the statement of Theorem 5.4 for the case of $\mathbb{R}$-groups.

Theorem. *Let $\mathbf{G}$ be a connected semisimple $\mathbb{R}$-group and let Γ_0 be a lattice in $\mathbf{G}(\mathbb{R})$. Denote by $\mathbf{G}^{\text{is}} \subset \mathbf{G}$ the (almost direct) product of all $\mathbb{R}$-isotropic factors of $\mathbf{G}$. Suppose* $\operatorname{rank}_{\mathbb{R}} \mathbf{G} \geq 2$, *the lattice Γ_0 is irreducible, and the subgroup $\mathbf{G}^{\text{is}}(\mathbb{R}) \cdot \Gamma_0$ is dense in $\mathbf{G}(\mathbb{R})$. Then*

(i) if N is a normal subgroup of Γ_0, then either $N \subset \mathscr{Z}(G)$ or Γ_0/N is finite;
(ii) $\Gamma_0/\mathscr{D}(\Gamma_0)$ is finite.

(5.6) The following theorem is a particular case of Theorem 5.5.

Theorem. *Let Γ be an irreducible lattice in G. Suppose that* $\operatorname{rank} G \geq 2$ *and for each $\alpha \in A$ the group $\mathbf{G}_\alpha$ has no k_α-anisotropic factors. Then*

(i) if N is a normal subgroup of Γ, then either $N \subset \mathscr{Z}(G)$ or Γ/N is finite;
(ii) $\Gamma/\mathscr{D}(\Gamma)$ is finite.

(5.7) Proposition. *Let Γ be an irreducible arithmetic lattice in G, l a field, $\mathbf{F}$ an algebraic l-group, and $\delta\colon \Gamma \to \mathbf{F}(l)$ a homomorphism. Assume* $\operatorname{rank} G \geq 2$. *Then*

(i) if $\operatorname{char} l = 0$, *then the l-group $\overline{\delta(\Gamma)}$ is semisimple;*
(ii) if $\operatorname{char} l \neq \operatorname{char} k_\alpha$ *for some (and hence for all) $\alpha \in A$, then the group $\delta(\Gamma)$ is finite.*

Proof. With K, $\mathbf{H}$, S, φ, Λ, and $\operatorname{rank}_S \mathbf{H}$ as in the proof of Proposition 5.3, we recall that (as was observed there) Λ is an S-arithmetic subgroup in $\mathbf{H}(K)$ and $\operatorname{rank}_S \mathbf{H} \geq 2$. Thus,

1) if $\operatorname{char} l = 0$, then the l-group $\overline{\delta(\varphi(\Lambda))}$ is semisimple (see Theorem VIII.3.10);
2) if $\operatorname{char} l \neq \operatorname{char} K$, then the group $\delta(\varphi(\Lambda))$ is finite (see Theorem IX.3.8(ii)(b)).

All that remains now is to observe that $\operatorname{char} K = \operatorname{char} k_\alpha$ for each $\alpha \in A$ and, since $\varphi(\Lambda)$ is of finite index in Γ (see 5.3), it follows that $\delta(\varphi(\Lambda))$ is of finite index in $\delta(\Gamma)$. Hence $\overline{\delta(\varphi(\Lambda))}$ is of finite index in $\overline{\delta(\Gamma)}$. (It may not be out of place to recall that if an algebraic group $\mathbf{F}$ contains a semisimple algebraic subgroup of finite index, then $\mathbf{F}$ is semisimple.) □

In view of remark 2 in 1.9, Theorem A in 1.9 and Proposition 5.7 imply the following

(5.8) Theorem. *Let $\Gamma \subset G$ be an irreducible lattice with property (QD) in G, l a field, $\mathbf{F}$ an algebraic l-group, and $\delta\colon \Gamma \to \mathbf{F}(l)$ a homomorphism. Suppose that* $\operatorname{rank} G \geq 2$. *Then*

(i) if $\operatorname{char} l = 0$, *then the l-group $\overline{\delta(\Gamma)}$ is semisimple;*
(ii) if $\operatorname{char} l \neq \operatorname{char} k_\alpha$ *for all $\alpha \in A$, then the group $\delta(\Gamma)$ is finite.*

(5.9) In view of Lemma VII.5.22, the following can be deduced from Theorem 5.8(i).

Corollary. *Let $\Gamma \subset G$ be an irreducible lattice with property (QD) in G. Assume* $\operatorname{rank} G \geq 2$. *Then $H^1(\Gamma, \rho) = 0$ for any representation ρ of the group Γ on a finite-dimensional vector space over a field of characteristic* 0.

(5.10) For the case of $\mathbb{R}$-groups Theorem 5.8 and Corollary 5.9 can be stated as follows.

Theorem. *With $\mathbf{G}$, Γ_0 and $\mathbf{G}^{\text{is}}$ as in the statement of Theorem 5.5, we assume* $\operatorname{rank}_{\mathbb{R}} \mathbf{G} \geq 2$, *the lattice Γ_0 is irreducible, and the subgroup $\mathbf{G}^{\text{is}}(\mathbb{R}) \cdot \Gamma_0$ is dense in* $\mathbf{G}(\mathbb{R})$.

(i) Let l be a field, $\mathbf{F}$ an algebraic l-group, and $\delta\colon \Gamma_0 \to \mathbf{F}(l)$ a homomorphism. Then
  (a) if $\operatorname{char} l = 0$, *then the l-group $\overline{\delta(\Gamma_0)}$ is semisimple;*
  (b) if $\operatorname{char} l \neq 0$, *then the group $\delta(\Gamma)$ is finite.*

(ii) $H^1(\Gamma_0, \rho) = 0$ for any representation ρ of the group Γ on a finite-dimensional vector space over a field of characteristic 0.

(5.11) As usual $H^q(\Lambda, \rho)$ denotes the q-th cohomology group of Λ with coefficients in ρ. One has the following

Theorem. (See [Bo-W], Chapter XIII, Propositions 3.6 and 3.7 and Lemma 3.2.) *Let Γ be an irreducible cocompact lattice in G and let ρ be a representation of Γ on a finite-dimensional vector space over a field k of characteristic* 0. *Suppose that for each $\alpha \in A$ the group $\mathbf{G}_\alpha$ has no k_α-anisotropic factors.*

(i) Let $k = \mathbb{C}$ and suppose that the representation ρ is unitary. Then
  (a) $H^q(\Gamma, \rho) = 0$ if $0 < q <$ $\operatorname{rank} G$;
  (b) if for each $\alpha \in A$ the field k_α is non-archimedean, then $H^q(\Gamma, \rho) = 0$ for $q \neq 0$ and $q \neq$ $\operatorname{rank} G$.

(ii) *Assume that* rank G ≥ 2 *and for each* $\alpha \in A$ *the field* k_α *is non-archimedean. Then* $H^q(\Gamma, \rho) = 0$ *whenever* $q \neq 0$ *and* $q \neq$ rank G. *(Here* ρ *is not assumed to be unitary.)*

Remarks.

(i) In [Bo-W] assertion (ii) was deduced from assertion (i) (b) with the help of the superrigidity theorem (namely Theorem VII.5.6). It was assumed in the context of (ii) that char $k_\alpha = 0$ for all $\alpha \in A$. But this restriction was merely dued to the fact that at the time when the book [Bo-W] was written, the superrigidity theorem was only known for the case in which char $k_\alpha = 0$ for all $\alpha \in A$. Furthermore, it was assumed in [Bo-W] that if among the fields k_α there is an archimedean one, then char $k_\alpha = 0$ for all $\alpha \in A$. But this restriction can also be dropped in view of Theorem 1.11.

(ii) For results on cohomology of discrete and arithmetic subgroups, in addition to [Bo-W], also see [B-K], [Bo 8], [Bo 9], [Bo 10], [Bo 11], [Bo-Se 2], [Ga 1], [Ga 2], [G-S 1], [G-S 2], [G-S 23], [Har 4], [Har 6], [Har 7], [Kaz 4], [L-S], [Mills 1], [Mills 3], [Mi-R], [Rag 3], [Rag 4], [Rag 5], [Rohlf], [R-S], [Schwe 1], [Schwe 2], [Ser 2], [So], [Stu] (The list is far from complete).

(5.12) Theorem. *Let* $\Gamma \subset G$ *be an irreducible lattice with property (QD) in G, k a local field of characteristic* 0, $\mathbf{F}$ *an algebraic k-group, and* $\delta\colon \Gamma \to \mathbf{F}(k)$ *a homomorphism. Suppose that* rank G ≥ 2.

(i) *If for each* $\alpha \in A$, *the fields* k_α *and* k *are of different type (in the sense of I.0.31), then the subgroup* $\delta(\Gamma)$ *is relatively compact in* $\mathbf{F}(k)$.

(ii) *Assume that the group* $\mathbf{G}_\alpha$ *is simply connected for each* $\alpha \in A$. *Then* δ *almost extends (in the sense of Definition V.3.5) to a continuous homomorphism* $\tilde{\delta}\colon G \to \mathbf{F}(k)$. *In other words, there exist a continuous homomorphism* $\delta\colon G \to \mathbf{F}(k)$ *and a homomorphism* $f\colon \Gamma \to \mathbf{F}(k)$ *such that* $\delta(\gamma) = \tilde{\delta}(\gamma)\cdot f(\gamma)$ *for all* $\gamma \in \Gamma$ *and the subgroup* $f(\Gamma)$ *is relatively compact in* $\mathbf{F}(k)$ *and commutes with* $\tilde{\delta}(G)$.

We shall not present detailed proof of this theorem. We only remark that Lemma VII.5.1 and Lemma 2.4 enable us to make a reduction to the case that the group $\overline{\delta(\Gamma)}$ is connected and for each $\alpha \in A$, $\mathbf{G}_\alpha$ has no k_α-anisotropic factors. But in that case it suffices to make use of Corollary VII.5.16, Theorem VII.5.13(b), and Theorem 5.8(i).

(5.13) It is very likely that the following is true.

Conjecture. *Let* Γ *be an irreducible arithmetic lattice in G, k a local field,* $\mathbf{F}$ *an algebraic k-group, and* $\delta\colon \Gamma \to \mathbf{F}(k)$ *a homomorphism. Suppose that* rank G ≥ 2 *and for each* $\alpha \in A$ *the group* $\mathbf{G}_\alpha$ *is simply connected. Then*

(i) *there is a subgroup* Γ_0 *of finite index in* Γ *(depending on* δ*) such that the restriction of* δ *to* Γ_0 *extends to a continuous homomorphism* $\tilde{\delta}\colon G \to \mathbf{F}(k)$;

(ii) *there exist a continuous homomorphism* $\tilde{\delta}\colon G \to \mathbf{F}(k)$ *and a homomorphism* $f\colon \Gamma \to \mathbf{F}(k)$ *such that* $\delta(\gamma) = \tilde{\delta}(\gamma) \cdot f(\gamma)$ *for all* $\gamma \in \Gamma$ *and the subgroup* $f(\Gamma)$ *is finite and commutes with* $\tilde{\delta}(G)$.

(5.14) *Remarks on Conjecture 5.13.*

(i) Assertion (ii) can easily be deduced from (i) with the help of Lemma VII.5.1.

(ii) If for each $\alpha \in A$ the group $\mathbf{G}_\alpha$ has no k_α-anisotropic factors, then Lemma II.2.3 and Proposition II.4.6(b) imply that the homomorphism $\tilde{\delta}$ is uniquely determined.

(iii) By applying Theorems (B) and (C) stated at the beginning of Chapter VIII and Proposition 5.3(ii), it is not hard to prove Conjecture 5.13 in any of the following cases:

(a) $\operatorname{char} k_\alpha = 0$ for some (and hence for all) $\alpha \in A$;

(b) $\operatorname{char} k \neq \operatorname{char} k_\alpha$;

(c) the k-group $\overline{\delta(\Gamma)}$ is reductive.

It thus remains to consider the case

(d) $\operatorname{char} k = \operatorname{char} k_\alpha \neq 0$ and the k-group $\overline{\delta(\Gamma)}$ is not reductive.

There is a hope that case (d) can be reduced to case (c) by applying Theorem VII.5.19, Corollary VIII.5.21, Proposition 5.3(ii), and the results of Prasad and Raghunathan on the second cohomology of semisimple groups over local fields (see [Pr-R 3]).

(5.15) In connection with Theorem 5.12, in the case where the group $\overline{\delta(\Gamma)}$ is connected and semisimple, we remark this theorem in fact asserts that the condition "for each $\alpha \in A$ the group $\mathbf{G}_\alpha$ has no k_α-anisotropic factors" in Theorem VII.5.13(b) can be replaced by the weaker one "Γ has property (QD) in G". It is not hard to show (using Lemma VII.5.1 and Lemma 2.4) that one can make the analogous replacement in the statements of other results of Sections 5 and 6 in Chapter VII. Furthermore, if we consider a lattice Γ_0 in $\mathbf{G}(\mathbb{R})$, where $\mathbf{G}$ is a connected semisimple $\mathbb{R}$-group, then the condition "$\mathbf{G}$ has no $\mathbb{R}$-anisotropic factors" ought to be replaced by the condition "the subgroup $\mathbf{G}^{\mathrm{is}}(\mathbb{R}) \cdot \Gamma_0$ is dense in $\mathbf{G}(\mathbb{R})$", where as before $\mathbf{G}^{\mathrm{is}}$ denotes the product of $\mathbb{R}$-anisotropic factors of $\mathbf{G}$.

6. Arithmeticity, Volume of Quotient Spaces, Finiteness of Factor Groups, and Superrigidity of Lattices in Semisimple Lie Groups

For any $\mathbb{R}$-group $\mathbf{F}$, let $\mathbf{F}(\mathbb{R})^0$ denote the identity component of the Lie group $\mathbf{F}(\mathbb{R})$. As in Section 5 we denote by $\bar{M}$ the Zariski closure of a subset M of an algebraic variety.

Throughout this section H denotes a connected semisimple Lie group. We set $\mathbf{G} = \overline{\operatorname{Ad} H}$, where Ad as usual denotes the adjoint representation. It is well known that $\mathbf{G}$ is a connected semisimple adjoint $\mathbb{R}$-group, $\operatorname{Ad} H = \mathbf{G}(\mathbb{R})^0$, $\operatorname{Ker} \operatorname{Ad} = \mathscr{Z}(H)$, and the isomorphism $H/\mathscr{Z}(H) \to \operatorname{Ad} H$ is a topological group isomorphism. Denote by $\mathbf{G}^{\text{is}} \subset \mathbf{G}$ the almost direct product of $\mathbb{R}$-isotropic almost $\mathbb{R}$-simple factors of $\mathbf{G}$ and by $H^{\text{is}} \subset H$ the identity component of the Lie group $\text{Ad}^{-1}(\mathbf{G}^{\text{is}}(\mathbb{R}))$. The subgroup H^{is} can be characterized as the minimal element in the class of connected normal subgroups $F \subset H$ with H/F compact. We remark that the group H is the almost direct product of the subgroup H^{is} and the maximal connected compact normal subgroup of H.

We set $\text{rank}\, H = \text{rank}_{\mathbb{R}}\, \mathbf{G}$ and call it the *rank* of the group H. It can easily be verified that $\text{rank}\, H$ coincides with the dimension of a maximal $\mathbb{R}$-split torus of H (a connected commutative subgroup $T \subset H$ is said to be an $\mathbb{R}$-*split torus* if for each $x \in T$, the transformation $\text{Ad}\, x$ is semisimple all of whose eigenvalues being real). The rank of H equals 0 if and only if $H^{\text{is}} = \{e\}$.

(6.1) Lemma. *Let Γ be a lattice in H.*

(a) The following conditions are equivalent:
- *(i) $\mathscr{Z}(H) \cdot \Gamma$ is a lattice in H;*
- *(ii) the subgroup $\mathscr{Z}(H) \cdot \Gamma$ is discrete in H;*
- *(iii) the subgroup $\mathscr{Z}(H) \cap \Gamma$ is of finite index in $\mathscr{Z}(H)$;*
- *(iv)* $\text{Ad}\, \Gamma$ *is a lattice in* $\mathbf{G}(\mathbb{R})$*;*
- *(v) the subgroup* $\text{Ad}\, \Gamma$ *is discrete in* $\mathbf{G}(\mathbb{R})$.

(b) If the subgroup $H^{\text{is}} \cdot \Gamma$ is dense in H, then $\mathscr{Z}(H) \cdot \Gamma$ is a lattice in H (see [Rag 5], Corollary 5.17).

We shall show (a). Conditions (i) and (ii) are equivalent, because any discrete subgroup containing a lattice is itself a lattice. Condition (iii) is satisfied if and only if Γ is of finite index in $\mathscr{Z}(H) \cdot \Gamma$. Thus, the equivalence (i) $\iff$ (iii) is a consequence of the fact that a subgroup $\Gamma' \subset H$ containing the lattice Γ is a lattice if and only if Γ is of finite index in Γ'. Finally, since $\text{Ker}\, \text{Ad} = \mathscr{Z}(H)$ and $\text{Ad}\, H = \mathbf{G}(\mathbb{R})^0$ are of finite index in $\mathbf{G}(\mathbb{R})$, the equivalences (i) $\iff$ (iv) and (ii) $\iff$ (v) hold. □

(6.2) Definition. *We say that a lattice $\Gamma \subset H$ is* reducible *if there exist connected infinite normal subgroups H' and H'' in H such that $H' \cap H'' \subset \mathscr{Z}(H)$, $H' \cdot H'' = H$, and the subgroup $(\Gamma \cap H') \cdot (\Gamma \cap H'')$ is of finite index in Γ; otherwise Γ is called* irreducible.

If $\text{Ad}\, \Gamma$ is a lattice in $\mathbf{G}(\mathbb{R})$, then it is easy to see that the irreducibility of the lattice $\Gamma \subset H$ is equivalent to that of $\text{Ad}\, \Gamma$ in the sense of III.5.9.

(6.3) Definition. *We say that an irreducible lattice $\Gamma \subset H$ is* arithmetic *if the following conditions are satisfied:*

(i) $\text{Ad}\, \Gamma$ *is an arithmetic lattice in* $\mathbf{G}(\mathbb{R})$ *(in the sense of the definitions given in Section 1);*

(ii) the subgroup $\mathscr{Z}(H) \cap \Gamma$ is of finite index in $\mathscr{Z}(H)$.

(6.4) *Remarks.*

(i) If Γ is a lattice in H and $\operatorname{Ad}\Gamma$ is an arithmetic lattice in $\mathbf{G}(\mathbb{R})$, then by remarks (vii) in 1.2 and (iv) in 1.6 the subgroup $\mathbf{G}^{\text{is}}(\mathbb{R}) \cdot \operatorname{Ad}\Gamma$ is dense in $\mathbf{G}(\mathbb{R})$. Hence, the subgroup $H^{\text{is}} \cdot \Gamma$ is dense in H. From this and Lemma 6.1 we deduce that (ii) in Definition 6.3 is a consequence of (i).

(ii) If $f\colon H \to H'$ is a continuous Lie group epimorphism and $\operatorname{Ker} f \subset \mathscr{Z}(H)$, then the arithmeticity of a lattice $\Gamma \subset H$ is equivalent to that of the lattice $f(\Gamma) \subset H'$.

(6.5) Theorem. *Let Γ be an irreducible lattice in H with $H^{\text{is}} \cdot \Gamma$ dense in H. Suppose that either* $\operatorname{rank} H \geq 2$ *or Γ is of infinite index in* $\operatorname{Comm}_H(\Gamma)$. *Then the lattice Γ is arithmetic.*

Proof. It follows from Lemma 6.1 that the subgroup $\mathscr{Z}(H) \cap \Gamma$ is of finite index in $\mathscr{Z}(H)$ and $\operatorname{Ad}\Gamma$ is a lattice in $\mathbf{G}(\mathbb{R})$. Since the lattice Γ is irreducible, it follows that $\operatorname{Ad}\Gamma$ is irreducible (see 6.2). Furthermore, since $H^{\text{is}} \cdot \Gamma$ is dense in H and the factor group $\mathbf{G}(\mathbb{R})/\mathbf{G}^{\text{is}}(\mathbb{R})$ is connected (see remark (vii) in 1.2). The subgroup $\mathbf{G}^{\text{is}}(\mathbb{R}) \cdot \operatorname{Ad}\Gamma$ is dense in $\mathbf{G}(\mathbb{R})$. Since $\mathscr{Z}(H) \cap \Gamma$ is of finite index in $\mathscr{Z}(H) = \operatorname{Ker}\operatorname{Ad}$, it follows that Γ is of infinite index in $\operatorname{Comm}_H(\Gamma)$ if and only if $\operatorname{Ad}\Gamma$ is of infinite index in $\operatorname{Comm}_{\mathbf{G}(\mathbb{R})}(\operatorname{Ad}\Gamma)$. All that remains now is to make use of Theorem 1.16. □

(6.6) The following Corollaries 6.7–6.12 are deduced with the help of Lemma 6.1 from the results of Section 4 which are indicated in brackets.

(6.7) Corollary. (See Corollary 4.5.) *Suppose there are two non-isomorphic simple factors of the complexification of the Lie algebra of H and H has no non-trivial compact factor groups. Then H does not contain irreducible lattices.*

(6.8) Corollary. (See Corollary 4.17.) *Let Γ be an irreducible lattice in H with $H^{\text{is}} \cdot \Gamma$ dense in H and* $\operatorname{rank} H \geq 2$. *Then there is a subgroup Γ_1 of finite index in Γ such that in some basis of the Lie algebra of H, the group* $\operatorname{Ad}\Gamma_1$ *consists of matrices whose entries are algebraic integers.*

(6.9) Corollary. (See Corollary 4.20.) *Let Γ be an irreducible non-cocompact lattice in H. Suppose that either* $\operatorname{rank} H \geq 2$ *or Γ is of finite index in* $\operatorname{Comm}_H(\Gamma)$. *Assume in addition that H has no non-trivial compact factor groups. Then the group* $\operatorname{Ad}\Gamma$ *of linear transformations of the Lie algebra of H is definable over $\mathbb{Z}$, and hence for each $\gamma \in \Gamma$ the coefficients of the characteristic polynomial of* $\operatorname{Ad}\gamma$ *are integers.*

(6.10) Corollary. (See assertion 4.21 (A).) *Suppose that* $\operatorname{rank} H \geq 2$ *and H has no non-trivial compact factor groups. Then there is a neighbourhood $W \subset H$ of the identity such that the intersection $W \cap \Gamma$ of W with any irreducible non-cocompact lattice $\Gamma \subset H$ consists of unipotent elements, i.e., of elements of the form* $\exp x$, *where x is a nilpotent element of the Lie algebra of H.*

(6.11) Corollary. (See Proposition 4.23.) *Let $\Gamma \subset H$ be an irreducible non-arithmetic lattice. Assume H has no non-trivial compact factor groups. Then there is a discrete subgroup $\Lambda \subset H$ containing Γ such that any subgroup of H commensurable with Γ is contained in Λ.*

(6.12) Corollary. (See Theorem 4.24.) *Let Γ be a lattice in H. Suppose that H has no non-trivial compact factor groups. Then there exists $c > 0$ such that $c\mu_H(\Gamma_1 \backslash H)$ is an integer for any subgroup Γ_1 of H commensurable with Γ.*

(6.13) *Remarks to Corollary 6.12.*

(i) Suppose that the group $\operatorname{Ad} H$ contains a compact Cartan subgroup and has no non-trivial compact factor groups. It then follows from the results of [Har 3] that
   (a) *there exists $c > 0$ such that $c\mu_H(\Gamma \setminus H)$ is an integer for any torsion free lattice $\Gamma \subset H$;*
   (b) *the number $\mu_H(\Gamma_1 \setminus H)/\mu(\Gamma_2 \setminus H)$ is rational for any two lattices Γ_1, Γ_2 in H.*

   We remark that (b) can be deduced from (a). For this apply Lemma 6.1, assertion 3.1(ii), and the fact that any finitely generated subgroup in $\mathbf{GL}_n(\mathbb{C})$ contains a torsion free subgroup of finite index (see [Rag 5], Theorem 6.11).

(ii) It is not clear if assertion (b) above is true for arbitrary H. This seems to be false even for $H = \mathbf{SL}_2(\mathbb{C})$. For further details we refer to ([Bo 12], 7.7), where in particular the connection between this and some number-theoretic problems is indicated.

(iii) Denote by $W(H)$ (resp. $W_{ar}(J)$) the set of the numbers $\mu_H(\Gamma \setminus H)$, where Γ is a lattice (resp. arithmetic lattice) in H. Then
   (1) the set $W(H)$ is discrete if no factor group of H is locally isomorphic to $\mathbf{SL}_2(\mathbb{R})$ or $\mathbf{SL}_2(\mathbb{C})$, and otherwise is not discrete (see [Bo 12], 8.3, [Th], Chapter V, and [WH 2]);
   (2) the set $W_{ar}(H)$ is discrete. (This follows from Theorem 8.2 in [Bo 12] and assertion (1).)

   We remark that for $H = \mathbf{SL}_2(\mathbb{R})$ the non-discreteness of the set $W(H)$ is due to the existence of torsion lattices (otherwise we would have a contradiction to remark (i) (a)).

   If $H = \mathbf{SL}_2(\mathbb{C})$, the set $\{\mu_H(\Gamma \setminus H) \mid \Gamma \text{ is a torsion free lattice in } H\}$ is not discrete.

(6.14) Theorem. *Let Γ be an irreducible lattice in H with $H^{\text{is}} \cdot \Gamma$ dense in H. Suppose that* $\operatorname{rank} H \geq 2$ *and the center $\mathscr{Z}(H)$ is finite. Then*

(i) for any normal subgroup $N \subset \Gamma$ either $N \subset \mathscr{Z}(H)$ or Γ/N is finite;
(ii) the factor group $\Gamma/D(\Gamma)$ is finite.

Proof. As in the proof of Theorem 6.5 we can verify that $\operatorname{Ad}\Gamma$ is an irreducible lattice in $\mathbf{G}(\mathbb{R})$ and the subgroup $\mathbf{G}^{\text{is}}(\mathbb{R}) \cdot \operatorname{Ad}\Gamma$ is dense in $\mathbf{G}(\mathbb{R})$. On the other hand, $\mathscr{Z}(G) = \{e\}$ and the kernel $\operatorname{Ker}\operatorname{Ad} = \mathscr{Z}(H)$ is finite. Therefore in view of Theorem 5.5 we have

(i) if N is a normal subgroup in Γ, then either $\operatorname{Ad} N = \{e\}$, and hence $N \subset \mathscr{Z}(H)$, or the factor group $\operatorname{Ad}\Gamma / \operatorname{Ad} N$ (and hence Γ / N) is finite;
(ii) the factor group $\operatorname{Ad}\Gamma / \mathscr{D}(\operatorname{Ad}\Gamma)$ (and hence $\Gamma / \mathscr{D}(\Gamma)$) is finite. □

(6.15) Theorem. *Let Γ be an irreducible lattice in H with $H^{\mathrm{is}} \cdot \Gamma$ dense in H. Suppose that* $\operatorname{rank} H \geq 2$ *and $\mathscr{Z}(H)$ is finite.*

(i) Let l be a field, $\mathbf{F}$ an algebraic l-group, and $\delta\colon \Gamma \to \mathbf{F}(l)$ a homomorphism. Then
(a) if $\operatorname{char} l = 0$, *the l-group $\overline{\delta(\Gamma)}$ is semisimple;*
(b) if $\operatorname{char} l \neq 0$, *the group $\delta(\Gamma)$ is finite.*

(ii) $H^1(\Gamma, \rho) = 0$ for any representation ρ of Γ on a finite-dimensional vector space over a field of characteristic 0.

Proof. In view of Lemma VII.5.22, (ii) is a consequence of (i) (a). We shall prove (i). We may assume $F = \overline{\delta(\Gamma)}$. Then $B \overset{\text{def}}{=} \delta(\Gamma \cap \mathscr{Z}(H)) \subset \mathscr{Z}(F)$, and hence the subgroup B is normal in F. Let $\pi\colon F \to F/B$ be the natural epimorphism. Clearly, $(\pi \circ \delta)(\Gamma \cap \mathscr{Z}(H)) = \{e\}$. Since $\mathscr{Z}(H)$ is finite, the subgroup $B \subset \mathbf{F}(l)$ is finite. Thus,

1) B (and hence $\mathbf{F}/B$) is an l-group;
2) if $\mathbf{F}/B$ is semisimple, then $\mathbf{F}$ is semisimple;
3) if $(\pi \circ \delta)(\Gamma)$ is finite, $\delta(\Gamma)$ is finite as well.

Hence, replacing $\mathbf{F}$ and δ by $\mathbf{F}/B$ and $\pi \circ \delta$, we may assume that $\delta(\Gamma \cap \mathscr{Z}(H)) = \{e\}$. Since $\operatorname{Ker} \operatorname{Ad} = \mathscr{Z}(H)$ one can then define a homomorphism $\delta'\colon \operatorname{Ad}\Gamma \to \mathbf{F}(l)$ with $\delta = \delta' \circ \operatorname{Ad}$. Obviously, $\delta(\Gamma) = \delta'(\operatorname{Ad}\Gamma)$. All that remains now is to make use of Theorem 5.10(i), observing first that $\operatorname{Ad}\Gamma$ is an irreducible lattice in $\mathbf{G}(\mathbb{R})$ and the subgroup $\mathbf{G}^{\mathrm{is}}(\mathbb{R}) \cdot \operatorname{Ad}\Gamma$ is dense in $\mathbf{G}(\mathbb{R})$ (see the proof of Theorem 6.5). □

(6.16) Theorem. *Let $\Gamma \subset H$ be a lattice, Λ a countable subgroup of* $\operatorname{Comm}_H(\Gamma)$ *containing Γ, k a local field, $\mathbf{F}$ a connected semisimple k-group, and $\delta\colon \Lambda \to \mathbf{F}(k)$ a homomorphism with $\delta(\Gamma)$ Zariski dense in $\mathbf{F}$. Suppose that the subgroup $H^{\mathrm{is}} \cdot \Gamma$ is dense in H and either* $\operatorname{rank} H \geq 2$ *and Γ is irreducible or Λ is dense in H. Then*

(a) if k is isomorphic neither to $\mathbb{R}$ nor to $\mathbb{C}$, i.e., k is non-archimedean, then the subgroup $\delta(\Gamma)$ is relatively compact in $\mathbf{F}(k)$;
(b) if k is archimedean, $\delta(\Gamma)$ is not relatively compact in $\mathbf{F}(k)$, and the group $\mathbf{F}$ is adjoint and k-simple, then δ extends uniquely to a continuous homomorphism $\tilde{\delta}\colon H \to \mathbf{F}(k)$;
(c) if $k = \mathbb{R}$ and $\mathbf{F}$ is adjoint and has no $\mathbb{R}$-anisotropic factors, then δ extends uniquely to a continuous homomorphism $\tilde{\delta}\colon H \to \mathbf{F}(\mathbb{R})$.

Proof. As in 6.5 and 6.14 we observe that $\operatorname{Ad}\Gamma$ is an irreducible lattice in $\mathbf{G}(\mathbb{R})$ and the subgroup $\mathbf{G}^{\mathrm{is}}(\mathbb{R}) \cdot \operatorname{Ad}\Gamma$ is dense in $\mathbf{G}(\mathbb{R})$. Since $\overline{\delta(\Gamma)} = \mathbf{F}$ and $\mathscr{Z}(\mathbf{F}) = \{e\}$, we further have that $\delta(\Gamma \cap \mathscr{Z}(H)) = \{e\}$. But $\operatorname{Ker}\operatorname{Ad} = \mathscr{Z}(H)$. So one can define a homomorphism $\delta'\colon \operatorname{Ad}\Lambda \to \mathbf{F}(k)$ with $\delta = \delta' \circ \operatorname{Ad}$. Now to prove (a) and (b) it suffices to apply Corollary VII.5.16 and Theorem VII.5.9 to the homomorphism δ' with the replacements in their statements which were mentioned in 5.15.

Furthermore, to prove the uniqueness of $\tilde{\delta}$, we observe that $\tilde{\delta}(\mathscr{Z}(H)) = \{e\}$ (because $\overline{\delta(\Gamma)} = \mathbf{F}$ and $\mathscr{Z}(\mathbf{F}) = \{e\}$). Assertion (c) is a consequence of (b) applied to homomorphisms of the form $p \circ \delta$, where p is the natural projection of $\mathbf{F}$ onto an $\mathbb{R}$-simple factor of $\mathbf{F}$. It should be noted here that if $\mathbf{D}$ is an $\mathbb{R}$-isotropic $\mathbb{R}$-simple group, then no subgroup $B \subset \mathbf{D}(\mathbb{R})$ which is Zariski dense in $\mathbf{D}$ is relatively compact in $\mathbf{D}(\mathbb{R})$. This fact is a consequence of the non-compactness of the group $\mathbf{D}(\mathbb{R})$ and the theorem on the algebraicity of a compact real linear group (see I.0.31). □

(6.17) *Remarks.*

(i) In view of Theorem 6.15 it is not necessary to assume that the k-group $\mathbf{F}$ in Theorem 6.16(a) is connected and semisimple.

(ii) If $\mathbf{B}$ is an algebraic $\mathbb{R}$-group and $D \subset \mathbf{B}(\mathbb{R})$ is a connected semisimple subgroup which is Zariski dense in $\mathbf{B}$, then $D = \mathbf{B}(\mathbb{R})^0$ (see [V-O], Chapter III, §7, Theorem 3). Thus, in Theorem 6.16(c) we have $\tilde{\delta}(H) = \mathbf{F}(\mathbb{R})^0$.

(iii) Suppose H has trivial center and no non-trivial compact factor groups, and let Γ be an irreducible lattice in H. If $B \subset H$ is a non-trivial proper normal subgroup and $p\colon H \to H/B$ the natural epimorphism, then the subgroup $p(\Gamma)$ is not discrete in H/B (see [Rag 5], Corollary 5.21). Therefore, in view of remark (ii), if in this context the subgroup $\delta(\Gamma)$ is discrete in $\mathbf{F}(\mathbb{R})$ and is different from $\{e\}$, then the homomorphism $\tilde{\delta}$ in Theorem 6.16 is an isomorphism of the group H onto $\mathbf{F}(\mathbb{R})^0$.

(6.18) The following assertion holds.

(A) *Let $\Lambda \subset H$ be a subgroup such that* $\operatorname{Ad}\Lambda$ *is an irreducible lattice in* $\mathbf{H}(\mathbb{R})$. *Suppose that* $\operatorname{rank} H \geq 2$ *and H has no non-trivial compact factor groups. Then the subgroup $\Lambda \cap \mathscr{Z}(H)$ is of finite index in $\mathscr{Z}(H)$.*

In the case where the lattice $\operatorname{Ad}\Lambda$ is cocompact, assertion (A) was proved in [Rag 10] during the course of the proof of Proposition 2.12. In this case the condition that $\operatorname{rank} H \geq 2$ is superfluous. If the lattice $\operatorname{Ad}\Lambda$ is not cocompact, assertion (A) can be deduced from the results of [De] and [Rag 10] (see [Rag 10] Theorems 1.4 and 2.1) and the centrality of the congruence kernel (see VIII.2.16), after observing that by Theorem 1.16 that $\operatorname{Ad}\Lambda$ is arithmetic. We remark that in view of Lemma 6.1, Theorem 6.14(ii), and the equality $\mathscr{D}(\Lambda \cdot \mathscr{Z}(H)) = \mathscr{D}(\Lambda)$, assertion (A) is equivalent to the following

(B) *Let Γ be an irreducible lattice in H. Assume* $\operatorname{rank} H \geq 2$ *and H has no non-trivial compact factor groups. Then $\Gamma/\mathscr{D}(\Gamma)$ is finite.*

In connection with (B) we state

(6.19) Proposition. *Suppose that* **G** *has no almost* $\mathbb{R}$*-simple factors of* $\mathbb{R}$*-rank* 1. *Then*

(i) H has property (T);

(ii) if Γ is a lattice in H, then Γ has property (T);

(iii) if Γ is a lattice in H, then $\Gamma/\mathscr{D}(\Gamma)$ is finite.

Assertion (iii) is a consequence of Theorem III.2.5 and assertion (ii). Assertion (ii) is in turn a consequence of Theorem III.2.12 and assertion (i). We omit the proof of (i). Note however that this proof is based on the methods developed in Chapter III and is in many respects similar to the proof of Theorem III.5.3.

(6.20) *Remarks.*

(i) Making use of assertion (A) in 6.18 it is not hard to show that the condition that the center $\mathscr{Z}(H)$ is finite can be dropped in the statements of Theorems 6.14 and 6.15.

(ii) Let $T(H)$ denote the intersection of all irreducible lattices $\Gamma \subset H$. It can easily be verified that $T(H) \subset \mathscr{Z}(H)$. On the other hand, the following assertion is plausible: if $\operatorname{rank} H \geq 2$ and H has no non-trivial compact factor groups, then $T(H)$ is of finite index in $\mathscr{Z}(H)$. For further results in this direction we refer to [De] and [Rag 10].

7. Applications to the Theory of Symmetric Spaces and Theory of Complex Manifolds

In this section for any Lie group F we denote its identity component by F^0.

(7.1) Let X be a simply connected symmetric Riemannian space of non-positive curvature (for the definitions and basic facts in the theory of symmetric spaces the reader is referred to [He]). We denote by $H(X)$ the isometry group of X. We remark that $H(X)^0$ is of finite index in $H(X)$. If X can not be decomposed into a direct product of symmetric spaces of positive dimension, we say that X is *irreducible.* The space X can be decomposed into a direct product of irreducible symmetric spaces $X_1, \ldots, X_n$ that will be called the *irreducible factors* of X. If none of the factors X_i, $1 \leq i \leq n$, is 1-dimensional, then $H(X)^0$ is a connected semisimple Lie group with trivial center which is decomposable into the direct product of the connected noncompact simple Lie groups $H(X_i)^0$.

The space X is representable in the form $X' \times X''$, where X' is Euclidean space and X'' has no 1-dimensional irreducible factors. Furthermore, $H(X)^0 = H(X')^0 \times H(X'')^0$. Thus, we may assume that $H(X)^0 = \mathbf{G}_X(\mathbb{R})^0$, where $\mathbf{G}_X$ is a connected $\mathbb{R}$-group. Furthermore,

1) $\mathbf{G}_X = \mathbf{G}_{X'} \times \mathbf{G}_{X''}$;
2) if X has no 1-dimensional irreducible factors, then $\mathbf{G}_X$ is a connected semisimple adjoint $\mathbb{R}$-group without $\mathbb{R}$-anisotropic factors;
3) if X is n-dimensional Euclidean space, then $\mathbf{G}_X$ is the semi-direct product of the special orthogonal group $\mathbf{SO}_n$ and the n-dimensional additive group. We remark that the groups $H(X)^0$ and $\mathbf{G}_X$ are semisimple if and only if X does not have 1-dimensional irreducible factors.

We shall say that the space X is *typewise homogeneous* if all simple factors of the complexification of the Lie algebra of $H(X)$ are isomorphic. If X has no

1-dimensional irreducible factors, then X is typewise homogeneous if and only if all absolutely simple factors of the semisimple $\mathbb{R}$-group $\mathbf{G}_X$ have common type (in the sense of I.0.27).

Let Γ be a discrete group of isometries of X such that the following equivalent conditions are satisfied:

1) the volume of the quotient space $\Gamma \setminus X$ is finite;
2) Γ is a lattice in $H(X)$.

We say that the group Γ is *irreducible* if X can not be decomposed into a direct product of symmetric spaces X' and X'' of lower dimension so that Γ is commensurable with a direct product of some discrete subgroups $\Gamma' \subset H(X')$ and $\Gamma'' \subset H(X'')$. If the group $H(X)$ is semisimple, i.e., if X has no 1-dimensional irreducible factors, then the irreducibility of Γ is equivalent to the condition that $\Gamma \cap H(X)^0$ is an irreducible (in the sense of Definition 6.2) lattice in $H(X)^0$.

We now consider a connected complete locally symmetric Riemannian space M of finite volume and non-positive curvature, fundamental group $\pi_1(M)$, and the simply connected covering $\tilde{M}$. We may assume that $M = \pi_1(M) \setminus \tilde{M}$, where $\pi_1(M)$ operates on $\tilde{M}$ by isometries. We shall say that M is *reducible* if some finite covering of M is a direct product (as a Riemannian manifold) of locally symmetric spaces of positive dimension; otherwise M is called *irreducible.* It can easily be verified that the irreducibility of M is equivalent to that of the isometry group $\pi_1(M) \subset H(\tilde{M})$.

(7.2) From what we have discussed in 7.1 and Corollary 6.7 we deduce the following

Corollary.
(i) Let X be a simply connected symmetric Riemannian space of non-positive curvature. Suppose that X has no 1-dimensional irreducible factors and is not typewise homogeneous. Then there exist no discrete irreducible group of isometries of X with quotient space of finite volume.
(ii) Let M be a connected complete locally symmetric Riemannian space of finite volume and non-positive curvature. Suppose that the simply connected covering $\tilde{M}$ of M has no 1-dimensional irreducible factors and is not typewise homogeneous. Then M is reducible.

(7.3) Let L^n and B^n denote, respectively, n-dimensional Lobachevsky space and n-dimensional complex ball equipped with Bergman metric. Both of these spaces are irreducible symmetric spaces of rank 1. The groups $H(L^n)$ and $H(B^n)$ are locally isomorphic, respectively, to $\mathbf{SO}_{n,1}(\mathbb{R})$ and $\mathbf{SU}_{n,1}(\mathbb{R})$, where $\mathbf{SO}_{n,1}$ and $\mathbf{SU}_{n,1}$ were defined in III.5.5. We recall some well-known facts on the groups $\mathbf{SO}_{n,1}$ and $\mathbf{SU}_{n,1}$. Let $n, m \in \mathbb{N}^+$, $n \geq 2$. Then

1) $\mathbf{SO}_{n,1}$ is absolutely almost simple for $n \neq 3$;
2) $\mathbf{SO}_{3,1}$ is $\mathbb{R}$-isogeneous to $R_{\mathbb{C}/\mathbb{R}}\mathbf{SO}_{2,1}$;
3) $\mathbf{SU}_{m,1}$ is absolutely almost simple;
4) $\mathbf{SO}_{n,1}$ and $\mathbf{SU}_{m,1}$ are not isogeneous;

5) $\mathbf{SU}_{n,1}$ and $\mathbf{SU}_{m,1}$ are not isogeneous for $n \neq m$;
6) $\mathbf{SO}_{n,1}$ and $\mathbf{SO}_{m,1}$ are not isogeneous for $n > m$ and $n \geq 4$.

From this, 7.1, and assertion (i) of 4.6 we deduce assertions (a) and (b) below. These assertions can be viewed as special cases of Corollary 7.2.

(a) *There exist no discrete irreducible group of isometries of X with quotient space of finite volume if one of the following conditions holds:*
   1) $X = L^n \times L^m$, $n > m \geq 2$, $n \geq 4$;
   2) $X = L^n \times B^m$, $n \geq 2$, $m \geq 1$;
   3) $X = B^n \times B^m$, $n \geq 1$, $m \geq 1$, $n \neq m$.

(b) *Let M be a connected complete locally symmetric Riemannian space of finite volume which is locally isometric to the direct product of symmetric spaces X_1 and X_2. Assume one of the following conditions is satisfied:*
   1) $X_1 = L^n$, $X_2 = L^m$, $n > m \geq 2$, $n \geq 4$;
   2) $X_1 = L^n$, $X_2 = B^m$, $n \geq 2$, $m \geq 1$;
   3) $X_1 = B^n$, $X_2 = B^m$, $n \geq 1$, $m \geq 1$, $n \neq m$.

   Then some finite covering of M is (as a Riemannian manifold) the direct product of two spaces which are likewise locally isomorphic to X_1 and X_2.

Remark. Suppose that either $X = L^n \times L^n$, $n \geq 2$, or $X = L^3 \times L^2$ or $X = B^n \times B^n$, $n \geq 1$. Then there is a discrete irreducible group Γ of isometries of X such that the quotient space $\Gamma \setminus X$ is compact. This is a consequence of Theorem C in 4.7 (in the cases where $X = L^n \times L^n$ and $X = L^3 \times L^2$, also see the remarks at the end of 4.6). Since Γ is finitely generated and has a faithful linear finite-dimensional representation, it follows that there exists a torsion free subgroup Γ_1 of finite index in Γ (see [Rag 5], Corollary 6.13). We set $M = \Gamma_1 \setminus X$. Then

1) M is a compact Riemannian manifold locally isometric to X;
2) if M' is an arbitrary finite covering of M, then M' can not be represented in the form of a direct product of locally symmetric spaces of positive dimension (In fact, even as a topological space, M can not be represented as a direct product of manifolds of positive dimension. It is not hard to prove this fact using a purely topological argument and the indecomposability of the group Γ in the sense of 2.2.).

(7.4) Let M be a connected complex manifold, $\pi_1(M)$ its fundamental group, and $\tilde{M}$ the simply connected covering of M. We can then assume that $M = \pi_1(M)\setminus\tilde{M}$, where $\pi_1(M)$ acts on $\tilde{M}$ by biholomorphic transformations. Suppose that M is a bounded domain in $\mathbb{C}^n$. Let us denote by ρ the Bergman metric on $\tilde{M}$. This metric is invariant under biholomorphic transformations and in particular under $\pi_1(M)$. So ρ induces a distance function on M which will also be called the Bergman metric. We can now restate assertion (b) of 7.3 for the case where $X_1 = B^n$ and $X_2 = B^m$.

(7.5) Proposition. *Let M be a connected complex manifold and let $n > m \geq 1$. Suppose that the simply connected covering of M, viewed as a complex manifold is isomorphic to $B^n \times B^m$ and M is of finite volume with respect to the Bergman* *metric. Then some finite covering M' of M can be represented in the form $M' = M_1 \times M_2$, where the simply connected covering of M_1 (resp. M_2), viewed as a complex manifold, is isomorphic to B^n (resp. B^m).*

Remark. Obviously, if M is compact, then

1) the condition that M is of finite volume with respect to the Bergman metric is automatically satisfied;
2) the manifolds M_1 and M_2 are compact.

(7.6) As in 7.4 let M be a connected complex manifold, $\pi_1(M)$ its fundamental group and $\tilde{M}$ the simply connected covering. By $F(\tilde{M})$ we denote the group of biholomorphic transformations of $\tilde{M}$. We may assume that $\pi_1(M) \subset F(\tilde{M})$ and $M = \pi_1(M) \setminus \tilde{M}$.

The projections of the product $M \times M$ onto the first and the second factor will be denoted, respectively, by p_1 and p_2. By a *non-singular correspondence for* M we mean a (non-singular) complex submanifold $Y \subset M \times M$ such that the restrictions to Y of the projections p_1 and p_2 are finite-sheeted coverings of M. Since $\tilde{M}$ is simply connected, non-singular correspondences for $\tilde{M}$ are precisely the graphs of transformations in $F(\tilde{M})$. On the other hand, if Y is a non-singular correspondence for M and φ is the natural projection $\tilde{M} \times \tilde{M} \to M \times M$, then the connected components of the manifold $\varphi^{-1}(Y)$ are non-singular correspondences for $\tilde{M}$. Therefore any non-singular correspondence for M is a set of the form

$$\Omega_g \stackrel{\text{def}}{=} \{(\pi(x), \pi(gx)) \mid x \in \tilde{M}\},$$

where $g \in F(\tilde{M})$ and $\pi\colon \tilde{M} \to M$ is the natural projection. Furthermore, since the maps $p_{1|\Omega_g}$ and $p_{2|\Omega_g}$ are finite-sheeted, $g \in \mathrm{Comm}_{F(\tilde{M})}(\pi_1(M))$. Conversely, if $g \in \mathrm{Comm}_{F(\tilde{M})}(\pi_1(M))$, then Ω_g is a non-singular correspondence for M. Clearly, $\Omega_g = \Omega_{g'}$ if and only if $g \in \pi_1(M)g'$. Thus, the following conditions are equivalent:

(i) there exist infinitely many non-singular correspondences for M;
(ii) $\pi_1(M)$ is of infinite index in $\mathrm{Comm}_{F(\tilde{M})}(\pi_1(M))$.

We now assume that $\tilde{M}$ is a bounded symmetric domain in $\mathbb{C}^n$. Then $F(\tilde{M})$ is a semisimple Lie group with finite number of connected components and its identity component $F(\tilde{M})^0$ has no compact factor groups (see [He], Chapter VIII). Furthermore, $\tilde{M}$ equipped with Bergman metric is an Hermitian symmetric space of non-positive curvature and the group $F(\tilde{M})^0$ coincides with the identity component of the isometry group of $\tilde{M}$. We shall say that M is an *arithmetic variety* if $\pi_1(M) \cap F(\tilde{M})^0$ is an arithmetic lattice in $F(\tilde{M})^0$. If M is of finite volume with respect to the Bergman metric, then $\pi_1(M) \cap F(\tilde{M})^0$ is a lattice in $F(\tilde{M})^0$. We say that M is *irreducible* if this lattice is irreducible. The following proposition is a straightforward consequence of Theorem 6.5 and the equivalence of (i) and (ii) above.

(7.7) Proposition. *Let M be a connected complex manifold. Suppose that there exist infinitely many non-singular correspondences for M. Assume in addition that the simply connected covering of M is a bounded symmetric domain in $\mathbb{C}^n$ and M*

is irreducible and of finite volume with respect to the Bergman metric. Then M is an arithmetic variety.

(7.8) If M is a compact Riemann surface of genus $g > 1$, then its simply connected covering space, viewed as a complex manifold, is isomorphic to the unit disk $\{z \in \mathbb{C} \mid |z| < 1\}$. So from Proposition 7.7 we deduce

Corollary. *Let M be a compact Riemann surface of genus $g \geq 2$. Suppose that there exist infinitely many non-singular correspondences for M. Then M is an arithmetic variety.*

(7.9) *Remark.* Corollary 7.8 can be generalized as follows. Let M be a Riemann surface obtained from a compact Riemann surface of genus g by deleting a finite set of p points. Suppose that there exist infinitely many non-singular correspondences for M. Assume in addition that either $g \geq 2$ or $g = 1$ and $p \geq 1$ or $g = 0$ and $p \geq 3$. Then M is an arithmetic variety.

(7.10) In what follows, for any symmetric Riemannian space X of non-positive curvature, we denote by $H(X)$ and $\mathbf{G}_X$ the same objects as in 7.1. By the *Zariski topology* on $H(X)^0 = \mathbf{G}_X(\mathbb{R})^0$ we mean the topology induced from the Zariski topology of $\mathbf{G}_X$.

From now until the end of this section, X will denote a simply connected symmetric Riemannian space of non-positive curvature without 1-dimensional irreducible factors. We recall that $H(X)^0$ has no non-trivial compact factor groups and is a connected semisimple Lie group with trivial center and $\mathbf{G}_X$ is a connected semisimple adjoint $\mathbb{R}$-group without $\mathbb{R}$-anisotropic factors.

Let $\operatorname{rank} X$ denote the dimension of a maximal flat totally geodesic submanifold in X (a Riemannian manifold is said to be *flat* if its curvature tensor equals 0). It is well known that $\operatorname{rank} X = \operatorname{rank} H(X)^0 = \operatorname{rank}_{\mathbb{R}} \mathbf{G}_X$.

(7.11) For any lattice $\Gamma \subset H(X)$ we denote by $v(\Gamma \backslash X)$ the volume of the quotient space $\Gamma \backslash X$. It can easily be verified that the ratio

$$\mu_{H(X)}(\Gamma \backslash H(X))/v(\Gamma \backslash X)$$

does not depend upon the lattice $\Gamma \subset H(X)$. Therefore from Corollary 6.12 we deduce the following

Corollary. *Let $\Gamma \subset H(X)$ be a discrete subgroup with $v(\Gamma \backslash X) < \infty$. Then there exists $c > 0$ such that $c \cdot v(\Gamma_1 \backslash X)$ is an integer for any subgroup $\Gamma_1 \subset H(X)$ commensurable with Γ.*

(7.12) Lemma. *Every lattice $\Gamma \subset H(X)^0 = \mathbf{G}_X(\mathbb{R})^0$ is Zariski dense in $\mathbf{G}_X$, and hence in $H(X)^0$.*

Proof. Since $\mathbf{G}_X(\mathbb{R})^0$ is of finite index in $\mathbf{G}_X(\mathbb{R})$, it follows that Γ is a lattice in $\mathbf{G}_X(\mathbb{R})$. So Γ is Zariski dense in $\mathbf{G}_X$ (see Corollary II.4.4). □

(7.13) Lemma. *Let $\Gamma \subset H(X)^0$ be a subgroup which is Zariski dense in $H(X)^0$. Then*

(i) if φ is a continuous automorphism of the group $H(X)^0$ and $\varphi(\gamma) = \gamma$ for all $\gamma \in \Gamma$, then $\varphi(h) = h$ for all $h \in H(X)^0$.
(ii) $\mathscr{Z}_{H(X)}(\Gamma) = \{e\}$, and hence $\mathscr{Z}(\Gamma) = \{e\}$.
(iii) if N is a finite subgroup of $H(X)$ and Γ normalizes N, then $N = \{e\}$.
(iv) if Γ normalizes a subgroup N of $H(X)$ and $N \cap H(X)^0 = \{e\}$, then $N = \{e\}$.

Proof. (i), (ii). Since $\mathbf{G}_X$ is a connected semisimple adjoint $\mathbb{R}$-group, by virtue of Proposition I.2.6.5 any continuous automorphism of the group $H(X)^0 = G_X(\mathbb{R})^0$ is rational (i.e. extends to an $\mathbb{R}$-automorphism of $\mathbf{G}_X$). Thus, the set $\{h \in H(X)^0 \mid \varphi(h) = h\}$ is closed in $H(X)^0$ in the Zariski topology. This proves (i). It is well known from the theory of symmetric spaces that the centralizer $\mathscr{Z}_{H(X)}(H(X)^0)$ is trivial. From this and (i) applied to the automorphisms of the form $\operatorname{Int} h|_{H(X)^0}$, $h \in H(X)$, we deduce (ii).

(iii) We set $\Gamma_N = \Gamma \cap \mathscr{Z}_{H(X)}(N)$. Since N is finite and Γ normalizes N, it follows that Γ_N is of finite index in Γ. But Γ is Zariski dense in $\mathbf{G}_X$. From this and (ii) we deduce that

$$N \subset \mathscr{Z}_{H(X)}(\Gamma_N) = \{e\}.$$

(iv) is a consequence of (iii) and the fact that $H(X)^0$ is of finite index in $H(X)$. □

(7.14) Theorem. *Let Γ be a discrete group of isometries of X such that the quotient space $\Gamma \backslash X$ has finite volume (i.e. Γ is a lattice in $H(X)$). Suppose that* $\operatorname{rank} X \geq 2$ *and Γ is irreducible. Then*

(i) If N is a normal subgroup of Γ, then either $N = \{e\}$ or Γ/N is finite.
(ii) $\Gamma/\mathscr{D}(\Gamma)$ is finite.

Proof. (i) We set $\Gamma_0 = \Gamma \cap H(X)^0$ and $N_0 = N \cap H(X)^0$. Since Γ is irreducible, Γ_0 is an irreducible lattice in $H(X)^0$. It then follows from Theorem 6.14(i) and the triviality of the center of Γ_0 that either $N_0 = \{e\}$ or Γ_0/N_0 is finite (see Lemma 7.12 and Lemma 7.13(ii)). But

1) if $N_0 = \{e\}$, then by Lemma 7.12 and Lemma 7.13(iv) we have $N = \{e\}$;
2) since $H(X)^0$ is of finite index in $H(X)$, the finiteness of the factor group Γ_0/N_0 is equivalent to that of Γ/N.

(ii) By virtue of Theorem 6.14(ii) the factor group $\Gamma_0/\mathscr{D}(\Gamma_0)$ is finite. On the other hand, since $H(X)^0$ is of finite index in $H(X)$, it follows that Γ_0 is of finite index in Γ. Thus, $\Gamma/\mathscr{D}(\Gamma)$ is finite. □

(7.15) Corollary. *Let M be a connected complete locally symmetric space of non-positive curvature and of finite volume, $\pi_1(M)$ its fundamental group, and $\tilde{M}$ the simply connected covering of M. Suppose that M is irreducible, $\tilde{M}$ has no 1-dimensional irreducible factors, and* $\operatorname{rank} \tilde{M} \geq 2$. *Then*

(i) If N is a normal subgroup of $\pi_1(M)$, then either $N = \{e\}$ or $\pi_1(M)/N$ is finite.
(ii) $\pi_1(M)/\mathscr{D}(\pi_1(M))$ is finite.
(iii) The cohomology group $H^1(M, \mathbb{Z})$ is finite, and hence $H^1(M, \mathbb{R}) = 0$.

Assertions (i) and (ii) follow from Theorem 7.14 and the last paragraph of 7.1. It is well known that for any arcwise connected space Y the group $H^1(Y,\mathbb{Z})$ is naturally isomorphic to $\pi_1(Y)/\mathscr{D}(\pi_1(Y))$. So (iii) is a consequence of (ii).

(7.16) *Remarks.*

(i) If rank $X = 1$ and Γ is a lattice in $H(X)$, then there is a subgroup Γ' of finite index in Γ containing a free non-commutative normal (in Γ') subgroup of infinite index. It is not hard to deduce this assertion from the results of 6.5 in [Gro].

(ii) With L^n and B^n as in 7.3, we have that for each $n \geq 2$ the groups $H(L^n)$ and $H(B^{n-1})$ contain lattices with infinite factor groups by their commutator subgroups (see [Mills 1] and [Kaz 4]). On the other hand, if X has no irreducible factors isometric to B^n or L^n and Γ is a lattice in $H(X)$, then Γ has property (T), and hence $\Gamma/\mathscr{D}(\Gamma)$ is finite (see Theorem III.2.5). This fact is a consequence of Theorems III.5.6 and III.2.12 and the description of the groups $H(B^n)$ and $H(L^n)$ presented in 7.3.

(7.17) Let $X_1,\ldots,X_n$ be the irreducible factors of the space X. Multiplying the metric on X_i, $1 \leq i \leq n$, by a constant $c_i > 0$ we obtain another metric on X. This metric is $H(X)^0$-invariant and X with this metric is a symmetric space (because $H(X)^0 = H(X_1)^0 \times \ldots \times H(X_n)^0$). This procedure of modification of the metric on X will be called the *renormalization*. If X is irreducible, then the renormalization reduces to the multiplication of the metric by a constant. Notice that all $H(X)^0$-invariant Riemannian metrics on X can be obtained by the renormalization of the initial metric.

Any isometry $h \in H(X)$ is defined by the isometries $h_i\colon X_i \to X_{\sigma_h(i)}$ $1 \leq i \leq n$, where σ_h is a permutation of the set $\{1,\ldots,n\}$. The renormalization defined by the constants $c_1,\ldots,c_n$ is said to be *h-invariant* if the following equivalent conditions are satisfied:

1) the renormalized metric is h-invariant;
2) $c_{\sigma_h(i)} = c_i$ for all i, $1 \leq i \leq n$.

We remark that if $\sigma_h = \mathrm{Id}$, e.g. if $h \in H(X)^0$, then any renormalized metric is h-invariant.

Let $\Gamma \subset H(X)$ be a discrete group of isometries of X. We call the renormalization *Γ-invariant* if it is h-invariant for each $h \in \Gamma$. Every Γ-invariant renormalization of a metric on X induces the *renormalization* of the metric on $\Gamma \backslash X$. If Γ is torsion free, then the set of renormalized metrics on $\Gamma \backslash X$ coincides with the set of Riemannian metrics invariant under locally geodesic symmetries of $\Gamma \backslash X$ (We recall that a locally geodesic symmetry with respect to a point p is the map $\exp ty \mapsto \exp(-ty)$ of some ball about p, where $t \mapsto \exp ty$ is a geodesic with tangent vector y passing through p.).

Now let M be a connected complete locally symmetric space of non-positive curvature such that the simply connected covering $\tilde{M}$ does not have 1-dimensional irreducible factors. Then representing M in the form $\pi_1(M) \backslash \tilde{M}$, we can define in the fashion indicated above the *renormalization* of the metric on M.

(7.18) Let Y and Y' be two connected Riemannian manifolds and let $\varphi\colon Y \to Y'$ be a differentiable map. We recall that φ is said to be an *isometric immersion* if φ leaves the Riemannian metric invariant. An isometric immersion φ is said to be an *isometric embedding* if the map $\varphi\colon Y \to \varphi(Y)$ is one-to one.

We shall say that φ is a *strict isometric immersion* (resp. *strict isometric embedding*) if the following conditions are satisfied:

(a) φ is an isometric immersion (resp. isometric embedding);
(b) φ transforms geodesics to geodesics.

We remark that for an isometric embedding φ condition (b) is equivalent to (b′) $\varphi(Y)$ is a totally geodesic submanifold of Y. It is easily seen that φ is a strict isometric immersion if and only if the following is satisfied: (c) for each point $y \in Y$ there is a neighbourhood $U_y \subset Y$ of y such that for all $u, v \in U_y$ the distance between u and v is equal to the distance between $\varphi(u)$ and $\varphi(v)$.

Now let Y be one of the Riemannian manifolds mentioned in 7.17, i.e., Y is either X or M. We say that a map $\varphi\colon Y \to Y'$ is an *r-isometry* (resp. *strict r-isometric immersion, strict r-isometric embedding*) if φ becomes an isometry (resp. strict isometric immersion, strict isometric embedding) after some renormalization of the metric on Y.

(7.19) Suppose we are given a homomorphism ϑ of a subgroup $\Gamma \subset H(X)$ to an isometry group of a metric space Y. A map $\varphi\colon X \to Y$ will be called *Γ-equivariant* if $\varphi(\gamma x) = \vartheta(\gamma)\varphi(x)$ for all $\gamma \in \Gamma$ and $x \in X$.

(7.20) Theorem. *Let $\Gamma \subset H(X)$ be a discrete group of isometries of the space X such that the quotient space $\Gamma \backslash X$ has finite volume. Let X' be a simply connected symmetric space of non-positive curvature and $\vartheta\colon \Gamma \to H(X')$ a homomorphism with $\vartheta(\Gamma)$ discrete in $H(X')$ and infinite. Suppose that* $\operatorname{rank} X \geq 2$ *and the group Γ is irreducible. Then*

(a) There is a Γ-equivariant strict r-isometric embedding $\varphi\colon X \to X'$.
(b) If the subgroup $\vartheta(\Gamma) \cap H(X')^0$ is Zariski dense in $\mathbf{G}_{X'}$, then there is a Γ-equivariant r-isometry of X onto X'.

In the proof we shall need the following

(7.21) Lemma. *Let Y be a simply connected symmetric space of non-positive curvature and $F \subset H(Y)$ a closed subgroup of the isometry group of Y. Assume F is a semisimple Lie group with finitely many connected components. Then*

(i) There is a point $y \in Y$ such that the orbit Fy is a connected totally geodesic submanifold of Y (see [Kar], [Most 2]).
(ii) If F acts transitively on Y, then $F^0 = H(Y)^0$.

We shall prove (ii). Let K be a maximal compact subgroup of F^0. If $K \backslash F^0$ is equipped with an F^0-invariant metric and F^0 acts effectively on $K \backslash F^0$, then F^0 coincides with the identity component of the group of isometries of $K \backslash F^0$ (see Theorems 4.1 in Chapter V and 1.1 in Chapter VI in [He]). On the other hand, since F acts transitively on Y, Y is connected, and F^0 is of finite index in F, then F^0 acts transitively on Y. Thus, it suffices to show that K coincides with the stabilizer $\{g \in F^0 \mid gy = y\}$ of some point $y \in Y$. But this is a consequence of Cartan's theorem on the existence of a fixed point of a compact group of isometries of a simply connected complete Riemannian manifold of non-positive curvature (see [He], Theorem 13.5 of Chapter I) and the fact that for each $y \in Y$ the stabilizer $\{h \in H(Y) \mid hy = y\}$ is compact. □

(7.22) *Proof of Theorem 7.20.* We first prove (b). Set

$$\Gamma_0 = \{\gamma \in \Gamma \cap H(X)^0 \mid \vartheta(\gamma) \in H(X')^0\}.$$

Since $H(X')^0$ is of finite index in $H(X')$, it follows that Γ_0 is of finite index in Γ, and hence $\vartheta(\Gamma_0)$ is of finite index in $\vartheta(\Gamma)$. On the other hand, $\Gamma \cap H(X)^0$ is an irreducible lattice in $H(X)^0$, the subgroup $\vartheta(\Gamma) \cap H(X')^0$ is Zariski dense in $\mathbf{G}_{X'}$, and the $\mathbb{R}$-group $\mathbf{G}_{X'}$ is connected. Therefore Γ_0 is an irreducible lattice in $H(X)^0$ and the subgroup $\vartheta(\Gamma_0)$ is Zariski dense in $\mathbf{G}_{X'}$. Then, in view of Theorem 6.15(i), $\mathbf{G}_{X'}$ is semisimple, and hence X' has no 1-dimensional irreducible factors. Now making use of Theorem 6.16(c) and remark (iii) in 6.17 we obtain that $\vartheta|_{\Gamma_0}$ extends to a continuous isomorphism of the group $H(X)^0$ onto the group $H(X')^0 = \mathbf{G}_{X'}(\mathbb{R})^0$. It is known from the theory of symmetric spaces that the space X (resp. X') is representable in the form of the quotient group of $H(X)^0$ (resp. $H(X')^0$) by a maximal compact subgroup and that all maximal compact subgroups in the groups $H(X)^0$ and $H(X')^0$ are conjugate. So the isomorphism $\tilde{\vartheta}$ induces an $H(X)^0$-equivariant diffeomorphism $\varphi\colon X \to X'$. Since all $H(X)^0$-invariant Riemannian metrics on X can be obtained by renormalization of the initial metric on X, φ is an r-isometry.

We will show that the r-isometry φ is Γ-equivariant. Pick $\gamma \in \Gamma$ and $\gamma_0 \in \Gamma_0$. Since the subgroups $H(X)^0$ and $H(X')^0$ are normal in $H(X)$ and $H(X')$, the subgroup Γ_0 is normal in Γ, and hence $\gamma\gamma_0\gamma^{-1} \in \Gamma_0$. On the other hand, since φ is $H(X)^0$-equivariant, and hence Γ_0-equivariant, $\varphi\gamma'\varphi^{-1} = \vartheta(\gamma')$ for all $\gamma' \in \Gamma_0$. Thus, we have

$$\begin{aligned}(1)\qquad &(\varphi\gamma\varphi^{-1})\vartheta(\gamma_0)(\varphi\gamma\varphi^{-1})^{-1} = \\ &= \varphi\gamma(\varphi^{-1}\vartheta(\gamma_0)\varphi)\gamma^{-1}\varphi^{-1} = \\ &= \varphi\gamma\gamma_0\gamma^{-1}\varphi^{-1} = \vartheta(\gamma\gamma_0\gamma^{-1}).\end{aligned}$$

Since ϑ is a homomorphism, we have

$$(2)\qquad \vartheta(\gamma)\vartheta(\gamma_0)\vartheta(\gamma)^{-1} = \vartheta(\gamma\gamma_0\gamma^{-1}).$$

It follows from (1) and (2) that the element $(\varphi\gamma\varphi^{-1})\vartheta(\gamma)^{-1}$ belongs to the centralizer $\mathscr{Z}_{H(X')}(\vartheta(\Gamma_0))$. But since the subgroup $\vartheta(\Gamma_0)$ is Zariski dense in $H(X')^0$ and X' has no 1-dimensional irreducible factors, Lemma 7.13 implies that $\mathscr{Z}_{H(X')}(\vartheta(\Gamma_0)) = \{e\}$. So $(\varphi\gamma\varphi^{-1})\vartheta(\gamma)^{-1} = e$ and hence $\varphi\gamma\varphi^{-1} = \vartheta(\gamma)$, which proves the Γ-equivariance of the r-isometry ϑ.

We now turn to the proof of (a). By embedding X' in a larger symmetric space, we may assume that X' has no 1-dimensional irreducible factors, and hence $\mathbf{G}_{X'}$ is a connected semisimple adjoint $\mathbb{R}$-group. Then any continuous automorphism of $H(X')^0 = \mathbf{G}_{X'}(\mathbb{R})^0$ is rational (see Proposition I.2.6.5). In particular, for each $h \in H(X')$ the automorphism $h_0 \mapsto hh_0h^{-1}$, $h_0 \in H(X')^0$, is continuous in the Zariski topology. Therefore for any subgroup $\Lambda \subset H(X')^0$ we have

$$(3) \qquad \mathcal{N}_{H(X')}(\Lambda) \subset \mathcal{N}_{H(X')}(\bar{\Lambda}),$$

where $\bar{\Lambda}$ denotes the closure of the subgroup Λ in $H(X')^0$ in the Zariski topology.

We denote by $\mathbf{B}$ the Zariski closure of the subgroup $\vartheta(\Gamma_0)$ in $\mathbf{G}_{X'}$. Since the subgroup $\vartheta(\Gamma_0)$ is normal in $\vartheta(\Gamma)$, it follows from (3) that $\vartheta(\Gamma)$ normalizes $\mathbf{B}(\mathbb{R})^0$. So $F \overset{\text{def}}{=} \vartheta(\Gamma) \cdot \mathbf{B}(\mathbb{R})^0$ is a subgroup. Since $\vartheta(\Gamma_0)$ is of finite index in $\vartheta(\Gamma)$ and $\mathbf{B}(\mathbb{R})^0$ is of finite index in $\mathbf{B}(\mathbb{R})$, it follows that $\mathbf{B}(\mathbb{R})^0$ is of finite index in F. On the other hand, since rank $X \geq 2$ and Γ_0 is an irreducible lattice in $H(X)^0$, by Theorem 6.15(i) the $\mathbb{R}$-group $\mathbf{B}$ is semisimple. Thus, F is a semisimple Lie group with finitely many connected components. Therefore there exists a point $x \in X'$ such that Fx is a connected totally geodesic submanifold in X' (see Lemma 7.21(i)). Then, since X' is a simply connected symmetric space of non-positive curvature, Fx is a simply connected symmetric space of non-positive curvature. We define a continuous homomorphism $q\colon F \to H(Fx)$ via the equality

$$(4) \qquad q(h)y = hy, \ h \in F, \ y \in Fx.$$

Since for each $z \in X'$ the stabilizer $\{h \in H(X') \mid hz = z\}$ is compact, the kernel $\operatorname{Ker} q$ is compact. From this and the fact that $\vartheta(\Gamma)$ is discrete and infinite we deduce that the subgroup $q(\vartheta(\Gamma))$ is discrete in $H(Fx)$ and infinite. Now, in view of assertion (b) and equality (4), it suffices to show that the subgroup $q(\vartheta(\Gamma)) \cap H(Fx)^0$ is Zariski dense in $\mathbf{G}_{Fx}$.

Since F operates transitively on Fx and $F^0 = \mathbf{B}(\mathbb{R})^0$, by Lemma 7.21(ii) we have

$$(5) \qquad q(\mathbf{B}(\mathbb{R})^0) = H(Fx)^0 = \mathbf{G}_{Fx}(\mathbb{R})^0.$$

We denote by $\mathbf{B}^0$ the identity component of $\mathbf{B}$. Since $\mathbf{B}$ is semisimple it follows from (5) and Proposition I.2.6.5 that the restriction of the homomorphism q to $\mathbf{B}(\mathbb{R})^0 = \mathbf{B}^0(\mathbb{R})^0$ extends to an $\mathbb{R}$-epimorphism $\tilde{q}\colon \mathbf{B}^0 \to \mathbf{G}_{Fx}$. On the other hand, since $\vartheta(\Gamma) \supset \vartheta(\Gamma_0)$, $\mathbf{B}$ is the Zariski closure of the subgroup $\vartheta(\Gamma_0) \subset \mathbf{B}(\mathbb{R})$ and $\mathbf{B}(\mathbb{R})^0$ is of finite index in $\mathbf{B}(\mathbb{R})$, it follows that the subgroup $\vartheta(\Gamma) \cap \mathbf{B}(\mathbb{R})^0$ is Zariski dense in $\mathbf{B}^0$. So the subgroup

$$q(\vartheta(\Gamma) \cap \mathbf{B}(\mathbb{R})^0) \subset q(\vartheta(\Gamma)) \cap H(Fx)^0$$

is Zariski dense in $\mathbf{G}_{Fx}$. This completes the proof. □

(7.23) Theorem. *Let M and M' be connected complete locally symmetric Riemannian spaces of non-positive curvature, $\Gamma = \pi_1(M)$, and $\Gamma' = \pi_1(M')$ their fundamental groups, and let $\tilde{M}$ and $\tilde{M}'$ be the simply connected coverings of M and M'. Let $f\colon M \to M'$ be a continuous map. Denote by $\vartheta\colon \Gamma \to \Gamma'$ the homomorphism* *of fundamental groups induced from f (the homomorphism ϑ is defined up to an inner automorphism of Γ). Suppose that*

(a) the space M is of finite volume and irreducible;

(b) rank $\tilde{M} \geq 2$ *and $\tilde{M}$ has no one-dimensional irreducible factors (we do not impose similar restrictions on $\tilde{M}'$);*

(c) f is not homotopic to a constant map or, equivalently, the homomorphism ϑ is non-trivial.

Then the following assertions are true

(i) The map f is homotopic to a strictly r-isometric immersion $\varphi\colon M \to M'$. Furthermore, if $\vartheta(\Gamma) = \Gamma'$, then φ is an embedding.

(ii) If M' is of finite volume and ϑ is an isomorphism, then f is homotopic to an r-isometry $\varphi\colon M \to M'$ or, equivalently, the isomorphism ϑ is induced from an r-isometry of the spaces M and M'.

Proof. (i) We may assume that $M = \Gamma \backslash \tilde{M}$ and $M' = \tilde{\Gamma} \backslash \tilde{M}'$, where Γ and Γ' act on $\tilde{M}$ and $\tilde{M}'$ respectively by isometries. Let $p\colon \tilde{M} \to M$ and $p'\colon \tilde{M}' \to M'$ be the natural projections. It is easily seen that there exists a Γ-equivariant continuous map $\tilde{f}\colon \tilde{M} \to \tilde{M}'$ such that $f \circ p = p' \circ \tilde{f}$. According to the Cartan fixed point theorem (mentioned in 7.21), any finite isometry group of $\tilde{M}'$ has a fixed point. But $\vartheta(\Gamma) \neq \{e\}$ and $\gamma x \neq x$ for all $x \in \tilde{M}'$ and $\gamma \in \Gamma'$, $\gamma \neq e$. Hence, the discrete subgroup $\vartheta(\Gamma)$ is infinite. So by Theorem 7.20(a) there is a Γ-equivariant strictly r-isometric embedding $\tilde{\varphi}\colon \tilde{M} \to \tilde{M}'$. We can then define a strictly r-isometric immersion $\varphi\colon M \to M'$ by setting $\varphi = p' \circ \tilde{\varphi} \circ p^{-1}$. If $\vartheta(\Gamma) = \Gamma'$, then the submanifold $\tilde{\varphi}(\tilde{M})$ is Γ-invariant and hence φ is an embedding. Since $\tilde{M}'$ is a complete simply connected manifold of non-positive curvature, any two points in $\tilde{M}'$ can be joined by exactly one geodesic. Therefore the Γ-equivariant maps $\tilde{f}$ and $\tilde{\varphi}$ are homotopic in the class of continuous Γ-equivariant maps. This implies that f is homotopic to φ.

(ii) The submanifold $\tilde{\varphi}(\tilde{M})$ is totally geodesic in $\tilde{M}'$ and $\vartheta(\Gamma)$-invariant. On the other hand, if Y and Y' are connected Riemannian manifolds and $\varphi\colon Y \to Y'$ is an isometric embedding with $\varphi(Y) = Y'$, then φ is an isometry. So it remains to prove that if $M' = \Gamma' \backslash \tilde{M}'$ is of finite volume, then any Γ'-invariant totally geodesic submanifold $L \subset \tilde{M}'$ coincides with $\tilde{M}'$. We leave the proof of this fact to the reader as an easy exercise. (Hint: consider the map $\pi\colon \tilde{M}' \to L$ sending $x \in \tilde{M}'$ to the base of the perpendicular from x to L). □

(7.24) Assertion (ii) of Theorem 7.23 is a weak version of the following strong rigidity theorem for locally symmetric spaces.

Theorem. (See [Most 7], Sections 1 and 24.) *Let M and M' be connected complete locally symmetric Riemannian spaces of finite volume and non-positive curvature. Let $\tilde{M}$ be the simply connected covering of M and $p\colon \tilde{M} \to M$ the natural projection. Suppose the following conditions are satisfied.*

(a) M has no one-dimensional irreducible factors;

(b) for any 2-dimensional irreducible factor Y of $\tilde{M}$ and each point $x \in M$ the set $\pi_Y(p^{-1}(x))$ is not discrete in Y, where $\pi_Y\colon \tilde{M} \to Y$ is the natural projection

(equivalently, M does not contain 2-dimensional totally geodesic submanifolds being locally factors of M).

Then any fundamental group isomorphism $\vartheta\colon \pi_1(M) \to \pi_1(M')$ is induced from an r-isometry φ. In particular, if the groups $\pi_1(M)$ and $\pi_1(M')$ are isomorphic, then the spaces M and M′ become isometric after a renormalization of the metric on M.

(7.25) Recall that $\mathbf{PSL}_2(\mathbb{R})$ denotes the factor group of $\mathbf{SL}_2(\mathbb{R})$ by its center. We provide a theorem which is a restatement of Theorem 7.24 in terms of Lie group theory.

Theorem. (See [Most 7], Theorem 24.2.) *Let H and H′ be connected semisimple Lie groups with trivial center and no non-trivial compact factor groups. Let Γ and Γ' be lattices in H and H′ respectively. Suppose that for any continuous epimorphism $\pi\colon H \to \mathbf{PSL}_2(\mathbb{R})$, the subgroup $\pi(\Gamma)$ is not discrete. Then any isomorphism $\vartheta\colon \Gamma \to \Gamma'$ extends to a continuous isomorphism $\tilde{\vartheta}\colon H \to H'$.*

(7.26) *Remarks.*

(i) It is not hard to prove the uniqueness of the r-isometry φ in Theorems 7.20(b), 7.23(ii), and 7.24, and also the uniqueness of the extension $\tilde{\vartheta}$ in Theorem 7.25. In Theorem 7.20(a) the embedding φ is not in general uniquely determined. However, if $\varphi_1, \varphi_2\colon X \to X'$ are Γ-equivariant strictly r-isometric embeddings, then there exists an isometry h of X' such that $\varphi_1 = h\varphi_2$ and $h \in \mathscr{Z}_{H(X')}(\vartheta(\Gamma))$. Similarly, with M, M', Γ, Γ', and ϑ as in the statement of Theorem 7.23, we assume $\vartheta(\Gamma) = \Gamma'$. Then if $\varphi_1, \varphi_2\colon M \to M'$ are homotopic strictly r-isometric embeddings there exists an isometry ω of M' such that $\varphi_1 = \omega \circ \varphi_2$.

(ii) In proving Theorem 7.24 one can easily make a reduction to the case where M is irreducible. So the difference between Theorem 7.24 and Theorem 7.23(ii) consists of the additional consideration in Theorem 7.24 of the case where $\operatorname{rank} M = 1$ (in Theorem 7.25 this corresponds to the case where $\operatorname{rank} H = 1$). The proof for this case is based on refined and profound arguments (see [Most 7], Sections 19–23). This proof utilises, in particular, the methods of the theory of quasiconformal mappings (not just ordinary quasi-conformal mappings but also quasi-conformal mappings over K, where K is a division algebra over $\mathbb{R}$). In the case where M is of constant negative curvature, another approach has been offered by Gromov (see [Th], 6.3); for this case see also [Mar 1].

(iii) In the statement of Theorem 7.24 the condition "for any continuous homomorphism $\pi\colon H \to \mathbf{PSL}_2(\mathbb{R})$ the subgroup $\pi(\Gamma)$ is not discrete" is necessary (the same is true for condition (b) in Theorem 7.24). Indeed, if $H = \mathbf{PSL}_2(\mathbb{R})$, then it is well known that H contains a continuous family of isomorphic lattices Γ_t, $0 \le t \le 1$, which are not conjugate by automorphisms in H.

(iv) Assertion (ii) of Theorem 7.23 can also easily be deduced from Theorem VII.7.5.

Appendices

A. Proof of the Multiplicative Ergodic Theorem

We let k, n, $\mathscr{W}$, $\|w\|(w \in \mathscr{W})$, and $\|B\|(B \in \mathrm{End}(\mathscr{W}))$ be as in Chapter V. We shall also use the notation and definitions introduced in V.2.0.

A.1 Ljapunov Cohomology and the Class Ω

A measurable function A on a measure space (X, μ) taking values in the group $\mathbf{GL}(\mathscr{W})$ will be called a *Ljapunov function* (with respect to a dynamical system $\{L^m\}$), provided that

(1) $$\lim_{n \to \pm\infty} (1/m) \ln \|A(L^m x)\| = \lim_{m \to \pm\infty} (1/m) \ln \|A(L^m x)^{-1}\| = 0$$

for almost all $x \in X$ (in measure μ). Two cocycles u and $\tilde{u}$ will be called *cohomologous* if there exists a measurable function C on X taking values in $\mathbf{GL}(\mathscr{W})$ such that

(2) $$C(L^m x)^{-1} u(m, x) C(x) = \tilde{u}(m, x)$$

for all $m \in \mathbf{Z}$ and almost all $x \in X$. We remark that in view of the equalities (2) in V.2.0, condition (2) above is equivalent to the following one:

(3) $$C(Lx)^{-1} w(x) C(x) = \tilde{w}(x),$$

where $w(x) = u(1, x)$ and $\tilde{w}(x) = \tilde{u}(1, x)$.

We then say that the function C *realizes the cohomology* of the cocycles u and $\tilde{u}$. A cohomology realized by a Ljapunov function C will be called a *Ljapunov cohomology*; in such a case the cocycles u and $\tilde{u}$ will be called *$\mathscr{L}$-cohomologous.*

In what follows the automorphism L is assumed to be ergodic. Let Ω denote the set of Borel (not necessarily integrable) cocycles u for which assertions (i) and (ii) of the multiplicative ergodic theorem hold. A simple but significant observation is that if a cocycle $\tilde{u}$ is $\mathscr{L}$-cohomologous to a cocycle $u \in \Omega$, then $\tilde{u} \in \Omega$. We leave the proof of this assertion to the reader. Just notice that

(a) $\mathscr{L}$-cohomologous cocycles have the same characteristic exponents;

(b) if two cocycles u and $\tilde{u}$ satisfy condition (2), then $\psi_i(x) = C(x)\tilde{\psi}_i(x)$ for almost all $x \in X$, where ψ_i (resp. $\tilde{\psi}_i$) is the characteristic map defined by the cocycle u (resp. $\tilde{u}$) and the characteristic exponent χ_i.

A.2 Reduction to the Case in Which the Cocycle u is Triangular

We fix a basis in $\mathscr{W}$ and identify $\mathbf{GL}(\mathscr{W})$ with $\mathbf{GL}_n(k)$. A cocycle u is called (lower) *triangular* if $u(m, x)$ is a lower triangular matrix for all $m \in \mathbb{Z}$ and $x \in X$. Now let u be an arbitrary integrable cocycle. The desired reduction would be complete if one could find an integrable triangular cocycle $\mathscr{L}$-cohomologous to the cocycle u, because Ljapunov cohomologies leave Ω invariant. But an arbitrary cocycle is not necessarily $\mathscr{L}$-cohomologous to a triangular one. However, we will show that over some extension of the dynamical system $\{L^m\}$ the cocycle u is $\mathscr{L}$-cohomologous to a triangular one.

We denote by $B \subset \mathbf{GL}(\mathscr{W}) = \mathbf{GL}_n(k)$ the group of lower triangular matrices and set $D = \mathbf{GL}_n(k)/B$ and $\hat{X} = X \times D$. We define the group $\{\hat{L}^m\}$ of Borel automorphisms of $\hat{X}$ by

$$\hat{L}^m(x, d) = (L^m x, u(m, x)d), \tag{4}$$

where $u(m, x)d$ is the translation of d by $u(m, x)$($\{\hat{L}^m\}$ is a group because u is a cocycle). Note that the Iwasawa decomposition implies that D is compact (see Theorem I.2.2.1). Let us consider the set F of (regular Borel) measures $\hat{\mu}$ on $\hat{X}$ with $\hat{\mu}(A \times D) = \mu(A)$ for any Borel set $A \subset X$. In other words, $F = \{\hat{\mu} \in \mathscr{M}(\hat{X}) \mid$ the image of $\hat{\mu}$ under the natural projection $\hat{X} \to X$ coincides with $\mu\}$. The automorphism $\hat{L}$ acts on measures by the formula $(\hat{L}(\hat{\mu}))(A) = \hat{\mu}(\hat{L}(A))$. Since F is compact in the weak*-topology (because D is compact) and convex, $\hat{L}(F) = F$, and $\hat{L}$ acts continuously on the space of measures (in the weak*-topology), it follows from the Tychonov fixed point theorem that there is an $\hat{L}$-invariant ergodic measure $\hat{\mu} \in F$. The automorphism $\hat{L}$ fails, in general, to be ergodic in $\hat{\mu}$. But if we take as $\hat{\mu}$ an extreme point of the convex compact set $F_0 = \{\mu' \in F \mid \hat{L}(\mu') = \mu'\}$, then $\hat{L}$ will be ergodic in measure $\hat{\mu}$. (Indeed, otherwise the measure $\hat{\mu}$ can be represented as a sum of two non-proportional $\hat{L}$-invariant measures μ_1 and μ_2. It then follows from the ergodicity of L that the images of the measures μ_1 and μ_2 under the natural projection of $X \times D$ onto X are proportional. Thus, the measures $c_1\mu_1$ and $c_2\mu_2$ belong to F_0, where $c_1 = \mu(X)/\mu_1(X \times D)$ and $c_2 = \mu(X)/\mu_2(X \times D)$. But this contradicts the fact that the measure $\hat{\mu} = \mu_1 + \mu_2$ is an extreme point of the set F_0 and μ_1 and μ_2 are not proportional.)

We now set

$$\hat{u}(m, x, d) = u(m, x), \tag{5}$$

$m \in \mathbb{Z}$, $x \in X$, $d \in D$. Since u is a cocycle with respect to $\{L^m\}$, it follows that $\hat{u}$ is a cocycle with respect to $\{\hat{L}^m\}$. Since the cocycle u is integrable, so is $\hat{u}$. The cocycle $\hat{u}(m, x, d)$ does not depend upon d. Therefore if $\hat{u} \in \Omega$, then $\chi^+(w, x, d)$ and $\chi^-(w, x, d)$ do not depend upon d, and this implies that $u \in \Omega$. But Ljapunov cohomologies leave Ω invariant (see A.1). Therefore to obtain the desired reduction it suffices to show that the cocycle $\hat{u}$ is $\mathscr{L}$-cohomologous to an integrable triangular cocycle. Since the space $D = \mathbf{GL}_n(k)/B$ is compact, by Theorem I.4.4.1 (on the existence of Borel sections) there exists a Borel map $f\colon D \to \mathbf{GL}_n(k)$ such that the set $f(D)$ is relatively compact in $\mathbf{GL}_n(k)$ and $\pi(f(d)) = d$ for all $d \in D$, where $\pi\colon \mathbf{GL}_n(k) \to D$ is the natural projection. It follows that

(a) $f(gd)^{-1}gf(d) \in B$ for all $g \in \mathbf{GL}_n(k)$ and $d \in D$ and
(b) $\sup_{g\in f(D)} \|g^{\pm 1}\| < \infty$.

Therefore, in view of (4) and (5), the function $C(x,d) = f(d)$ realizes the desired cohomology.

A.3 The Classes Ψ and Ψ_0

We denote by $u_{ts}(m,x)$ (resp. $u_{ts}(x)$) the entry of the matrix $u(m,x)$ (resp. $w(x)$) in t-th row and s-th column. Let Ψ be the set of measurable triangular cocycles u such that

(a) $w(x) = u(1,x)$ is a Ljapunov function and
(b) $\ln |w_{tt}| \in L_1(x,\mu)$ for all t, $1 \leq t \leq n$.

If $f \in L_1(X,\mu)$, then the Birkhoff individual ergodic theorem (see Theorem I.4.6.5) applied to the function $f(x) - f(Lx)$ implies that f is a Ljapunov function. From this we easily obtain that any integrable triangular cocycle belongs to Ψ. Therefore, in view of the reduction presented in A.2, it suffices to show that

$$\Psi \subset \Omega. \tag{6}$$

Let $u \in \Psi$. Since L is ergodic, the Birkhoff individual ergodic theorem (see Theorem I.4.6.5), the property of the cocycle u being triangular, and equalities (2) in V.2.0 imply that for any t, $1 \leq t \leq n$, there exists $\lambda_t(u)$ such that

$$\lim_{m\to\pm\infty} (1/m) \ln |u_{tt}(m,x)| = \lambda_t(u) \tag{7}$$

for almost all $x \in X$. We set $\Psi_0 = \{u \in \Psi \mid u_{ts}(m,x) \equiv 0$ whenever $1 \leq s < t \leq n$ and $\lambda_t(u) \neq \lambda_s(u)\}$. We will show that

$$\Psi_0 \subset \Omega. \tag{8}$$

Let $u \in \Psi_0$. Then, expressing the entries of the matrices $u(m,x) = w(L^{m-1}x)u(m-1,x)$ and $u(-m,x) = w(L^{-m}x)^{-1}u(-m+1,x)$ via the entries of the matrices $w(L^{m-1}x)$, $u(m-1,x)$, $w(L^{-m}x)^{-1}$, and $u(-m+1,x)$ and using the fact that w is a Ljapunov function, one can prove by induction on $t-s$ that for all t and s, $1 \leq s < t \leq n$, and almost all $x \in X$ the values of $|u_{ts}(m,x)/u_{tt}(m,x)|$ (and hence the values of $|u_{ts}(m,x)/u_{ss}(m,x)|$) increase, as functions of m, slower, as $m \to \pm\infty$, than any exponential function $\exp c|m|$, $c > 0$. From this and (7) we deduce that $u \in \Omega$. Furthermore, the characteristic exponents of the cocycle u are the numbers $\lambda_1(u), \ldots, \lambda_n(u)$, and the values of the characteristic maps at any point are most easily described as the characteristic subspaces of the linear transformation given by the diagonal matrix

$$\begin{pmatrix} \tilde{\lambda}_1(u) & & & \\ & \tilde{\lambda}_2(u) & & 0 \\ & 0 & \ddots & \\ & & & \tilde{\lambda}_n(u) \end{pmatrix},$$

where $\tilde{\lambda}_i(u) \in k$ and $\tilde{\lambda}_i(u) \neq \tilde{\lambda}_j(u)$ if and only if $\lambda_i(u) \neq \lambda_j(u)$, $1 \leq i, j \leq n$, verifying (8). On the other hand (see A.1), a cocycle $\mathscr{L}$-cohomologous to a cocycle belonging to Ω is likewise in Ω. Therefore to establish (6) it suffices to show that any cocycle from Ψ is $\mathscr{L}$-cohomologous to a cocycle from Ψ_0. In A.4 and A.5 below we shall prove a bit stronger assertion. Namely, we show that

(∗) *for each cocycle $u \in \Psi$ there exist a cocycle $\tilde{u} \in \Psi_0$ and a Ljapunov function C such that*
 (a) *C realizes the cohomology of the cocycles u and $\tilde{u}$ and*
 (b) *$C(x)$ are lower triangular matrices with ones in the diagonal.*

A.4 Proof of (∗) in Case $n = 2$

If $\lambda_1(u) = \lambda_2(u)$, then $u \in \Psi_0$. So we can assume that $\lambda_1(u) \neq \lambda_2(u)$. Then, replacing if necessary L by L^{-1}, we may assume that $\lambda_1(u) < \lambda_2(u)$. Since $u(m+1, x) = w(L^m x) u(m, x)$, we have

$$(9) \qquad \frac{u_{21}(m+1,x)}{u_{22}(m+1,x)} = \frac{w_{21}(L^m x)}{w_{22}(L^m x)} \cdot \frac{u_{11}(m,x)}{u_{22}(m,x)} + \frac{u_{21}(m,x)}{u_{22}(m,x)} .$$

Since w is a Ljapunov function taking values in a group of triangular matrices, it follows from (7) and (9) that for almost all $x \in X$

$$(10) \qquad \limsup_{m \to +\infty} \frac{1}{m} \ln \left| \frac{u_{21}(m+1,x)}{u_{22}(m+1,x)} - \frac{u_{21}(m,x)}{u_{22}(m,x)} \right| \leq \lambda_1(u) - \lambda_2(u) .$$

It follows from (10) and the inequality $\lambda_1(u) < \lambda_2(u)$ that for almost all $x \in X$ the following limit exist:

$$(11) \qquad c(x) \stackrel{\text{def}}{=} \lim_{m \to +\infty} (u_{21}(m,x)/u_{22}(m,x)).$$

Furthermore,

$$(12) \qquad \limsup_{m \to +\infty} \frac{1}{m} \ln \left| \frac{u_{21}(m,x)}{u_{22}(m,x)} - c(x) \right| \leq \lambda_1(u) - \lambda_2(u) .$$

We denote by y_x the vector $(1, -c(x))$. Then (7) and (12) imply that for almost all $x \in X$,

$$(13) \qquad \lim_{m \to +\infty} (1/m) \ln \|u(m,x) y_x\| = \lambda_1(u).$$

We set $Y_x = \{ay_x \mid a \in k\}$. Since $\lambda_2(u) > \lambda_1(u)$, it follows from (13) that for almost all $x \in X$, we have

$$Y_x = \{0\} \cup \{y \in k^2 - \{0\} \mid \lim_{m \to +\infty} (1/m) \ln \|u(m,x)y\| = \lambda_1(u)\}.$$

From this and the equality $u(m+r,x) = u(m, L^r x)u(r,x)$ we deduce that

$$u(r,x)Y_x = Y_{L^r x} \tag{14}$$

for each $r \in \mathbb{Z}$ and almost all $x \in X$. We set $C(x) = \begin{pmatrix} 1 & 0 \\ -c(x) & 1 \end{pmatrix}$. It then follows from (14) that C realizes the cohomology of the cocycles u and $\tilde{u} \in \Psi_0$, where $\tilde{u}(m,x) = \begin{pmatrix} u_{11}(m,x) & 0 \\ 0 & u_{22}(m,x) \end{pmatrix}$. We will show that C is a Ljapunov function. This is equivalent to the condition that for almost all $x \in X$,

$$\lim_{r \to +\infty} (1/r) \ln^+ |c(L^r x)| = 0. \tag{15}$$

Since w is a Ljapunov function taking values in a group of triangular matrices, it follows from (7) and (9) that for almost all $x \in X$

$$\lim_{m \to -\infty} \sup \frac{1}{|m|} \ln \left| \frac{u_{21}(m+1,x)}{u_{22}(m+1,x)} - \frac{u_{21}(m,x)}{u_{22}(m,x)} \right| \leq \lambda_2(u) - \lambda_1(u) ,$$

and hence

$$\lim_{m \to -\infty} \sup (1/|m|) \ln |u_{21}(m,x)/u_{22}(m,x))| \leq \lambda_2(u) - \lambda_1(u). \tag{16}$$

Since $Y_x = \{ay_x | a \in k\}$, $y_x = (1, -c(x))$, and u is a triangular cocycle, it follows from equality (14) that for almost all $x \in X$ and each $r \in \mathbb{Z}$,

$$c(L^r x) = \left(c(x) - \frac{u_{21}(r,x)}{u_{22}(r,x)} \right) \cdot \frac{u_{22}(r,x)}{u_{11}(r,x)} . \tag{17}$$

Now equality (15), as $r \to +\infty$, is a consequence of (7) and (12) and, as $r \to -\infty$, is a consequence of (7) and (16).

A.5 Proof of (*) for arbitrary *n*

A function $A = (a_{ij})$ on X taking values in $\mathbf{GL}_n(k)$ will be called (t,s)-*elementary*, where $1 \leq t \leq n$, $1 \leq s \leq n$, $t \neq s$, if

1) $a_{ii}(x) \equiv 1$ for each i, $1 \leq i \leq n$, and
2) $a_{ij}(x) \equiv 0$ whenever $i \neq j$ and the (ordered) pair (i,j) is different from (t,s).

A cohomology realized by a (t,s)-elementary function will be called (t,s)-*elementary*. For $i \in \mathbb{N}^+$, we further set

$\Phi_i = \{u \in \Psi \mid u_{ts}(m,x) \equiv 0$, whenever $1 \leq s < t \leq n$, $t - s < i$, and $\lambda_t(u) \neq \lambda_s(u)\}$.

Clearly, $\Phi_1 = \Psi$ and $\Phi_n = \Psi_0$. We set

$$U_{ts}(m,x) = \begin{pmatrix} u_{ss}(m,x) & 0 \\ u_{ts}(m,x) & u_{tt}(m,x) \end{pmatrix} \in \mathbf{GL}_2(k),$$

where $m \in \mathbb{N}^+$, $x \in X$, and $1 \leq s < t \leq n$. Let $u \in \Phi_i$, $1 \leq s < t \leq n$, $t - s = i$, and $\lambda_t(u) \neq \lambda_s(u)$. Then for $s < q < t$, either $u_{qs}(m,x) \equiv 0$ or $u_{tq}(m,x) \equiv 0$. From this and the fact that u is a triangular cocycle we deduce that U_{ts} is a cocycle. But in case $n = 2$ assertion (*) was already established. Therefore there exists a (t,s)-elementary Ljapunov cohomology after the action of which $u_{ts}(m,x) \equiv 0$. On the other hand, since u is a triangular cocycle, the indicated cohomology can alter only the entries u_{tr} and u_{qs}, where $r \leq s$ and $q \geq t$. Thus, applying (t,s)-elementary Ljapunov cohomologies ($1 \leq s < t \leq n$, $t - s = i$, and $\lambda_t(u) \neq \lambda_s(u)$) to $u \in \psi_i$ we can attain the inclusion $u \in \psi_{i+1}$. Using this we establish (*) by induction on $n - i$ for any cocycle $u \in \psi_i$. All that remains now is to recall that $\Phi_1 = \Psi$, completing the proof. □

B. Free Discrete Subgroups of Linear Groups

In this appendix using the methods of the paper [Ti 3], we shall prove Theorem V.5.6. As in Section 3 of Chapter VI, k will denote a local field, $\mathscr{W}$ a finite-dimensional vector space over k, and $\pi\colon \mathscr{W} - \{0\} \to \mathbf{P}(\mathscr{W})$ the natural projection. We shall assume $\mathbf{GL}(\mathscr{W})$ acts naturally on $\mathbf{P}(\mathscr{W})$ namely, $gx = \pi(g\pi^{-1}(x))$, $g \in \mathbf{GL}(\mathscr{W})$, $x \in \mathbf{P}(\mathscr{W})$.

If X is a subset of an algebraic variety M, then the topology on X induced from the Zariski topology of M will be called the *Zariski topology* of X.

(B.1) Let $g \in \mathbf{GL}(\mathscr{W})$. We denote by $\Omega(g)$ the set of characteristic exponents of the transformation g and by $\mathscr{W}_a(g) \subset \mathscr{W}$ the characteristic subspace of g corresponding to $a \in \Omega(g)$ (for the definitions we refer to II.1). Let $a_0 = \max\{a \mid a \in \Omega(g)\}$ be the maximal characteristic exponent of g. We set

$$A(g) = \pi(\mathscr{W}_{a_0}(g)) \text{ and } A'(g) = \pi\Big(\bigoplus_{a\in\Omega(g)-\{a_0\}} \mathscr{W}_a(g)\Big).$$

From Proposition II.1.2 we deduce

(1) Lemma. *The transformation g attracts $\mathbf{P}(\mathscr{W}) - A'(g)$ towards $A(g)$, i.e., for any open subset $U \subset \mathbf{P}(\mathscr{W})$ and any compact set $K \subset \mathbf{P}(\mathscr{W}) - A'(g)$, there exists $m = m(g, U, K) \in \mathbb{N}^+$ such that $g^r K \subset U$ for all $r \geq m$.*

As in VI.3.5 we say that g is *proximal* if the linear subspace $\mathscr{W}_{a_0}(g)$ is one-dimensional or, equivalently, if $A(g)$ is a singleton.

(2) Lemma. (See [Ti 3], Lemma 3.11.) *Let F be a subgroup of $\mathbf{GL}(\mathscr{W})$ which is connected in the Zariski topology and leaves invariant no proper nontrivial linear* *subspace of $\mathscr{W}$. Suppose F contains at least one semisimple proximal element g_0. Then the set $X = \{x \in F \mid x$ and x^{-1} are proximal$\}$ is Zariski dense in F.*

Remark. Modifying slightly the proof of Lemma 2 given in [Ti 3], one can show that the assumption of semisimplicity of g_0 is superfluous.

(B.2) We assume k is equipped with an absolute value extended to any finite field extension of k.

(3) Lemma. *Let F be a subgroup of $\mathbf{GL}(\mathscr{W})$ and let X be a Zariski dense and open subset of F. Suppose* $\mathrm{End}(\mathscr{W})$ *is spanned by F and F is not relatively compact in $\mathbf{GL}(\mathscr{W})$. Then the set of traces $\{\mathrm{Tr}\, X\}$is not relatively compact in k, and hence there exists $x \in X$ such that at least one of the eigenvalues of the transformation x is greater than 1 in absolute value.*

Lemma 3 is a particular case of Lemma 2.6 in [Ti 3].

(4) Lemma. *Let $\mathbf{H}$ be a connected semisimple k-group and let F be a subgroup of $\mathbf{H}(k)$ which is Zariski dense in $\mathbf{H}$. Suppose F is not relatively compact in the topology of $\mathbf{H}(k)$ induced from that of the field k. Then there exist $x \in F$, $r > 1$, and a rational k-irreducible r-dimensional representation ρ defined over k of the group $\mathbf{H}$ such that the transformation $\rho(x)$ is semisimple and proximal.*

Lemma 4 is a strengthening of Lemma VI.4.5, because we claim that $\rho(x)$ is not only proximal but also semisimple. Lemma 4 can be deduced from Lemma 3 in the same way as Lemma VI.4.5 was deduced from Lemma VI.4.4. To this end it suffices to observe that in any semisimple algebraic group $\mathbf{G}$ the set of semisimple elements contains a Zariski dense and open subset of $\mathbf{G}$.

We will need the following

(5) Lemma. (See [Ti 3], Lemma 3.11.) *Let $F \subset \mathbf{GL}(\mathscr{W})$ be a subgroup leaving invariant no proper non-trivial linear subspace of $\mathscr{W}$ and let $\mathscr{W}_1$, $\mathscr{W}_2$ be two linear subspaces of $\mathscr{W}$ with $\mathscr{W}_1 \neq \{0\}$ and $\mathscr{W}_2 \neq \mathscr{W}$. Then the set $\{g \in F \mid g\mathscr{W}_1 \not\subset \mathscr{W}_2\}$ is non-empty and Zariski open in F.*

(B.3) We now proceed the proof of Theorem V.5.6. Let $\mathbf{H}^0$ be the Zariski connected component of the identity in $\mathbf{H}$. Since $\mathbf{H}/\mathbf{H}^0$ is finite, replacing $\mathbf{H}$ by $\mathbf{H}^0$ and Λ by $\Lambda \cap \mathbf{H}^0$, we may assume that $\mathbf{H}$ is connected. Then by Lemma 4 there exist $\lambda \in \Lambda$, $r > 1$, and a rational k-irreducible r-dimensional representation ρ defined over k of the group $\mathbf{H}$ such that the transformation $\rho(\lambda)$ is semisimple and proximal. Making use of Lemma 2 we deduce that the set $X \overset{\mathrm{def}}{=} \{x \in \rho(\Lambda \mid x$ and x^{-1} are proximal$\}$ is Zariski dense in $\rho(\Lambda)$. For $x \in X$, we set $B(x) = A(x) \cup A(x^{-1})$, $B'(x) = A'(x) \cup A'(x^{-1})$, and denote by $W \simeq k^r$ the space of the representation ρ. It is easily seen that $B'(x) \neq \mathbf{P}(\mathscr{W})$. On the other hand, $x \in X$, $r > 1$, and the group $\mathbf{H}$ (and hence the group Λ) is connected in the Zariski topology. Therefore it follows from Lemma 5 that there exists $h \in \rho(\Lambda)$ such that

$$B(x) \cap hB'(x) = B'(x) \cap hB(x) = \varnothing. \tag{1}$$

We set $g_1 = x$ and $g_2 = hxh^{-1}$. It is easily seen that $A(g_2) = hA(x)$ and $A'(g_2) = hA'(x)$. Therefore (1) implies that for $i = 1, 2$,

$$(A(g_i) \cup A(g_i^{-1})) \cap (A'(g_{3-i}) \cup A'(g_{3-i}^{-1})) = \varnothing,$$

and hence there exist neighbourhoods U_1, U_1', U_2, U_2' of the points $A(g_1)$, $A(g_1^{-1})$, $A(g_2)$, $A(g_2^{-1})$ and a point $p \in \mathbf{P}(\mathscr{W})$ such that

$$(\bar{U}_i \cup \bar{U}_i) \cap (A'(g_{3-i}) \cup A(g_{3-i}^{-1})) = \varnothing, \ i = 1, 2, \tag{2}$$

and

$$p \notin \bigcup_{i=1,2} \bar{U}_i \cup \bar{U}_i' \cup A'(g_i) \cup A'(g_i^{-1}), \tag{3}$$

where $\bar{U}_i$ and $\bar{U}_i'$ are the closures of U_i and U_i' respectively in the topology induced from that of the field k. Then in view of Lemma 1 there exists $m \in \mathbb{N}^+$ such that for $i = 1, 2$ and all $r \in \mathbb{Z}$, $|r| \geq m$, we have

$$g_i^r(\bar{U}_{3-i} \cup \bar{U}_{3-i}' \cup \{p\}) \subset U_i \cup U_i'. \tag{4}$$

We set $\tilde{g}_i = g_i^m$. Then by induction on s we obtain from (4) that

$$\tilde{g}_{i_s}^{r_s} \cdot \tilde{g}_{i_{s-1}}^{r_{s-1}} \cdot \ldots \cdot \tilde{g}_{i_1}^{r_1}(p) \in U_{i_s} \cup U_{i_s}'$$

for all $s \in \mathbb{N}^+$, $i_l = 1, 2, i_l \neq i_{l-1}$, $r_l \in \mathbb{Z} - \{0\}$. From this and (3) we deduce that $\tilde{g}_1$ and $\tilde{g}_2$ generate a free subgroup. But $\tilde{g}_1, \tilde{g}_2 \in \rho(\Lambda)$. Thus Λ contains a free discrete subgroup. □

C. Examples of Non-Arithmetic Lattices

It is a well known fact that the group $\mathbf{SL}_2(\mathbb{R})$ contains non-arithmetic lattices. Indeed, there exist only countably many mutually non-conjugate arithmetic lattices. On the other hand, in $\mathbf{SL}_2(\mathbb{R})$ there are continuous families $\{\Gamma_t\}$ of mutually non-conjugate lattices. Thus, the existence of non-arithmetic lattices is due to the fact that in $\mathbf{SL}_2(\mathbb{R})$ there are lattices which are not locally rigid. In this connection, for some time it seemed likely that any locally rigid lattice in a connected semisimple Lie group should be arithmetic. But according to the Selberg-Weil local rigidity theorem (see [Rag 5], Theorem 6.7 and Corollary 7.66) if G is a connected semisimple Lie group without non-trivial compact factor groups and which is not locally isomorphic to $\mathbf{SL}_2(\mathbb{R})$, then any irreducible cocompact lattice $\Gamma \subset G$ is locally rigid. (We remark that if in addition G is not locally isomorphic to $\mathbf{SL}_2(\mathbb{C})$, then the analogous assertion is also valid for non-cocompact lattices.) So the conjecture arised that if G is a connected semisimple Lie group without non-trivial compact factor groups and which is not locally isomorphic to $\mathbf{SL}_2(\mathbb{R})$, then any irreducible lattice $\Gamma \subset G$ is arithmetic. Theorem IX.6.5 asserts that the conjecture is "almost always" true, namely, in case $\operatorname{rank} G > 1$. However, there are simple rank 1 Lie groups non-locally isomorphic to $\mathbf{SL}_2(\mathbb{R})$ and which contain non-arithmetic lattices. The first examples of this kind have been constructed by Makarov and Vinberg. They have shown (see [Mak 2] and [Vi 1]) that there exist non-arithmetic lattices, both cocompact and non-cocompact, in the groups of motions of 3, 4, and 5-dimensional Lobachevsky spaces (here and in what follows, by an arithmetic lattice in a semisimple Lie group G with finitely many connected components, we mean a lattice whose intersection with the identity component of G is an arithmetic lattice in the sense of Definition IX.6.3). Recently, using hybrids of hyperbolic manifolds, Gromov and Piatetski-Shapiro [Gro-P] constructed non-arithmetic lattices in n-dimensional Lobachevsky space for any $n \geq 3$.

It should be noted that, according to a recent result of Lubotzky (see [Lu 2]), if k is a non-archimedean local field and $\mathbf{G}$ is a connected almost k-simple k-group of k-rank 1, then $\mathbf{G}(k)$ contains an uncountable number of conjugacy classes of cocompact lattices and in particular non-arithmetic lattices. If $\operatorname{char} k > 0$, then the same holds for non-arithmetic lattices.

In Sections 1 and 2 of this appendix we present in short the papers [Vi 1] and [Gro-P] mentioned above. In Section 3 some results are given on discrete subgroups of $\mathbf{SU}_{n,1}(\mathbb{R})$ generated by complex reflections, where $\mathbf{SU}_{n,1}$ was defined in III.5.5 (we recall that the group $\mathbf{SU}_{n,1}(\mathbb{R})$ is locally isomorphic to the group of motions of n-dimensional complex ball equipped with the Bergmann metric).

1. Discrete Groups Generated by Reflections in Lobachevsky Spaces

(1.1) Fundamental Polyhedron of a Discrete Group Generated by Reflections. Let S be either n-dimensional Euclidean space E^n or n-dimensional Lobachevsky space Λ^n. By a *convex polyhedron* in S we mean the intersection of a finite number of closed half-spaces with non-empty interior and by a *reflection* we mean a hyperplane reflection.

Let Γ be a discrete group of motions of S generated by a finite number of reflections. The hyperplanes of all the reflections from Γ partition S into convex polyhedra, named *Γ-cells.* Each Γ-cell is a fundamental domain for Γ and the reflections relative to the faces of this Γ-cell generate Γ. A convex polyhedron P is a Γ-cell for a group Γ of the type in question if and only if all dihedral angles of P are of form π/n with n an integer. Such convex polyhedra will be called *Coxeter polyhedra.* Clearly, a Γ-cell P is compact (resp. is of finite volume) if and only if Γ is a cocompact (resp. non-cocompact) lattice in the group of motions of S.

(1.2) *C*-polyhedra in Euclidean Spaces. To describe a convex polyhedron $P \subset E^n$ consider unit vectors e_i orthogonal to the bounding hyperplanes H_i of P and directed outwards. The Gram matrix $(a_{ij} = \langle e_i, e_j \rangle)$ of the vector system $\{e_i\}$ will be called the *Gram matrix of P*.

A polyhedron P is said to be *decomposable* if E^n can be represented in the form of a direct product $E_1 \times \ldots \times E_k$ of Euclidean spaces E_i so that $P = P_1 \times \ldots \times P_k$, where P_i is a convex polyhedron in E_i. The Gram matrix A of P is in this case a direct sum of the Gram matrices A_i of P_i.

A polyhedron P is called a *C-polyhedron* if all the angles formed by the bounding hyperplanes of P do not exceed $\pi/2$ and P does not contain any straight line.

The Gram matrix A of a C-polyhedron in E^n is a non-negative definite symmetric matrix of rank n with $a_{ii} = 1$ and $a_{ij} \leq 0$ $(i \neq j)$. The matrices of such a kind will be called *C-matrices*. It is well known (see [Cox]) that C-matrices of rank n which are not decomposable into a direct sum can be of two types:

(a) *Indecomposable C^+-matrices*, which are positive definite matrices of order n. Each of these matrices is the Gram matrix of a unique (up to isometry) convex polyhedron in E^n which is a simplicial angle.
(b) *Indecomposable C^0-matrices*, which are non-negative definite matrices of order $n+1$. Each of these matrices is the Gram matrix of a unique (up to similarity) convex polyhedron in E^n which is a simplex.

By *C^+-matrices* (resp. *C^0-matrices*) we mean those C-matrices decomposable into a direct sum of indecomposable C^+- (resp. C^0-)matrices, i.e. C-matrices being the Gram matrices of simplicial angles (resp. of direct products of simplices).

(1.3) C^--polyhedra in Lobachevsky Spaces. Let $E^{n,1}$ be $(n+1)$-dimensional vector space over $\mathbb{R}$ equipped with a non-degenerate bilinear symmetric form $\langle\ ,\ \rangle$ of non-negative inertia index 1. Let us consider the set

$$V = \{x \in E^{n,1} \mid \langle x, x \rangle < 0\}.$$

This set consists of two connected components that will be denoted by V_+ and V_-. We can identify the points of n-dimensional Lobachevsky space Λ^n with rays (from the origin) in $E^{n,1}$ lying in the cone V_+. The motions of Λ^n in this model are the transformations induced from the linear transformations of $E^{n,1}$ leaving the form $\langle , \rangle$ and the set V_+ invariant.

A linear subspace of $E^{n,1}$ is called *hyperbolic* if the bilinear form induced on it from $\langle , \rangle$ is non-degenerate and indefinite. To any s-dimensional plane $\Pi \subset \Lambda^n$ there corresponds an $(s+1)$-dimensional hyperbolic subspace in E^{n+1} that will be denoted by $\hat{\Pi}$. To any half-space in Λ^n there corresponds the half-space in $E^{n,1}$ bounded by a hyperbolic subspace of co-dimension 1. A convex polyhedron $P \subset \Lambda^n$ is, by definition, the intersection of several half-spaces in Λ^n. Denote by $\hat{P}$ the intersection of the corresponding half-spaces in $E^{n,1}$. This intersection is a polyhedral angle with its vertex at the origin. Let us agree to exclude the origin from $\hat{P}$. The polyhedron P is defined uniquely by $\hat{P}$ as the union of rays lying in $\hat{P} \cap V_+$. The boundedness of the polyhedron P in Λ^n is equivalent to the inclusion $\hat{P} \subset V_+$, and the finiteness of its volume is equivalent to the inclusion $\hat{P} \subset \bar{V}_+$, where $\bar{V}_+$ is the closure of V_+. We remark that in the latter case the boundary of V_+ contains only certain edges of the angle $\hat{P}$, i.e. those which correspond to so called "vertices at infinity" of the polyhedron P.

Pick in $E^{n,1}$ the vectors e_i perpendicular to the hyperplanes $\hat{H}_i$ of the n-dimensional faces of the angle $\hat{P}$ and directed outwards. We then normalize them by the condition $\langle e_i, e_i \rangle = 1$; this is possible, because $\hat{H}_i$ are hyperbolic subspaces, and hence $\langle e_i, e_i \rangle > 0$.

The Gram matrix $A = (a_{ij} = \langle e_i, e_j \rangle)$ of the vector system $\{e_i\}$ will be called the *Gram matrix of the polyhedron* P. It can easily be shown that if the hyperplanes H_i and H_j of two faces of P form an angle α, then $a_{ij} = -\cos\alpha$, and if they do not meet, then $a_{ij} \leq -1$. Therefore if all the angles between the bounding hyperplanes of P do not exceed $\pi/2$, then all non-diagonal entries of A are negative or equal to 0.

A convex polyhedron $P \subset \Lambda^n$ is said to be a *C^--polyhedron* if P is of finite volume and all the angles between the bounding hyperplanes of P do not exceed $\pi/2$. The Gram matrix of any C^--polyhedron $P \subset \Lambda^n$ is indecomposable of rank $n+1$ (see [Vi 1], Lemmas 1 and 2). It is easily seen that if the Gram matrix A of a vector system in $E^{n,1}$ is of rank $n+1$, then the negative inertia index of A equals 1. From what we have discussed in this and the former paragraphs we deduce that the Gram matrix $A = (a_{ij})$ of any C^--polyhedron $P \subset \Lambda^n$ satisfies the following conditions:

(L1) The rank of A equals $n+1$ and the negative inertia index of A equals 1.
(L2) $a_{ii} = 1$ and $a_{ij} \leq 0 (i \neq j)$.
(L3) A is indecomposable.

Suppose A is symmetric matrix satisfying conditions (L1)–(L3). Then (see [Vi 1], Lemma 4 or [Vi 12], Theorem 2.5) A is the Gram matrix of a convex polyhedron $P \subset \Lambda^n$. We now state a criterion for a matrix A to be the Gram matrix of a C^--polyhedron. Recall that a submatrix of a matrix A which is symmetric relative to the principal diagonal of A is called a *principal submatrix* of A.

(1) Theorem. (See [Vi 1], Theorem 1.) *For a symmetric matrix $A = (a_{ij})$ to be the Gram matrix of a C^--polyhedron P in Λ^n it is necessary and sufficient that A satisfies* (L1)–(L3) *and the following conditions:*

(L4) *A has at least one principal submatrix which is either a C^+-matrix of rank n or a C^0-matrix of rank $n-1$ (such submatrices will be called nodal).*
(L5) *For any nodal submatrix B_1 of A and any C^+-submatrix B of B_1 of rank $n-1$ there exists a nodal submatrix B_2 of A containing B and different from B_1.*

We clarify the meaning of conditions (L4) and (L5). Let P be a convex polyhedron in Λ^n and A its Gram matrix. Then there is the natural map $F \mapsto A(F)$ of the set of all faces of P (including vertices at infinity) to the set of principal submatrices of A (to a face F we assign the matrix $A(F)$ of the system of those vectors e_i orthogonal to F). Furthermore (see [Vi 1], Lemmas 3 and 5),

1) if A_1 is a principal C^+-submatrix of A of rank $n-s$, then $A_1 = A(F)$, where F is an ordinary s-dimensional face of P;
2) if A_1 is a principal C^0-submatrix of A of rank $n-1$, then $A_1 = A(F)$, where F is a vertex at infinity;

3) if P is a C^--polyhedron and a face F of P is different from P, then $A(F)$ is either a C^+-matrix or a C^0-matrix.

So property (L4) means that P has vertices (possibly at infinity) and property (L5) means that any edge of P going out one vertex goes to another vertex. We remark also that a C^--polyhedron P is bounded if and only if A does not contain principal C^0-submatrices.

Symmetric matrices A satisfying condition (L1)–(L5) will be called *C^--matrices.*

(1.4) Coxeter Diagrams. As was observed in 1.1, a convex polyhedron P is a Γ-cell for some group Γ generated by reflections if and only if P is a Coxeter polyhedron, i.e., if all dihedral angles of P are of form π/n with n an integer. Therefore to find all (up to conjugacy) discrete groups of motions of Euclidean space generated by reflections it suffices to consider all C-matrices (see 1.2) which have non-diagonal entries of the form

(1) $$a_{ij} = -\cos(\pi/m_{ij}),$$

where m_{ij} can take the values 2, 3, ..., ∞. This was done by Coxeter in [Cox]. For the description of matrices of this kind he proposed the diagrams which can be constructed as follows: we take as many vertices as the order of a matrix and join the vertices v_i and v_j by a $(m_{ij}-2)$-multiple line or by the ordinary one with index m_{ij}.

With this notation, indecomposable C^+-matrices satisfying condition (1) can be described by the following diagrams (the subscript equals the rank of a matrix).

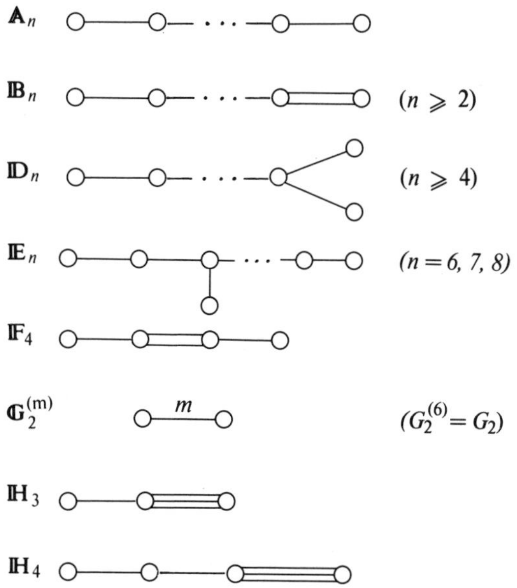

Indecomposable C^0-matrices satisfying condition (1) can be described by the following diagrams (the subscript equals the rank of a matrix).

$\tilde{\mathbb{A}}_n$ $(n \geqslant 2)$

$\tilde{\mathbb{B}}_n$ $(n \geqslant 3)$

$\tilde{\mathbb{C}}_n$ $(n \geqslant 2)$

$\tilde{\mathbb{D}}_n$ $(n \geqslant 4)$

$\tilde{\mathbb{E}}_6$

$\tilde{\mathbb{E}}_7$

$\tilde{\mathbb{E}}_8$

$\tilde{\mathbb{F}}_4$

$\tilde{\mathbb{G}}_2$

$\mathbb{I}$

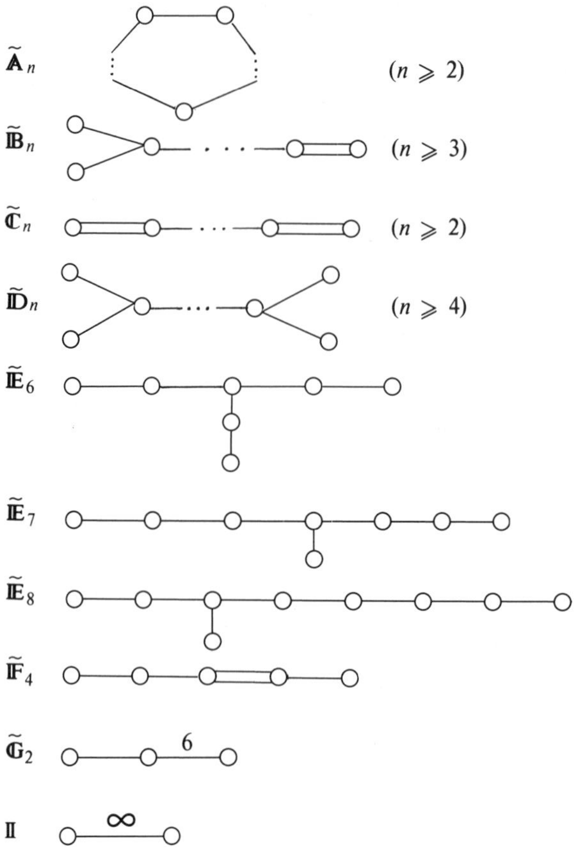

Similarly, to find all discrete groups of motions of Lobachevsky space generated by a finite number of reflections with finite volume of the fundamental domain it suffices to choose from among all C^--matrices (see 1.3) those having non-diagonal entries of the form

(2) $$a_{ij} = -\cos(\pi/m_{ij}) \text{ or } a_{ij} < -1.$$

For such C^--matrices, nodal submatrices are of the Coxeter type. Therefore to check (L4) and (L5) we need not any calculations. It suffices to make use of the tables above and observe that the Coxeter diagram of a direct sum of C-matrices is the disconnected union of the Coxeter diagrams of the summands.

For the description of C^--matrices satisfying condition (2) we shall use Coxeter diagrams with the additional stipulation that in case $a_{ij} < -1$ the vertices v_i and v_j are joined by a dotted line with index $-a_{ij}$. By the Coxeter diagram of a Coxeter polyhedron and a lattice generated by reflections we mean the Coxeter diagram of the corresponding C^--matrix.

(1.5) Arithmeticity Criterion for Lattices Generated by Reflections.

(2) Theorem. (See [Vi 1], Theorem 2.) *Let Γ be a lattice in the group of motions of n-dimensional Lobachevsky space Λ^n which is generated by a finite number of reflections. Choose a Γ-cell $P \subset \Lambda^n$. Denote by $A = (a_{ij})$ the Gram matrix of the polyhedron P, by $\tilde{K}$ the field generated by the entries a_{ij}, and by K the field generated by the numbers of the form*

$$(1) \qquad 2^m a_{i_1 i_2} a_{i_2 i_3} \cdot \ldots \cdot a_{i_{m-1} i_m} a_{i_m i_1}.$$

For Γ to be arithmetic it is necessary and sufficient that the following two conditions are satisfied.

(a) $\tilde{K}$ is a totally real algebraic number field and for any morphism $\sigma\colon \tilde{K} \to \mathbb{R}$ which is not the identity on K, the matrix ${}^\sigma A = (\sigma(a_{ij}))$ is non-negative definite.

(b) The algebraic numbers (1) *are algebraic integers.*

We remark (see [Vi 1], §6, Remark 1) that if Γ is non-cocompact, then the arithmeticity of Γ is equivalent to the property that the numbers (1) are rational integers.

(1.6) Examples. (See [Vi 1], III.) In this section we present some examples of non-arithmetic lattices in the groups of motions of Lobachevsky spaces Λ^n. Moreover, non-cocompact lattices are constructed in dimensions 3, 4, and 5 and cocompact ones in dimensions 3 and 4. We consider only the case in which a lattice Γ is generated by finitely many reflections relative to the faces of a fundamental polyhedron $P \subset \Lambda^n$. The proofs, in view of Theorems 1 and 2, can be reduced to elementary calculations.

a) The most simple is the construction of a group Γ whose fundamental polyhedron is a simplex. To this end it suffices to draw any connected diagram satisfying the following conditions.
   1) The number of vertices equals $n + 1$.
   2) The diagram is not contained in Coxeter's table given in 1.4.
   3) Any subdiagram of n vertices is either a union of several diagrams from the first Coxeter table or contained in the second one.

In this context to subdiagrams of the first type in condition 3) there correspond ordinary vertices of the simplex P and to subdiagrams of the second type there correspond vertices at infinity.

The groups Γ with bounded associated simplex P exist only for $n \leq 4$. In Λ^3 there are 9 such groups and 5 in Λ^4. Among them, for $n > 2$, there is only one non-arithmetic group. Its diagram is the following:

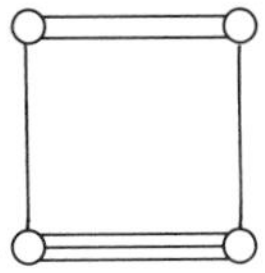

There exist only a few groups Γ whose fundamental domain is an unbounded simplex. Such groups exist only for $n \leq 9$. Among them there are exactly 7 non-arithmetic groups for $n > 2$, namely, 6 in Λ^3 and 1 in Λ^5. They have the following diagrams.

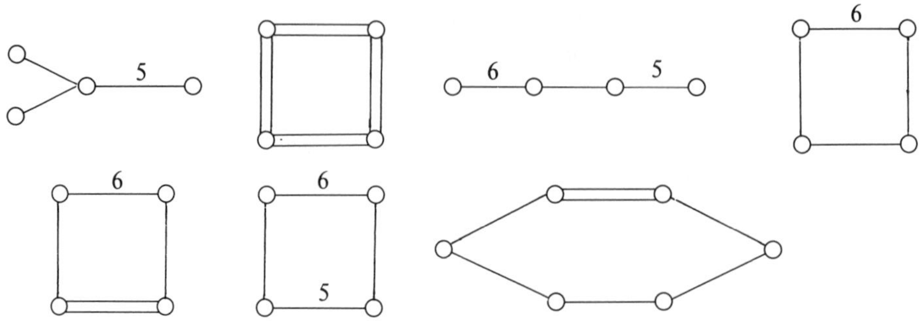

We remark that the simplex having the diagram

is the fundamental domain of the (finite) group of motions of the regular dodecahedron with vertices at infinity in the 3-dimensional Lobachevsky space.

b) The first example of a non-arithmetic lattice in a simple non-compact Lie group which is not locally isomorphic to $\mathbf{SL}_2(\mathbb{R})$ is due to Makarov (see [Mak 2]). He considered the case in which $n = 3$ and the fundamental polyhedron $P \subset \Lambda^3$ has the combinatorial type of the triangular prism defined by the diagram

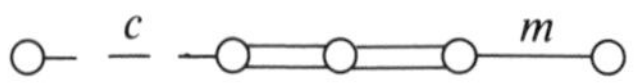

$$\left(m \geq 5,\ c = \frac{\cos(\pi/m)}{\sqrt{\cos(2\pi/m)^{1/2}}}\right).$$

It was established in [Mak 2] that for m sufficiently large the corresponding group Γ_m is not arithmetic. The proof given in [Mak 2] is based on the fact that for any simple non-compact Lie group G the orders of the elements of finite order belonging to non-cocompact arithmetic lattices $\Gamma \subset G$ are bounded by a constant depending only upon G. It follows from Theorem 2 that the group Γ'_m is not arithmetic for all $m \neq 6$.

Similar but bounded polyhedra in Λ^3 have been considered in [Mak 1] (the arithmeticity problem was not considered in that paper). The following are the diagrams of one series of such polyhedra:

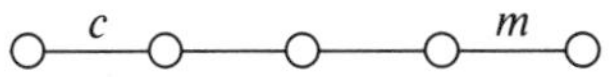

$$\left(m \geq 7,\ c = \sqrt{\frac{3\cos(2\pi/m)-1}{4\cos(2\pi/m)-2}}\right).$$

The corresponding groups are arithmetic for $m = 7, 8, 9, 10, 14$ and non-arithmetic for the other values of m.

c) We present two examples of non-arithmetic lattices in the motion of Λ^4. These are the lattices corresponding to the polyhedra with diagrams

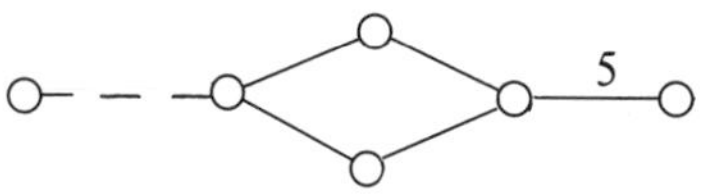

and

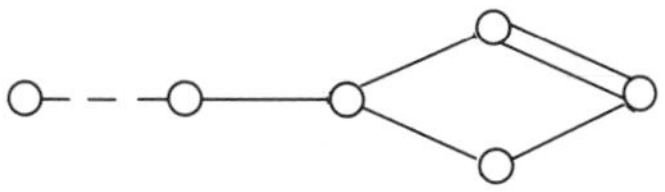

These polyhedra have the combinatorial type of a 4-dimensional prism having a tetrahedron as its base. The first of these polyhedra is unbounded and the second one is bounded.

(1.7) *Remarks.*

(i) Examples of cocompact lattices generated by reflections in the groups of motions of Λ^6 and Λ^7 have been constructed in [Bu]. It is an open question if there exist bounded Coxeter polyhedra in Lobachevsky spaces of dimension ≥ 8. But in Λ^n, for $n \geq 30$, there are no bounded Coxeter polyhedra (see [Vi 10]).

(ii) There are examples of unbounded Coxeter polyhedra which are the Γ-cells of arithmetic lattices Γ in Λ^n for $n \leq 19$ (see [Vi 6], [Vi 9], and [V-K]). The most simple example of such a polyhedron in a Lobachevsky space of higher dimension is the polyhedron in Λ^{17} with diagram

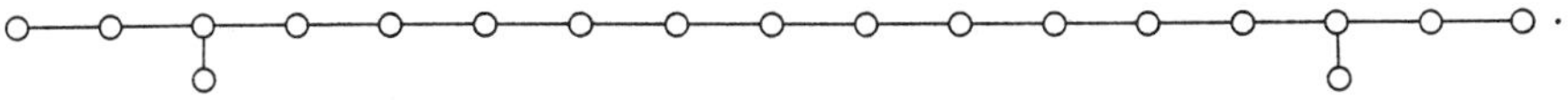

Its completion is combinatorically a pyramid over the direct product of two 8-dimensional simplices. We note also that

a) there are no Coxeter polyhedra of finite volume in Lobachevsky spaces of dimension greater than 995 (see [Hov] and [Pro]) and

b) there are no arithmetic lattices generated by reflections in the group of motions of Λ^n for $n \geq 30$ (see [Vi 10]).

(iii) For applications of the theory of discrete groups generated by reflections in algebraic geometry we refer to [N 4] and [Vi 9].

(iv) In [An 1] and [An 2] an almost exhaustive classification of lattices generated by reflections was obtained in the group of motions of Λ^3. Namely, in these papers some simple necessary and sufficient conditions have been given for the existence in Λ^3 of a convex polyhedron of finite volume and of a given combinatorial type with two given dihedral angles not exceeding $\pi/2$ (for instance, being of form π/n with n an integer).

(v) The existence of non-arithmetic lattices in the group of motions of Λ^3 is also a consequence of assertions (1) and (2) of IX.6.13 (iii) (on the volumes of quotient spaces).

2. "Hybrids" of Hyperbolic Manifolds and Examples of Non-Arithmetic Lattices in Lobachevsky Spaces

In this section $\mathrm{PO}(n,1)$ denotes the projective orthogonal group $O(n,1)/\{+1,-1\}$. Recall that $\mathrm{PO}(n,1)$ is the isometry group of the Lobachevsky space Λ^n.

(2.1) "Hybrids" of Hyperbolic Manifolds. Let Γ_1 and Γ_2 be two lattices in $\mathrm{PO}(n,1)$. Assume that the subgroups Γ_1 and Γ_2 have no torsion. Then the quotient spaces $V_i = \Gamma_i \setminus \Lambda^n$ are hyperbolic manifolds (i.e. complete Riemannian manifolds of constant negative curvature) and Γ_i is the fundamental group of V_i for $i = 1,2$. We assume there exist connected submanifolds $V_1^+ \subset V_1$ and $V_2^+ \subset V_2$ of dimension n with boundaries $\partial V_1^+ \subset V_1$ and $\partial V_2^+ \subset V_2$ such that

(a) Every connected component M of the hypersurface $\partial V_i^+ \subset V_i$ is totally geodesic in V_i, $i = 1,2$. That is, the universal covering of M is a hyperplane in the universal covering Λ^n of V_i. In particular, M is an $(n-1)$-dimensional hyperbolic manifold.

(b) The manifolds ∂V_1^+ and ∂V_2^+ are isometric.

Now we produce the *hybrid manifold* V by gluing together V_1^+ and V_2^+ via an isometry between ∂V_1^+ and ∂V_2^+. The manifold V carries a natural metric of constant negative curvature coming from those on V_1^+ and V_2^+. If V_1 and V_2 are compact then V is complete (because in this case there exists ε such that every $x \in V$ has a neighbourhood isometric to an ε-ball in L^n). If $n = 2$ and the manifolds V_1 and V_2 are not compact, then V could be non-complete. If $n \geq 3$ and $\partial V_1^+ = \partial V_2^+$ has finite volume, then V is always complete (see 2.10 in [Gro-P]; it is possible to prove that for $n \geq 3$ the condition "$\partial V_1^+ = \partial V_2^+$ has finite volume" is always satisfied).

From now on we assume that V is complete. Then the universal covering of V is isomorphic to Λ^n and the fundamental group Γ of V is a lattice in $\mathrm{PO}(n,1)$. Note that if the subgroups Γ_1 and Γ_2 are cocompact (i.e. if V_1 and V_2 are compact) then also Γ is cocompact.

There are natural homomorphisms of fundamental groups

$$\alpha_i\colon \pi_1(V_i^+) \to \pi_1(V_i) = \Gamma_i \text{ and}$$
$$\alpha_i'\colon \pi_1(V_i^+) \to \pi_1(V) = \Gamma \text{ for } 1,2.$$

The submanifold V_i^+ has convex (in fact, totally geodesic) boundary and so every class in $\pi_1(V_i^+)$ is represented by a *geodesic* loop in V_i^+. Such a loop is contractible neither in V nor in V_i, because these manifolds are complete and of negative curvature. This implies that α_i and α'_i are monomorphisms.

Let us fix a hyperbolic hyperplane $\Lambda^{n-1} \subset \Lambda^n$ and a connected component M of $\partial V_1^+ = \partial V_2^+$. Then there exist a discrete subgroup F of the isometry group $\mathrm{PO}(n-1, 1)$ of Λ^{n-1} and an isomorphism $\beta\colon \pi_1(M) \to F$ such that β induces an isometry of M onto $F \backslash \Lambda^{n-1}$. One can assume that α_i and α'_i agree on $\pi_1(M)$ with β. Let $\tilde{V}_i^+$ and $\tilde{M} \subset \tilde{V}_i^+$ denote the universal coverings of V_i^+ and M respectively. If f_1 and f_2 are two isometric immersions of $\tilde{V}_i^+$ into Λ^n which agree on $\tilde{M}$ and map $\tilde{M}$ onto Λ^{n-1} then f_1 is equal to f_2 (up to the orthogonal reflection of Λ^n in Λ^{n-1}). This and the equality $\alpha_{i|\pi_1(M)} = \alpha'_{i|\pi_1(M)}$ allows us to assume that $\alpha' = \alpha'_i$.

(2.2) Density Criterion for Hyperbolic Manifolds with Boundary. Let W^+ be a connected n-dimensional manifold of constant negative curvature with non-empty totally geodesic boundary ∂W^+ having finitely many connected components. Assume W^+ is complete as a metric space and $\mathrm{Vol}\, W^+ < \infty$.

(1) Lemma. *Suppose that the (image of the) fundamental group of every component of ∂W^+ has finite index in the fundamental group of W^+. Then $n = 2$ and W^+ is simply connected. It follows that W^+ is isometric to a k-gon in Λ^2 with vertices at infinity.*

Proof. The finite index condition shows that the universal covering $\tilde{W}^+$ of W^+ also has finitely many boundary components. Then one may assume without loss of generality that the deck transformation group maps every component into itself. Let ∂_0 be one of the components of $\partial \tilde{W}^+$ and let $\bar{\partial}_i \subset \partial_0$ be the normal projections of the remaining components ∂_i, $i = 1, \ldots, k$, to ∂_0. The condition $\mathrm{Vol}\, W^+ < \infty$ implies that $\bigcup_{i=1}^k \bar{\partial}_i \subset \partial_0$ is a subset of full measure. Hence, $n = 2$, and the action of deck transformations is trivial. □

(2) Lemma. *If* $\mathrm{Vol}\, \partial W^+ < \infty$, *then the fundamental group* $\pi_1(W^+)$ *of* W^+ *is Zariski dense in* $\mathrm{PO}(n, 1)^0$ *(here and below* F^0 *denotes the connected component of* e *in a group* F*).*

Proof. Since $\mathrm{Vol}\, \partial W^+ < \infty$, the Zariski closure $\bar{\Gamma}^+ \subset \mathrm{PO}(n, 1)$ of Γ^+ contains $\mathrm{PO}(n-1, 1)$ (Borel density theorem), where $\mathrm{PO}(n-1, 1) \subset \mathrm{PO}(n, 1)$ is identified with the isometry group of the space Λ^{n-1} serving as the universal covering of each component of ∂W^+. By Lemma 1, $\dim \bar{\Gamma}^+ > \dim \mathrm{PO}(n-1, 1)$ because the algebraic group $\bar{\Gamma}^+$ has at most finitely many connected components. It follows that $\bar{\Gamma}^+ = \mathrm{PO}(n, 1)^0$ since $\mathrm{PO}(n-1, 1)^0$ is a *maximal* connected subgroup in $\mathrm{PO}(n, 1)$. □

(2.3) Arithmetic Subgroups in $O(n, 1)$. Let $K \subset \mathbb{R}$ be a number field and F be a non-singular quadratic form in $n + 1$ variables with coefficients in K. Denote by $\Gamma(F)$ the group of F-orthogonal matrices with entries from the ring of integers in K. Suppose that K is totally real and that the form ${}^\sigma F$ is positive definite for any nontrivial embedding $\sigma\colon K \to \mathbb{R}$. Then $\Gamma(F)$ is an arithmetic subgroup in (some conjugate of) the orthogonal group $O(n,1)$.

Take a prime ideal $\mathfrak{p}$ in the ring of integers of K and define the *congruence subgroup* $\Gamma_{\mathfrak{p}}(F) \subset \Gamma(F)$ by

$$\Gamma_{\mathfrak{p}}(F) = \{\gamma \in \Gamma(F) \mid \gamma \equiv \mathrm{Id} \pmod{\mathfrak{p}}\}.$$

If $|K/\mathfrak{p}|$ is sufficiently large, then $\Gamma_{\mathfrak{p}}(F)$ has no torsion and the action of $\Gamma_{\mathfrak{p}}(F)$ on Λ^n is free.

(2.4) Examples of Non-Arithmetic Lattices Corresponding to Hybrids. As in 2.1, let us fix a hyperbolic hyperplane $\Lambda^{n-1} \subset \Lambda^n$. Let F_0 be either the quadratic form

$$x_1^2 + \ldots + x_{n-1}^2 - x_n^2 \text{ or } x_1^2 + \ldots + x_{n-1}^2 - \sqrt{2}x_n^2.$$

Consider the forms $F_1 = x_0^2 + F_0$ and $F_2 = 3x_0^2 + F_0$. Let us consider congruence subgroups $\Gamma_{\mathfrak{p}}(F_0)$, $\Gamma_{\mathfrak{p}}(F_1)$ and $\Gamma_{\mathfrak{p}}(F_2)$. For every k, let $\pi\colon O(k,1) \to \mathrm{PO}(k,1)$ be the natural epimorphism and let

$$\Gamma_0' = \pi(\Gamma_{\mathfrak{p}}(F_0)) \subset \mathrm{PO}(n-1,1) = \mathrm{Is}\Lambda^{n-1} \text{ and}$$
$$\Gamma_i' = \pi(\Gamma_{\mathfrak{p}}(F_i)) \subset \mathrm{PO}(n,1) = \mathrm{Is}\Lambda^n$$

It is easy to show that

(a) the subgroup of Γ_i', $i = 1,2$ which stabilizes Λ^{n-1} induces the group Γ_0' on Λ^{n-1},

(b) Γ_1' is not commensurable to Γ_2'.

It can be shown (see 2.8 in [Gro-P]) that for sufficiently large $\mathfrak{p}$ the following holds:

(c) the canonical map $\varphi_i\colon V_0' \overset{\text{def}}{=} \Gamma_0' \backslash \Lambda^{n-1} \to V_i' \overset{\text{def}}{=} \Gamma_i' \backslash \Lambda^n$ is a proper embedding and $\varphi_i(V_0')$ does not bound in V_i' for $i = 1,2$.

Now there exists an obvious double covering V_i of V_i' such that the lift of $\varphi_i(V_0')$ to V_i consists of two disjoint copies of $\varphi_i(V_0)$ which do bound some connected submanifold $V_i^+ \subset V_i$. In other words, the boundary ∂V_i^+ is the union of two copies of V_0'. Now we can consider the "hybrid" V of V_1^+ and V_2^+. The manifold V is the quotient space $\Gamma \backslash \Lambda^n$, where Γ is a lattice in $\mathrm{PO}(n,1)$. Using the equality fact at the end of 2.1, one can show that (after the conjugation of Γ by an element of $\mathrm{PO}(n,1)$) the intersection $\Gamma \cap \Gamma_i$ contains the subgroup $\alpha_i(\pi_1(\partial V_i^+))$. But according to Lemma 2, the subgroup $\alpha_i(\pi_1(\partial V_i^+))$ is Zariski dense in $\mathrm{PO}(n,1)$. Thus, $\Gamma \cap \Gamma_i$ is Zariski dense in $\mathrm{PO}(n,1)^0$ for $i = 1,2$. On the other hand, it is not difficult to prove that if the intersection of two arithmetic subgroups of $\mathrm{PO}(n,1)^0$ is Zariski dense in $\mathrm{PO}(n,1)^0$, then these subgroups are commensurable. This and property (b) imply that the lattice Γ is not arithmetic.

Remark. If $F_0 = x_1^2 + \ldots + x_{n-1}^2 - x_n^2$, then Γ is not cocompact and if $F_0 = x_1^2 + \ldots + x_{n-1}^2 - \sqrt{2}x_n^2$, then Γ is cocompact.

3. Examples of Non-Arithmetic Lattices in $SU_{n,1}(\mathbb{R})$

(3.1) Complex Reflections. Let V be n-dimensional vector space over $\mathbb{C}$. By a *complex reflection* (or a *$\mathbb{C}$-reflection*) in V we mean a linear map $R\colon V \to V$ with $n-1$ eigenvalues equal 1. A complex reflection can be written in the form $x \mapsto x + \beta(x)e$, where $e \in V$, β is a linear function, and $1+\beta(e)$ is a root of unity different from 1.

Let H be a non-degenerate Hermitian form on V which is not necessarily positive definite. For $v, w \in V$, we denote by $\langle v, w\rangle$ the value $H(v, w)$. Given $p \in \mathbb{N}^+$ and $e \in V$ with $\langle e, e\rangle = 1$, we denote by $R_{e,p}$ the complex reflection

$$x \mapsto x + (\xi - 1)\langle x, e\rangle e, \;\; \xi = \exp(2\pi i/p).$$

Clearly, $R_{e,p}$ leaves H invariant, is of order p, and leaves the points of the set $e^\perp \stackrel{\text{def}}{=} \{x \in V \mid \langle x, e\rangle = 0\}$ fixed.

(3.2) Groups Generated by Complex Reflections. (see [Most 9] and [Most 10], 2.2 and 2.3)

Finite groups generated by complex reflections were classified by Shepherd and Todd and the diagrams for their description were introduced by Coxeter. These diagrams serve equally well for the description of infinite groups generated by complex reflections. The diagram consists of points (vertices) and lines (edges) to which we assign positive integers. A point is represented, for convenience, by encircling the integer assigned to it. To each diagram D

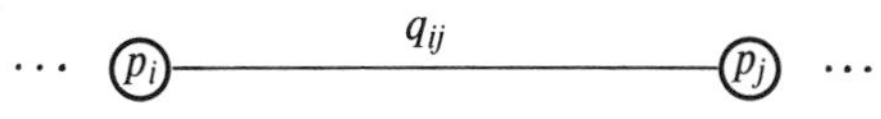

we assign the family of groups generated by complex reflections in $\mathbb{C}^n$ as follows.

Let $e_1, \ldots, e_n$ be a basis in $\mathbb{C}^n$. We define an Hermitian form H by setting

$$\langle e_i, e_j\rangle = -\alpha_{ij}\varphi_{ij}, \;\; i \neq j, \;\; \langle e_i, e_i\rangle = 1, \tag{1}$$

where

$$\alpha_{ij} = \left(\frac{\cos(\pi/p_1 - \pi/p_2) + \cos 2\pi/q_{ij}}{2\sin\pi/p_1 \cdot \sin\pi/p_2}\right)^{1/2}$$

and $|\varphi_{ij}| = 1$. Now for each i, $1 \leq i \leq n$, one can define the complex reflection $R_i = R_{e_i,p_i}$ (see 2.1). The group Γ_{ij} generated by R_i and R_j is finite if and only if the following condition is satisfied

$$(1/p_i) + (1/p_j) + (2/q_{ij}) \geq 1 \text{ and } q_{ij} \text{ is even if } p_i \neq p_j \tag{2}$$

We assume below that (2) is satisfied. In the group Γ_{ij} the following relation holds

$$(R_iR_j)^{q_{ij}/2} = (R_jR_i)^{q_{ij}/2}.$$

Furthermore, if q_{ij} is odd, this equality means that

$$R_iR_j \ldots R_i = R_jR_i \ldots R_j$$

(in both sides of this equality there are q_{ij} factors). Notice that in view of our comments in 2.1, R_i is of order p_i.

We denote by Γ the group generated by the reflections R_i, $1 \leq i \leq n$. This group leaves the form H invariant. We remark that the group depends not only on the diagram D but also on the numbers φ_{ij}. On the other hand, the group Γ_{ij} does not depend (up to isomorphism) on the choice of the numbers φ_{ij}. More generally, if the graph of the diagram D is a tree, then the group Γ does not depend (up to isomorphism) on the choice of φ_{ij}. To verify this it suffices to replace e_i by $\varphi_i e_i$, $1 \leq i \leq n$, $|\varphi_i| = 1$. If the graph of D has a cycle $i_1 i_2 \ldots i_k$, then the product $\varphi_{i_1 i_2} \cdot \varphi_{i_2 i_3} \cdot \ldots \cdot \varphi_{i_k i_1}$ is invariant under the above-mentioned replacements. So it turns out that two groups having common diagram need not be isomorphic. The set $\{\varphi_{ij} \mid i \neq j,\ 1 \leq i, j \leq n\}$ is called a *phase shift* of the form H. Two phase shifts define isomorphic groups Γ if (but not only if) their products through all closed cycles coincide.

(3.3) Groups $\Gamma(p,t)$. (see [Most 9] and [Most 10]) Let $p = 3, 4, 5$ and $t \in \mathbb{Q}$, $|t| < 3((1/2) - (1/p))$. Let us consider the diagram

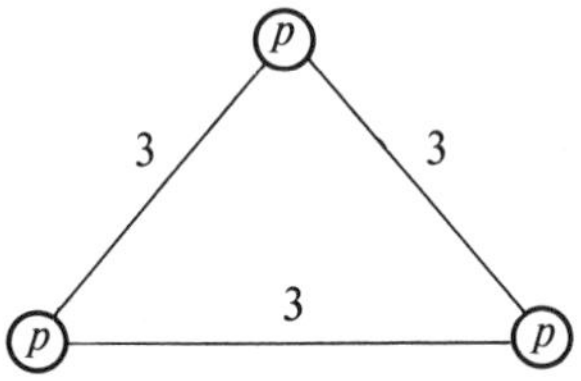

and define a phase shift by setting $\varphi_{ij} = \varphi$, $1 \leq i, j \leq 3$, where $\varphi^3 = \exp \pi i t$. We denote the corresponding group by $\Gamma(p,t)$. It can be verified by direct computations that under the restrictions indicated above the Hermitian form defined in 2.2 has signature (2,1). So we may assume that $\Gamma(p,t)$ is a subgroup in $\mathbf{SU}_{2,1}(\mathbb{R})$. The subgroup is discrete only for a finite number of values of the parameters. If Γ is discrete it is a lattice in $\mathbf{SU}_{2,1}(\mathbb{R})$.

(1) Theorem. (See [Most 10], Theorem 17.3.) *In* $\mathbf{SU}_{2,1}(\mathbb{R})$ *there are non-arithmetic lattices of the form* $\Gamma(p,t)$. *There exist only seven such lattices. They correspond to the following values of the parameters*

$$(p,t) = (3, 5/42),\ (3, 1/12),\ (3, 1/30),\ (4, 3/20),$$
$$(4, 1/12),\ (5, 1/5),\ (5, 11/30),$$

Non-cocompact lattices of the form $\Gamma(p,t)$ *are arithmetic.*

Since $\Gamma(p,t) \subset \mathbf{SU}_{2,1}(\mathbb{R})$, it follows that $\Gamma(p,t)$ operates naturally on 2-dimensional complex ball $\mathbb{C}h^2$.

(2) Theorem. (See [Most 10], §19.) *Assume* $|t| < ((1/2) - (1/p))$. *Then there exist a discrete holomorphic action of the group* $\Gamma(p,t)$ *on a 2-dimensional complex manifold* $Y(p,t)$ *and a* $\Gamma(p,t)$*-invariant holomorphic map* $\pi\colon\ Y(p,t) \to \mathbb{C}h^2$.

In [M-S] with the help of Theorem 2 the first known examples of compact Riemannian manifolds M of negative curvature non-diffeomorphic to a locally symmetric space were constructed. Furthermore, M is an algebraic surface and has negative curvature relative to a Kähler-metric.

(3.4) *Remarks.*

(i) The main difficulty in proving Theorem 1 is the check of discreteness of the groups $\Gamma(p,t)$ for indicated values of the parameters.

(ii) In the papers [Schw 2], [Schw 3], and [Hö] ramified coverings of 2-dimensional complex projective space are studied and applied to the construction of cocompact lattices in $\mathbf{SU}_{2,1}(\mathbb{R})$ generated by complex reflections. In this the Yau criterion for the universal covering of 2-dimensional Kähler manifold to be a complex ball is used.

(iii) In [Most 11] and [D-M] another approach to the construction of discrete subgroups has been proposed. This is based on the study of the monodromy groups of hypergeometric equations and has been applied to the construction of non-arithmetic lattices, both cocompact and non-cocompact, in $\mathbf{SU}_{2,1}(\mathbb{R})$ and $\mathbf{SU}_{3,1}(\mathbb{R})$.

(3.5) Non-Standard Homomorphisms. Let $\mathbf{G}$ be a connected absolutely simple $\mathbb{R}$-isotropic $\mathbb{R}$-group, Γ a lattice in $\mathbf{G}(\mathbb{R})$, k a local field, $\mathbf{H}$ an absolutely almost simple k-group. A homomorphism $\tau\colon\ \Gamma \to \mathbf{H}(k)$ will be called *standard* if the subgroup $\tau(\Gamma)$ is Zariski dense in $\mathbf{H}$ and either $\tau(\Gamma)$ is relatively compact in $\mathbf{H}(k)$ or k is archimedean and τ extends to a $\mathbb{C}$-epimorphism $\tilde{\tau}\colon\ \mathbf{G} \to \mathbf{H}$. We denote by L the field generated by the entries of the matrices in Γ. It is not hard to show that Γ is arithmetic if and only if, for any embedding of L in a local field l, the homomorphism $\sigma^0\colon\ \Gamma \to {}^{\sigma}\mathbf{G}(l)$ is standard (this follows, for instance, from the analysis of the proof of the arithmeticity theorems in IX.2). So any non-arithmetic lattice in $\mathbf{G}(\mathbb{R})$ has non-standard homomorphisms. For the non-arithmetic lattices mentioned in 2.3 a bit stronger assertion holds (see [Most 10], Section 22). Let $\Gamma = \Gamma(p,t)$ be a non-arithmetic lattice in $\mathbf{SU}_{2,1}(\mathbb{R})$. Then there exists an automorphism σ of the field $\mathbb{C}$ such that the homomorphism $\sigma^0\colon\ \mathrm{Ad}\,\Gamma \to {}^{\sigma}\mathbf{G}(\mathbb{C})$ is not standard, where $\mathbf{G} = \mathrm{Ad}\,\mathbf{SU}_{2,1}$.

The existence of non-standard homomorphisms is in sharp contrast to the superrigidity theorem for $\mathbb{R}$-rank > 1 groups (see Theorem VII.5.9).

We indicate one more example of non-standard homomorphisms. Let $\Gamma_1 = \Gamma(5,-1/10)$ and $\Gamma_2 = \Gamma(5,7/10)$. Then (see [Most 10], §22) Γ_1 and Γ_2 are arithmetic lattices in $\mathbf{SU}_{2,1}(\mathbb{R})$ and there exists an epimorphism $\rho\colon\ \Gamma_1 \to \Gamma_2$ with infinite kernel. The construction of such an epimorphism ρ presented in [Most 10] is based on the study of generators and defining relations of the groups Γ_1 and Γ_2.

Historical and Bibliographical Notes

The Borel-Wang density theorem presented in Chapter II was established by Borel in [Bo 1] for all archimedean k and by Wang in [WS 8] in the general case. At present, there are various generalizations and proofs of this theorem (see [Da 3], [Fu 7], [Mosk], and [Rag 6]). The approach to the proof of the density theorems presented in Sections 2 and 4 of Chapter II is close to the one proposed in [Da 3] and [Fu 7].

Lemma 3.2 of Chapter II is actually a modification of Mautner's lemma proved by Prasad (see [Pr 2], Lemma 2). In the case that the fields k_α are archimedean, Proposition II.3.3 is a particular case of the Moore-Mautner theorem (see [Moo 1], Theorem 1) and in case $\operatorname{char} A = 1$ was in fact established by Prasad in the process of proving Proposition 2 in [Pr 2]. The proof given in this book is a modification of the proofs given in [Moo 1] and [Pr 2]. Lemma II.3.4 is due to Kirillov (see [D-K], the proof of Lemma 5, or [G-G-P], the proof of the lemma in §2 of the appendix of Chapter II). Proposition II.3.9 and Corollary II.3.10 were established in §2.3 of [Mar 8].

Theorem II.5.1 was proved in [Mar 8] and [Pr 5] (the proof given in this book is a version of the one given in [Mar 8]). For the results of Section 6 in Chapter II (see [Mar 8], [Pr 5], and Section 1 in [Pr 6] (for the strong approximation theorem given in 6.8, see also [Kn 3], [Pl 2], [Pl 3], and [Pl 4]). In the case where the fields k_α are archimedean, Theorem II.7.2 is actually a special case of the Mostow-Moore-Mautner theorem (see [Most 5], Lemma 4) and, in case $\operatorname{card} A = 1$ and $\operatorname{char} k_\alpha = 0$ for all $\alpha \in A$, was essentially proved by Dani (see [Da 1], Corollary 3.2). The proof of this theorem given in this book is a modification of the proofs in [Da 1] and [Most 5]. We refer the reader who is interested in the ergodicity problem of flows on the homogeneous spaces of arbitrary real Lie groups to [B-M]. Questions involving the interrelationships between the theory of discrete subgroups and ergodic theory have been studied in [Da 1,2,4-8], [Rat 1-3], [Sul 1], and many other papers.

As was remarked at the beginning of Chapter III, property (T) was introduced by Kazhdan in [Kaz 1]. The main results concerning this property were given in that paper. However the exposition there is very concise. More detailed expositions have been given in [D-K] and [WS 4]. The case of semisimple algebraic groups over fields of characteristic 2 has been considered in [WS 18], where several other results on property (T) of nonsemisimple groups have also been obtained. The implication (ii) $\Rightarrow$ (i) in Theorem III.2.12 was established in [WS 11]. The proof of Watatani's theorem given in Section 3 of Chapter III, asserting that the groups with property (T) are not amalgams, is actually the same as in [Wat]. In III.5.12 we present the results of the papers [Mar 14] and [Sul 2]. These two papers differ one from another by the choice of dense subgroups with property (T).

It should be noted that the forthcoming book [H-V] is specially devoted to property (T).

In Chapter IV we have presented the substance of the papers [Mar 11] and [Mar 12]. The proof of Theorem IV.1.3 is a modification of that of Theorem III.1.2.3 in [D-S]. Theorem IV.1.5 is deduced from Theorem IV.1.3 in the same way as in Section 7 of [Br], the usual density point theorem is deduced from the Vitali covering theorem. Proposition IV.2.4 was in fact established in [Li].

The approach to the proof of the existence of equivariant measurable maps based on an application of the multiplicative ergodic theorem presented in Chapter V was proposed for the first time in [Mar 7]. Another approach (based on boundary theory) presented in Chapter VI was offered by Furstenberg (see [Fu 6] and [Fu 8]). For other approaches, see VII.5.26.

An indication of the proof of Theorem VII.5.9 in case $\Lambda' = \Gamma'$ was given in [Mar 7] and complete details offered in [Mar 9]. The method developed in [Mar 7] and [Mar 9] can immediately be applied to the proof of Theorem VII.5.6 in case $\operatorname{char} k_\alpha = 0$ for all $\alpha \in A$. The approach to the proof of the superrigidity theorems presented in Chapter VII includes some ideas and arguments due to E. Vishik. Repeating the statement made in [Mar 7] the author would particularly like to mention that while proving the superrigidity theorems he was under the influence of Mostow's remarkable ideas on the extension of isomorphisms of lattices to compactifications of symmetric spaces.

Theorem VII.7.5, in the case where the lattice Γ' is cocompact, was first established by Mostow in [Most 1]. Developing Mostow's method, Prasad proved Theorem VII.7.5 for non-cocompact lattices Γ' of $\mathbb{Q}$-rank 1 (see [Pr 1]) and Theorem VII.7.1 for Γ' cocompact (see [Pr 6]). Theorem VII.7.5 in case Γ' is non-cocompact was established by the author in the paper entitled "Non-cocompact lattices in semisimple algebraic groups" and published in the same transactions as [Mar 6]. The proof given in this paper is based on the study of unipotent elements in lattices and therefore can not be carried over the case in which Γ' is cocompact.

Theorem VIII (A) was first established in 2.4 of [Mar 12]. The majority of the results of VIII.2, though in somewhat weaker form, were established in that paper. Theorem VIII (B) in a slightly different form was stated in [Mar 7]. In case $\operatorname{char} k = 0$, Propositions VIII.3.20 and VIII.3.22 were established in [Vi 4]; several other results on definition rings of linear groups were also obtained in this paper.

The main results of Chapter IX (i.e. Theorems A and B in 1.9) in case $\operatorname{char} k_\alpha = 0$ for all $\alpha \in A$ and $\mathbf{G} - \mathbf{G}^{\mathrm{is}}$ were stated (in a slightly different form) in [Mar 7].

Theorem IX.1.16 under the assumption "$\operatorname{rank}_{\mathbb{R}} \mathbf{G} \geq 2$ and $\mathbf{G} = \mathbf{G}^{\mathrm{is}}$" was first established in [Mar 7] and [Mar 9] and had been conjectured by Selberg in case Γ_0 is non-cocompact (see [Sel 2], [Sel 3]) and by Piatetski-Shapiro in the general case (see [Pi 2], [Pi 3]). Various results, essentially for the case of non-cocompact lattices, have been obtained by these two authors while aiming at the proof of this conjecture (see [Pi 1], [Pi 2], [Pi 3], [Sel 1], [Sel 3]). For Γ_0 non-cocompact, Theorem IX.1.16 under the assumption mentioned above was established in [Mar 3] and [Mar 5] by a method based on a study of unipotent elements of Γ_0 and which is therefore inapplicable to cocompact lattices.

A slightly weaker result on arithmeticity of non-cocompact lattices in real semisimple groups was announced by the author in the paper "On the arithmeticity of discrete subgroups" (Soviet Math. Dokl., 10 (1969), 900–902). Following this, a substantial progress towards a proof of Theorem IX.1.16, for non-cocompact Γ_0, was also made by Raghunathan (see [Rag 6]), whose method is based again on a study of unipotent elements of Γ_0.

For certain groups G, under the assumption $\operatorname{char} k_\alpha = 0$ for all $\alpha \in A$, Theorem A of IX.1.9 was conjectured (though in somewhat inaccurate form) by Piatetski-Shapiro (see [Pi 2]). For the case of $\mathbb{R}$-groups Theorem B of IX.1.9 was conjectured, at least essentially, in [P-S] and [Pi 2] and so was Proposition IX.7.7. Theorem IX.1.9 B was proved by Kazhdan in [Kaz 2] for non-cocompact lattices in $\mathbf{SL}_2(\mathbb{R})$ consisting of matrices with algebraic integral entries.

Corollary IX.4.17 for the case $\mathbf{G} = \mathbf{G}^{\mathrm{is}}$ was conjectured in various forms by Selberg (see [Sel 1] and [Sel 3]). Corollary IX.7.2 yields a positive answer to Selberg's question of whether there exist symmetric spaces without irreducible discrete groups of motions of finite covolume.

References

[A-M] Alon, N.; Milman, V.D.: λ_1, isoperimetric inequalities for graphs and superconcentrators. J. Combinatorial Theory. **38** (1985) 73–88

[An 1] Andreev, E.M.: Convex polyhedra in Lobachevsky space. Math. USSR Sb. **10** (1970) 413–440

[An 2] Andreev, E.M.: Convex polyhedra of finite volume in Lobachevsky space. Math. USSR Sb. **12** (1970) 255–259

[A-W] Akemann, C.A.; Walter, M.E.: Unbounded negative definite functions. Can. J. Math. **33** (1981) 862–871

[Be 1] Behr, H.: Zur starken Approximation in algebraischen Gruppen über globalen Körpern. J. Reine Angew. Math. **229** (1968) 107–116

[Be 2] Behr, H.: Endliche Erzeugbarkeit arithmetischer Gruppen über Funktionenkörpern. Invent. math. **7** (1969) 1–32

[Be 3] Behr, H.: Chevalley groups of rank 2 over $\mathbb{F}_q(t)$ are not finitely presentable. Proc. Symp. "Homological and combinatorial techniques in group theory" (Durham 1977). London Math. Soc., Lecture Notes Ser. 36, 213–224

[Be 4] Behr, H.: Finite presentability of arithmetic groups over global function fields. Proc. Edinburgh Math. Soc. **30** (1987) 23–39

[Bi] Billingsley, P.: Ergodic theory and information. Wiley, New York 1965

[B-K] Bernstein, I.N.; Kazhdan, D.A.: The one-dimensional cohomology of discrete subgroups. Funct. Anal. Appl. **4** (1970) 1–4

[B-M] Brezin, J.; Moore, C.C.: Flows on homogeneous spaces: a new look. Amer. J. Math. **103** (1981) 571–613

[B-M-S] Bass, H.; Milnor, J.; Serre, J.-P.: Solution of the congruence subgroup problem for $\mathbf{SL}_n$ and $\mathbf{Sp}_{2n}$. Publ. Math. IHES **33** (1967) 59–137

[Bo 1] Borel, A.: Density properties for certain subgroups of semisimple groups without compact components. Ann. Math. **72** (1960) 179–188

[Bo 2] Borel, A.: Compact Clifford-Klein forms of symmetric spaces. Topology **2** (1963) 111–122

[Bo 3] Borel, A.: Some finiteness properties of adèle groups over number fields. Publ. Math. IHES **16** (1963) 1–30

[Bo 4] Borel, A.: Density and maximality of arithmetic subgroups. J. Reine Angew. Math. **224** (1966) 78–89

[Bo 5] Borel, A.: Ensembles fondamentaux pour les groups arithmétiques et formes automorphes. Faculté des Sciences de Paris, cours mimeographié, 1967

[Bo 6] Borel, A.: Linear algebraic groups. Benjamin, New York 1969

[Bo 7] Borel, A.: Some metric properties of arithmetic quotients of symmetric spaces and an extension theorem. J. Differential Geometry, **6** (1972) 543–560

[Bo 8] Borel, A.: Cohomologie de certains groupes discrets et laplacien p-adique. Seminaire Bourbaki, Exp. 437, 1973/74, Springer Lecture Notes, No. 431

[Bo 9] Borel, A.: Cohomology of arithmetic groups. Proc. Int. Congress Math. (Vancouver 1974), vol. 1 (1975), pp. 435–442

[Bo 10] Borel, A.: Cohomologie de sous-groupes discrets et représentations de groupes semi-simples. Astérisque **32–33** (1976) 73–112

[Bo 11] Borel, A.: Stable and L^2-cohomology of arithmetic groups. Bull. Amer. Math. Soc., New Ser. **3** (1980) 1025–1027

[Bo 12] Borel, A.: Commensurability classes and volumes of hyperbolic 3-manifolds. Ann. Scuola Norm. Sup. Pisa Cl. Sci (4) **8** (1981) 1–23

[Bo-Hard] Borel, A.; Harder, G.: Existence of discrete cocompact subgroups of reductive groups over local fields. J. Reine Angew. Math. **298** (1978) 53–64

[Bo-Hari] Borel, A.; Harish-Chandra: Arithmetic subgroups of algebraic groups. Ann. Math. **75** (1962) 485–535

[Boro 1] Borovoj, M.V.: Abstract simplicity for some anisotropic algebraic groups over number fields. Dokl. Akad. Nauk SSSR **283** (1985) 794–797 (in Russian)

[Boro 2] Borovoj, M.V.: Abstract simplicity of groups of type $\mathbb{D}_n$ over number fields. Usp. Mat. Nauk, V. 43, **5** (263) 179–180 (in Russian)

[Bo-Se 1] Borel, A.; Serre, J.-P.: Corners and arithmetic groups. Comment. Math. Helv. **48** (1973) 436–491

[Bo-Se 2] Borel, A.; Serre, J.-P.: Cohomologie d'immeubles et de groupes S-arithmétiques. Topology **15** (1967) 211–232

[Bo-Sp] Borel, A.; Springer, T.A.: Rationality properties of linear algebraic groups II. Tohoku Math. J. **20** (1968) 443–497

[Bo-T 1] Borel, A.; Tits, J.: Groupes réductifs. Publ. Math. IHES **27** (1965) 55–150

[Bo-T 2] Borel, A.; Tits, J.: Eléments unipotents et sous-groupes paraboliques de groupes réductifs. I. Invent. math. **12** (1971) 95–104

[Bo-T 3] Borel, A.; Tits, J.: Compléments à l'article "Groups réductifs". Publ. Math. IHES **41** (1972) 253–276

[Bo-T 4] Borel, A.; Tits, J.: Homomorphismes "abstraits" de groupes algébriques simples. Ann. Math. **97** (1973) 499–571

[Bou 1] Bourbaki, N.: Topologie Générale. Hermann, Paris: Ch. I–II (1951), Ch. III–IV (1951), Ch. V–VIII (1955)

[Bou 2] Bourbaki, N.: Espaces Vectorielles Topologiques. Hermann, Paris: Ch. I–II (1953), Ch. III–V (1955), Fascicule de résultats 1955

[Bou 3] Bourbaki, N.: Intégration. Ch. VI–VIII, Hermann, Paris 1960

[Bou 4] Bourbaki, N.: Variétés Différentielles et Analytiques. Fascicule des résultats, Hermann, Paris: Paragr. 1–7 (1967), Paragr. 8–15 (1971)

[Bou 5] Bourbaki, N.: Groupes et Algèbres de Lie. Hermann, Paris: Ch. I (1971), Ch. II–III (1972), Ch. IV–VI (1968)

[Bo-W] Borel, A.; Wallach, N.: Continuous cohomology, discrete subgroups and representations of reductive groups. Ann. Math. Studies, No. 94, Princeton Univ. Press, Princeton 1980

[Br] Brudno, A.A.: Theory of functions of real variables. Nauka, Moscow 1971 (in Russian)

[B-T 1] Bruhat, F.; Tits, J.: Groupes réductifs sur un corps local, I. Publ. Math. IHES **41** (1972) 5–252

[B-T 2] Bruhat, F.; Tits, J.: Groupes réductifs sur un corps local, II. Publ. Math. IHES **60** (1984) 5–184

[B-T 3] Bruhat, F.; Tits, J.: Groupes algébriques simples sur un corps local: cohomologie galoisienne, décompositions d''Iwasawa et de Cartan. C.R. Acad. Sci. Paris. Série A, **263** (1966) 867–869

[Bu] Bugayenko, V.O.: Groups of automorphisms of unimodular hyperbolic quadratic forms over the ring $\mathbb{Z}[(\sqrt{5}+1)/2]$. Moscow Univ. Math. Bull. **39** (1984) 6–14

[C 1] Chevalley, C.: Theory of Lie groups, I. Princeton University Press, Princeton 1946

[C 2] Chevalley, C.: Théorie des groupes de Lie, II: groupes algébriques. Hermann, Paris 1951

[C 3] Chevalley, C.: Théorie des groupes de Lie, III: théorèmes généraux sur les algèbres de Lie. Hermann, Paris 1955

[Ch 1] Chernousov, V.I.: On the projective simplicity of algebraic groups split over a quadratic extension of a number field. Dokl. Akad. Nauk SSSR **296** (1987) 1301–1405 (in Russian)

[Ch 2] Chernousov, V.I.: On the structure of the groups of rational points of algebraic groups of type $\mathbb{D}_n$. Dokl. Akad. Nauk SSSR **31** (1987) 593–596 (in Russian)

[Ch 3] Chernousov, V.I.: On the projective simplicity for some groups of rational points over number fields. Izv. Akad. Nauk SSSR, Ser. math. **53** (1989) 398–410 (in Russian)

[Ch-W] Choda, M.; Watatani, Y.: Fixed point algebra and property (T). Math. Japonica **27** (1982) 263–266

[Cox] Coxeter, H.S.M.: Discrete groups generated by reflections. Ann. Math. **35** (1934) 588–621

[Co-W] Connes, A.; Weiss, B.: Property (T) and asymptotically invariant sequences. Israel. J. Math. **37** (1980) 209–210

[Da 1] Dani, S.G.: Kolmogorov automorphisms on homogeneous spaces. Amer. J. Math. **98** (1976) 119–163

[Da 2] Dani, S.G.: On invariant measures, minimal sets, and a lemma of Margulis. Invent. math. **51** (1979) 239–260

[Da 3] Dani, S.G.: A simple proof of Borel's density theorem. Math. Zeit. **174** (1980) 81–94

[Da 4] Dani, S.G.: Invariant measures and minimal sets of horospherical flows. Invent. math. **64** (1981) 357–385

[Da 5] Dani, S.G.: Continuous equivariant images of lattice actions on boundaries. Ann. Math. **119** (1984) 111–119

[Da 6] Dani, S.G.: On orbits of unipotent flows on homogeneous spaces. Ergod. Theor. and Dynam. Sys. **4** (1984) 25–34

[Da 7] Dani, S.G.: On orbits of unipotent flows on homogeneous spaces, II. Ergod. Theor. and Dynam. Sys. **6** (1986) 167–182

[Da 8] Dani, S.G.: Orbits of horospherical flows. Duke Math. Z. **53** (1986) 177–188

[De] Deligne, P.: Extensions centrales non-résiduellement finies de groupes arithmétiques. C.R. Acad. Sci. Paris. Série A, **287** (1978) 203–208

[Die] Dieudonné, J.: La géométrie des groupes classiques. Springer, New York 1971

[Dix] Dixmier, J.: Les C^*-algèbres et leurs représentations. Gauthier-Villars, Paris 1969

[D-K] Delaroche, C.; Kirillov, A.: Sur les relations entre l'espace dual d'un groupe et la structure de ses sous-groupes fermés. Séminaire Bourbaki, vol. 1967/68, exposé 343, Benjamin, New York 1969

[D-M] Deligne, P.; Mostow, G.D.: Monodromy of hypergeometric functions and non-lattice integral monodromy. Publ. Math. IHES **63** (1986) 5–89

[Dr] Drinfel'd, V.G.: Finitely additive measures on S^2 and S^3 invariant with respect to rotations. Funct. Anal. Appl. **18** (1984) 245–246

[D-R] Del Junco, A.; Rosenblatt, J.: Counterexamples in ergodic theory and number theory. Math. Ann. **245** (1979) 185–197

[D-S] Dunford, N.; Schwartz, T.: Linear operators, Part I: general theory. Interscience publishers, New York 1958

[Fa] Faddeev, L.D.: The eigen-function expansion of Laplace's operator on the fundamental domain of a discrete group on Lobachevsky plane. Trudy Moskov. Mat. Obšč. **17** (1967) 323–350 (in Russian)

[Fe] Fell, J.M.G.: Weak containment and induced representations of groups. Can. J. Math. **14** (1962) 237–268

[F-G] Feldman, J.; Greenleaf, F.P.: Existence of Borel transversals in groups. Pacific J. Math. **25** (1968) 455–461

[F-K] Furstenberg, H.; Kesten, H.: Products of random matrices. Ann. Math. Statistics **31** (1960) 457–469

[Fu 1] Furstenberg, H.: Non-commuting random products. Trans. Amer. Math. Soc. **108** (1963) 377–428

[Fu 2] Furstenberg, H.: Poisson boundaries and envelopes of discrete groups. Bull. AMS **73** (1967) 350–356

[Fu 3] Furstenberg, H.: The unique ergodicity of the horocycle flow, in: Recent advances in topological dynamics. Springer Lecture Notes No. 318

[Fu 4] Furstenberg, H.: Random walks and discrete subgroups of Lie groups, in: Advances in Probability and related topics, v. 1, Marcel Dekker, New York 1973

[Fu 5] Furstenberg, H.: Boundary theory and stochastic processes on homogeneous spaces, in: Harmonic analysis on homogeneous spaces. (Symposia on Pure and Applied Math.), Williamstown, Mass. 1972, Proceedings, vol. **26** (1973) 193–229

[Fu 6] Furstenberg, H.: Equivariant measurable maps and representations of lattice subgroups. Hebrew University, Jerusalem 1975, preprint

[Fu 7] Furstenberg, H.: A note on Borel's density theorem. Proc. AMS **55** (1976) 209–212

[Fu 8] Furstenberg, H.: Rigidity and cocycles for ergodic actions of semisimple groups (after G.A. Margulis and R. Zimmer). Seminaire Bourbaki, No. 559, 1979/80

[Ga 1] Garland, H.: p-adic curvature and the cohomology of discrete subgroups of p-adic groups. Ann. Math. (2) **97** (1973) 375–423

[Ga 2] Garland, H.: On the cohomology of discrete subgroups of p-adic groups. Proc. In. Congress Math. Vancouver 1974, v. **1** (1975) 449–453

[G-G-P] Gel'fand, I.M.; Graev, M.I.; Piatetski-Shapiro, I.I.: Representation theory and automorphic functions. Fiz. Mat. Giz., Moscow 1966 (in Russian)

[Go] Godement, R.: Les fonctions de type positif et la théorie des groupes. Trans. AMS **63** (1948) 1–84

[G-R] Garland, H.; Raghunathan, M.S.: Fundamental domains for lattices in R-rank 1 semisimple Lie groups. Ann. Math. **92** (1970) 279–326

[Gre] Greenleaf, F.: Invariant means on topological groups. van Nostrand, New York 1969

[Gro] Gromov, M.: Hyperbolic manifolds, groups, and actions, in: Riemann surfaces and related topics, ed. I. Kra and B. Maskit, Annals of Math. Studies, No. 97. Princeton University Press 1981

[Gro-P] Gromov, M.; Piatetski-Shapiro, I.: Non-arithmetic groups in Lobachevsky spaces. Publ. Math. IHES (1988)

[G-S 1] Grunewald, F.; Schwermer, J.: A non-vanishing theorem for the cuspidal cohomology of $\mathbf{SL}_2$ over imaginary quadratic integers. Math. Ann. **258** (1981) 183–200

[G-S 2] Grunewald, F.; Schwermer, J.: Free non-abelian quotients of $\mathbf{SL}_2$ over orders of imaginary quadratic number fields. J. Algebra **69** (1981) 298–304

[G-S 3] Grunewald, F.; Schwermer, J.: Arithmetic quotients of hyperbolic 3-space, cusp forms and link complements. Duke Math. J. **48**, No. 2 (1981) 351–358

[Gu] Guivarc'h, Y.: Quelques propriétés asymptotiques des produits de matrices aléatoires. Springer Lecture Notes, No. **774** (1980) 178–250

[G-V] Gel'fand, I.M.; Vilenkin, N.J.: Some applications of harmonic analysis. Equipped Hilbert spaces (Generalized functions No. 4), Gosud. Izdat. Fiz. Mat. Lit., Moscow 1961 (in Russian)

[Hal 1] Halmos, P.R.: Measure Theory. New York 1950

[Hal 2] Halmos, P.R.: Lectures on ergodic theory. Math. Soc. of Japan, Tokyo 1956

[Har 1] Harder, G.: Eine Bemerkung zum schwachen Approximationsatz. Arch. d. Math. **19** (1968) 465–471

[Har 2] Harder, G.: Minkowskische Reduktionstheorie über Funktionenkörpern. Invent. math. **7** (1969) 33–54

[Har 3] Harder, G.: A Gauss-Bonnet formula for discrete arithmetically defined groups. An. Sc. d. Ec. Norm. Sup. 4 Serie t. **4** (1971) 409–455

[Har 4] Harder, G.: On the cohomology of discrete arithmetically defined groups, in: Proc. Int. Colloq. on discrete subgroups of Lie groups and applications to moduli (Bombay 1973). Oxford University Press 1975, 129–160

[Har 5] Harder, G.: Über die Galoiskohomologie halbeinfacher algebraischer Gruppen, III. J. Reine Angew. Math. **274/275** (1975) 125–138

[Har 6] Harder, G.: Die Kohomologie S-arithmetischer Gruppen über Funktionenkörpern. Invent. math. **42** (1977) 135–175

[Har 7] Harder, G.: Eisenstein cohomology of arithmetic groups. The case $\mathbf{GL}_2$. Invent. math. **89** (1987) 37–118

[He] Helgason, S.: Differential Geometry and Symmetric Spaces. Academic Press, New York 1962

[Ho] Hopf, E.: Ergodentheorie. Ergebnisse der Mathematik, vol. 2. J. Springer, Berlin 1937

[Hov] Hovanskiĭ, A.G.: Hyperplane sections of polyhedra, toroidal manifolds, and discrete groups in Lobachevsky space. Funct. Anal. Appl. (1986) 50–61 (in Russian)

[Hö] Höfer, T.: Ballquotienten als verzweigte Überlegungen der projektiven Ebene. Diss. Universität Bonn, Bonn 1985

[Hul] Hulanicki, A.: Means and Følner conditions on locally compact groups. Studia Math. **27** (1966) 87–104

[Hum 1] Humphreys, J.E.: Linear Algebraic Groups. Springer, New York 1975

[Hum 2] Humphreys, J.E.: Arithmetic groups. Springer Lecture Notes, No. **789** (1980)

[H-V] Harpe, P.; Valette, A.: La propriété (T) de Kazhdan pour les groupes localement compacts. To appear

[J-M] Jonson, D.; Millson, J.: Deformation spaces associated to compact hyperbolic manifolds. Preprint

[J-W-W] James, D.; Waterhouse, W.; Weisfeiler, B.: Abstract homomorphisms of algebraic groups: problems and bibliography. Communications in Algebra **9** (1981) 95–114

[Kaz 1] Kazhdan, D.A.: On the connection between the dual space of a group and the structure of its closed subgroups. Funct. Anal. Appl. **1** (1967) 63–65

[Kaz 2] Kazhdan, D.A.: Construction of Γ-rational groups for certain discrete subgroups Γ of the group $\mathbf{SL}(2,\mathbb{R})$. Funct. Anal. Appl. **2** (1968) 34–37

[Kaz 3] Kazhdan, D.A.: On arithmetic varieties. Proc. of the Summer School of the Bolyai Janos Math. Soc. on Lie groups and their representations, Budapest 1971. Akademia Kiado, Budapest 1975, pp. 151–217

[Kaz 4] Kazhdan, D.A.: Some applications of the Weil representation, Mimeographed notes.

[Kar] Karpelevich, F.I.: Transitivity surfaces of a semisimple subgroup of the motion group of a symmetric space. Dokl. Akad. Nauk SSSR **93** (1953) 401–404 (in Russian)

[Ki] Kirillov, A.A.: Basic representation theory. Nauka, Moscow 1972 (in Russian)

[Kn 1] Kneser, M.: Orthogonale Gruppen über algebraischen Zahlkörpern. J. Reine Angew. Math. **196** (1956) 213–220

[Kn 2] Kneser, M.: Schwache Approximation in algebraischen Gruppen. Colloque sur la théorie des groupes algébriques. CBRM Brussels (1962) 41–52

[Kn 3] Kneser, M.: Strong approximation I, II. Proc. Symp. Pure Math. **9** (1966) 99–103, 187–196

[Kn 4] Kneser, M.: Normal subgroups of integral orthogonal groups. Algebraic K-theory and its geometric applications. Springer Lecture-Notes No. **108** (1969) 67–71

[Kn 5] Kneser, M.: Normalteiler ganzzahliger Spingruppen. J. Reine Angew. Math. 311/312 (1979) 191–214

[Ko] Kostant, B.: On the existence and irreducibility of certain series of representations. Bull. AMS **75** (1969) 627–642

[K-S-F] Kornfeld, I.P.; Sinai, J.G.; Fomin, S.V.: Ergodic Theory. Gosud. Izd. Fiz. Mat. Lit., Moscow 1980 (in Russian)

[Ku] Kurosh, A.G.: Group Theory. Nauka, Moscow 1967 (in Russian)

[La] Lachaud, G.: Spectral analysis of automorphic forms on rank one groups by perturbation methods, in: Harmonic Analysis of Homogeneous spaces. Proc. Symp. Pure Math. **26** (1973) 441–450

[Lan 1] Lang, S.: Algebra. Addison-Wesley, Reading, Mass. 1965

[Lan 2] Lang, S.: $\mathbf{SL}_2(\mathbb{R})$. Addison-Wesley, Reading, Mass. 1975

[Li] Lind, D.A.: Translation invariant sigma algebras of groups. Proc. AMS **42** (1974) 218–221

[Lo] Loève, M.: Probability Theory. D. Van Nostrand Comp. Inc., Princeton New Jersey 1960

[L-R] Losert, V.; Rindler, H.: Almost invariant sets. Bull. London Math. Soc. **13** (1981) 145–148

[L-S] Lee, R.; Schwermer, J.: Cohomology of arithmetic subgroups of $\mathbf{SL}_3$ at infinity. J. Reine Angew. Math. **330** (1982) 100–131

[Lu 1] Lubotzky, A.: Group presentation, p-adic analytic groups and lattices in $\mathbf{SL}_2(\mathbb{C})$. Ann. Math. (2) **118** (1983) 115–130

[Lu 2] Lubotzky, A.: Trees and discrete subgroups of Lie groups over local fields. Bull. AMS **20** (1989) 23–30

[Macd] Macdonald, I.G.: Spherical functions on a group of p-adic type. Univ. of Madras, India, 1971

[Mack] Mackey, G.W.: Induced representations of locally compact groups. Ann. Math. **55** (1952) 101–139

[Mak 1] Makarov, V.S.: A class of decompositions of the Lobachevsky space. Soviet Math. Dokl. **6** (1965) 400–401

[Mak 2] Makarov, V.S.: On a certain class of discrete groups of Lobachevsky space having an infinite fundamental region of finite measure. Soviet Math. Dokl. **7** (1966) 328–331

[Mak 3] Makarov, V.S.: On a certain class of 2-dimensional Fedorov groups. Math. USSR Izv. **1** (1968) 515–524

[Mal] Mal'cev, A.I.: On a class of homogeneous spaces, AMS Transl. **39** (1951)

[Mar 1] Margulis, G.A.: The isometry of closed manifolds of constant negative curvature with the same fundamental group. Soviet Math. Dokl. **11** (1970) 722–723

[Mar 2] Margulis, G.A.: Explicit construction of expanders. Probl. Inform. Transmission (1975) 325–332 (in Russian)

[Mar 3] Margulis, G.A.: Arithmeticity of discrete subgroups. Russ. Math. Surv. **29** (1974) 107–156

[Mar 4] Margulis, G.A.: Arithmeticity and finite-dimensional representations of uniform lattices. Funct. Anal. Appl. **8** (1974) 258–259

[Mar 5] Margulis, G.A.: Arithmeticity of non-uniform lattices in weakly noncompact groups. Funct. Anal. Appl. **9** (1975) 31–38

[Mar 6] Margulis, G.A.: On the action of unipotent groups in the space of lattices, in: Proc. of the Summer School of the Bolyai Janos Math. Soc. on Lie groups and their representations, Budapest 1971, Akademia Kiado, Budapest 1975, 365–370

[Mar 7] Margulis, G.A.: Discrete groups of motions of manifolds of non-positive curvature. Proc. Int. Congress Math. (Vancouver 1974), AMS Transl. **109** (1977) 33–45

[Mar 8] Margulis, G.A.: Cobounded subgroups of algebraic groups over local fields. Funct. Anal. Appl. **11** (1977) 119–128

[Mar 9] Margulis, G.A.: Arithemticity of irreducible lattices in semisimple groups of rank greater than 1, appendix to Russian translation of M. Raghunathan, Discrete subgroups of Lie groups. Mir, Moscow 1977 (in Russian). Invent. math. **76** (1984) 93–120

[Mar 10] Margulis, G.A.: Factor groups of discrete subgroups. Soviet Math. Dokl. **19** (1978) 1145–1149

[Mar 11] Margulis, G.A.: Quotient groups of discrete subgroups and measure theory. Funct. Anal. Appl. **12** (1978) 295–305

[Mar 12] Margulis, G.A.: Finiteness of quotient groups of discrete subgroups. Funct. Anal. Appl. **13** (1979) 178–187

[Mar 13] Margulis, G.A.: On the multiplicative group of a quaternion algebra over a global field. Soviet Math. Dokl. **21** (1980) 780–784

[Mar 14] Margulis, G.A.: Some remarks on invariant means. Monatshefte für Math. **90** (1980) 233–235

[Mar 15] Margulis, G.A.: On the decomposition of discrete subgroups into amalgams. Selecta Mathematica Sovetica **1** (1981) 197–213

[Mar 16] Margulis, G.A.: Finitely-additive invariant measures on Euclidean spaces. Ergodeor. Th. and Dynam. Sys. **2** (1982) 383–396

[Mar 17] Margulis, G.A.: Free totally discontinuous groups of affine transformations. Soviet Math. Dokl. **28** (1983) 435–439

[Mar 18] Margulis, G.A.: Complete affine locally flat manifolds with free fundamental group. Zapiski Naučn. Sem. Leningrad Otd. Mat. Inst. Steklov **134** (1984) 190–205 (in Russian)

[Ma-R] Margulis, G.A.; Rohlfs, J.: On the proportionality of covolumes of discrete subgroups. Math. Ann. **275** (1986) 197–205

[Mat] Matsumoto, H.: Sur les sous-groupes arithmétiques des groupes semi-simples deployés. Ann. Scient. E. Norm. Sup., 4-th Series, **2** (1969) 1–62

[Mel] Mezzljakov, Ju.I.: A survey of latest results on automorphisms of classical groups, in: Collection "Isomorphisms of classical groups", Mir, Moscow 1976, 250–259 (in Russian)

[Me2] Mezzljakov, Ju.I.: Theory of isomorphisms of classical groups in 1976–1980, in: Collection "Isomorphisms of classical groups over integral rings", Mir, Moscow 1980, 252–258 (in Russian)

[Milli] Millionshchikov, V.M.: A criterion for the stability of the probable spectrum of linear system of differential equations with recurrent coefficients, and a criterion for the almost reducibility of systems with almost periodic coefficients. Math. USSR Sb. **7** (1969) 171–193

[Mills 1] Millson, J.: On the first Betti number of a constant negatively curved manifold. Ann. Math. **104** (1976) 235–247

[Mills 2] Millson, J.: Real vector bundles with discrete structural group. Topology **18** (1979) 83–89

[Mills 3] Millson, J.: Cycles and harmonic forms on locally symmetric spaces. Preprint

[Mi-R] Millson, J.; Raghunathan, M.S.: Geometric construction of cohomology for arithmetic groups I. Papers dedicated to the memory of V.K. Patodi Indian Acad. Sci. Bangalore 1980

[Moo 1] Moore, C.C.: Ergodicity of flows on homogeneous spaces. Amer. J. Math. **88** (1966) 154–178

[Moo 2] Moore, C.C.: The Mautner phenomenon for general unitary representations. Pacific J. Math. **86** (1980) 155–169

[Mosk] Moskowitz, M.: On the density theorem of Borel and Furstenberg. Arkch. Math. **16** (1978) 11–27

[Most 1] Mostow, G.D.: Self-adjoint groups. Ann. Math. **62** (1955) 44–55

[Most 2] Mostow, G.D.: Some new decomposition theorems for semi-simple groups. Memoirs AMS **14** (1955) 31–54

[Most 3] Mostow, G.D.: Fundamental groups of homogeneous spaces. Ann. Math. **66** (1957) 249–255

[Most 4] Mostow, G.D.: Quasi-conformal mappings in *n*-space and the rigidity of hyperbolic space forms. Publ. Math. IHES **34** (1968) 53–104

[Most 5] Mostow, G.D.: Intersections of discrete subgroups with Cartan subgroups. J. Indian Math. Soc. **34** (1970) 203–214

[Most 6] Mostow, G.D.: The rigidity of locally symmetric spaces. Proc. Int. Congr. Math. **2** (1970) 187–197

[Most 7] Mostow, G.D.: Strong rigidity of locally symmetric spaces. Ann. Math. Studies, No. 78, Princeton University Press, Princeton, N.J. 1973

[Most 8] Mostow, G.D.: Discrete subgroups of Lie groups. Adv. Math. **16** (1975) 112–123

[Most 9] Mostow, G.D.: Existence of a non-arithmetic lattice in $\mathbf{SU}(2,1)$. Proc. Nat. Acad. Sci. USA **75** (1978) 3029–3033

[Most 10] Mostow, G.D.: On a remarkable class of polyhedra in complex hyperbolic space. Pacific J. Math. **86** (1980) 171–276

[Most 11] Mostow, G.D.: Existence of nonarithmetic monodromy groups. Proc. Nat. Acad. Sci. USA **78** (1981) 5948–5950

[M-S] Mostow, G.D.; Siu, Y.T.: A compact Kaehler surface of negative curvature not covered by the ball. Ann. Math. **112** (1980) 321–360

[M-T] Mostow, G.D.; Tamagawa, T.: On the compactness of the arithmetically defined homogeneous spaces. Ann. Math. **76** (1962) 440–463

[N 1] Nikulin, V.V.: On factor groups of the automorphism groups of hyperbolic forms modulo subgroups generated by 2-reflections. Soviet Math. Dokl. **20** (1979) 1156–1158

[N 2] Nikulin, V.V.: On arithmetic groups generated by reflections in Lobachevsky spaces. Math. USSR Izv. **16** (1981) 573–601

[N 3] Nikulin, V.V.: On the classification of arithmetic groups generated by reflections in Lobachevsky spaces. Math. USSR Izv. **18** (1982) 99–123

[N 4] Nikulin, V.V.: On the factor groups of the automorphism groups of hyperbolic forms by the subgroups generated by 2-reflections. Applications to algebraic geometry, in Sovremennye problemy matematiki 18, Moscow VINITI 1981, 3–114 (in Russian)

[O] Oseledec, V.I.: A multiplicative ergodic theorem. Trans. Moscow Math. Soc. **19** (1968) 197–231

[Pi 1] Piatetski-Shapiro, I.I.: Discrete groups of analitic automorphisms of a polycilinder and automorphic functions. Dokl. Akad. Nauk SSSR **124** (1959) 760–763 (in Russian)

[Pi 2] Piatetski-Shapiro, I.I.: Automorphic forms and arithmetic groups. Proc. Int. Congr. Math. (Moscow 1966), Mir, Moscow 1968, 232–247 (in Russian), AMS Transl. (2) **70** (1968) 185–201

[Pi 3] Piatetski-Shapiro, I.I.: Discrete subgroups of Lie groups. Trans. Moscow Math. Soc. **18** (1968) 1–18

[Pl 1] Platonov, V.P.: Linear algebraic group theory and torsion groups. Izv. Akad. Nauk SSSR, Ser. math. **30** (1966) 573–620 (in Russian)

[Pl 2] Platonov, V.P.: The problem of strong approximation and the Kneser-Tits hypothesis for algebraic groups. Math. USSR Izv. **3** (1970) 1139–1147; ibid. **4** (1970) 784–786

[Pl 3] Platonov, V.P.: Arithmetic and structural problems in linear algebraic groups. Proc. Int. Congress Math. (Vancouver 1974), AMS Transl. **109** (1977)

[Pl 4] Platonov, V.P.: Arithmetic theory of algebraic groups. Usp. Mat. Nauk. v. 37, **3** (255) (1982) 3–54 (in Russian)

[Pl-J] Platonov, V.P.; Janchevskiĭ, V.I.: On Harder's conjecture. Soviet Math. Dokl. **16** (1975) 424–427

[Pl-R 1] Platonov, V.P.; Rapinchuk, A.S.: On the group of rational points of 3-dimensional groups. Dokl. Akad. Nauk SSSR **247** (1979) 279–282 (in Russian)

[Pl-R 2] Platonov, V.P.; Rapinchuk, A.S.: The multiplicative structure of division rings over number fields and the Hasse norm principle. Proc. Steklov Inst. Math. **165** (1985) 187–205

[Pr 1] Prasad, G.: Strong rigidity of $\mathbb{Q}$-rank 1 lattices. Invent. math. **21** (1973) 255–286

[Pr 2] Prasad, G.: Triviality of certain automorphisms of semi-simple groups over local fields. Math. Ann. **218** (1975) 219–227

[Pr 3] Prasad, G.: Discrete subgroups isomorphic to lattices in Lie groups. Amer. J. Math., v. **98** (1976) 241–261, 853–863

[Pr 4] Prasad, G.: Non-vanishing of the first cohomology. Bull. Soc. Math. France **105** (1977) 415–418

[Pr 5] Prasad, G.: Strong approximation for semi-simple groups over function fields. Ann. Math. **105** (1977) 553–572

[Pr 6] Prasad, G.: Lattices in semi-simple groups over local fields. Studies in algebra and number theory. Adv. Math. Suppl. Studies **6** (1979)

[Pr 7] Prasad, G.: Elementary proof of a theorem of Bruhat-Tits-Rousseau and a theorem of Tits. Bull. Soc. Math. France **110** (1982) 197–202

[Pro] Prokhorov, M.N.: The non-existence of discrete groups of reflections with non-compact fundamental polyhedron of finite volume in Lobachevsky spaces of higher dimension. Izv. AN SSSR Ser. math. **50** (1986) 320–332 (in Russian)

[Pr-R 1] Prasad, G.; Raghunathan, M.S.: Cartan subgroups and lattices in semi-simple groups. Ann. Math. **96** (1972) 296–317

[Pr-R 2] Prasad, G.; Raghunathan, M.S.: On the congruence subgroup problem: determination of the "metaplectic kernel". Invent. math. **71** (1983) 21–42

[Pr-R 3] Prasad, G.; Raghunathan, M.S.: Topological central extension of semi-simple groups over local fields. Ann. Math. **119** (1984) 141–201, 203–268

[Pr-R 4] Prasad, G.; Raghunathan, M.S.: On the Kneser-Tits problem. Comment. Math. Helv. **60** (1985) 107–121

[P-S] Piatetski-Shapiro, I.I.; Šafarevič, I.R.: Galois theory of transcendental extensions and uniformization. AMS Transl. (2) **69** (1968) 111–145

[Rag 1] Raghunathan, M.S.: On the first cohomology of discrete subgroups of semi-simple Lie groups. Amer. J. Math. **87** (1965) 102–139

[Rag 2] Raghunathan, M.S.: Vanishing theorems for cohomology groups associated to discrete subgroups of semi-simple Lie groups. Osaka J. Math. **3** (1966) 243–256. Corrections, Osaka J. Math. **16** (1979) 295–299

[Rag 3] Raghunathan, M.S.: Cohomology of arithmetic subgroups of algebraic groups, I. Ann. Math. **86** (1967) 409–424

[Rag 4] Raghunathan, M.S.: Cohomology of arithmetic subgroups of algebraic groups, II. Ann. Math. **87** (1968) 279–304

[Rag 5] Raghunathan, M.S.: Discrete subgroups of Lie groups. Springer, New York 1972

[Rag 6] Raghunathan, M.S.: Discrete groups and $\mathbb{Q}$-structures on semisimple Lie groups. Proc. Int. Coll. on Discrete subgroups of Lie groups and appl. to moduli, Oxford University Press 1975, 225–321

[Rag 7] Raghunathan, M.S.: On the congruence subgroup problem. Publ. Math. IHES **46** (1976) 107–161

[Rag 8] Raghunathan, M.S.: A proof of Oseledec's multiplicative ergodic theorem. Israel J. Math. **32** (1979) 356–362

[Rag 9] Raghunathan, M.S.: Arithmetic lattices in semisimple groups. Proc. Indian Acad. Sci. (Math. Sci.) **91** (1982) 133–138

[Rag 10] Raghunathan, M.S.: Torsion in cocompact lattices in $\mathbf{SO}(2, n)$. Math. Ann. **266** (1984) 403–419

[Rag 11] Raghunathan, M.S.: On the group of norm 1 elements in a division algebra. Math. Ann. **279** (1988) 457–484

[Rag 12] Raghunathan, M.S.: On the congruence subgroup problem, II. Invent. math. **85** (1986) 73–117

[Rag 13] Raghunathan, M.S.: Discrete subgroups of algebraic groups over local fields of positive characteristics. To appear in Proceedings of Indian Academy of Sciences

[Rap 1] Rapinchuk, A. S.: Multiplicative arithmetic of division algebras over number fields and the metaplectic problem. Izv. Akad. Nauk SSSR, Ser. math. **51** (1987) 1033–1065 (in Russian)

[Rap 2] Rapinchuk, A.S.: The congruence subgroup problem for algebraic groups and strong approximation in affine manifolds. Dokl. Akad. Nauk SSSR **32** (1988) 581–584 (in Russian)

[Rat 1] Ratner, M.: Rigidity of horocycle flows. Ann. Math. **115** (1983) 597–614

[Rat 2] Ratner, M.: Horocycle flows, joinings, and rigidity of products. Ann. Math. **118** (1983) 277–313

[Rat 3] Ratner, M.: Ergodic theory in hyperbolic space. Preprint

[Rohlf] Rohlfs, J.: On the cuspidal cohomology of the Bianchi modular groups. Preprint

[Rohli 1] Rohlin, V.A.: On the fundamental ideas of measure theory. AMS Transl. (1) **10** (1962) 1–54

[Rohli 2] Rohlin, V.A.: Lectures on entropic transformation theory with invariant measure. Usp. Mat. Nauk **22** (1967) 3–56 (in Russian)

[Rosenb] Rosenblatt, J.: Uniqueness of invariant means for measure preserving transformations. Trans. AMS **265** (1981) 623–636

[Rosenl] Rosenlicht, M.: Questions of rationality for solvable algebraic groups over non-perfect fields. Annali di Mat. IV **61** (1963) 97–120

[R-S] Rohlfs, J.; Speh, B.: Representations with cohomology in the direct spectrum of subgroups of $\mathbf{SO}(n,1)(\mathbb{Z})$ and Lefschetz numbers. Ann. Sci. Ec. Norm. Super., IV, Sér. **20** (1987) 89–136

[Ru] Ruelle, D.: Ergodic theory of differential dynamical systems. Preprint IHES 1978

[Schm] Schmidt, K.: Amenability, Kazhdan's property T, strong ergodicity and invariant means for ergodic group actions. Ergod. Theor. and Dynam. Syst. **1** (1981) 223–226

[Scho] Schoenberg, I.J.: Metric spaces and positive definite functions. Trans. AMS **44** (1938) 522–536

[Schw 1] Schwartzmann, O.V.: On discrete arithmetic subgroups of complex Lie groups. Mat. Sb. **77** (1968) 541–544 (in Russian)

[Schw 2] Schwartzmann, O.V.: Discrete groups of reflections in complex ball. Funct. Anal. Appl. **18** (1984) 88–89 (in Russian)

[Schw 3] Schwartzmann, O.V.: Discrete groups of reflections in complex ball. Questions of group theory and homological algebra. Proc. Jaroslavl' (1985) 61–77 (in Russian)

[Schwe 1] Schwermer, J.: Sur la cohomologie de $\mathbf{SL}_n(\mathbb{Z})$ a l'infini et les séries d'Eisenstein. C.R. Acad. Sci. Paris **289** (1979) 413–415

[Schwe 2] Schwermer, J.: Kohomologie arithmetisch definierter Gruppen und Eisensteinreihen. Springer Lecture Notes, No. 988, 1983

[Sel 1] Selberg, A.: On discontinuous groups in higher dimensional symmetric spaces. Contributions to Function Theory (Int. Colloq. Bombay, 1960). Tata Inst. Fund. Res., Bombay 1960, 147–164

[Sel 2] Selberg, A.: Discontinuous groups and harmonic analysis. Proc. Int. Congr. Math. (Stockholm 1962), Inst. Mittag-Leffler, Ojursholm, 1963, 177–189

[Sel 3] Selberg, A.: Recent development in the theory of discontinuous groups of motions of symmetric spaces. Proc. 15th Scandinavian Math. Congr. (Oslo 1968), Springer Lecture Notes No. **118** (1970) 99–120

[Ser 1] Serre, J.-P.: Le problème de groupes de congruence pour $\mathbf{SL}_2$. Ann. Math. **92** (1970) 489–527

[Ser 2] Serre, J.-P.: Cohomologie des groupes discrets. Prospects in Math., Ann. Math. Studies **70** (1971) 77–169

[Ser 3] Serre, J.-P.: Arbres, amalgames, $\mathbf{SL}_2$. Soc. Math. France, Asterisque **46** (1977)

[Sha] Shafarevich, I.R.: Foundations of Algebraic Geometry. Nauka, Moscow 1973 (in Russian)

[Shil] Shilov, G.E.: Mathematical analysis. Fiz. Mat. Giz., Moscow 1960 (in Russian)

[Shir] Shirakawa, H.: An example of infinite measure preserving ergodic geodesic flow on a surface with constant negative curvature. Comment. Math. Univ. Sacti Pauli **31** (1982) 163–182

[So] Soulé, C.: The cohomology of $\mathbf{SL}_3(\mathbb{Z})$. Topology **17** (1978) 1–22

[Spa] Spatzier, R.J.: On lattices acting on boundaries of semisimple groups. Ergod. Theor. and Dynam. Sys. **1** (1981) 489–494

[Spr] Springer, T.A.: Reductive groups, in: Automorphic forms, representations and L-functions. Proc. Symp. Pure Math. v. 33, Part 1, 3–27

[Ste] Steinberg, R.: Lectures on Chevalley groups. Notes prepared by John Faulkner and Robert Wilson, Yale University 1967

[Stu] Stuhler, U.: Homological properties of certain arithmetic groups in function field case. Invent. math. **57** (1980) 269–281

[Sul 1] Sullivan, D.: On the ergodic theory at infinity of an arbitrary discrete group of hyperbolic motions, in Proc. of the Stony Brook Conf. on Kleinian groups and Riemann surfaces, New York 1978

[Sul 2] Sullivan, D.: For $n > 3$ there is only one finitely additive rotationally invariant measure on the n-sphere defined on all Lebesgue measurable subsets. Bull. AMS **4** (1981) 121–123

[Sup] Suprunenko, D.A.: Matrix groups. Nauka, Moscow 1972 (in Russian)

[Th] Thurston, W.: The geometry and topology of 3-manifolds. Mimeographed notes, Princeton University

[Ti 1] Tits, J.: Algebraic and abstract simple groups. Ann. Math. **80** (1964) 313–329

[Ti 2] Tits, J.: Classification of algebraic semisimple groups, algebraic groups, and discontinuous groups. Symposium Colorado, Boulder 1965, AMS Proc. Symp. Pure Math. IX

[Ti 3] Tits, J.: Free subgroups in linear groups. J. Algebra **20** (1972) 250–270

[Ti 4] Tits, J.: Travaux de Margulis sur les sous-groupes discretes de groupes de Lie. Springer Lecture Notes, No. 577 (Seminaire Bourbaki 1975/76), Springer 1977, exposè 482, 114–130

[To 1] Tomanov, G.M.: Sur la structure des groupes algébriques simples des type $\mathbb{D}_n$ definis sur des corps de nombres. C.R. Acad. Sci. Paris **306** (1988) 647–650

[To 2] Tomanov, G.M.: On the congruence-subgroup problem for groups of type $\mathbb{G}_2$. C.R. de l'Ac. Bulg. des Sciences **42** (1989) No. 6

[To 3] Tomanov, G.M.: Projective simplicity of groups of rational points of simply connected algebraic groups over number fields. Banach Center Publ. (Topics in Algebra) **26** (1989)

[To 4] Tomanov, G.M.: On the congruence-subgroup problem for some anisotropic algebraic groups over number fields. J. Reine Angew. Math. **402** (1989) 138–152

[To 5] Tomanov, G.M.: Remarques sur la structure des groupes algébriques définis sur des corps de nombres. C.R. Acad. Sci. Paris **310** (1990) 33–36

[T-W] Tong, G.L.; Wang, S.P.: Harmonic forms dual to geodesic cycles in quotients of $\mathbf{SU}(p, 1)$. Math. Ann. **258** (1982) 289–318

[Va 1] Vaserštein, L.N.: On the group $\mathbf{SL}_2$ over Dedekind rings of arithmetic type. Math. USSR Sb. **18** (1972) 321–332

[Va 2] Vaserštein, L.N.: Structure of classical arithmetic groups of rank greater than 1. Matem. Sb. **91** (1973) 445–470 (in Russian)

[Ve] Venkataramana, T.N.: On superrigidity and arithmeticity of lattices in semi-simple groups over local fields of arbitrary characteristic. Invent. math. **92**, no. 2 (1988) 255–306

[Vi 1] Vinberg, E.B.: Discrete groups generated by reflections in Lobachevsky spaces. Math. USSR Sb. **1** (1967) 429–444

[Vi 2] Vinberg, E.B.: Compact Lie groups. Moscow Univ. Press, Moscow 1967 (in Russian)

[Vi 3] Vinberg, E.B.: Some examples of crystallographic groups in Lobachevsky spaces. Math. USSR Sb. **7** (1969) 617–622

[Vi 4] Vinberg, E.B.: Rings of definition of dense subgroups of semisimple linear groups. Math. USSR Izv. **5** (1971) 45–55

[Vi 5] Vinberg, E.B.: Discrete linear groups generated by reflections. Math. USSR Izv. **5** (1971) 1083–1119

[Vi 6] Vinberg, E.B.: Some arithmetic discrete groups in Lobachevsky spaces, in Proc. Int. Coll. on Discrete subgrops of Lie groups and appl. to moduli (Bombay 1973). Oxford University Press (1975) 323–348

[Vi 7] Vinberg, E.B.: Properties of the root decomposition of a semisimple Lie algebra over an algebraically nonclosed field. Funct. Anal. Appl. **9** (1975) 17–21

[Vi 8] Vinberg, E.B.: Absence of crystallographic groups of reflections in Lobachevsky spaces of large dimension. Funct. Anal. Appl. **15** (1981) 128–130

[Vi 9] Vinberg, E.B.: Two almost algebraic K3 surfaces. Math. Ann. **265** (1983) 1–21

[Vi 10] Vinberg, E.B.: Absence of crystallographic groups of reflections in Lobachevsky spaces of large dimension. Trans. Moscow Math. Soc. **47** (1984) 68–102 (in Russian)

[Vi 11] Vinberg, E.B.: Discrete groups of reflections in Lobachevsky spaces. Proc. Int. Congr. Math., Warsaw 1983, 585–592

[Vi 12] Vinberg, E.B.: Hyperbolic reflection groups. Russ. Math. Surv. **40** (1985) 31–75

[V-K] Vinberg, E.B.; Kaplinskaja, I.M.: On the groups $O_{18,1}(\mathbb{Z})$ and $O_{19,1}(\mathbb{Z})$. Soviet Math. Dokl. **19** (1978) 194–197

[V-O] Vinberg, E.B.; Oniščik, A.L.: Seminaire on algebraic groups and Lie groups. Moscow Univ. Press, Moscow 1969 (in Russian)

[Wae 1] van der Waerden, B.L.: Algebra I. Springer, New York 1971

[Wae 2] van der Waerden, B.L.: Algebra II. Springer, New York 1967

[Wat] Watatani, Y.: Property (T) of Kazhdan implies property (FA) of Serre. Math. Jap. **27** (1982) 97–103

[Weil 1] Weil, A.: On discrete subgroups of Lie groups I, II. Ann. Math. (2) **27** (1960) 369–384; ibid. (2) **75** (1962) 578–602

[Weil 2] Weil, A.: Adeles and algebraic groups. Inst. for Advanced Study, Princeton, N.J. 1961

[Weil 3] Weil, A.: Remarks on the cohomology of groups. Ann. Math. **80** (1964) 149–157

[Weil 4] Weil, A.: Basic number theory. Springer, New York 1967

[Weis 1] Weisfeiler, B.: On abstract monomorphisms of k-forms of **PGL**(2). J. Algebra **57** (1979) 522–543

[Weis 2] Weisfeiler, B.: Abstract homomorphisms of big subgroups of algebraic groups. Notre Dame Math. Lect. University of Notre Dame Press **10** (1982) 135–181

[Wh] Whitney, H.: Elementary structure of real algebraic varieties. Ann. Math. **66** (1957) 545–556

[WH 1] Wang, H.C.: On a maximality property of discrete subgroups with fundamental domain of finite measure. Amer. J. Math. **89** (1967) 124–132

[WH 2] Wang, H.C.: in: Topics on totally discontinuous groups in symmetric spaces. W. Boothly (ed.) M. Dekker 1972, pp. 460–487

[WS 1] Wang, S.P.: Limit of lattices in a Lie groups. Trans. AMS **133** (1968) 519–526

[WS 2] Wang, S.P.: On a theorem of representation of lattices. Proc. AMS **23** (1969) 583–587

[WS 3] Wang, S.P.: On a conjecture of Chabauty. Proc. AMS **23** (1969) 569–572

[WS 4] Wang, S.P.: The dual space of semi-simple Lie groups. Amer. J. Math. **91** (1969) 921–937

[WS 5] Wang, S.P.: On the centralizer of a lattice. Proc. AMS **21** (1969) 21–23

[WS 6] Wang, S.P.: On the limit of subgroups in a group. Amer. J. Math. **92** (1970) 708–724

[WS 7] Wang, S.P.: Some properties of lattices in a Lie group. Illinois J. Math. **14** (1970) 35–39

[WS 8] Wang, S.P.: On density properties of S-subgroups of locally compact groups. Ann. Math. **94** (1971) 325–329

[WS 9] Wang, S.P.: On a lemma of Mahler. Amer. J. Math. 95, No. **4** (1973) 703–712

[WS 10] Wang, S.P.: On the first cohomology group of discrete groups with property (T). Proc. AMS **42** (1974) 621–624

[WS 11] Wang, S.P.: On isolated points in the dual spaces of locally compact groups. Math. Ann. **218** (1975) 19–34

[WS 12] Wang, S.P.: On subgroups with property (P) and maximal discrete subgroups. Amer. J. Math. **97** (1975) 404–414

[WS 13] Wang, S.P.: Homogeneous spaces with finite invariant measure. Amer. J. Math. **98** (1976) 311–324

[WS 14] Wang, S.P.: On density properties of certain subgroups of locally compact groups. Duke Math. J. **43** (1976) 561–578; Correction Duke Math. J. **45** (1978) 953

[WS 15] Wang, S.P.: On L-subgroups of locally compact groups. Adv. Math. **28** (1978) 89–100

[WS 16] Wang, S.P.: On density properties of certain subgroups with boundedness conditions. Monatsh. Math. **89** (1980) 141–162

[WS 17] Wang, S.P.: A note on free subgroups in linear groups. J. Algebra **71** (1981) 232–234

[WS 18] Wang, S.P.: On the Mautner phenomenon and groups with property (T). Amer. J. Math. **104** (1982) 1191–1210

[Zi 1] Zimmer, R.L.: Strong rigidity for ergodic actions of semisimple Lie groups. Annals of Math. **112** (1980) 511–529

[Zi 2] Zimmer, R.L.: Orbit equivalence and rigidity for ergodic actions of Lie groups. Ergod. Theor. and Dynam. Sys. **1** (1981) 237–253

[Zi 3] Zimmer, R.L.: Equivariant images of projective space under the action of $\mathbf{SL}(n, \mathbb{Z})$. Ergod. Theor. and Dynam. Sys. **1** (1981) 519–622

[Zi 4] Zimmer, R.L.: Ergodic theory, group representations, and rigidity. Bull. AMS **6** (1982) 383–416

[Zi 5] Zimmer, R.L.: On the Mostow rigidity theorem and measurable foliations by hyperbolic space. Israel J. Math. **43** (1982) 281–290

[Zi 6] Zimmer, R.L.: Volume preserving actions of lattices in semisimple groups on compact manifolds. Publ. Math. IHES **59** (1984) 5–33

[Zi 7] Zimmer, R.L.: Kazhdan groups acting on compact manifolds. Invent. math. **75** (1984) 425–436

[Zi 8] Zimmer, R.L.: Ergodic theory and semisimple groups. Birkhäuser, Boston 1985

[Z-S] Zariski, O.; Samuel, P.: Commutative Algebra I, II. D. Van Nostrand Comp. Inc. Princeton, New Jersey 1958, 1960

Subject Index